Surveys in Differential Geometry

Vol. 1: Lectures given in 1990
edited by S.-T. Yau and H. Blaine Lawson

Vol. 2: Lectures given in 1993
edited by C.C. Hsiung and S.-T. Yau

Vol. 3: Lectures given in 1996
edited by C.C. Hsiung and S.-T. Yau

Vol. 4: Integrable systems
edited by Chuu Lian Terng and Karen Uhlenbeck

Vol. 5: Differential geometry inspired by string theory
edited by S.-T. Yau

Vol. 6: Essays on Einstein manifolds
edited by Claude LeBrun and McKenzie Wang

Vol. 7: Papers dedicated to Atiyah, Bott, Hirzebruch, and Singer
edited by S.-T. Yau

Vol. 8: Papers in honor of Calabi, Lawson, Siu, and Uhlenbeck
edited by S.-T. Yau

Vol. 9: Eigenvalues of Laplacians and other geometric operators
edited by A. Grigor'yan and S-T. Yau

Vol. 10: Essays in geometry in memory of S.-S. Chern
edited by S.-T. Yau

Vol. 11: Metric and comparison geometry
edited by Jeffrey Cheeger and Karsten Grove

Vol. 12: Geometric flows
edited by Huai-Dong Cao and S.-T. Yau

Vol. 13: Geometry, analysis, and algebraic geometry
edited by Huai-Dong Cao and S.-T.Yau

Vol. 14: Geometry of Riemann surfaces and their moduli spaces
edited by Lizhen Ji, Scott A. Wolpert, and S.-T. Yau

Vol. 15: Perspectives in mathematics and physics: Essays dedicated to Isadore Singer's 85th birthday
edited by Tomasz Mrowka and S.-T. Yau

Vol. 16: Geometry of special holonomy and related topics
edited by Naichung Conan Leung and S.-T. Yau

Vol. 17: In Memory of C. C. Hsiung: Lectures given at the JDG Symposium, Lehigh University, June 2010
edited by Huai-Dong Cao and S.-T. Yau

Volume XVII

Surveys in Differential Geometry

In Memory of C. C. Hsiung

Lectures given at the JDG Symposium,
Lehigh University, June 2010

edited by

Huai-Dong Cao
and Shing-Tung Yau

International Press
www.intlpress.com

Series Editor: Shing-Tung Yau

Surveys in Differential Geometry, Vol. 17 (2012)
In Memory of C. C. Hsiung: Lectures given at the JDG Symposium, Lehigh University, June 2010

Volume Editors:
Huai-Dong Cao (Lehigh University)
Shing-Tung Yau (Harvard University)

ISBN 978-1-57146-237-4

Printed in the United States of America.

16 15 14 13 12 1 2 3 4 5 6 7 8 9

Preface

On May 6, 2009, Professor Chuan-Chih Hsiung passed away. I was given this sad news by his daughter Nancy.

Professor Hsiung had joined Lehigh University in 1952. He officially retired in 1984 but thereafter remained active mathematically—and within the department—for many years.

Besides his substantial and important works in the subject of differential geometry, Professor Hsiung made a tremendously important contribution to mathematics by founding the *Journal of Differential Geometry* in 1967. At that time, he had the foresight and vision to understand that the subject of geometry would grow and become one of the most crucial elements in the development of mathematics.

Note that prior to this no journal has been dedicated exclusively to geometry. Professor Hsiung had the skill to persuade the provost of Lehigh of the journal's importance to the University and to gain his support for the journal. Professor Hsiung set a high standard for the *Journal of Differential Geometry* by inviting first-class mathematicians to its editorial board and by selecting only the best papers.

In the early days of the journal, Professor Hsiung consulted with my thesis advisor, Professor Shiing-Shen Chern, whose advice regarding the journal's development proved invaluable. In 1980 Professors Chern and Hsiung together convinced me to become the journal's editor. We saw the development of geometry influenced by many of the papers which thereafter appeared in the journal.

My collaboration with Professor Hsiung over thirty years' time was pleasant and productive, and I am sad to have lost such a good friend.

I am grateful to Lehigh University for giving this conference, the International Symposium on Geometry and Topology (Bethlehem, Penn., May 2010), in honor of Professor Hsiung. And I am grateful to the many distinguished geometers who presented talks at the conference, and who contributed articles to this volume of the *Surveys in Differential Geometry* book series.

In this volume, we have articles in geometric analysis by Christian Bär and Werner Ballmann on boundary value problems for elliptic differential operators of the first order; by Simon Brendle on rigidity phenomena involving scalar curvature; by Simon Donaldson on stability, birational transformations, and the Kähler-Einstein problem; by D. H. Phong, Jian Song, and Jacob Sturm concerning complex Monge-Ampère equations; by Weimin Sheng, Neil S. Trudinger, and Xu-Jia Wang on the k-Yamabe problem; by S.-T. Yau on topics of geometric analysis. We also have articles in geometry by Gergely Berczi and Frances Kirwan on a geometric construction for invariant jet differentials; by B. Bramham and H. Hofer concerning symplectic dynamics; by Dominic Joyce on C-infinity schemes and C-infinity algebraic geometry; by Kenji Fukaya, Yong-Geun Oh, Hiroshi Ohta, Kaoru Ono on Lagrangian Floer theory on compact toric manifolds; and by Paul Seidel concerning Fukaya categories.

Ten years ago, Huai-Dong Cao came to Lehigh University to lead the geometry group there. He certainly has been the key to managing the *Journal of Differential Geometry.* I am grateful to him for organizing the conference and for editing this volume.

Shing-Tung Yau
Harvard University

Acknowledgments

The editors are very grateful to the Sponsors, and to the members of the Scientific Committee and the Local Organizing Committee.

Sponsors

Journal of Differential Geometry
Department of Mathematics of Lehigh University
Office of International Affairs of Lehigh University
College of Arts and Sciences of Lehigh University
National Science Foundation

Scientific Committee

Huai-Dong Cao (Lehigh University)
Sylvain Cappell (New York University)
Shiu-Yuen Cheng (University of Sciences and Technology, Hong Kong)
Simon Donaldson (Imperial College, UK)
Gerhard Huisken (Albert Einstein Institute/Max-Planck Institute, Germany)
H. Blaine Lawson (Stoney Brook University)
Jun Li (Stanford University)
Richard Schoen (Stanford University)
Neil Trudinger (Australia National University, Australia)
Claire Voisin (Institut de mathématiques de Jussieu, France)
S.-T. Yau (Harvard University)

Local Organizing Committee at Lehigh University

Huai-Dong Cao
Donald M. Davis
David L. Johnson (Chair)
Terry Napier
Robert Neel
Xiaofeng Sun
Susan Szczepanski
Steven H. Weintraub

Contents

Boundary value problems for elliptic differential operators of first order

Christian Bär and Werner Ballmann

ABSTRACT. We study boundary value problems for linear elliptic differential operators of order one. The underlying manifold may be noncompact, but the boundary is assumed to be compact. We require a symmetry property of the principal symbol of the operator along the boundary. This is satisfied by Dirac type operators, for instance.

We provide a selfcontained introduction to (nonlocal) elliptic boundary conditions, boundary regularity of solutions, and index theory. In particular, we simplify and generalize the traditional theory of elliptic boundary value problems for Dirac type operators. We also prove a related decomposition theorem, a general version of Gromov and Lawson's relative index theorem and a generalization of the cobordism theorem.

CONTENTS

2010 *Mathematics Subject classification*: 35J56, 58J05, 58J20, 58J32.

Key words and phrases. Elliptic differential operators of first order, elliptic boundary conditions, completeness, coercivity at infinity, boundary regularity, index theory, decomposition theorem, relative index theorem, cobordism theorem.

1. Introduction

In their attempt to generalize Hirzebruch's signature theorem to compact manifolds with boundary, Atiyah, Patodi, and Singer arrived at a boundary condition which is nonlocal in nature and involves the spectrum of an associated selfadjoint differential operator on the boundary. In fact, they obtained an index theorem for a certain class of first order elliptic differential operators on compact manifolds with boundary [**APS**, Theorem 3.10]. For the standard differential operators of first order encountered in Riemannian geometry, their assumption means that a sufficiently small collar about the boundary is cylindrical, that is, isometric to the Riemannian product of an interval times the boundary. In many applications, this is a completely satisfactory assumption. They also discuss the L^2-theory for the natural extension of the operator to the noncompact manifold which is obtained by extending the cylinder beyond the boundary to a one-sided infinite cylinder. The work of Atiyah, Patodi, and Singer lies at the heart of many investigations concerning boundary value problems and L^2-index theory for first order elliptic differential operators, and this includes the present article.

The original motivation for our present studies came from the relative index theorem of Gromov and Lawson, see [**GL**, Thm. 4.18] or Theorem 1.21 below. After the decomposition theorem in [**BW**, Thm. 23.3] and [**BL1**, Thm. 4.3] was used in [**BB1**] and [**BB2**] to obtain index theorems for first order geometric differential operators on certain noncompact Riemannian manifolds, we observed that the decomposition theorem could also be used for a short proof of the relative index theorem. The drawback of this argument is, however, that the proof of the decomposition theorem in the above references [**BW**] and [**BL1**] involves a heavy technical machinery so that, as a whole, the proof of the relative index theorem would not be simplified. Therefore, our first objective is a simplification of the theory of boundary value problems for (certain) first order elliptic differential operators, and the main result of this endeavor is formulated in Theorem 1.12 (together with Addendum 1.13) and Theorem 1.15 below. Another objective is the L^2-index theory for noncompact manifolds, see Theorem 1.18. We arrive, finally, at simple proofs of the decomposition and relative index theorems.

To formulate our main results, we start by fixing the setup for our investigations. We consider Hermitian vector bundles E, F over a manifold M with compact boundary ∂M and a differential operator D from E to F. We do not equip M with a Riemannian metric but we assume that M is endowed with a smooth volume element μ. Then functions can be integrated over M and the spaces $L^2(M,E)$ and $L^2(M,F)$ of square-integrable sections of E and F are defined. By D_{cc} we denote D, considered as an unbounded operator from $L^2(M,E)$ to $L^2(M,F)$ with domain $C^\infty_{cc}(M,E)$, the space of smooth sections of E with compact support in the interior of M[†].

We denote by D^* the formal adjoint of D. The *maximal extension* $D_{\max}$ of D is the adjoint of D^*_{cc} in the sense of functional analysis. That is, $\operatorname{dom}(D_{\max})$ is the space of all $\Phi \in L^2(M,E)$ such that there is a section $\Psi \in L^2(M,F)$ with $(\Phi, D^*\Xi)_{L^2(M)} = (\Psi, \Xi)_{L^2(M)}$ for all $\Xi \in C^\infty_{cc}(M,F)$. Then we set $D_{\max}\Phi := \Psi$. The *graph norm* $\|\cdot\|_{D_{\max}}$ of $D_{\max}$, defined by

$$\|\Phi\|_D^2 := \|\Phi\|^2_{L^2(M)} + \|D_{\max}\Phi\|^2_{L^2(M)},$$

turns $\operatorname{dom}(D_{\max})$ into a Hilbert space. Clearly, $C^\infty_c(M,E) \subset \operatorname{dom}(D_{\max})$.

DEFINITION 1.1. We say that D is *complete* if the subspace of compactly supported sections in $\operatorname{dom}(D_{\max})$ is dense in $\operatorname{dom}(D_{\max})$ with respect to the graph norm of $D_{\max}$.

By definition, completeness holds if M is compact. In the noncompact case, completeness signifies that square integrability is a decent boundary condition at infinity for D. In practice, completeness can often be checked using the following theorem:

THEOREM 1.2. *Suppose that M carries a complete Riemannian metric with respect to which the principal symbol σ_D of D satisfies an estimate*

$$|\sigma_D(\xi)| \le C(\operatorname{dist}(x, \partial M)) \cdot |\xi|$$

for all $x \in M$ and $\xi \in T^*_x M$, where $C : [0,\infty) \to \mathbb{R}$ is a positive monotonically increasing continuous function with

$$\int_0^\infty \frac{dr}{C(r)} = \infty.$$

Then D and D^* are complete.

This theorem applies, for instance, if C is a constant.

[†]Throughout the article, the indices c and cc indicate compact support in M and in the interior of M, respectively. In the index at the end of the article, the reader finds much of the standard and all of the nonstandard notation.

Remark 1.3. Note that we do not assume that μ is the volume element induced by the Riemannian metric. In fact, the principal symbols of D and D^* do not depend on the choice of μ, although D^* does.

Fix a vector field T of M along ∂M pointing into the interior of M and let τ be the *associated one-form* along ∂M, that is, $\tau(T) = 1$ and $\tau|_{T\partial M} = 0$. Set $\sigma_0 := \sigma_D(\tau)$. Note that σ_0 is invertible if D is elliptic.

Definition 1.4. We say that D is *boundary symmetric* (with respect to T) if D is elliptic and

$$(1) \qquad \sigma_0(x)^{-1} \circ \sigma_D(\xi) : E_x \to E_x \quad \text{and} \quad \sigma_D(\xi) \circ \sigma_0(x)^{-1} : F_x \to F_x$$

are skew-Hermitian, for all $x \in \partial M$ and $\xi \in T_x^*\partial M$. Here we identify ξ with its extension to T_xM which satisfies $\xi(T) = 0$.

It is somewhat hidden in this definition, but will become clear in the text, that D is boundary symmetric if and only if D^* is boundary symmetric. Clearly, boundary symmetry of D does not depend on the choice of μ.

We will see in Lemma 4.1 that the boundary symmetry of D implies existence of essentially selfadjoint elliptic first order differential operators A on $E|_{\partial M}$ and $\tilde{A}$ on $F|_{\partial M}$ with symbols given by $\sigma_0^{-1} \circ \sigma_D$ and $(\sigma_D \circ \sigma_0^{-1})^*$, respectively. We will call such operators *adapted* (with respect to the choice of T). Note that we may add zero-order terms to adapted operators without violating the requirement on their principal symbols. Lemma 4.1 also shows that, in a collar about ∂M, D and D^* can be represented in the normal form

$$(2) \qquad D = \sigma_t \left(\frac{\partial}{\partial t} + A + R_t \right) \quad \text{and} \quad D^* = -\sigma_t^* \left(\frac{\partial}{\partial t} + \tilde{A} + \tilde{R}_t \right),$$

where t is an inward coordinate in the collar with $\partial/\partial t = T$ along $\partial M = t^{-1}(0)$ and where, for $0 \le t < r$,

(i) $\sigma(t,x) = \sigma_t(x) = \sigma_D(dt|_{(t,x)})$ for all $x \in \partial M$;
(ii) R_t and $\tilde{R}_t$ are differential operators of $E|_{\partial M}$ and $F|_{\partial M}$ of order at most one, respectively, whose coefficients depend smoothly on t and such that R_0 and $\tilde{R}_0$ are of order zero.

Since $dt = \tau$ along ∂M, the notation σ_t is in peace with the notation $\sigma_0 = \sigma_D(\tau)$ further up.

To put the normal form (2) into perspective, we mention that Atiyah, Patodi, and Singer consider the case where σ does not depend on t and where $R_t = \tilde{R}_t = 0$.

Standard Setup 1.5. Our standard setup, which we usually assume, is as follows:

- M is a manifold with compact boundary ∂M (possibly empty);
- μ is a smooth volume element on M;

- E and F are Hermitian vector bundles over M;
- D is an elliptic differential operator from E to F of order one;
- T is an interior vector field along ∂M;
- D is boundary symmetric with respect to T;
- A and $\tilde{A}$ are adapted differential operators on $E|_{\partial M}$ and $F|_{\partial M}$ for D and D^*, respectively.

EXAMPLE 1.6. Let M be a Riemannian manifold with boundary and μ be the associated Riemannian volume element. Then Dirac type operators over M are elliptic and boundary symmetric with respect to the interior unit normal field of ∂M, compare Example 4.3 (a).

For $I \subset \mathbb{R}$ and $s \in \mathbb{R}$, we denote by $H^s_I(A)$ the closed subspace of the Sobolev space $H^s(\partial M, E)$ spanned by the eigenspaces of A with eigenvalue in I. We denote by Q_I the (spectral) projection of $H^s(\partial M, E)$ onto $H^s_I(A)$ with kernel $H^s_{\mathbb{R}\setminus I}(A)$. For $a \in \mathbb{R}$, we let

$$\check{H}(A) := H^{1/2}_{(-\infty,a)}(A) \oplus H^{-1/2}_{[a,\infty)}(A). \tag{3}$$

Since, for any finite interval I, the space $H^s_I(A)$ is finite-dimensional and contained in $C^\infty(\partial M, E)$, different choices of a lead to the same topological vector space.

THEOREM 1.7. *Assume the Standard Setup 1.5 and that D and D^* are complete. Then*

(1) $C^\infty_c(M, E)$ is dense in $\operatorname{dom}(D_{\max})$;
(2) the trace map $\mathscr{R}$ on $C^\infty_c(M, E)$, $\mathscr{R}\Phi := \Phi|_{\partial M}$, extends to a surjective bounded linear map $\mathscr{R} : \operatorname{dom}(D_{\max}) \to \check{H}(A)$;
(3) $\Phi \in \operatorname{dom}(D_{\max})$ is in $H^{k+1}_{\mathrm{loc}}(M, E)$ if and only if $D\Phi$ is in $H^k_{\mathrm{loc}}(M, F)$ and $Q_{[0,\infty)}(\mathscr{R}\Phi)$ is in $H^{k+1/2}(\partial M, E)$, for any integer $k \geq 0$.

REMARKS 1.8. (a) Completeness is irrelevant for Assertions (2) and (3), and a similar remark applies to the boundary regularity results below; compare Remark 3.5 below.

(b) In the case where $\partial M = \emptyset$, Assertion (1) of Theorem 1.7 says that $D_{\max}$ is equal to the closure of D_{cc}. If D is *formally selfadjoint*, i.e., if $D = D^*$, this means that D is *essentially selfadjoint*.

(c) Assertion (2) of Theorem 1.7 implies that, as a topological vector space, $\check{H}(A)$ does not depend on the choice of A (as long as its symbol is given by $\sigma_0^{-1} \circ \sigma_D$). Carron pointed out to us a direct proof of this fact: If A' is a further operator of the required kind, then the difference $\delta = Q_{[a,\infty)} - Q'_{[a,\infty)}$ of the corresponding spectral projections is a pseudo-differential operator of order -1, thus δ maps $H^{-1/2}(\partial M, E)$ to $H^{1/2}(\partial M, E)$.

Theorem 1.7 implies that, for any closed subspace $B \subset \check{H}(A)$, the restriction $D_{B,\max}$ of $D_{\max}$ to

$$\operatorname{dom}(D_{B,\max}) := \{\Phi \in \operatorname{dom}(D_{\max}) \mid \mathscr{R}(\Phi) \in B\} \tag{4}$$

is a closed extension of D_{cc}. We will see in Proposition 7.2 that any closed operator between D_{cc} and $D_{\max}$ is of this form. In particular, the *minimal extension* $D_{\min}$ of D, that is, the closure of D_{cc}, is given by the Dirichlet boundary condition, that is, by $D_{0,\max}$. We arrive at the following

DEFINITION 1.9. A *boundary condition* for D is a closed subspace of $\check{H}(A)$.

As we already mentioned above, we have

$$D_{cc} \subset D_{\min} \subset D_{B,\max} \subset D_{\max} \tag{5}$$

for any boundary condition $B \subset \check{H}(A)$, explaining the terminology *minimal* and *maximal* extension.

We will see in Subsection 7.2 that, for any boundary condition B, the adjoint operator of $D_{B,\max}$ is the closed extension $D^*_{B^{\mathrm{ad}},\max}$ of D^*_{cc} with boundary condition

$$B^{\mathrm{ad}} = \{\psi \in \check{H}(\tilde{A}) \mid \beta(\varphi,\psi) = 0 \text{ for all } \varphi \in B\}, \tag{6}$$

where β denotes the natural extension of the sesqui-linear form

$$\beta(\varphi,\psi) := (\sigma_0\varphi,\psi)_{L^2(\partial M)} \tag{7}$$

on $C^\infty(\partial M,E) \times C^\infty(\partial M,F)$ to $\check{H}(A) \times \check{H}(\tilde{A})$, compare Lemma 6.3.

DEFINITION 1.10. A boundary condition B is said to be

(i) *elliptic* if

$$\operatorname{dom}(D_{B,\max}) \subset H^1_{\mathrm{loc}}(M,E) \quad \text{and} \quad \operatorname{dom}(D^*_{B^{\mathrm{ad}},\max}) \subset H^1_{\mathrm{loc}}(M,F);$$

(ii) *m-regular*, where $m \geq 0$ is an integer, if

$$D_{\max}\Phi \in H^k_{\mathrm{loc}}(M,F) \implies \Phi \in H^{k+1}_{\mathrm{loc}}(M,E),$$
$$D^*_{\max}\Psi \in H^k_{\mathrm{loc}}(M,E) \implies \Psi \in H^{k+1}_{\mathrm{loc}}(M,F)$$

for all $\Phi \in \operatorname{dom}(D_{B,\max})$, $\Psi \in \operatorname{dom}(D^*_{B^{\mathrm{ad}},\max})$, and $0 \leq k \leq m$.

(iii) *∞-regular* if it is m-regular for all integers $m \geq 0$.

We recall that the implications in (ii) hold in the interior of M for all integers $k \geq 0$, by the standard interior regularity theory for elliptic differential operators.

REMARKS 1.11. (a) By definition, we have $H^0_{\mathrm{loc}}(M,E)=L^2_{\mathrm{loc}}(M,E)$, and similarly for F. Hence a boundary condition is elliptic if and only if it is 0-regular.

(b) By Theorem 1.7, Remark 1.8, and the above, a boundary condition B is elliptic if and only if

$$B\subset H^{1/2}(\partial M,E) \quad \text{and} \quad B^{\mathrm{ad}}\subset H^{1/2}(\partial M,F).$$

To state our main theorem on boundary regularity, we introduce the notation $U^s := U\cap H^s(\partial M,E)$, for any $U\subset \cup_{s'\in\mathbb{R}}H^{s'}(\partial M,E)$ and $s\in\mathbb{R}$.

THEOREM 1.12. *Assume the Standard Setup 1.5 and that D and D^* are complete. Let $B\subset \check{H}(A)$ be a boundary condition. Then B is elliptic if and only if, for some (and then any) $a\in\mathbb{R}$, there are L^2-orthogonal decompositions*

$$L^2_{(-\infty,a)}(A)=V_-\oplus W_- \quad \textit{and} \quad L^2_{[a,\infty)}(A)=V_+\oplus W_+ \tag{8}$$

such that

(i) W_- and W_+ are finite-dimensional and contained in $H^{1/2}(\partial M,E)$;
(ii) there is a bounded linear map $g:V_-\to V_+$ with

$$g(V_-^{1/2})\subset V_+^{1/2} \quad \textit{and} \quad g^*(V_+^{1/2})\subset V_-^{1/2}$$

such that

$$B=W_+\oplus\{v+gv\mid v\in V_-^{1/2}\}. \tag{9}$$

ADDENDUM 1.13. *Assume the Standard Setup 1.5 and that D and D^* are complete. Let B be an elliptic boundary condition as in Theorem 1.12. Then*

$$B^{\mathrm{ad}}=(\sigma_0^{-1})^*\left(W_-\oplus\{v-g^*v\mid v\in V_+^{1/2}\}\right). \tag{10}$$

REMARK 1.14. For boundary conditions B and B^{ad} as in Theorem 1.12 and Addendum 1.13, $\sigma_0^*(B^{\mathrm{ad}})$ is the L^2-orthogonal complement of B in $H^{1/2}(\partial M,E)$, by the orthogonality of the decomposition (8), (9), and (10).

THEOREM 1.15. *Assume the Standard Setup 1.5 and that D and D^* are complete. Let B be an elliptic boundary condition as in Theorem 1.12 and $m\geq 0$ be an integer. Then B is m-regular if and only if*

(i) $W_\pm\subset H^{m+1/2}(\partial M,E)$;
(ii) $g(V_-^{m+1/2})\subset V_+^{m+1/2}$ and $g^*(V_+^{m+1/2})\subset V_-^{m+1/2}$.

EXAMPLES 1.16. (a) The boundary condition

$$B_{\mathrm{APS}}:=H^{1/2}_{(-\infty,0)}(A) \tag{11}$$

of Atiyah, Patodi, and Singer is ∞-regular. More generally, given any $a \in \mathbb{R}$,

$$B(a) = H^{1/2}_{(-\infty,a)}(A) \tag{12}$$

is an ∞-regular boundary condition with

$$B(a)^{\mathrm{ad}} = (\sigma_0^{-1})^* \left(H^{1/2}_{[a,\infty)}(A)\right). \tag{13}$$

Boundary conditions of this type are called *generalized Atiyah-Patodi-Singer boundary conditions*. From the point of view of Theorems 1.12, they are the most basic examples of elliptic or ∞-regular boundary conditions.

(b) Elliptic boundary value problems in the classical sense of Lopatinsky and Shapiro are ∞-regular, see Corollary 7.22.

(c) *Chiral boundary conditions*: Let G be a chirality operator of E along ∂M, that is, G is a field of unitary involutions of E along ∂M which anti-commutes with all $\sigma_A(\xi)$, $\xi \in T^*\partial M$. Let $E^\pm$ be the subbundle of E over ∂M which consists of the eigenbundle of G for the eigenvalue ± 1. Then the (local) boundary condition requiring that $\mathscr{R}\Phi$ be a section of $E^\pm$ is ∞-regular, see Corollary 7.23.

Suppose for example that M is Riemannian, that D is a Dirac operator in the sense of Gromov and Lawson, and that T is the interior unit normal vector field along ∂M. Then Clifford multiplication with iT defines a field G of involutions as above. This type of boundary condition goes under the name *MIT bag condition.* If M is even dimensional, there is another choice for G, namely Clifford multiplication with the complex volume form, compare [**LM**, Ch. I, § 5].

(d) *Transmission conditions*: For convenience assume $\partial M = \emptyset$. Let N be a closed and two-sided hypersurface in M. Cut M along N to obtain a manifold M' whose boundary $\partial M'$ consists of two disjoint copies N_1 and N_2 of N. There are natural pull-backs μ', E', F', and D' of μ, E, F, and D from M to M'. Assume that there is an interior vector field T along $N = N_1$ such that D' is boundary symmetric with respect to T along N_1. (If, for instance, D is of Dirac type, we may choose T to be the interior normal vector field of M' along N_1.) Then D' is also boundary symmetric with respect to the interior vector field $-T$ of M' along $N = N_2$. The *transmission condition*

$$B := \left\{(\varphi, \psi) \in H^{1/2}(N, E) \oplus H^{1/2}(N, E) \mid \varphi = \psi\right\}$$

reflects the fact that H^1_{loc}-sections of E have a well-defined trace along N. The natural pull-back to M' yields a 1-1 correspondence between H^1_{loc}-sections of E and H^1_{loc}-sections of E' with boundary values in B. This boundary condition is basic for the Fredholm theory of D in the case where M is noncompact. We will see in Example 7.28 that B is an ∞-regular boundary condition for D'.

DEFINITION 1.17. We say that D is *coercive at infinity* if there is a compact subset $K \subset M$ and a constant C such that

$$\|\Phi\|_{L^2(M)} \leq C\|D\Phi\|_{L^2(M)}, \tag{14}$$

for all smooth sections Φ of E with compact support in $M \backslash K$.

As for completeness, coercivity at infinity is automatic if M is compact.

THEOREM 1.18. *Assume the Standard Setup 1.5 and that D and D^* are complete. Let $B \subset \check{H}(A)$ be an elliptic boundary condition.*

Then D is coercive at infinity if and only if $D_{B,\max}$ has finite-dimensional kernel and closed image. In this case

$$\begin{aligned} \operatorname{ind} D_{B,\max} &= \dim\ker D_{B,\max} - \dim\ker D^*_{B^{\mathrm{ad}},\max} \\ &= \operatorname{ind} D_{B(a)} + \dim W_+ - \dim W_- \in \mathbb{Z} \cup \{-\infty\}, \end{aligned}$$

where we choose the representation of B as in Theorem 1.12.

In particular, $D_{B,\max}$ is a Fredholm operator if and only if D and D^* are coercive at infinity.

Given an elliptic or regular boundary condition B as in Theorem 1.12 above, we obtain a continuous one-parameter family B_s, $1 \geq s \geq 0$, of elliptic or regular boundary conditions by substituting sg for g. Then $B_1 = B$ and $B_0 = W_+ \oplus V_-^{1/2}$. The proof of the second index formula above relies on the constancy of $\operatorname{ind} D_{B_t,\max}$ under this deformation, given that D is coercive at infinity. It is clear that, in $H^{1/2}_{(-\infty,a)}(A) \oplus H^{1/2}_{[a,\infty)}(A)$, any sum $W_- \oplus W_+$ of subspaces W_- and W_- of dimensions $k_- := \dim W_- < \infty$ and $k_+ := \dim W_+ < \infty$ can be deformed into any other such sum as long as the latter has the same pair of dimensions k_- and k_+. We conclude that for fixed a, the pair (k_-, k_+) of dimensions is a complete invariant† for the homotopy classes of elliptic boundary conditions, and similarly for regular boundary conditions.

Suppose now for convenience that $\partial M = \emptyset$ is empty, and let $M = M' \cup M''$ be a decomposition of M into two pieces with common boundary $N = \partial M' = \partial M''$, a compact hypersurface. Let E and F be Hermitian vector bundles over M and D a first-order elliptic differential operator from E to F. Denote by E', F', and D' the restrictions of E, F, and D to M', and analogously for M''. Assume that D' is boundary symmetric with respect to an interior normal field T of M' along N. Then D'' is boundary symmetric with respect to $-T$. Choose an adapted operator A for D' as in the Standard Setup 1.5. Then $-A$ is an adapted operator for D'' as in the Standard Setup 1.5 and will be our preferred choice. With respect to this choice, the boundary condition of Atiyah, Patodi, and Singer for D'' is given by $H^{1/2}_{(-\infty,0)}(-A) = H^{1/2}_{(0,\infty)}(A)$.

†The pair (k_-, k_+) depends on the choice of a, though.

THEOREM 1.19 (Decomposition Theorem). *Let $M = M' \cup M''$ and notation be as above. Assume that D' is boundary symmetric with respect to an interior normal field T of M' along N and choose an adapted operator A as above.*

Then D and D^* are complete and coercive at infinity if and only if D' and D'' and their formal adjoints are complete and coercive at infinity. In this case, $D_{\max}$, $D'_{B',\max}$, and $D_{B'',\max}$ are Fredholm operators and their indices satisfy

$$\operatorname{ind} D_{\max} = \operatorname{ind} D'_{B',\max} + \operatorname{ind} D''_{B'',\max},$$

where $B' = B(a) = H^{1/2}_{(-\infty,a)}(A)$ and $B'' = H^{1/2}_{[a,\infty)}(A)$ or, more generally, where B' is elliptic and B'' is the L^2-orthogonal complement of B' in $H^{1/2}(A)$.

DEFINITION 1.20. For $i = 1, 2$, let M_i be manifolds, E_i and F_i Hermitian vector bundles over M_i, and $D_i : C^\infty(M_i, E_i) \to C^\infty(M_i, F_i)$ be differential operators. Let $K_i \subset M_i$ be closed subsets. Then we say that D_1 *outside* K_1 *agrees with* D_2 *outside* K_2 if there are a diffeomorphism $f : M_1 \setminus K_1 \to M_2 \setminus K_2$ and smooth fiberwise linear isometries $\mathscr{I}_E : E_1|_{M_1\setminus K_1} \to E_2|_{M_2\setminus K_2}$ and $\mathscr{I}_F : F_1|_{M_1\setminus K_1} \to F_2|_{M_2\setminus K_2}$ such that

$$\begin{array}{ccc} E_1|_{M_1\setminus K_1} & \xrightarrow{\mathscr{I}_E} & E_2|_{M_2\setminus K_2} \\ \downarrow & & \downarrow \\ M_1\setminus K_1 & \xrightarrow{f} & M_2\setminus K_2 \end{array} \qquad \begin{array}{ccc} F_1|_{M_1\setminus K_1} & \xrightarrow{\mathscr{I}_F} & F_2|_{M_2\setminus K_2} \\ \downarrow & & \downarrow \\ M_1\setminus K_1 & \xrightarrow{f} & M_2\setminus K_2 \end{array}$$

commute and

$$\mathscr{I}_F \circ (D_1\Phi) \circ f^{-1} = D_2(\mathscr{I}_E \circ \Phi \circ f^{-1}) \tag{15}$$

for all smooth sections Φ of E_1 over $M_1 \setminus K_1$.

If M_i are Riemannian and D_i are Dirac type operators, then f is necessarily an isometry because the principal symbol of a Dirac type operator determines the Riemannian metric on the underlying manifold via the Clifford relations (31) and (32).

To each Dirac type operator D over M there is an associated 1-density α_D on M, the *index density*, see [**BGV**, Ch. 4]. At any point $x \in M$ the value of $\alpha_D(x)$ can be computed in local coordinates from the coefficients of D and their derivatives at x.

We are now ready to state a general version of Gromov and Lawson's relative index theorem:

THEOREM 1.21 (Relative Index Theorem). *Let M_1 and M_2 be complete Riemannian manifolds without boundary. Let $D_i : C^\infty(M_i, E_i) \to C^\infty(M_i, F_i)$ be Dirac type operators which agree outside compact subsets $K_1 \subset M_1$ and $K_2 \subset M_2$.*

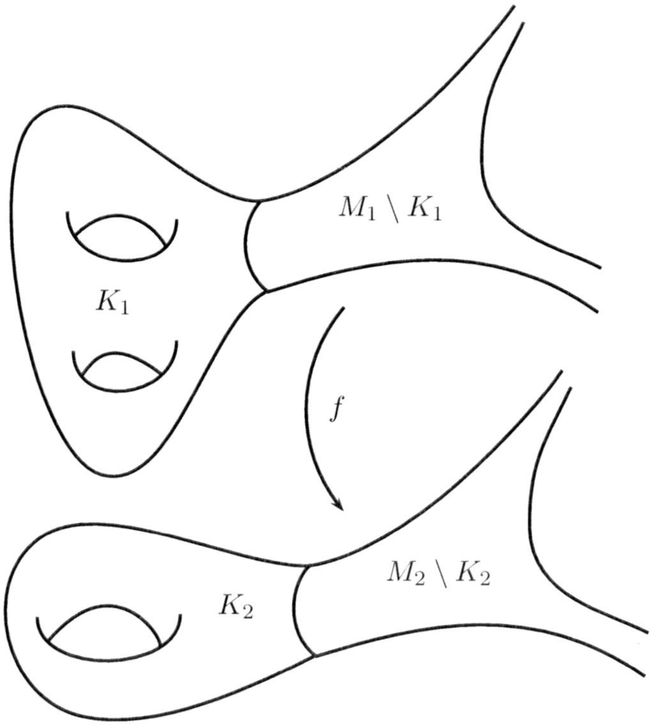

FIGURE 1

Then $D_{1,\max}$ is a Fredholm operator if and only if $D_{2,\max}$ is a Fredholm operator. In this case,

$$\operatorname{ind} D_{1,\max} - \operatorname{ind} D_{2,\max} = \int_{K_1} \alpha_{D_1} - \int_{K_2} \alpha_{D_2}.$$

The reason for the restriction to Dirac type operators in Theorem 1.21 is mostly for convenience. We need that the operators are boundary symmetric along an auxiliary hypersurface $N_i \subset M_i$. This is automatic for Dirac type operators. Moreover, we need the local version of the Atiyah-Singer index theorem.

Finally, we show

THEOREM 1.22 (Cobordism Theorem). *Let M be a complete and connected Riemannian manifold, $E \to M$ be a Hermitian vector bundle, and $D : C^\infty(M,E) \to C^\infty(M,E)$ be a formally selfadjoint differential operator of Dirac type. Let A be an adapted boundary operator for D with respect to the interior unit normal vector field, and let $E = E^+ \oplus E^-$ be the splitting into the eigensubbundles of the involution $i\sigma_0$ for the eigenvalues ± 1. With* *respect to this splitting, we write*

$$A = \begin{pmatrix} A_{++} & A^- \\ A^+ & A_{--} \end{pmatrix}.$$

Then, if D is coercive at infinity,

$$\operatorname{ind} A^+ = \operatorname{ind} A^- = 0.$$

Originally, the cobordism theorem was formulated for compact manifolds M with boundary and showed the cobordism invariance of the index. This played an important role in the original proof of the Atiyah-Singer index theorem, compare e.g. [**Pa**, Ch. XVII] and [**BW**, Ch. 21]. In this case, one can also derive the cobordism invariance from Roe's index theorem for partitioned manifolds [**R, Hi**]. We replace compactness of the bordism by the weaker assumption of coercivity of D. Our proof is comparatively simple and makes no use of the Calderón projector.

Bibliographical Notes: Theorem 1.2 extends results of Wolf [**Wo**] and Chernoff [**Ch**] to operators which are not necessarily essentially self-adjoint and live on manifolds with possibly nonempty boundary. In the case where σ_D is uniformly bounded, our proof consists of an adaptation of the argument in the proof of Thm. II.5.7 in [**LM**]. The seemingly weaker assumption of Chernoff on the growth of σ_D as stated in 1.2 follows from an appropriate conformal change of the underlying Riemannian metric of the manifold, see Section 3.

Theorem 1.7 extends Propositions 2.30 and 5.7 of [**BBC**] (where the higher regularity part $k > 0$ is not discussed). The higher regularity parts of Theorem 1.7 and Theorem 1.15 generalize the Regularity Theorem 6.5 of [**BL2**]. Our proof is rather elementary. It is worth mentioning that one of the main points in [**BBC**] is the low regularity of the given data, whereas we assume throughout that the data are smooth.

In the articles [**BF1**], [**BF2**], and [**BFO**] of Booss-Bavnbek et al., the space $\operatorname{dom} D_{\max}/\operatorname{dom} D_{\min}$ of abstract boundary values of $D_{\max}$ is investigated and identified with the space $\check{H}(A)$ as in (3), see Proposition 7.15 of [**BF2**]. Conversely, it is immediate from Theorem 1.7 and Corollary 6.6 that $\check{H}(A)$ is topologically isomorphic to $\operatorname{dom} D_{\max}/\operatorname{dom} D_{\min}$.

A preliminary and unpublished version of our Theorem 1.12 was taken up in [**BBC**], where it is proved for Dirac operators in the sense of Gromov and Lawson. Conversely, the presentation and results in [**BBC**] have influenced our discussion. In particular, in the text we interchange definition and characterization of elliptic boundary conditions, compare Subsection 7.3. In contrast to [**BBC**], our arguments do not involve the Calderón projection and its regularity properties.

Similar, but more special boundary conditions have been considered in [**BC**]. The emphasis there lies on Dirac type operators with coefficients of low regularity. Interior and boundary regularity of solutions and certain Fredholm properties are derived.

The first part of Theorem 1.18 generalizes a result of Anghel [**An**, Thm. 2.1] to manifolds with boundary and operators which are not necessarily essentially selfadjoint. Variants of the index formulas in the second part of Theorem 1.18 are also contained in [**BW**] and [**BBC**]. Theorem 1.21 corresponds to Theorem 3.1 of [**Ca**]. At the cost of introducing a bit more of (elementary) functional analysis, we could replace our assumption on the Fredholmness by the weaker assumption of non-parabolicity used in [**Ca**].

There is a vast literature on elliptic boundary value problems based on pseudo-differential techniques, see e.g. [**B, BLZ, G2, RS, S1, S2**] and the references therein. Not all of the boundary conditions which we consider can be treated with pseudo-differential operators. Our main objective, however, is to provide a more elementary and simplified approach using standard functional analysis only. A further reference for an elementary approach to boundary value problems of Dirac type operators is [**FS**]; however, the proof of the main result in [**FS**], Theorem 4, is not complete.

PREREQUISITES: The reader should be familiar with basic differential geometric concepts such as manifolds and vector bundles. The functional analysis of Hilbert and Banach spaces, selfadjoint operators and their spectrum is assumed. From the field of partial differential equations, we only require knowledge about linear differential operators, principal symbols, and the standard interior elliptic regularity theory. No previous knowlegde of boundary value problems is necessary.

STRUCTURE OF THE ARTICLE: Most of this is clear from the table of contents. We just give references to the places, where the results stated in the introduction are proved: Theorem 1.2 in Section 3, Theorem 1.7 in Section 6, Theorem 1.12 and Addendum 1.13 in Subsection 7.3, Theorem 1.15 in Subsection 7.4, Theorem 1.18 in Subsections 8.1 and 8.3, the decomposition and relative index theorems 1.19 and 1.21 in Subsection 8.4 and, finally, the cobordism theorem in Subsection 8.5.

2. Preliminaries

2.1. Measured manifolds. We will consider differential operators which live on manifolds with boundary (possibly empty). In general, the manifolds will not be Riemannian, but we will assume that they are equipped with a smooth *volume element*. By this we mean a nowhere-vanishing smooth one-density. If M is oriented, this is essentially the same thing as a nowhere vanishing n-form, where $n = \dim(M)$. The volume element is needed to define the integral of functions over M.

DEFINITION 2.1. A *measured manifold* is a pair consisting of a manifold M (possibly with boundary) and a smooth volume element μ on M.

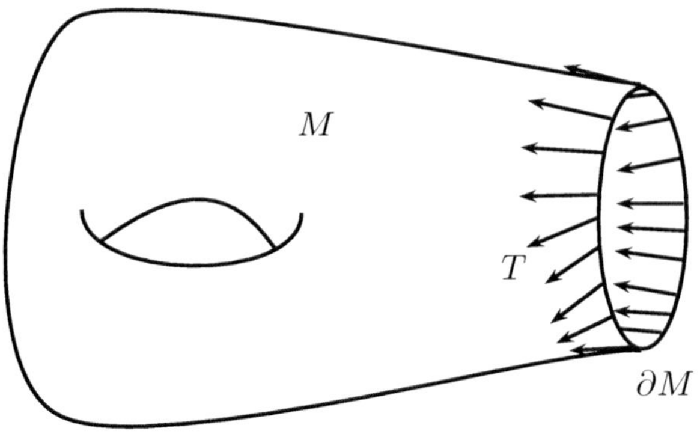

FIGURE 2

Let M be a manifold with nonempty boundary ∂M. Let T be a smooth vector field on M along ∂M pointing into the interior of M. In particular, T does not vanish anywhere.

At each $x \in \partial M$ there is a unique $\tau(x) \in T_x^* M$ such that

$$\langle \tau(x), T(x) \rangle = 1 \quad \text{and} \quad \tau(x)|_{T_x \partial M} = 0. \tag{16}$$

Here $\langle \cdot, \cdot \rangle$ denotes the natural evaluation of 1-forms on tangent vectors. We call τ the *one-form associated to* T.

If μ is a smooth volume element on M, then ∂M inherits a smooth volume element ν such that along ∂M we have $\mu = |\tau| \otimes \nu$, i.e.,

$$\nu(X_1, \dots, X_{n-1}) = \mu(T, X_1, \dots, X_{n-1}) \tag{17}$$

for all vector fields $X_1, \dots, X_{n-1}$ on ∂M. This turns ∂M into a measured manifold.

REMARK 2.2. If M is a Riemannian manifold with boundary, then the Riemannian volume element μ turns M canonically into a measured manifold. The natural choice for T is the interior unit normal vector field along ∂M. The induced volume element ν on ∂M then coincides with the volume element of the induced Riemannian metric on ∂M.

Conversely, given a smooth volume element μ on M and an interior vector field T along ∂M, one can always find a Riemannian metric on M inducing μ and T in this way.

NOTATION 2.3. Throughout this article we will write

$$Z_I := I \times \partial M$$

where $I \subset \mathbb{R}$ is any interval. We think of Z_I as a cylinder over ∂M.

LEMMA 2.4. *Let (M, μ) be a measured manifold with compact boundary and let τ be the one-form associated to an interior vector field T.*

Then there is a neighborhood U about ∂M in M, a constant $r > 0$, and a diffeomorphism $\Psi = (t, \psi) : U \to Z_{[0,r)}$ such that

(i) $\partial M = t^{-1}(0)$,
(ii) $\psi|_{\partial M} = \mathrm{id}_{\partial M}$,
(iii) $d\Psi(T) = \partial/\partial t$ *along* ∂M,
(iv) $\tau = dt$ *along* ∂M,
(v) $\Psi_*(\mu) = |dt| \otimes \nu$ *on* $Z_{[0,r)}$.

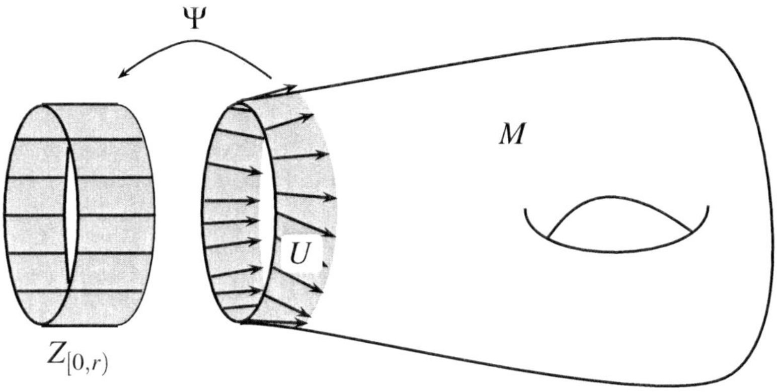

FIGURE 3

PROOF. Extend T to a smooth vector field T_1 without zeros in a neighborhood of ∂M in M. Solve for a smooth real function f on a possibly smaller neighborhood such that

$$0 = \mathrm{div}(f \cdot T_1) = df(T_1) + f \cdot \mathrm{div}(T_1) \tag{18}$$

with $f_{|\partial M} = 1$. Note that the divergence is defined because M carries a smooth volume element. Since (18) is an ordinary differential equation along the integral curves of T_1, there is a unique solution f.

The vector field fT_1 is a smooth extension of T, which we denote again by T. Let Φ be the flow of T. Since ∂M is compact, there is a neighborhood U of ∂M in M and an $r > 0$ such that

$$Z_{[0,r)} \to U, \quad (t, x) \mapsto \Phi_t(x),$$

is a diffeomorphism. Let Ψ be the inverse of this diffeomorphism. Properties (i), (ii), and (iii) are clear by construction and they imply (iv). Since T is divergence free, its flow preserves μ. This shows (v). □

DEFINITION 2.5. A diffeomorphism as in Lemma 2.4 will be called *adapted to* (M, μ) and T.

We will often identify a neighborhood U of the boundary with the cylinder $Z_{[0,r)}$ using an adapted diffeomorphism. Property (v) is not really necessary in our reasoning below, but it simplifies the exposition.

2.2. Vector bundles and differential operators. Let (M, μ) be a measured manifold (possibly with boundary) and $E \to M$ be a Hermitian vector bundle over M.

The space of smooth sections of E is denoted by $C^\infty(M, E)$. Here smoothness means smoothness up to the boundary, i.e., in local coordinates all derivatives have continuous extensions to the boundary. Smooth sections with compact support form the space $C_c^\infty(M, E)$. Note that the support of $\Phi \in C_c^\infty(M, E)$ may intersect the boundary ∂M. The space of smooth sections with compact support contained in the interior of M is denoted by $C_{cc}^\infty(M, E)$. Obviously, we have

$$C_{cc}^\infty(M, E) \subset C_c^\infty(M, E) \subset C^\infty(M, E).$$

The L^2-scalar product of $\Phi, \Psi \in C_c^\infty(M, E)$ is defined by

$$(\Phi, \Psi)_{L^2(M)} := \int_M \langle \Phi, \Psi \rangle \, d\mu,$$

the corresponding norm by

$$\|\Phi\|^2_{L^2(M)} = (\Phi, \Phi)_{L^2(M)} = \int_M |\Phi|^2 d\mu.$$

Here $\langle \cdot, \cdot \rangle$ and $|\cdot|$ denote the Hermitian product and norm of E. The completion of $C_c^\infty(M, E)$ with respect to $\|\cdot\|_{L^2(M)}$ is denoted by $L^2(M, E)$. Its elements are called *square integrable* sections. An alternative, but equivalent, definition of $L^2(M, E)$ would be the space of all measurable sections with finite L^2-norm modulo sections vanishing almost everywhere. Measurable sections whose restrictions to compact subsets have finite L^2-norm are called *locally square integrable*. The space of locally square integrable sections modulo sections vanishing almost everywhere is denoted by $L^2_{\mathrm{loc}}(M, E)$. If M is compact, then, of course, $L^2(M, E) = L^2_{\mathrm{loc}}(M, E)$. The L^2-norm and hence $L^2(M, E)$ depend on the volume element μ, while $L^2_{\mathrm{loc}}(M, E)$ does not because, over compact subsets of M, any two smooth volume elements on M can be bounded by each other.

Let $E \to M$ and $F \to M$ be Hermitian vector bundles over M and D be a (linear) differential operator from E to F. Associated with D, there is a unique differential operator

$$D^* : C^\infty(M, F) \to C^\infty(M, E),$$

called the *formal adjoint* of D, such that

$$\int_M \langle D\Phi, \Psi \rangle \, d\mu = \int_M \langle \Phi, D^*\Psi \rangle \, d\mu \tag{19}$$

for all sections $\Phi \in C^\infty_{cc}(M,E)$ and $\Psi \in C^\infty_{cc}(M,F)$. Locally, D^* is obtained from D by integration by parts. Clearly, we have $D^{**} = D$.

Let D_c and D_{cc} be the operator D, considered as an unbounded operator on the Hilbert space $L^2(M,E)$ with domain $\mathrm{dom}(D_c) = C^\infty_c(M,E)$ and $\mathrm{dom}(D_{cc}) = C^\infty_{cc}(M,E)$, respectively, and similarly for D^*. Note that $\mathrm{dom}(D_c)$ and $\mathrm{dom}(D_{cc})$ are dense in $L^2(M,E)$ and that $\mathrm{dom}(D_{cc})$ is contained in $\mathrm{dom}(D_c)$. We write $D_{cc} \subset D_c$ to express the latter fact.

Suppose that, as additional data, we are given a Riemannian metric g on M and a Hermitian connection ∇ on E. For any $\Phi \in C^\infty(M,E)$ and integer $k \geq 0$, we then get the kth covariant derivative

$$\nabla^k \Phi \in C^\infty(M, \underbrace{T^*M \otimes \cdots \otimes T^*M}_{k \text{ times}} \otimes E),$$

where both, the Levi-Civita connection of g and ∇, are used in the definition of the higher covariant derivatives of Φ. Together with g, the Hermitian metric on E induces a metric on $T^*M \otimes \cdots \otimes T^*M \otimes E$ so that we obtain the formal adjoint $(\nabla^k)^*$ of the differential operator ∇^k.

For a section $\Phi \in L^2_{\mathrm{loc}}(M,E)$, we call $\Psi \in L^2_{\mathrm{loc}}(M, T^*M \otimes \cdots \otimes T^*M \otimes E)$ the *k-th weak covariant derivative* of Φ if

$$(\Psi, \Xi)_{L^2(M)} = (\Phi, (\nabla^k)^* \Xi)_{L^2(M)}$$

for all $\Xi \in C^\infty_{cc}(M, T^*M \otimes \cdots \otimes T^*M \otimes E)$. If Ψ exists, it is uniquely determined and we write $\Psi = \nabla^k \Phi$ (instead of $\Psi = \nabla^k_{\max} \Phi$). The space of $\Phi \in L^2_{\mathrm{loc}}(M,E)$, whose weak covariant derivatives up to order k exist in $L^2_{\mathrm{loc}}(M,E)$, is denoted by $H^k_{\mathrm{loc}}(M,E)$. We have the inclusions

$$C^\infty(M,E) \subset H^k_{\mathrm{loc}}(M,E) \subset H^{k-1}_{\mathrm{loc}}(M,E) \subset \cdots \subset H^0_{\mathrm{loc}}(M,E) = L^2_{\mathrm{loc}}(M,E)$$

and, by the Sobolev embedding theorem,

$$\bigcap_{k=0}^{\infty} H^k_{\mathrm{loc}}(M,E) = C^\infty(M,E).$$

The space of $\Phi \in L^2(M,E)$, whose weak covariant derivatives up to order k exist in $L^2(M,E)$, is denoted by $H^k(M,E)$. It is a Hilbert space with respect to the *Sobolev norm* $\|\cdot\|_{H^k(M)}$ defined by

$$\|\Phi\|^2_{H^k(M)} = \|\Phi\|^2_{L^2(M)} + \|\nabla\Phi\|^2_{L^2(M)} + \cdots + \|\nabla^k\Phi\|^2_{L^2(M)}. \tag{20}$$

If M is noncompact, then $H^k(M,E)$ depends on the choice of g and ∇ (given the smooth volume element μ and the Hermitian vector bundle E). If M is compact, then any two H^k-norms (for the same k) are equivalent so that $H^k(M,E)$ is independent of these choices.

For M compact (possibly with boundary), the *Rellich embedding theorem* [**Au**, Theorem 2.34, p. 55] says in particular that the embedding

$$H^{k+1}(M,E) \hookrightarrow H^k(M,E)$$

is compact; in other words, bounded sequences in $H^{k+1}(M,E)$ subconverge in $H^k(M,E)$.

We collect some of the spaces introduced so far in the following table:

TABLE 1

space of	notation
smooth sections	$C^\infty(M,E)$
smooth sections with compact support	$C_c^\infty(M,E)$
smooth sections with compact support contained in the interior of M	$C_{cc}^\infty(M,E)$
square integrable sections	$L^2(M,E)$
locally square integrable sections	$L^2_{\mathrm{loc}}(M,E)$
sections in $L^2_{\mathrm{loc}}(M,E)$ with first k weak derivatives in $L^2_{\mathrm{loc}}(M,E)$	$H^k_{\mathrm{loc}}(M,E)$

We can restrict the bundle E to ∂M and consider the corresponding spaces such as $C^\infty(\partial M,E)$, $L^2(\partial M,E)$ etc. Further spaces of sections will be defined as needed.

Suppose from now on that D is a differential operator from E to F of order one and denote the *principal symbol* of D by σ_D. For any $x \in M$, $\sigma_D(x) : T_x^*M \to \mathrm{Hom}(E_x,F_x)$ is a linear map which is characterized by the property that

$$D(f\Phi) = fD\Phi + \sigma_D(df)\Phi, \tag{21}$$

for all $f \in C^\infty(M)$ and $\Phi \in C^\infty(M,E)$. For all $\xi \in T^*M$, we have†

$$\sigma_{D^*}(\xi) = -\sigma_D(\xi)^*. \tag{22}$$

Equation (19) holds if the supports of Φ and Ψ are compact and contained in the interior of M. In case they meet the boundary, there is an additional boundary term involving the principal symbol of D:

LEMMA 2.6 (Green's formula). *Let (M,μ) be a measured manifold with boundary and let τ be the one-form associated to an interior vector field. Then*

$$\int_M \langle D\Phi,\Psi\rangle\, d\mu - \int_M \langle \Phi, D^*\Psi\rangle\, d\mu = -\int_{\partial M} \langle \sigma_D(\tau)\Phi,\Psi\rangle\, d\nu$$

for all $\Phi \in C^\infty(M,E)$ and $\Psi \in C^\infty(M,F)$ such that $\mathrm{supp}(\Phi)\cap\mathrm{supp}(\Psi)$ *is compact.*

PROOF. Choose a Riemannian metric on M inducing μ and T as in Remark 2.2. Now the lemma follows from the standard Green's formula for Riemannian manifolds, see e.g. [**Ta**, Prop. 9.1, p. 160]. □

†In order to get rid of the sign in (22), a factor i is often included in the definition of the principal symbol.

We say that D is *elliptic* if $\sigma_D(\xi)$ is invertible for each nonzero covector ξ. It the boundary of M is empty and D is elliptic (of order one), then

$$D_{\max}\Phi \in H^k_{\mathrm{loc}}(M,F) \implies \Phi \in H^{k+1}_{\mathrm{loc}}(M,E) \tag{23}$$

for all $\Phi \in \mathrm{dom}(D_{\max})$, by *interior elliptic regularity theory*, see e.g. [**LM**, Ch. III, § 5].

3. Completeness

Let (M,μ) be a measured manifold with compact boundary. Let $E, F \to M$ be Hermitian vector bundles and $D : C^\infty(M,E) \to C^\infty(M,F)$ be a differential operator of first order. We start by generalizing Equation (21) to Lipschitz functions and sections in the maximal domain of D:

LEMMA 3.1. *Let $\chi : M \to \mathbb{R}$ be a Lipschitz function with compact support and $\Phi \in \mathrm{dom}(D_{\max})$. Then $\chi\Phi \in \mathrm{dom}(D_{\max})$ and*

$$D_{\max}(\chi\Phi) = \sigma_D(d\chi)\Phi + \chi D_{\max}\Phi.$$

PROOF. Suppose $\Psi \in H^1_{\mathrm{loc}}(M,F)$ has compact support in the interior of M. Let K be a compact subset in the interior of M which contains the support of Ψ in its interior. Then there is a sequence of Ψ_j in $C^\infty_{cc}(M,F)$ with supports in K which converge to Ψ in the H^1-norm (over K). In particular, $D^*\Psi$ is well defined and $\lim_{n\to\infty} D^*\Psi_n = D^*\Psi$ with respect to the L^2-norm. We conclude that, for any $\Phi \in \mathrm{dom}(D_{\max})$,

$$(\Phi, D^*\Psi) = \lim_{n\to\infty}(\Phi, D^*\Psi_n) = \lim_{n\to\infty}(D_{\max}\Phi, \Psi_n) = (D_{\max}\Phi, \Psi).$$

Assume now that $\chi : M \to \mathbb{R}$ is a Lipschitz function with compact support, and let $\Phi \in \mathrm{dom}(D_{\max})$. For any $\Psi \in C^\infty_{cc}(M,F)$ we have $\chi\Psi \in H^1_{\mathrm{loc}}(M,F)$, the support of $\chi\Psi$ is compact and contained in the interior of M, and $D^*(\chi\Psi) = \sigma_{D^*}(d\chi)\Psi + \chi D^*\Psi$ in $L^2(M,F)$, as we see by approximating χ in the H^1-norm by smooth functions with compact support. Hence

$$\begin{aligned}(\chi\Phi, D^*\Psi)_{L^2(M)} &= (\Phi, \chi D^*\Psi)_{L^2(M)} \\ &= (\Phi, D^*(\chi\Psi) - \sigma_{D^*}(d\chi)\Psi)_{L^2(M)} \\ &= (D_{\max}\Phi, \chi\Psi)_{L^2(M)} + (\sigma_D(d\chi)\Phi, \Psi)_{L^2(M)} \\ &= (\chi D_{\max}\Phi + \sigma_D(d\chi)\Phi, \Psi)_{L^2(M)}.\end{aligned}$$ □

Recall from Definition 1.1 that D is complete if and only if the space of compactly supported sections in $\mathrm{dom}(D_{\max})$ is dense in $\mathrm{dom}(D_{\max})$.

REMARK 3.2. If $\partial M = \emptyset$ and D is elliptic, then D is complete if and only if the minimal and maximal extensions of D on $C^\infty_c(M,E) = C^\infty_{cc}(M,E)$ coincide. Namely, if the extensions coincide, then $C^\infty_{cc}(M,E)$ is dense in $\mathrm{dom}(D_{\max})$. Conversely, since D is elliptic, $\mathrm{dom}(D_{\max})$ is contained in $H^1_{\mathrm{loc}}(M,E)$, by interior elliptic regularity. Furthermore, any compactly supported H^1-section can be approximated by smooth sections with support contained in a fixed compact domain in the H^1-norm and hence in the graph norm of D.

THEOREM 3.3. *Let (M,μ) be a measured manifold with compact boundary and $D : C^\infty(M,E) \to C^\infty(M,F)$ be a differential operator of first order. Suppose that there exists a constant $C > 0$ and a complete Riemannian metric on M with respect to which*

$$|\sigma_D(\xi)| \le C\,|\xi|$$

for all $x \in M$ and $\xi \in T_x^*M$. Then D and D^* are complete.

PROOF OF THEOREM 3.3. Let $r : M \to \mathbb{R}$ be the distance function from the boundary, $r(x) = \operatorname{dist}(x, \partial M)$. Then r is a Lipschitz function with Lipschitz constant 1. Choose $\rho \in C^\infty(\mathbb{R},\mathbb{R})$ so that $0 \le \rho \le 1$, $\rho(t) = 0$ for $t \ge 2$, $\rho(t) = 1$ for $t \le 1$, and $|\rho'| \le 2$. Set

$$\chi_m(x) := \rho\left(\frac{r(x)}{m}\right).$$

Then χ_m is a Lipschitz function and we have almost everywhere

$$|d\chi_m(x)| \le \frac{2}{m}.$$

Moreover, $\{\chi_m\}_m$ is a uniformly bounded sequence of functions converging pointwise to 1.

Now let $\Phi \in \operatorname{dom}(D_{\max})$. Then $\|\chi_m\Phi - \Phi\|_{L^2(M)} \to 0$ as $m \to \infty$ by Lebesgue's theorem. Furthermore, $\chi_m\Phi$ has compact support and $\chi_m\Phi \in \operatorname{dom}(D_{\max})$ by Lemma 3.1. Since

$$\begin{aligned}
&\|D_{\max}(\chi_m\Phi) - D_{\max}\Phi\|_{L^2(M)} \\
&\quad \le \|(1-\chi_m)D_{\max}\Phi\|_{L^2(M)} + \|\sigma_D(d\chi_m)\Phi\|_{L^2(M)} \\
&\quad = \|(1-\chi_m)D_{\max}\Phi\|_{L^2(M)} + \frac{2C}{m}\,\|\Phi\|_{L^2(M)} \to 0
\end{aligned}$$

as $m \to \infty$, we conclude that $\chi_m\Phi \to \Phi$ in the graph norm of $D_{\max}$.

The same discussion applies to D^* because $|\sigma_{D^*}(\xi)| = |-\sigma_D(\xi)^*| = |\sigma_D(\xi)|$, and the theorem is proved. □

EXAMPLE 3.4. If D is of Dirac type with respect to a complete Riemannian metric, see Example 4.3 (a), then D is complete. Namely, by the Clifford relations (31) we have $\sigma_D(\xi)^*\sigma_D(\xi) = |\xi|^2 \cdot \mathrm{id}$ and hence $|\sigma_D(\xi)| = |\xi|$.

The assumption in Theorem 3.3 that the principal symbol of D is uniformly bounded can be weakened to the condition considered by Chernoff in [**Ch**, Theorems 1.3 and 2.2]:

PROOF OF THEOREM 1.2. Choose a smooth function $f: M \to \mathbb{R}$ with

$$C(\operatorname{dist}(x, \partial M)) \le f(x) \le 2\,C(\operatorname{dist}(x, \partial M))$$

for all $x \in M$. Let g denote the complete Riemannian metric as in the assumptions of Theorem 1.2. Then $h := f^{-2}g$ is also complete because the h-length of a curve $c\colon [0, \infty) \to M$, starting in ∂M and parametrized by arc-length with respect to g, is estimated by

$$\begin{aligned} L_h(c) &= \int_0^\infty \frac{|c'(t)|_g}{f(c(t))}\,dt \\ &\ge \frac{1}{2}\int_0^\infty \frac{|c'(t)|_g}{C(\operatorname{dist}(c(t), \partial M))}\,dt \\ &\ge \frac{1}{2}\int_0^\infty \frac{1}{C(t)}\,dt \\ &= \infty. \end{aligned}$$

With respect to h, the principal symbol σ_D is uniformly bounded as required in Theorem 3.3. □

REMARK 3.5. It will be convenient to assume further on that D is complete. However, in questions concerning boundary regularity, assuming completeness is somewhat artificial because it is a property "at infinity". But, in such questions, we can always pass to a complete differential operator on vector bundles over the cylinder $Z_{[0,\infty)}$ which coincides with the given operator in a neighborhood of the boundary. In this sense, assuming completeness causes no loss of generality when studying the operator near the boundary.

4. Normal form

Throughout this section, let (M, μ) be a given measured manifold with compact boundary. Let T be an interior vector field along ∂M and τ be the associated one-form. Identify a neighborhood U of the boundary ∂M with a cylinder $Z_{[0,r)}$ via an adapted diffeomorphism as in Lemma 2.4.

Let $E, F \to M$ be Hermitian vector bundles. Identify the restrictions of E and F to $Z_{[0,r)}$ as Hermitian vector bundles with the pull-back of their restriction to ∂M with respect to the canonical projection onto ∂M along the family of t-lines (t, x), $0 \le t < r$ and $x \in \partial M$. This can be achieved by using parallel transport along the t-lines with respect to Hermitian connections on E and F. Then sections of E and F over $Z_{[0,r)}$ can be viewed as t-dependent sections of E and F over ∂M. Using this identification we have, by (v) of Lemma 2.4,

$$\int_{Z_{[0,r)}} |\Phi(p)|^2 d\mu(p) = \int_0^r \int_{\partial M} |\Phi(t,x)|^2 d\nu(x)\,dt$$

for any $\Phi \in L^2(Z_{[0,r)}, E)$ or $\Phi \in L^2(Z_{[0,r)}, F)$.

Let $D: C^\infty(M,E) \to C^\infty(M,F)$ be an elliptic differential operator of first order and set

$$\sigma := \sigma_D(dt). \tag{24}$$

By (iv) of Lemma 2.4, we have $\sigma_{(0,x)} = \sigma_D(\tau(x))$ for each $x \in \partial M$. Since D is elliptic, $\sigma_{(t,x)} : E_x \to F_x$ is an isomorphism for each $(t,x) \in Z_{[0,r)}$. We usually suppress the x-dependence in the notation and write σ_t instead of $\sigma_{(t,x)}$.

The main point about boundary symmetric operators as in Definition 1.4 is that, in coordinates adapted to (M,μ) and T as in Lemma 2.4, D and D^* admit good normal forms near the boundary:

LEMMA 4.1 (Normal form). *Let $D: C^\infty(M,E) \to C^\infty(M,F)$ be an elliptic differential operator of first order. If D is boundary symmetric, then there are formally selfadjoint elliptic differential operators*

$$A: C^\infty(\partial M,E) \to C^\infty(\partial M,E) \quad \text{and} \quad \tilde A: C^\infty(\partial M,F) \to C^\infty(\partial M,F)$$

such that, over $Z_{[0,r)}$,

$$D = \sigma_t\left(\frac{\partial}{\partial t} + A + R_t\right), \tag{25}$$

$$D^* = -\sigma_t^*\left(\frac{\partial}{\partial t} + \tilde A + \tilde R_t\right). \tag{26}$$

The remainders

$$R_t: C^\infty(\partial M,E) \to C^\infty(\partial M,E) \quad \text{and} \quad \tilde R_t: C^\infty(\partial M,F) \to C^\infty(\partial M,F)$$

are families of differential operators of order at most one whose coefficients depend smoothly on $t \in [0,r)$. They satisfy an estimate

$$\|R_t\Phi\|_{L^2(\partial M)} \le C\left(t\|A\Phi\|_{L^2(\partial M)} + \|\Phi\|_{L^2(\partial M)}\right), \tag{27}$$

$$\|\tilde R_t\Psi\|_{L^2(\partial M)} \le C\left(t\|\tilde A\Psi\|_{L^2(\partial M)} + \|\Psi\|_{L^2(\partial M)}\right), \tag{28}$$

for all $\Phi \in C^\infty(\partial M,E)$ and $\Psi \in C^\infty(\partial M,F)$.

PROOF. For $x \in \partial M$, identify $T_x^*\partial M$ with the subspace of $\xi \in T_x^*M$ such that $\xi(T) = 0$. In this sense, the principal symbol of the desired operator A is given by

$$\sigma_A(\xi) = \sigma_0(x)^{-1} \circ \sigma_D(\xi), \tag{29}$$

by (25), (27), and the definition of σ. Since D is boundary symmetric, $\sigma_0(x)^{-1} \circ \sigma_D(\xi)$ is skew-Hermitian for all $x \in \partial M$ and $\xi \in T_x^*\partial M$. Hence we can choose a formally selfadjoint differential operator $A: C^\infty(\partial M,E) \to C^\infty(\partial M,E)$ of order one with principal symbol as required by (29). Since the principal symbol is composed of invertible symbols, A is elliptic.

Over $Z_{[0,r)}$, we have

$$D = \sigma_t \left(\frac{\partial}{\partial t} + \mathscr{D}_t \right),$$

where $\mathscr{D}_t : C^\infty(\partial M, E) \to C^\infty(\partial M, E)$ is a family of elliptic differential operators of order one whose coefficients depend smoothly on $t \in [0, r)$. Hence

$$R_t := \mathscr{D}_t - A$$

is a family of differential operators of order at most one whose coefficients depend smoothly on t. Since $\mathscr{D}_0$ and A have the same principal symbol, R_0 is of order 0. Since ∂M is a closed manifold, we conclude that

$$\|R_t \Phi\|_{L^2(\partial M)} \leq C' \big(t \|\Phi\|_{H^1(\partial M)} + \|\Phi\|_{L^2(\partial M)} \big)$$

for some constant C'. Now A is elliptic of order one. Hence, by standard elliptic estimates, the H^1-norm is bounded by the graph norm of A; that is, we have an estimate as claimed in (27). This finishes the proof of the assertions concerning D.

By (22), the principal symbol of the desired operator $\tilde{A}$ is given by

$$\begin{aligned} \sigma_{\tilde{A}}(\xi) &= \sigma_{D^*}(\tau(x))^{-1} \circ \sigma_{D^*}(\xi) \\ &= (\sigma_D(\tau(x))^*)^{-1} \circ \sigma_D(\xi)^* \\ &= (\sigma_0(x)^*)^{-1} \circ \sigma_D(\xi)^* \\ &= \big(\sigma_D(\xi) \circ \sigma_0(x)^{-1}\big)^*, \end{aligned} \tag{30}$$

which is also skew-Hermitian. Thus the analogous arguments as above show the assertions concerning D^*. □

REMARKS 4.2. (a) Conversely, it is immediate that an elliptic differential operator D with a normal form as in Lemma 4.1 is boundary symmetric. In particular, D is boundary symmetric if and only if D^* is boundary symmetric. The latter is also obvious from Equation (22) (as we see from the end of the above proof).

(b) The operators A and $\tilde{A}$ in Lemma 4.1 are not unique. One can add symmetric zero-order terms to them by paying with a corresponding change of the remainder-terms R_t and $\tilde{R}_t$. Equations (27) and (28) will still be valid for the modified remainder terms.

We conclude this section with a few examples.

EXAMPLES 4.3. (a) (Dirac type operators) We say that the operator D is of *Dirac type* if M carries a Riemannian metric $\langle \cdot, \cdot \rangle$ such that the principal symbol of D satisfies the Clifford relations

$$\sigma_D(\xi)^* \sigma_D(\eta) + \sigma_D(\eta)^* \sigma_D(\xi) = 2 \langle \xi, \eta \rangle \cdot \mathrm{id}_{E_x}, \tag{31}$$

$$\sigma_D(\xi) \sigma_D(\eta)^* + \sigma_D(\eta) \sigma_D(\xi)^* = 2 \langle \xi, \eta \rangle \cdot \mathrm{id}_{F_x}, \tag{32}$$

for all $x \in M$ and $\xi, \eta \in T_x^*M$. They easily imply that D is elliptic and boundary symmetric with respect to the interior normal field of the given Riemannian metric. If D is of Dirac type, then so is D^*.

The class of Dirac type operators contains in particular Dirac operators on Dirac bundles as in [**LM**, Ch. II, § 5]. The classical Dirac operator on a spin manifold is an important special case.

(b) A somewhat artificial example of a boundary symmetric operator which is not of Dirac type can be constructed as follows. Let g and g' be two Riemannian metrics on a manifold M with spin structure. Let $D : C^\infty(M, E) \to C^\infty(M, F)$ and $D' : C^\infty(M, E') \to C^\infty(M, F')$ be the corresponding Dirac operators acting on spinors. Assume that g and g' are conformal along the boundary, i.e., $g' = fg$ for some smooth positive function f on ∂M. Let T be an interior normal vector field along ∂M, perpendicular to ∂M for g and g'. Then

$$\begin{pmatrix} D & 0 \\ 0 & D' \end{pmatrix} + V : C^\infty(M, E \oplus E') \to C^\infty(M, F \oplus F')$$

is boundary symmetric (with respect to T) but in general not of Dirac type. Here V may be an arbitrary zero-order term.

(c) More importantly, let D be a Dirac type operator. If one changes D in the interior of M in such a way that it remains an elliptic first-order operator, then D is still boundary symmetric.

5. The model operator

Throughout this section, assume the Standard Setup 1.5, identify a neighborhood U of the boundary ∂M with a cylinder $Z_{[0,r)}$ via an adapted diffeomorphism as in Lemma 2.4, and fix a normal form for D and D^* as in Lemma 4.1. Consider A as an unbounded operator in the Hilbert space $L^2(\partial M, E)$ with domain $\mathrm{dom}(A) = C^\infty(\partial M, E)$. The *model operator* associated to D and A is the operator

$$D_0 := \sigma_0 \left(\frac{\partial}{\partial t} + A \right) \tag{33}$$

on the half-infinite cylinder $Z_{[0,\infty)}$. Here $\sigma_0(x) = \sigma_D(\tau(x))$, compare Lemma 4.1. The coefficients of D_0 do not depend on t. With respect to the product measure $\mu_0 := dt \otimes \nu$ on $Z_{[0,\infty)}$, we have

$$D_0^* = -\sigma_0^* \Big(\frac{\partial}{\partial t} - (\sigma_0^*)^{-1} A \sigma_0^* \Big) \quad \text{and} \quad (\sigma_0^{-1} D_0)^* = -\Big(\frac{\partial}{\partial t} - A \Big). \tag{34}$$

We will keep the above setup and will use the abbrevation $\Phi' := \partial\Phi/\partial t$.

Lemma 5.1. *For any* $\Phi \in C_c^\infty(Z_{[0,\infty)}, E)$, *we have*

$$\|\sigma_0^{-1} D_0 \Phi\|^2_{L^2(Z_{[0,\infty)})} = \|\Phi'\|^2_{L^2(Z_{[0,\infty)})} + \|A\Phi\|^2_{L^2(Z_{[0,\infty)})} - (A\Phi_0, \Phi_0)_{L^2(\partial M)},$$

where Φ_0 *denotes the restriction of* Φ *to* $\{0\} \times \partial M$.

PROOF. We fix $t \in [0, \infty)$ and integrate over ∂M:

$$\begin{aligned}
&\|\sigma_0^{-1} D_0 \Phi\|^2_{L^2(\partial M)} \\
&\quad = (\Phi' + A\Phi, \Phi' + A\Phi)_{L^2(\partial M)} \\
&\quad = \|\Phi'\|^2_{L^2(\partial M)} + \|A\Phi\|^2_{L^2(\partial M)} + (\Phi', A\Phi)_{L^2(\partial M)} + (A\Phi, \Phi')_{L^2(\partial M)} \\
&\quad = \|\Phi'\|^2_{L^2(\partial M)} + \|A\Phi\|^2_{L^2(\partial M)} + (A\Phi', \Phi)_{L^2(\partial M)} + (A\Phi, \Phi')_{L^2(\partial M)} \\
&\quad = \|\Phi'\|^2_{L^2(\partial M)} + \|A\Phi\|^2_{L^2(\partial M)} + (A\Phi, \Phi)'_{L^2(\partial M)}.
\end{aligned}$$

Here we used that A does not depend on t and that it is formally selfadjoint. Now we integrate this identity with respect to $t \in [0, \infty)$. Since Φ vanishes for sufficiently large t, the last term gives a boundary contribution only for $t = 0$. □

LEMMA 5.2. *There exist constants $r, C > 0$ such that*

$$\|(D - D_0)\Phi\|_{L^2(Z_{[0,\rho]})} \le C(\rho \, \|D_0\Phi\|_{L^2(Z_{[0,\rho]})} + \|\Phi\|_{L^2(Z_{[0,\rho]})})$$

for all $0 < \rho < r$ and all $\Phi \in C^\infty(Z_{[0,\infty)}, E)$ with support in $Z_{[0,\rho]}$.

PROOF. We have, by Lemma 4.1,

$$D - D_0 = \sigma_t \left(\frac{\partial}{\partial t} + A + R_t \right) - \sigma_0 \left(\frac{\partial}{\partial t} + A \right) = (\sigma_t - \sigma_0)\sigma_0^{-1} D_0 + \sigma_t R_t.$$

Since $(\sigma_t - \sigma_0)\sigma_0^{-1} = \mathrm{O}(t)$ and σ is uniformly bounded, it remains to estimate $R_t\Phi$. By (27) it suffices to estimate

$$t\|A\Phi\|_{L^2(\partial M)} = \|A(t\Phi)\|_{L^2(\partial M)}.$$

Now $t\Phi$ is in $C_c^\infty(Z_{[0,\infty)}, E)$ and vanishes at $t = 0$. Therefore, by Lemma 5.1,

$$\|\sigma_0^{-1} D_0(t\Phi)\|^2_{L^2(Z_{[0,\infty)})} = \|(t\Phi)'\|^2_{L^2(Z_{[0,\infty)})} + \|A(t\Phi)\|^2_{L^2(Z_{[0,\infty)})}.$$

Hence

$$\begin{aligned}
\|A(t\Phi)\|_{L^2(\partial M)} &\le \|\sigma_0^{-1} D_0(t\Phi)\|_{L^2(Z_{[0,\infty)})} \\
&\le C' \big(\|t D_0 \Phi\|_{L^2(Z_{[0,\infty)})} + \|\Phi\|_{L^2(Z_{[0,\infty)})} \big).
\end{aligned}$$

The asserted inequality follows. □

Since ∂M is compact without boundary, the minimal and maximal extensions of the operator A coincide. Hence A is essentially selfadjoint in the Hilbert space $L^2(\partial M, E)$. For any $s \in \mathbb{R}$, the positive operator $(\mathrm{id} + A^2)^{s/2}$ is defined by functional calculus.

DEFINITION 5.3. For any $s \in \mathbb{R}$, we define the *Sobolev H^s-norm* on $C^\infty(\partial M, E)$ by

$$\|\varphi\|^2_{H^s(\partial M)} := \|(\mathrm{id} + A^2)^{s/2} \varphi\|^2_{L^2(\partial M)}.$$

We denote by $H^s(\partial M, E)$ the completion of $C^\infty(\partial M, E)$ with respect to this norm.

By standard elliptic estimates, this norm is equivalent to the Sobolev norms defined in (20) if $s \in \mathbb{N}$. It is a nice feature of Definition 5.3 that it makes sense for all $s \in \mathbb{R}$. The values $s = 1/2$ and $s = -1/2$ will be of particular importance. Let

$$-\infty \leftarrow \cdots \leq \lambda_{-2} \leq \lambda_{-1} \leq \lambda_0 \leq \lambda_1 \leq \lambda_2 \leq \cdots \rightarrow +\infty$$

be the spectrum of A with each eigenvalue being repeated according to its (finite) multiplicity, and fix a corresponding L^2-orthonormal basis φ_j, $j \in \mathbb{Z}$, of eigensections of A. Then, for $\varphi = \sum_{j=-\infty}^{\infty} a_j \varphi_j$, one has

$$\|\varphi\|^2_{H^s(\partial M)} = \sum_{j=-\infty}^{\infty} |a_j|^2 (1 + \lambda_j^2)^s.$$

FACTS 5.4. The following facts are basic in our considerations:

(i) $H^0(\partial M, E) = L^2(\partial M, E)$;

(ii) if $s < t$, then $\|\varphi\|_{H^s(\partial M)} \leq \|\varphi\|_{H^t(\partial M)}$ and, by the Rellich embedding theorem, the induced embedding $H^t(\partial M, E) \hookrightarrow H^s(\partial M, E)$ is compact;

(iii) by the Sobolev embedding theorem, $\bigcap_{s \in \mathbb{R}} H^s(\partial M, E) = C^\infty(\partial M, E)$;

(iv) for all $s \in \mathbb{R}$, the L^2-product $(\varphi, \psi) = \int_{\partial M} \langle \varphi, \psi \rangle \, d\nu$, where $\varphi, \psi \in C^\infty(\partial M, E)$, extends to a perfect pairing

$$H^s(\partial M, E) \times H^{-s}(\partial M, E) \rightarrow \mathbb{C}$$

and therefore renders $H^s(\partial M, E)$ and $H^{-s}(\partial M, E)$ as pairwise dual;

(v) for all $k \geq 1$, the restriction map $\mathscr{R} : C_c^\infty(M, E) \rightarrow C^\infty(\partial M, E)$, $\mathscr{R}(\Phi) := \Phi|_{\partial M}$, extends by the *trace theorem* [**Ad**, Thm. 5.22] to a continuous linear map

$$\mathscr{R} : H^k_{\mathrm{loc}}(M, E) \rightarrow H^{k-1/2}(\partial M, E).$$

For $I \subset \mathbb{R}$, let Q_I be the spectral projection of the selfadjoint operator A,

$$Q_I : \sum_{j=-\infty}^{\infty} a_j \varphi_j \mapsto \sum_{\lambda_j \in I} a_j \varphi_j.$$

Then Q_I is an orthogonal projection in $L^2(\partial M, E)$ and

$$Q_I(H^s(\partial M, E)) \subset H^s(\partial M, E)$$

for all $s \in \mathbb{R}$. In particular, $Q_I(C^\infty(\partial M, E)) \subset C^\infty(\partial M, E)$. We abbreviate

$$H^s_I(A) := Q_I(H^s(\partial M, E)).$$

Fix $\Lambda \in \mathbb{R}$ and define

$$\|\varphi\|^2_{\check{H}(A)} := \|Q_{(-\infty,\Lambda]}\varphi\|^2_{H^{1/2}(\partial M)} + \|Q_{(\Lambda,\infty)}\varphi\|^2_{H^{-1/2}(\partial M)}. \tag{35}$$

This norm is, up to equivalence, independent of the choice of Λ. Namely, if $\Lambda_1 < \Lambda_2$, then the corresponding $\check{H}$-norms coincide on the L^2-orthogonal complement of $Q_{[\Lambda_1,\Lambda_2]}(C^\infty(\partial M, E))$. Now, the latter space is finite-dimensional so that any two norms on it are equivalent, hence the claim.

The completion of $C^\infty(\partial M, E)$ with respect to $\|\cdot\|_{\check{H}(A)}$ will be denoted $\check{H}(A)$. In other words,

$$\check{H}(A) = H^{1/2}_{(-\infty,\Lambda]}(A) \oplus H^{-1/2}_{(\Lambda,\infty)}(A). \tag{36}$$

Similarly, we set

$$\|\varphi\|^2_{\hat{H}(A)} := \|Q_{(-\infty,\Lambda]}\varphi\|^2_{H^{-1/2}(\partial M)} + \|Q_{(\Lambda,\infty)}\varphi\|^2_{H^{1/2}(\partial M)} \tag{37}$$

and

$$\hat{H}(A) := H^{-1/2}_{(-\infty,\Lambda]}(A) \oplus H^{1/2}_{(\Lambda,\infty)}(A). \tag{38}$$

While $H^s(\partial M, E)$ is independent of A, the definitions of $H^s_I(A)$, $\check{H}(A)$, and $\hat{H}(A)$ do depend on A. We have

$$\hat{H}(A) = \check{H}(-A). \tag{39}$$

The L^2-product on $C^\infty(\partial M, E)$ uniquely extends to a perfect pairing

$$\check{H}(A) \times \hat{H}(A) \to \mathbb{C}. \tag{40}$$

Hence $\hat{H}(A)$ is canonically isomorphic to the dual space of $\check{H}(A)$ and conversely. Let

$$H^{\text{fin}}(A) := \Big\{\varphi = \sum_{j=-\infty}^{\infty} a_j\varphi_j \,\Big|\, a_j = 0 \text{ for all but finitely many } j\Big\} \tag{41}$$

be the space of "finite Fourier series". We have the inclusions

$$\begin{aligned} H^{\text{fin}}(A) &\subset C^\infty(\partial M, E) \\ &\subset H^{1/2}(\partial M, E) \subset \left\{ \begin{matrix} L^2(\partial M, E) \\ \hat{H}(A) \\ \check{H}(A) \end{matrix} \right\} \subset H^{-1/2}(\partial M, E). \end{aligned}$$

The space $H^{\text{fin}}(A)$ is dense in any of these spaces. Sections Φ in

$$L^2(Z_{[0,\infty)}, E) = L^2([0,\infty), L^2(\partial M, E))$$

can be developed in the basis $(\varphi_j)_j$ with coefficients depending on t,

$$\Phi(t,x) = \sum_{j=-\infty}^{\infty} a_j(t)\varphi_j(x).$$

We fix a constant $r > 0$ and a smooth cut-off function $\chi : \mathbb{R} \to \mathbb{R}$ with

$$\chi(t) = 1 \text{ for all } t \leq r/3 \quad \text{and} \quad \chi(t) = 0 \text{ for all } t \geq 2r/3. \tag{42}$$

We define, for $\varphi \in H^{\text{fin}}(A)$, a smooth section $\mathscr{E}\varphi$ of E over $Z_{[0,\infty)}$ by

$$(\mathscr{E}\varphi)(t) := \chi(t) \cdot \exp(-t|A|)\varphi. \tag{43}$$

In other words, for $\varphi(x) = \sum_{j=-\infty}^{\infty} a_j \varphi_j(x)$ we have

$$(\mathscr{E}\varphi)(t,x) = \chi(t) \sum_{j=-\infty}^{\infty} a_j \cdot \exp(-t|\lambda_j|) \cdot \varphi_j(x).$$

We obtain a linear map $\mathscr{E} : H^{\text{fin}}(A) \to C_c^\infty(Z_{[0,\infty)}, E)$.

LEMMA 5.5. *There is a constant $C = C(\chi, A) > 0$ such that*

$$\|\mathscr{E}\varphi\|_{D_0}^2 \leq C\|\varphi\|_{\check{H}(A)}^2$$

for all $\varphi \in H^{\text{fin}}(A)$.

PROOF. Without loss of generality we choose $\Lambda = 0$ in (36), the definition of $\check{H}(A)$. Since the eigenspaces of A are pairwise L^2-orthogonal, it suffices to consider the Fourier coefficients of $\mathscr{E}\varphi$ separately. We see that

$$D_0 \exp(-t|A|)Q_{(0,\infty)}\varphi = \sigma_0 \left(\frac{\partial}{\partial t} + A\right) \exp(-tA)Q_{(0,\infty)}\varphi = 0,$$

and hence that

$$D_0(\mathscr{E}Q_{(0,\infty)}\varphi) = \chi' \sigma_0 \exp(-tA)Q_{(0,\infty)}\varphi.$$

It follows that the graph norm $\|\mathscr{E}Q_{(0,\infty)}\varphi\|_{D_0}$ can be bounded from above by $\|\exp(-tA)Q_{(0,\infty)}\varphi\|_{L^2(Z_{[0,r)})}$. Now there is some $\varepsilon > 0$ such that $|\lambda| \geq \varepsilon > 0$ for all nonzero eigenvalues λ of A. Hence, for $\varphi = \sum_j a_j \varphi_j$, we have

$$\begin{aligned}
\|\exp(-t|A|)Q_{(0,\infty)}\varphi\|_{L^2(Z_{[0,r)})}^2 &= \sum_{\lambda_j \geq \varepsilon} |a_j|^2 \int_0^r e^{-2t\lambda_j}\, dt \\
&= \frac{1}{2} \sum_{\lambda_j \geq \varepsilon} |a_j|^2 \cdot \lambda_j^{-1}(1 - e^{-2r\lambda_j}) \\
&\leq \frac{1}{2} \sum_{\lambda_j \geq \varepsilon} |a_j|^2 \cdot \lambda_j^{-1} \\
&\leq \frac{(1+\varepsilon^{-2})^{1/2}}{2} \sum_{\lambda_j \geq \varepsilon} |a_j|^2 \cdot (1+\lambda_j^2)^{-1/2} \\
&= \frac{(1+\varepsilon^{-2})^{1/2}}{2} \|Q_{(0,\infty)}\varphi\|_{H^{-1/2}(\partial M,E)}^2 \\
&= \frac{(1+\varepsilon^{-2})^{1/2}}{2} \|Q_{(0,\infty)}\varphi\|_{\check{H}(A)}^2.
\end{aligned}$$

The claimed inequality for $\|\mathscr{E}Q_{(0,\infty)}\varphi\|_{D_0}^2$ follows. The estimate for $\|\mathscr{E}Q_{(-\infty,0]}\varphi\|_{D_0}^2$ follows from similar considerations, where now

$$(D_0\exp(-t|A|)Q_{(-\infty,0]}\varphi)(t) = 2\sigma_0 A\exp(-t|A|)Q_{(-\infty,0]}\varphi$$

and hence

$$(D_0\mathscr{E}Q_{(-\infty,0]}\varphi)(t) = \sigma_0(2\chi A + \chi')\exp(tA)Q_{(-\infty,0]}\varphi.$$

Thus the graph norm $\|\mathscr{E}Q_{(-\infty,0]}\varphi\|_{D_0}$ can be bounded from above by

$$\|(\mathrm{id} + |A|)\exp(-t|A|)Q_{(-\infty,0]}\varphi\|_{L^2(Z_{[0,r)})}.$$

Now we have

$$\begin{aligned}
&\|(\mathrm{id} + |A|)\exp(-t|A|)Q_{(-\infty,0]}\varphi\|_{L^2(Z_{[0,r)})}^2 \\
&\quad = \sum_{\lambda_j\le 0}\int_0^r |a_j|^2(1+|\lambda_j|)^2 e^{2t\lambda_j}\,dt \\
&\quad = r\sum_{\lambda_j=0}|a_j|^2 + \frac{1}{2}\sum_{\lambda_j\le-\varepsilon}|a_j|^2(1+|\lambda_j|)^2|\lambda_j|^{-1}(1-e^{2r\lambda_j}) \\
&\quad \le r\sum_{\lambda_j=0}|a_j|^2 + \frac{1}{2}\sum_{\lambda_j\le-\varepsilon}|a_j|^2(1+|\lambda_j|)^2|\lambda_j|^{-1} \\
&\quad \le r\sum_{\lambda_j=0}|a_j|^2 + C_1(\varepsilon)\sum_{\lambda_j\le-\varepsilon}|a_j|^2(1+|\lambda_j|^2)^{1/2} \\
&\quad \le C_2(\varepsilon,r)\sum_{\lambda_j\le 0}|a_j|^2(1+|\lambda_j|^2)^{1/2} \\
&\quad = C_2(\varepsilon,r)\|Q_{(-\infty,0]}\varphi\|_{H^{1/2}(\partial M,E)}^2 \\
&\quad = C_2(\varepsilon,r)\|Q_{(-\infty,0]}\varphi\|_{\check{H}(A)}^2.
\end{aligned}$$

□

REMARK 5.6. With the same methods one can show that $\|\mathscr{E}\varphi\|_{D_0}$ can also be bounded from below by $\|\varphi\|_{\check{H}(A)}$.

6. The maximal domain

Throughout this Section, assume the Standard Setup 1.5, identify a neighborhood U of the boundary ∂M with a cylinder $Z_{[0,r)}$ via an adapted diffeomorphism as in Lemma 2.4, and fix a normal form for D and D^* as in Lemma 4.1. Assume furthermore that D and D^* are complete.

6.1. Regularity properties of the maximal domain. For some of our assertions concerning estimates over the cylinder $Z_{[0,r)}$, it will be convenient to consider the operator $\sigma^{-1}D$ instead of D. Its formal adjoint is given by

$$(\sigma^{-1}D)^* = -\left(\frac{\partial}{\partial t} - A - R_t^*\right) \tag{44}$$

so that we can let $-A$ take over the role of $\tilde{A}$ in the normal form of $\sigma^{-1}D$ as in Lemma 4.1. Then we have $\check{H}(-A) = \hat{H}(A)$.

LEMMA 6.1. *For $\varphi \in H^{\mathrm{fin}}(A)$, the section $\mathscr{E}\varphi$ of E over $Z_{[0,r)}$ belongs to the maximal domain of D. Moreover, there is a constant $C > 0$ such that*

$$\|\mathscr{E}\varphi\|_D \le C\|\varphi\|_{\check{H}(A)} \quad \text{and} \quad \|\mathscr{E}\varphi\|_{(\sigma^{-1}D)^*} \le C\|\varphi\|_{\hat{H}(A)}.$$

PROOF. The section $\mathscr{E}\varphi$ belongs to the maximal domain of D because $\mathscr{E}\varphi \in C_c^\infty(Z_{[0,r)}, E) \subset C_c^\infty(M, E) \subset \mathrm{dom}(D_{\max})$. The first estimate has been shown in Lemma 5.5 with the model operator D_0 instead of D. By the definition of D_0, we have

$$\begin{aligned}
\|D(\mathscr{E}\varphi)\|_{L^2(M)} &= \|D(\mathscr{E}\varphi)\|_{L^2(Z_{[0,r)})} \\
&\le C_1 \cdot \|\sigma_0\sigma_t^{-1}D(\mathscr{E}\varphi)\|_{L^2(Z_{[0,r)})} \\
&= C_1 \cdot \|(D_0 + \sigma_0 R_t)(\mathscr{E}\varphi)\|_{L^2(Z_{[0,r)})} \\
&\le C_1 \cdot \|D_0(\mathscr{E}\varphi)\|_{L^2(Z_{[0,r)})} + C_2 \cdot \|R_t(\mathscr{E}\varphi)\|_{L^2(Z_{[0,r)})} \\
&\le C_1 \cdot \|\mathscr{E}\varphi\|_{D_0} + C_2 \cdot \|R_t(\mathscr{E}\varphi)\|_{L^2(Z_{[0,r)})}.
\end{aligned}$$

It remains to estimate $\|R_t(\mathscr{E}\varphi)\|_{L^2(Z_{[0,r)})}$. By (27), we get

$$\begin{aligned}
&\|R_t(\mathscr{E}\varphi)\|^2_{L^2(Z_{[0,r)})} \\
&\le C_3 \int_0^r \left(\|tA(\mathscr{E}\varphi)\|^2_{L^2(\partial M)} + \|\mathscr{E}\varphi\|^2_{L^2(\partial M)}\right) dt \\
&= C_3 \int_0^r \chi(t)^2 \left(\|tA\exp(-t|A|)\varphi\|^2_{L^2(\partial M)} + \|\exp(-t|A|)\varphi\|^2_{L^2(\partial M)}\right) dt.
\end{aligned}$$

If $A\varphi_j = \lambda_j\varphi_j$, then we compute, substituting $\tilde{t} = t \cdot |\lambda_j|$,

$$\begin{aligned}
&\int_0^r \chi(t)^2 \left(\|tA\exp(-t|A|)\varphi_j\|^2_{L^2(\partial M)} + \|\exp(-t|A|)\varphi_j\|^2_{L^2(\partial M)}\right) dt \\
&\le \int_0^r \exp(-2t|\lambda_j|)(t^2|\lambda_j|^2 + 1)\|\varphi_j\|^2_{L^2(\partial M)}\, dt \\
&= \int_0^{r|\lambda_j|} \exp(-2\tilde{t})(\tilde{t}^2 + 1)\|\varphi_j\|^2_{L^2(\partial M)} \frac{d\tilde{t}}{|\lambda_j|}
\end{aligned}$$

$$\le \frac{\|\varphi_j\|^2_{L^2(\partial M)}}{|\lambda_j|} \int_0^\infty \exp(-2\tilde{t})(\tilde{t}^2+1)\, d\tilde{t}$$
$$= \frac{3}{4} \cdot \||\lambda_j|^{-1/2}\varphi_j\|^2_{L^2(\partial M)}.$$

Expanding φ in an orthonormal eigenbasis for A, this shows

$$\int_0^r \chi(t)^2 \left(\|tA\exp(-t|A|)\varphi\|^2_{L^2(\partial M)} + \|\exp(-t|A|)\varphi\|^2_{L^2(\partial M)} \right) dt$$
$$\le C_4 \|\varphi\|^2_{H^{-1/2}(\partial M)} \le C_4 \|\varphi\|^2_{\check{H}(A)}.$$

This concludes the proof of the first inequality. For the proof of the second, we recall from (44) that $-A$ is an adapted boundary operator for $(\sigma^{-1}D)^*$ and that $\check{H}(-A) = \hat{H}(A)$. □

LEMMA 6.2. *There is a constant* $C > 0$ *such that, for all* $\Phi \in C_c^\infty(Z_{[0,r)}, E)$,

$$\|\Phi_{|\partial M}\|_{\check{H}(A)} \le C \|\Phi\|_D.$$

PROOF. Since the pairing between $\check{H}(A)$ and $\hat{H}(A)$ introduced in (40) is perfect and since $H^{\text{fin}}(A)$ is dense in $\hat{H}(A)$, we have

$$\|\Phi_{|\partial M}\|_{\check{H}(A)} = \sup\{|(\Phi_{|\partial M}, \psi)| \mid \psi \in H^{\text{fin}}(A), \|\psi\|_{\hat{H}(A)} = 1\}.$$

Now Lemma 2.6 and Lemma 6.1 imply that, for any $\psi \in H^{\text{fin}}(A)$ with $\|\psi\|_{\hat{H}(A)} = 1$,

$$\begin{aligned}
|(\Phi_{|\partial M}, \psi)| &= \left|(\sigma^{-1}D\Phi, \mathscr{E}\psi)_{L^2(M)} - (\Phi, (\sigma^{-1}D)^*(\mathscr{E}\psi))_{L^2(M)}\right| \\
&\le \|\sigma^{-1}D\Phi\|_{L^2(M)} \|\mathscr{E}\psi\|_{L^2(M)} + \|\Phi\|_{L^2(M)} \|(\sigma^{-1}D)^*(\mathscr{E}\psi)\|_{L^2(M)} \\
&\le C' \cdot \|\Phi\|_D \cdot \|\mathscr{E}\psi\|_{(\sigma^{-1}D)^*} \\
&\le C \cdot \|\Phi\|_D \cdot \|\psi\|_{\hat{H}(A)} \\
&= C \cdot \|\Phi\|_D.
\end{aligned}$$
□

LEMMA 6.3. *Over* ∂M, *the homomorphism field* $(\sigma_0^{-1})^* : E \to F$ *induces an isomorphism* $\hat{H}(A) \to \check{H}(\tilde{A})$. *Here* $\tilde{A}$ *is an adapted boundary operator for* D^*. *In particular, the sesquilinear form*

$$\beta : \check{H}(A) \times \check{H}(\tilde{A}) \to \mathbb{C}, \quad \beta(\varphi, \psi) := -(\sigma_0\varphi, \psi),$$

is a perfect pairing of topological vector spaces.

PROOF. For $\varphi \in C^\infty(\partial M, E)$, we have $(\sigma_0^{-1})^*\varphi \in C^\infty(\partial M, F)$ and

$$\begin{aligned}
\|(\sigma_0^{-1})^*\varphi\|^2_{\check{H}(\tilde{A})} &\le C_1 \|(\sigma^{-1})^*\mathscr{E}\varphi\|^2_{D^*} \\
&= C_1\big(\|D^*(\sigma^{-1})^*\mathscr{E}\varphi\|^2_{L^2(Z_{[0,r)})} + \|(\sigma^{-1})^*\mathscr{E}\varphi\|^2_{L^2(Z_{[0,r)})}\big) \\
&\le C_2\big(\|(\sigma^{-1}D)^*\mathscr{E}\varphi\|^2_{L^2(Z_{[0,r)})} + \|\mathscr{E}\varphi\|^2_{L^2(Z_{[0,r)})}\big) \\
&= C_2 \|\mathscr{E}\varphi\|^2_{(\sigma^{-1}D)^*} \\
&\le C_3 \|\varphi\|^2_{\hat{H}(A)}
\end{aligned}$$

by Lemmas 6.2 (for D^*) and 6.1, respectively.

Conversely, for $\psi \in C^\infty(\partial M, F)$, we have $\sigma_0^*\psi \in C^\infty(\partial M, E)$. Furthermore, if $\tilde{\mathscr{E}}$ denotes the extension operator defined in (43) associated to the boundary operator $\tilde{A}$, we obtain

$$\begin{aligned}
\|\sigma_0^*\psi\|^2_{\hat{H}(A)} &= \|(\sigma^*\tilde{\mathscr{E}}\psi)|_{\partial M}\|^2_{\check{H}(-A)} \\
&\le C_4 \|\sigma^*\tilde{\mathscr{E}}\psi\|^2_{(\sigma^{-1}D)^*} \\
&= C_4\big(\|(\sigma^{-1}D)^*\sigma^*\tilde{\mathscr{E}}\psi\|^2_{L^2(Z_{[0,r)})} + \|\sigma^*\tilde{\mathscr{E}}\psi\|^2_{L^2(Z_{[0,r)})}\big) \\
&\le C_5\big(\|D^*\tilde{\mathscr{E}}\psi\|^2_{L^2(Z_{[0,r)})} + \|\tilde{\mathscr{E}}\psi\|^2_{L^2(Z_{[0,r)})}\big) \\
&= C_5 \|\tilde{\mathscr{E}}\psi\|^2_{D^*} \\
&\le C_6 \|\psi\|^2_{\check{H}(\tilde{A})}
\end{aligned}$$

by Lemma 6.2 applied to the operator $(\sigma^{-1}D)^*$ and Lemma 6.1 applied to the operator D^*.

Since $C^\infty(\partial M, E)$ and $C^\infty(\partial M, F)$ are dense in $\hat{H}(A)$ and $\check{H}(A)$, respectively, and since $(\sigma_0^*)^{-1} = (\sigma_0^{-1})^*$, we conclude that $(\sigma_0^{-1})^*$ induces a topological isomorphism between the Hilbert spaces $\hat{H}(A)$ and $\check{H}(A)$ as asserted. The second claim is now an immediate consequence since the corresponding pairing between $\check{H}(A)$ and $\hat{H}(A)$ is perfect. □

Consider the intersection

$$H^1_D(M,E) := \operatorname{dom}(D_{\max}) \cap H^1_{\mathrm{loc}}(E,F). \tag{45}$$

The H^1_D-norm on $H^1_D(M,E)$ is defined by

$$\|\Phi\|^2_{H^1_D(M)} := \|\chi\Phi\|^2_{H^1(M)} + \|\Phi\|^2_{L^2(M)} + \|D\Phi\|^2_{L^2(M)}, \tag{46}$$

where χ is as in (42), and turns $H^1_D(M,E)$ into a Hilbert space. Since the support of χ as a function on M is compact, the specific choice of H^1-norm is irrelevant and leads to equivalent H^1_D-norms. The H^1_D-norm is stronger than the graph norm for D; it controls in addition H^1-regularity near the boundary. The completeness of D is responsible for the following properties of $H^1_D(M,E)$.

LEMMA 6.4. *(i)* $C_c^\infty(M,E)$ *is dense in* $H^1_D(M,E)$*;*
(ii) $C_{cc}^\infty(M,E)$ *is dense in* $\{\Phi \in H^1_D(M,E) \mid \Phi|_{\partial M} = 0\}$.

PROOF. (i) We have to show that any given $\Phi \in H^1_D(M,E)$ can be approximated by compactly supported smooth sections in the H^1_D-norm. Let $\chi \in C_c^\infty(M,\mathbb{R})$ be the cut-off function used in the definition of $\|\cdot\|_{H^1_D(M)}$, that is, the function from (42). Choose a second cut-off function $\chi_2 \in C_c^\infty(M,\mathbb{R})$ with $\chi_2 \equiv 1$ on the support of χ and a third cut-off function $\chi_3 \in C_c^\infty(M,\mathbb{R})$ with $\chi_3 \equiv 1$ on the support of χ_2.

Then $\chi_3\Phi \in H^1_D(M,E)$ because $\Phi \in H^1_{\mathrm{loc}}(M,E)$ and $\chi_3\Phi$ has compact support. Therefore $\chi_3\Phi$ can be approximated by smooth compactly supported sections in any H^1-norm, hence also in the H^1_D-norm.

It remains to approximate $(1-\chi_3)\Phi \in H^1_D(M,E)$. Since D is complete, there exists $\Phi_0 \in \mathrm{dom}(D_{\max})$ with compact support such that

$$\|(1-\chi_3)\Phi - \Phi_0\|_D < \epsilon$$

for any given $\varepsilon > 0$. Since χ_2 vanishes on the support of $1-\chi_3$, we have

$$\begin{aligned}
\|(1-\chi_3)\Phi - (1-\chi_2)\Phi_0\|_{H^1_D(M)} &= \|(1-\chi_2)((1-\chi_3)\Phi - \Phi_0)\|_{H^1_D(M)} \\
&= \|(1-\chi_2)((1-\chi_3)\Phi - \Phi_0)\|_D \\
&\le C \cdot \|(1-\chi_3)\Phi - \Phi_0\|_D \\
&< C \cdot \varepsilon.
\end{aligned}$$

Hence it suffices to approximate $(1-\chi_2)\Phi_0$. Since $(1-\chi_2)\Phi_0$ has compact support, this is possible exactly like for $\chi_3\Phi$.

(ii) Suppose that $\Phi \in H^1_D(M,E)$ vanishes along ∂M. We have to show that Φ can be approximated by sections from $C_{cc}^\infty(M,E)$. As in the first part of the proof, we may assume that the support of Φ is contained in a neighborhood of ∂M, say in $Z_{r/2}$. For $n \in \mathbb{N}$ sufficiently large, define sections Φ_n of E over $Z_{[0,r)}$ by

$$\Phi_n(t,x) = \begin{cases} 0 & \text{for } 0 \le t \le 1/n, \\ \Phi(t-1/n, x) & \text{for } 1/n \le t < r. \end{cases}$$

Then Φ_n has compact support in $Z_{[1/n,r]}$, $\Phi_n \in H^1(Z_{[0,r)},E)$, and $\Phi_n \to \Phi$ in $H^1(Z_{[0,r)},E)$, therefore also in $H^1_D(M,E)$. Since Φ_n has compact support

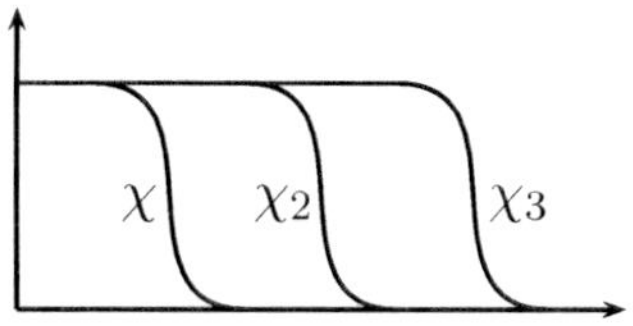

FIGURE 4

in $Z_{(0,r)}$, it can be approximated in $H^1_D(M,E)$ by smooth sections of E with compact support in $Z_{(0,r)}$. □

LEMMA 6.5. *There is a constant $C>0$ such that*

$$\|\Phi\|_D \le \|\Phi\|_{H^1_D(M)} \le C\,\|\Phi\|_D$$

for all $\Phi \in C^\infty_c(M,E)$ with $Q_{(0,\infty)}(\Phi_{|\partial M}) = 0$.

PROOF. We show that the H^1_D-norm is bounded from above by the graph norm, the converse inequality being clear. We write $\Phi = \Phi_1 + \Phi_2$ where $\Phi_1 = \chi\Phi \in C^\infty_c(Z_{[0,r)}, E)$ and $\Phi_2 = (1-\chi)\Phi \in C^\infty_{cc}(M,E)$ has support disjoint from $Z_{r/3}$. On the space of Φ_2's the graph norm and the H^1_D-norm are equivalent. Therefore we can assume that $\Phi = \Phi_1$ has compact support in the closure of $Z_{2r/3}$.

Since $Q_{(0,\infty)}\Phi_{|\partial M} = 0$, we have $(\Phi_{|\partial M}, A\Phi_{|\partial M})_{L^2(\partial M)} \le 0$ and hence

$$\begin{aligned}
\|D_0\Phi\|^2_{L^2(Z_{[0,r)})} &= \|\sigma_0(\Phi' + A\Phi)\|^2_{L^2(Z_{[0,r)})}\\
&\ge C_1\|\Phi' + A\Phi\|^2_{L^2(Z_{[0,r)})}\\
&= C_1\{\|\Phi'\|^2_{L^2(Z_{[0,r)})} + \|A\Phi\|^2_{L^2(Z_{[0,r)})} - (\Phi_{|\partial M}, A\Phi_{|\partial M})_{L^2(\partial M)}\}\\
&\ge C_1\{\|\Phi'\|^2_{L^2(Z_{[0,r)})} + \|A\Phi\|^2_{L^2(Z_{[0,r)})}\}, \qquad (47)
\end{aligned}$$

where we use Lemma 5.1 to pass from line 2 to line 3. Thus

$$\begin{aligned}
\|\Phi\|^2_{H^1_D(M)} &\le C_2\,\{\|\Phi\|^2_{L^2(Z_{[0,r)})} + \|\Phi'\|^2_{L^2(Z_{[0,r)})} + \|A\Phi\|^2_{L^2(Z_{[0,r)})}\}\\
&\le C_3\,\{\|\Phi\|^2_{L^2(Z_{[0,r)})} + \|D_0\Phi\|^2_{L^2(Z_{[0,r)})}\}\\
&\le C_4\,\|\Phi\|^2_D.
\end{aligned}$$

The first inequality follows from the ellipticity of A, the second from (47), and the third from Lemma 5.2. □

COROLLARY 6.6. *Assume the Standard Setup 1.5 and that D and D^* are complete. Then* $\operatorname{dom} D_{\min} = \{\Phi \in H^1_D(M,E) \mid \Phi|_{\partial M} = 0\}$.

PROOF. Sections in $C^\infty_{cc}(M,E)$ vanish along ∂M, hence satisfy the boundary condition $Q_{(0,\infty)}(\Phi_{|\partial M}) = 0$ required in Lemma 6.5, and hence $\operatorname{dom} D_{\min}$ is contained in $\{\Phi \in H^1_D(M,E) \mid \Phi|_{\partial M} = 0\}$. Now Lemma 6.4 (ii) concludes the proof. □

THEOREM 6.7. *Assume the Standard Setup 1.5 and that D and D^* are complete. Then we have:*

(i) $C^\infty_c(M,E)$ *is dense in* $\operatorname{dom}(D_{\max})$ *with respect to the graph norm.*
(ii) *The trace map* $C^\infty_c(M,E) \to C^\infty(\partial M,E)$, $\Phi \mapsto \Phi_{|\partial M}$, *extends uniquely to a surjective bounded linear map* $\mathscr{R} : \operatorname{dom}(D_{\max}) \to \check{H}(A)$.
(iii) $H^1_D(M,E) = \{\Phi \in \operatorname{dom}(D_{\max}) \mid \mathscr{R}\Phi \in H^{1/2}(\partial M,E)\}$.

The corresponding statements hold for $\operatorname{dom}((D^*)_{\max})$. *Furthermore, for all sections* $\Phi \in \operatorname{dom}(D_{\max})$ *and* $\Psi \in \operatorname{dom}((D^*)_{\max})$, *we have*

$$(D_{\max}\Phi, \Psi)_{L^2(M)} - (\Phi, (D^*)_{\max}\Psi)_{L^2(M)} = -(\sigma_0 \mathscr{R}\Phi, \mathscr{R}\Psi)_{L^2(\partial M)}. \tag{48}$$

Proof. (i) Extend (M, μ) smoothly to a larger measured manifold $(\tilde{M}, \tilde{\mu})$ with $\partial\tilde{M} = \emptyset$. Do it in such a way that E and F extend smoothly to Hermitian vector bundles $\tilde{E}$ and $\tilde{F}$ over $\tilde{M}$ and D to an elliptic operator $\tilde{D} : C^\infty(\tilde{M}, \tilde{E}) \to C^\infty(\tilde{M}, \tilde{F})$. This is possible, since we may choose $\tilde{M} \setminus M$ to be diffeomorphic to the product $\partial M \times (-1, 0)$.

Let D_c be D with domain $\operatorname{dom}(D_c) = C_c^\infty(M, E)$. We have to show that the closure of D_c equals $D_{\max}$. Let $\Psi \in L^2(M, F)$ be in the domain of the adjoint operator $(D_c)^{\mathrm{ad}}$. Extend Ψ and $(D_c)^{\mathrm{ad}}\Psi$ by the trivial section 0 on $\tilde{M} \setminus M$ to sections $\tilde{\Psi} \in L^2(\tilde{M}, \tilde{F})$ and $\tilde{\Xi} \in L^2(\tilde{M}, \tilde{E})$, respectively. Let $\tilde{\Phi} \in C_c^\infty(\tilde{M}, \tilde{E}) = C_{cc}^\infty(\tilde{M}, \tilde{E})$. Then the restriction Φ of $\tilde{\Phi}$ to M is in $\operatorname{dom}(D_c)$ and hence, since $\tilde{\Psi} = \tilde{\Xi} = 0$ on $\tilde{M} \setminus M$,

$$\begin{aligned}(\tilde{D}\tilde{\Phi}, \tilde{\Psi})_{L^2(\tilde{M})} &= (D_c\Phi, \Psi)_{L^2(M)} \\ &= (\Phi, (D_c)^{\mathrm{ad}}\Psi)_{L^2(M)} \\ &= (\tilde{\Phi}, \tilde{\Xi})_{L^2(\tilde{M})}.\end{aligned}$$

Thus $\tilde{\Psi}$ is a weak solution of $\tilde{D}^*\tilde{\Phi} = \tilde{\Xi} \in L^2(\tilde{M}, \tilde{E})$. By interior elliptic regularity theory, $\tilde{\Psi} \in H^1_{\mathrm{loc}}(\tilde{M}, \tilde{F})$. It follows that Ψ is in $H^1_{D^*}(M, F)$ and that $\Psi_{|\partial M} = 0$. By Corollary 6.6, $\Psi \in \operatorname{dom}((D^*)_{\min})$, the domain of the minimal extension of $(D^*)_{cc}$. Hence $(D_c)^{\mathrm{ad}} \subset (D^*)_{\min}$, and therefore the closure $\overline{D}_c$ of D_c satisfies

$$\overline{D}_c = ((D_c)^{\mathrm{ad}})^{\mathrm{ad}} \supset ((D^*)_{\min})^{\mathrm{ad}} = D_{\max} \supset \overline{D}_c.$$

Thus $\overline{D}_c = D_{\max}$ as asserted.

(ii) By Lemma 6.2, the trace map $C_c^\infty(M, E) \to C^\infty(\partial M, E)$, $\Phi \mapsto \Phi_{|\partial M}$, extends to a bounded linear map $\operatorname{dom}(D_{\max}) \to \check{H}(A)$. By (i), this extension is unique. By Lemma 6.1, the map $\mathscr{E} : H^{\mathrm{fin}}(A) \to C_c^\infty(M, E)$ extends to a bounded linear map $\mathscr{E} : \check{H}(A) \to \operatorname{dom}(D_{\max})$ with $\mathscr{R} \circ \mathscr{E} = \mathrm{id}$. This proves surjectivity.

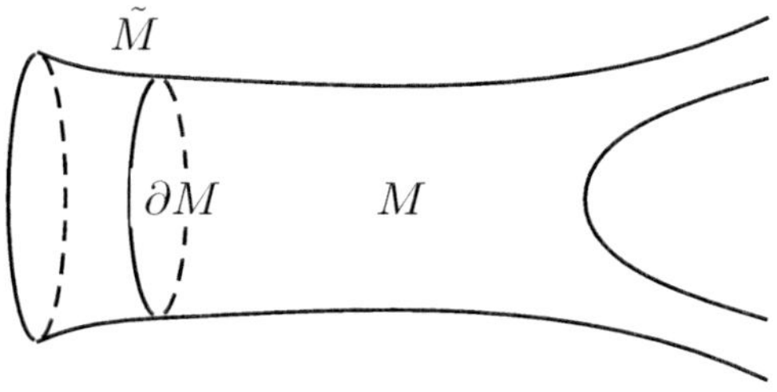

Figure 5

(iii) The inclusion $H^1_D(M,E)\subset\{\Phi\in\mathrm{dom}(D_{\max})\mid \mathscr{R}\Phi\in H^{1/2}(\partial M)\}$ is clear from the definition of $H^1_D(M,E)$ and the standard trace theorem. To show the converse inclusion, let $\Phi\in\mathrm{dom}(D_{\max})$ with $\mathscr{R}\Phi\in H^{1/2}(\partial M)$. Put $\varphi:=Q_{(0,\infty)}\mathscr{R}\Phi\in H^{1/2}(\partial M)$. Expanding φ with respect to an eigenbasis of A, one easily sees that $\mathscr{E}\varphi\in H^1_D(M,E)$. In particular, $\Phi-\mathscr{E}\varphi\in \mathrm{dom}(D_{\max})$ and $Q_{(0,\infty)}\mathscr{R}(\Phi-\mathscr{E}\varphi)=0$. Now (i) and Lemma 6.5 imply $\Phi-\mathscr{E}\varphi\in H^1_D(M,E)$. Thus

$$\Phi=\mathscr{E}\varphi+(\Phi-\mathscr{E}\varphi)\in H^1_D(M,E).$$

The asserted formula for integration by parts holds for all $\Phi,\Psi\in C^\infty_c(M,E)$, hence for all $\Phi\in\mathrm{dom}(D_{\max})$ and all $\Psi\in\mathrm{dom}((D^*)_{\max})$, by (i), (ii), and Lemma 6.3. □

6.2. Higher regularity. Fix $\rho>0$. For $g\in L^2([0,\rho],\mathbb{C})$ and $t\in[0,\rho]$ set

$$(R_\lambda g)(t):=\begin{cases}\int_0^t g(s)\,e^{\lambda(s-t)}ds & \text{if } \lambda\ge 0,\\ -\int_t^\rho g(s)\,e^{\lambda(s-t)}ds & \text{if } \lambda<0.\end{cases} \tag{49}$$

For $\lambda>0$, we compute for $f=R_\lambda g$:

$$\begin{aligned}\|f\|^2_{L^2([0,\rho])}&=\int_0^\rho\left|\int_0^t g(s)e^{\lambda(s-t)}ds\right|^2dt\\ &\le\int_0^\rho\left(\int_0^t|g(s)|^2e^{\lambda(s-t)}ds\right)\left(\int_0^t e^{\lambda(s-t)}ds\right)dt\\ &\le\frac{1}{\lambda}\int_0^\rho\int_s^\rho|g(s)|^2e^{\lambda(s-t)}\,dt\,ds\\ &=\frac{1}{\lambda^2}\int_0^\rho|g(s)|^2\,ds\\ &=\frac{1}{\lambda^2}\|g\|^2_{L^2([0,\rho])}.\end{aligned}$$

There is a similar computation in the case $\lambda<0$. We obtain

$$\|f\|^2_{L^2([0,\rho])}\le\begin{cases}\frac{1}{\lambda^2}\|g\|^2_{L^2([0,\rho])} & \text{for } \lambda\ne 0,\\ \frac{\rho^2}{2}\|g\|^2_{L^2([0,\rho])} & \text{for } \lambda=0.\end{cases} \tag{50}$$

Now f satisfies

$$f'+\lambda f=g. \tag{51}$$

Hence

$$\|f'\|^2_{L^2([0,\rho])}\le\begin{cases}4\|g\|^2_{L^2([0,\rho])} & \text{for } \lambda\ne 0,\\ \|g\|^2_{L^2([0,\rho])} & \text{for } \lambda=0.\end{cases}$$

In conclusion,

$$\|f\|^2_{H^1([0,\rho])} \leq \begin{cases} (4+\frac{1}{\lambda^2})\|g\|^2_{L^2([0,\rho])} & \text{for } \lambda \neq 0, \\ (1+\frac{\rho^2}{2})\|g\|^2_{L^2([0,\rho])} & \text{for } \lambda = 0. \end{cases} \tag{52}$$

REMARKS 6.8. (a) The above considerations are an adaption and small simplification of the corresponding discussion in [**APS**, Sec. 2].

(b) Note that the bound in (52) is independent of ρ for $\lambda \neq 0$ and depends only on an upper bound on ρ for $\lambda = 0$.

From now on assume that $\rho < r$ where $r < \infty$ is as in Lemma 2.4. We apply the above discussion to sections of E over the cylinder $Z_{[0,\rho]} = [0,\rho] \times \partial M$. We choose an orthonormal Hilbert basis of $L^2(\partial M, E)$ consisting of eigensections φ_j of A with corresponding eigenvalues λ_j as in Section 5. For

$$\Psi \in L^2(Z_{[0,\rho]}, F) = L^2([0,\rho], L^2(\partial M, F)), \quad \sigma_0^{-1}\Psi(t,x) = \sum_{j=-\infty}^{\infty} g_j(t)\varphi_j(x),$$

we set

$$S_0\Psi := \sum_{j=-\infty}^{\infty} (R_{\lambda_j} g_j)\varphi_j. \tag{53}$$

Estimates (50) and (52) show that $\Phi = S_0\Psi$ satisfies

$$\begin{aligned} \|\Phi\|^2_{H^1(Z_{[0,\rho]})} &\leq C_1 \left(\|\Phi\|^2_{L^2(Z_{[0,\rho]})} + \|\Phi'\|^2_{L^2(Z_{[0,\rho]})} + \|A\Phi\|^2_{L^2(Z_{[0,\rho]})} \right) \\ &= C_1 \sum_{j=-\infty}^{\infty} \left(\|R_{\lambda_j} g_j\|^2_{H^1([0,\rho])} + \lambda_j^2 \|R_{\lambda_j} g_j\|^2_{L^2([0,\rho])} \right) \\ &\leq C_1 C_2 \sum_{j=-\infty}^{\infty} \|g_j\|^2_{L^2([0,\rho])} \\ &= C_1 C_2 \|\sigma_0^{-1}\Psi\|^2_{L^2(Z_{[0,\rho]})} \\ &\leq C_1 C_2 C_3 \|\Psi\|^2_{L^2(Z_{[0,\rho]})}. \end{aligned}$$

Here C_1 depends on the specific choice of H^1-norm, C_2 on r (which is an upper bound for ρ) and a lower bound for the modulus of the nonzero λ_j, and C_3 on an upper bound for σ_0^{-1}. Differentiating (51) repeatedly, we get

$$\|\Phi\|^2_{H^{m+1}(Z_{[0,\rho]})} \leq C_4 \|\Psi\|^2_{H^m(Z_{[0,\rho]})}, \tag{54}$$

where C_4 depends on the specific choice of H^k-norms, on m, on r, on a lower bound for the nonzero $|\lambda_j|$, and on an upper bound for the derivatives of order up to m of σ_0^{-1}. By definition, C_4 is a bound for

$$S_0 : H^m(Z_{[0,\rho]}, F) \to H^{m+1}(Z_{[0,\rho]}, E).$$

Moreover, with the model operator D_0 as in (33), we have

$$D_0 S_0 \Psi = \Psi \tag{55}$$

and, by definition, $S_0\Psi$ satisfies the boundary condition

$$Q_{[0,\infty)}(S_0\Psi)(0) = 0 \quad \text{and} \quad Q_{(-\infty,0)}(S_0\Psi)(\rho) = 0. \tag{56}$$

By Theorem 6.7 (ii), the boundary values of elements in the maximal domain of D_0 over $Z_{[0,\rho]}$ constitute the space $\check{H}(A) \oplus \hat{H}(A)$, where $\check{H}(A)$ is responsible for the left part of the boundary, $\{0\} \times \partial M$, and $\hat{H}(A)$ for the right, $\{\rho\} \times \partial M$. Set

$$B_0 := H^{1/2}_{(-\infty,0)}(A) \oplus H^{1/2}_{[0,\infty)}(A) \tag{57}$$

and, for $m \geq 1$,

$$H^m(Z_{[0,\rho]}, E; B_0) := \{f \in H^m(Z_{[0,\rho]}, E) \mid (f(0), f(\rho)) \in B_0\}.$$

By (56), S_0 is a bounded operator $H^m(Z_{[0,\rho]}, F) \to H^{m+1}(Z_{[0,\rho]}, E; B_0)$.

Proposition 6.9. *Let S_0 be as in* (53) *and B_0 as in* (57). *Then, for all $m \geq 0$, the model operator*

$$D_0 : H^{m+1}(Z_{[0,\rho]}, E; B_0) \to H^m(Z_{[0,\rho]}, F)$$

is an isomorphism with inverse S_0. Furthermore, for elements Φ in the maximal domain of D_0 satisfying $Q_{(-\infty,0)}(\Phi(\rho)) = 0$, we have

$$\Phi - S_0 D_0 \Phi = \exp(-tA)(Q_{[0,\infty)}\Phi(0)).$$

Proof. Surjectivity of D_0 follows from (55) and (56). The Fourier coefficients f_j of an element in the kernel of D_0 satisfy $f_j' + \lambda_j f_j = 0$, thus the boundary condition B_0, i.e. $f_j(0) = 0$ or $f_j(\rho) = 0$, implies that they vanish. This shows injectivity and proves the first assertion.

Let Φ be in the maximal domain of D_0 with $Q_{(-\infty,0)}(\Phi(\rho)) = 0$. Since $D_0(\Phi - S_0 D_0 \Phi) = 0$, the Fourier coefficients f_j of $\Phi - S_0 D_0 \Phi$ satisfy $f_j' + \lambda_j f_j = 0$. From $Q_{(-\infty,0)}(\Phi(\rho)) = 0$ and $Q_{(-\infty,0)}(S_0 D_0 \Phi(\rho)) = 0$ we have $f_j(\rho) = 0$ for all $\lambda_j < 0$. Hence $f_j \equiv 0$ whenever $\lambda_j < 0$. Thus

$$(\Phi - S_0 D_0 \Phi)(t, x) = \sum_{\lambda_j \geq 0} e^{-t\lambda_j} f_j(0) \varphi_j(x). \tag{58}$$

On the other hand,

$$\begin{aligned} \exp(-tA)(Q_{[0,\infty)}\Phi(0)) &= \exp(-tA)(Q_{[0,\infty)}(\Phi - S_0 D_0 \Phi)(0)) \\ &= \exp(-tA)\Big(\sum_{\lambda_j \geq 0} f_j(0)\varphi_j\Big) \\ &= \sum_{\lambda_j \geq 0} e^{-t\lambda_j} f_j(0)\varphi_j. \end{aligned} \tag{59}$$

Equating (58) and (59) concludes the proof. □

We return from the model operator D_0 to the original operator D.

LEMMA 6.10. *Let (M, μ) be a measured manifold with compact boundary and let D and A be as in the Standard Setup 1.5. Let $m \geq 0$. If $\rho > 0$ is sufficiently small, then*

(i) D_0 and D have the same maximal domain when regarded as unbounded operators from $L^2(Z_{[0,\rho]}, E)$ to $L^2(Z_{[0,\rho]}, F)$.

(ii) $(D - \sigma_0 R_0) : H^{m+1}(Z_{[0,\rho]}, E; B_0) \to H^m(Z_{[0,\rho]}, F)$ is an isomorphism.

PROOF. By Lemma 5.2, $D - D_0$ is relatively D_0-bounded with constant $C\rho$. For ρ so small that $C\rho < 1$ the first assertion follows from [**Ka**, Thm. 1.1, p. 190].

If ρ is small enough, we have by Lemma 4.1, for given $\varepsilon > 0$,

$$\|(D_0 - (D - \sigma_0 R_0))(\Phi)\|_{H^m(Z_{[0,\rho]}, E)} \leq \varepsilon \|\Phi\|_{H^{m+1}(Z_{[0,\rho]}, E)}$$

for all $\Phi \in H^{m+1}(M, E)$. Since ρ is small, the norm of the inverse $S_0 : H^m(Z_{[0,\rho]}, F) \to H^{m+1}(Z_{[0,\rho]}, E; B_0)$ of D_0 is bounded by a constant C independent of ρ, see (54). Thus if $\varepsilon < 1/C$, then

$$(D - \sigma_0 R_0) : H^{m+1}(Z_{[0,\rho]}, E; B_0) \to H^m(Z_{[0,\rho]}, F)$$

is also an isomorphism. □

THEOREM 6.11. *Assume the Standard Setup 1.5 and that D and D^* are complete. Then, for any $m \geq 0$,*

$$\begin{aligned} &\operatorname{dom}(D_{\max}) \cap H^{m+1}_{\mathrm{loc}}(M, E) \\ &\quad = \{\Phi \in \operatorname{dom}(D_{\max}) \mid D\Phi \in H^m_{\mathrm{loc}}(M, F) \text{ and } Q_{[0,\infty)}(\mathscr{R}\Phi) \\ &\quad \in H^{m+1/2}(\partial M, E)\}. \end{aligned}$$

PROOF. The case $m = 0$ is assertion (iii) in Theorem 6.7, so that we may assume $m > 0$ in the following.

Clearly, if $\Phi \in \operatorname{dom}(D_{\max}) \cap H^{m+1}_{\mathrm{loc}}(M, E)$, then $D\Phi \in H^m_{\mathrm{loc}}(M, F)$ and $\mathscr{R}\Phi \in H^{m+1/2}(\partial M, E)$. A fortiori, $Q_{[0,\infty)}(\mathscr{R}\Phi) \in H^{m+1/2}(\partial M, E)$.

Conversely, let $\Phi \in \operatorname{dom}(D_{\max})$ satisfy $D\Phi \in H^m_{\mathrm{loc}}(M, F)$ and $Q_{[0,\infty)}(\mathscr{R}\Phi) \in H^{m+1/2}(\partial M, E)$. By interior elliptic regularity, we may assume that Φ has support in $Z_{[0,\rho)}$, where ρ is as in Lemma 6.10. By induction on m, we may also assume that $\Phi \in H^m_{\mathrm{loc}}(M, E)$, and then $\sigma_0 R_0 \in H^m(Z_{[0,\rho]}, F)$.

Since now $\Phi|_{\{\rho\} \times \partial M} = 0$ and Φ is in the maximal domain of D_0 over $Z_{[0,\rho]}$ by Lemma 6.10 (i), we can apply Proposition 6.9 and get

$$\Phi = S_0 D_0 \Phi + \exp(-tA)(Q_{[0,\infty)}(\mathscr{R}\Phi)) =: \Phi_0 + \Phi_1.$$

Since $Q_{[0,\infty)}(\mathscr{R}\Phi) \in H^{m+1/2}(\partial M, E)$, we have $\Phi_1 \in H^{m+1}(Z_{[0,\rho]}, E)$. Furthermore, $\Phi_0 = S_0 D_0 \Phi \in H^1(Z_{[0,\rho]}, E; B_0)$ and

$$(D - \sigma_0 R_0)\Phi_0 = (D - \sigma_0 R_0)\Phi - (D - \sigma_0 R_0)\Phi_1 \in H^m(Z_{[0,\rho]}, F).$$

By Lemma 6.10 (ii) we have $\Phi_0 \in H^{m+1}(Z_{[0,\rho]}, E; B_0)$ and hence

$$\Phi = \Phi_0 + \Phi_1 \in H^{m+1}(Z_{[0,\rho]}, E). \qquad \Box$$

7. Boundary value problems

Assume the Standard Setup 1.5 and that D and D^* are complete.

7.1. Boundary conditions. We will use the following notation. For any subset $U \subset \bigcup_{s' \in \mathbb{R}} H^{s'}(A)$ and any $s \in \mathbb{R}$, we let

$$U^s := U \cap H^s(A), \quad \check{U} := U \cap \check{H}(A), \quad \hat{U} := U \cap \hat{H}(A).$$

DEFINITION 7.1. A closed linear subspace $B \subset \check{H}(A)$ will be called a *boundary condition* for D.

Later we will study the equation $D\Phi = \Psi$ with given $\Psi \in L^2(M,F)$ subject to the boundary condition $\varphi := \mathscr{R}\Phi \in B$ for $\Phi \in \mathrm{dom}(D_{\max})$. This explains the terminology; compare also Proposition 7.2 below.

For a boundary condition $B \subset \check{H}(A)$, we consider the operators $D_{B,\max}$ and D_B with domains

$$\begin{aligned} \mathrm{dom}(D_{B,\max}) &= \{\Phi \in \mathrm{dom}(D_{\max}) \mid \mathscr{R}\Phi \in B\}, \\ \mathrm{dom}(D_B) &= \{\Phi \in H^1_D(M,E) \mid \mathscr{R}\Phi \in B\} \\ &= \{\Phi \in \mathrm{dom}(D_{\max}) \mid \mathscr{R}\Phi \in H^{1/2}(\partial M, E)\}, \end{aligned}$$

and similarly for the formal adjoint D^*. By (ii) of Theorem 6.7, $\mathrm{dom}(D_{B,\max})$ is a closed subspace of $\mathrm{dom}(D_{\max})$. Since the trace map extends to a bounded linear map $\mathscr{R}: H^1_D(M,E) \to H^{1/2}(\partial M, E)$ and $H^{1/2}(\partial M,E) \hookrightarrow \check{H}(A)$ is a continuous embedding, $\mathrm{dom}(D_B)$ is also a closed subspace of $H^1_D(M,E)$.

In particular, $D_{B,\max}$ is a closed operator. Conversely, Proposition 1.50 of [**BBC**] reads as follows:

PROPOSITION 7.2. *Any closed extension of D between D_{cc} and $D_{\max}$ is of the form $D_{B,\max}$, where $B \subset \check{H}(A)$ is a closed subspace.*

PROOF. Let $\bar{D} \subset D_{\max}$ be a closed extension of D_{cc}. Then $\bar{D}$ extends the minimal closed extension $D_{\min} = D_{0,\max}$ of D_{cc}, that is,

$$\mathrm{dom}(D_{\min}) = \{\Phi \in \mathrm{dom}(D_{\max}) \mid \mathscr{R}(\Phi) = 0\} \subset \mathrm{dom}(\bar{D}),$$

compare Corollary 6.6. It follows that

$$\mathrm{dom}\,\bar{D} = \{\Phi \in \mathrm{dom}(D_{\max}) \mid \mathscr{R}(\Phi) \in B\},$$

where $B \subset \check{H}(A)$ is the space of all φ such that there exists a $\Phi \in \mathrm{dom}\,\bar{D}$ with $\mathscr{R}(\Phi) = \varphi$. In particular, $\mathscr{E}(\varphi)$ is contained in $\mathrm{dom}\,\bar{D}$, for any $\varphi \in B$. Now let (φ_j) be a sequence in B converging to φ in $\check{H}(A)$. Then $(\mathscr{E}(\varphi_j))$ converges to $\mathscr{E}(\varphi)$ in $\mathrm{dom}(D_{\max})$, by Lemma 6.1. Since $\bar{D}$ is closed, we have $\mathscr{E}(\varphi) \in \mathrm{dom}\,\bar{D}$. It follows that $\varphi \in B$ and hence that B is closed in $\check{H}(A)$. $\Box$

LEMMA 7.3. *Let B be a boundary condition. Then $B \subset H^{1/2}(\partial M, E)$ if and only if $D_B = D_{B,\max}$, and then there exists a constant $C > 0$ such that*

$$\|\Phi\|^2_{H^1_D(M)} \le C \cdot \left(\|\Phi\|^2_{L^2(M)} + \|D\Phi\|^2_{L^2(M)} \right)$$

for all $\Phi \in \mathrm{dom}(D_B)$.

PROOF. If $D_{B,\max} = D_B$ then $B \subset H^{1/2}(\partial M, E)$ by (iii) of Theorem 6.7. Conversely, if $B \subset H^{1/2}(\partial M, E)$, then

$$\begin{aligned} \mathrm{dom}(D_{B,\max}) &= \{\Phi \in \mathrm{dom}(D_{\max}) \mid \mathscr{R}\Phi \in B\} \\ &\subset \{\Phi \in \mathrm{dom}(D_{\max}) \mid \mathscr{R}\Phi \in H^{1/2}(\partial M, E)\} \\ &= H^1_D(M, E), \end{aligned}$$

again by (iii) of Theorem 6.7, and therefore $D_{B,\max} = D_B$.

Suppose now that $D_{B,\max} = D_B$. Since $\mathrm{dom}(D_{B,\max})$ is a closed subspace of $\mathrm{dom}(D_{\max})$ and $\mathrm{dom}(D_B)$ is a closed subspace of $H^1_D(M, E)$, we conclude that the H^1_D-norm and the graph norm for D are equivalent on $\mathrm{dom}(D_{B,\max}) = \mathrm{dom}(D_B)$. This shows the asserted inequality. □

7.2. Adjoint boundary condition. For any boundary condition B, we have $D_{cc} \subset D_{B,\max}$. Hence the L^2-adjoint operators satisfy

$$(D_{B,\max})^{\mathrm{ad}} \subset (D_{cc})^{\mathrm{ad}} = (D^*)_{\max}. \tag{60}$$

From (48), we conclude that

$$\begin{aligned} &\mathrm{dom}((D_{B,\max})^{\mathrm{ad}}) \\ &\quad = \{\Psi \in \mathrm{dom}((D^*)_{\max}) \mid (\sigma_0 \mathscr{R}\Phi, \mathscr{R}\Psi) = 0 \text{ for all } \Phi \in \mathrm{dom}(D_{B,\max})\}. \end{aligned} \tag{61}$$

Now for any $\varphi \in B$ there is a $\Phi \in \mathrm{dom}(D_{B,\max})$ with $\mathscr{R}\Phi = \varphi$. Therefore

$$(D_{B,\max})^{\mathrm{ad}} = (D^*)_{B^{\mathrm{ad}},\max} \tag{62}$$

with

$$B^{\mathrm{ad}} := \{\psi \in \check{H}(\tilde{A}) \mid (\sigma_0 \varphi, \psi) = 0 \text{ for all } \varphi \in B\}. \tag{63}$$

By Lemma 6.3, B^{ad} is a closed subspace of $\check{H}(\tilde{A})$. In other words, it is a boundary condition for D^*.

DEFINITION 7.4. We call B^{ad} the boundary condition *adjoint to B*.

The perfect pairing between $H^{1/2}(\partial M, E)$ and $H^{-1/2}(\partial M, E)$ as in (iv) on page 26 and the analogous perfect pairing between $\check{H}(A)$ and $\hat{H}(A)$ as in (40) coincide on the intersection $H^{1/2}(\partial M, E) \times \hat{H}(A)$. For a boundary condition B which is contained in $H^{1/2}(\partial M, E)$, it follows that

$$\sigma_0^*(B^{\mathrm{ad}}) = B^0 \cap \hat{H}(A), \tag{64}$$

where the superscript 0 indicates the annihilator in $H^{-1/2}(\partial M, F)$.

7.3. Elliptic boundary conditions.

DEFINITION 7.5. A linear subspace $B \subset H^{1/2}(\partial M, E)$ is called an *elliptic boundary condition* for A (or for D) if there is an L^2-orthogonal decomposition

$$L^2(\partial M, E) = V_- \oplus W_- \oplus V_+ \oplus W_+$$

such that

(i) W_- and W_+ are finite-dimensional and contained in $H^{1/2}(\partial M, E)$;
(ii) $V_- \oplus W_- \subset L^2_{(-\infty,a]}(A)$ and $V_+ \oplus W_+ \subset L^2_{[-a,\infty)}(A)$ for some $a \geq 0$;
(iii) there is a bounded (with respect to $\|\cdot\|_{L^2(\partial M)}$) linear map $g : V_- \to V_+$ with

$$g(V_-^{1/2}) \subset V_+^{1/2} \quad \text{and} \quad g^*(V_+^{1/2}) \subset V_-^{1/2}$$

such that

$$B = W_+ \oplus \{v + gv \mid v \in V_-^{1/2}\}.$$

EXAMPLE 7.6. If we put $V_- = L^2_{(-\infty,0)}(A)$, $V_+ = L^2_{[0,\infty)}(A)$, $W_- = W_+ = 0$, and $g = 0$, then we have

$$B = B_{\mathrm{APS}} := H^{1/2}_{(-\infty,0)}(A).$$

This is the well-known *Atiyah-Patodi-Singer boundary condition* as introduced in [**APS**].

We will use the notation $\Gamma(g) := \{v + gv \mid v \in V_-\}$ to denote the graph of g and similarly for the restriction of g to $V_-^{1/2}$, $\Gamma(g)^{1/2} := \{v + gv \mid v \in V_-^{1/2}\}$.

LEMMA 7.7. *Let $B \subset H^{1/2}(\partial M, E)$ be an elliptic boundary condition. Then*

$$V_- \oplus V_+ = \Gamma(g) \oplus \Gamma(-g^*),$$

where both decompositions are L^2-orthogonal.

PROOF. The decomposition $V_- \oplus V_+$ is L^2-orthogonal by assumption. With respect to the splitting $V_- \oplus V_+$ we have

$$\Gamma(g) = \left\{ \begin{pmatrix} v_- \\ gv_- \end{pmatrix} \mid v_- \in V_- \right\} \quad \text{and} \quad \Gamma(-g^*) = \left\{ \begin{pmatrix} -g^* v_+ \\ v_+ \end{pmatrix} \mid v_+ \in V_+ \right\}.$$

By the definition of g^*, the decomposition on the right hand side is L^2-orthogonal. An arbitrary element of $\Gamma(g) + \Gamma(-g^*)$ is given by

$$\begin{pmatrix} v_- - g^* v_+ \\ gv_- + v_+ \end{pmatrix} = \begin{pmatrix} \mathrm{id} & -g^* \\ g & \mathrm{id} \end{pmatrix} \begin{pmatrix} v_- \\ v_+ \end{pmatrix}.$$

Since

$$\begin{pmatrix} \mathrm{id} & g^* \\ -g & \mathrm{id} \end{pmatrix} \begin{pmatrix} \mathrm{id} & -g^* \\ g & \mathrm{id} \end{pmatrix} = \begin{pmatrix} \mathrm{id} & -g^* \\ g & \mathrm{id} \end{pmatrix} \begin{pmatrix} \mathrm{id} & g^* \\ -g & \mathrm{id} \end{pmatrix} = \begin{pmatrix} \mathrm{id} + g^*g & 0 \\ 0 & \mathrm{id} + gg^* \end{pmatrix}$$

we see that $\begin{pmatrix} \mathrm{id} & -g^* \\ g & \mathrm{id} \end{pmatrix}$ is an isomorphism with inverse

$$\begin{pmatrix} \mathrm{id} & g^* \\ -g & \mathrm{id} \end{pmatrix} \begin{pmatrix} \mathrm{id} + g^* g & 0 \\ 0 & \mathrm{id} + g g^* \end{pmatrix}^{-1}.$$

The decomposition follows. □

REMARK 7.8. With respect to the splitting $V_+ \oplus V_-$, the orthogonal projection onto $\Gamma(g)$ is given by

$$\begin{pmatrix} \mathrm{id} & -g^* \\ g & \mathrm{id} \end{pmatrix} \begin{pmatrix} v_- \\ v_+ \end{pmatrix} \mapsto \begin{pmatrix} \mathrm{id} & 0 \\ g & 0 \end{pmatrix} \begin{pmatrix} v_- \\ v_+ \end{pmatrix},$$

hence by the matrix

$$\begin{pmatrix} \mathrm{id} & 0 \\ g & 0 \end{pmatrix} \begin{pmatrix} \mathrm{id} & -g^* \\ g & \mathrm{id} \end{pmatrix}^{-1}. \tag{65}$$

LEMMA 7.9. *Let $B \subset H^{1/2}(\partial M, E)$ be an elliptic boundary condition. Then*

(i) the spaces $B, V_\pm^{1/2}$, and $W_\pm$ are closed in $H^{1/2}(\partial M, E)$ and

$$H^{1/2}(\partial M, E) = V_-^{1/2} \oplus W_- \oplus V_+^{1/2} \oplus W_+;$$

(ii) the spaces $V_\pm^{1/2}$ are dense in $V_\pm$;

(iii) the map g restricts to a continuous (w. r. t. $\|\cdot\|_{H^{1/2}(\partial M)}$) linear map $g : V_-^{1/2} \to V_+^{1/2}$, and similarly for g^*;

(iv) the L^2-orthogonal projections

$$\pi_\pm : L^2(\partial M, E) \twoheadrightarrow V_\pm \subset L^2(\partial M, E)$$

restrict to continuous projections

$$\pi_\pm^{1/2} : H^{1/2}(\partial M, E) \twoheadrightarrow V_\pm^{1/2} \subset H^{1/2}(\partial M, E).$$

PROOF. We start by proving (iv). Since $V_+ \oplus W_+ \subset L^2_{[-a,\infty)}(A)$ is the orthogonal complement of $V_- \oplus W_-$ in $L^2(\partial M, E)$, we have

$$L^2_{(-\infty,-a)}(A) \subset V_- \oplus W_- \subset L^2_{(-\infty,a]}(A).$$

Hence the orthogonal complement F of $L^2_{(-\infty,-a)}(A)$ in $V_- \oplus W_-$ is contained in $L^2_{[-a,a]}(A)$. In particular, F is finite-dimensional and contained in $H^{1/2}(\partial M, E)$. Thus the orthogonal projection $\pi_F : L^2(\partial M, E) \twoheadrightarrow F \subset L^2(\partial M, E)$ restricts to a continuous projection $H^{1/2}(\partial M, E) \twoheadrightarrow F \subset H^{1/2}(\partial M, E)$, and similarly for the orthogonal projection π_{W_-} onto W_-. Since the orthogonal projection $\pi_- : L^2(\partial M, E) \twoheadrightarrow V_- \subset L^2(\partial M, E)$ is given by $\pi_- = Q_{(-\infty,-a)} + \pi_F - \pi_{W_-}$, it restricts to a continuous projection $H^{1/2}(\partial M, E) \twoheadrightarrow V_-^{1/2} \subset H^{1/2}(\partial M, E)$ as asserted. The case V_+ is analogous.

Clearly, (iv) implies (ii) and shows that $V_\pm^{1/2}$ is closed in $H^{1/2}(\partial M,E)$. Since they are finite-dimensional, W_+ and W_- are also closed in $H^{1/2}(\partial M,E)$. Moreover, we obtain the decomposition of $H^{1/2}(\partial M,E)$ as asserted in (i).

We now show (iii). Let $\varphi_j \to \varphi$ in $V_-^{1/2}$ and $g\varphi_j \to \psi$ in $V_+^{1/2}$ as $j \to \infty$. Then $\varphi_j \to \varphi$ in V_- and, since g is continuous on V_-, $g\varphi_j \to g\varphi$ in V_+. Thus $\psi = g\varphi$. This shows that the graph $\Gamma(g)^{1/2}$ of $g : V_-^{1/2} \to V_+^{1/2}$ is closed. The closed graph theorem implies (iii).

Since W_+ is closed in $H^{1/2}(\partial M,E)$ and $\Gamma(g)^{1/2}$ is closed in $V_-^{1/2} \oplus V_+^{1/2}$, it follows that B is closed in $H^{1/2}(\partial M,E)$ as well. This concludes the proof of (i) and of the lemma. □

Lemma 7.10. *Let $B \subset H^{1/2}(\partial M,E)$ be an elliptic boundary condition. Then there exists a constant $C>0$ such that*

$$\|\varphi\|_{H^{1/2}(\partial M)} \le C \cdot \|\varphi\|_{\check{H}(A)}$$

for all $\varphi \in B$.

Since we always have $\|\cdot\|_{\check{H}(A)} \le \|\cdot\|_{H^{1/2}(\partial M)}$, Lemma 7.10 says that on B the $H^{1/2}$ and $\check{H}$-norms are equivalent. In particular, B is also closed in $\check{H}(A)$ and hence a boundary condition in the sense of Definition 7.1.

Proof. We apply Proposition A.3 with $X = B$ (with the $H^{1/2}$-norm), $Y = H^{1/2}_{(-\infty,\Lambda]}(A)$, and $Z = H^{-1/2}(\partial M,E)$. The linear map

$$L : X = B \hookrightarrow H^{1/2}(\partial M,E) \xrightarrow{Q_{(-\infty,\Lambda]}} H^{1/2}_{(-\infty,\Lambda]}(A) = Y$$

is bounded and the inclusion

$$K : X = B \hookrightarrow H^{1/2}(\partial M,E) \hookrightarrow H^{-1/2}(\partial M,E) = Z$$

is compact. Since K is injective, $\ker K \cap \ker L = \{0\}$. We need to show that the kernel of L is finite-dimensional and that its range is closed. Then the implication (i) ⇒ (iv) of Proposition A.3 yields the desired inequality.

Without loss of generality we assume that $\Lambda \ge a$, where Λ is as in (35) and a as in Definition 7.5. Let $w + v + gv \in \ker(L)$, $w \in W_+$, $v \in V_-$. Then

$$0 = Q_{(-\infty,\Lambda]}(w + v + gv) = Q_{(-\infty,\Lambda]}(w) + v + Q_{[-a,\Lambda]}(gv),$$

hence

$$\begin{aligned} v &= -Q_{(-\infty,\Lambda]}(w) - Q_{[-a,\Lambda]}(gv) \\ &\in Q_{(-\infty,\Lambda]}(W_+) + Q_{[-a,\Lambda]}(H^{1/2}(\partial M,E)) =: F. \end{aligned}$$

Thus $\ker(L) \subset W_+ + \Gamma(g|_F)$ which is finite-dimensional. Here $\Gamma(g|_F) = \{x + gx \mid x \in F\}$ denotes the graph of g restricted to the finite-dimensional space F.

The image of L is given by

$$\operatorname{im}(L) = Q_{(-\infty,\Lambda]}(B) = \underbrace{Q_{(-\infty,\Lambda]}(W_+)}_{\text{finite dimensional}} + \underbrace{\Gamma(Q_{(-\infty,\Lambda]} \circ g : V_-^{1/2} \to Y)}_{\text{closed}}$$

which is closed because g (and hence $Q_{(-\infty,\Lambda]} \circ g$) is $H^{1/2}$–bounded by Lemma 7.9(iii). □

THEOREM 7.11. *Assume the Standard Setup 1.5. Then, for a linear subspace $B \subset H^{1/2}(\partial M, E)$, the following are equivalent:*

(i) B is closed in $\check{H}(A)$ and $B^{\mathrm{ad}} \subset H^{1/2}(\partial M, F)$.

(ii) For any $a \in \mathbb{R}$, we can choose orthogonal decompositions

$$L^2_{(-\infty,a)}(A) = V_- \oplus W_- \quad \textit{and} \quad L^2_{[a,\infty)}(A) = V_+ \oplus W_+$$

and $g : V_- \to V_+$ as in Definition 7.5 such that

$$B = W_+ \oplus \{v + gv \mid v \in V_-^{1/2}\}.$$

(iii) B is an elliptic boundary condition.

Moreover, for an elliptic boundary condition B, the adjoint boundary condition $B^{\mathrm{ad}} \subset \check{H}(\tilde{A}) = (\sigma_0^*)^{-1}(\hat{H}(A))$ is also elliptic (for D^*) with

$$\sigma_0^*(B^{\mathrm{ad}}) = W_- \oplus \{u - g^*u \mid u \in V_+^{1/2}\}.$$

REMARK 7.12. In [**BBC**], the point of departure for elliptic boundary conditions is Property (i). The equivalence between (i) and (ii) was already observed there. Since our setup is slightly different and more general, we repeat the argument.

PROOF. It is trivial that (ii) implies (iii). We show that both (i) and the assertion on the adjoint boundary condition are implied by (iii). By Lemma 7.10, B is closed in $\check{H}(A)$. Denote the closure of $V_\pm$ with respect to $\|\cdot\|_{H^{-1/2}(\partial M)}$ by $V_\pm^{-1/2}$. Then we have

$$\hat{H}(A) = V_-^{-1/2} \oplus W_- \oplus V_+^{1/2} \oplus W_+.$$

By (64), an element

$$v_- + w_- + v_+ + w_+ \in V_-^{-1/2} \oplus W_- \oplus V_+^{1/2} \oplus W_+$$

lies in $\sigma_0^*(B^{\mathrm{ad}})$ if and only if

$$(\tilde{w}_+ + \tilde{v}_- + g\tilde{v}_-, v_- + w_- + v_+ + w_+) = 0$$

for all $\tilde{w}_+ \in W_+, \tilde{v}_- \in V_-^{1/2}$, i.e. if and only if

$$(\tilde{w}_+, w_+) + (\tilde{v}_-, v_-) + (g\tilde{v}_-, v_+) = 0$$

for all $\tilde{w}_+ \in W_+, \tilde{v}_- \in V_-^{1/2}$, i.e., if and only if it is of the form

$$-g^* v_+ + w_- + v_+, \quad \text{where } w_- \in W_-,\ v_+ \in V_+^{1/2}.$$

Hence

$$\sigma_0^*(B^{\mathrm{ad}}) = W_- \oplus \{v_+ - g^* v_+ \mid v_+ \in V_+^{1/2}\}$$

as asserted. In particular, $B^{\mathrm{ad}} \subset (\sigma_0^*)^{-1}(H^{1/2}(\partial M, E)) = H^{1/2}(\partial M, F)$.

It remains to show that (i) implies (ii). Fix $a \in \mathbb{R}$. Consider the spaces W_+, $V_-^{1/2}$, $V_+^{1/2}$, V_+, and $V_+^{-1/2}$ as in Lemma 7.13 below. In particular,

$$B = W_+ \oplus \{v + gv \mid v \in V_-^{1/2}\}. \tag{66}$$

By (64), $C := \sigma_0^*(B^{\mathrm{ad}}) = B^0 \cap \hat{H}(A)$. In particular, C is closed in $\hat{H}(A) = \check{H}(-A)$. By assumption, $C \subset H^{1/2}(\partial M, E)$, hence we can apply Lemma 7.13 with $-A$ instead of A and $-a-\varepsilon$ instead of a, where $\varepsilon > 0$ is chosen so small that A has no eigenvalues in $[a-\varepsilon, a)$. We obtain subspaces

(a) $W_- = C \cap H^{1/2}_{[-a+\varepsilon,\infty)}(-A) = C \cap H^{1/2}_{(-\infty,a-\varepsilon]}(A) = C \cap H^{1/2}_{(-\infty,a)}(A)$;
(b) $U_+^{1/2} = Q_{[a,\infty)}(C)$;
(c) $U_-^{-1/2}$, the annihilator of W_- in $H^{-1/2}_{(-\infty,a)}(A)$;
(d) $U_- = U_-^{-1/2} \cap L^2(\partial M, E)$ and $U_-^{1/2} = U_-^{-1/2} \cap H^{1/2}(\partial M, E)$.

Moreover,

$$C = W_- \oplus \{u + hu \mid u \in U_+^{1/2}\}. \tag{67}$$

We have

$$\begin{aligned} W_- &= C \cap H^{1/2}_{(-\infty,a)}(A) \\ &= B^0 \cap \hat{H}(A) \cap H^{1/2}_{(-\infty,a)}(A) \\ &= B^0 \cap \hat{H}(A) \cap H^{-1/2}_{(-\infty,a)}(A), \end{aligned}$$

where we use that $B^0 \cap \hat{H}(A) \subset H^{1/2}(\partial M, E)$ to pass from the second to the third line. Now $\hat{H}(A) \cap H^{-1/2}_{(-\infty,a)}(A) = H^{-1/2}_{(-\infty,a)}(A)$ by the definition of $\hat{H}(A)$. We conclude that

$$\begin{aligned} W_- &= B^0 \cap H^{-1/2}_{(-\infty,a)}(A) \\ &= \{x \in H^{-1/2}_{(-\infty,a)}(A) \mid (x, w + v + gv) = 0 \text{ for all } w \in W_+,\ v \in V_-^{1/2}\} \\ &= \{x \in H^{-1/2}_{(-\infty,a)}(A) \mid (x, v) = 0 \text{ for all } v \in V_-^{1/2}\}. \end{aligned}$$

Hence W_- is the annihilator of $V_-^{1/2}$ in $H^{-1/2}_{(-\infty,a)}(A)$. By Lemma 7.13, W_- is also the annihilator of $U_-^{1/2}$ in $H^{-1/2}_{(-\infty,a)}(A)$. Thus

$$U_-^{1/2} = V_-^{1/2}.$$

By interchanging the roles of B and C, we also get

$$U_+^{1/2} = V_+^{1/2}.$$

We set

$$V_- := U_- \quad \text{and} \quad V_-^{-1/2} := U_-^{-1/2}$$

and get by Lemma 7.13 (iii)

$$H^{-1/2}(\partial M, E) = H^{-1/2}_{(-\infty,a)}(A) \oplus H^{-1/2}_{[a,\infty)}(A) = W_- \oplus V_-^{-1/2} \oplus W_+ \oplus V_+^{-1/2}$$

and similarly for $L^2(\partial M, E)$ and $H^{1/2}(\partial M, E)$.

It follows that the annihilators of B and C in $H^{-1/2}(\partial M, E)$ are given by

$$B^0 = W_- \oplus \{u - g'u \mid u \in V_+^{-1/2}\},$$
$$C^0 = W_+ \oplus \{v - h'v \mid u \in V_-^{-1/2}\},$$

where $g' : V_+^{-1/2} \to V_-^{-1/2}$ is the dual map of g and similarly for h. Furthermore, we get that the restriction of $-h'$ to $V_-^{1/2}$ equals g and the restriction of $-g'$ to $V_-^{1/2}$ coincides with h. By interpolation, $-h'$ restricts to a continuous linear map $V_- \to V_+$, again denoted by g, and $-g'$ restricts to $-g^*$. This shows that (i) implies (ii). □

LEMMA 7.13. *Let $B \subset \check{H}(A)$ be a boundary condition which is contained in $H^{1/2}(\partial M, E)$. Let $a \in \mathbb{R}$. Define*

(a) $W_+ := B \cap H^{1/2}_{[a,\infty)}(A)$;

(b) $V_-^{1/2} := Q_{(-\infty,a)}(B)$;

(c) $V_+^{-1/2}$ *to be the annihilator of* W_+ *in* $H^{-1/2}_{[a,\infty)}(A)$;

(d) $V_+ := V_+^{-1/2} \cap L^2(\partial M, E)$ *and* $V_+^{1/2} := V_+^{-1/2} \cap H^{1/2}(\partial M, E)$.

Then

(i) W_+ *is finite-dimensional and equals the annihilator of* $V_+^{1/2}$ *in* $H^{-1/2}_{[a,\infty)}(A)$;

(ii) $V_-^{1/2}$ *is a closed subspace of* $H^{1/2}_{(-\infty,a)}(A)$;

(iii) $H^{-1/2}_{[a,\infty)}(A) = W_+ \oplus V_+^{-1/2}$,
$L^2_{[a,\infty)}(A) = W_+ \oplus V_+$,
$H^{1/2}_{[a,\infty)}(A) = W_+ \oplus V_+^{1/2}$,
where the second decomposition is L^2-orthogonal;

(iv) the perfect pairing between $H^{1/2}(\partial M, E)$ and $H^{-1/2}(\partial M, E)$ restricts to a perfect pairing between $V_+^{1/2}$ and $V_+^{-1/2}$;

(v) there exists a continuous linear map $g : V_-^{1/2} \to V_+^{1/2}$ such that

$$B = W_+ \oplus \{v + gv \mid v \in V_-^{1/2}\}.$$

PROOF. Since B is closed in $\check{H}(A)$ and contained in $H^{1/2}(\partial M, E)$, there is an estimate

$$\begin{aligned} \|\varphi\|_{H^{1/2}(\partial M)} &\le C\|\varphi\|_{\check{H}(A)} \\ &\le C'(\|Q_{(-\infty,a)}\varphi\|_{H^{1/2}(\partial M)} + \|Q_{[a,\infty)}\varphi\|_{H^{-1/2}(\partial M)}) \end{aligned} \tag{68}$$

for all $\varphi \in B$. Since $B \subset H^{1/2}(\partial M, E)$ is closed, the linear map

$$Q_{[a,\infty)} : B \to H^{1/2}_{[a,\infty)}(A) \hookrightarrow H^{-1/2}_{[a,\infty)}(A)$$

is compact by Rellich's theorem. It follows from Proposition A.3 that $Q_{(-\infty,a)} : B \to H^{1/2}_{(-\infty,a)}(A)$ has finite-dimensional kernel W_+ and closed image $V_-^{1/2}$. This shows (ii) and the first part of (i).

Since $W_+ \subset H^{1/2}_{[a,\infty)}(A)$ is finite-dimensional, Lemma A.4 implies (iii), (iv), and the second part of (i).

Let G be the L^2-orthogonal complement of W_+ in B. Then $B = W_+ \oplus G$ and $Q_{(-\infty,a)} : G \to V_-^{1/2}$ is an isomorphism. Compose its inverse $V_-^{1/2} \to G$ with $Q_{[a,\infty)} : G \to H^{1/2}_{[a,\infty)}(A)$ to obtain a continuous linear map $g : V_-^{1/2} \to H^{1/2}_{[a,\infty)}(A)$. Since G is L^2-orthogonal to W_+ and since V_+ is the L^2-orthogonal complement of W_+ in $L^2_{[a,\infty)}(A)$ by (iii), the map g takes values in $V_+^{1/2}$. In conclusion,

$$B = W_+ \oplus \{v + gv \mid v \in V_-^{1/2}\},$$

where g is now considered as a continuous linear map $V_-^{1/2} \to V_+^{1/2}$. □

7.4. Boundary regularity.

LEMMA 7.14. *Let $B \subset H^{1/2}(\partial M, E)$ be an elliptic boundary condition. Let $s \ge 1/2$. Then the following are equivalent:*

(i) There exist $V_\pm$, $W_\pm$, and g for B as in Definition 7.5 such that $W_+ \subset H^s(\partial M, E)$ and $g(V_-^s) \subset V_+^s$.
(ii) For all $V_\pm$, $W_\pm$, and g for B as in Definition 7.5, we have that $W_+ \subset H^s(\partial M, E)$ and $g(V_-^s) \subset V_+^s$.
(iii) For all $\varphi \in B$ with $Q_{(-\infty,0)}(\varphi) \in H^s(\partial M, E)$, we have $\varphi \in H^s(\partial M, E)$.

PROOF. Clearly, (ii) implies (i). We show that (i) implies (iii). Let $V_\pm$, $W_\pm$, and g be as in (i). Let $\varphi \in B$ with $Q_{(-\infty,0)}(\varphi) \in H^s(\partial M, E)$. We have to show $\varphi \in H^s(\partial M, E)$. Write $\varphi = w_+ + v_- + gv_-$ with $w_+ \in W_+$ and $v_- \in V_-$. Since $V_+ \subset L^2_{[-a,\infty)}(A)$ for some $a \ge 0$, we have

$$Q_{(-\infty,-a)}(\varphi) = Q_{(-\infty,-a)}(w_+ + v_-)$$

and thus, using $V_- \subset L^2_{(-\infty,a]}(A)$,

$$\begin{aligned} v_- &= Q_{(-\infty,a]}(v_-) \\ &= Q_{(-\infty,-a)}(v_-) + Q_{[-a,a]}(v_-) \\ &= Q_{(-\infty,-a)}(\varphi) - Q_{(-\infty,-a)}(w_+) + Q_{[-a,a]}(v_-) \\ &= Q_{(-\infty,0)}(\varphi) - Q_{[-a,0)}(\varphi) - Q_{(-\infty,-a)}(w_+) + Q_{[-a,a]}(v_-). \end{aligned}$$

Since for bounded intervals I the projection Q_I takes values in smooth sections and since $w_+ \in H^s(\partial M, E)$, all four terms on the right hand side are contained in $H^s(\partial M, E)$. Hence $v_- \in H^s(\partial M, E)$. By the assumption on g, we also have $gv_- \in H^s(\partial M, E)$ and therefore $\varphi \in H^s(\partial M, E)$.

It remains to show that (iii) implies (ii). Let $V_\pm$, $W_\pm$, and g be for B as in Definition 7.5. Since $W_+ \subset L^2_{[-a,\infty)}(A)$, we have for $w_+ \in W_+$ that

$$Q_{(-\infty,0)}(w_+) = Q_{[-a,0)}(w_+) \in C^\infty(\partial M, E).$$

By (iii), $w_+ \in H^s(\partial M, E)$. Hence $W_+ \subset H^s(\partial M, E)$.

Now let $v_- \in V_-^s$. Then $v_- + gv_- \in B$ and

$$Q_{(-\infty,0)}(v_- + gv_-) = Q_{(-\infty,0)}(v_-) + Q_{[-a,0)}(gv_-) \in H^s(\partial M, E).$$

Again by (iii), $v_- + gv_- \in H^s(\partial M, E)$ and therefore $gv_- \in H^s(\partial M, E)$. This shows $g(V_-^s) \subset V_+^s$. □

DEFINITION 7.15. An elliptic boundary condition $B \subset H^{1/2}(\partial M, E)$ is called *$(s+1/2)$-regular* if the assertions in Lemma 7.14 hold for B and for B^{ad}. If B is $(s+1/2)$-regular for all $s \geq 1/2$, then B is called *∞-regular.*

REMARKS 7.16. (a) An elliptic boundary condition $B \subset H^{1/2}(\partial M, E)$ is $(s+1/2)$-regular if and only if for some (or equivalently all) $V_\pm$, $W_\pm$, and g as in Definition 7.5 we have $W_+ \cup W_- \subset H^s(\partial M, E)$ and $g(V_-^s) \subset V_+^s$ as well as $g^*(V_+^s) \subset V_-^s$. In this case, g and g^* are continuous with respect to the H^s-norm.

(b) If B is $(s+1/2)$-regular, then so is B^{ad}.

(c) By definition, every elliptic boundary condition is 1-regular. By interpolation one sees that if B is $(s+1/2)$-regular, then B is also t-regular for all $1 \leq t \leq s+1/2$.

THEOREM 7.17 (Higher boundary regularity). *Assume the Standard Setup 1.5 and that D and D^* are complete. Let $m \in \mathbb{N}$ and $B \subset H^{1/2}(\partial M, E)$ be an m-regular elliptic boundary condition for D.*

Then, for all $0 \leq k < m$ and $\Phi \in \mathrm{dom}(D_{\max})$, we have

$$\mathscr{R}(\Phi) \in B \quad \textit{and} \quad D_{\max}\Phi \in H^k_{\mathrm{loc}}(M, F) \implies \Phi \in H^{k+1}_{\mathrm{loc}}(M, E).$$

PROOF. Since B is an m-regular elliptic boundary condition, Theorem 7.11, Lemma 7.14, and (c) in Remark 7.16 above imply that we have an orthogonal decomposition

$$H_{(-\infty,0)}(A) = V_- \oplus W_- \quad \text{and} \quad H_{[0,\infty)}(A) = V_+ \oplus W_+$$

with $W_\pm \subset H^{m-1/2}(\partial M, E)$ of finite dimension and a bounded linear map $g : V_- \to V_+$ with $g(V_-^{k+1/2}) \subset V_+^{k+1/2}$ and $g^*(V_+^{k+1/2}) \subset V_-^{k+1/2}$ for all $0 \le k < m$ such that

$$B = W_+ \oplus \{v + gv \mid v \in V_-^{1/2}\},$$
$$\sigma_0^*(B^{\mathrm{ad}}) = W_- \oplus \{u - g^*u \mid u \in V_+^{1/2}\}.$$

It is no loss of generality to enlarge B by extending g along W_- by 0. Hence we may assume that $V_- = H_{(-\infty,0)}(A)$ and $W_- = \{0\}$.

We identify a small collar about the boundary with $Z_{[0,\rho]} = [0,\rho] \times \partial M$, where $0 < \rho < r$ and r is as in Lemma 2.4. Assume first that we have constant coefficients $D = D_0$ over $Z_{[0,\rho]}$.

At the "right end" $t = \rho$ we impose the boundary condition $B_\rho = H_{[0,\infty)}(A)$, at the "left end" $t = 0$ we impose B. Since $V_- = H_{(-\infty,0)}(A)$, we have for Φ with $\mathscr{R}(\Phi) = \Phi(0) \in B$,

$$Q_{[0,\infty)}(\mathscr{R}(\Phi)) = gQ_{(-\infty,0)}(\mathscr{R}(\Phi)) + P_{W_+}(\mathscr{R}(\Phi)),$$

where $P_{W_+} : L^2(\partial M, E) \to W_+$ is the orthogonal projection. Using this and Proposition 6.9 one easily checks that the maps

$$\mathscr{F}_{B,0} : H^{k+1}(Z_{[0,\rho]}, E; B \oplus B_\rho) \to H^k(Z_{[0,\rho]}, F) \oplus W_+,$$
$$\mathscr{F}_{B,0}(\Phi) := (D_0\Phi, P_{W_+}\mathscr{R}(\Phi)),$$

and

$$\mathscr{G}_{B,0} : H^k(Z_{[0,\rho]}, F) \oplus W_+ \to H^{k+1}(Z_{[0,\rho]}, E; B \oplus B_\rho),$$
$$\mathscr{G}_{B,0}(\Psi, \varphi) := S_0\Psi + \exp(-tA)(gQ_{(-\infty,0)}(\mathscr{R}(S_0\Psi)) + \varphi),$$

are inverse to each other for all $k \in \mathbb{N}$. Hence $\mathscr{F}_{B,0}$ is an isomorphism for all $k \in \mathbb{N}$.

Passing now to variable coefficients, we compare with constant coefficients. We still let B_ρ on the right side of $Z_{[0,\rho]}$ be defined with respect to A. Arguing as in the proof of Lemma 6.10, we find that

$$\mathscr{F}_B : H^{k+1}(Z_{[0,\rho]}, E; B \oplus B_\rho) \to H^k(Z_{[0,\rho]}, E) \oplus W_+,$$
$$\mathscr{F}_B(\Phi) := ((D - \sigma_0 R_0)\Phi, P_{W_+}\mathscr{R}(\Phi)),$$

is also an isomorphism for $0 \le k < m$ and ρ small enough. Denote the inverse of $\mathscr{F}_B$ by $\mathscr{G}_B$.

Suppose now that $\mathscr{R}(\Phi) \in B$ and $D\Phi \in H^k_{\mathrm{loc}}(M, F)$. We only need to show that Φ is H^{k+1} near ∂M. Multiplying by a smooth cut-off function, we can assume that the support of Φ lies inside $Z_{[0,\rho]}$ and does not intersect the right part of the boundary of $Z_{[0,\rho]}$. Then $\Phi \in H^1(Z_{[0,\rho]}, E; B \oplus B_\rho)$ by Theorem 6.7 (iii) because $B \subset H^{1/2}(\partial M, E)$. By induction on k, we can also assume that $\Phi \in H^k(Z_{[0,\rho]}, E)$. Then

$$\mathscr{F}_B(\Phi) = ((D - \sigma_0 R_0)\Phi, P_{W_+}\mathscr{R}(\Phi)) \in H^k(Z_{[0,\rho]}, F) \oplus W_+$$

and therefore

$$\Phi = \mathscr{G}_B(\mathscr{F}_B(\Phi)) \in H^{k+1}(Z_{[0,\rho]}, E; B \oplus B_\rho). \qquad \square$$

COROLLARY 7.18. *Assume the Standard Setup 1.5 and that D and D^* are complete. Let $B \subset H^{1/2}(\partial M, E)$ be an ∞-regular elliptic boundary condition for D.*

Then each $\Phi \in \mathrm{dom}(D_{\max})$ with $D_{\max}\Phi = 0$ and $\mathscr{R}\Phi \in B$ is smooth up to the boundary.

PROOF. By standard elliptic regularity theory, Φ is smooth in the interior of M. Theorem 7.17 shows that $\Phi \in H^m_{\mathrm{loc}}(M, E)$ for all m. The Sobolev embedding theorem now implies that Φ is smooth up to the boundary. $\square$

7.5. Local and pseudo-local boundary conditions.

DEFINITION 7.19. We say that $B \subset H^{1/2}(\partial M, E)$ is a *local boundary condition* if there is a subbundle $E' \subset E_{|\partial M}$ such that

$$B = H^{1/2}(\partial M, E').$$

More generally, we call B *pseudo-local* if there is a classical pseudo-differential operator P of order 0 acting on sections of E over ∂M inducing an orthogonal projection on $L^2(\partial M, E)$ such that

$$B = P(H^{1/2}(\partial M, E)).$$

THEOREM 7.20. *Assume the Standard Setup 1.5 and that D and D^* are complete. Let P be a classical pseudo-differential operator of order zero acting on section of E over ∂M. Suppose that P induces an orthogonal projection in $L^2(\partial M, E)$. Then the following are equivalent:*

(i) *$B = P(H^{1/2}(\partial M, E))$ is an elliptic boundary condition for D.*
(ii) *For some (and then all) $a \in \mathbb{R}$, $P - Q_{[a,\infty)} : L^2(\partial M, E) \to L^2(\partial M, E)$ is a Fredholm operator.*
(iii) *For some (and then all) $a \in \mathbb{R}$, $P - Q_{[a,\infty)} : L^2(\partial M, E) \to L^2(\partial M, E)$ is an elliptic classical pseudo-differential operator of order zero.*
(iv) *For all $\xi \in T^*_x\partial M \setminus \{0\}$, $x \in \partial M$, the principal symbol $\sigma_P(\xi) : E_x \to E_x$ restricts to an isomorphism from the sum of the eigenspaces to the negative eigenvalues of $i\sigma_A(\xi)$ onto its image $\sigma_P(\xi)(E_x)$.*

REMARKS 7.21. (a) Variants of the equivalence of (i) with (ii) are contained in [**Gi**, pp. 75–77], [**BL2**, Thm. 5.6], and [**BBC**, Thm. 1.95]. A special case of the equivalence of (ii) with (iv) is contained in [**BL2**, Thm. 7.2].

(b) Not every elliptic boundary condition is pseudo-local, i.e., the orthogonal projection onto an elliptic boundary condition B is not always given by a pseudo-differential operator. For example, let $V_- = L^2_{(-\infty,0]}(A)$, $W_- = 0$, and $g = 0$. Choose $\varphi \in H^{1/2}_{(0,\infty)}(A)$ which is not smooth, put $W_+ = \mathbb{C}\cdot\varphi$ and let V_+ be the orthogonal complement of W_+ in $L^2_{(0,\infty)}(A)$. Each eigensection for A to a positive eigenvalue which occurs in the spectral decomposition of φ is a smooth section which is mapped by P to a nontrivial multiple of φ. If P were pseudo-differential it would map smooth sections to smooth sections.

PROOF OF THEOREM 7.20. The equivalence of (ii) and (iii) follows from Theorems 19.5.1 and 19.5.2 in [**Hö**].

It is known that the spectral projection $Q := Q_{[a,\infty)}$, $a \in \mathbb{R}$, is a classical pseudo-differential operator of order zero and that its principal symbol $i\sigma_Q(\xi)$, $\xi \in T^*\partial M \setminus \{0\}$, is the orthogonal projection onto the sum of the eigenspaces of positive eigenvalues of $i\sigma_A(\xi)$, see [**APS**, p. 48] together with [**Se**] or [**BW**, Prop. 14.2]. Both, $i\sigma_P(\xi)$ and $i\sigma_Q(\xi)$, are orthogonal projections. Hence $i\sigma_{P-Q}(\xi) = i\sigma_P(\xi) - i\sigma_Q(\xi)$ is an isomorphism if and only if $i\sigma_P(\xi)$ induces an isomorphism between its image and the orthogonal complement of the image of $i\sigma_Q(\xi)$. This shows the equivalence of (iii) and (iv).

It remains to show the equivalence of (i) with the other conditions. We show that (iii) implies (i). We check the first characterization in Theorem 7.11. First we observe that if $\varphi_j \in B$ converge in $H^{1/2}(\partial M, E)$ to an element φ, then $\varphi_j = P(\varphi_j) \to P(\varphi) \in B$, hence B is closed in $H^{1/2}(\partial M, E)$. Since $P-Q$ is elliptic of order zero, it has a parametrix R, that is, R is a classical pseudo-differential operator of order zero such that $R(P-Q) = \mathrm{id} + S$, where S is a smoothing operator. For $\varphi \in B$ we have

$$\begin{aligned}
\|\varphi\|_{H^{1/2}(\partial M)} &\le \|(\mathrm{id}+S)\varphi\|_{H^{1/2}(\partial M)} + \|S\varphi\|_{H^{1/2}(\partial M)} \\
&= \|R(P-Q)\varphi\|_{H^{1/2}(\partial M)} + \|S\varphi\|_{H^{1/2}(\partial M)} \\
&\le C_1 \cdot \Big(\|(P-Q)\varphi\|_{H^{1/2}(\partial M)} + \|\varphi\|_{H^{-1/2}(\partial M)}\Big) \\
&= C_1 \cdot \Big(\|(\mathrm{id}-Q)\varphi\|_{H^{1/2}(\partial M)} + \|\varphi\|_{H^{-1/2}(\partial M)}\Big) \\
&\le C_2 \cdot \|\varphi\|_{\check H(A)}.
\end{aligned}$$

Thus the $H^{1/2}$ and $\check H$–norms are equivalent on B, hence B is closed in $\check H(A)$.

Now let $\psi \in \sigma_0^*(B^{\mathrm{ad}})$. Then $\psi \in \hat H(A)$, and we have for all $\varphi \in H^{1/2}(A)$:

$$0 = (P\varphi, \psi) = (\varphi, P\psi).$$

Hence $P\psi = 0$ and thus $(P - Q)(\psi) = -Q(\psi) \in H^{1/2}(A)$ since $\psi \in \hat{H}(A)$. Therefore

$$\psi = (\mathrm{id} + S)\psi - S\psi = \underbrace{R(P-Q)\psi}_{\in H^{1/2}} - \underbrace{S\psi}_{\in C^\infty} \in H^{1/2}(\partial M, E).$$

Thus $\sigma_0^*(B^{\mathrm{ad}}) \subset H^{1/2}(\partial M, E)$. Hence $B^{\mathrm{ad}} \subset H^{1/2}(\partial M, F)$ and (i) follows.

Finally, we show that (i) implies (ii). Let $L^2(\partial M, E) = V_- \oplus W_- \oplus V_+ \oplus W_+$ and $B = P(H^{1/2}(\partial M, E)) = W_+ \oplus \Gamma(g)^{1/2}$, hence $P(L^2(\partial M, E)) = W_+ \oplus \Gamma(g)$. Since the sum of a Fredholm operator and a finite-rank-operator is again a Fredholm operator and since W_+ and W_- are finite-dimensional, we can assume without loss of generality that $W_+ = W_- = 0$. Let $a \in \mathbb{R}$ such that $V_- \subset L^2_{(-\infty,a]}(A)$ and $V_+ \subset L^2_{[-a,\infty)}(A)$. Since $L^2_{[-a,a]}(A)$ is finite-dimensional, we can furthermore assume that $a = 0$, $V_- = L^2_{(-\infty,0)}(A)$, and $V_+ = L^2_{[0,\infty)}(A)$. With respect to the splitting $L^2(\partial M, E) = V_- \oplus V_+$ we have

$$Q = \begin{pmatrix} 0 & 0 \\ 0 & \mathrm{id} \end{pmatrix} \quad \text{and, by (65),} \quad P = \begin{pmatrix} \mathrm{id} & 0 \\ g & 0 \end{pmatrix} \begin{pmatrix} \mathrm{id} & -g^* \\ g & \mathrm{id} \end{pmatrix}^{-1},$$

hence

$$P - Q = \begin{pmatrix} \mathrm{id} & 0 \\ 0 & -\mathrm{id} \end{pmatrix} \begin{pmatrix} \mathrm{id} & -g^* \\ g & \mathrm{id} \end{pmatrix}^{-1}$$

is an isomorphism. □

COROLLARY 7.22. *Assume the Standard Setup 1.5 and that D and D^* are complete. Let $E' \subset E_{|\partial M}$ be a subbundle and let $P : E_{|\partial M} \to E'$ be the fiberwise orthogonal projection.*

If $(D, \mathrm{id} - P)$ is an elliptic boundary value problem in the classical sense of Lopatinsky and Shapiro, then $B = H^{1/2}(\partial M, E')$ is a local elliptic boundary value condition in the sense of Definition 7.5.

PROOF. We use a boundary defining function t as in Lemma 2.4. Let $(D, \mathrm{id} - P)$ be an elliptic boundary value problem in the classical sense of Lopatinsky and Shapiro, see e. g. [**Gi**, Sec. 1.9]. This means that the rank of E' is half of that of E and that, for any $x \in \partial M$, any $\eta \in T_x^*\partial M \setminus \{0\}$, and any $\varphi \in (E'_x)^\perp$, there is a unique solution $f : [0, \infty) \to E_x$ to the ordinary differential equation

$$\left(i\sigma_A(\eta) + \frac{d}{dt} \right) f(t) = 0 \tag{69}$$

subject to the boundary conditions

$$(\mathrm{id} - P)f(0) = \varphi \quad \text{and} \quad \lim_{t\to\infty} f(t) = 0.$$

Recall from Definition 1.4 that $i\sigma_A(\eta)$ is Hermitian, hence diagonalizable with real eigenvalues. The solution to (69) is given by $f(t) = \exp(-it\sigma_A(\eta))\varphi$. The condition $\lim_{t\to\infty} f(t) = 0$ is therefore equivalent to φ lying in the sum of the eigenspaces to the positive eigenvalues of $i\sigma_A(\eta)$. This shows criterion (iv) of Theorem 7.20. □

As a direct consequence of Theorem 7.20 (iv) we obtain

COROLLARY 7.23. *Assume the Standard Setup 1.5 and that D and D^* are complete. Let $E_{|\partial M} = E' \oplus E''$ be a decomposition such that $\sigma_A(\xi)$ interchanges E' and E'' for all $\xi \in T^*\partial M$.*

Then $B' := H^{1/2}(\partial M, E')$ and $B'' := H^{1/2}(\partial M, E'')$ are local elliptic boundary conditions for D. □

This corollary applies, in particular, if A itself interchanges sections of E' and E''.

Solutions to elliptic equations under pseudo-local elliptic boundary conditions have optimal regularity properties. Namely, we have

PROPOSITION 7.24. *Every pseudo-local elliptic boundary condition is ∞-regular.*

PROOF. Let B be a pseudo-local elliptic boundary condition, and let $s \geq 1/2$. We show that B is $(s + 1/2)$-regular by checking criterion (iii) of Lemma 7.14. Let $\varphi \in B$ with $Q_{(-\infty,0)}(\varphi) \in H^s(\partial M, E)$. Since B is pseudo-local, there is a classical pseudo-differential operator P of order 0 inducing an orthogonal projection on $L^2(\partial M, E)$ such that $B = P(H^{1/2}(\partial M, E))$ and such that $P - Q_{[0,\infty)} : L^2(\partial M, E) \to L^2(\partial M, E)$ is elliptic, see criterion (iii) in Theorem 7.20. Now, since $P\varphi = \varphi$,

$$(P - Q_{[0,\infty)})(\varphi) = \varphi - Q_{[0,\infty)}(\varphi) = Q_{(-\infty,0)}(\varphi) \in H^s(\partial M, E).$$

Ellipticity of $P - Q_{[0,\infty)}$ implies that $\varphi \in H^s(\partial M, E)$. □

7.6. Examples. We start with examples of well-known local boundary conditions of great geometric significance. They are all of the form described in Corollary 7.23.

EXAMPLE 7.25 (Differential forms). Let M carry a Riemannian metric g and let

$$E = \bigoplus_{j=0}^{n} \Lambda^j T^*M \otimes_{\mathbb{R}} \mathbb{C} = \Lambda^* T^*M \otimes_{\mathbb{R}} \mathbb{C}$$

be the complexification of the sum of the form bundles over M. The operator is given by $D = d + d^*$, where d denotes exterior differentiation and d^* is its formal adjoint with respect to the volume element of g.

Let T be the interior unit normal vector field along the boundary ∂M and let τ be the associated unit conormal one-form. Then σ_D is symmetric with respect to T along the boundary. For each $x \in \partial M$ and j we have a canonical identification

$$\Lambda^j T_x^* M = \Lambda^j T_x^* \partial M \oplus \tau(x) \wedge \Lambda^{j-1} T_x^* \partial M, \quad \varphi = (\varphi)^{\tan} + \tau(x) \wedge (\varphi)^{\mathrm{nor}}.$$

The local boundary condition corresponding to the subbundle $E' := \Lambda^* M \otimes_{\mathbb{R}} \mathbb{C} \subset E_{|\partial M}$ is called the *absolute boundary condition*,

$$B_{\mathrm{abs}} = \{\varphi \in H^{1/2}(\partial M, E) \mid (\varphi)^{\mathrm{nor}} = 0\},$$

while $E'' := \tau \wedge \Lambda^* \partial M \otimes_{\mathbb{R}} \mathbb{C} \subset E_{|\partial M}$ yields the *relative boundary condition*,

$$B_{\mathrm{rel}} = \{\varphi \in H^{1/2}(\partial M, E) \mid (\varphi)^{\tan} = 0\}.$$

Both boundary conditions are known to be elliptic in the classical sense, see e. g. [**Gi**, Lemma 4.1.1]. Indeed, for $\xi \in T^*\partial M$, $\sigma_D(\xi)$ leaves the subbundles E' and E'' invariant while $\sigma_D(\tau)$ interchanges them. Hence, by (29), $\sigma_A(\xi)$ interchanges E' and E''.

Example 7.26 (Chirality conditions). Let M be a Riemannian spin manifold and D be the Dirac operator acting on spinor fields, i.e., sections of the spinor bundle $E = F = \Sigma M$. We say that a morphism $G : \Sigma M|_{\partial M} \to \Sigma M|_{\partial M}$ is a *boundary chirality operator* if G is an orthogonal involution of E such that

$$\begin{aligned} &\sigma_D(\tau) \circ G = G \circ \sigma_D(\tau) \quad \text{and} \\ &\sigma_D(\xi) \circ G = -G \circ \sigma_D(\xi), \quad \text{for all } \xi \in T^*\partial M. \end{aligned}$$

Now let G be a chirality operator of E as above and E' and E'' be the subbundles of E given by the ± 1-eigenspaces of G. The displayed properties of G show that $\sigma_A(\xi) = \sigma_D(\tau)^{-1} \circ \sigma_D(\xi)$ interchanges the two subbundles. Corollary 7.23 then implies that both subbundles give rise to local elliptic boundary conditions for D.

For instance, Clifford multiplication with i times the exterior unit normal field yields a chirality operator along ∂M. The resulting boundary condition is sometimes called the *MIT-bag condition*.

Chirality conditions have been used to show the positivity of the ADM mass on asymptotically flat manifolds [**GHHP, He**] (using an idea of Witten [**Wi**]). Eigenvalue estimates for the Dirac operator under selfadjoint chirality boundary conditions have been derived in [**HMZ, HMR**].

We now discuss examples of pseudo-local boundary conditions.

Example 7.27 (Generalized Atiyah-Patodi-Singer boundary conditions). In Example 7.6 we have seen that the Atiyah-Patodi-Singer boundary conditions are elliptic in the sense of Definition 7.5. More generally, fix $a \in \mathbb{R}$ and put $V_- = L^2_{(-\infty,a)}(A)$, $V_+ = L^2_{[a,\infty)}(A)$, $W_- = W_+ = 0$, and $g = 0$. Then we have

$$B = B(a) := H^{1/2}_{(-\infty,a)}(A).$$

These elliptic boundary conditions are known as *generalized Atiyah-Patodi-Singer boundary conditions*. As remarked in the proof of Theorem 7.20 these boundary conditions are pseudo-local by [**APS**, p. 48] together with [**Se**] or by [**BW**, Prop. 14.2].

EXAMPLE 7.28 (Transmission conditions). Let (M, μ) be a measured manifold. For the sake of simplicity, assume that the boundary of M is empty, even though this is not really necessary. Let $N \subset M$ be a compact hypersurface with trivial normal bundle. Cut M along N to obtain a measured manifold M' with boundary. The boundary $\partial M'$ consists of two copies N_1 and N_2 of N. We may write $M' = (M \setminus N) \sqcup N_1 \sqcup N_2$.

Let $E, F \to M$ be Hermitian vector bundles and $D : C^\infty(M, E) \to C^\infty(M, F)$ a linear elliptic differential operator of first order. We get induced bundles $E' \to M'$ and $F' \to M'$ and an elliptic operator $D' : C^\infty(M', E') \to C^\infty(M', F')$. For $\Phi \in H^1_{\mathrm{loc}}(M, E)$ we get $\Phi' \in H^1_{\mathrm{loc}}(M', E')$ such that $\Phi'|_{N_1} = \Phi'|_{N_2}$. We use this as a boundary condition for D' on M'. We set

$$B := \left\{ (\varphi, \varphi) \in H^{1/2}(N_1, E) \oplus H^{1/2}(N_2, E) \mid \varphi \in H^{1/2}(N, E) \right\}.$$

Here we used the canonical identification

$$H^{1/2}(N_1, E) = H^{1/2}(N_2, E) = H^{1/2}(N, E).$$

Now we assume that D' is boundary symmetric with respect to some interior vector field T along $N = N_1$. This corresponds to a condition on the principal symbol of D along N. It is satisfied e.g. if D is of Dirac type. Let $A = A_0 \oplus -A_0$ be an adapted boundary operator for D'. Here A_0 is a selfadjoint

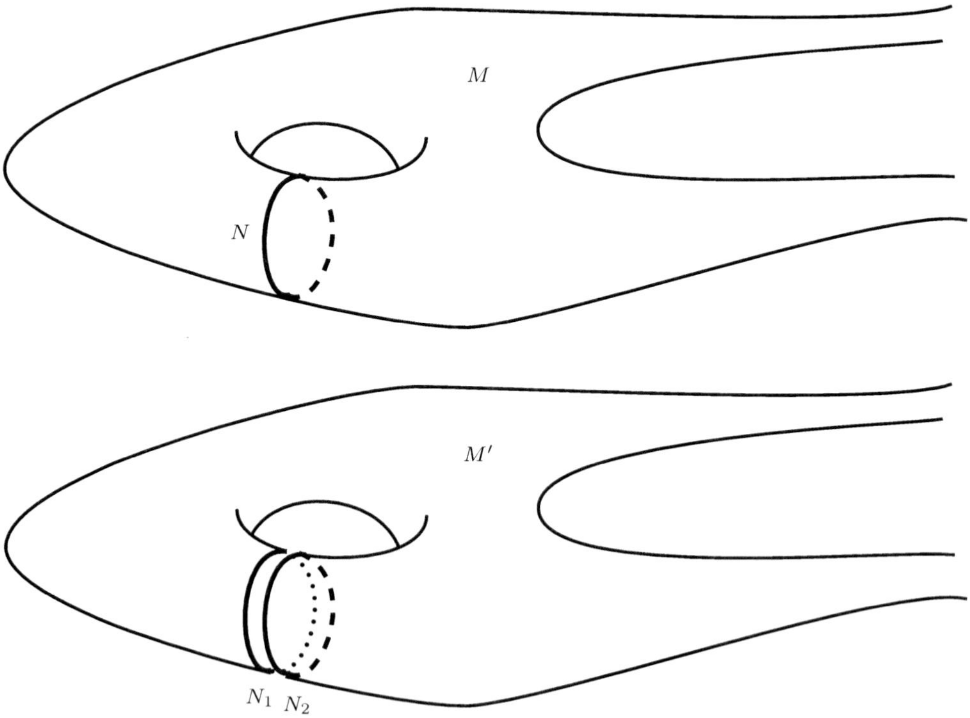

FIGURE 6

elliptic operator on $C^\infty(N,E) = C^\infty(N_1,E')$ and similarly $-A_0$ on $C^\infty(N,E) = C^\infty(N_2,E')$. The sign is due to the opposite relative orientations of N_1 and N_2 in M'.

To see that B is an elliptic boundary condition, put

$$\begin{aligned}
V_+ &:= L^2_{(0,\infty)}(A_0 \oplus -A_0) = L^2_{(0,\infty)}(A_0) \oplus L^2_{(-\infty,0)}(A_0),\\
V_- &:= L^2_{(-\infty,0)}(A_0 \oplus -A_0) = L^2_{(-\infty,0)}(A_0) \oplus L^2_{(0,\infty)}(A_0),\\
W_+ &:= \{(\varphi,\varphi) \in \ker(A_0) \oplus \ker(A_0)\},\\
W_- &:= \{(\varphi,-\varphi) \in \ker(A_0) \oplus \ker(A_0)\},
\end{aligned}$$

and

$$g : V_-^{1/2} \to V_+^{1/2}, \quad g = \begin{pmatrix} 0 & \mathrm{id} \\ \mathrm{id} & 0 \end{pmatrix}.$$

With these choices B is of the form required in Definition 7.5. We call these boundary conditions *transmission conditions*. Clearly, they are ∞-regular.

If M has a nonempty boundary and N is disjoint from ∂M, let us assume that we are given an elliptic boundary condition for ∂M. Then the same discussion applies if one keeps the boundary condition on ∂M and adapts B on $\partial M' = \partial M \sqcup N_1 \sqcup N_2$ accordingly.

8. Index theory

Throughout this section, assume the Standard Setup 1.5 and that D and D^* are complete. In addition to completeness, we introduce a second property of D at infinity: coercivity, see Definition 8.2 below. This property will be crucial for the Fredholm property of D so that we can speak of its index.

REMARK 8.1. Recall that we have $D_B = D_{B,\max}$ with domain

$$H^1_D(M,E;B) := \{\Phi \in H^1_D(M,E) \mid \mathscr{R}\Phi \in B\}$$

if B is an elliptic boundary condition for D, compare Definition 1.10 and Lemma 7.3. In particular, if ∂M is empty, then $D = D_{\max}$ with domain $H^1_D(M,E)$. This explains the difference in notation between the introduction and this section.

8.1. The Fredholm property.

DEFINITION 8.2. We say that D is *coercive at infinity* if there is a compact subset $K \subset M$ and a constant C such that

$$\|\Phi\|_{L^2(M)} \le C\|D\Phi\|_{L^2(M)}, \tag{70}$$

for all smooth sections Φ of E with compact support in $M\backslash K$.

EXAMPLES 8.3. (a) By definition, if M is compact, then D is coercive at infinity.

(b) Coercivity at infinity of a Dirac type operator $D: C^\infty(M,E) \to C^\infty(M,F)$ can be shown using a *Weitzenböck formula*. This is a formula of the form

$$D^*D = \nabla^*\nabla + \mathscr{K}$$

where ∇ is a connection on E and $\mathscr{K}$ is a symmetric endomorphism field on E. Now if, outside a compact subset $K \subset M$, all eigenvalues of $\mathscr{K}$ are bounded below by a constant $c > 0$, then we have, for all $\Phi \in C^\infty_{cc}(M,E)$ with support disjoint from K,

$$\|D\Phi\|^2_{L^2(M)} = \|\nabla\Phi\|^2_{L^2(M)} + (\mathscr{K}\Phi, \Phi)_{L^2(M)} \geq c\|\Phi\|^2_{L^2(M)}.$$

Hence D is coercive at infinity.

LEMMA 8.4. *The operator D is coercive at infinity if and only if there is no sequence (Φ_n) in $C^\infty_{cc}(M,E)$ such that*

$$\|\Phi_n\|_{L^2(M)} = 1 \quad \text{and} \quad \lim_{n\to\infty} \|D\Phi_n\|_{L^2(M)} = 0$$

and such that, for any compact subset $K \subset M$, we have

$$\operatorname{supp}\Phi_n \cap K = \emptyset$$

for all sufficiently large n.

PROOF. Choose an exhaustion of M by compact sets $K_1 \subset K_2 \subset \cdots \subset M$. If D is not coercive at infinity, then for each n there exists Φ_n with $\operatorname{supp}\Phi_n \cap K_n = \emptyset$ and $\|\Phi_n\|_{L^2(M)} > n\|D\Phi_n\|_{L^2(M)}$. Normalizing Φ_n, we obtain a sequence as in Lemma 8.4.

Conversely, a sequence as in Lemma 8.4 directly violates the condition in Definition 8.2. □

THEOREM 8.5. *Assume the Standard Setup 1.5 and that D and D^* are complete. Let $B \subset H^{1/2}(\partial M, E)$ be an elliptic boundary condition for D.*

Then D is coercive at infinity if and only if

$$D_B : H^1_D(M,E;B) \to L^2(M,F)$$

has finite-dimensional kernel and closed image. In this case

$$\operatorname{ind} D_B = \dim\ker D_B - \dim\ker(D^*)_{B^{\mathrm{ad}}} \in \mathbb{Z} \cup \{-\infty\}.$$

PROOF. Suppose first that D is coercive at infinity. We want to apply Proposition A.3. Let (Φ_n) be a bounded sequence in $H^1_D(M,E;B)$ such that $D\Phi_n \to \Psi \in L^2(M,F)$. We have to show that (Φ_n) has a convergent subsequence in $H^1_D(M,E)$.

Passing to a subsequence if necessary, we can assume, by the Rellich embedding theorem, that there is a section $\Phi \in L^2_{\mathrm{loc}}(M,E)$ such that $\Phi_n \to \Phi$ in $L^2_{\mathrm{loc}}(M,E)$. Choose a compact subset $K \subset M$ and a constant C as in

(70). Let $\chi \in C_c^\infty(M,\mathbb{R})$ be a cut-off function which is equal to 1 on K and $0 \le \chi \le 1$ everywhere. Set $K' := \operatorname{supp}(\chi)$. By (70),

$$\begin{aligned}
\|\Phi_n - \Phi_m\|_{L^2(M)} &\le \|\chi(\Phi_n - \Phi_m)\|_{L^2(M)} + \|(1-\chi)(\Phi_n - \Phi_m))\|_{L^2(M)} \\
&\le \|\Phi_n - \Phi_m\|_{L^2(K')} + C\|D((1-\chi)(\Phi_n - \Phi_m)))\|_{L^2(M)} \\
&\le \|\Phi_n - \Phi_m\|_{L^2(K')} + C\| - \sigma_D(d\chi)(\Phi_n - \Phi_m)\|_{L^2(M)} \\
&\quad + C\|(1-\chi)(D\Phi_n - D\Phi_m)\|_{L^2(M)} \\
&\le C'\|\Phi_n - \Phi_m\|_{L^2(K')} + C\|D\Phi_n - D\Phi_m\|_{L^2(M)} \to 0.
\end{aligned}$$

It follows that (Φ_n) is a Cauchy sequence in $L^2(M,E)$, hence $\Phi \in L^2(M,E)$ and $\Phi_n \to \Phi$ in $L^2(M,E)$.

Since B is an elliptic boundary condition, the H^1_D-norm and the graph norm of D are equivalent on $H^1_D(M,E;B)$. Now (Φ_n) and $(D\Phi_n)$ converge in $L^2(M,E)$ and $L^2(M,F)$, respectively. Hence (Φ_n) converges in the graph norm of D, and therefore in $H^1_D(M,E)$. By the implication (iii) $\Rightarrow$ (i) of Proposition A.3, D_B has finite-dimensional kernel and closed image.

Suppose now that D is not coercive at infinity. Let (Φ_n) be a sequence as in Lemma 8.4. By the assumption on the support of Φ_n, $\Phi_n \rightharpoonup 0$ in $L^2(M,E)$. Since, on the other hand, $\|\Phi_n\|_{L^2(M)} = 1$ for all n, no subsequence converges in $L^2(M,E)$. In particular, no subsequence converges in $H^1_D(M,E)$. Hence D_B does not satisfy criterion (iii) in Proposition A.3, and hence D_B has infinite-dimensional kernel or nonclosed image. □

COROLLARY 8.6. *Assume the Standard Setup 1.5 and that D and D^* are complete and coercive at infinity. Let $B \subset H^{1/2}(\partial M,E)$ be an elliptic boundary condition for D. Then*

$$D_B : H^1_D(M,E;B) \to L^2(M,F)$$

is a Fredholm operator and

$$\operatorname{ind} D_B = \dim\ker D_B - \dim\ker(D^*)_{B^{\mathrm{ad}}} \in \mathbb{Z}. \qquad \square$$

The following corollary is immediate from Proposition A.1.

COROLLARY 8.7. *Assume the Standard Setup 1.5 and that D and D^* are complete and coercive at infinity. Let $B \subset H^{1/2}(\partial M,E)$ be an elliptic boundary condition for D. Let $\check{C}$ and C be closed complements of B in $\check{H}(A)$ and in $H^{1/2}(\partial M,E)$, respectively. Let $\check{P} : \check{H}(A) \to \check{H}(A)$ and $P : H^{1/2}(\partial M,E) \to H^{1/2}(\partial M,E)$ be the projections with kernel B and image $\check{C}$ and C, respectively. Then*

$$\check{L} : \operatorname{dom}(D_{\max}) \to L^2(M,F) \oplus \check{C}, \quad \check{L}\Phi = (D_{\max}\Phi, \check{P}\mathscr{R}\Phi),$$

and

$$L : H^1_D(M,E) \to L^2(M,F) \oplus C, \quad L\Phi = (D\Phi, P\mathscr{R}\Phi),$$

are Fredholm operators with the same index as $D_{B,\max} = D_B$. □

COROLLARY 8.8. *Assume the Standard Setup 1.5 and that D and D^* are complete and coercive at infinity. Let $B_1 \subset B_2 \subset H^{1/2}(\partial M, E)$ be elliptic boundary conditions for D.*

Then $\dim(B_2/B_1)$ *is finite and*

$$\operatorname{ind}(D_{B_2}) = \operatorname{ind}(D_{B_1}) + \dim(B_2/B_1).$$

PROOF. Choose closed complements C_1 and C_2 in $H^{1/2}(\partial M, E)$ of B_1 and B_2, respectively, such that $C_2 \subset C_1$. Then we have the following commutative diagram:

$$\begin{array}{ccc} & & L^2(M,E) \oplus C_1 \\ & \overset{(D,Q_1)}{\nearrow} & \\ H^1_D(M,E) & & \Big\downarrow \mathrm{id}\oplus\pi \\ & \underset{(D,Q_2)}{\searrow} & \\ & & L^2(M,E) \oplus C_2 \end{array}$$

where Q_j is $\mathscr{R}$ composed with the projection onto C_j as in Corollary 8.7 and $\pi : C_1 \to C_2$ is the projection. Since (D, Q_1) and (D, Q_2) are Fredholm operators, $\mathrm{id} \oplus \pi$ is one too. In particular,

$$\dim(B_2/B_1) = \dim(C_1/C_2) = \operatorname{ind}(\pi) = \operatorname{ind}(\mathrm{id} \oplus \pi)$$

is finite. Moreover,

$$\begin{aligned} \operatorname{ind}(D_{B_1}) &= \operatorname{ind}(D, Q_1) \\ &= \operatorname{ind}(D, Q_2) - \operatorname{ind}(\mathrm{id} \oplus \pi) \\ &= \operatorname{ind}(D_{B_2}) - \dim(B_2/B_1). \end{aligned}$$

□

EXAMPLE 8.9. Assume the Standard Setup 1.5 and that D and D^* are complete and coercive at infinity. Then we have, for generalized Atiyah-Patodi-Singer boundary conditions $B(a)$ and $B(b)$ with $a < b$,

$$\operatorname{ind} D_{B(b)} = \operatorname{ind} D_{B(a)} + \dim L^2_{[a,b)}(A).$$

8.2. Deformations of boundary conditions.

DEFINITION 8.10. A family of boundary conditions $B_s \subset H^{1/2}(\partial M, E)$, $0 \le s \le 1$, is called a *continuous family of boundary conditions* if there exist isomorphisms

$$k_s : B_0 \to B_s,$$

where $k_0 = \mathrm{id}$ and where the map $[0,1] \to \mathscr{L}(B_0, H^{1/2}(\partial M, E))$, $s \mapsto k_s$, is continuous. Here $\mathscr{L}(B_0, H^{1/2}(\partial M, E))$ denotes the Banach space of bounded operators from B_0 to $H^{1/2}(\partial M, E)$ equipped with the operator norm.

The following auxiliary isomorphisms allow us to consider families of operators between fixed spaces, i.e., spaces independent of the deformation parameter. This will be useful since it will allow us to apply the standard fact that the index of a continuous family of Fredholm operators has constant index.

LEMMA 8.11. *The map*

$$J_E : H^1_D(M,E;0) \oplus H^{1/2}(\partial M,E) \to H^1_D(M,E),$$
$$J_E(\Phi,\varphi) = \Phi + \mathscr{E}\varphi,$$

is an isomorphism. If $B \subset H^{1/2}(\partial M,E)$ *is an elliptic boundary condition, then the restriction*

$$J_{E,B} : H^1_D(M,E;0) \oplus B \to H^1_D(M,E;B)$$

is again an isomorphism.

PROOF. For $\varphi \in H^{1/2}(\partial M,E)$, we have $\mathscr{E}\varphi \in \mathrm{dom}(D_{\max})$ by Lemma 6.1. By Theorem 6.7 (iii), $\mathscr{E}\varphi \in H^1_D(M,E)$. Thus J_E maps indeed to $H^1_D(M,E)$. The inverse map of J_E is given by

$$K_E : H^1_D(M,E) \to H^1_D(M,E;0) \oplus H^{1/2}(\partial M,E),$$
$$K_E(\Psi) = (\Psi - \mathscr{E}\mathscr{R}\Psi, \mathscr{R}\Psi).$$

Clearly, $J_E(\Phi,\varphi) \in H^1_D(M,E;B)$ if and only if

$$\varphi = \mathscr{R}(J_E(\Phi,\varphi)) \in B.$$

Hence J_E restricts to an isomorphism between $H^1_D(M,E;0) \oplus B$ and $H^1_D(M,E;B)$. □

THEOREM 8.12. *Assume the Standard Setup 1.5 and that* D *and* D^* *are complete and coercive at infinity. Let* B_s *be a continuous family of elliptic boundary conditions for* D.

Then, for all $0 \le s \le 1$,

$$\mathrm{ind}\, D_{B_s} = \mathrm{ind}\, D_{B_0}.$$

PROOF. Let the $k_s : B_0 \to B_s$ be as in Definition 8.10. Consider the commutative diagram of isomophisms:

$$\begin{array}{ccc}
H^1_D(M,E;0) \oplus B_0 & \xrightarrow{J_{E,B_0}} & H^1_D(M,E;B_0) \\
{\scriptstyle \mathrm{id}\oplus k_s}\downarrow & & \downarrow{\scriptstyle K_s} \\
H^1_D(M,E;0) \oplus B_s & \xrightarrow{J_{E,B_s}} & H^1_D(M,E;B_s),
\end{array}$$

where $K_s(\Psi) = \Psi + \mathscr{E}\big(k_s(\mathscr{R}(\Psi)) - \mathscr{R}(\Psi)\big)$. Then the composition

$$H^1_D(M,E;B_0) \xrightarrow{K_s} H^1_D(M,E;B_s) \xrightarrow{D_{B_s}} L^2(M,F)$$

is a continuous family of operators. Here continuity refers to the operator norm in $\mathscr{L}(H^1_D(M,E;B_0),L^2(M,F))$. Thus the index of $D_{B_s}\circ K_s$ is constant. Since K_s is an isomorphism, $\operatorname{ind} D_{B_s} = \operatorname{ind}(D_{B_s}\circ K_s)$. $\square$

EXAMPLE 8.13. Writing an elliptic boundary condition B_1 as in Definition 7.5, we let $B_0 := W_+ \oplus V_-^{1/2}$ and obtain a continuous deformation by keeping $V_\pm$ and $W_\pm$ constant and replacing the map $g : V_- \to V_+$ by sg, $0 \le s \le 1$. In other words,

$$k_s : B_0 \to B_s, \quad k_s(\varphi + \psi) = \varphi + \psi + sg\psi,$$

where $\varphi \in W_+$ and $\psi \in V_-$. This allows us to reduce index computations to the case $g = 0$.

8.3. Fredholm pairs. Recall the generalized Atiyah-Patodi-Singer boundary conditions $B(a)$, $a \in \mathbb{R}$, from Example 1.16 (a) or Example 7.27. The next result illustrates the central role of this kind of boundary condition.

THEOREM 8.14. *Assume the Standard Setup 1.5 and that D and D^* are complete and coercive at infinity.*

Let $B \subset H^{1/2}(\partial M, E)$ be an elliptic boundary condition for D. Let $V_\pm$ and $W_\pm$ be for B as in Theorem 7.11 (ii). *In other words, $L^2_{(-\infty,a)}(A) = V_- \oplus W_-$ and $B = W_+ \oplus \Gamma(g)^{1/2}$, where $g : V_- \to V_+$. Then we have*

$$\operatorname{ind} D_B = \operatorname{ind} D_{B(a)} + \dim W_+ - \dim W_-.$$

PROOF. As explained in Example 8.13, we can assume without loss of generality that $g = 0$, i.e., $B = W_+ \oplus V_-^{1/2}$. Consider one further boundary condition,

$$B' := W_- \oplus W_+ \oplus V_-^{1/2} = H^{1/2}_{(-\infty,a)}(A) \oplus W_+ = B(a) \oplus W_+.$$

Applying Corollary 8.8 twice we conclude

$$\begin{aligned}\operatorname{ind}(D_B) &= \operatorname{ind}(D_{B'}) - \dim W_- \\ &= \operatorname{ind}(D_{B(a)}) + \dim W_+ - \dim W_-.\end{aligned}$$ $\square$

We recall the notion of Fredholm pairs of subspaces. Let H be a Hilbert space and let $X, Y \subset H$ be closed subspaces. Then (X,Y) is called a *Fredholm pair* if $X \cap Y$ is finite-dimensional and $X + Y$ has finite codimension. In particular, $X + Y$ is closed.* Intuitively, constituting a Fredholm pair means that X and Y are complements in H up to finite-dimensional errors. The number

$$\operatorname{ind}(X,Y) := \dim(X \cap Y) - \dim(H/(X+Y)) \in \mathbb{Z} \tag{71}$$

*Note that, in general, the sum of two closed subspaces in a Hilbert space need not be closed.

is called the *index* of the pair (X, Y). If (X, Y) is a Fredholm pair, then the orthogonal complements also form a Fredholm pair $(X^\perp, Y^\perp)$ and one has

$$\operatorname{ind}(X^\perp, Y^\perp) = -\operatorname{ind}(X, Y). \tag{72}$$

See [**Ka**, Ch. IV, §4] for a detailed discussion. Under the assumptions of Theorem 8.14, the pair $(L^2_{[a,\infty)}, \bar{B})$, where $\bar{B}$ denotes the L^2-closure of B, is a Fredholm pair and the index formula says that

$$\operatorname{ind} D_B - \operatorname{ind} D_{B(a)} = \operatorname{ind}(L^2_{[a,\infty)}, \bar{B}) = -\operatorname{ind}(L^2_{(-\infty,a)}, \bar{B}^\perp) \tag{73}$$

THEOREM 8.15. *Assume the Standard Setup 1.5 and that D and D^* are complete and coercive at infinity.*

Let $B_1, B_2 \subset H^{1/2}(\partial M, E)$ be elliptic boundary conditions for D. Let g_1 and g_2 be maps for B_1 and B_2 respectively, as in Theorem 7.11 (ii) *for the same number $a \in \mathbb{R}$.*

If the L^2-operator norms satisfy

$$\|g_1\| \cdot \|g_2\| < 1, \tag{74}$$

then the L^2-closures $(\bar{B}_1, \bar{B}_2^\perp)$ form a Fredholm pair in $L^2(M, E)$ and

$$\operatorname{ind}(D_{B_1}) - \operatorname{ind}(D_{B_2}) = \operatorname{ind}(\bar{B}_1, \bar{B}_2^\perp). \tag{75}$$

PROOF. In the notation of Theorem 7.5 we have

$$\bar{B}_1 = W_{1,+} \oplus \Gamma(g_1)$$

where $g_1 : V_{1,-} \to V_{1,+}$ and, by Lemma 7.7,

$$\bar{B}_2^\perp = W_{2,-} \oplus \Gamma(-g_2^*).$$

A general element of $\bar{B}_1 \cap \bar{B}_2^\perp$ is therefore of the form

$$w_{1,+} + v_{1,-} + g_1 v_{1,-} = w_{2,-} + v_{2,+} - g_2^* v_{2,+} \tag{76}$$

where $w_{1,+} \in W_{1,+}$, $w_{2,-} \in W_{2,-}$, $v_{1,-} \in V_{1,-}$, and $v_{2,+} \in V_{2,+}$. Projecting to $L^2_{(-\infty,a)}(A)$ and to $L^2_{[a,\infty)}(A)$, we see that (76) is equivalent to the two equations

$$v_{1,-} = w_{2,-} - g_2^* v_{2,+}, \tag{77}$$

$$v_{2,+} = w_{1,+} + g_1 v_{1,-}. \tag{78}$$

We claim that the linear map $\bar{B}_1 \cap \bar{B}_2^\perp \to W_{1,+} \oplus W_{2,-}$ mapping the element in (76) to $(w_{1,+}, w_{2,-})$ is injective. Namely, if $w_{1,+} = w_{2,-} = 0$, then (77) and (78) say

$$v_{2,+} = g_1 v_{1,-} = -g_1 g_2^* v_{2,+},$$

hence

$$(\mathrm{id} + g_1 g_2^*) v_{2,+} = 0.$$

Here we have extended g_1 to $L^2_{(-\infty,a)}(A)$ by setting it $=0$ on $W_{1,-}$ and similarly for g_2^*. Since

$$\|g_1 g_2^*\| \leq \|g_1\| \cdot \|g_2^*\| = \|g_1\| \cdot \|g_2\| < 1, \tag{79}$$

we see that $\mathrm{id} + g_1 g_2^*$ is an isomorphism on $L^2_{[a,\infty)}(A)$. Hence $v_{2,+} = 0$ and, by (77), also $v_{1,-} = 0$. This implies

$$\dim(\bar{B}_1 \cap \bar{B}_2^\perp) \leq \dim(W_{1,+} \oplus W_{2,-}) < \infty.$$

Next we show that $\bar{B}_1 + \bar{B}_2^\perp$ is closed in $L^2(M,E)$. Since $W_{1,+}$ and $W_{2,-}$ are finite-dimensional, it suffices to show that $\Gamma(g_1) + \Gamma(-g_2^*)$ is closed. Let $u_k + g_1 u_k + v_k - g_2^* v_k$ converge in $L^2(M,E)$ as $k \to \infty$ with $u_k \in V_{1,-}$ and $v_k \in V_{2,+}$. Projecting onto $L^2_{(-\infty,a)}(A)$ and onto $L^2_{[a,\infty)}(A)$ we see that

$$u_k - g_2^* v_k \to x_- \in L^2_{(-\infty,a)}(A) \quad \text{and}$$
$$v_k + g_1 u_k \to x_+ \in L^2_{[a,\infty)}(A).$$

From

$$g_1(u_k - g_2^* v_k) = v_k + g_1 u_k - (\mathrm{id} + g_1 g_2^*) v_k$$

we have

$$v_k = (\mathrm{id} + g_1 g_2^*)^{-1}(v_k + g_1 u_k - g_1(u_k - g_2^* v_k)).$$

This shows that (v_k) converges with

$$\lim_{k\to\infty} v_k = (\mathrm{id} + g_1 g_2^*)^{-1}(x_+ - g_1 x_-).$$

Then (u_k) also converges and $u_k + g_1 u_k + v_k - g_2^* v_k$ converges to an element in $\Gamma(g_1) + \Gamma(-g_2^*)$.

Now that we know that $\bar{B}_1 + \bar{B}_2^\perp$ is closed, we compute

$$\frac{L^2(M,E)}{\bar{B}_1 + \bar{B}_2^\perp} \cong (\bar{B}_1 + \bar{B}_2^\perp)^\perp = \bar{B}_1^\perp \cap \bar{B}_2 \hookrightarrow W_{2,+} \oplus W_{1,-}.$$

Hence $\dim(L^2(M,E)/(\bar{B}_1 + \bar{B}_2^\perp)) < \infty$ and $(\bar{B}_1, \bar{B}_2^\perp)$ is a Fredholm pair.

It remains to prove (75). If we replace g_1 and g_2 by sg_1 and sg_2, respectively, with $s \in [0,1]$, then (74) clearly remains valid, so that the corresponding boundary conditions form Fredholm pairs for all $s \in [0,1]$. Since the index of D subject to these boundary conditions remains constant by Theorem 8.12 and the index of the Fredholm pairs by [**Ka**, Thm. 4.30, p. 229], we may without loss of generality assume that $g_1 = g_2 = 0$.

In this case (77) and (78) tell us that

$$\bar{B}_1 \cap \bar{B}_2^\perp = (W_{1,+} \cap V_{2,+}) \oplus (W_{2,-} \cap V_{1,-}) \tag{80}$$

and similarly

$$\bar{B}_2 \cap \bar{B}_1^\perp = (W_{2,+} \cap V_{1,+}) \oplus (W_{1,-} \cap V_{2,-}). \tag{81}$$

Let $\tilde{\perp}$ denote the orthogonal complement in $L^2_{(\infty,a)}(A)$ rather than in $L^2(M,E)$. Then

$$\begin{aligned}
\dim(W_{2,-}\cap V_{1,-}) &= \dim\left(\left(W_{2,-}^{\tilde{\perp}} + V_{1,-}^{\tilde{\perp}}\right)^{\tilde{\perp}}\right) \\
&= \dim\left((V_{2,-} + W_{1,-})^{\tilde{\perp}}\right) \\
&= \dim\left(\left(V_{2,-} \oplus \frac{W_{1,-}}{W_{1,-}\cap V_{2,-}}\right)^{\tilde{\perp}}\right) \\
&= \dim(W_{2,-}) - \dim(W_{1,-}) + \dim(W_{1,-}\cap V_{2,-}). \qquad (82)
\end{aligned}$$

Similarly, one sees

$$\dim(W_{2,+}\cap V_{1,+}) - \dim(W_{1,+}\cap V_{2,+}) = \dim(W_{2,+}) - \dim(W_{1,+}). \qquad (83)$$

Equations (80)–(83) combine to give

$$\operatorname{ind}(\bar{B}_1, \bar{B}_2^{\perp}) = \dim(W_{1,+}) + \dim(W_{2,-}) - \dim(W_{1,-}) - \dim(W_{2,+}). \qquad (84)$$

Theorem 8.14 yields

$$\operatorname{ind} D_{B_1} = \operatorname{ind} D_{B_{\mathrm{APS}}} + \dim W_{1,+} - \dim W_{1,-} \qquad (85)$$

and

$$\operatorname{ind} D_{B_2} = \operatorname{ind} D_{B_{\mathrm{APS}}} + \dim W_{2,+} - \dim W_{2,-}. \qquad (86)$$

Subtracting (85) and (86) and inserting (84) concludes the proof. □

REMARK 8.16. Assumption (74) is sharp. For instance, we can choose B_1 and B_2 such that they have the same $W_\pm$ and $V_\pm$, namely $W_\pm = 0$, $V_- = L^2_{(-\infty,0)}(A)$, and $V_+ = L^2_{[0,\infty)}(A)$. For g_1 we choose a unitary isomorphism $g_1 : V_- \to V_+$ which is also continuous with respect to the $H^{1/2}$-norm. For g_2 we choose $g_2 = -g_1$. Then

$$\bar{B}_2^{\perp} = \Gamma(-g_2^*) = \Gamma(g_1^*) = \Gamma(g_1^{-1}) = \Gamma(g_1) = \bar{B}_1.$$

Thus $(\bar{B}_1, \bar{B}_2^{\perp})$ does not form a Fredholm pair. In this case, $\|g_1\| = \|g_2\| = 1$.

If one of the two boundary conditions is a generalized Atiyah-Patodi-Singer boundary condition $B(b)$, $b \in \mathbb{R}$, then the corresponding g vanishes so that assumption (74) holds without any restriction on the other boundary condition. In this case, Theorem 8.15 reduces to Theorem 8.14.

8.4. Relative index theory. Throughout this subsection, assume the Standard Setup 1.5 and that D and D^* are complete and coercive at infinity. For convenience assume also that M is connected and $\partial M = \emptyset$.

For what follows, compare Example 7.28. Let N be a closed and two-sided hypersurface in M. Cut M along N to obtain a manifold M', possibly connected, whose boundary $\partial M'$ consists of two disjoint copies N_1 and N_2 of N, see Figure 6 on page 56. There are natural pull-backs μ', E', F', and D' of μ, E, F, and D from M to M'. Assume that there is an interior vector field T along $N = N_1$ such that D' is boundary symmetric with respect to T along N_1. Then D' is also boundary symmetric with respect to the interior vector field $-T$ of M' along $N = N_2$. Choose an adapted operator A for D' along N_1 as in the Standard Setup 1.5. Then $-A$ is an adapted operator for D' along N_2 as in the Standard Setup 1.5 and will be used in what follows.

Theorem 8.17 (Splitting Theorem). *Let M, M', and notation be as above. Assume that D' is boundary symmetric with respect to an interior vector field T of M' along N and choose an adapted operator A as above.*

Then, D and D^* are complete and coercive at infinity if and only if D' and $(D')^*$ are complete and coercive at infinity. In this case, D and $D'_{B_1 \oplus B_2}$ are Fredholm operators with

$$\operatorname{ind} D = \operatorname{ind} D'_{B_1 \oplus B_2},$$

where $B_1 = B(a) = H^{1/2}_{(-\infty,a)}(A)$ and $B_2 = H^{1/2}_{[a,\infty)}(A)$, considered as boundary conditions along N_1 and N_2, respectively. More generally, we may choose any elliptic boundary condition $B_1 \subset H^{1/2}(N, E)$ and its L^2-orthogonal complement $B_2 \subset H^{1/2}(N, E)$.

Remarks 8.18. (a) If M is a complete Riemannian manifold and D is of Dirac type, then M' is also a complete Riemannian manifold and D' is of Dirac type as well. Completeness of the metrics implies completeness of D and D', see Example 3.4. Moreover, D' is boundary symmetric with respect to the interior unit normal field T of M' along N. Finally, if a curvature condition as in Example 8.3 (b) ensures coercivity at infinity of D and D', then the index formula of Theorem 8.17 applies.

(b) The Decomposition Theorem 1.19 is the special case of Theorem 8.17 where N decomposes M into two components, compare Remark 8.1.

Proof of the Splitting Theorem 8.17. The first assertion, namely that D and D^* are complete and coercive at infinity if and only if D' and $(D')^*$ are complete and coercive at infinity, is immediate from Definition 1.1, Definition 1.17 (resp. 8.2), and the compactness of N.

Assume now that D and D^* are complete and coercive at infinity. Then D and $D'_{B_1 \oplus B_2}$ are Fredholm operators, by what we just said and Theorem 1.18.

Under the canonical identifications $E|_{N_1} = E|_N = E|_{N_2}$, the transmission condition from Example 7.28 reads

$$B = \{(\varphi, \psi) \in H^{1/2}(N_1, E) \oplus H^{1/2}(N_2, E) \mid \varphi = \psi\}.$$

Recall that B is an elliptic boundary condition. Furthermore, with respect to the canonical pull-back of sections from E to E', we have

$$\operatorname{dom} D = H^1_D(M, E) = H^1(M', E'; B) = \operatorname{dom} D'_B$$

and

$$\operatorname{ind} D = \operatorname{ind} D'_B, \tag{87}$$

It follows from the discussion in Example 7.28 that B is homotopic to the boundary condition $W_+ \oplus V_-^{1/2}$ (that is, $g = 0$) with

$$\begin{aligned} V_- &= L^2_{(-\infty,0)}(A) \oplus L^2_{(-\infty,0)}(-A) = L^2_{(-\infty,0)}(A) \oplus L^2_{(0,\infty)}(A) \\ W_+ &= \{(\varphi, \varphi) \mid \varphi \in \ker A\}. \end{aligned}$$

We note that here $V_-^{1/2} = B(0) = H^{1/2}_{(-\infty,0)}(A) \oplus H^{1/2}_{(0,\infty)}(A)$, hence

$$\begin{aligned} \operatorname{ind} D'_B &= \operatorname{ind} D'_{B(0)} + \dim W_+ \\ &= \operatorname{ind} D'_{B(0)} + \dim \ker A \\ &= \operatorname{ind} D'_{H^{1/2}_{(-\infty,0)}(A) \oplus H^{1/2}_{[0,\infty)}(A)}, \end{aligned} \tag{88}$$

by applying Corollary 8.8 twice. If B_1 is an elliptic boundary condition,

$$B_1 = W_{1,+} \oplus \Gamma(g)^{1/2},$$

where the notation is as in Theorem 1.12 with $a = 0$, then

$$B_2 = W_{1,-} \oplus \Gamma(-g^*)^{1/2}$$

is the L^2-orthogonal complement of B_1 in $H^{1/2}(N, E)$. By Example 8.13, we may assume that $g = 0$. Then

$$B_1 = W_{1,+} \oplus V_{1,-}^{1/2} \quad \text{and} \quad B_2 = W_{1,-} \oplus V_{1,+}^{1/2} \tag{89}$$

with

$$V_{1,-}^{1/2} \oplus W_{1,-} = H^{1/2}_{(-\infty,0)}(A) \quad \text{and} \quad V_{1,+}^{1/2} \oplus W_{1,+} = H^{1/2}_{[0,\infty)}(A). \tag{90}$$

Applying Corollary 8.8 to (89) and (90), respectively, we get

$$\begin{aligned} \operatorname{ind} D'_{B_1 \oplus B_2} &= \operatorname{ind} D'_{V_{1,-}^{1/2} \oplus V_{1,+}^{1/2}} + \dim W_{1,+} + \dim W_{1,-} \\ &= \operatorname{ind} D'_{H^{1/2}_{(-\infty,0)}(A) \oplus H^{1/2}_{[0,\infty)}(A)}. \end{aligned} \tag{91}$$

The claimed index formula is now immediate from (87), (88), and (91). □

Let M_1 and M_2 be complete Riemannian manifolds without boundary. Let $D_i : C^\infty(M_i, E_i) \to C^\infty(M_i, F_i)$ be Dirac type operators which agree outside closed subsets $K_i \subset M_i$ and choose f and $\mathscr{I}_E$ as in Definition 1.20. For $i = 1, 2$, choose a decomposition $M_i = M_i' \cup M_i''$ such that $N_i = M_i' \cap M_i''$ is a compact hypersurface in M_i, K_i is contained in the interior of M_i', $f(M_1'') = M_2''$, and $f(N_1) = N_2$. Denote the restriction of D_i to M_i' by D_i'. The following result is a general version of the Φ-relative index theorem of Gromov and Lawson [**GL**, Thm. 4.35].

THEOREM 8.19. *Under the above assumptions, let $B_1 \subset H^{1/2}(N_1, E_1)$ and $B_2 \subset H^{1/2}(N_2, E_2)$ be elliptic boundary conditions which correspond to each other under the identifications given by f and $\mathscr{I}_E$ from Definition 1.20. Assume that D_1 and D_2 and their formal adjoints are coercive at infinity.*

Then D_1, D_2, D'_{1,B_1}, and D'_{2,B_2} are Fredholm operators such that

$$\operatorname{ind} D_1 - \operatorname{ind} D_2 = \operatorname{ind} D'_{1,B_1} - \operatorname{ind} D'_{2,B_2}.$$

PROOF. The Decomposition Theorem implies that D_1, D_2, D'_{1,B_1}, and D'_{2,B_2} are Fredholm operators.

Let T be the normal vector field along N_1 pointing into M_1'. Since D_1 is of Dirac type, D_1 is boundary symmetric with respect to T. Let A be an adapted operator on $E_1|_N$. By the Decomposition Theorem, we have

$$\operatorname{ind} D_1 = \operatorname{ind} D'_{1,B'} + \operatorname{ind} D''_{1,B''}, \tag{92}$$

where $B' = B_1$ and B'' is the L^2-orthogonal complement of B_1 in $H^{1/2}(\partial M, E_1)$.

Use f and $\mathscr{I}_E$ from Definition 1.20 to identify $M_1 \backslash K_1$ with $M_2 \backslash K_2$ and E_1 over $M_1 \backslash K_1$ with E_2 over $M_2 \backslash K_2$. Identify N_1 and M_1'' with their images N_2 and M_2'' under f. Then, with the same choice of T and A as above, we obtain a corresponding index formula

$$\operatorname{ind} D_2 = \operatorname{ind} D'_{2,B'} + \operatorname{ind} D''_{2,B''}, \tag{93}$$

where $B' = B_2$ and B'' correspond to the above B' and B'' under the chosen identifications. Now $\operatorname{ind} D''_{1,B''} = \operatorname{ind} D''_{2,B''}$ since D_1 outside K_1 agrees with D_2 outside K_2, and therefore the asserted formula follows by subtracting (93) from (92). □

PROOF OF THE RELATIVE INDEX THEOREM 1.21. The first assertion of Theorem 1.21, namely that D_1 is a Fredholm operator if and only if D_2 is a Fredholm operator, is immediate from Theorem 1.18 and the compactness of K_1 and K_2.

Assume now that D_1 and D_2 are Fredholm operators. Choose a compact hypersurface $N = N_1$ in $M_1 \backslash K_1$ which decomposes M_1 into $M_1 = M_1' \cup M_1''$, where M_1' is compact and K_1 is contained in the interior of M_1', and choose a corresponding decomposition of M_2. Then we get

$$\operatorname{ind} D_1 - \operatorname{ind} D_2 = \operatorname{ind} D'_{1,B_1} - \operatorname{ind} D'_{2,B_2}, \tag{94}$$

in the setup of and the index fromula in by Theorem 8.19.

Now choose a compact Riemannian manifold X with boundary N equipped with a Dirac type operator such that gluing along N yields a smooth closed Riemannian manifold $\tilde{M}_1 = M'_1 \cup_N X$ together with a smooth extension† $\tilde{D}_1$ of D_1. Since D_1 outside K_1 agrees with D_2 outside K_2, $\tilde{M}_2 = M'_2 \cup_N X$ is also a smooth closed Riemannian manifold and comes with a smooth extension $\tilde{D}_2$ of D_2. As above, we get

$$\operatorname{ind} \tilde{D}_1 - \operatorname{ind} \tilde{D}_2 = \operatorname{ind} D'_{1,B_1} - \operatorname{ind} D'_{2,B_2}. \tag{95}$$

Now, the Atiyah-Singer index theorem for Dirac type operators on closed manifolds gives

$$\operatorname{ind} \tilde{D}_i = \int_{\tilde{M}_i} \alpha_{\tilde{D}_i}, \tag{96}$$

where $\alpha_{\tilde{D}_i}$ denotes the index density of $\tilde{D}_i$. We apply (94) – (96), the fact that $\alpha_{\tilde{D}_1} = \alpha_{\tilde{D}_2}$ on $M_1 \backslash K_1 = M_2 \backslash K_2$, and that $\alpha_{\tilde{D}_i} = \alpha_{D_i}$ on K_i, and obtain

$$\begin{aligned}\operatorname{ind} D_1 - \operatorname{ind} D_2 &= \operatorname{ind} \tilde{D}_1 - \operatorname{ind} \tilde{D}_2 \\ &= \int_{\tilde{M}_1} \alpha_{\tilde{D}_1} - \int_{\tilde{M}_2} \alpha_{\tilde{D}_2} \\ &= \int_{K_1} \alpha_{D_1} - \int_{K_2} \alpha_{D_2}.\end{aligned}$$

□

REMARK 8.20. In Theorems 8.19 and 1.21, it is also possible to deal with the situation that M_1 and M_2 have compact boundary and elliptic boundary conditions B_1 and B_2 along their boundaries are given. One then chooses the hypersurface $N = N_i$ such that it does not intersect the boundary of M_i and such that the boundary of M_i is contained in M'_i. The same arguments as above yield

$$\operatorname{ind} D_{1,B_1} - \operatorname{ind} D_{2,B_2} = \operatorname{ind} D'_{1,B_1 \oplus B'_1} - \operatorname{ind} D'_{2,B_2 \oplus B'_2},$$

where B'_i and B'_2 are elliptic boundary condition along N_1 and N_2 which correspond to each other under the identifications given by f and $\mathscr{I}_E$ as in Definition 1.20.

Using Theorem 8.14, one can reduce to the case of Atiyah-Patodi-Singer boundary conditions. If the domains M'_i are compact, one can then express the indices on the right hand side as integrals over the index densities plus boundary contributions. There are two kinds of boundary contribution: the

†Such an extension exists. The manifold X can be chosen to be diffeomorphic to M'_1, for instance.

eta invariant of the adapted boundary operator and the boundary integral of the transgression form, compare [**APS**, Thm. 3.10] and [**G1**, Cor. 5.3].

Since the boundary contributions along N_1 and N_2 agree, they cancel each other when taking the difference $\operatorname{ind} D'_{1,B_1\oplus B'_1} - \operatorname{ind} D'_{2,B_2\oplus B'_2}$. This observation leads to another proof of the relative index formula (in the case where $\partial M_i = \emptyset$), not using the auxiliary manifold X, but the local index theorem for compact manifolds with boundary.

8.5. The cobordism theorem. Assume that $D : C^\infty(M,E) \to C^\infty(M,E)$ is a formally selfadjoint operator of Dirac type over a complete Riemannian manifold M with compact boundary ∂M. Then $\sigma_D(\xi)$ is skew-Hermitian, for any $\xi \in T^*M$.

Let τ be the interior unit conormal field along ∂M. Then, by the Clifford relations (31) and (32), $i\sigma_0 = i\sigma_D(\tau)$ is a field of unitary involutions of $E|_{\partial M}$ which anticommutes with $\sigma_A(\xi) = \sigma_0^{-1} \circ \sigma_D(\xi)$, for all $\xi \in T^*\partial M$. The eigenspaces of $i\sigma_0$ for the eigenvalues ± 1 split $E|_{\partial M}$ into a sum of fiber-wise perpendicular subbundles $E^\pm$. By Corollary 7.23 and Proposition 7.24, $B^\pm := H^{1/2}(\partial M, E^\pm)$ are ∞-regular elliptic boundary conditions for D.

With respect to the splitting $E|_{\partial M} = E^+ \oplus E^-$, any adapted boundary operator takes the form

$$A = \begin{pmatrix} A_{++} & A^- \\ A^+ & A_{--} \end{pmatrix}.$$

Since the differential operators $A^\pm$ are elliptic of order one, they define Fredholm operators $A^\pm : H^1(\partial M, E^\pm) \to L^2(\partial M, E^\mp)$. The cobordism theorem is concerned with the index of theses operators.

Since $\sigma_A(\xi)$ interchanges the bundles E^+ and E^-, the operators A_{++} and A_{--} are of order zero. Thus we may, without loss of generality, assume that $A_{++} = A_{--} = 0$. Moreover, $Z = (A^-)^* - A^+$ is also of order zero. Since the addition of a zero order term does not change the index of an operator of order one, we may replace A^+ by $A^+ + Z$ and arrive at the normal form for A which we use from now on,

$$A = \begin{pmatrix} 0 & A^- \\ A^+ & 0 \end{pmatrix}, \tag{97}$$

where $(A^+)^* = A^-$. The adapted operator A is still not uniquely determined by these requirements; we have the freedom to replace A^+ by $A^+ + V$ and A^- by $A^- + V^*$, where $V : E^+ \to E^-$ is any zero-order term.

In the notation of Theorems 1.12 and 7.11, the boundary conditions B^+ and B^- can be written as

$$B^+ = W_+ \oplus \Gamma(g)^{1/2} \quad \text{and} \quad B^- = W_- \oplus \Gamma(-g)^{1/2},$$

where

$$W_\pm = \ker A^\pm, \quad V_- = L^2_{(-\infty,0)}(A), \quad V_+ = L^2_{(0,\infty)}(A),$$

and $gx = i\sigma_0 x$. Since $\sigma_0^* = \sigma_0^{-1} = -\sigma_0$, we have $g^* = g$ and hence

$$B^- = \sigma_0(B^-) = (\sigma_0^{-1})^*(B^-)$$

is the adjoint boundary condition of B^+.

LEMMA 8.21. *Let A, B^+, and B^- be chosen as above and assume that each connected component of M has a nonempty boundary. Then*

$$\ker D_{B^+} = \ker D_{B^-} = 0.$$

PROOF. Let $\Phi \in \ker D_{B^\pm,\max}$. By (48), we have

$$\begin{aligned} 0 &= (D_{\max}\Phi, \Phi)_{L^2(M)} - (\Phi, D_{\max}\Phi)_{L^2(M)} \\ &= -(\sigma_0 \mathscr{R}\Phi, \mathscr{R}\Phi)_{L^2(\partial M)} \\ &= \pm i \|\mathscr{R}\Phi\|^2_{L^2(\partial M)}, \end{aligned}$$

and hence $\mathscr{R}\Phi = 0$.

Now extend M, E, and D beyond the boundary ∂M to a larger manifold $\tilde{M}$ (without boundary) equipped with a bundle $\tilde{E}$ and Dirac type operator $\tilde{D}$ as in the proof of Theorem 6.7. Moreover, extend Φ to a section of $\tilde{E}$ by setting $\tilde{\Phi} = 0$ on $\tilde{M} \backslash M$. For any test section $\Psi \in C^\infty_{cc}(\tilde{M}, \tilde{E})$ we have, again by (48),

$$\begin{aligned} (\tilde{\Phi}, \tilde{D}^*\Psi)_{L^2(\tilde{M})} &= (\Phi, D^*\Psi)_{L^2(M)} \\ &= (D_{\max}\Phi, \Psi)_{L^2(M)} + (\sigma_0 \mathscr{R}\Phi, \mathscr{R}\Psi)_{L^2(\partial M)} \\ &= 0. \end{aligned}$$

Thus $\tilde{\Phi}$ is a weak (and, by elliptic regularity theory, smooth) solution of $\tilde{D}\tilde{\Phi} = 0$. Hence $\tilde{\Phi}$ is a solution to the Laplace type equation $\tilde{D}^2\tilde{\Phi} = 0$ which vanishes on an nonempty open subset of $\tilde{M}$. The unique continuation theorem of Aronszajn [**Ar**] implies $\tilde{\Phi} = 0$ and hence $\Phi = 0$. □

We have

$$\operatorname{ind} A^+ = \dim W_+ - \dim W_- = -\operatorname{ind} A^- \tag{98}$$

since A^- is the adjoint operator of A^+.

PROOF OF THEOREM 1.22. By (62), D_{B^-} is the adjoint of D_{B^+}. Lemma 8.21 therefore yields

$$-\operatorname{ind} D_{B^-} = \operatorname{ind} D_{B^+} = \dim\ker D_{B^+} - \dim\ker D_{B^-} = 0.$$

Without loss of generality we may assume $g = 0$, see Example 8.13. Then Corollary 8.8 yields

$$\begin{aligned} \operatorname{ind} D_{B^+} &= \operatorname{ind} D_{B_{\mathrm{APS}}} + \dim W_+, \\ \operatorname{ind} D_{B^-} &= \operatorname{ind} D_{B_{\mathrm{APS}}} + \dim W_-. \end{aligned}$$

We obtain that $\dim W_+ = \dim W_-$. Equation (98) concludes the proof. □

Appendix A. Some functional analytic facts

We collect some functional analytic facts, which have been used in the main body of the text, and which are not easily found in the standard literature.

PROPOSITION A.1. *Let H be a Hilbert space, let E and F be Banach spaces and let $L : H \to E$ and $P : H \to F$ be bounded linear maps. We assume that $P : H \to F$ is onto.*

Then the following hold:

(i) *The kernel of $L_{|\ker(P)} : \ker(P) \to E$ and the kernel of $L \oplus P : H \to E \oplus F$ have equal dimension.*
(ii) *The range of $L_{|\ker(P)} : \ker(P) \to E$ is closed if and only if the range of $L \oplus P : H \to E \oplus F$ is closed.*
(iii) *The cokernel of $L_{|\ker(P)} : \ker(P) \to E$ and the cokernel of $L \oplus P : H \to E \oplus F$ have equal dimension.*
(iv) *$L_{|\ker(P)} : \ker(P) \to E$ is Fredholm of index k if and only if $L \oplus P : H \to E \oplus F$ is Fredholm of index k.*

PROOF. Write $L_1 := L_{|\ker(P)}$, $L_2 := L_{|\ker(P)^\perp}$, and $P_2 := P_{|\ker(P)^\perp}$. With respect to the splittings $H = \ker(P) \oplus \ker(P)^\perp$ and $E \oplus F$ the operator $L \oplus P$ takes the matrix form

$$L \oplus P = \begin{pmatrix} L_1 & L_2 \\ 0 & P_2 \end{pmatrix}.$$

By the open mapping theorem $P_2 : \ker(P)^\perp \to F$ is an isomorphism (with a bounded inverse), so that $\begin{pmatrix} \mathrm{id}_{\ker(P)} & 0 \\ 0 & P_2^{-1} \end{pmatrix}$ is an isomorphism. Moreover, $\begin{pmatrix} \mathrm{id}_E & -L_2P_2^{-1} \\ 0 & \mathrm{id}_F \end{pmatrix}$ is an isomorphism with inverse $\begin{pmatrix} \mathrm{id}_E & L_2P_2^{-1} \\ 0 & \mathrm{id}_F \end{pmatrix}$. Therefore the kernel of $L \oplus P$ has the same dimension as the kernel of

$$\begin{pmatrix} \mathrm{id}_E & -L_2P_2^{-1} \\ 0 & \mathrm{id}_F \end{pmatrix} \begin{pmatrix} L_1 & L_2 \\ 0 & P_2 \end{pmatrix} \begin{pmatrix} \mathrm{id}_{\ker(P)} & 0 \\ 0 & P_2^{-1} \end{pmatrix} = \begin{pmatrix} L_1 & 0 \\ 0 & \mathrm{id}_F \end{pmatrix}$$

which is the same as the dimension of the kernel of L_1. The other statements follow similarly. □

REMARK A.2. The proposition also holds and the proof works without change if H is only a Banach space and one assumes that $\ker(P)$ has a closed complement.

PROPOSITION A.3. *Let X and Y be Banach spaces and $L : X \to Y$ be a bounded linear map. Then the following are equivalent:*

(i) *The operator L has finite-dimensional kernel and closed image.*

(ii) There is a Banach space Z, a compact linear map $K: X \to Z$, and a constant C such that

$$\|x\|_X \leq C \cdot (\|Kx\|_Z + \|Lx\|_Y),$$

for all $x \in X$. In particular, $\ker K \cap \ker L = \{0\}$.

(iii) Every bounded sequence (x_n) in X such that (Lx_n) converges in Y has a convergent subsequence in X.

Moreover, these equivalent conditions imply

(iv) For any Banach space Z and compact linear map $K: X \to Z$ such that $\ker K \cap \ker L = \{0\}$, there is a constant C such that

$$\|x\|_X \leq C(\|Kx\|_Z + \|Lx\|_Y)$$

for all $x \in X$.

The equivalence of (i) and (iii) is Proposition 19.1.3 in [**Hö**].

PROOF OF PROPOSITION A.3. We show (i) $\Rightarrow$ (ii). Assume (i). Since $\ker L$ is finite-dimensional, it has a closed complement U in X. Since the image of L is closed, $L|_U : U \to \operatorname{im} L$ is an isomorphism of Banach spaces. Moreover, since U is closed and $\ker L$ is finite-dimensional, the projection $K: X \to \ker L$ along U is a compact operator. It follows that

$$F: X \to \ker L \oplus \operatorname{im} L, \quad Fx = (Kx, Lx),$$

is an isomorphism of topological vector spaces, hence we have (ii) with $Z = \ker L$.

Next we show (ii) $\Rightarrow$ (iii). Assume (ii), and let K be a compact linear map as assumed in (ii). Let (x_n) be a bounded sequence in X such that (Lx_n) converges in Y. Since K is compact, we may assume, by passing to a subsequence if necessary, that (Kx_n) converges in Z. The estimate in (ii) then implies that (x_n) is a Cauchy sequence in X, hence (iii).

Now we show (iii) $\Rightarrow$ (i). Assume (iii), and let (x_n) be a bounded sequence in $\ker L$. Then (x_n) subconverges in $\ker L$, by (iii). It follows that $\dim \ker L < \infty$.

Let U be a complement of $\ker L$ in X. Let (x_n) be a sequence in X such that $Lx_n \to y \in Y$. We have to show that y lies in the image of L. For this, it is sufficient to show that (x_n) subconverges. Without loss of generality we assume that $x_n \in U$, for all n.

If $\|x_{n_k}\|_X \to \infty$ for some subsequence (x_{n_k}), then $u_k := x_{n_k}/\|x_{n_k}\|_X$ has norm 1 and $Lu_k \to 0$. By (iii), the sequence of u_k has a convergent subsequence. The limit u is a unit vector in U with $Lu = 0$. This is a contradiction, because U is complementary to $\ker L$. Hence we may assume that the sequence of (x_n) is bounded. But then it subconverges, by (iii).

Finally, we show (ii) $\Rightarrow$ (iv). Assume (ii) and let $K_0 : X \to Z_0$ be a compact linear map as in (ii). Let $K: X \to Z$ be any compact linear map such that $\ker K \cap \ker L = \{0\}$. If the assertion does not hold, then there is a sequence (x_n) of unit vectors in X such that $\|Kx_n\| + \|Lx_n\| \to 0$. This implies, in particular, that $\|Lx_n\| \to 0$ and hence that $\|K_0 x_n\| \geq \delta$ for some $\delta > 0$. Now K_0 is compact, hence, up to passing to a subsequence, $(K_0 x_n)$ is a Cauchy sequence. Since $Lx_n \to 0$, this implies that (x_n) is a Cauchy sequence, by (ii). If $x := \lim x_n \in X$, then x is a unit vector with $Kx = Lx = 0$. This contradicts $\ker K \cap \ker L = \{0\}$. □

LEMMA A.4. *Let B_+ and B_- be Banach spaces, let $(\cdot,\cdot) : B_+ \times B_- \to \mathbb{C}$ be a perfect pairing. Let H be a Hilbert space and let there be continuous imbeddings $B_+ \subset H \subset B_-$. Assume that the restriction of $(\cdot,\cdot)$ to $B_+ \times H$ coincides with the restriction of the scalar product on H.*

Let $W \subset B_+$ be a finite-dimensional subspace and let V_- be the annihilator of W in B_-, i. e. $V_- = \{x \in B_- \mid (w,x) = 0 \ \forall x \in W\}$. Put $V_0 := V_- \cap H$ and $V_+ := V_- \cap B_+$.

Then

$$B_- = W \oplus V_-, \quad H = W \oplus V_0, \quad B_+ = W \oplus V_+.$$

Moreover, the second decomposition is orthogonal and W is the annihilator of V_+ in B_-. The pairing $(\cdot,\cdot)$ restricts to a perfect pairing of V_+ and V_-.

PROOF. For $w \in W \cap V_-$ we have $0 = (w,w) = \|w\|_H^2$, thus $w = 0$. This shows $W \cap V_- = 0$ and hence also $W \cap H = W \cap V_+ = 0$. Since the codimension of the annihilator of W in any space can at most be $\dim(W)$ we conclude $B_- = W \oplus V_-$ and similarly for H and B_+.

Orthogonality of the second decomposition is clear because the restrictions of the pairing and of the scalar product coincide. Clearly, W is contained in the annihilator of V_+ in B_-. Conversely, let $v = w + v_- \in B_- = W \oplus V_-$ be in the annihilator of V_+. Then for all $v_+ \in V_+$ we have $0 = (v_+, w + v_-) = (v_+, v_-)$, thus $v_- = 0$, i. e. $v = w \in W$.

Nondegeneracy of the pairing of V_+ and V_- also follows easily. □

Acknowledgments

We would like to thank Bernd Ammann, Bernhelm Booss-Bavnbek, and Gilles Carron for helpful remarks and suggestions. We also gratefully acknowledge the support by the Max Planck Institute for Mathematics, the Hausdorff Center for Mathematics, the Sonderforschungsbereich 647, and the Erwin Schrödinger Institute for Mathematical Physics. W. B. thanks the University of Potsdam for its hospitality.

References

[Ad] R. Adams: *Sobolev spaces.* Academic Press, New York, 1975.

[An] N. Anghel: An abstract index theorem on non-compact Riemannian manifolds. *Houston J. Math.* 19 (1993), 223–237.

[Ar] N. Aronszajn: A unique continuation theorem for solutions of elliptic partial differential equations or inequalities of second order. *J. Math. pur. appl., IX. Sér.* 36 (1957), 235–249.

[APS] M. Atiyah, V. Patodi, and I. Singer: Spectral asymmetry and Riemannian Geometry. I. *Math. Proc. Cambridge Philos. Soc.* 77 (1975), 43–69.

[Au] T. Aubin: *Some nonlinear problems in Riemannian geometry.* Springer-Verlag, Berlin, 1998

[BB1] W. Ballmann and J. Brüning: On the spectral theory of surfaces with cusps. In: *Geometric analysis and partial differential equations*, 13–37, Springer, Berlin, 2003.

[BB2] W. Ballmann and J. Brüning: On the spectral theory of manifolds with cusps. *J. Math. Pures Appl.* 80 (2001), 593–625.

[BBC] W. Ballmann, J. Brüning, and G. Carron: Regularity and index theory for Dirac-Schrödinger systems with Lipschitz coefficients. *J. Math. Pures Appl.* 89 (2008), 429–476.

[BC] R. Bartnik and P. Chruściel: Boundary value problems for Dirac-type equations. *J. Reine Angew. Math.* 579 (2005), 13–73.

[BGV] N. Berline, E. Getzler, and M. Vergne: *Heat kernels and Dirac operators.* Springer-Verlag, Berlin, 1992.

[BF1] B. Booss-Bavnbek and K. Furutani: The Maslov index: a functional analytical definition and the spectral flow formula. *Tokyo J. Math.* 21 (1998), 1–34.

[BF2] B. Booss-Bavnbek and K. Furutani: Symplectic functional analysis and spectral invariants. Geometric aspects of partial differential equations (Roskilde, 1998), 53–83, *Contemp. Math.* 242, Amer. Math. Soc., Providence, RI, 1999.

[BFO] B. Booss-Bavnbek, K. Furutani, and N. Otsuki: Criss-cross reduction of the Maslov index and a proof of the Yoshida-Nicolaescu theorem. *Tokyo J. Math.* 24 (2001), 113–128.

[BLZ] B. Booss-Bavnbek, M. Lesch, and C. Zhu: The Calderón projection: New definition and applications. *J. Geom. Phys.* 59 (2009), 784–826.

[BW] B. Booss–Bavnbek and K. Wojciechowski: *Elliptic Boundary Problems for Dirac Operators.* Birkhäuser 1993.

[B] L. Boutet de Monvel: Boundary problems for pseudo-differential operators. *Acta Math.* 126 (1971), 11–51.

[BL1] J. Brüning and M. Lesch: Spectral theory of boundary value problems for Dirac type operators. *Geometric aspects of partial differential equations* (Roskilde, 1998), 203–215, Contemp. Math., 242, Amer. Math. Soc., Providence, RI, 1999.

[BL2] J. Brüning and M. Lesch: On boundary value problems for Dirac type operators. 1. Regularity and self-adjointness. *J. Funct. Anal.* 185 (2001), 1–62.

[Ca] G. Carron: Un Théorème de l'indice relatif. *Pacific J. Math.* 198 (2001), 8–107.

[Ch] P. Chernoff: Essential Self-Adjointness of Powers of Generators of Hyperbolic Equations. *J. Funct. Anal.* 12 (1973), 401–414.

[FS] S. Farinelli and G. Schwarz: On the spectrum of the Dirac operator under boundary conditions. *J. Geom. Phys.* 28 (1998), 67–84.

[GHHP] G. W. Gibbons, S. W. Hawking, G. T. Horowitz, and M. J. Perry: Positive mass theorems for black holes. *Commun. Math. Phys.* 88 (1983), 295–308.

[Gi] P. Gilkey: *Invariance theory, the heat equation, and the Atiyah-Singer index theorem.* Publish or Perish, Wilmington, USA, 1984.

[GL] M. Gromov and H.B. Lawson: Positive scalar curvature and the Dirac operator on complete Riemannian manifolds. *Inst. Hautes Études Sci. Publ. Math.* 58 (1983), 83–196.

[G1] G. Grubb: Heat operator trace expansions and index for general Atiyah-Patodi-Singer boundary problems. *Commun. Part. Diff. Eq.* 17 (1992), 2031–2077.

[G2] G. Grubb: *Functional calculus of pseudodifferential boundary problems.* Second edition. Progress in Mathematics, 65. Birkhäuser, Boston, 1996.

[He] M. Herzlich: The positive mass theorem for black holes revisited. *J. Geom. Phys.* 26 (1998), 97–111.

[Hi] N. Higson: A note on the cobordism invariance of the index. *Topology* 30 (1991), 439–443.

[HMR] O. Hijazi, S. Montiel, and A. Roldán: Eigenvalue Boundary Problems for the Dirac Operator. *Commun. Math. Phys.* 231 (2002), 375–390.

[HMZ] O. Hijazi, S. Montiel, and X. Zhang: Eigenvalues of the Dirac Operator on Manifolds with Boundary. *Commun. Math. Phys.* 221 (2001), 255–265.

[Hö] L. Hörmander: *The analysis of linear partial differential operators. III Pseudo-differential operators.* Springer-Verlag, Berlin, 1985.

[Ka] T. Kato: *Perturbation Theory for Linear Operators.* Second Ed., Springer-Verlag, 1986.

[LM] H. B. Lawson and M.-L. Michelsohn: *Spin Geometry.* Princeton University Press, Princeton, USA, 1989.

[Pa] R. S. Palais: *Seminar on the Atiyah-Singer index theorem.* With contributions by M. F. Atiyah, A. Borel, E. E. Floyd, R. T. Seeley, W. Shih, and R. Solovay. Princeton University Press, Princeton, 1965.

[RS] S. Rempel and B.-W. Schulze: *Index theory of elliptic boundary problems.* Reprint of the 1982 edition, North Oxford Academic Publishing, London, 1985.

[R] J. Roe: Partitioning noncompact manifolds and the dual Toeplitz problem. *Operator algebras and applications*, Vol. 1, 187–228, London Math. Soc. Lecture Note Ser., 135, Cambridge Univ. Press, Cambridge, 1988.

[S1] B.-W. Schulze: An algebra of boundary value problems not requiring Shapiro-Lopatinskij conditions. *J. Funct. Anal.* 179 (2001), 374–408.

[S2] B.-W. Schulze: Toeplitz operators, and ellipticity of boundary value problems with global projection conditions. *Aspects of boundary problems in analysis and geometry*, 342–429, Oper. Theory Adv. Appl. 151, Birkhäuser, Basel, 2004.

[Se] R. T. Seeley: Complex powers of an elliptic operator. *Proc. Sympos. Pure Math.* 10 (1967), 288–307.

[Ta] M. E. Taylor: *Partial differential equations. I. Basic theory.* Springer-Verlag, New York, 1996.

[Wi] E. Witten: A new proof of the positive energy theorem. *Commun. Math. Phys.* 80 (1981), 381–402.

[Wo] J. A. Wolf: Essential self-adjointness for the Dirac operator and its square. *Indiana Univ. Math. J.* 22 (1972/73), 611–640.

Index

Universität Potsdam, Institut für Mathematik, Am Neuen Palais 10, Haus 8, 14469 Potsdam, Germany
E-mail address: baer@math.uni-potsdam.de

Max Planck Institute for Mathematics, Vivatsgasse 7, 53111 Bonn, Germany
E-mail address: hwbllmnn@mpim-bonn.mpg.de

A geometric construction for invariant jet differentials

Gergely Berczi and Frances Kirwan

1. Introduction

The action of the reparametrization group $\mathbb{G}_k$, consisting of k-jets of germs of biholomorphisms of $(\mathbb{C}, 0)$, on the bundle $J_k = J_k T^*X$ of k-jets at 0 of germs of holomorphic curves $f : \mathbb{C} \to X$ in a complex manifold X has been a focus of investigation since the work of Demailly [**5**] which built on that of Green and Griffiths [**13**]. Here $\mathbb{G}_k$ is a non-reductive complex algebraic group which is the semi-direct product $\mathbb{G}_k = \mathbb{U}_k \rtimes \mathbb{C}^*$ of its unipotent radical $\mathbb{U}_k$ with $\mathbb{C}^*$; it has the form

$$\mathbb{G}_k \cong \left\{ \begin{pmatrix} \alpha_1 & \alpha_2 & \alpha_3 & \cdots & \alpha_k \\ 0 & \alpha_1^2 & \cdots & & \\ 0 & 0 & \alpha_1^3 & \cdots & \\ \cdot & \cdot & \cdot & \cdot & \cdot \\ 0 & 0 & 0 & \cdots & \alpha_1^k \end{pmatrix} : \alpha_1 \in \mathbb{C}^*, \alpha_2, \ldots, \alpha_k \in \mathbb{C} \right\}$$

where the entries above the leading diagonal are polynomials in $\alpha_1, \ldots, \alpha_k$, and $\mathbb{U}_k$ is the subgroup consisting of matrices of this form with $\alpha_1 = 1$. The bundle of Demailly-Semple jet differentials of order k over X has fibre at $x \in X$ given by the algebra $\mathcal{O}((J_k)_x)^{\mathbb{U}_k}$ of $\mathbb{U}_k$-invariant polynomial functions on the fibre $(J_k)_x = (J_k T^*X)_x$ of $J_k T^*X$. More generally following [**25**] we can replace $\mathbb{C}$ with $\mathbb{C}^p$ for $p \geq 1$ and consider the bundle $J_{k,p}T^*X$ of k-jets at 0 of holomorphic maps $f : \mathbb{C}^p \to X$ and the reparametrization group $\mathbb{G}_{k,p}$ consisting of k-jets of germs of biholomorphisms of $(\mathbb{C}^p, 0)$; then $\mathbb{G}_{k,p}$ is the semi-direct product of its unipotent radical $\mathbb{U}_{k,p}$ and the complex reductive group $\mathrm{GL}(p)$, while its subgroup $\mathbb{G}'_{k,p} = \mathbb{U}_{k,p} \rtimes \mathrm{SL}(p)$ (which equals $\mathbb{U}_{k,p}$ when $p = 1$) fits into an exact sequence $1 \to \mathbb{G}'_{k,p} \to \mathbb{G}_{k,p} \to \mathbb{C}^* \to 1$. The generalized Demailly-Semple algebra is then $\mathcal{O}((J_{k,p})_x)^{\mathbb{G}'_{k,p}}$.

This work was supported by the Engineering and Physical Sciences Research Council [grant numbers GR/T016170/1,EP/G000174/1].

The Demailly-Semple algebras $\mathcal{O}(J_k)^{\mathbb{U}_k}$ and their generalizations have been studied for a long time. The invariant jet differentials play a crucial role in the strategy devised by Green, Griffiths [**13**], Bloch [**4**], Demailly [**5, 6**], Siu [**28, 29, 30**] and others to prove Kobayashi's 1970 hyperbolicity conjecture [**19**] and the related conjecture of Green and Griffiths in the special case of hypersurfaces in projective space. This strategy has been recently used successfully by Diverio, Merker and Rousseau in [**7**] and then by the first author in [**1**] to give effective lower bounds for the degrees of generic hypersurfaces in $\mathbb{P}_n$ for which the Green-Griffiths conjecture holds.

In particular it has been a long-standing problem to determine whether the algebras of invariants $\mathcal{O}((J_{k,p})_x)^{\mathbb{G}'_{k,p}}$ and bi-invariants $\mathcal{O}((J_{k,p})_x)^{\mathbb{G}'_{k,p}\times U_{n,x}}$ (where $U_{n,x}$ is a maximal unipotent subgroup of $\mathrm{GL}(T_xX)\cong\mathrm{GL}(n)$) are finitely generated as graded complex algebras, and if so to provide explicit finite generating sets. In [**20**] Merker showed that when $p=1$ and both k and $n=\dim X$ are small then these algebras are finitely generated, and for $p=1$ and all k and n he provided an algorithm which produces finite sets of generators when they exist. In this paper we will describe methods inspired by [**2**] and the approach of [**9**] to non-reductive geometric invariant theory (GIT) to prove the finite generation of $\mathcal{O}((J_k)_x)^{\mathbb{U}_k}$ for all n and $k\geq 2$ (from which the finite generation of the corresponding bi-invariants follows). In fact we will show that $\mathbb{U}_k$ is a Grosshans subgroup of $\mathrm{SL}(k)$, so that the algebra $\mathcal{O}(\mathrm{SL}(k))^{\mathbb{U}_k}$ is finitely generated and hence every linear action of $\mathbb{U}_k$ which extends to a linear action of $\mathrm{SL}(k)$ has finitely generated invariants. We will also give a geometric description of a finite set of generators for $\mathcal{O}(\mathrm{SL}(k))^{\mathbb{U}_k}$, and a geometric description of the associated affine variety

$$\mathrm{SL}(k)//\mathbb{U}_k=\mathrm{Spec}(\mathcal{O}(SL(k))^{\mathbb{U}_k})$$

which leads to a geometric description of the affine variety

$$(J_k)_x//\mathbb{U}_k=\mathrm{Spec}(\mathcal{O}((J_k)_x)^{\mathbb{U}_k})$$

as a GIT quotient

$$((J_k)_x\times(\mathrm{SL}(k)//\mathbb{U}_k))//\mathrm{SL}(k)$$

by the reductive group $\mathrm{SL}(k)$, in the sense of classical geometric invariant theory [**23**]. Similarly we expect that if $p>1$ and k is sufficiently large (depending on p) then $\mathbb{G}'_{k,p}$ is a subgroup of $\mathrm{SL}(\mathrm{sym}^{\leq k}(p))$, where

$$\mathrm{sym}^{\leq k}(p)=\sum_{i=1}^{k}\dim\mathrm{Sym}^i\mathbb{C}^p,$$

such that the algebra $\mathcal{O}(\mathrm{SL}(\mathrm{sym}^{\leq k}(p)))^{\mathbb{G}'_{k,p}}$ is finitely generated, and thus that the algebra and $\mathcal{O}((J_{k,p})_x)^{\mathbb{G}'_{k,p}}$ is also finitely generated, and we have a geometric description of the associated affine variety

$$(J_{k,p})_x//\mathbb{G}'_{k,p}.$$

The layout of this paper is as follows. §2 reviews the reparametrization groups $\mathbb{G}_k$ and $\mathbb{G}_{k,p}$ and their actions on jet bundles and jet differentials over a complex manifold X. Next §3 reviews some of the results of [**9**] on non-reductive geometric invariant theory. In §4 we recall from [**2**] a geometric description of the quotients by $\mathbb{U}_k$ and $\mathbb{G}_k$ of open subsets of $(J_k)_x$, and in §5 this is used to find explicit affine and projective embeddings of these quotients and explicit embeddings of $\mathrm{SL}(k)/\mathbb{U}_k$. In §6 we see that the complement of $\mathrm{SL}(k)/\mathbb{U}_k$ in its closure for a suitable embedding in an affine space has codimension at least two. In §7 we conclude that $\mathbb{U}_k$ is a Grosshans subgroup of $\mathrm{SL}(k)$ when $k \geq 2$, so that $\mathcal{O}(SL(k))^{\mathbb{U}_k}$ and $\mathcal{O}((J_k)_x)^{\mathbb{U}_k}$ are finitely generated, and provide a geometric description of a finite set of generators of $\mathcal{O}(\mathrm{SL}(k))^{\mathbb{U}_k}$. Finally §8 and §9 discuss how to extend the results of §6 and §7 to the action of $\mathbb{G}_{k,p}$ on the jet bundle $J_{k,p} \to X$ of k-jets of germs of holomorphic maps from $\mathbb{C}^p$ to X for $p > 1$.

2. Jets of curves and jet differentials

Let X be a complex n-dimensional manifold and let k be a positive integer. Green and Griffiths in [**13**] introduced the bundle $J_k \to X$ of k-jets of germs of parametrized curves in X; its fibre over $x \in X$ is the set of equivalence classes of germs of holomorphic maps $f : (\mathbb{C}, 0) \to (X, x)$, with the equivalence relation $f \sim g$ if and only if the derivatives $f^{(j)}(0) = g^{(j)}(0)$ are equal for $0 \leq j \leq k$. If we choose local holomorphic coordinates $(z_1, \ldots, z_n)$ on an open neighbourhood $\Omega \subset X$ around x, the elements of the fibre $J_{k,x}$ are represented by the Taylor expansions

$$f(t) = x + tf'(0) + \frac{t^2}{2!}f''(0) + \cdots + \frac{t^k}{k!}f^{(k)}(0) + O(t^{k+1})$$

up to order k at $t = 0$ of $\mathbb{C}^n$-valued maps

$$f = (f_1, f_2, \ldots, f_n)$$

on open neighbourhoods of 0 in $\mathbb{C}$. Thus in these coordinates the fibre is

$$J_{k,x} = \left\{(f'(0), \ldots, f^{(k)}(0)/k!)\right\} = (\mathbb{C}^n)^k,$$

which we identify with $\mathbb{C}^{nk}$. Note, however, that J_k is not a vector bundle over X, since the transition functions are polynomial, but not linear.

Let $\mathbb{G}_k$ be the group of k-jets at the origin of local reparametrizations of $(\mathbb{C}, 0)$

$$t \mapsto \varphi(t) = \alpha_1 t + \alpha_2 t^2 + \cdots + \alpha_k t^k, \quad \alpha_1 \in \mathbb{C}^*, \alpha_2, \ldots, \alpha_k \in \mathbb{C},$$

in which the composition law is taken modulo terms t^j for $j > k$. This group acts fibrewise on J_k by substitution. A short computation shows that this is a linear action on the fibre:

$$\begin{aligned} f\circ\varphi(t) &= f'(0)\cdot(\alpha_1 t+\alpha_2 t^2+\cdots+\alpha_k t^k)\\ &+\frac{f''(0)}{2!}\cdot(\alpha_1 t+\alpha_2 t^2+\cdots+\alpha_k t^k)^2\\ &+\cdots+\frac{f^{(k)}(0)}{k!}\cdot(\alpha_1 t+\alpha_2 t^2+\cdots+\alpha_k t^k)^k \text{ (modulo } t^{k+1})\end{aligned}$$

so the linear action of φ on the k-jet $(f'(0), f''(0)/2!, \ldots, f^{(k)}(0)/k!)$ is given by the following matrix multiplication:

$$(1)\qquad (f'(0), f''(0)/2!, \ldots, f^{(k)}(0)/k!) \cdot \begin{pmatrix} \alpha_1 & \alpha_2 & \alpha_3 & \cdots & \alpha_k \\ 0 & \alpha_1^2 & 2\alpha_1\alpha_2 & \cdots & \alpha_1\alpha_{k-1}+\cdots+\alpha_{k-1}\alpha_1 \\ 0 & 0 & \alpha_1^3 & \cdots & 3\alpha_1^2\alpha_{k-2}+\cdots \\ \cdot & \cdot & \cdot & \cdot & \cdot \\ 0 & 0 & 0 & \cdots & \alpha_1^k \end{pmatrix}$$

where the matrix has general entry

$$(G_k)_{i,j} = \sum_{s_1\geq 1,\ldots,s_i\geq 1,\ s_1+\cdots+s_i=j} \alpha_{s_1}\ldots\alpha_{s_i}$$

for $i, j \leq k$.

There is an exact sequence of groups:

$$(2)\qquad 1 \to \mathbb{U}_k \to \mathbb{G}_k \to \mathbb{C}^* \to 1,$$

where $\mathbb{G}_k \to \mathbb{C}^*$ is the morphism $\varphi \to \varphi'(0) = \alpha_1$ in the notation used above, and

$$\mathbb{G}_k = \mathbb{U}_k \rtimes \mathbb{C}^*$$

is a semi-direct product. With the above identification, $\mathbb{C}^*$ is the subgroup of $\mathbb{G}_k$ consisting of diagonal matrices satisfying $\alpha_2 = \cdots = \alpha_k = 0$ and $\mathbb{U}_k$ is the unipotent radical of $\mathbb{G}_k$, consisting of matrices of the form above with $\alpha_1 = 1$. The action of $\lambda \in \mathbb{C}^*$ on k-jets is thus described by

$$\lambda\cdot(f'(0), f''(0)/2!, \ldots, f^{(k)}(0)/k!) = (\lambda f'(0), \lambda^2 f''(0)/2!, \ldots, \lambda^k f^{(k)}(0)/k!)$$

Let $\mathcal{E}_{k,m}^n$ denote the vector space of complex valued polynomial functions

$$Q(u_1, u_2, \ldots, u_k)$$

of $u_1 = (u_{1,1}, \ldots, u_{1,n}), \ldots, u_k = (u_{k,1}, \ldots, u_{k,n})$ of weighted degree m with respect to this $\mathbb{C}^*$ action, where $u_i = f^{(i)}(0)/i!$; that is, such that

$$Q(\lambda u_1, \lambda^2 u_2, \ldots, \lambda^k u_k) = \lambda^m Q(u_1, u_2, \ldots, u_k).$$

Thus elements of $\mathcal{E}^n_{k,m}$ have the form

$$Q(u_1, u_2, \ldots, u_k) = \sum_{|i_1|+2|i_2|+\cdots+k|i_k|=m} u_1^{i_1} u_2^{i_2} \ldots u_k^{i_k},$$

where $i_1 = (i_{1,1}, \ldots, i_{1,n}), \ldots, i_k = (i_{k,1}, \ldots, i_{k,n})$ are multi-indices of length n. There is an induced action of $\mathbb{G}_k$ on the algebra $\bigoplus_{m\geq 0} \mathcal{E}^n_{k,m}$. Following Demailly (see [**5**]), we denote by $E^n_{k,m}$ (or $E_{k,m}$) the Demailly-Semple bundle whose fibre at x consists of the $\mathbb{U}_k$-invariant polynomials on the fibre of J_k at x of weighted degree m, i.e those which satisfy

$$\begin{aligned}
&Q((f\circ\varphi)'(0), (f\circ\varphi)''(0)/2!, \ldots, (f\circ\varphi)^{(k)}(0)/k!) \\
&\quad = \varphi'(0)^m \cdot Q(f'(0), f''(0)/2!, \ldots, f^{(k)}(0)/k!),
\end{aligned}$$

and we let $E^n_k = \oplus_m E^n_{k,m}$ denote the Demailly-Semple bundle of graded algebras of invariants.

We can also consider higher dimensional holomorphic surfaces in X, and therefore we fix a parameter $1 \leq p \leq n$, and study germs of maps $\mathbb{C}^p \to X$.

Again we fix the degree k of our map, and introduce the bundle $J_{k,p} \to X$ of k-jets of maps $\mathbb{C}^p \to X$. The fibre over $x \in X$ is the set of equivalence classes of germs of holomorphic maps $f : (\mathbb{C}^p, 0) \to (X, x)$, with the equivalence relation $f \sim g$ if and only if all derivatives $f^{(j)}(0) = g^{(j)}(0)$ are equal for $0 \leq j \leq k$.

We need a description of the fibre $J_{k,p,x}$ in terms of local coordinates as in the case when $p = 1$. Let $(z_1, \ldots, z_n)$ be local holomorphic coordinates on an open neighbourhood $\Omega \subset X$ around x, and let $(u_1, \ldots, u_p)$ be local coordinates on $\mathbb{C}^p$. The elements of the fibre $J_{k,p,x}$ are $\mathbb{C}^n$-valued maps

$$f = (f_1, f_2, \ldots, f_n)$$

on $\mathbb{C}^p$, and two maps represent the same jet if their Taylor expansions around $\mathbf{z} = 0$

$$f(\mathbf{z}) = x + \mathbf{z}f'(0) + \frac{\mathbf{z}^2}{2!}f''(0) + \cdots + \frac{\mathbf{z}^k}{k!}f^{(k)}(0) + O(\mathbf{z}^{k+1})$$

coincide up to order k. Note that here

$$f^{(i)}(0) \in \mathrm{Hom}\,(\mathrm{Sym}^i\mathbb{C}^p, \mathbb{C}^n)$$

and in these coordinates the fibre is a finite-dimensional vector space

$$J_{k,p,x} = \Big\{(f'(0), \ldots, f^{(k)}(0)/k!)\Big\} \cong \mathbb{C}^{n\binom{k+p-1}{k-1}}.$$

Let $\mathbb{G}_{k,p}$ be the group of k-jets of germs of biholomorphisms of $(\mathbb{C}^p, 0)$. Elements of $\mathbb{G}_{k,p}$ are represented by holomorphic maps

$$\begin{aligned}\mathbf{u} \to \varphi(\mathbf{u}) &= \Phi_1\mathbf{u} + \Phi_2\mathbf{u}^2 + \cdots + \Phi_k\mathbf{u}^k \\ &= \sum_{\mathbf{i}\in\mathbb{Z}^p\setminus 0} a_{i_1\dots i_p}u_1^{i_1}\dots u_p^{i_p}, \quad \Phi_1 \text{ is non-degenerate}\end{aligned} \tag{3}$$

where $\Phi_i \in \mathrm{Hom}\,(\mathrm{Sym}^i\mathbb{C}^p, \mathbb{C}^p)$. The group $\mathbb{G}_{k,p}$ admits a natural fibrewise right action on $J_{k,p}$, by reparametrizing the k-jets of holomorphic p-discs. A computation similar to that in **[2]** shows that

$$\begin{aligned}f\circ\varphi(\mathbf{u}) = f'(0)\Phi_1\mathbf{u} &+ \left(f'(0)\Phi_2 + \frac{f''(0)}{2!}\Phi_1^2\right)\mathbf{u}^2 + \cdots \\ &+ \sum_{i_1+\cdots+i_l=d}\frac{f^{(l)}(0)}{l!}\Phi_{i_1}\dots\Phi_{i_l}\mathbf{u}^l.\end{aligned}$$

This defines a linear action of $\mathbb{G}_{k,p}$ on the fibres $J_{k,p,x}$ of $J_{k,p}$ with the matrix representation given by

$$\begin{pmatrix} \Phi_1 & \Phi_2 & \Phi_3 & \dots & \Phi_k \\ 0 & \Phi_1^2 & \Phi_1\Phi_2 & \dots & \\ 0 & 0 & \Phi_1^3 & \dots & \\ \cdot & \cdot & \cdot & \cdot & \cdot \\ & & & & \Phi_1^k \end{pmatrix}, \tag{4}$$

where

- $\Phi_i \in \mathrm{Hom}\,(\mathrm{Sym}^i\mathbb{C}^p, \mathbb{C}^p)$ is a $p\times\dim(\mathrm{Sym}^i\mathbb{C}^p)$-matrix, the ith degree component of the map Φ, which is represented by a map $(\mathbb{C}^p)^{\otimes i} \to \mathbb{C}^p$;
- $\Phi_{i_1}\dots\Phi_{i_l}$ is the matrix of the map $\mathrm{Sym}^{i_1+\cdots+i_l}(\mathbb{C}^p) \to \mathrm{Sym}^l\mathbb{C}^p$, which is represented by

$$\sum_{\sigma\in\mathcal{S}_l}\Phi_{i_1}\otimes\cdots\otimes\Phi_{i_l} : (\mathbb{C}^p)^{\otimes i_1}\otimes\cdots\otimes(\mathbb{C}^p)^{\otimes i_l} \to (\mathbb{C}^p)^{\otimes l};$$

- the (l, m) block of $\mathbb{G}_{k,p}$ is $\sum_{i_1+\cdots+i_l=m}\phi_{i_1}\dots\Phi_{i_l}$. The entries in these boxes are indexed by pairs (τ,μ) where $\tau\in\binom{p+l-1}{l-1}, \mu\in\binom{p+m-1}{m-1}$ correspond to bases of $\mathrm{Sym}^l(\mathbb{C}^p)$ and $\mathrm{Sym}^m(\mathbb{C}^p)$.

EXAMPLE 2.1. *For $p = 2, k = 3$, using the standard basis*

$$\{e_i, e_ie_j, e_ie_je_k : 1 \le i \le j \le k \le 2\}$$

of $(J_{3,2})_x$, *we get the following* 9×9 *matrix for a general element of* $\mathbb{G}_{3,2}$*:*

$$\begin{pmatrix}
\alpha_{10} & \alpha_{01} & \alpha_{20} & \alpha_{11} & \alpha_{02} & \alpha_{30} & a_{21} & \alpha_{12} & \alpha_{03} \\
\beta_{10} & \beta_{01} & \beta_{20} & \beta_{11} & \beta_{02} & \beta_{30} & \beta_{21} & \beta_{12} & \beta_{03} \\
0 & 0 & \alpha_{10}^2 & \alpha_{10}\alpha_{01} & \alpha_{01}^2 & \alpha_{10}\alpha_{20} & \alpha_{10}\alpha_{11}+\alpha_{01}\alpha_{20} & \alpha_{10}\alpha_{02}+\alpha_{11}\alpha_{01} & \alpha_{01}\alpha_{02} \\
0 & 0 & \alpha_{10}\beta_{10} & \alpha_{10}\beta_{01}+\alpha_{01}\beta_{10} & \alpha_{01}\beta_{01} & \alpha_{10}\beta_{20}+\alpha_{20}\beta_{10} & P & Q & \alpha_{01}\beta_{02}+\alpha_{02}\beta_{01} \\
0 & 0 & \beta_{10}^2 & \beta_{10}\beta_{01} & \beta_{01}^2 & \beta_{10}\beta_{20} & \beta_{10}\beta_{11}+\beta_{20}\beta_{01} & \beta_{01}\beta_{11}+\beta_{02}\beta_{10} & \beta_{01}\beta_{02} \\
0 & 0 & 0 & 0 & 0 & \alpha_{10}^3 & \alpha_{10}^2\alpha_{01} & \alpha_{10}\alpha_{01}^2 & \alpha_{01}^3 \\
0 & 0 & 0 & 0 & 0 & \alpha_{10}^2\beta_{10} & \alpha_{10}\alpha_{10}\beta_{01} & \alpha_{10}\alpha_{01}\beta_{01} & \alpha_{01}\beta_{01}^2 \\
0 & 0 & 0 & 0 & 0 & \alpha_{10}\beta_{10}^2 & \alpha_{10}\beta_{10}\beta_{01} & \alpha_{10}\beta_{01}\beta_{01} & \alpha_{01}\beta_{01}^2 \\
0 & 0 & 0 & 0 & 0 & \beta_{10}^3 & \beta_{10}^2\beta_{01} & \beta_{10}\beta_{01}^2 & \beta_{01}^3
\end{pmatrix} \tag{5}$$

where

$$P=\alpha_{10}\beta_{11}+\alpha_{11}\beta_{10}+\alpha_{20}\beta_{01}+\alpha_{01}\beta_{20} \quad \text{and}$$
$$Q=\alpha_{01}\beta_{11}+\alpha_{11}\beta_{01}+\alpha_{02}\beta_{10}+\alpha_{10}\beta_{02}.$$

This is a subgroup of the standard parabolic $P_{2,3,4}\subset GL(9)$. *The diagonal blocks are the representations* $\mathrm{Sym}^i\mathbb{C}^2$ *for* $i=1,2,3$ *of* $GL(2)$, *where* $\mathbb{C}^2$ *is the standard representation of* $GL(2)$.

In general the linear group $\mathbb{G}_{k,p}$ is generated along its first p rows; that is, the parameters in the first p rows are independent, and all the remaining entries are polynomials in these parameters. The assumption on the parameters is that the determinant of the smallest diagonal $p\times p$ block is nonzero; for the $p=2, k=3$ example above this means that

$$\det\begin{pmatrix}\alpha_{10} & \alpha_{01}\\ \beta_{10} & \beta_{01}\end{pmatrix}\neq 0.$$

The parameters in the $(1,m)$ block are indexed by a basis of $\mathrm{Sym}^m(\mathbb{C}^p)\times\mathbb{C}^p$, so they are of the form α^l_ν where $\nu\in\binom{p+m-1}{m-1}$ is an m-tuple and $1\leq l\leq p$. An easy computation shows that:

PROPOSITION 2.2. *The polynomial in the* (l,m) *block and entry indexed by*

$$\tau=(\tau[1],\ldots,\tau[l])\in\binom{p+l-1}{l-1}$$

and $\nu \in \binom{p+m-1}{m-1}$ *is*

$$(\mathbb{G}_{k,p})_{\tau,\nu} = \sum_{\nu_1+\cdots+\nu_l=\nu} \alpha_{\nu_1}^{\tau[1]}\alpha_{\nu_2}^{\tau[2]}\ldots\alpha_{\nu_l}^{\tau[l]} \tag{6}$$

Note that $\mathbb{G}_{k,p}$ is an extension of its unipotent radical $\mathbb{U}_{k,p}$ by $GL(p)$; that is, we have an exact sequence

$$1 \to \mathbb{U}_{k,p} \to \mathbb{G}_{k,p} \to GL(p) \to 1,$$

and $\mathbb{G}_{k,p}$ is the semi-direct product $\mathbb{U}_{k,p} \rtimes GL(p)$. Here $\mathbb{G}_{k,p}$ has dimension $p \times \mathrm{sym}^{\leq k}(p)$ where $\mathrm{sym}^{\leq k}(p) = \dim(\oplus_{i=1}^{k}\mathrm{Sym}^i\mathbb{C}^p)$, and is a subgroup of the standard parabolic subgroup $P_{p,\mathrm{sym}^2(p),\ldots,\mathrm{sym}^k(p)}$ of $GL(\mathrm{sym}^{\leq k}(p))$ where $\mathrm{sym}^i(p) = \dim(\mathrm{Sym}^i\mathbb{C}^p)$. We define $\mathbb{G}'_{k,p}$ to be the subgroup of $\mathbb{G}_{k,p}$ which is the semi-direct product

$$\mathbb{G}'_{k,p} = \mathbb{U}_{k,p} \rtimes SL(p)$$

(so that $\mathbb{G}'_{k,p} = \mathbb{U}_{k,p}$ when $p = 1$) fitting into the exact sequence

$$1 \to \mathbb{U}_{k,p} \to \mathbb{G}'_{k,p} \to SL(p) \to 1.$$

The action of the maximal torus $(\mathbb{C}^*)^p \subset GL(p)$ of the Levi subgroup of $\mathbb{G}_{k,p}$ is

$$(\lambda_1,\ldots,\lambda_p)\cdot f^{(i)} = \left(\lambda_1^i\frac{\partial^i f}{\partial u_1^i},\ldots,\lambda_1^{i_1}\cdots\lambda_p^{i_p}\frac{\partial^i f}{\partial u_1^{i_1}\cdots\partial u_p^{i_p}}\cdots\lambda_p^i\frac{\partial^i f}{\partial u_p^i}\right) \tag{7}$$

We introduce the *Green-Griffiths* vector bundle $E_{k,p,m}^{GG} \to X$, whose fibres are complex-valued polynomials

$$Q(f'(0), f''(0)/2!, \ldots, f^{(k)}(0)/k!)$$

on the fibres of $J_{k,p}$, having weighted degree $(m,\ldots,m)$ with respect to the action (7) of $(\mathbb{C}^*)^p$. That is, for $Q \in E_{k,p,m}^{GG}$

$$\begin{aligned}Q(\lambda f'(0), \lambda f''(0)/2!, \ldots, \lambda f^{(k)}(0)/k!) &= \lambda_1^m\cdots\lambda_p^m\\ Q(f'(0), f''(0)/2!, \ldots, f^{(k)}(0)/k!)&\end{aligned}$$

for all $\lambda \in \mathbb{C}^p$ and $(f'(0), f''(0)/2!, \ldots, f^{(k)}(0)/k!) \in J_{k,p,m}$.

DEFINITION 2.3. *The generalized Demailly-Semple bundle* $E_{k,p,m} \to X$ *over* X *has fibre consisting of the* $\mathbb{G}'_{k,p}$*-invariant jet differentials of order* k *and weighted degree* $(m,\ldots,m)$*; that is, the complex-valued polynomials* $Q(f'(0), f''(0)/2!, \ldots, f^{(k)}(0)/k!)$ *on the fibres of* $J_{k,p}$ *which transform under any reparametrization* $\phi \in \mathbb{G}_{k,p}$ *of* $(\mathbb{C}^p, 0)$ *as*

$$Q(f\circ\phi) = (J_\phi)^m Q(f)\circ\phi,$$

where $J_\phi = \det \Phi_1$ *denotes the Jacobian of* ϕ *at 0. The generalized Demailly-Semple bundle of algebras* $E_{k,p} = \oplus_{m\geq 0} E_{k,p,m}$ *is the associated graded algebra of* $\mathbb{G}'_{k,p}$*-invariants, whose fibre at* $x \in X$ *is the generalized Demailly-Semple algebra* $\mathcal{O}((J_{k,p})_x)^{\mathbb{G}'_{k,p}}$.

The determination of a suitable generating set for the invariant jet differentials when $p=1$ is important in the longstanding strategy to prove the Green-Griffiths conjecture. It has been suggested in a series of papers [**13, 5, 27, 20, 7, 21**] that the Schur decomposition of the Demailly-Semple algebra, together with good estimates of the higher Betti numbers of the Schur bundles and an asymptotic estimation of the Euler charactristic, should result in a positive lower bound for the global sections of the Demailly-Semple jet differential bundle.

3. Geometric invariant theory

Suppose now that Y is a complex quasi-projective variety on which a linear algebraic group G acts. For geometric invariant theory (GIT) we need a linearization of the action; that is, a line bundle L on Y and a lift $\mathcal{L}$ of the action of G to L. Usually L is ample, and hence (as it makes no difference for GIT if we replace L with $L^{\otimes k}$ for any integer $k>0$) we can assume that for some projective embedding $Y \subseteq \mathbb{P}^n$ the action of G on Y extends to an action on $\mathbb{P}^n$ given by a representation $\rho: G \to GL(n+1)$, and take for L the hyperplane line bundle on $\mathbb{P}^n$.

For classical GIT developed by Mumford [**23**] (cf. also [**8, 22, 24, 26**]) we require the complex algebraic group G to be reductive. Let Y be a projective complex variety with an action of a complex reductive group G and linearization $\mathcal{L}$ with respect to an ample line bundle L on Y. Then $y \in Y$ is *semistable* for this linear action if there exists some $m>0$ and $f \in H^0(Y, L^{\otimes m})^G$ not vanishing at y, and y is *stable* if also the action of G on the open subset

$$Y_f := \{x \in Y \mid f(x) \neq 0\}$$

is closed with all stabilizers finite. Y^{ss} has a projective categorical quotient $Y^{ss} \to Y//G$, which restricts on the set of stable points to a geometric quotient $Y^s \to Y^s/G$ (see [**23**] Theorem 1.10). The morphism $Y^{ss} \to Y//G$ is surjective, and identifies $x, y \in Y^{ss}$ if and only if the closures of the G-orbits of x and y meet in Y^{ss}; moreover each point in $Y//G$ is represented by a unique closed G-orbit in Y^{ss}. There is an induced action of G on the homogeneous coordinate ring

$$\hat{\mathcal{O}}_L(Y) = \bigoplus_{k\geq 0} H^0(Y, L^{\otimes k})$$

of Y. The subring $\hat{\mathcal{O}}_L(Y)^G$ consisting of the elements of $\hat{\mathcal{O}}_L(Y)$ left invariant by G is a finitely generated graded complex algebra because G is reductive, and the GIT quotient $Y//G$ is the projective variety $\mathrm{Proj}(\hat{\mathcal{O}}_L(Y)^G)$ [**23**].

The subsets Y^{ss} and Y^s of Y are characterized by the following properties (see [**23**, Chapter 2] or [**24**]).

PROPOSITION 3.1. *(Hilbert-Mumford criteria) (i) A point $x \in Y$ is semistable (respectively stable) for the action of G on Y if and only if for every $g \in G$ the point gx is semistable (respectively stable) for the action of a fixed maximal torus of G.*
(ii) A point $x \in Y$ with homogeneous coordinates $[x_0 : \ldots : x_n]$ in some coordinate system on $\mathbb{P}^n$ is semistable (respectively stable) for the action of a maximal torus of G acting diagonally on $\mathbb{P}^n$ with weights $\alpha_0, \ldots, \alpha_n$ if and only if the convex hull

$$\mathrm{Conv}\{\alpha_i : x_i \neq 0\}$$

contains 0 (respectively contains 0 in its interior).

Similarly if a complex reductive group G acts linearly on an affine variety Y then we have a GIT quotient

$$Y//G = \mathrm{Spec}(\mathcal{O}(Y)^G)$$

which is the affine variety associated to the finitely generated algebra $\mathcal{O}(Y)^G$ of G-invariant regular functions on Y. In this case $Y^{ss} = Y$ and the inclusion $\mathcal{O}(Y)^G \hookrightarrow \mathcal{O}(Y)$ induces a morphism of affine varieties $Y \to Y//G$.

Now suppose that H is any complex linear algebraic group, with unipotent radical $U \trianglelefteq H$ (so that $R = H/U$ is reductive and H is isomorphic to the semi-direct product $U \rtimes R$), acting linearly on a complex projective variety Y with respect to an ample line bundle L. Then $\mathrm{Proj}(\hat{\mathcal{O}}_L(Y)^H)$ is not in general well-defined as a projective variety, since the ring of invariants

$$\hat{\mathcal{O}}_L(Y)^H = \bigoplus_{k \geq 0} H^0(Y, L^{\otimes k})^H$$

is not necessarily finitely generated as a graded complex algebra, and so it is not obvious how GIT might be generalised to this situation (cf. [**9, 11, 10, 14, 15, 18**]). However in some cases it is known that $\hat{\mathcal{O}}_L(Y)^U$ is finitely generated, which implies that

$$\hat{\mathcal{O}}_L(Y)^H = \left(\bigoplus_{k \geq 0} H^0(Y, L^{\otimes k})^U \right)^{H/U}$$

is finitely generated and hence the *enveloping quotient* in the sense of [**9**] is given by the associated projective variety

$$Y//H = \mathrm{Proj}(\hat{\mathcal{O}}_L(Y)^H).$$

Similarly if Y is affine and H acts linearly on Y with $\mathcal{O}(Y)^H$ finitely generated, then we have the enveloping quotient

$$Y//H = \mathrm{Spec}(\mathcal{O}(Y)^H).$$

There is a morphism
$$q: Y^{ss} \to Y/\!/H,$$
from an open subset Y^{ss} of Y (where $Y^{ss} = Y$ when Y is affine), which restricts to a geometric quotient
$$q: Y^s \to Y^s/H$$
for an open subset $Y^s \subset Y^{ss}$. However in contrast with the reductive case, the morphism $q: Y^{ss} \to Y/\!/H$ is not in general surjective; indeed the image of q is not in general a subvariety of $Y/\!/H$, but is only a constructible subset.

If there is a complex reductive group G containing the unipotent radical U of H such that the algebra $\mathcal{O}(G)^U$ is finitely generated and the action of U on Y extends to a linear action of G, then
$$\mathcal{O}(Y)^U \cong (\mathcal{O}(Y) \otimes \mathcal{O}(G)^U)^G$$
is finitely generated and hence so is
$$\mathcal{O}(Y)^H = (\mathcal{O}(Y)^U)^{H/U}$$
(or if Y is projective with an ample linearisation L then $\hat{\mathcal{O}}_L(Y)^U$ is finitely generated and hence so is $\hat{\mathcal{O}}_L(Y)^H$). In this situation we say that U is a Grosshans subgroup of G (cf. [**16, 17**]). Then geometrically G/U is a quasi-affine variety with $\mathcal{O}(G/U) \cong \mathcal{O}(G)^U$, and it has a canonical affine embedding as an open subvariety of the affine variety
$$G/\!/U = \mathrm{Spec}(\mathcal{O}(G)^U)$$
with complement of codimension at least two. Moreover if a linear action of U on an affine variety Y extends to a linear action of G then
$$Y/\!/U \cong (Y \times G/\!/U)/\!/G$$
(and a corresponding result is true if Y is projective). Conversely if we can find an embedding of G/U as an open subvariety of an affine variety Z with complement of codimension at least two, then
$$\mathcal{O}(G)^U \cong \mathcal{O}(Z)$$
is finitely generated and $G/\!/U \cong Z$.

Suppose that U is a unipotent group with a reductive group R of automorphisms of U given by a homomorphism $\phi: R \to \mathrm{Aut}(U)$ such that R contains a central one-parameter subgroup $\lambda: \mathbb{C}^* \to R$ for which the weights of the induced $\mathbb{C}^*$ action on the Lie algebra $\mathfrak{u}$ of U are all nonzero. Then we can form the semi-direct product
$$\hat{U} = \mathbb{C}^* \ltimes U \subseteq R \ltimes U$$
given by $\mathbb{C}^* \times U$ with group multiplication
$$(z_1, u_1).(z_2, u_2) = (z_1 z_2, (\lambda(z_2^{-1})(u_1))u_2).$$

The groups $\mathbb{G}_k = \mathbb{U}_k \rtimes \mathbb{C}^*$ and $\mathbb{G}_{k,p} = \mathbb{U}_{k,p} \rtimes \mathrm{GL}(p)$ which act on the fibres of the jet bundles J_k and $J_{k,p}$ are of this form. We will use this structure to study the Demailly-Semple algebras of invariant jet differentials E_k^n and $E_{k,p}^n$ and prove

THEOREM 3.2. *The fibres $\mathcal{O}((J_k)_x)^{\mathbb{U}_k}$ and $\mathcal{O}((J_{k,p})_x)^{\mathbb{G}'_{k,p}}$ of the bundles E_k^n and $E_{k,p}^n$ are finitely generated graded complex algebras.*

Thus we have non-reductive GIT quotients

$$(J_k)_x//\mathbb{U}_k = \mathrm{Spec}(\mathcal{O}((J_k)_x)^{\mathbb{U}_k})$$

and

$$(J_{k,p})_x//\mathbb{G}'_{k,p} = \mathrm{Spec}(\mathcal{O}((J_{k,p})_x)^{\mathbb{G}'_{k,p}})$$

and we would like to understand them geometrically. There is a crucial difference here from the case of reductive group actions, even though the invariants are finitely generated: when H is a non-reductive group we cannot describe $Y//H$ geometrically as Y^{ss} modulo some equivalence relation. Instead our aim is to use methods inspired by [**2**] to study these geometric invariant theoretic quotients and the associated algebras of invariants.

Here a crucial ingredient would be to find an open subset W of $(J_{k,p})_x$ with a geometric quotient $W/\mathbb{G}'_{k,p}$ embedded as an open subset of an affine variety Z such that the complement of $W/\mathbb{G}'_{k,p}$ in Z has (complex) codimension at least two, and the complement of W in $(J_{k,p})_x$ has codimension at least two. For then we would have

$$\mathcal{O}((J_{k,p})_x) = \mathcal{O}(W)$$

and

$$\mathcal{O}((J_{k,p})_x)^{\mathbb{G}'_{k,p}} = \mathcal{O}(W)^{\mathbb{G}'_{k,p}} = \mathcal{O}(W/\mathbb{G}'_{k,p}) = \mathcal{O}(Z),$$

and it follows that $\mathcal{O}((J_{k,p})_x)^{\mathbb{G}'_{k,p}}$ is finitely generated since Z is affine, and that

$$Z = \mathrm{Spec}(\mathcal{O}(Z)) = \mathrm{Spec}(\mathcal{O}((J_{k,p})_x)^{\mathbb{G}'_{k,p}}) = ((J_{k,p})_x)//\mathbb{G}'_{k,p}.$$

Similarly if we can find a complex reductive group G containing $\mathbb{G}'_{k,p}$ as a subgroup, and an embedding of $G/\mathbb{G}'_{k,p}$ as an open subset of an affine variety Z with complement of codimension at least two, then $\mathcal{O}(G)^{\mathbb{G}'_{k,p}}$ is finitely generated. It follows as above that if Y is any affine variety on which G acts linearly then

$$\mathcal{O}(Y)^{\mathbb{G}'_{k,p}} \cong (\mathcal{O}(Y) \otimes \mathcal{O}(G)^{\mathbb{G}'_{k,p}})^G$$

is finitely generated, and hence so is $\mathcal{O}(Y)^{\mathbb{G}_{k,p}} = (\mathcal{O}(Y)^{\mathbb{G}'_{k,p}})^{\mathbb{C}^*}$, and similarly $\hat{\mathcal{O}}_L(Y)^{\mathbb{G}'_{k,p}}$ and $\hat{\mathcal{O}}_L(Y)^{\mathbb{G}_{k,p}}$ are finitely generated if Y is any projective variety wtih an ample line bundle L on which G acts linearly.

We can use the ideas of [**2**] to look for suitable affine varieties Z as above, and in particular to prove

THEOREM 3.3. *$\mathbb{G}'_{k,p}$ is a subgroup of the special linear group* $\mathrm{SL}(\mathrm{sym}^{\leq k}p)$ *where*

$$\mathrm{sym}^{\leq k}p = \sum_{i=1}^{k} \dim \mathrm{Sym}^i \mathbb{C}^p = \begin{pmatrix} k+p-1 \\ k-1 \end{pmatrix}$$

such that the algebra of invariants $\mathcal{O}(\mathrm{SL}(\mathrm{sym}^{\leq k}p))^{\mathbb{G}'_{k,p}}$ is finitely generated, and every linear action of $\mathbb{G}'_{k,p}$ or $\mathbb{G}_{k,p}$ on an affine or projective variety (with an ample linearisation) which extends to a linear action of $\mathrm{GL}(\mathrm{sym}^{\leq k}p)$ *has finitely generated invariants.*

Theorem 3.2 is an immediate consequence of this theorem, since the action of $\mathbb{G}_{k,p}$ on $(J_{k,p})_x$ extends to an action of the general linear group $\mathrm{GL}(\mathrm{sym}^{\leq k}p)$. Moreover we will find a geometric description of

$$\mathrm{SL}(\mathrm{sym}^{\leq k}p)//\mathbb{G}'_{k,p} \cong \mathrm{Spec}(\mathcal{O}(\mathrm{SL}(\mathrm{sym}^{\leq k}p))^{\mathbb{G}'_{k,p}})$$

and thus a geometric description of

$$(J_{k,p})_x//\mathbb{G}'_{k,p} \cong ((J_{k,p})_x \times \mathrm{SL}(\mathrm{sym}^{\leq k}p)//\mathbb{G}'_{k,p})//\mathrm{SL}(\mathrm{sym}^{\leq k}p).$$

4. A description via test curves

In [**2**] the action of $\mathbb{G}_k$ on jet bundles is studied using an idea coming from global singularity theory. The construction goes as follows.

If u, v are positive integers, let $J_k(u,v)$ denote the vector space of k-jets of holomorphic maps $(\mathbb{C}^u, 0) \to (\mathbb{C}^v, 0)$ at the origin; that is, the set of equivalence classes of maps $f : (\mathbb{C}^u, 0) \to (\mathbb{C}^v, 0)$, where $f \sim g$ if and only if $f^{(j)}(0) = g^{(j)}(0)$ for all $j = 1, \ldots, k$.

With this notation, the fibres of J_k are isomorphic to $J_k(1,n)$, and the group $\mathbb{G}_k$ is simply $J_k(1,1)$ with the composition action on itself.

If we fix local coordinates $z_1, \ldots, z_u$ at $0 \in \mathbb{C}^u$ we can again identify the k-jet of f, using derivatives at the origin, with $(f'(0), f''(0)/2!, \ldots, f^{(k)}(0)/k!)$, where $f^{(j)}(0) \in \mathrm{Hom}(\mathrm{Sym}^j \mathbb{C}^u, \mathbb{C}^v)$. This way we get an identification

$$J_k(u,v) = \oplus_{j=1}^{k} \mathrm{Hom}(\mathrm{Sym}^j \mathbb{C}^u, \mathbb{C}^v).$$

We can compose map-jets via substitution and elimination of terms of degree greater than k; this leads to the composition maps

$$\begin{aligned} &J_k(v,w) \times J_k(u,v) \to J_k(u,w), \\ (8) \qquad &(\Psi_2, \Psi_1) \mapsto \Psi_2 \circ \Psi_1 \text{ modulo terms of degree} > k. \end{aligned}$$

When $k = 1$, $J_1(u,v)$ may be identified with u-by-v matrices, and (8) reduces to multiplication of matrices.

The k-jet of a curve $(\mathbb{C}, 0) \to (\mathbb{C}^n, 0)$ is simply an element of $J_k(1,n)$. We call such a curve φ *regular* if $\varphi'(0) \neq 0$. Let us introduce the notation $J_k^{\mathrm{reg}}(1,n)$ for the set of regular curves:

$$J_k^{\mathrm{reg}}(1,n) = \left\{\gamma \in J_k(1,n); \gamma'(0) \neq 0\right\}.$$

Note that if $n>1$ then the complement of $J_k^{\rm reg}(1,n)$ in $J_k(1,n)$ has codimension at least two. Let $N\geq n$ be any integer and define

$$\Upsilon_k=\{\Psi\in J_k(n,N):\exists\gamma\in J_k^{\rm reg}(1,n):\Psi\circ\gamma=0\}$$

to be the set of those k-jets which take at least one regular curve to zero. By definition, Υ_k is the image of the closed subvariety of $J_k(n,N)\times J_k^{\rm reg}(1,n)$ defined by the algebraic equations $\Psi\circ\gamma=0$, under the projection to the first factor. If $\Psi\circ\gamma=0$, we call γ a *test curve* of Ψ.

This term originally comes from global singularity theory, where this is called the test curve model of A_k-singularities. In global singularity theory singularities of polynomial maps $f:(\mathbb{C}^n,0)\to(\mathbb{C}^m,0)$ are classified by their local algebras, and

$$\Sigma_k=\{f\in J_k(n,m):\mathbb{C}[x_1,\dots,x_n]/\langle f_1,\dots,f_m\rangle\simeq\mathbb{C}[t]/t^{k+1}\}$$

is called a Morin singularity, or A_k-singularity. The test curve model of Gaffney [**12**] tells us that

$$\overline{\Sigma_k}=\overline{\Upsilon_k}$$

in $J_k(n,m)$.

A basic but crucial observation is the following. If γ is a test curve of $\Psi\in\Upsilon_k$, and $\varphi\in J_k^{\rm reg}(1,1)=G_k$ is a holomorphic reparametrization of $\mathbb{C}$, then $\gamma\circ\varphi$ is, again, a test curve of Ψ:

$$\mathbb{C}\xrightarrow{\ \varphi\ }\mathbb{C}\xrightarrow{\ \gamma\ }\mathbb{C}^n\xrightarrow{\ \Psi\ }\mathbb{C}^N \tag{9}$$

$$\Psi\circ\gamma=0\ \Rightarrow\ \Psi\circ(\gamma\circ\varphi)=0.$$

In fact, we get all test curves of Ψ in this way from a single γ if the following open dense property holds: the linear part of Ψ has 1-dimensional kernel. Before stating this more precisely in Proposition 4.3 below, let us write down the equation $\Psi\circ\gamma=0$ in coordinates in an illustrative case. Let $\gamma=(\gamma',\gamma'',\dots,\gamma^{(k)})\in J_k^{\rm reg}(1,n)$ and $\Psi=(\Psi',\Psi'',\dots,\Psi^{(k)})\in J_k(n,N)$ be the k-jets. Using the chain rule, the equation $\Psi\circ\gamma=0$ reads as follows for $k=4$:

$$\begin{aligned}(10)\qquad &\Psi'(\gamma')=0,\\ &\tfrac{1}{2!}\Psi'(\gamma'')+\Psi''(\gamma',\gamma')=0,\\ &\tfrac{1}{3!}\Psi'(\gamma''')+\tfrac{2}{2!}\Psi''(\gamma',\gamma'')+\Psi'''(\gamma',\gamma',\gamma')=0,\\ &\tfrac{1}{4!}\Psi'(\gamma'''')+\tfrac{2}{3!}\Psi''(\gamma',\gamma''')+\tfrac{1}{2!2!}\Psi''(\gamma'',\gamma'')+\tfrac{3}{2!}\Psi'''(\gamma',\gamma',\gamma'')\\ &\quad+\Psi''''(\gamma',\gamma',\gamma',\gamma')=0.\end{aligned}$$

DEFINITION 4.1. *To simplify our formulas we introduce the following notation for a partition $\tau=[i_1\dots i_l]$ of the integer $i_1+\dots+i_l$:*

- *the length: $|\tau|=l$,*
- *the sum: $\sum\tau=i_1+\dots+i_l$,*

- *the number of permutations:* $\mathrm{perm}(\tau)$ *is the number of different sequences consisting of the numbers* $i_1, \ldots, i_l$ *(e.g.* $\mathrm{perm}([1,1,1,3]) = 4$*),*
- $\gamma_\tau = \prod_{j=1}^{l} \gamma^{(i_j)} \in \mathrm{Sym}^l \mathbb{C}^n$ *and* $\Psi(\gamma_\tau) = \Psi^l(\gamma^{(i_1)}, \ldots, \gamma^{(i_l)}) \in \mathbb{C}^N$.

LEMMA 4.2. *Let* $\gamma = (\gamma', \gamma'', \ldots, \gamma^{(k)}) \in J_k^{\mathrm{reg}}(1,n)$ *and* $\Psi = (\Psi', \Psi'', \ldots, \Psi^{(k)}) \in J_k(n,N)$ *be* k*-jets. Then the equation* $\Psi \circ \gamma = 0$ *is equivalent to the following system of* k *linear equations with values in* $\mathbb{C}^N$*:*

$$\sum_{\tau \in \Pi[m]} \frac{\mathrm{perm}(\tau)}{\prod_{i \in \tau} i!} \Psi(\gamma_\tau) = 0, \quad m = 1, 2, \ldots, k, \tag{11}$$

where $\Pi[m]$ *denotes the set of all partitions of* m.

For a given $\gamma \in J_k^{\mathrm{reg}}(1,n)$ let $\mathcal{S}_\gamma$ denote the set of solutions of (11); that is,

$$\mathcal{S}_\gamma = \{\Psi \in J_k(n,N); \Psi \circ \gamma = 0\}.$$

The equations (11) are linear in Ψ, hence

$$\mathcal{S}_\gamma \subset J_k(n,N)$$

is a linear subspace of codimension kN. Moreover, the following holds:

PROPOSITION 4.3. *([**2**], Proposition 4.4)*

(i) *For* $\gamma \in J_k^{\mathrm{reg}}(1,n)$*, the set of solutions* $\mathcal{S}_\gamma \subset J_k(n,N)$ *is a linear subspace of codimension* kN.

(ii) *Set*

$$J_k^o(n,N) = \left\{\Psi \in J_k(n,N) | \dim \ker(\Psi') = 1\right\}.$$

For any $\gamma \in J_k^{\mathrm{reg}}(1,n)$*, the subset* $\mathcal{S}_\gamma \cap J_k^o(n,N)$ *of* $\mathcal{S}_\gamma$ *is dense.*

(iii) *If* $\Psi \in J_k^o(n,N)$*, then* Ψ *belongs to at most one of the spaces* $\mathcal{S}_\gamma$*. More precisely,*

$$\text{if } \gamma_1, \gamma_2 \in J_k^{\mathrm{reg}}(1,n), \quad \Psi \in J_k^o(n,N) \text{ and } \Psi \circ \gamma_1 = \Psi \circ \gamma_2 = 0,$$

then there exists $\varphi \in J_k^{\mathrm{reg}}(1,1)$ *such that* $\gamma_1 = \gamma_2 \circ \varphi$.

(iv) *Given* $\gamma_1, \gamma_2 \in J_k^{\mathrm{reg}}(1,n)$*, we have* $\mathcal{S}_{\gamma_1} = \mathcal{S}_{\gamma_2}$ *if and only if there is some* $\varphi \in J_k^{\mathrm{reg}}(1,1)$ *such that* $\gamma_1 = \gamma_2 \circ \varphi$.

By the second part of Proposition 4.3 we have a well-defined map

$$\nu : J_k^{\mathrm{reg}}(1,n) \to \mathrm{Grass}(\mathrm{codim} = kN, J_k(n,N)), \quad \gamma \mapsto \mathcal{S}_\gamma$$

to the Grassmannian of codimension-kN subspaces in $J_k(n,N)$. From the last part of Proposition 4.3 it follows that:

PROPOSITION 4.4. *([2]) ν is $\mathbb{G}_k$-invariant on the $J_k^{\mathrm{reg}}(1,1)$-orbits, and the induced map on the orbits*

$$\bar{\nu} : J_k^{\mathrm{reg}}(1,n)/\mathbb{G}_k \hookrightarrow \mathrm{Grass}(\mathrm{codim} = kN, J_k(n,N)) \tag{12}$$

is injective.

5. Embedding into the flag of equations

In this section we will recast the embedding (12) of $J_k^{\mathrm{reg}}(1,n)/\mathbb{G}_k$ given by Proposition 4.4 into a more useful form, still following [2]. Let us rewrite the linear system $\Psi \circ \gamma = 0$ associated to $\gamma \in J_k^{\mathrm{reg}}(1,n)$ in a dual form. The system is based on the standard composition map (8):

$$J_k(n,N) \times J_k(1,n) \longrightarrow J_k(1,N),$$

which, via the identification $J_k(n,N) = J_k(n,1) \otimes \mathbb{C}^N$, is derived from the map

$$J_k(n,1) \times J_k(1,n) \longrightarrow J_k(1,1)$$

via tensoring with $\mathbb{C}^N$. Observing that composition is linear in its first argument, and passing to linear duals, we may rewrite this correspondence in the form

$$\phi : J_k(1,n) \longrightarrow \mathrm{Hom}\,(J_k(1,1)^*, J_k(n,1)^*). \tag{13}$$

If $\gamma = (\gamma', \gamma'', \dots, \gamma^{(k)}) \in J_k(1,n) = (\mathbb{C}^n)^k$ is the k-jet of a curve, we can put $\gamma^{(j)} \in \mathbb{C}^n$ into the jth column of an $n \times k$ matrix, and

- identify $J_k(1,n)$ with $\mathrm{Hom}\,(\mathbb{C}^k, \mathbb{C}^n)$;
- identify $J_k(n,1)^*$ with $\mathrm{Sym}^{\leq k}\mathbb{C}^n = \oplus_{l=1}^k \mathrm{Sym}^l\mathbb{C}^n$;
- identify $J_k(1,1)^*$ with $\mathbb{C}^k$.

Using these identifications, we can recast the map ϕ in (13) as

$$\phi_k : \mathrm{Hom}\,(\mathbb{C}^k, \mathbb{C}^n) \longrightarrow \mathrm{Hom}\,(\mathbb{C}^k, \mathrm{Sym}^{\leq k}\mathbb{C}^n), \tag{14}$$

which may be written out explicitly as follows

$$\begin{aligned} &(\gamma', \gamma'', \dots, \gamma^{(k)}) \\ &\quad \longmapsto \left(\gamma', \gamma'' + (\gamma')^2, \dots, \sum_{i_1+i_2+\dots+i_s=d} \frac{1}{i_1! \dots i_s!} \gamma^{(i_1)}\gamma^{(i_2)} \dots \gamma^{(i_s)} \right). \end{aligned}$$

The set of solutions $\mathcal{S}_\gamma$ is the linear subspace orthogonal to the image of $\phi_k(\gamma', \dots \gamma^{(k)})$ tensored by $\mathbb{C}^N$; that is,

$$\mathcal{S}_\gamma = \mathrm{im}(\phi_k(\gamma))^\perp \otimes \mathbb{C}^N \subset J_k(n,N).$$

Consequently, it is straightforward to take $N = 1$ and define

$$\mathcal{S}_\gamma = \mathrm{im}(\phi_k(\gamma)) \in \mathrm{Grass}(k, \mathrm{Sym}^{\leq k}\mathbb{C}^n). \tag{15}$$

Moreover, let $B_k \subset GL(k)$ denote the Borel subgroup consisting of upper triangular matrices and let

$$\begin{aligned}\mathrm{Flag}_k(\mathbb{C}^n) &= \mathrm{Hom}\,(\mathbb{C}^k, \mathrm{Sym}^{\leq k}\mathbb{C}^n)/B_k \\ &= \{0 = F_0 \subset F_1 \subset \cdots \subset F_k \subset \mathbb{C}^n,\ \dim F_l = l\}\end{aligned}$$

denote the full flag of k-dimensional subspaces of $\mathrm{Sym}^{\leq k}\mathbb{C}^n$. In addition to (15) we can analogously define

$$\mathcal{F}_\gamma = (\mathrm{im}(\phi(\gamma^1)) \subset \mathrm{im}(\phi(\gamma^2)) \subset \cdots \subset \mathrm{im}(\phi(\gamma^k))) \in \mathrm{Flag}_k(\mathrm{Sym}^{\leq k}\mathbb{C}^n). \tag{16}$$

Using these definitions Proposition 4.3 implies the the following version of Proposition 4.4, which does not contain the parameter N.

PROPOSITION 5.1. *The map ϕ in (14) is a $\mathbb{G}_k$-invariant algebraic morphism*

$$\phi : J_k^{\mathrm{reg}}(1,n) \to \mathrm{Hom}\,(\mathbb{C}^k, \mathrm{Sym}^{\leq k}\mathbb{C}^n),$$

which induces

- *an injective map on the $\mathbb{G}_k$-orbits to the Grassmannian:*
$$\phi^{Gr} : J_k^{\mathrm{reg}}(1,n)/\mathbb{G}_k \hookrightarrow \mathrm{Grass}(k, \mathrm{Sym}^{\leq k}\mathbb{C}^n)$$
defined by $\phi^{Gr}(\gamma) = \mathcal{S}_\gamma$;
- *an injective map on the $\mathbb{G}_k$-orbits to the flag manifold:*
$$\phi^{Flag} : J_k^{\mathrm{reg}}(1,n)/G_k \hookrightarrow \mathrm{Flag}_k(\mathrm{Sym}^{\leq k}\mathbb{C}^n)$$
defined by $\phi^{Flag}(\gamma) = \mathcal{F}_\gamma$.

In addition,

$$\phi^{Gr} = \phi^{Flag} \circ \pi_k$$

where $\pi_k : \mathrm{Flag}(k, \mathrm{Sym}^{\leq k}\mathbb{C}^n) \to \mathrm{Grass}_k(\mathrm{Sym}^{\leq k}\mathbb{C}^n)$ is the projection to the k-dimensional subspace.

Composing ϕ^{Gr} with the Plücker embedding

$$\mathrm{Grass}(k, \mathrm{Sym}^{\leq k}\mathbb{C}^n) \hookrightarrow \mathbb{P}(\wedge^k \mathrm{Sym}^{\leq k}\mathbb{C}^n)$$

we get an embedding

$$\phi^{\mathrm{Proj}} : J_k^{\mathrm{reg}}(1,n)/\mathbb{G}_k \hookrightarrow \mathbb{P}(\wedge^k(\mathrm{Sym}^{\leq k}\mathbb{C}^n)). \tag{17}$$

The image

$$\phi^{Gr}(J_k^{\mathrm{reg}}(1,n))/\mathbb{G}_k \subset \mathrm{Grass}(k, \mathrm{Sym}^{\leq k}\mathbb{C}^n)$$

is a $GL(n)$-orbit in $\mathrm{Grass}(k, \mathrm{Sym}^{\leq k}\mathbb{C}^n)$, and therefore a nonsingular quasi-projective variety. Its closure is, however, a highly singular subvariety of $\mathrm{Grass}(k, \mathrm{Sym}^{\leq k}\mathbb{C}^n)$, which when $k \leq n$ is a finite union of $GL(n)$ orbits.

Definition 5.2. *Recall that we can identify* $J_k(1,n)$ *with* $\mathrm{Hom}(\mathbb{C}^k,\mathbb{C}^n)$ *and then*

$$J_k^{\mathrm{reg}}(1,n)=\{\rho\in \mathrm{Hom}(\mathbb{C}^k,\mathbb{C}^n):\rho(e_1)\neq 0\}.$$

Let

$$J_k^{\mathrm{nondeg}}(1,n)=\{\rho\in \mathrm{Hom}(\mathbb{C}^k,\mathbb{C}^n):\mathrm{rank}\rho=\max\{k,n\}\}$$

and let

$$X_{n,k}=\phi^{\mathrm{Proj}}(J_k^{\mathrm{nondeg}}(1,n)),\ \ Y_{n,k}=\phi^{\mathrm{Proj}}(J_k^{\mathrm{reg}}(1,n)),$$

so that if $n\le k$ *then*

$$X_{n,k}\subset Y_{n,k}\subset \mathrm{Grass}(n,\mathrm{Sym}^{\le k}\mathbb{C}^n)\subset \mathbb{P}(\wedge^k(\mathrm{Sym}^{\le k}\mathbb{C}^n)).$$

It is clear that $J_k^{\mathrm{nondeg}}(1,n)$ is an open subset of $J_k^{\mathrm{reg}}(1,n)$. If we identify the elements of $J_k(1,n)$ with $n\times k$ matrices whose columns are the derivatives of the map germs $f=(f',\dots,f^{(n)}):\mathbb{C}\to\mathbb{C}^n$, then $J_k^{\mathrm{nondeg}}(1,n)$ is the set of such matrices of maximal rank and $J_k^{\mathrm{reg}}(1,n)$ consists of the matrices with nonzero first column.

Definition 5.3. *Let* $e_1,\dots,e_n$ *be the standard basis of* $\mathbb{C}^n$*; then*

$$\{e_{i_1,i_2,\dots,i_s}=e_{i_1}\dots e_{i_s}:1\le i_1\le\cdots\le i_s\le n,1\le s\le k\}$$

is a basis of $\mathrm{Sym}^{\le k}\mathbb{C}^n$*, and*

$$\{e_{\varepsilon_1}\wedge\cdots\wedge e_{\varepsilon_n}:\varepsilon_l\in\Pi_{\le n}\}$$

is a basis of $\mathbb{P}(\wedge^n(\mathrm{Sym}^{\le k}\mathbb{C}^n))$*, where*

$$\Pi_{\le n}=\{(i_1,i_2,\dots,i_s):1\le i_1\le\cdots\le i_s\le n,1\le s\le k\}.$$

The corresponding coordinates of $x\in\mathrm{Sym}^{\le k}\mathbb{C}^n$ *will be denoted by* $x_{\varepsilon_1,\varepsilon_2,\dots,\varepsilon_d}$*. Let* $A_{n,k}\subset\mathbb{P}(\wedge^k(\mathrm{Sym}^{\le k}\mathbb{C}^n))$ *consist of the points whose projection to* $\wedge^k(\mathbb{C}^n)$ *is nonzero. This is the subset where* $x_{i_1,i_2,\dots,i_k}\neq 0$ *for some* $1\le i_1\le\cdots\le i_k\le n$*.*

Remark 5.4. If $n=k$ then $A_{n,n}\subset\mathbb{P}(\wedge^k(\mathrm{Sym}^{\le k}\mathbb{C}^n))$ is the affine chart where $x_{1,2,\dots,n}\neq 0$.

Let us take a closer look at the space $\mathrm{Grass}(n,\mathrm{Sym}^{\le k}\mathbb{C}^n)$, which has an induced $\mathrm{GL}(n)$ action coming from the $\mathrm{GL}(n)$ action on $\mathrm{Sym}^{\le k}\mathbb{C}^n$. Since ϕ^{Proj} is a $\mathrm{GL}(n)$-equivariant embedding, we conclude that

Lemma 5.5. *(i) For* $k\le n$ $X_{n,k}$ *is the* $\mathrm{GL}(n)$ *orbit of*

$$\mathbf{z}=\phi^{\mathrm{Proj}}(e_1,\dots,e_k)=\left[e_1\wedge(e_2+e_1^2)\wedge\cdots\wedge\left(\sum_{i_1+\cdots i_s=k}e_{i_1}\dots e_{i_s}\right)\right]\tag{18}$$

in $\mathbb{P}(\wedge^k(\mathrm{Sym}^{\leq k}\mathbb{C}^n))$*. For arbitrary* $g \in \mathrm{GL}(n)$ *with column vectors* $v_1, \ldots, v_n$ *the action is given by*

$$g \cdot \mathbf{z} = \phi^{\mathrm{Proj}}(g) = \phi^{\mathrm{Proj}}(v_1, \ldots, v_n)$$
$$= \left[v_1 \wedge (v_2 + v_1^2) \wedge \cdots \wedge \left(\sum_{i_1 + \cdots + i_s = n} v_{i_1} \ldots v_{i_s} \right) \right].$$

(ii) *For* $k \leq n$ $Y_{n,k}$ *is a finite union of* $\mathrm{GL}(n)$ *orbits.*
(iii) *For* $k > n$ *the images* $X_{n,k}$ *and* $Y_{n,k}$ *are* $\mathrm{GL}(n)$*-invariant quasi-projective varieties with no dense* $\mathrm{GL}(n)$ *orbit.*

Lemma 5.6. *If* $k \leq n$ *then*
(i) $A_{n,k}$ *is invariant under the* $\mathrm{GL}(n)$ *action on* $\mathbb{P}(\wedge^k(\mathrm{Sym}^{\leq k}\mathbb{C}^n))$*.*
(ii) $X_{n,k} \subset A_{n,k}$*; however,* $Y_{n,k} \not\subseteq A_{n,k}$*.*

Proof. To prove the first part take a lift

$$\tilde{z} = \tilde{z}^1 \oplus \tilde{z}^2 \in \mathrm{Hom}\,(\mathbb{C}^n, \mathrm{Sym}^{\leq k}\mathbb{C}^n)$$

of $z \in \mathrm{Grass}(n, \mathrm{Sym}^{\leq k}\mathbb{C}^n)$, where

$$z^1 \in \mathrm{Hom}\,(\mathbb{C}^n, \mathbb{C}^n) \quad \text{and} \quad z^2 \in \mathrm{Hom}\,(\mathbb{C}^n, \oplus_{i=2}^n \mathrm{Sym}^i(\mathbb{C}^n))$$

Then $z \in A_{n,k}$ if and only if $x_{1,2,\ldots,n}(z) = \det(\tilde{z}^1) \neq 0$, which is preserved by the $\mathrm{GL}(n)$ action. For the second part note that for $(v_1, \ldots, v_k) \in J_k^{\mathrm{nondeg}}(1,n)$ we have $v_1 \wedge \cdots \wedge v_k \neq 0$ so by definition $\phi^{\mathrm{Proj}}(v_1, \ldots, v_k) \in A_{n,k}$. On the other hand

$$\phi^{\mathrm{Proj}}(e_1, 0, \ldots, 0) = e_1 \wedge e_1^2 \wedge \cdots \wedge e_1^k \in Y_{n,k} \setminus A_{n,k}. \qquad \square$$

When $k = n$ we have

Lemma 5.7. $X_{k,k} \cong \mathrm{GL}(k)/\mathbb{G}_k$ *is embedded in the affine space* $A_{k,k} \subset \mathbb{P}(\wedge^k \mathrm{Sym}^{\leq k}\mathbb{C}^k)$ *as the* $\mathrm{GL}(k)$ *orbit of* $[e_1 \wedge (e_2 + e_1^2) \wedge \cdots \wedge (\sum_{i_1+\cdots+i_s=k} e_{i_1} \ldots e_{i_s})]$.

6. Affine embeddings of $\mathrm{SL}(k)/\mathbb{U}_k$

In the last section we embedded $\mathrm{GL}(k)/\mathbb{G}_k$ in the affine space $A_{k,k} \subset \mathbb{P}(\wedge^k(\mathrm{Sym}^{\leq k}\mathbb{C}^k))$ as the $\mathrm{GL}(k)$ orbit of

$$\left[e_1 \wedge (e_2 + e_1^2) \wedge \cdots \wedge \left(\sum_{i_1+\cdots+i_s=k} e_{i_1} \ldots e_{i_s} \right) \right] \in \mathbb{P}(\wedge^k(\mathrm{Sym}^{\leq k}\mathbb{C}^k)).$$

Equivalently we have

$$\mathrm{SL}(k)/\mathrm{SL}(k) \cap \mathbb{G}_k = \mathrm{SL}(k)/\mathbb{U}_k \rtimes F_k$$

embedded in $\wedge^k(\mathrm{Sym}^{\leq k}\mathbb{C}^k)$ as the SL(k) orbit of

$$p_k = e_1 \wedge (e_2 + e_1^2) \wedge \cdots \wedge \left(\sum_{i_1+\cdots+i_s=k} e_{i_1} \dots e_{i_s} \right),$$

where $\mathrm{SL}(k) \cap \mathbb{G}_k$ is the semi-direct product $\mathbb{U}_k \rtimes F_k$ of $\mathbb{U}_k$ by the finite group F_k of ℓ_kth roots of unity in $\mathbb{C}$ for $\ell_k = 1+\cdots+k = \binom{k+1}{2}$, embedded in SL($k$) as

$$\epsilon \mapsto \begin{pmatrix} \epsilon & 0 & \dots & 0 \\ 0 & \epsilon^2 & \dots & 0 \\ & & \ddots & \\ 0 & 0 & \dots & \epsilon^k \end{pmatrix} \in \mathrm{SL}(k).$$

In this section we will look for affine embeddings of $\mathrm{SL}(k)/\mathbb{U}_k$ in spaces of the form

$$W_{k,K} = \wedge^k(\mathrm{Sym}^{\leq k}\mathbb{C}^k) \otimes (\mathbb{C}^k)^{\otimes K}$$

for suitable K and study their closures.

LEMMA 6.1. *Let $K = M(1+2+\cdots+k)+1 = \binom{k+1}{2}M+1$ where $M \in \mathbb{N}$. Then the point*

$$p_k \otimes e_1^{\otimes K} \in \wedge^k(\mathrm{Sym}^{\leq k}\mathbb{C}^k) \otimes (\mathbb{C}^k)^{\otimes K}$$

where

$$p_k = e_1 \wedge (e_2 + e_1^2) \wedge \cdots \wedge \left(\sum_{i_1+\cdots+i_s=k} e_{i_1} \dots e_{i_s} \right) \in \wedge^k(\mathrm{Sym}^{\leq k}\mathbb{C}^k)$$

has stabiliser $\mathbb{U}_k$ in SL(k).

PROOF. By Proposition 5.1 the stabiliser of

$$[p_k] \in \mathbb{P}(\wedge^k(\mathrm{Sym}^{\leq k}\mathbb{C}^k)) \cong \mathbb{P}(\wedge^k(\mathrm{Sym}^{\leq k}\mathbb{C}^k) \otimes (\mathbb{C}e_1)^{\otimes K}) \subseteq \mathbb{P}(W_{k,K})$$

in $GL(k)$ is $\mathbb{G}_k = \mathbb{U}_k \rtimes \mathbb{C}^*$, so the stabiliser of

$$p_k \otimes e_1^{\otimes K} \in \wedge^k(\mathrm{Sym}^{\leq k}\mathbb{C}^k) \otimes (\mathbb{C}^k)^{\otimes K}$$

is contained in $\mathbb{G}_k$. Moreover by the proof of Proposition 5.1 the stabiliser of $p_k \otimes e_1^{\otimes K}$ contains $\mathbb{U}_k$. Finally

$$\begin{pmatrix} z & 0 & \dots & 0 \\ 0 & z^2 & \dots & 0 \\ & & \ddots & \\ 0 & 0 & \dots & z^k \end{pmatrix} \in \mathbb{C}^* \subseteq \mathbb{G}_k$$

acts on $p_k \otimes e_1^{\otimes K}$ as multiplication by

$$z^{1+2+\cdots+k+K} = z^{(M+1)(1+2+\cdots+k)+1}$$

and has determinant 1 if and only if $z^{1+2+\cdots+k} = 1$, so it lies in $\mathrm{SL}(k)$ and fixes $p_k \otimes e_1^{\otimes K}$ if and only if $z = 1$. □

We will prove

THEOREM 6.2. *If $k \geq 4$ and $K = M(1+2+\cdots+k)+1$ where $M \in \mathbb{N}$ is sufficiently large, then the orbit of $p_k \otimes e_1^{\otimes K}$ where*

$$p_k = e_1 \wedge (e_2 + e_1^2) \wedge \cdots \wedge \left(\sum_{i_1+\cdots+i_s=k} e_{i_1} \ldots e_{i_s} \right) \in \wedge^k(\,\mathrm{Sym}^{\leq k}\mathbb{C}^k)$$

under the natural action of $\mathrm{SL}(k)$ on

$$W_{k,K} = \wedge^k(\,\mathrm{Sym}^{\leq k}\mathbb{C}^k) \otimes (\mathbb{C}^k)^{\otimes K}$$

is isomorphic to $\mathrm{SL}(k)/\mathbb{U}_k$, and its complement in its closure $\overline{\mathrm{SL}(k)(p_k \otimes e_1^{\otimes K})}$ in $W_{k,K}$ has codimension at least two.

This theorem has an immediate corollary.

COROLLARY 6.3. *If $k \geq 2$ then $\mathbb{U}_k$ is a Grosshans subgroup of $\mathrm{SL}(k)$, so that every linear action of $\mathbb{U}_k$ which extends to a linear action of $\mathrm{SL}(k)$ has finitely generated invariants.*

PROOF. This follows directly from Theorem 6.2 when $k \geq 4$. When $k = 2$ and $k = 3$ it is already known (cf. [**27**]). □

The remainder of this section will be devoted to proving Theorem 6.2.

It follows directly from Lemma 6.1 that the $\mathrm{SL}(k)$-orbit of $p_k \otimes e_1^{\otimes K}$ in $W_{k,K} = \wedge^k(\,\mathrm{Sym}^{\leq k}\mathbb{C}^k) \otimes (\mathbb{C}^k)^{\otimes K}$ is isomorphic to $\mathrm{SL}(k)/\mathbb{U}_k$.

Recall that

$$\mathbb{U}_k = \left\{ \begin{pmatrix} 1 & \alpha_2 & \alpha_3 & \cdots & \alpha_k \\ 0 & 1 & 2\alpha_2 & \cdots & 2\alpha_{k-1} + \cdots \\ 0 & 0 & 1 & \cdots & 3\alpha_{k-2} + \cdots \\ \cdot & \cdot & \cdot & \cdot & \cdot \\ 0 & 0 & \cdots & 1 & (k-1)\alpha_2 \\ 0 & 0 & 0 & \cdots & 1 \end{pmatrix} : \alpha_2, \ldots, \alpha_k \in \mathbb{C} \right\}$$

so that $\mathbb{U}_k$ is generated along its last column as well as along its first row.

Let $B_k \subset \mathrm{SL}(k)$ denote the standard Borel subgroup of $\mathrm{SL}(k)$ which stabilises the filtration $\mathbb{C}e_1 \subset \mathbb{C}e_1 \oplus \mathbb{C}e_2 \subset \cdots \mathbb{C}^k$. Then $B_k = B_{k-1} \cdot \mathbb{U}_k$ where the Borel subgroup B_{k-1} of $\mathrm{GL}(k-1) = \mathrm{GL}(\mathbb{C}e_1 \oplus \mathbb{C}e_2 \oplus \cdots \oplus \mathbb{C}e_{k-1})$ is embedded diagonally in $\mathrm{SL}(k)$ via

$$A \mapsto \begin{pmatrix} A & 0 \\ 0 & (\det A)^{-1} \end{pmatrix}.$$

Since $\mathbb{U}_k$ stabilises p_k and e_1 we have

$$\overline{B_k(p_k \otimes e_1^{\otimes K})} = \overline{B_{k-1}(p_k \otimes e_1^{\otimes K})},$$

and since $\mathrm{SL}(k)/B_k$ is projective we have

$$\overline{\mathrm{SL}(k)(p_k \otimes e_1^{\otimes K})} = \mathrm{SL}(k)\overline{B_k(p_k \otimes e_1^{\otimes K})} = \mathrm{SL}(k)\overline{B_{k-1}(p_k \otimes e_1^{\otimes K})}.$$

Since the closure $\overline{\mathrm{SL}(k)(p_k \otimes e_1^{\otimes K})}$ of the $\mathrm{SL}(k)$-orbit of $p_k \otimes e_1^{\otimes K}$ in $W_{k,K}$ is the union of finitely many $\mathrm{SL}(k)$-orbits, to prove Theorem 6.2 it suffices to prove

LEMMA 6.4. *Suppose that $k \geq 4$ and a and b are strictly positive integers with b/a large enough and that x lies in the closure in*

$$(\wedge^k(\operatorname{Sym}^{\leq k}\mathbb{C}^k))^{\otimes a} \otimes (\mathbb{C}^k)^{\otimes b}$$

of the orbit $B_k(p_k^{\otimes a} \otimes e_1^{\otimes b})$ of $p_k^{\otimes a} \otimes e_1^{\otimes b}$ under the natural action of the Borel subgroup B_k of $\mathrm{SL}(k)$. Then either $x \in B_k(p_k^{\otimes a} \otimes e_1^{\otimes b})$ or the stabiliser of x in $\mathrm{SL}(k)$ has dimension at least $k+1$.

We will split the proof of this lemma into two parts. Let T_k denote the standard maximal torus of $\mathrm{SL}(k)$ consisting of the diagonal matrices in $\mathrm{SL}(k)$. Lemma 6.4 follows immediately from Lemmas 6.5 and 6.6 below.

LEMMA 6.5. *Suppose that $k \geq 4$ and a and b are strictly positive integers with b/a large enough and that x lies in the closure $\overline{T_k(p_k^{\otimes a} \otimes e_1^{\otimes b})}$ in*

$$(\wedge^k(\operatorname{Sym}^{\leq k}\mathbb{C}^k))^{\otimes a} \otimes (\mathbb{C}^k)^{\otimes b}$$

of the orbit $T_k(p_k^{\otimes a} \otimes e_1^{\otimes b})$ of $p_k^{\otimes a} \otimes e_1^{\otimes b}$ under the natural action of the maximal torus T_k of $\mathrm{SL}(k)$. Then either $x \in T_k(p_k^{\otimes a} \otimes e_1^{\otimes b})$ or the stabiliser of x in $\mathrm{SL}(k)$ has dimension at least $k+1$.

LEMMA 6.6. *Suppose that $k \geq 2$ and a and b are strictly positive integers and that x lies in the closure in*

$$(\wedge^k(\operatorname{Sym}^{\leq k}\mathbb{C}^k))^{\otimes a} \otimes (\mathbb{C}^k)^{\otimes b}$$

of the orbit $B_k(p_k^{\otimes a} \otimes e_1^{\otimes b})$ of $p_k^{\otimes a} \otimes e_1^{\otimes b}$ under the natural action of the Borel subgroup B_k of $\mathrm{SL}(k)$. Then either $x \in B_k\overline{T_k(p_k^{\otimes a} \otimes e_1^{\otimes b})}$ or the stabiliser of x in $\mathrm{SL}(k)$ has dimension at least $k+1$.

We will start with the proof of Lemma 6.6.

PROOF. We have

$$x \in \overline{B_k(p_k^{\otimes a} \otimes e_1^{\otimes b})} = \overline{B_{k-1}(p_k^{\otimes a} \otimes e_1^{\otimes b})}$$

as above, so there is a sequence of matrices

$$b^{(m)} = \begin{pmatrix} b_{11}^{(m)} & b_{12}^{(m)} & \dots & b_{1k-1}^{(m)} & 0 \\ 0 & b_{22}^{(m)} & \dots & b_{2k-1}^{(m)} & 0 \\ & & \ddots & & \\ 0 & 0 & \dots & 0 & b_{kk}^{(m)} \end{pmatrix} \in B_{k-1} \subset \mathrm{SL}(k)$$

such that $b^{(m)}(p_k^{\otimes a} \otimes e_1^{\otimes b}) \to x$ as $m \to \infty$. Now expanding the wedge product in the definition of p_k we get

$$b^{(m)}(p_k^{\otimes a}) = \left(e_1 \wedge \cdots \wedge e_n + \cdots + (b_{11}^{(m)})^{1+2+\cdots+k} e_1 \otimes e_1^2 \otimes \cdots \otimes e_1^k\right)^{\otimes a}$$

while

$$b^{(m)}(e_1^{\otimes b}) = (b_{11}^{(m)})^b e_1^{\otimes b},$$

so by considering the coefficient of $(e_1 \wedge \cdots \wedge e_n)^{\otimes a} \otimes e_1^{\otimes b}$ we see that $(b_{11}^{(m)})^b$ tends to a limit in $\mathbb{C}$ as $m \to \infty$. Thus, by replacing the sequence $(b^{(m)})$ with a subsequence if necessary, we can assume that

$$b_{11}^{(m)} \to b_{11}^{(\infty)} \in \mathbb{C}$$

as $m \to \infty$.

First suppose that $k = 2$. Then $\mathrm{Sym}^{\leq k}\mathbb{C}^k = \mathbb{C}^2 \oplus \mathrm{Sym}^2\mathbb{C}^2$ and

$$(\wedge^k(\mathrm{Sym}^{\leq k}\mathbb{C}^k))^{\otimes a} \otimes (\mathbb{C}^k)^{\otimes b} = (\wedge^2(\mathbb{C}^2 \oplus \mathrm{Sym}^2\mathbb{C}^2))^{\otimes a} \otimes (\mathbb{C}^2)^{\otimes b}$$

and

$$p_k = e_1 \wedge (e_2 + e_1^2),$$

so if

$$b^{(m)} = \begin{pmatrix} b_{11}^{(m)} & b_{12}^{(m)} \\ 0 & b_{22}^{(m)} \end{pmatrix} \in \mathrm{SL}(2)$$

then $b_{11}^{(m)} b_{22}^{(m)} = 1$ and

$$b^{(m)}(p_2^{\otimes a} \otimes e_1^{\otimes b}) = (b_{11}^{(m)})^b (e_1 \wedge (e_2 + (b_{11}^{(m)})^3 e_1^2)))^{\otimes a} \otimes e_1^{\otimes b}$$
$$\to x = (b_{11}^{(\infty)})^b (e_1 \wedge (e_2 + (b_{11}^{(\infty)})^3 e_1^2)))^{\otimes a} \otimes e_1^{\otimes b}$$

as $m \to \infty$. If $b_{11}^{(\infty)} \neq 0$ then $x \in \mathrm{SL}(2)((p_2^{\otimes a} \otimes e_1^{\otimes b})$, while if $b_{11}^{(\infty)} = 0$ then $x = 0$ is fixed by $\mathrm{SL}(2)$ which has dimension $3 = k+1$.

Now suppose that $k > 2$, and assume first that $b_{11}^{(\infty)} \neq 0$. We have that

$$b^{(m)}(p_k^{\otimes a} \otimes e_1^{\otimes b}) = (b_{11}^{(m)})^b (b^{(m)} p_k)^{\otimes a}) \otimes e_1^{\otimes b} \to x$$

and $b_{11}^{(m)} \to b_{11}^{(\infty)} \in \mathbb{C} \setminus \{0\}$ as $m \to \infty$, so by replacing the sequence $(b^{(m)})$ with a subsequence if necessary, we can assume that

$$(b_{11}^{(m)})^{b/a} b^{(m)} p_k \to p_k^{\infty} \in \wedge^k(\mathrm{Sym}^{\leq k}\mathbb{C}^k)$$

as $m \to \infty$, where

$$b^{(m)}p_k = b_{11}^{(m)}e_1 \wedge (b_{22}^{(m)}e_2 + (b_{11}^{(m)})^2 e_1^2) \wedge \cdots \wedge (b_{ii}^{(m)}e_i + b_{i-1i}^{(m)}e_{i-1} + \cdots + b_{1i}^{(m)}e_1 + \sum_{s=2}^{i-1} \sum_{i_1+\cdots+i_s=i} (b_{i_1i_1}^{(m)}e_{i_1} + \cdots + b_{1i_1}^{(m)}e_1) \cdots \times (b_{i_si_s}^{(m)}e_{i_s} + \cdots + b_{1i_s}^{(m)}e_1) + (b_{11}^{(m)})^i e_1^i) \wedge \cdots \tag{19}$$

Looking at the coefficient of

$$e_1 \wedge e_1^2 \wedge \cdots \wedge e_1^{i-1} \wedge e_j \wedge e_1^{i+1} \wedge \cdots \wedge e_1^k$$

when $1 \leq j \leq i \leq k$, we see that

$$(b_{11}^{(m)})^{1+2+\cdots+(i-1)+(i+1)+\cdots+k} b_{ji}^{(m)}$$

tends to a limit in $\mathbb{C}$ as $m \to \infty$, and so since $b_{11}^{(\infty)} \neq 0$

$$b_{ji}^{(m)} \to b_{ji}^{(\infty)} \in \mathbb{C}.$$

Also $b_{11}^{(m)} b_{22}^{(m)} \cdots b_{kk}^{(m)} = 1$ for all m, so $b_{11}^{(\infty)} b_{22}^{(\infty)} \cdots b_{kk}^{(\infty)} = 1$, so $b^{(m)} \to b^{(\infty)} \in \mathrm{SL}(k)$. Therefore

$$x = b^{(\infty)}(p_k^{\otimes a} \otimes e_1^{\otimes b})$$

lies in the orbit of $p_k^{\otimes a} \otimes e_1^{\otimes b}$ as required.

So it remains to consider the case when $b_{11}^{(\infty)} = 0$. If $p_k^\infty = 0$ then its stabiliser is $\mathrm{SL}(k)$ which has dimension $k^2 - 1 \geq k+1$, so we can assume that $p_k^\infty \neq 0$. Recall that then

$$(b_{11}^{(m)})^{b/a} b^{(m)} p_k \to p_k^\infty \in \wedge^k(\mathrm{Sym}^{\leq k}\mathbb{C}^k)$$

and

$$[b^{(m)}p_k] \to [p_k^\infty] \in \mathbb{P}(\wedge^k(\mathrm{Sym}^{\leq k}\mathbb{C}^k))$$

as $m \to \infty$, where

$$b^{(m)}p_k = b_{11}^{(m)}e_1 \wedge (b_{22}^{(m)}e_2 + (b_{11}^{(m)})^2 e_1^2) \wedge \cdots \wedge (b_{ii}^{(m)}e_i + b_{i-1i}^{(m)}e_{i-1} + \cdots + b_{1i}^{(m)}e_1 + \sum_{s=2}^{i-1} \sum_{i_1+\cdots+i_s=i} (b_{i_1i_1}^{(m)}e_{i_1} + \cdots + b_{1i_1}^{(m)}e_1) \cdots \times (b_{i_si_s}^{(m)}e_{i_s} + \cdots + b_{1i_s}^{(m)}e_1) + (b_{11}^{(m)})^i e_1^i) \wedge \cdots$$

By replacing the sequence $(b^{(m)})$ with a subsequence if necessary, we can assume that

$$[b_{ii}^{(m)}e_i + b_{i-1i}^{(m)}e_{i-1} + \cdots + b_{1i}^{(m)}e_1] \to [c_{ii}^{(\infty)}e_i + c_{i-1i}^{(\infty)}e_{i-1} + \cdots + c_{1i}^{(\infty)}e_1] \in \mathbb{P}(\mathbb{C}^k)$$

as $m \to \infty$ for $2 \leq i \leq k$, which implies that

$$[(b^{(m)}_{i_1 i_1} e_{i_1} + \cdots + b^{(m)}_{1 i_1} e_1) \dots (b^{(m)}_{i_s i_s} e_{i_s} + \cdots + b^{(m)}_{1 i_s} e_1)]$$
$$\to [(c^{(\infty)}_{i_1 i_1} e_{i_1} + \cdots + c^{(\infty)}_{1 i_1} e_1) \dots (c^{(\infty)}_{i_s i_s} e_{i_s} + \cdots + c^{(\infty)}_{1 i_s} e_1)] \in \mathbb{P}(\mathrm{Sym}^i \mathbb{C}^k)$$

whenever $i_1 + \cdots + i_s = i \in \{2, \dots, k\}$, and hence that

$$p_k^\infty \in \wedge^k (\mathrm{Sym}^{\leq k} D)$$

where D is the span in $\mathbb{C}^k$ of

$$\{e_1\} \cup \{c^{(\infty)}_{ii} e_i + c^{(\infty)}_{i-1 i} e_{i-1} + \cdots + c^{(\infty)}_{1i} e_1 : 2 \leq i \leq k\}.$$

Moreover since $b^{(m)} \in B_{k-1}$ we have $b^{(m)}_{jk} = 0$ if $j < k$ so

$$[c^{(\infty)}_{kk} e_k + c^{(\infty)}_{k-1 k} e_{k-1} + \cdots + c^{(\infty)}_{1k} e_1] = [e_k]$$

so $e_k \in D$.

Note that $b^{(m)} \in B_{k-1}$ and B_{k-1} normalises the maximal unipotent subgroup U_k of B_k which contains the stabiliser $\mathbb{U}_k$ of p_k. Therefore for each m there is a $(k-1)$-dimensional subgroup of U_k which stabilises $b^{(m)} p_k$, and it follows that there is a $(k-1)$-dimensional subgroup $\mathbb{U}_k^\infty$ of U_k which stabilises p_k^∞. In addition by **[3]** Theorem 6.4 if p_k^∞ does not lie in $\mathrm{SL}(k) p_k$ then it is stabilised by a nontrivial one-parameter subgroup $\lambda^\infty : \mathbb{C}^* \to \mathrm{SL}(k)$ of $\mathrm{SL}(k)$. Moreover if $D \neq \mathbb{C}^k$ then there is some $j \in \{2, \dots, k-1\}$ such that e_j is not in D, and then there is an automorphism of $\mathbb{C}^k$ which fixes every element of D and sends e_j to $e_j + e_k$. This automorphism is independent of $\mathbb{U}_k^\infty$ (since $\mathbb{U}_k^\infty \subseteq U_k$) and the one-parameter subgroup λ^∞ of $\mathrm{SL}(k)$ fixing p_k^∞, so the stabiliser of p_k^∞ in $\mathrm{SL}(k)$ has dimension at least

$$\dim \mathbb{U}_k^\infty + 2 = k + 1.$$

Thus we can assume that $D = \mathbb{C}^k$, and hence $c^{(\infty)}_{ii} \neq 0$ for $2 \leq i \leq k$, so that

$$\frac{b^{(m)}_{ji}}{b^{(m)}_{ii}} \to \frac{c^{(\infty)}_{ji}}{c^{(\infty)}_{ii}} \in \mathbb{C}$$

as $m \to \infty$. Then by applying an element of B_{k-1} to p_k^∞ we can assume that

$$[c^{(\infty)}_{ii} e_i + c^{(\infty)}_{i-1 i} e_{i-1} + \cdots + c^{(\infty)}_{1i} e_1] = [e_i]$$

or equivalently that

$$[b^{(m)}_{ii} e_i + b^{(m)}_{i-1 i} e_{i-1} + \cdots + b^{(m)}_{1i} e_1] \to [e_i]$$

as $m \to \infty$ for $2 \leq i \leq k$, and hence that

$$[(b^{(m)}_{i_1 i_1} e_{i_1} + \cdots + b^{(m)}_{1 i_1} e_1) \dots (b^{(m)}_{i_s i_s} e_{i_s} + \cdots + b^{(m)}_{1 i_s} e_1)] \to [e_{i_1} \cdots e_{i_s}] \in \mathbb{P}(\mathrm{Sym}^i \mathbb{C}^k)$$

whenever $i_1+\cdots+i_s=i\in\{2,\ldots,k\}$. Now by again replacing the sequence $(b^{(m)})$ with a subsequence if necessary, we can assume that

$$[b_{ii}^{(m)}e_i+b_{i-1i}^{(m)}e_{i-1}+\cdots+b_{1i}^{(m)}e_1+\sum_{s=2}^{i-1}\sum_{i_1+\cdots+i_s=i}(b_{i_1i_1}^{(m)}e_{i_1}+\cdots+b_{1i_1}^{(m)}e_1]$$

$$\to[d_i^\infty]\in\mathbb{P}(\operatorname{Sym}^{\leq k}\mathbb{C}^k)$$

where

$$d_i^\infty=\gamma_i^{(\infty)}e_i+\sum_{s=2}^{i}\sum_{i_1+\cdots+i_s=i}\gamma_{i_1\ldots i_s}^{(\infty)}e_{i_1}\cdots e_{i_s}\in\operatorname{Sym}^{\leq k}\mathbb{C}^k\setminus\{0\}$$

for some $\gamma_{i_1\ldots i_s}^{(\infty)}\in\mathbb{C}$. In addition $\{d_i^\infty:1\leq i\leq k\}$ is linearly independent so that

$$[p_k^\infty]=[d_i^\infty\wedge\cdots\wedge d_k^\infty]\in\mathbb{P}(\wedge^k(\operatorname{Sym}^{\leq k}\mathbb{C}^k))$$

and $p_k^\infty=\lim_{m\to\infty}t^{(m)}p_k$ where $t^{(m)}$ is the diagonal matrix with entries $b_{11}^{(m)},\ldots,b_{kk}^{(m)}$.

Thus we can assume that $p_k^\infty\in\overline{T_kp_k}$ where T_k is the standard maximal torus in $\mathrm{SL}(k)$, which completes the proof of Lemma 6.6. □

It therefore remains to prove Lemma 6.5. We can continue with the notation above and use the following standard result:

LEMMA 6.7. *Let T be an algebraic torus acting on the projective variety Z, and $z\in Z$. Then $y\in\overline{Tz}$ if and only if there is $\tau\in T$, and a one-parameter subgroup $\lambda:\mathbb{C}^*\to T$ such that $\tau y\in\overline{\lambda(\mathbb{C}^*)z}$.*

Hence we may assume without loss of generality that there is a one-parameter subgroup

$$t\mapsto\lambda(t)=\begin{pmatrix}t^{\lambda_1}&0&\cdots&&0\\0&t^{\lambda_2}&0&\cdots&0\\&&\cdots&&\\0&\cdots&&0&t^{\lambda_k}\end{pmatrix}$$

of $\mathrm{SL}(k)$ such that $\lambda_1>0$ and $t^{\lambda_1b/a}\lambda(t)p_k\to p_k^\infty$ as $t\to0$. Therefore

$$p_k^\infty=\lim_{t\to0}t^{\lambda_1b/a}e_1\wedge(e_2+t^{2\lambda_1-\lambda_2}e_1^2)\wedge\cdots\wedge$$
$$\times\left(e_k+\sum_{s=2}^{k}\sum i_1+\cdots+i_s=kt^{\lambda_{i_1}+\cdots+\lambda_{i_s}-\lambda_k}e_{i_1}\cdots e_{i_s}\right)$$

where $\lambda_1 + \cdots + \lambda_k = 0$. We are assuming that $p_k^\infty \neq 0$ so

$$[p_k^\infty] = \lim_{t\to 0} \Bigg[e_1 \wedge (e_2 + t^{2\lambda_1 - \lambda_2} e_1^2) \wedge \cdots \wedge \\ \times \left(e_k + \sum_{s=2}^{k} \sum i_1 + \cdots + i_s = k t^{\lambda_{i_1} + \cdots + \lambda_{i_s} - \lambda_k} e_{i_1} \cdots e_{i_s} \right) \Bigg].$$

If $\lambda_{i_1} + \cdots + \lambda_{i_s} < \lambda_j$ for some $j \in \{2, \ldots, k-1\}$ and $s \geq 2$ and $i_1, \ldots, i_s \geq 1$ such that $i_1 + \cdots + i_s = j$, then $[p_k^\infty]$ is independent of e_j and so as above the stabiliser of p_k^∞ in $\mathrm{SL}(k)$ has dimension at least $k+1$. So we can assume that

$$\lambda_{i_1} + \cdots + \lambda_{i_s} \geq \lambda_j \tag{20}$$

for any $j \in \{2, \ldots, k-1\}$ and $s \geq 2$ and $i_1, \ldots, i_s \geq 1$ such that $i_1 + \cdots + i_s = j$, and in particular that $\lambda_j \leq j\lambda_1$ for each $j \in \{2, \ldots, k-1\}$. Let

$$\rho_j = j\lambda_1 - \lambda_j \tag{21}$$

for $j \in \{1, \ldots, k-1\}$; then $\rho_1 = 0$ and $\rho_j \geq 0$ and

$$\rho_{i_1} + \cdots + \rho_{i_s} \leq \rho_j$$

for any $j \in \{2, \ldots, k-1\}$ and $s \geq 2$ and $i_1, \ldots, i_s \geq 1$ such that $i_1 + \cdots + i_s = j$. In addition looking at the coefficient of

$$e_1 \wedge e_2 \wedge \cdots \wedge e_{k-1} \wedge e_{i_1} \cdots e_{i_s}$$

where $i_1 + \cdots + i_s = k$, we find that

$$\begin{aligned} 0 &\leq \lambda_1 b/a + \lambda_{i_1} + \cdots + \lambda_{i_s} - \lambda_k \\ &= \lambda_1 (b/a + k(k+1)/2) - (\rho_{i_1} + \cdots + \rho_{i_s} + \rho_2 + \cdots + \rho_{k-1}), \end{aligned}$$

and since $p_k^\infty \neq 0$ there is some $i_1, \ldots, i_s$ with $i_1 + \cdots + i_s = k$ and

$$\lambda_1 b/a + \lambda_{i_1} + \cdots + \lambda_{i_s} = \lambda_k \tag{22}$$

or equivalently

$$\lambda_1 (b/a + k(k+1)/2) = \rho_{i_1} + \cdots + \rho_{i_s} + \rho_2 + \cdots + \rho_{k-1}.$$

Thus

$$(23)\quad p_k^\infty = \lim_{t\to 0} e_1 \wedge (e_2 + t^{2\lambda_1-\lambda_2}e_1^2) \wedge \cdots \wedge$$
$$\times \left(e_{k-1} + \sum_{s=2}^{k-1} \sum_{i_1+\cdots+i_s=k-1} t^{\lambda_{i_1}+\cdots+\lambda_{i_s}-\lambda_{k-1}} e_{i_1}\cdots e_{i_s} \right)$$
$$\wedge \left(t^{\lambda_1 b/a} \sum_{s=2}^{k} \sum_{i_1+\cdots+i_s=k} t^{\lambda_{i_1}+\cdots+r_{i_s}-r_k} e_{i_1}\cdots e_{i_s} \right.$$
$$= e_1 \wedge \cdots \wedge \left(e_{k-1} + \sum_{s=2}^{k-1} \sum_{i_1+\cdots+i_s=k-1:\rho_{i_1}+\cdots+\rho_{i_s}=\rho_{k-1}} e_{i_1}\cdots e_{i_s} \right)$$
$$\wedge \left(\sum_{s=2}^{k} \sum_{\substack{i_1+\cdots+i_s=k\\ \lambda_1(b/a+k(k+1)/2)=\rho_{i_1}+\cdots+\rho_{i_s}+\rho_2+\cdots+\rho_{k-1}}} e_{i_1}\cdots e_{i_s} \right)$$

is independent of e_k and hence is fixed by the automorphisms of $\mathbb{C}^k$ which fix $e_1,\ldots,e_{k-1}$ and send e_k to e_k+e_j for $j\in\{1,\ldots,k-1\}$, as well as by the one-parameter subgroup

$$\lambda(t) = \begin{pmatrix} t^{\lambda_1} & 0 & \cdots & & 0 \\ 0 & t^{\lambda_2} & 0 & \cdots & 0 \\ & & \cdots & & \\ 0 & \cdots & & 0 & t^{\lambda_k} \end{pmatrix}$$

of T_k. Thus to complete the proof of Lemma 6.5 and hence of Theorem 6.2, it suffices to find an additional one-dimensional stabiliser, which will be done in the rest of this section.

Letting

$$\mathbf{z} = [p_k] = \left[e_1 \wedge (e_2 + e_1^2) \wedge \cdots \wedge \left(\sum_{i_1+\cdots+i_s=k} e_{i_1}\ldots e_{i_s} \right) \right]$$

as at (18) we have

$$\lambda(t)\mathbf{z} = \left[t^{\lambda_1} e_1 \wedge (t^{\lambda_2}e_2 + t^{2\lambda_1}e_1^2) \wedge \cdots \wedge \left(\sum_{i_1+\cdots+i_s=k} t^{\lambda_{i_1}+\cdots+\lambda_{i_s}} e_{i_1}\ldots e_{i_s} \right) \right]$$
$$= [t^{\lambda_1+\cdots+\lambda_k}(e_1\wedge\cdots\wedge e_k) + t^{\lambda_1+2\lambda_1+\lambda_3+\cdots+\lambda_k}$$
$$\times (e_1 \wedge e_1^2 \wedge e_3 \wedge \cdots \wedge e_k) + \cdots].$$

The generic term in this expression is

$$t^{\lambda_{\varepsilon_1}+\lambda_{\varepsilon_2}+\cdots\lambda_{\varepsilon_k}}(\mathbf{e}_{\varepsilon_1}\wedge\cdots\wedge\mathbf{e}_{\varepsilon_k}),\quad \Sigma(\varepsilon_i)=i$$

where

$$\lambda_\tau = \sum_{i \in \tau} \lambda_i \quad \text{and} \quad \mathbf{e}_\tau = \Pi_{i \in \tau} e_i \quad \text{if } \tau = (i_1, \ldots, i_s). \tag{24}$$

Definition 6.8. *For any one-parameter subgroup λ as above let*

- $m_\lambda = \min_{\substack{(\varepsilon_1, \ldots \varepsilon_k) \\ 1 \le \Sigma(\varepsilon_i) \le k}} (\lambda_{\varepsilon_1} + \lambda_{\varepsilon_2} + \cdots \lambda_{\varepsilon_k})$,
- $\mathbf{z}_\lambda = [\sum_{1 \le \Sigma\varepsilon \le k, \lambda_\varepsilon = m_\lambda} \mathbf{e}_\varepsilon]$,
- $m_\lambda[i] = \min_{\Sigma(\varepsilon) = i} \lambda_\varepsilon$ *for* $1 \le i \le k$,
- $\mathbf{z}_\lambda[i] = [\sum_{\Sigma\varepsilon = i, \lambda_\varepsilon = m_\lambda[i]} \mathbf{e}_\varepsilon]$.

Let $\mathcal{O}_\lambda$ denote the $\mathrm{SL}(k)$*-orbit of* $\mathbf{z}_\lambda$.

It is clear that the one-parameter subgroup $\tilde{\lambda}(t) = (t, t^2, \ldots, t^k)$ stabilises $\mathbf{z}$, where $\mathbf{z}$ is defined as at (18), and therefore $\mathbf{z} = \mathbf{z}_{\tilde{\lambda}}$ and its $\mathrm{SL}(k)$-orbit is equal to its $\mathrm{GL}(k)$-orbit.

We need a more precise description of the orbit structure of the closure of the orbit $\mathcal{O}_0 = \mathcal{O}_{\tilde{\lambda}}$. Since $\tilde{\lambda}_i = i\tilde{\lambda}_1$ for $i = 1, \ldots, k$, for $\lambda \ne \tilde{\lambda}$ we have a smallest index $\sigma \in \{2, \ldots, k\}$ with $\lambda_\sigma \ne \sigma\lambda_1$.

Definition 6.9. *We call $\sigma = Head(\lambda)$ the head of $\lambda = (\lambda_1, \ldots, \lambda_n)$ if*

$$\lambda_i = i\lambda_1 \quad \textit{for } i < \sigma \quad \textit{and} \quad \lambda_\sigma \ne \sigma\lambda_1.$$

If $\lambda_\sigma < \sigma\lambda_1$ then we call λ regular *; otherwise we call λ* degenerate.

We will say that a one-parameter subgroup λ is *maximal* if the closure of the orbit $\mathrm{GL}(k) \cdot \mathbf{z}_\lambda$ is a maximal boundary component of the closure of the orbit of $\mathbf{z}$.

Definition 6.10. *Fix $0 < \varepsilon < 1$ and $2 \le \sigma \le k$. Let $\lambda^\sigma = (\lambda_1^\sigma, \ldots, \lambda_k^\sigma)$ and $\mu^\sigma = (\mu_1^\sigma, \ldots, \mu_k^\sigma)$ be the following one-parameter subgroups of* $\mathrm{GL}(k)$*:*

$$\lambda_i^\sigma = i - \left\lfloor \frac{i}{\sigma} \right\rfloor \varepsilon \quad \textit{for } 1 \le i \le k, \tag{25}$$

$$\mu_i^\sigma = \begin{cases} i & \textit{for } i \ne \sigma, i \le k, \\ \sigma + \varepsilon & \textit{for } i = \sigma. \end{cases} \tag{26}$$

It is easy to see that $\mathrm{Head}(\lambda^\sigma) = \mathrm{Head}(\mu^\sigma) = \sigma$, and λ^σ is regular, whereas μ^σ is degenerate.

Definition 6.11. *Let λ be a 1-parameter subgroup. We call*

$$\sharp\{i : \mathbf{z}_\lambda[i] = e_i\}$$

the toral dimension of the limit point $\mathbf{z}_\lambda$.

Lemma 6.12. *If the* $\mathrm{SL}(k)$*-orbit of p_k^∞ has codimension one in $\overline{\mathrm{SL}(k)p_k}$, then $[p_k^\infty]$ lies in the orbit of one of $\mathbf{z}_{\lambda^2}, \ldots, \mathbf{z}_{\lambda^k}$ or $\mathbf{z}_{\mu^2}, \ldots, \mathbf{z}_{\mu^{k-1}}$.*

PROOF. We can assume that $[p_k^\infty] = \mathbf{z}_\lambda$ for some one-parameter subgroup λ. First suppose that λ is a regular one-parameter subgroup with $\mathrm{Head}(\lambda) = \sigma$ and $[p_k^\infty] = \mathbf{z}_\lambda$. Without loss of generality we can assume that

$$\lambda_i = i \quad \text{for } i < \sigma \quad \text{and} \quad \lambda_\sigma = \sigma - \varepsilon.$$

We will call $d(i) = \lfloor \frac{i}{\sigma} \rfloor$ the defect of i and $d(\tau) = d(i_1) + \cdots + d(i_s)$ the defect of $\tau = (i_1, \ldots, i_s)$, so that when $i \le \sigma$ we have $d(i)\epsilon = \rho_i$ as defined at (21). Since

$$\lambda_{(j,\underbrace{\sigma, \ldots, \sigma}_{m})} = j + m(\sigma - \varepsilon) \quad \text{for } 1 \le j \le \sigma - 1, \ m \ge 0,$$

we have

$$m_\lambda[i] \le i - d(i)\varepsilon \quad \text{for } 1 \le i \le k. \tag{27}$$

If $\lambda_s < s - d(s)\varepsilon$ for $s > i$ and s is the smallest index with this property then $m_\lambda[s] = \lambda_s$ and $\mathbf{z}_\lambda[s] = e_s$, so

$$\mathbf{z}_\lambda[1] = e_1, \ \mathbf{z}_\lambda[\sigma] = e_\sigma, \ \mathbf{z}_\lambda[s] = e_s,$$

while $\mathbf{z}_\lambda$ is independent of e_k by (23), so $[p_k^\infty]$ is fixed by a three-dimensional torus in $\mathrm{SL}(k)$ and thus p_k^∞ is fixed by a two-dimensional torus in $\mathrm{SL}(k)$ as well as a unipotent subgroup of dimension $k - 1$. So we can assume that $\lambda_i \ge i - d(i)\varepsilon$ for $1 \le i \le k$, and therefore

$$m_\lambda[i] = i - d(i)\varepsilon \quad \text{for } 1 \le i \le k.$$

So

$$\mathbf{e}_\tau \notin \mathbf{z}_\lambda[i] \quad \text{if } d(\tau) > d(i). \tag{28}$$

On the other hand the distinguished 1-parameter subgroup λ^σ is defined as $\lambda_i^\sigma = i - d(i)\varepsilon$, and therefore

$$\mathbf{z}_{\lambda^\sigma}[i] = \sum_{\Sigma(\tau)=i, d(\tau)=d(i)} \mathbf{e}_\tau. \tag{29}$$

Comparing (28) and (29) we conclude

$$\mathbf{z}_\lambda[i] \subset \mathbf{z}_{\lambda^\sigma}[i] \quad \text{for } 1 \le i \le k.$$

Now let μ be a degenerate 1-parameter subgroup with $\mathrm{Head}(\mu) = \sigma$. Without loss of generality we can assume again that

$$\mu_i = i \quad \text{for } i < \sigma \quad \text{and} \quad \mu_\sigma = \sigma + \varepsilon.$$

Since

$$\mu_{(\underbrace{1, \ldots 1}_{i})} = i \quad \text{for } 1 \le i \le k$$

we have

$$m_\mu[i] \le i. \tag{30}$$

Again, $\mu_s < s$ cannot happen for $s > \sigma$ since in that case $\mathbf{z}_\mu[s] = e_s$ would hold and the codimension of $\mathrm{SL}(k)p_k^\infty$ would be at least two. So $\mu_s \geq s$ and therefore $\mu_\tau \geq \Sigma(\tau)$ with strict inequality if $\sigma \in \tau$. Therefore

$$\mathbf{e}_\tau \notin \mathbf{z}_\mu[i] \quad \text{if } \sigma \in \tau. \tag{31}$$

On the other hand μ^σ satisfies equality in (30), and

$$\mathbf{z}_{\mu^\sigma}[i] = \sum_{\Sigma(\tau)=i, \sigma \notin \tau} \mathbf{e}_\tau. \tag{32}$$

Comparing (31) and (32) we get

$$\mathbf{z}_\mu[i] \subset \mathbf{z}_{\mu^\sigma}[i] \quad \text{for } 1 \leq i \leq k,$$

and so it remains to consider the possibility that $[p_k^\infty] = \mathbf{z}_{\mu^k}$. But by (22) there is some $i_1, \ldots, i_s$ with $i_1 + \cdots + i_s = k$ and

$$\lambda_1 b/a + \lambda_{i_1} + \cdots + \lambda_{i_s} = \lambda_k$$

and hence $\lambda_k > \lambda_{i_1} + \cdots + \lambda_{i_s}$. Thus $[p_k^\infty]$ cannot be equal to $\mathbf{z}_{\mu^k}$ because the coefficient of $e_1 \wedge e_1^2 \ldots \wedge e_1^k$ is nonzero for $\mathbf{z}_{\mu^k}$ but zero for $[p_k^\infty]$, and the result follows. □

We summarize our information about the maximal boundary components in

PROPOSITION 6.13. *We have* $\mathbf{z}_{\lambda^\sigma} = \wedge_{i=1}^k \mathbf{z}_{\lambda^\sigma}[i]$, *where* $\mathbf{z}_{\lambda^\sigma}[i] = \oplus_{\Sigma(\tau)=i, d(\tau)=d(i)} \mathbf{e}_\tau$, *and* $\mathbf{z}_{\mu^\sigma} = \wedge_{i=1}^k \mathbf{z}_{\mu^\sigma}[i]$ *where* $\mathbf{z}_{\mu^\sigma}[i] = \oplus_{\Sigma(\tau)=i, \sigma \notin \tau} \mathbf{e}_\tau$.

REMARK 6.14. Since the one-parameter subgroup $\tilde{\lambda}(t) = (t, t^2, \ldots, t^k)$ of $\mathrm{GL}(k)$ stabilises $T_k\mathbf{z}$, it follows from Lemma 6.12 that it is enough to prove the codimension-at-least-two property we require only for the one-parameter subgroups $\tilde{\lambda}^\sigma$ (for $2 \leq s \leq k$) and $\tilde{\mu}^\sigma$ (for $2 \leq s \leq k-1$) of $\mathrm{SL}(k)$ given by

$$\tilde{\lambda}^\sigma(t) = (\lambda^\sigma(t)\tilde{\lambda}(t)^{q_\sigma})^{n_\sigma}$$

and

$$\tilde{\mu}^\sigma(t) = (\mu^\sigma(t)\tilde{\lambda}(t)^{r_\sigma})^{m_\sigma}$$

for suitable $q_\sigma, r_\sigma \in \mathbb{Q}$ and $n_\sigma, m_\sigma \in \mathbb{Z}$. But we observed at (20) that the property is satisfied by a one-parameter subgroup λ of $\mathrm{SL}(k)$ if $\lambda_{i_1} + \cdots + \lambda_{i_s} < \lambda_j$ for any $j \in \{2, \ldots, k-1\}$ such that $i_1 + \cdots + i_s = j$, so it is enough to consider the one-parameter subgroups $\tilde{\lambda}^\sigma$ for $2 \leq s \leq k$.

6.1. The limit of the stabilisers. In order to prove Lemma 6.5, it now suffices by Remark 6.14 to find a k-dimensional unipotent subgroup of the stabiliser $G_{\mathbf{z}_{\lambda^\sigma}}$ of $\mathbf{z}_{\lambda^\sigma}$ in $\mathrm{GL}(k)$ for each σ when $\mathbf{z}_{\lambda^\sigma} = [p_k^\infty]$, since we know that p_k^∞ is fixed by a one-parameter subgroup of the maximal torus T_k of $\mathrm{SL}(k)$, and any unipotent group which stabilises $\mathbf{z}_{\lambda^\sigma} = [p_k^\infty]$ also stabilises p_k^∞.

In this subsection we will study the limits $\lim G_{\lambda^\sigma(t)\mathbf{z}}$ of the stabiliser groups for the one-parameter subgroups λ^σ for $2 \le \sigma \le k$, and use this to prove Lemma 6.5, which together with Lemma 6.6 will complete the proof of Theorem 6.2.

PROPOSITION 6.15. *$G^\sigma = \lim_{t\to 0} G_{\lambda^\sigma(t)\mathbf{z}} \subset GL(k)$ is a k-dimensional subgroup of $G_{\mathbf{z}_{\lambda^\sigma}}$ which contains a $k-1$-dimensional subgroup of the maximal unipotent subgroup U_k of* $\mathrm{SL}(k)$.

PROOF. Consider the stabilizer

$$G_{\lambda^\sigma(t)\mathbf{z}} = \lambda^\sigma(t)^{-1} G_{\mathbf{z}} \lambda^\sigma(t).$$

Recall that

$$G_{\mathbf{z}} = \left\{ \left(\begin{array}{cccccc} \alpha_1 & \alpha_2 & \alpha_3 & \cdots & \alpha_k \\ 0 & \alpha_1^2 & 2\alpha_1\alpha_2 & \cdots & 2\alpha_1\alpha_{n-1} + \cdots \\ 0 & 0 & \alpha_1^3 & \cdots & 3\alpha_1^2\alpha_{k-2} + \cdots \\ 0 & 0 & 0 & \cdots & \cdot \\ \cdot & \cdot & \cdot & \cdots & \alpha_1^d \end{array} \right) \right\}$$

where the polynomial in the (i,j) entry is

$$p_{i,j}(\alpha) = \sum_{a_1+a_2+\cdots+a_i=j} \alpha_{a_1}\alpha_{a_2}\cdots\alpha_{a_i}.$$

Therefore, the (i,j) entry of the stabilizer of $\lambda^s(t)\mathbf{z}$ is

$$(G_{\lambda^\sigma(t)\mathbf{z}})_{i,j} = t^{\lambda_i^\sigma - \lambda_j^\sigma} p_{i,j}(\alpha) \tag{33}$$

If ε is small enough then $\lambda_1^\sigma < \lambda_2^\sigma < \cdots < \lambda_k^\sigma$, and we define the positive number

$$n_i^\sigma = \max_{1 \le j \le n-i+1} (\lambda_{j+i-1}^\sigma - \lambda_j^\sigma), \quad i = 1, \ldots, k. \tag{34}$$

Note that by definition $n_1^\sigma = 0$ for all σ.

LEMMA 6.16. *Under the substitution*

$$\beta_i^\sigma = t^{-n_i^\sigma} \alpha_i^\sigma$$

we have

$$G_{\lambda^\sigma(t)\mathbf{z}}(\beta_1, \ldots, \beta_k) \in GL(\mathbb{C}[\beta_1, \ldots, \beta_k][t]),$$

so the entries are polynomials in t with coefficients in $\mathbb{C}[\beta_1, \ldots, \beta_k]$.

PROOF. Compute the substitution as follows:

$$(G_{\lambda^\sigma(t)\mathbf{z}})_{i,j} = t^{\lambda_i^\sigma - \lambda_j^\sigma} \sum_{a_1+a_2+\cdots+a_i=j} \alpha_{a_1}\alpha_{a_2}\ldots\alpha_{a_i} = \tag{35}$$

$$= \sum_{a_1+\cdots a_i=j} t^{\lambda_i^\sigma-\lambda_j^\sigma} t^{n_{a_1}^\sigma+n_{a_2}^\sigma+\cdots+n_{a_i}^\sigma}\beta_{a_1}\beta_{a_2}\ldots\beta_{a_i}. \tag{36}$$

By definition

$$n_{a_1}^\sigma \geq \lambda_{i+a_1-1}^\sigma - \lambda_i^\sigma;\ n_{a_2}^\sigma \geq \lambda_{i+a_1+a_2-2}^\sigma - \lambda_{i+a_1-1}^\sigma;\ \ldots\ ;$$
$$n_{a_j}^\sigma \geq \lambda_{i+a_1+\cdots+a_i-i}^\sigma - \lambda_{i+a_1+\cdots+a_{i-1}-(i-1)}^\sigma.$$

Adding up these inequalites and using $a_1+\cdots+a_i=j$ we get an alternating sum on the left cancelling up to

$$n_{a_1}^\sigma + \cdots + n_{a_i}^\sigma \geq \lambda_j^\sigma - \lambda_i^\sigma.$$

Substituting this into (35) we get

$$(G_{\lambda^\sigma(t)\mathbf{z}})_{i,j} = \sum_{a_1+\cdots a_i=j} t^{\lambda_i^\sigma-\lambda_j^\sigma} t^{n_{a_1}^\sigma+n_{a_2}^\sigma+\cdots+n_{a_i}^\sigma}\beta_{a_1}\beta_{a_2}\ldots\beta_{a_i} \in \mathbb{C}[\beta_1,\ldots,\beta_k][t]. \tag{37}$$

This proves Lemma 6.16. □

As a corollary we get the existence of

$$G^\sigma = \lim_{t\to 0} G_{\lambda^\sigma(t)\mathbf{z}}(\beta_1,\ldots,\beta_k) \in GL(\mathbb{C}[\beta_1,\ldots,\beta_k]).$$

To prove that $\dim G^\sigma = k$ and complete the proof of Proposition 6.15, for $1 \leq i \leq k$ choose $\theta(i)$ such that

$$n_i^\sigma = \lambda_{\theta(i)+i-1} - \lambda_{\theta(i)} \tag{38}$$

holds. Then

$$p_{\theta(i),\theta(i)+i-1}(\beta_1,\ldots,\beta_k) = \sum_{a_1+\cdots+a_{\theta(i)}=\theta(i)+i-1} t^{n_{a_1}^\sigma+\cdots+n_{a_{\theta(i)}}^\sigma}\beta_{a_1}\ldots\beta_{a_{\theta(i)}} \tag{39}$$

so

$$(G^\sigma)_{\theta(i),\theta(i)+i-1} = \lim_{t\to 0} t^{-n_i^\sigma} p_{\theta(i),\theta(i)+i-1}(\beta_1,\ldots,\beta_k) = \lim_{t\to 0}(t^{n_i^\sigma}\beta_1^{\theta(i)-1}\beta_i + \cdots)$$

$$= \beta_1^{\theta(i)-1}\beta_i + q_{\theta(i),\theta(i)+i-1} \tag{40}$$

where

$$q_{\theta(i),\theta(i)+i-1} \in \mathbb{C}[\beta_1,\ldots,\beta_k][t].$$

It follows that the elements $\frac{d}{dt}A^\sigma(t(e_1+e_i)1) \in \mathrm{Lie}(G^\sigma)$ are independent, where $t(e_1+e_i) = (t,0,\ldots,0,t,0,\ldots,0)$ with the t's are in the 1st and ith position if $i>1$ but interpreted as $(2t,0,\ldots,0)$ if $i=1$. This completes the proof of Proposition 6.15. □

In order to prove Lemma 6.5, it now suffices to find an extra one-dimensional unipotent subgroup of the stabiliser $G_{\mathbf{z}_{\lambda^\sigma}}$ of $\mathbf{z}_{\lambda^\sigma}$ for each σ when $\mathbf{z}_{\lambda^\sigma} = [p_k^\infty]$, since we know that p_k^∞ is fixed by a one-parameter subgroup of the maximal torus T_k of $\mathrm{SL}(k)$ and by a $k-1$-dimensional unipotent subgroup of $G^\sigma = \lim_{\theta \to 0} G_{\lambda^\sigma(t)\mathbf{z}}$ which is contained in the standard maximal unipotent subgroup U_k of $\mathrm{SL}(k)$. It turns out that we have to distinguish three cases here.

Case 1: $\sigma = k$.

PROOF. Let $T_\zeta \in GL(k)$ denote the transformation

$$T_\zeta(e_i) = e_i \text{ for } i \neq k-1; \quad T_\zeta(e_{k-1}) = e_{k-1} + \zeta e_k \text{ for } \zeta \in \mathbb{C}.$$

Since e_{k-1} does not occur just in $\mathbf{z}_{\lambda^\sigma}[k-1]$, T_ζ stabilises p_k^∞. This gives us a subgroup of $\mathrm{SL}(k)$ of dimension at least $k+1$ which stabilises p_k^∞, because T_ζ is unipotent but not upper triangular if $\zeta \neq 0$. □

Case 2: $\sigma < k$ and $k \neq -1 \bmod \sigma$.

PROOF. Let T be the transformation

$$T(e_i) = e_i \quad \text{for } i \neq k; \quad T(e_k) = e_k + \zeta e_\sigma. \tag{41}$$

Since e_k occurs only in $\mathbf{z}_{\lambda^\sigma}[k]$, and $\mathbf{z}_{\lambda^\sigma}[\sigma] = \sigma$, we have

$$\begin{aligned} T \cdot \mathbf{z}_{\lambda^\sigma} &= \mathbf{z}_{\lambda^\sigma}(e_1, \dots, e_{k-1}, e_k + \zeta e_\sigma) \\ &= \mathbf{z}_{\lambda^\sigma}[1] \wedge \cdots \wedge \mathbf{z}_{\lambda^\sigma}[\sigma-1] \wedge e_\sigma \wedge \mathbf{z}_{\lambda^\sigma}[\sigma+1] \wedge \cdots \wedge \mathbf{z}_{\lambda^\sigma}[k]) \\ &\quad + \zeta \cdot \mathbf{z}_{\lambda^\sigma}[1] \wedge \cdots \wedge \mathbf{z}_{\lambda^\sigma}[\sigma-1] \\ &\quad \wedge e_\sigma \wedge \mathbf{z}_{\lambda^\sigma}[\sigma+1] \wedge \cdots \wedge \mathbf{z}_{\lambda^\sigma}[k-1] \wedge e_\sigma = \mathbf{z}_{\lambda^\sigma}, \end{aligned} \tag{42}$$

so $T \in G_{\mathbf{z}_{\lambda^\sigma}}$.

It is slightly harder task to show that $T \notin G^\sigma = \lim_{\theta \to 0} G_{\lambda^\sigma(t)\mathbf{z}}$. First, we compute n_i for $i = k - \sigma$. We claim that for $k \neq -1 \bmod \sigma$

$$n_{k-\sigma+1} = \lambda_k^\sigma - \lambda_\sigma^\sigma = \lambda_{k-\sigma+1}^\sigma - \lambda_1^\sigma. \tag{43}$$

Indeed,

$$\lambda_{j+k-\sigma-1} - \lambda_j = \cdots \leq \lambda_k^\sigma - \lambda_\sigma^\sigma = \lambda_{k-\sigma+1}^\sigma - \lambda_1^\sigma$$

This means that we can choose $\theta(k-\sigma+1) = \sigma$ in (38) and substitute into (40)

$$(G^\sigma)_{\sigma,k} = \beta_1^{\sigma-1}\beta_{k-\sigma+1} + q_{\sigma,k}(\beta_1, \dots, \beta_k), \tag{44}$$

where $q_{\sigma,k}(\beta_1, \dots, \beta_k)$ is a polynomial, whose monomials $\beta_{i_1}^{b_1} \dots \beta_{i_\sigma}^{b_\sigma}$ satisfy

$$i_1 b_1 + \cdots + i_\sigma b_\sigma = k. \tag{45}$$

Moreover, we can also choose $\theta(k-\sigma+1) = 1$, by (43), and then (40) gives us

$$(G^\sigma)_{1,k-\sigma+1} = \beta_{k-\sigma+1}. \tag{46}$$

Suppose now that $T \in G^\sigma$, that is

$$T = G^\sigma(\beta_1, \dots, \beta_k) \quad \text{for some } \beta_1 \in \mathbb{C}^*, \beta_2, \dots, b_k \in \mathbb{C}. \tag{47}$$

Let $(T)_{i,j}$ denote the (i,j) entry of T. Then

$$(T)_{\sigma,k} = \zeta, \quad (T)_{i,j} = 0 \quad \text{for } i \neq j, \ (T)_{i,i} = 1.$$

Comparing the $(1,1)$ and $(1, k-\sigma+1)$ entries of T and G^σ we get

$$\beta_1 = 1, \quad \beta_{\delta-\sigma+1} = 0. \tag{48}$$

Choose $\theta(i)$ for $i = 2, \dots, k$ as in (38) and let $\theta(k-\sigma+1) = \sigma$. Since all off-diagonal entries of T but the (σ, k) are zero, (47) forces the following equations

$$\beta_i + q_{\theta(i),\theta(i)+i-1} = 0 \quad \text{for } i \neq k - \sigma + 1, \tag{49}$$

$$\beta_{k-\sigma+1} + q_{\sigma,k} = \zeta. \tag{50}$$

By (48), these are $k-1$ polynomial equations in $k-2$ variables, and the Jacobian at 0 is the origin, so we have finitely many solutions near the origin. Therefore, for some ζ, it follows that T is not in G^σ. □

Case 3: $\sigma < k$ and $d = -1 \bmod \sigma$.

PROOF. This case works very similarly to the previous one. Suppose $k - 1 > \sigma$, that is, if $k = c\sigma - 1$ where $c \geq 2$ (this holds because $k \geq \sigma$), the condition is that $c\sigma - 2 > \sigma$, which is true for all $k \geq 4$.

Let T be the transformation

$$T(e_i) = e_i \text{ for } i \neq k, k-1 \ ; \ T(e_{k-1}) = e_{k-1} + \zeta e_\sigma \ ; \ T(e_k) = e_k + \zeta e_\sigma \tag{51}$$

First we check again that $T \in G_{\mathbf{z}_{\lambda^\sigma}}$. We have

$$\begin{aligned} \mathbf{z}_{\lambda^\sigma}[\sigma] &= e_\sigma \ ; \\ \mathbf{z}_{\lambda^\sigma}[\sigma+1] &= e_{\sigma+1} + e_1 e_\sigma \ ; \\ \mathbf{z}_{\lambda^\sigma}[k] &= e_k + \sum_{i=1}^{k-1} e_i e_{k-i} \ . \end{aligned}$$

An easy computation shows that

$$\begin{aligned} T \cdot \mathbf{z}_{\lambda^\sigma} &= \mathbf{z}_{\lambda^\sigma}(e_1, \dots, e_{k-2}, e_{k-1} + \zeta e_\sigma, e_k + \zeta e_{\sigma+1}) \\ &= \mathbf{z}_{\lambda^\sigma}[1] \wedge \dots \wedge \mathbf{z}_{\lambda^\sigma}[k-2] \wedge (\mathbf{z}_{\lambda^\sigma}[k-1] \\ &\quad + \zeta \mathbf{z}_{\lambda^\sigma}[\sigma]) \wedge (\mathbf{z}_{\lambda^\sigma}[k] + \zeta \mathbf{z}_{\lambda^\sigma}[\sigma+1] \\ &= \mathbf{z}_{\lambda^\sigma}[1] \wedge \dots \wedge \mathbf{z}_{\lambda^\sigma}[k] = \mathbf{z}_{\lambda^\sigma}. \end{aligned} \tag{52}$$

Now we prove that $T \notin G^\sigma$ in a similar way to the second case above. Since $k - 1 \neq -1 \bmod \sigma$ we can substitute $k-1$ instead of k in (43):

$$n_{k-\sigma} = \lambda^\sigma_{k-1} - \lambda^\sigma_\sigma = \lambda^\sigma_{k-\sigma} - \lambda^\sigma_1. \tag{53}$$

Moreover, we also get the extra equation

$$n_{k-\sigma} = \lambda_k^\sigma - \lambda_{\sigma+1}^\sigma, \tag{54}$$

and similarly to (44) and (46) it follows that

$$(G^\sigma)_{\sigma,k-1} = \beta_1^{\sigma-1}\beta_{k-\sigma} + q_{\sigma,k-1}(\beta_1,\dots,\beta_k); \tag{55}$$

$$(G^\sigma)_{\sigma+1,k} = \beta_1^{\sigma}\beta_{k-\sigma} + q_{\sigma+1,k}(\beta_1,\dots,\beta_k); \tag{56}$$

$$(G^\sigma)_{1,k-\sigma} = \beta_{k-\sigma}. \tag{57}$$

Since T differs from the identity matrix only by the entries

$$(T)_{\sigma,k-1} = (T)_{\sigma+1,k} = \zeta,$$

the equality

$$T = G^\sigma(\beta_1,\dots,\beta_k)$$

forces $\beta_{k-\sigma} = 0, \beta_1 = 1$ and the analogue of (49) ,(50):

$$\beta_i + q_{\theta(i),\theta(i)+i-1} = 0 \quad \text{for } i \neq k-\sigma \tag{58}$$

$$\beta_{k-\sigma} + q_{\sigma,k-1} = \zeta \tag{59}$$

$$\beta_{k-\sigma} + q_{\sigma+1,k} = \zeta \tag{60}$$

which are, again, $k+1$ nondegenerate polynomial equations in $k-1$ variables, such that for some ζ there is no solution. □

We have now proved Lemma 6.5, which together with Lemma 6.6 completes the proof of Theorem 6.2.

7. Geometric description of Demailly-Semple invariants

As an immediate consequence of Corollary 6.3, we can now prove Theorem 3.3 in the case when $p = 1$.

THEOREM 7.1. *If $k \geq 2$ then $\mathbb{G}_k' = \mathbb{U}_k$ is a Grosshans subgroup of the special linear group $SL(k)$, so that $\mathcal{O}(\mathrm{SL}(k)^{\mathbb{U}_k})^{\mathrm{SL}(k)}$ is a finitely generated complex algebra and moreover every linear action of $\mathbb{U}_k$ or $\mathbb{G}_k$ on an affine or projective variety Y (with respect to an ample linearisation) which extends to a linear action of $GL(k)$ has finitely generated invariants.*

In particular we have the special case of Theorem 3.2 when $p = 1$.

THEOREM 7.2. *The fibre $\mathcal{O}((J_k)_x)^{\mathbb{U}_k}$ of the bundle E_k^n is a finitely generated graded complex algebra.*

PROOF. We have

$$\mathcal{O}((J_k)_x)^{\mathbb{U}_k} \cong (\mathcal{O}((J_k)_x) \otimes \mathcal{O}(\mathrm{SL}(k)^{\mathbb{U}_k})^{\mathrm{SL}(k)}$$

which is finitely generated because $\mathcal{O}(\mathrm{SL}(k)^{\mathbb{U}_k})^{\mathrm{SL}(k)}$ is finitely generated and $\mathrm{SL}(k)$ is reductive. □

Theorem 6.2 also allows us to describe the algebra $\mathcal{O}(SL(k))^{\mathbb{U}_k}$. In §6 we constructed an embedding of $SL(k)/\mathbb{U}_k$ in the affine space $\wedge^k(\mathrm{Sym}^{\leq k}\mathbb{C}^k)\otimes(\mathbb{C}^k)^{\otimes K}$ for suitable large K, and in Theorem 6.2 we proved that the boundary components of the closure $\overline{\mathrm{SL}(k)(p_k\otimes e_1^{\otimes K})}$ of its image have codimension at least two. Thus we obtain the following corollary of Theorem 6.2:

THEOREM 7.3. *(i) If $k\geq 4$ then the canonical affine completion*

$$\mathrm{SL}(k)//\mathbb{U}_k=\mathrm{Spec}(\mathcal{O}(\mathrm{SL}(k))^{\mathbb{U}_k})$$

of $\mathrm{SL}(k)/\mathbb{U}_k$ is isomorphic to the closure $\overline{\mathrm{SL}(k)(p_k\otimes e_1^{\otimes K})}$ of the orbit $\mathrm{SL}(k)(p_k\otimes e_1^{\otimes K})\cong\mathrm{SL}(k)/\mathbb{U}_k$ of $p_k\otimes e_1^{\otimes K}$ in $\wedge^k(\mathrm{Sym}^{\leq k}\mathbb{C}^k)\otimes(\mathbb{C}^k)^{\otimes K}$ where $K=M(1+2+\cdots+k)+1$ for any strictly positive integer M;

(ii) The algebra

$$\mathcal{O}(SL(k))^{\mathbb{U}_k}$$

is generated by the Plücker coordinates on $\mathbb{P}(\wedge^k(\mathrm{Sym}^{\leq k}\mathbb{C}^k))$, which can be expressed as

$$\{\Delta_{\mathbf{i}_1,\ldots,\mathbf{i}_s}:s\leq k\},$$

where $\mathbf{i}_j$ denotes a multi-index corresponding to basis elements of $\mathrm{Sym}^{\leq k}(\mathbb{C}^k)$, and $\Delta_{\mathbf{i}_1,\ldots,\mathbf{i}_s}$ is the corresponding minor of $\phi(f'\ldots,f^{(k)})\in\mathrm{Hom}(\mathbb{C}^k,\mathrm{Sym}^{\leq k}(\mathbb{C}^k))$, together with the coordinates $f_1',\ldots,f_k'$ of f'.

It follows immediately from this theorem that the non-reductive GIT quotient

$$(J_k)_x//\mathbb{U}_k=\mathrm{Spec}(\mathcal{O}((J_k)_x)^{\mathbb{U}_k})$$

is isomorphic to the reductive GIT quotient

$$((J_k)_x\times\overline{\mathrm{SL}(k)(p_k\otimes e_1^{\otimes K})})//\mathrm{SL}(k).$$

This can be identified with the quotient of the open subset $((J_k)_x\times\overline{\mathrm{SL}(k)(p_k\otimes e_1^{\otimes K})})^{ss}$ of $\mathrm{SL}(k)$-semistable points of $(J_k)_x\times\overline{\mathrm{SL}(k)(p_k\otimes e_1^{\otimes K})}$ by the equivalence relation $\sim$ such that $y\sim z$ if and only if the closures of the $\mathrm{SL}(k)$-orbits of y and z intersect in $((J_k)_x\times\overline{\mathrm{SL}(k)(p_k\otimes e_1^{\otimes K})})^{ss}$. Equivalently it can be identified with the closed $\mathrm{SL}(k)$-orbits in $((J_k)_x\times\overline{\mathrm{SL}(k)(p_k\otimes e_1^{\otimes K})})^{ss}$. Since $\overline{\mathrm{SL}(k)(p_k\otimes e_1^{\otimes K})}$ is the union of finitely many $\mathrm{SL}(k)$-orbits, with stabilisers $H_1=\mathbb{U}_k,H_2,\ldots,H_s$, say, we can stratify $(J_k)_x//\mathbb{U}_k$ so that the stratum corresponding to H_j is identified with the H_j-orbits in $(J_k)_x$ such that the corresponding $\mathrm{SL}(k)$-orbit in $(J_k)_x\times\overline{\mathrm{SL}(k)(p_k\otimes e_1^{\otimes K})}$ is semistable and closed in $((J_k)_x\times\overline{\mathrm{SL}(k)(p_k\otimes e_1^{\otimes K})})^{ss}$.

EXAMPLE 7.4. *When $k=2$ we have*

$$J_2^{\mathrm{reg}}(1,2)=\{(f_1',f_2',f_1'',f_2'')\in(\mathbb{C}^2)^2;(f_1',f_2')\neq(0,0)\},$$

and fixing a basis $\{e_1, e_2\}$ *of* $\mathbb{C}^2$ *and the induced basis* $\{e_1, e_2, e_1^2, e_1e_2, e_2^2\}$ *of* $\mathbb{C}^2 \oplus \mathrm{Sym}^2\mathbb{C}^2$, *the map* $\phi : J_2(1,2) = \mathrm{Hom}(\mathbb{C}^2, \mathbb{C}^2) \to \mathrm{Hom}(\mathbb{C}^2, \mathrm{Sym}^{\leq 2}\mathbb{C}^2)$ *of* (14) *is given by*

$$(f_1', f_2', f_1'', f_2'') \mapsto \begin{pmatrix} f_1' & f_2' & 0 & 0 & 0 \\ \frac{1}{2!}f_1'' & \frac{1}{2!}f_2'' & (f_1')^2 & f_1'f_2' & (f_2')^2 \end{pmatrix}.$$

The 2×2 *minors of this* 2×5 *matrix are* $(f_1')^3$, $(f_1')^2 f_2'$, $f_1'(f_2')^2, (f_2')^3$ *and*

$$\Delta_{[1,2]} = f_1' f_2'' - f_1'' f_2'.$$

On $SL(2)$ *we have* $\Delta_{[1,2]} = 1$ *and the algebra of invariants* $\mathcal{O}(SL(2))^{\mathbb{U}_2}$ *is generated by* f_1' *and* f_2', *as expected since* $\mathrm{SL}(2)/\mathbb{U}_2 \cong \mathbb{C}^2 \setminus \{0\}$ *and its canonical affine completion* $\mathrm{SL}(2)/\!/\mathbb{U}_2$ *is* $\mathbb{C}^2$.

EXAMPLE 7.5. *When* $k = 3$ *the finite generation of the Demailly-Semple algebra* $\mathcal{O}((J_k)_x)^{\mathbb{U}_k}$ *was proved by Rousseau in* [**27**]. *We have*

$$\begin{aligned} J_3^{\mathrm{reg}}(1,3) = \{&(f_1', f_2', f_3', f_1'', f_2'', f_3'', f_1''', f_2''', f_3''') \\ &\in (\mathbb{C}^3)^3; (f_1', f_2', f_3') \neq (0,0,0)\}, \end{aligned}$$

and if we fix a basis $\{e_1, e_2, e_3\}$ *of* $\mathbb{C}^3$ *and the induced basis*

$$\{e_1, e_2, e_3, e_1^2, e_1e_2, e_2^2, e_1e_3, e_2e_3, e_3^2, e_1^3, e_1^2e_2, \ldots, e_3^3\}$$

of $\mathbb{C}^3 \oplus \mathrm{Sym}^2\mathbb{C}^3 \oplus \mathrm{Sym}^3\mathbb{C}^3$, *the map* $\phi : \mathrm{Hom}(\mathbb{C}^3, \mathbb{C}^3) \to \mathrm{Hom}(\mathbb{C}^3, \mathrm{Sym}^{\leq 3}\mathbb{C}^3)$ *in* (14) *sends*

$$(f_1', f_2', f_3', f_1'', f_2'', f_3'', f_1''', f_2''', f_3''')$$

to a 3×19 *matrix, whose first* 9 *columns (corresponding to* $\mathrm{Sym}^{\leq 2}\mathbb{C}^3$*) are*

$$\begin{pmatrix} f_1' & f_2' & f_3' & 0 & 0 & 0 & 0 & 0 & 0 \\ \frac{1}{2!}f_1'' & \frac{1}{2!}f_2'' & \frac{1}{2!}f_3'' & (f_1')^2 & f_1'f_2' & (f_2')^2 & f_1'f_3' & f_2'f_3' & (f_3')^2 \\ \frac{1}{3!}f_1''' & \frac{1}{3!}f_2''' & \frac{1}{3!}f_3''' & f_1'f_1'' & f_1'f_2'' + f_1''f_2' & f_2'f_2'' & f_1'f_3'' + f_3'f_1'' & f_2'f_3'' + f_2''f_3' & f_3'f_3'' \end{pmatrix},$$

and the remaining 10 *columns (corresponding to* $\mathrm{Sym}^3\mathbb{C}^3$*) are*

$$\begin{pmatrix} 0 & 0 & 0 & 0 & 0 & 0 & 0 & 0 & 0 & 0 \\ 0 & 0 & 0 & 0 & 0 & 0 & 0 & 0 & 0 & 0 \\ (f_1')^3 & (f_1')^2f_2' & f_1'(f_2')^2 & (f_2')^3 & f_1'(f_3')^2 & (f_1')^2f_3' & (f_2')^2f_3' & f_2'(f_3')^2 & (f_3')^3 & f_1'f_2'f_3' \end{pmatrix}.$$

The 3×3 *minors of this matrix together with* f_1', f_2', f_3' *generate the algebra of invariants* $\mathcal{O}(SL(3))^{\mathbb{U}_3}$.

8. Generalized Demailly-Semple jet bundles

The aim of this section is to extend the earlier constructions for $p=1$ to generalized Demailly-Semple invariant jet differentials when $p>1$.

Let X be a compact, complex manifold of dimension n. We fix a parameter $1\le p\le n$, and study the maps $\mathbb{C}^p\to X$. Recall that as before we fix the degree k of the map, and introduce the bundle $J_{k,p}\to X$ of k-jets of maps $\mathbb{C}^p\to X$, so that the fibre over $x\in X$ is the set of equivalence classes of germs of holomorphic maps $f:(\mathbb{C}^p,0)\to(X,x)$, with the equivalence relation $f\sim g$ if and only if all derivatives $f^{(j)}(0)=g^{(j)}(0)$ are equal for $0\le j\le k$. Recall also that $\mathbb{G}_{k,p}$ is the group of k-jets of germs of biholomorphisms of $(\mathbb{C}^p,0)$, which has a natural fibrewise right action on $J_{k,p}$ with the matrix representation given by

$$G_{k,p}=\begin{pmatrix}\Phi_1 & \Phi_2 & \Phi_3 & \dots & \Phi_k\\ 0 & \Phi_1^2 & \Phi_1\Phi_2 & \dots & \\ 0 & 0 & \Phi_1^3 & \dots & \\ \cdot & \cdot & \cdot & \cdot & \cdot\\ & & & & \Phi_1^k\end{pmatrix}, \tag{61}$$

for $G_{k,p}\in\mathbb{G}_{p,k}$ where $\Phi_i\in\operatorname{Hom}(\operatorname{Sym}^i\mathbb{C}^p,\mathbb{C}^p)$ and $\det\Phi_1\neq0$. Recall also that $\mathbb{G}_{k,p}$ is generated along its first p rows, in the sense that the parameters in the first p rows are independent, and all the remaining entries are polynomials in these parameters. The parameters in the $(1,m)$ block are indexed by a basis of $\operatorname{Sym}^m(\mathbb{C}^p)\times\mathbb{C}^p$, so they are of the form α_ν^l where $\nu\in\binom{p+m-1}{m-1}$ is an m-tuple and $1\le l\le p$, and the polynomial in the (l,m) block and entry indexed by $\tau=(\tau[1],\dots,\tau[l])\in\binom{p+l-1}{l-1}$ and $\nu\in\binom{p+m-1}{m-1}$ is given by

$$(G_{k,p})_{\tau,\nu}=\sum_{\nu_1+\dots+\nu_l=\nu}\alpha_{\nu_1}^{\tau[1]}\alpha_{\nu_2}^{\tau[2]}\dots\alpha_{\nu_l}^{\tau[l]}. \tag{62}$$

Recall also that $\mathbb{G}_{k,p}=\mathbb{U}_{k,p}\rtimes\mathrm{GL}(p)$ is an extension of its unipotent radical $\mathbb{U}_{k,p}$ by $\mathrm{GL}(p)$, and that the generalized Demailly-Semple jet bundle $E_{k,p,m}\to X$ of invariant jet differentials of order k and weighted degree $(m,\dots,m)$ consists of the jet differentials which transform under any reparametrization $\phi\in\mathbb{G}_{k,p}$ of $(\mathbb{C}^p,0)$ as

$$Q(f\circ\phi)=(J_\phi)^mQ(f)\circ\phi,$$

where $J_\phi=\det\Phi_1$ denotes the Jacobian of ϕ, so that $E_{k,p}=\oplus_{m\ge0}E_{k,p,m}$ is the graded algebra of $\mathbb{G}'_{k,p}$-invariants where $\mathbb{G}'_{k,p}=\mathbb{U}_{k,p}\rtimes\mathrm{SL}(p)$.

8.1. Geometric description for $p>1$. As in the case when $p=1$ our goal is to prove that $\mathbb{G}'_{k,p}$ is a Grosshans subgroup of $\mathrm{SL}(\mathrm{sym}^{\le k}(p))$ where $\mathrm{sym}^{\le k}(p)=\sum_{i=1}^k\dim\operatorname{Sym}^i\mathbb{C}^p$ by finding a suitable embedding of the quotient $\mathrm{SL}(\mathrm{sym}^{\le k}(p))/\mathbb{G}'_{k,p}$.

REMARK 8.1. In [**25**] Pacienza and Rousseau generalize the inductive process given in [**5**] of constructing a smooth compactification of the Demailly-Semple jet bundles. Using the concept of a directed manifold, they define a bundle $X_{k,p} \to X$ with smooth fibres, and the effective locus $Z_{k,p} \subset X_{k,p}$, and a holomorphic embedding $J_{k,p}^{reg}/\mathbb{G}_{k,p} \hookrightarrow Z_{k,p}$ which identifies $J_{k,p}^{reg}/\mathbb{G}_{k,p}$ with $Z_{k,p}^{reg} = X_{k,p}^{reg} \cap Z_{k,p}$, so that $Z_{k,p}$ is a relative compactification of $J_{k,p}/\mathbb{G}_{k,p}$. We choose a different approach, generalizing the test curve model, resulting in a holomorphic embedding of $J_{k,p}/\mathbb{G}_{k,p}$ into a partial flag manifold and a different compactification, which is a singular subvariety of the partial flag manifold, such that the invariant jet differentials of degree divisible by $\mathrm{sym}^{\leq k}(p)$ are given by polynomial expressions in the Plücker coordinates.

Fix $x \in X$ and an identification of T_xX with $\mathbb{C}^n$; then let $J_k(p,n) = J_{k,p,x}$ as defined in §2. Let

$$J_k^{\mathrm{reg}}(p,n) = \{\gamma \in J_k(p,n) : \Gamma_1 \text{ is non-degenerate}\}$$

where γ is represented by

$$\mathbf{u} \mapsto \gamma(\mathbf{u}) = \Gamma_1\mathbf{u} + \Gamma_2\mathbf{u}^2 + \cdots + \Gamma_k\mathbf{u}^k$$

with $\Gamma_i \in \mathrm{Hom}(\mathrm{Sym}^i\mathbb{C}^p, \mathbb{C}^p)$. Let $N \geq n$ be any integer and define

$$\Upsilon_{k,p} = \left\{\Psi \in J_k(n,N) : \exists\gamma \in J_k^{\mathrm{reg}}(p,n) : \Psi \circ \gamma = 0\right\}.$$

REMARK 8.2. The global singularity theory description of $\Upsilon_{k,p}$ is

$$\begin{aligned}\Upsilon_{k,p} \doteq \{p = (p_1, \ldots, p_N) \in J_k(n,N) : \mathbb{C}[z_1, \ldots, z_n]/\langle p_1, \ldots, p_N\rangle \\ \cong \mathbb{C}[x,y]/\langle z_1, \ldots, z_n\rangle^{k+1}\}.\end{aligned}$$

Note, again, as in the $p = 1$ case, that if $\gamma \in J_k^{\mathrm{reg}}(p,n)$ is a test surface of $\Psi \in \Upsilon_{k,p}$, and $\varphi \in \mathbb{G}_k$ is a holomorphic reparametrization of $\mathbb{C}^p$, then $\gamma \circ \varphi$ is, again, a test surface of Ψ:

$$\mathbb{C}^p \xrightarrow{\ \varphi\ } \mathbb{C}^p \xrightarrow{\ \gamma\ } \mathbb{C}^n \xrightarrow{\ \Psi\ } \mathbb{C}^N \tag{63}$$

$$\Psi \circ \gamma = 0 \ \Rightarrow \ \Psi \circ (\gamma \circ \varphi) = 0$$

EXAMPLE 8.3. *Let $k = 2, p = 2$ and let $\Psi(\mathbf{z}) = \Psi'\mathbf{z} + \Psi''\mathbf{z}^2$ for $\mathbf{z} \in \mathbb{C}^n$, and*

$$\gamma(u_1,u_2) = \gamma_{10}u_1 + \gamma_{01}u_2 + \gamma_{20}u_1^2 + \gamma_{11}u_1u_2 + \gamma_{02}u_2^2, \quad \gamma_{ij} \in \mathbb{C}^n.$$

Then $\Psi \circ \gamma = 0$ has the form

$$\begin{aligned}&\Psi'(\gamma_{10}) = 0; \quad \Psi'(\gamma_{01}) = 0 \\ &\Psi'(\gamma_{20}) + \Psi''(\gamma_{10},\gamma_{10}) = 0,; \quad \Psi'(\gamma_{11}) + 2\Psi''(\gamma_{10},\gamma_{01}) = 0,; \\ &\Psi'(\gamma_{01}) + \Psi''(\gamma_{01},\gamma_{01}) = 0,\end{aligned} \tag{64}$$

We introduce

$$\mathcal{S}_\gamma = \{\Psi \in J_k(n,N) : \Psi \circ \gamma = 0\}$$

and the following analogue of $J_k^o(1,n)$:

$$J_k^o(n,N) = \{\Psi \in J_k(n,N) : \dim\ker\Psi = p\}.$$

The proof of the following proposition is analogous to that of Proposition 4.7 in [**2**], and we omit the details. We use the notation

$$\mathrm{sym}^i(p) = \dim(\mathrm{Sym}^i\mathbb{C}^p);$$

$$\mathrm{sym}^{\leq k}(p) = \dim(\mathbb{C}^p \oplus \mathrm{Sym}^2\mathbb{C}^p \oplus \cdots \oplus \mathrm{Sym}^k\mathbb{C}^p) = \sum_{i=1}^k \mathrm{sym}^i p.$$

PROPOSITION 8.4. *(i) If $\gamma \in J_k^{\mathrm{reg}}(p,n)$ then $\mathcal{S}_\gamma \subset J_k(n,N)$ is a linear subspace of codimension $N\mathrm{sym}^{\leq k}(p)$.*

(ii) For any $\gamma \in J_k^{\mathrm{reg}}(p,n)$, the subset $\mathcal{S}_\gamma \cap J_k^o(n,N)$ of $\mathcal{S}_\gamma$ is dense.

(iii) If $\Psi \in J_k^o(n,N)$, then Ψ belongs to at most one of the spaces $\mathcal{S}_\gamma$. More precisely, if $\gamma_1, \gamma_2 \in J_k^{\mathrm{reg}}(p,n)$, $\Psi \in J_k^o(n,N)$ and $\Psi\circ\gamma_1 = \Psi\circ\gamma_2 = 0$, then there exists $\varphi \in J_k^{\mathrm{reg}}(p,p)$ such that $\gamma_1 = \gamma_2 \circ \varphi$.

(iv) Given $\gamma_1, \gamma_2 \in J_k^{\mathrm{reg}}(1,n)$, we have $\mathcal{S}_{\gamma_1} = \mathcal{S}_{\gamma_2}$ if and only if there is some $\varphi \in J_k^{\mathrm{reg}}(1,1)$ such that $\gamma_1 = \gamma_2 \circ \varphi$.

With the notation

$$\Upsilon_{k,p} = \Upsilon_{k,p} \cap J_k^o(n,N),$$

we deduce from Proposition 8.4 the following

COROLLARY 8.5. *$\Upsilon^0_{k,p}$ is a dense subset of $\Upsilon_{k,p}$, and $\Upsilon^0_{k,p}$ has a fibration over the orbit space $J_k^{\mathrm{reg}}(p,n)/J_k^{\mathrm{reg}}(p,p) = J_k^{\mathrm{reg}}(p,n)/\mathbb{G}_{k,p}$ with linear fibres.*

REMARK 8.6. In fact, Proposition 8.4 says a bit more, namely that $\Upsilon^0_{k,p}$ is fibrewise dense in $\Upsilon_{k,p}$ over $J_k^{\mathrm{reg}}(p,n)/\mathbb{G}_{k,p}$, but we will not use this stronger statement.

By the first part of Proposition 8.4 the assignment $\gamma \to \mathcal{S}_\gamma$ defines a map

$$\nu : J_k^{\mathrm{reg}}(p,n) \to \mathrm{Grass}(kN, J_k(n,N))$$

which, by the fourth part, descends to the quotient

$$\bar{\nu} : J_k^{\mathrm{reg}}(p,n)/\mathbb{G}_{k,p} \hookrightarrow \mathrm{Grass}(kN, J_k(n,N)) \tag{65}$$

(cf. Proposition 4.4). Next, we want to rewrite this embedding in terms of the identifications introduced in §5. So we

- identify $J_k(p,n)$ with $\mathrm{Hom}\,(\mathbb{C}^{\mathrm{sym}^1 p} \oplus \cdots \oplus \mathbb{C}^{\mathrm{sym}^k p}, \mathbb{C}^n) = \mathrm{Hom}\,(\mathbb{C}^{\mathrm{sym}^{\leq k}(p)}, \mathbb{C}^n)$ where $\mathrm{sym}^j p = \dim\mathrm{Sym}^j\mathbb{C}^p$ and $\mathrm{sym}^{\leq k}(p) = \sum_{j=1}^k \mathrm{sym}^j p$;
- identify $J_k(n,1)^*$ with $\mathrm{Sym}^{\leq k}\mathbb{C}^n = \oplus_{l=1}^k \mathrm{Sym}^l\mathbb{C}^n$.

We think of an element v of $\mathrm{Hom}\,(\mathbb{C}^{\mathrm{sym}^{\leq k}(p)}, \mathbb{C}^n)$ as an $n \times \mathrm{sym}^{\leq k}(p)$ matrix, with column vectors in $\mathbb{C}^n$. These columns correspond to basis elements of $\mathbb{C}^{\mathrm{sym}^1 p} \oplus \cdots \oplus \mathbb{C}^{\mathrm{sym}^k p}$, and the columns in the ith component are indexed by i-tuples $1 \leq t_1 \leq t_2 \leq \cdots \leq t_i \leq p$, or equivalently by

$$(e_{t_1} + e_{t_2} + \cdots + e_{t_i}) \in \mathbb{Z}^p_{\geq 0}$$

where $e_j = (0, \ldots, 1, \ldots, 0)$ with 1 in the jth place, giving us

$$v = (v_{10,\ldots 0}, v_{01\ldots 0}, \ldots, v_{0\ldots 0k}) \in \mathrm{Hom}\,(\mathbb{C}^{\mathrm{sym}^{\leq k}(p)}, \mathbb{C}^n).$$

The elements of $J_k^{\mathrm{reg}}(p,n)$ correspond to matrices whose first p columns are linearly independent. When $n \geq \mathrm{sym}^{\leq k}(p)$ there is a smaller dense open subset $J_k^{\mathrm{nondeg}}(p,n) \subset J_k^{\mathrm{reg}}(p,n)$ consisting of the $n \times \mathrm{sym}^{\leq k}(p)$ matrices of rank $\mathrm{sym}^{\leq k}(p)$.

Define the following map, whose components correspond to the equations in (64):

$$(66) \qquad \phi : \mathrm{Hom}\,(\mathbb{C}^{\mathrm{sym}^{\leq k}(p)}, \mathbb{C}^n) \to \mathrm{Hom}\,(\mathbb{C}^{\mathrm{sym}^{\leq k}(p)}, \mathrm{Sym}^{\leq k}\mathbb{C}^n)$$
$$(v_{10,\ldots 0}, v_{01\ldots 0}, \ldots, v_{0\ldots 0k}) \mapsto (\ldots, \textstyle\sum_{\mathbf{s}_1+\mathbf{s}_2+\cdots+\mathbf{s}_j=\mathbf{s}} v_{\mathbf{s}_1} v_{\mathbf{s}_2} \cdots v_{\mathbf{s}_j}, \ldots),$$

where on the right hand side $\mathbf{s} \in \mathbb{Z}^p_{\geq 0}$.

EXAMPLE 8.7. *If $k = p = 2$ then ϕ is given by*

$$\phi(v_{10}, v_{01}, v_{20}, v_{11}, v_{02}) = (v_{10}, v_{01}, v_{20} + v_{10}^2, v_{11} + 2v_{10}v_{01}, v_{02} + v_{01}^2).$$

Let $P_{k,p} \subset \mathrm{GL}_{\mathrm{sym}^{\leq k}(p)}$ denote the standard parabolic subgroup with Levi subgroup

$$\mathrm{GL}(\mathrm{sym}^1 p) \times \cdots \times \mathrm{GL}(\mathrm{sym}^k p),$$

where $\mathrm{sym}^j p = \dim \mathrm{Sym}^j \mathbb{C}^p$ and $\mathrm{sym}^{\leq k}(p) = \sum_{j=1}^k \mathrm{sym}^j p$. Then (65) has the following reformulation, analogous to Proposition 5.1.

PROPOSITION 8.8. *The map ϕ in (66) is a $\mathbb{G}_{k,p}$-invariant algebraic morphism*

$$\phi : J_k^{\mathrm{reg}}(p,n) \to \mathrm{Hom}\,(\mathbb{C}^{\mathrm{sym}(p)}, \mathrm{Sym}^{\leq k}\mathbb{C}^n)$$

which induces an injective map ϕ^{Grass} on the $\mathbb{G}_{k,p}$-orbits:

$$\phi^{\mathrm{Grass}} : J_k^{\mathrm{reg}}(p,n)/\mathbb{G}_{k,p} \hookrightarrow \mathrm{Grass}_{\mathrm{sym}^{\leq k}(p)}(\mathrm{Sym}^{\leq k}\mathbb{C}^n)$$

and

$$\phi^{\mathrm{Flag}} : J_k^{\mathrm{reg}}(p,n)/\mathbb{G}_{k,p} \hookrightarrow \mathrm{Flag}_{\mathrm{sym}^1(p),\ldots,\mathrm{sym}^k(p)}(\mathrm{Sym}^{\leq k}\mathbb{C}^n)$$
$$\hookrightarrow \mathrm{Hom}\,(\mathbb{C}^{\mathrm{sym}(p)}, \mathrm{Sym}^{\leq k}\mathbb{C}^n)/P_{k,p}.$$

Composition with the Plücker embedding gives

$$\phi^{\mathrm{Proj}} = \mathrm{Pluck} \circ \phi^{\mathrm{Grass}} : J_k^{\mathrm{reg}}(p,n)/\mathbb{G}_{k,p} \hookrightarrow \mathbb{P}(\wedge^{\mathrm{sym}^{\leq k}(p)} \mathrm{Sym}^{\leq k}\mathbb{C}^n).$$

As in the case when $p=1$, we introduce the following notation

$$X_{n,k,p} = \phi^{\mathrm{Proj}}(J_k^{\mathrm{reg}}(p,n)),$$
$$Y_{n,k,p} = \phi^{\mathrm{Proj}}(J_k^{\mathrm{nondeg}}(p,n)) \subset \mathbb{P}(\wedge^{\mathrm{sym}^{\leq k}}(\mathrm{Sym}^{\leq k}\mathbb{C}^n)).$$

DEFINITION 8.9. *Let $n \geq \mathrm{sym}^{\leq k}(p) = \mathrm{sym}^1(p) + \cdots + \mathrm{sym}^k(p)$. Then the open subset of $\mathbb{P}(\wedge^{\mathrm{sym}^{\leq k}(p)}(\mathrm{Sym}^{\leq k}\mathbb{C}^n))$ where the projection to $\wedge^{\mathrm{sym}^{\leq k}(p)}\mathbb{C}^n$ is nonzero is denoted by $A_{n,k,p}$.*

Since ϕ^{Grass} and ϕ^{Proj} are $\mathrm{GL}(n)$-equivariant, and for $n \geq \mathrm{sym}^{\leq k}(p)$ the action of $\mathrm{GL}(n)$ is transitive on $\mathrm{Hom}^{\mathrm{nondeg}}(\mathbb{C}^{\mathrm{sym}^{\leq k}(p)}, \mathbb{C}^n)$, we have

LEMMA 8.10. (*i*) *If $n \geq \mathrm{sym}^{\leq k}(p)$ then $X_{n,k,p}$ is the $\mathrm{GL}(n)$ orbit of*
(67)

$$\mathbf{z} = \phi^{\mathrm{Proj}}(e_1, \ldots, e_{\mathrm{Sym}^{\leq k}(p)}) = \left[\wedge_{j_1+\cdots+j_p \leq k} \sum_{\mathbf{i}_1+\cdots+\mathbf{i}_s = (j_1,\ldots,j_p)} e_{\mathbf{i}_1} \cdots e_{\mathbf{i}_s} \right]$$

in $\mathbb{P}(\wedge^{\mathrm{sym}^{\leq k}(p)}(\mathrm{Sym}^{\leq k}\mathbb{C}^n))$.

(*ii*) *If $n \geq \mathrm{sym}^{\leq k}(p)$ then $X_{n,k,p}$ and $Y_{n,k,p}$ are finite unions of $\mathrm{GL}(n)$ orbits.*

(*iii*) *For $k > n$ the images $X_{n,k,p}$ and $Y_{n,k,p}$ are $\mathrm{GL}(n)$-invariant quasi-projective varieties, though they have no dense $\mathrm{GL}(n)$ orbit.*

Similar statements hold for the closure of the image in the Grassmannian

$$\mathrm{Grass}_{\mathrm{sym}^{\leq k}(p)}(\mathrm{Sym}^{\leq k}\mathbb{C}^n)$$

(or equivalently in the projective space $\mathbb{P}(\wedge^{\mathrm{sym}^{\leq} k(p)}(\mathrm{Sym}^{\leq k}\mathbb{C}^n))$).

LEMMA 8.11. *Let $n \geq \mathrm{sym}^{\leq k}(\mathbb{C}^n)$; then*

(*i*) *$A_{n,k,p}$ is invariant under the $GL(n)$ action on $\mathbb{P}(\wedge^{\mathrm{sym}^{\leq} k(p)}(\mathrm{Sym}^{\leq k}\mathbb{C}^n))$;*

(*ii*) *$X_{n,k,p} \subset A_{n,k,p}$, although $Y_{n,k,p} \not\subseteq A_{n,k,p}$;*

(*iii*) *$\overline{X}_{n,k,p}$ is the union of finitely many $GL(n)$-orbits.*

9. Affine embeddings of $\mathrm{SL}(\mathrm{sym}^{\leq k}p)/\mathbb{G}_{k,p}$

In this section we study the case when $n = \mathrm{sym}^{\leq k}p$ and so $\mathrm{GL}(n) \subset J_k^{\mathrm{reg}}(p,n)$. In the previous section we embedded $J_k^{\mathrm{reg}}(p,n)/\mathbb{G}_{k,p}$ in the affine space $A_{n,k,p} \subset \mathbb{P}(\wedge^n \mathrm{Sym}^{\leq k}\mathbb{C}^n)$, which can be restricted to $\mathrm{GL}(n)$ to give us an embedding

$$\mathrm{GL}(n)/\mathbb{G}_{k,p} \hookrightarrow \mathbb{P}(\wedge^n \mathrm{Sym}^{\leq k}\mathbb{C}^n)$$

as the GL(n) orbit of

$$\left[\cdots \wedge \sum_{|\mathbf{s}|=j} \sum_{\mathbf{s}_1+\mathbf{s}_2+\cdots+\mathbf{s}_j=\mathbf{s}} e_{\mathbf{s}_1} e_{\mathbf{s}_2} \ldots e_{\mathbf{s}_j} \wedge \cdots \right].$$

Equivalently we have $\mathrm{SL}(n)/(\mathrm{SL}(n) \cap \mathbb{G}_{k,p}) = \mathrm{SL}(n)/\mathbb{G}'_{k,p} \rtimes F_{k,p}$ embedded in $\wedge^k(\mathrm{Sym}^{\leq k}\mathbb{C}^k)$ as the SL(k) orbit of

$$p_{k,p} = \cdots \wedge \sum_{|\mathbf{s}|=j} \sum_{\mathbf{s}_1+\mathbf{s}_2+\cdots+\mathbf{s}_j=\mathbf{s}} e_{\mathbf{s}_1} e_{\mathbf{s}_2} \ldots e_{\mathbf{s}_j} \wedge \cdots,$$

where $\mathrm{SL}(n) \cap \mathbb{G}_{k,p}$ is the semi-direct product $\mathbb{G}'_{k,p} \rtimes F_{k,p}$ of $\mathbb{G}'_{k,p}$ by the finite group $F_{k,p}$ of $l_{k,p}$th roots of unity in $\mathbb{C}$ for $l_{k,p} = \sum_{i=1}^k i\mathrm{sym}^i p$. In analogy with §6 we can consider an embedding of $\mathrm{SL}(n)/\mathbb{G}'_{k,p}$ in

$$\wedge^n(\mathrm{Sym}^{\leq k}\mathbb{C}^n) \otimes (\wedge^p(\mathbb{C}^n))^{\otimes K}$$

for suitable K and its closure in this affine space. We expect the following result generalising Theorem 6.2.

CONJECTURE 9.1. *Let $K = M(\sum_{i=1}^k i\mathrm{sym}^i p)+1$ where $M \in \mathbb{N}$. Then the point*

$$p_{k,p} \otimes (e_1 \wedge \cdots \wedge e_p)^{\otimes K} \in \wedge^n(\mathrm{Sym}^{\leq k}\mathbb{C}^n) \otimes (\wedge^p(\mathbb{C}^n))^{\otimes K}$$

where

$$p_{k,p} = \cdots \wedge \sum_{|\mathbf{s}|=j} \sum_{\mathbf{s}_1+\mathbf{s}_2+\cdots+\mathbf{s}_j=\mathbf{s}} e_{\mathbf{s}_1} e_{\mathbf{s}_2} \ldots e_{\mathbf{s}_j} \wedge \cdots$$

has stabiliser $\mathbb{G}'_{k,p}$ in $\mathrm{SL}(n)$, and the closure of its $\mathrm{SL}(n)$ orbit

$$\overline{\mathrm{SL}(n)(p_{k,p} \otimes (e_1 \wedge \cdots \wedge e_p)^{\otimes K})}$$

is the union of the orbit of $p_{k,p} \otimes (e_1 \wedge \cdots \wedge e_p)^{\otimes K}$ and finitely many other $\mathrm{SL}(n)$-orbits, all of which have codimension at least two if k is large enough (depending on p) and M is sufficiently large (depending on k and p).

The proof of Conjecture 9.1 should be similar to that of Theorem 6.2, with the rôle of the Borel subgroup B_k of SL(k) played by the standard parabolic subgroup $P \subset \mathrm{SL}(n)$ which stabilises the filtration

$$0 \subset \mathbb{C}^p = \mathbb{C}e_1 \oplus \cdots \oplus \mathbb{C}e_p \subset \mathbb{C}^p \oplus \mathrm{Sym}^2\mathbb{C}^p \subset \cdots \subset \mathbb{C}^p \oplus \mathrm{Sym}^2\mathbb{C}^p \oplus \cdots \oplus \mathrm{Sym}^k\mathbb{C}^p = \mathbb{C}^n.$$

It follows immediately from Conjecture 9.1 that we would have

COROLLARY 9.2. *If $p \geq 1$ and k is large enough (depending on p) then the reparametrisation group $\mathbb{G}'_{k,p}$ is a subgroup of the special linear group* $\mathrm{SL}(\mathrm{sym}^{\leq k}p)$, *where*

$$\mathrm{sym}^{\leq k}p = \sum_{i=1}^{k} \dim \mathrm{Sym}^i \mathbb{C}^p = \binom{k+p-1}{k-1},$$

such that the algebra of invariants

$$\mathcal{O}(\mathrm{SL}(\mathrm{sym}^{\leq k}p))^{\mathbb{G}'_{k,p}}$$

is finitely generated, so that every linear action of $\mathbb{G}_{k,p}$ or $\mathbb{G}'_{k,p}$ on an affine or projective variety (with respect to an ample linearisation) which extends to a linear action of $GL(\mathrm{sym}^{\leq k}p)$ has finitely generated invariants.

In particular we would have

COROLLARY 9.3. *If $p \geq 1$ and k is large enough (depending on p) then the fibres $\mathcal{O}((J_{k,p})_x)^{\mathbb{G}'_{k,p}}$ of the bundle $E^n_{k,p}$ are finitely generated graded complex algebras.*

We would also obtain geometric descriptions of the associated affine varieties

$$\mathrm{Spec}(\mathcal{O}(\mathrm{SL}(\mathrm{sym}^{\leq k}p))^{\mathbb{G}'_{k,p}})$$

and $\mathrm{Spec}(\mathcal{O}((J_{k,p})_x)^{\mathbb{G}'_{k,p}})$ generalising those in §7.

Acknowledgments

We are indebted to Damiano Testa, who called our attention to the importance of the group $\mathbb{G}_k$ in the Green-Griffiths problem. We would also like to thank Brent Doran for helpful discussions.

The first author warmly thanks Andras Szenes, his former PhD supervisor, for his patience and their joint work from which this paper has grown.

References

[1] G. Bérczi, Thom polynomials and the Green-Griffiths conjecture, arxiv:1011. 4710.

[2] G. Bérczi, A. Szenes, Thom polynomials of Morin singularities, arXiv:math/0608285, *Annals of Math.*, to appear.

[3] D. Birkes, Orbits of linear algebraic groups, *Annals of Math.* 93 (1971) 459–475.

[4] A. Bloch, Sur les systmes de fonctions uniformes satisfaisant l'équation dune variété algébrique dont l'irrégularité dépasse la dimension, *J. de Math.* 5 (1926), 19–66.

[5] J.-P. Demailly, Algebraic criteria for Kobayashi hyperbolic projective varieties and jet differentials, *Proc. Sympos. Pure Math. 62 (1982), Amer. Math. Soc., Providence, RI*, 1997, 285–360.

[6] J.-P. Demailly, J. El-Goul, Hyperbolicity of generic surfaces of high degree in projective 3-space, *Amer. J. Math.* 122 (2000), 515–546.
[7] S. Diverio, J. Merker, E. Rousseau, Effective algebraic degeneracy, *Invent. Math.* 180(2010) 161–223.
[8] I. Dolgachev, Lectures on invariant theory, *London Mathematical Society Lecture Note Series 296, Cambridge University Press*, 2003.
[9] B. Doran and F. Kirwan, Towards non-reductive geometric invariant theory, *Pure Appl. Math. Q.* 3 (2007), 61–105.
[10] A. Fauntleroy, Categorical quotients of certain algebraic group actions, *Illinois Journal Math.* 27 (1983), 115–124.
[11] A. Fauntleroy, Geometric invariant theory for general algebraic groups, *Compositio Mathematica* 55 (1985), 63–87.
[12] T. Gaffney, The Thom polynomial of P^{1111}, *Singularities, Part 1, Proc. Sympos. Pure Math.*, 40, (1983), 399–408.
[13] M. Green, P. Griffiths, Two applications of algebraic geometry to entire holomorphic mappings, *The Chern Symposium 1979. (Proc. Intern. Sympos., Berkeley, California, 1979)* 41–74, Springer, New York, 1980.
[14] G.-M. Greuel and G. Pfister, Geometric quotients of unipotent group actions, *Proc. London Math. Soc.* (3) 67 (1993) 75–105.
[15] G.-M. Greuel and G. Pfister, Geometric quotients of unipotent group actions II, *Singularities (Oberwolfach 1996), 27–36, Progress in Math. 162, Birkhauser, Basel* 1998.
[16] F. Grosshans, Algebraic homogeneous spaces and invariant theory, *Lecture Notes in Mathematics, 1673, Springer-Verlag, Berlin*, 1997.
[17] F. Grosshans, The invariants of unipotent radicals of parabolic subgroups, *Invent. Math.* 73 (1983), 1–9.
[18] F. Kirwan, Quotients by non-reductive algebraic group actions, 'Moduli Spaces and Vector Bundles', L. Brambila-Paz, S. Bradlow, O. Garcia-Prada, S. Ramanan (editors), London Mathematical Society Lecture Note Series **359**, Cambridge University Press 2009.
[19] S. Kobayashi, Hyperbolic complex spaces, *Grundlehren der Mathematischen Wissenschaften* 318, Springer Verlag, Berlin, 1998.
[20] J. Merker, Applications of computational invariant theory to Kobayashi hyperbolicity and to Green-Griffiths algebraic degeneracy, *Journal of Symbolic Computations*, 45 (2010), 986–1074.
[21] J. Merker, Jets de Demailly-Semple dordres 4 et 5 en dimension 2, *Int. J. Contemp. Math. Sciences*, 3 (2008) no. 18. 861–933.
[22] S. Mukai, An introduction to invariants and moduli, *Cambridge University Press* 2003.
[23] D. Mumford, J. Fogarty and F. Kirwan, Geometric invariant theory, *3rd edition, Springer*, 1994.
[24] P. E. Newstead, Introduction to moduli problems and orbit spaces, *Tata Institute Lecture Notes, Springer*, 1978.
[25] G. Pacienza, E. Rousseau, Generalized Demailly-Semple jet bundles and holomorphic mappings into complex manifolds, arXiv:0810.4911.
[26] V. Popov, E. Vinberg, Invariant theory, Algebraic geometry IV, *Encyclopedia of Mathematical Sciences v. 55*, 1994.
[27] E. Rousseau, Etude des jets de Demailly-Semple en dimension 3, *Ann. Inst. Fourier (Grenoble)* 56 (2006), no. 2, 397–421.
[28] Y.-T. Siu, Some recent transcendental techniques in algebraic and complex geometry. *Proceedings of the International Congress of Mathematicians, Vol. I (Beijing, 2002), 439–448, Higher Ed. Press, Beijing*, 2002.

[29] Y.-T. Siu, Hyperbolicity in complex geometry, *The legacy of Niels Henrik Abel, Springer, Berlin*, 2004, 543–566.
[30] Y.-T. Siu, S.-K. Yeung, Hyperbolicity of the complement of a generic smooth curve of high degree in the complex projective plane, *Invent. Math.* 124, (1996), 573–618.

MATHEMATICAL INSTITUTE, OXFORD OX1 3BJ, UK

First Steps Towards a Symplectic Dynamics

B. Bramham and H. Hofer

Contents

1. What Should Symplectic Dynamics be?

Many interesting physical systems have mathematical descriptions as finite-dimensional or infinite-dimensional Hamiltonian systems. According to A. Weinstein, [**76**], Lagrange was the first to notice that the dynamical systems occurring in the mathematical description of the motion of the planets can be written in a particular form, which we call today a Hamiltonian system. Poincaré who started the modern theory of dynamical systems and symplectic geometry developed a particular viewpoint combining geometric and dynamical systems ideas in the study of Hamiltonian systems. After

Poincaré the field of dynamical systems and the field of symplectic geometry developed separately. Both fields have rich theories and the time seems ripe to develop the common core with highly integrated ideas from both fields. Given the state of both fields this looks like a promising undertaking. Though it is difficult to predict what "Symplectic Dynamics" ultimately will be, it is not difficult to give examples which show how dynamical systems questions and symplectic ideas come together in a nontrivial way.

Assume we have a cylinder, where the flow on the boundary is standard. The flow-lines enter in a standard way on the left and leave in a standard way on the right, see Figure 1. Here is the question: What must happen in the tube, assuming there are no rest points, so that not all flow-lines entering on the left will leave on the right?

It is easy to modify the flow by introducing a pair of periodic orbits with the desired properties, see Figure 2.

One could ask next whether we can achieve the desired effect without periodic orbits. That turns out to be a very hard question and is closely related to the Seifert conjecture, [**71**]. It was solved in the category of smooth vector fields by K. Kuperberg in 1993, [**54**]. One can modify the flow smoothly (even real analytically), without rest points and periodic orbits, so that not all orbits go through. In the category of volume-preserving flows, G. Kuperberg, [**53**], showed that the same holds on the C^1-level. Modulo the question of whether G. Kuperberg's example can be made smooth it seems that there cannot be any interesting additional contribution. However, it is precisely here where things become even more interesting and where we obtain a first glimpse of a "Symplectic Dynamics".

Fix on a compact three-manifold M (perhaps with boundary) a volume form Ω. A vector field X is volume preserving provided $L_X\Omega = 0$. By the Cartan homotopy formula this means that

$$0 = i_X d\Omega + di_X\Omega = d(i_X\Omega).$$

Let us assume for the moment that $H^1(M) = H^2(M) = 0$. Then we find a 1-form Γ_0 with

$$i_X\Omega = d\Gamma_0$$

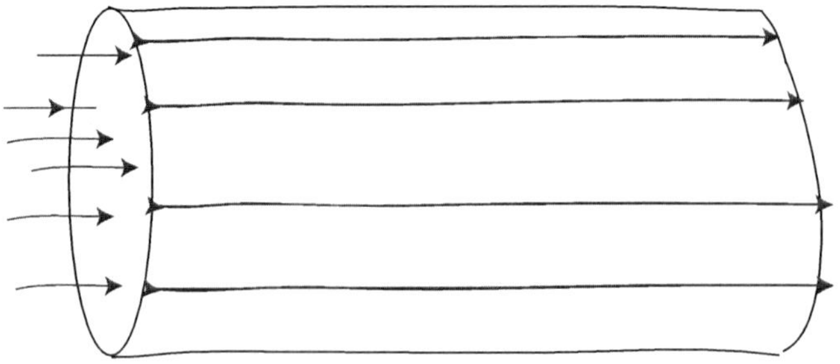

FIGURE 1. A flow tube with a flow being standard at the boundary.

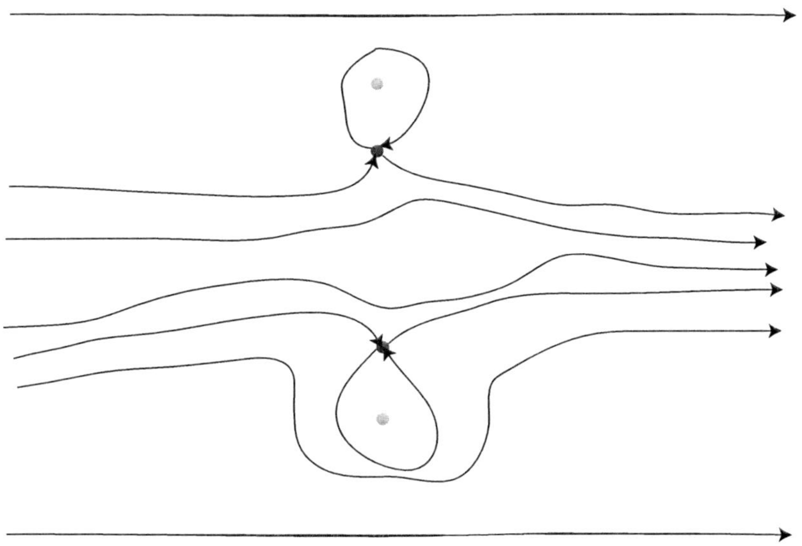

FIGURE 2. Two periodic orbits are introduced in order to have a flow without rest points, but not all flow lines entering left leave on the right.

and any 1-form Γ with that property can be written as

$$\Gamma = \Gamma_0 + dh$$

for a smooth map h.

Consider the collection $\mathcal{V}$ of all smooth nowhere vanishing Ω-preserving vector fields. Among these there is the interesting subset $\mathcal{V}^*$ consisting of those vector fields X for which there exists a Γ with $d\Gamma = i_X\Omega$, and so that $\Gamma(X) > 0$. Observe that if $X' \in \mathcal{V}$ is close to X we still have that $i_{X'}\Omega = d\Gamma'$ for a Γ' close to Γ and consequently $\Gamma'(X') > 0$ (by the compactness of M). So we see that $\mathcal{V}^*$ is open in $\mathcal{V}$, in for example the C^1-topology, provided $H^1(M) = H^2(M) = 0$.

DEFINITION 1.1. Let (M, Ω) be a closed three-manifold equipped with a volume form. An Ω-preserving vector field X is called *Reeb-like* provided there exist a one-form Γ satisfying $d\Gamma = i_X\Omega$ and $\Gamma(X) > 0$ at all points of M.

Assume that X is Reeb-like, so that $\Gamma(X) > 0$ for some Γ with $d\Gamma = i_X\Omega$. Note that this implies that $\Gamma \wedge d\Gamma$ is a volume form. In other words Γ is a contact form.

Define a positive function f by

$$f = \frac{1}{\Gamma(X)}.$$

Then $Y = fX$ satisfies $\Gamma(Y) = 1$ and $d\Gamma(Y, .) = i_Y d\Gamma = f i_X i_X \Omega = 0$. In particular

$$L_Y(\Gamma \wedge d\Gamma) = 0.$$

So Y is Reeb-like for the modified volume form $\Gamma \wedge d\Gamma$, but satisfies the stronger condition $\Gamma(Y) = 1$. Observe that Y and X have the same unparameterized flow lines. So for many questions one can study Y rather than X.

DEFINITION 1.2. Let M be a compact three-manifold. A *Reeb vector field* on M is a vector field for which there exists a contact form λ with $\lambda(X) = 1$ and $d\lambda(X, .) = 0$.

In [**23**] it was shown that if for a Reeb-like vector field not all orbits pass through, then there exists a periodic orbit. More precisely the method of proof in this paper shows the following result, where D is the closed unit disk in $\mathbb{R}^2$.

THEOREM 1.3 (Eliashberg-Hofer). *Let $Z = [0, 1] \times D$ with coordinates (z, x, y) equipped with a contact form λ which near $z = 0$ or $z = 1$ and $x^2 + y^2 = 1$ has the form $dz + xdy$, so that close to the boundary the associated Reeb vector field is given by $(1, 0, 0)$. Then, if not all entering orbits go through, there has to be a periodic orbit inside Z.*

In other words, complicated Reeb dynamics produces periodic orbits. But even much more is true as we shall see. There is a holomorphic curve theory, in the spirit of [**36**], related to the dynamics of Reeb-like vector fields, see [**41**]. The holomorphic curves allow to quantify the complexity of the dynamics in terms of periodic orbits and relations between them. The latter are again expressed in terms of holomorphic curves. Symplectic field theory (SFT), [**22**], uses the same ingredients to derive contact and symplectic invariants. However, it is possible to shift the focus onto the dynamical aspects. The already strongly developed SFT gives an idea of the possible richness of the theory one might expect. This is precisely the key observation which indicates that there should be a field accurately described as "Symplectic Dynamics" with ideas and techniques based on the close relationship between dynamics and associated holomorphic curve theories, as they occur in symplectic geometry and topology. Our paper describes some of the observations.

2. Holomorphic Curves

In the first subsection we introduce the holomorphic curve theory associated to a contact form on a three-dimensional manifold. This can also be done in higher dimensions. However, we shall restrict ourselves to low dimensions. Here the results which can be obtained look the strongest.

2.1. Contact Forms and Holomorphic Curves. Consider a three-manifold M equipped with a contact form λ. The Reeb vector field associated to λ is denoted by X and, as previously explained, defined by

$$i_X\lambda = 1 \quad \text{and} \quad i_X d\lambda = 0.$$

There is another piece of data associated to λ. Namely the *contact structure* ξ defined as the kernel bundle associated to λ. The form $d\lambda$ defines on the fibers of $\xi \to M$ a symplectic structure. Consequently, λ gives us a canonical way to split the tangent space TM of M into a line bundle L with preferred section X and a symplectic vector bundle $(\xi, d\lambda)$:

$$TM \equiv (L, X) \oplus (\xi, d\lambda).$$

We can pick a complex structure J for ξ, so that $d\lambda(h, Jh) > 0$ for $h \neq 0$. Then we can extend J to an $\mathbb{R}$-invariant almost complex structure $\tilde{J}$ on $\mathbb{R}\times M$ by requiring that the standard tangent vector $(1, 0)$ at $(a, m) \in \mathbb{R}\times M$ is mapped to $(0, X(m))$. At this point we have equipped $\mathbb{R} \times M$ with an $\mathbb{R}$-invariant almost complex structure that couples the Reeb vector field with the $\mathbb{R}$-direction. We will refer to such as an almost complex structure *compatible* with λ.

It is natural to ask about the existence of holomorphic maps from Riemann surfaces with image in $\mathbb{R} \times M$ and raise the question whether the geometry of these curves reflect in some way the dynamics of X, since $\tilde{J}$ couples the Reeb vector field with the $\mathbb{R}$-direction. That in fact turns out to be true and was used by the second author to prove certain cases of the Weinstein conjecture, [**41**]. This approach was in part motivated by Gromov's pseudoholomorphic curve theory for symplectic manifolds. However, the extension for contact manifolds is by no means straight forward, since the compactness issues for solution spaces are tricky, see [**41, 10**].

The Weinstein conjecture was formulated in [**75**] and stipulates that on a closed manifold a Reeb vector field has a periodic orbit. The first breakthrough came in [**74**] followed by [**27**]. In [**42**] it was shown that this conjecture can sometimes be solved if holomorphic spheres are present and this paper was the starting point of linking the Weinstein conjecture to

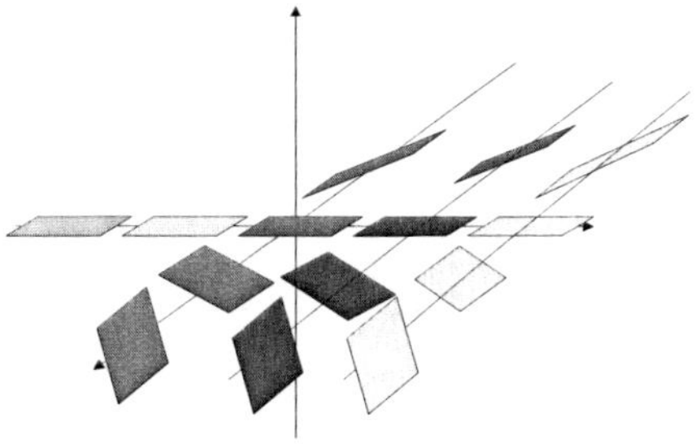

FIGURE 3. A contact structure.

Gromov-Witten theory, before it really existed, which was completed in [**63**]. The conjecture in dimension three was recently settled by Taubes, [**73**], and uses a relationship between Seiberg-Witten theory and holomorphic curve theory. The conjecture in higher dimensions is open. For example it is not known if every Reeb vector field on S^5 has a periodic orbit. The only result in higher dimensions, which proves the existence of periodic orbits for a class of Reeb vector fields on every closed manifold (which admits a Reeb vector field), is given in [**2**]. Let us begin with a detailed discussion of the holomorphic curve theory.

As it turns out one should study tuples $(S, j, \Gamma, \tilde{u})$ with (S, j) being a closed Riemann surface, Γ a finite set of punctures, and $\tilde{u} := (a, u) : S \backslash \Gamma \to \mathbb{R} \times M$ a smooth map with non-removable singularities at Γ satisfying the first order elliptic system

$$T\tilde{u} \circ j = \tilde{J} \circ T\tilde{u}.$$

This is a nonlinear Cauchy-Riemann-type equation. It also turns out to be useful to consider two tuples $(S, j, \Gamma, \tilde{u})$ and $(S', j', \Gamma', \tilde{u}')$ equivalent if there exists a biholomorphic map $\phi : (S, j) \to (S', j')$ with $\phi(\Gamma) = \Gamma'$ and $\tilde{u}' \circ \phi = \tilde{u}$. We denote an equivalence class by $[S, j, \Gamma, \tilde{u}]$. Note that in symplectic field theory we consider a somewhat different equivalence also incorporating the natural $\mathbb{R}$-action on $\mathbb{R} \times M$.

In a first step let us show that the dynamics of X can be viewed as a part of the theory. Given a solution $x : \mathbb{R} \to M$ of $\dot{x} = X(x)$ we can consider $[S^2, i, \{\infty\}, \tilde{u}]$ with $\tilde{u}(s+it) = (s, x(t))$. Here (S^2, i) is the standard Riemann sphere and $S^2 \backslash \{\infty\}$ is identified with $\mathbb{C}$ with coordinates $s + it$. Observe that if y is another solution of $\dot{y} = X(y)$ with $y(0) = x(t_0)$, then $[S^2, i, \{\infty\}, \tilde{v}]$ with $\tilde{v}(s + it) = (s + c, y(t))$ is the same class. Indeed take $\phi(s + it) = (s + c) + i(t + t_0)$ which defines a biholomorphic map $S^2 \to S^2$ fixing ∞. Then

$$\begin{aligned} \tilde{u} \circ \phi(s + it) &= \tilde{u}((s + c) + i(t + t_0)) = (s + c, x(t + t_0)) \\ &= (s + c, y(t)) = \tilde{v}(s + it). \end{aligned}$$

Hence

$$[S^2, i, \{\infty\}, \tilde{v}] = [S^2, i, \{\infty\}, \tilde{u}].$$

We call this particular type of class an *orbit plane*, or a *plane over a Reeb orbit.*

If an orbit x is periodic, say $x(t + T) = x(t)$, then it also gives us the class

$$[S^2, i, \{0, \infty\}, \tilde{u}]$$

where $S^2 \backslash \{0, \infty\}$ can be identified with $\mathbb{R} \times (\mathbb{R}/\mathbb{Z})$, and

$$\tilde{u}(s, [t]) = (Ts, x(Tt)).$$

We call this an *orbit cylinder*, or a *cylinder over a periodic Reeb orbit.*

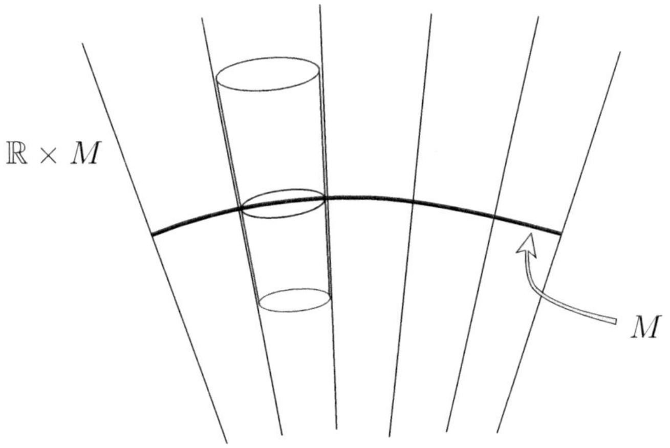

FIGURE 4. A cylinder over a periodic orbit.

That there is an interesting theory, which still has to be explored much further, comes from the fact that there are many holomorphic curves which interrelate these simple building blocks. When M is compact, these are the curves satisfying a finite energy condition.

2.2. Notions of Energy. Let us introduce two important quantities, which are certain energy measurements.

DEFINITION 2.1. Let $[S, j, \Gamma, \tilde{u}]$ be a pseudoholomorphic curve. We assume that the punctures are non-removable. Then we say it is a *finite energy curve* provided

$$E(\tilde{u}) := \sup_{\varphi} \int_{S \setminus \Gamma} \tilde{u}^* d(\varphi\lambda) < \infty.$$

Here the supremum is taken over the collection Σ of all smooth maps $\varphi : \mathbb{R} \to [0,1]$ with $\varphi'(s) \geq 0$.

If we compute the energy E of a cylinder over a T-periodic orbit we obtain the identity

$$E = T.$$

However, for the energy of a plane over a Reeb orbit we find

$$E = \infty.$$

There is another useful energy which can be introduced.

DEFINITION 2.2. The *$d\lambda$-energy* is defined by

$$E_{d\lambda}(\tilde{u}) = \int_{S \setminus \Gamma} u^* d\lambda.$$

This energy turns out to be 0 in both the previous cases. However, there are in general many interesting holomorphic curves which have a positive $d\lambda$-energy. These are in fact the curves used to establish relations between the periodic orbits. The easiest examples are finite energy planes.

DEFINITION 2.3. A *finite energy plane* is an equivalence class $[S^2, j, \{\infty\}, \tilde{u}]$ for which ∞ is (as usual) not removable and $0 < E < \infty$.

We will see in Theorem 2.5 that finite energy planes behave very differently to the (infinite) energy planes over orbits that we just encountered. Interesting properties of finite energy planes were used in [**41**] to prove cases of the Weinstein conjecture, [**75**].

It takes some analysis to show that if $\tilde{u} = (a, u) : \mathbb{C} \to \mathbb{R} \times M$ represents a finite energy plane, then its $\mathbb{R}$-component a is proper. That means that $a(z) \to \infty$ for $|z| \to \infty$. Then, as a consequence of Stokes' theorem one easily verifies

LEMMA 2.4. *For a finite energy plane $\tilde{u}$ we have the equality*

$$E(\tilde{u}) = E_{d\lambda}(\tilde{u}).$$

The finite energy planes have some nice properties. For example they detect contractible periodic orbits of the Reeb vector field.

THEOREM 2.5. *Assume that $\tilde{u} := (a, u) : \mathbb{C} \to \mathbb{R} \times M$ is smooth and satisfies the differential equation*

$$T\tilde{u} \circ i = \tilde{J} \circ T\tilde{u}.$$

Assume further that $\tilde{u}$ is nonconstant and $E(\tilde{u}) < \infty$. Then $T := E(\tilde{u}) \in (0, \infty)$ and for every sequence $r_k \to \infty$ there exists a subsequence r_{k_j} and a solution $x : \mathbb{R} \to M$ of

$$\dot{x} = X(x) \quad \text{and} \quad x(0) = x(T)$$

so that in addition

$$\lim_{j\to\infty} u(r_{k_j} \cdot e^{2\pi i t}) = x(Tt) \quad \text{in } C^\infty(\mathbb{R}/\mathbb{Z}, M).$$

In other words, non-constant solutions on the 1-punctured Riemann sphere are related to periodic orbits for the Reeb vector field. The period in fact being the quantity

$$T = \int_{\mathbb{C}} u^* d\lambda = E_{d\lambda}(\tilde{u}).$$

The main idea in [**41**] was to show that for the nonlinear Cauchy-Riemann problem associated to the Reeb vector field of an overtwisted contact form there always exists a finite energy plane, showing the existence of a periodic orbit.

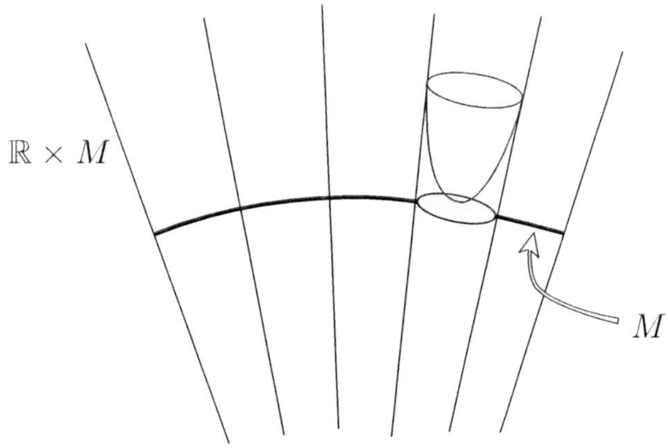

FIGURE 5. A finite energy plane asymptotic to a cylinder over a periodic orbit, and the projected disk in the contact manifold.

2.3. Global Systems of Surfaces of Section. Assume that λ is a contact form on the closed three-manifold M. Suppose we have fixed a compatible almost complex structure $\tilde{J}$ on $\mathbb{R} \times M$ as described before. If $[S, j, \Gamma, \tilde{u}]$ is an equivalence class of solutions associated to the nonlinear Cauchy-Riemann equation, we can associate to it its *image* F defined by

$$F_{[S,j,\Gamma,\tilde{u}]} = \tilde{u}(S \backslash \Gamma).$$

This definition does not depend on the representative we have picked. We call $[S, j, \Gamma, \tilde{u}]$ an *embedded solution* provided the map

$$\tilde{u} : S \backslash \Gamma \to \mathbb{R} \times M$$

is an embedding. Let us also observe that for a given solution $[S, j, \Gamma, \tilde{u}]$ and real constant $c \in \mathbb{R}$ we obtain another solution $[S, j, \Gamma, \tilde{u}]_c$ defined by

$$[S, j, \Gamma, \tilde{u}]_c := [S, j, \Gamma, \tilde{u}_c],$$

where $(a, u)_c = (a + c, u)$ and $\tilde{u} = (a, u)$. Observe that the image of $[S, j, M, \tilde{u}]_c$ is the image of $[S, j, \Gamma, \tilde{u}]$ shifted by c via the obvious $\mathbb{R}$-action on $\mathbb{R} \times M$.

DEFINITION 2.6. Let λ be a contact form on the three-manifold M and $\tilde{J}$ a compatible almost complex structure. A *finite energy foliation* $\tilde{\mathcal{F}}$ associated to this data is a smooth foliation of $\mathbb{R} \times M$ by the images of embedded curves, having finite energy, with the property that if F is a leaf, each F_c is also a leaf*.

*We will sometimes emphasize this last property by referring to a finite energy foliation as being "ℝ-invariant". In this article we will not consider finite energy foliations without this invariance under the ℝ-translations.

Let us observe that if we drop the requirement of finite energy we always have the following object.

DEFINITION 2.7. Let λ be a contact form on the three-manifold M. The *vertical foliation* $\tilde{\mathcal{F}}^{\nu}(M, \lambda)$ is defined to be the foliation of $\mathbb{R} \times M$ whose leaves take the form $\mathbb{R} \times \phi(\mathbb{R})$ over all Reeb trajectories $\phi : \mathbb{R} \to M$.

Note that the leaves of the vertical foliation are pseudoholomorphic for any almost complex structure compatible with λ, and they are invariant under $\mathbb{R}$-translations. However, the vertical foliation is only a finite energy foliation when every Reeb orbit is periodic. This simple observation will be crucial in Section 3.5.

Finite energy foliations, when they exist, have important consequences for the Reeb flow. Recall that a *surface of section** for a flow on a three-manifold is an embedded surface, possibly with boundary, with the property that the flow is transverse to the interior of the surface while each boundary circle is a periodic orbit, called a binding or spanning orbit. Poincaré used this notion to great effect, constructing such surfaces locally "by hand". It was observed in [**44**] that a finite energy foliation gives rise to a filling of the entire three-manifold by surfaces of section, simply by projecting the leaves down via the projection map

$$\mathrm{pr} : \mathbb{R} \times M \to M.$$

The resulting filling $\mathcal{F}$ of the three-manifold was therefore called in [**44, 45**] a global system of surfaces of section, which we loosely define as follows.

DEFINITION 2.8. Let M be a three-manifold with a nowhere vanishing vector field X having a globally defined flow. A *global system of surfaces of section* for this data is a finite collection of periodic orbits $\mathcal{P}$ of the flow, called the spanning orbits, and a smooth foliation of the complement

$$M \backslash \mathcal{P}$$

by embedded punctured Riemann surfaces $\mathcal{S}$, such that each leaf in $\mathcal{S}$ converges to a spanning orbit at each of its punctures, and such that the closure of each leaf in M is a surface of section for the flow.

An adapted open book associated to a contact three-manifold [**35**] provides a familiar example of a global system of surfaces of section, but in contrast to the situation we describe here, one only knows *there exists* a Reeb flow making the leaves surfaces of section. A further distinction, is that in an open book all leaves lie in a single S^1 family, in particular they all have the same collection of spanning orbits.

*Contrast this definition with that of a global surface of section which has the additional property that every orbit, other than the bindings or spanning orbits, hits the surface in forward and backward time.

PROPOSITION 2.9 (Hofer-Wysocki-Zehnder,[**46, 45**]). *Suppose that a three-manifold M equipped with a contact form λ admits an associated finite energy foliation $\tilde{\mathcal{F}}$ (with respect to some almost complex structure compatible with λ). Then the projection of the leaves down to M is a singular foliation*

$$\mathcal{F} := \{ \operatorname{pr}(F) \mid F \in \tilde{\mathcal{F}} \}$$

with the structure of a global system of surfaces of section for the Reeb flow.

Note that in this context, where the dynamics comes from a Reeb vector field, each surface of section in $\mathcal{F}$ comes naturally equipped with an area form of finite volume, preserved by the flow. Indeed, that the Reeb vector field is transverse to the interior of a surface $S \in \mathcal{F}$ implies that $d\lambda$ restricts to a non-degenerate 2-form on S, while the embeddedness of the leaf up to the boundary yields that the total volume, i.e. the $d\lambda$-energy of the corresponding holomorphic curve, is finite. In fact, by Stokes theorem the area of each leaf is equal to the sum of the periods of the positive punctures minus the sum of the periods of the negative punctures.

Proposition 2.9 can be seen as follows. If F is a leaf in $\tilde{\mathcal{F}}$, then either F is a cylinder over a periodic orbit, or it is disjoint from each of its $\mathbb{R}$-translates, F_c, $c \neq 0$. In the former case the projection of F down to M is just the periodic orbit it spans. In the latter case, the leaf F must be nowhere tangent to $(1, 0)$ in $T(\mathbb{R} \times M)$, and since $(1, 0)$ is coupled by $\tilde{J}$ with the Reeb vector field X, the leaf is transverse to the complex line $\mathbb{R}(1, 0) \oplus \mathbb{R}X$. This amounts to the projection of the leaf being transverse to X in M. A variety of necessary and sufficient conditions for the projection of a curve to be embedded are given in [**72**].

A global system of surfaces of section $\mathcal{F}$ inherits a certain amount of other structure from the finite energy foliation $\tilde{\mathcal{F}}$. In particular, if the contact form has only non-degenerate periodic orbits, there are only two possibilities for the local behavior near each spanning orbit, depending on the parity of its Conley-Zehnder index. Local cross-sections are illustrated in Figure 6*.

For a global illustration of a global system of surfaces of section on S^3 see Figures 7 and 8, and on a solid torus see 10, 11, 12, 13.

It is not clear at all if in any given situation a finite energy foliation exists. However, it turns out, that quite often they do. The first such result appeared in [**44**] and was generalized further in [**46**]. These papers study tight Reeb flows on S^3. According to a classification result every positive tight contact form λ on S^3 is, after a smooth change of coordinates, of the form $f\lambda_0$, where $f : S^3 \to (0, \infty)$ is a smooth map and $\lambda_0 = \frac{1}{2}[q \cdot dp - p \cdot dq]$ is the standard contact form on S^3 whose associated contact structure is the

*In general, the picture on the left of Figure 6 could happen at a spanning orbit having even parity Conley-Zehnder index if the orbit has a constraint in the form of an asymptotic "weight". But in all the examples in this paper, there are only weights on odd index orbits so this doesn't happen.

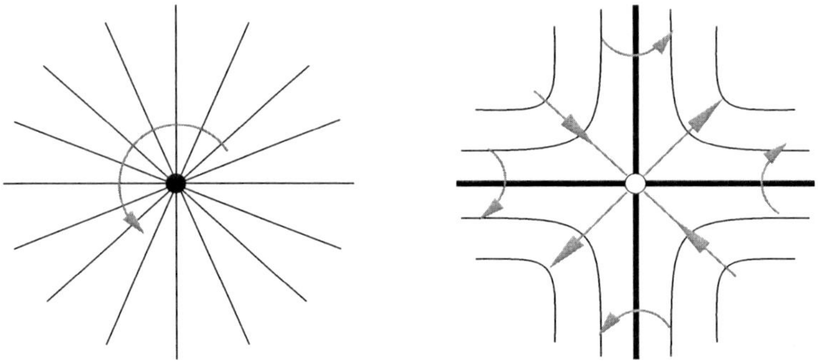

FIGURE 6. Illustrating the two forms of behavior possible in a neighborhood of a spanning orbit in a global system of surfaces of section $\mathcal{F}$ that arises as the projection of a finite energy foliation $\tilde{\mathcal{F}}$, for a non-degenerate contact form. On the left all surfaces which enter the neighborhood converge to the spanning orbit. This occurs when the spanning orbit has odd parity Conley-Zehnder index (see also the footnote on previous page). In the second case precisely four leaves entering a neighborhood of the orbit connect to it. This occurs whenever the spanning orbit has even Conley-Zehnder index. The Reeb trajectories are transversal to the page and to the leaves in the direction of the arrows. In an open book the picture is as on the left at every spanning orbit.

line bundle of complex lines in TS^3, where S^3 is seen as the unit sphere in $\mathbb{C}^2$ and $q + ip$ are the coordinates. The flow lines of X on S^3 associated to $f\lambda_0$ are conjugated to the Hamiltonian flow on the energy surface

$$N = \{\sqrt{f(z)}z \mid |z| = 1\},$$

where we indentify $\mathbb{C}^2$ with $\mathbb{R}^4$ via $q + ip \to (q_1, p_1, q_2, p_2)$, and the latter has the standard symplectic form $\omega = dq_1 \wedge dp_1 + dq_2 \wedge dp_2$. So we can formulate the results in terms of star-shaped energy surfaces in $\mathbb{R}^4$, i.e. energy surfaces bounding domains which are star-shaped with respect to 0.

THEOREM 2.10 (Hofer-Wysocki-Zehnder,[**44**]). *Assume that N bounds a strictly convex domain containing zero and is equipped with the contact form $\lambda_0|N$. For a generic admissible complex multiplication J on the associated contact structure there exists a finite energy foliation with precisely one leaf which is a cylinder over a periodic orbit and all other leaves are finite energy planes asymptotic to it.*

After projecting down to the 3-manifold N, each plane-like leaf gives rise to a disk-like surface of section with boundary the spanning orbit, and finite volume. The convexity implies that the generalized Conley-Zehnder index of the spanning orbit is at least 3. This implies that the orbits nearby intersect all the leaves of the projected foliation transversally enough that all orbits, besides the spanning orbit, hit each leaf in forwards and backwards time. Thus, fixing any leaf, we obtain a well defined return map, which is an area-preserving diffeomorphism of the open disk. By an important result of Franks we obtain that the disk map, if it has at least two periodic points must have infinitely many periodic orbits. Hence we obtain the following corollary.

COROLLARY 2.11 (Hofer-Wysocki-Zehnder). *On an energy surface in $\mathbb{R}^4$ which bounds a strictly convex bounded domain, we have either precisely two geometrically distinct periodic orbits or infinitely many.*

We would like to emphasize that no kind of genericity is assumed.

When we go to the most general case, namely that of an energy surface bounding a starshaped domain, we in general still need some genericity

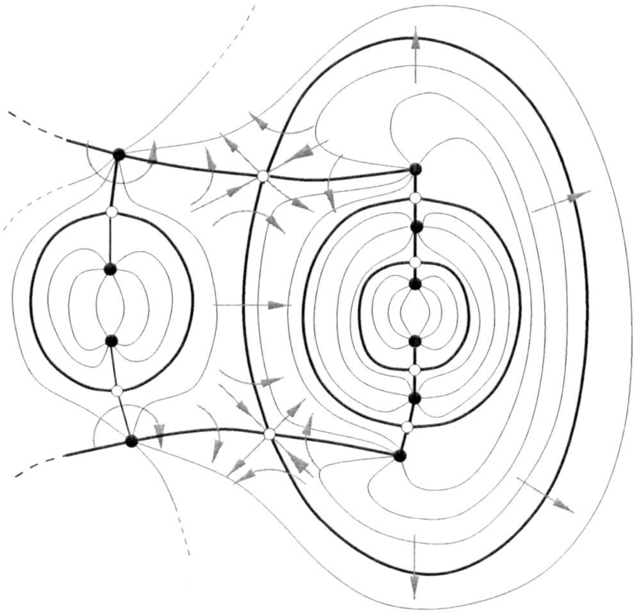

FIGURE 7. A global system of surfaces of section of S^3, obtained as the projection of a finite energy foliation of $\mathbb{R} \times S^3$. The 3-sphere is viewed as $\mathbb{R}^3 \cup \{\infty\}$, and the figure shows the trace of the surfaces of section cut by a plane. The dots represent the spanning periodic orbits; they are perpendicular to the page and two dots belong to the same periodic orbit. The white dots represent periodic orbits of index 2 and the black dots periodic orbits of index 3. The leaves are disk-like and annuli-like. The rigid surfaces are represented by bold curves. The grey arrows indicate the Reeb flow.

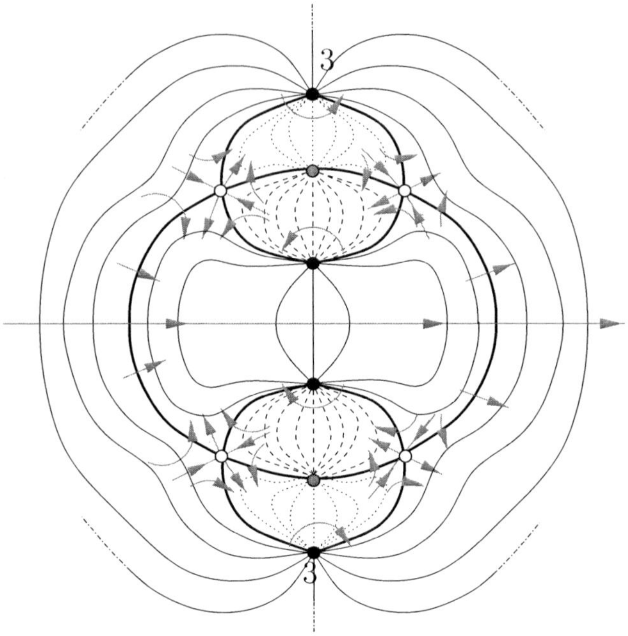

FIGURE 8. A second example, illustrating a finite energy foliation associated to S^3. The grey dots represent a periodic orbit of index 1, the white dots periodic orbits of index 2 and the black dots periodic orbits of index 3. The leaves are again disk-like and annuli-like surfaces. The rigid surfaces are represented by bold curves. Dotted and dashed surfaces represent families of annuli-like surfaces connecting periodic orbits of index 3 with periodic orbits of index 1. Thin curves represent disk-like surfaces asymptotic to periodic orbits with index 3. The grey arrows picture the flow of the Reeb vector field.

assumption*. For example assuming that all periodic orbits are non-degenerate and the stable and unstable manifolds of hyperbolic orbits are transversal where they intersect. This can always be obtained by a C^∞-small perturbation of the energy surface. Alternatively we may consider generic contact forms $f\lambda_0$ on S^3, so that the associated star-shaped energy surface has the previously described genericity properties.

THEOREM 2.12 (Hofer-Wysocki-Zehnder,[**46**]). *Let $\lambda = f\lambda_0$ be a generic contact form on S^3. Then for a generic complex multiplication on ξ with associated $\mathbb{R}$-invariant almost complex structure $\tilde{J}$ on $\mathbb{R} \times S^3$, there exists*

*This is only a technical assumption and one should be able to remove it. However, it might not be so easy to draw the same conclusions about the dynamics of the Reeb vector field.

an associated finite energy foliation. Besides finitely many cylinders over periodic orbits the other leaves are parameterized by punctured finite energy spheres with precisely one positive puncture, but which can have several negative punctures. The asymptotic limits at the punctures are simply covered.

The precise result also contains some more technical information about the Conley-Zehnder indices of the periodic spanning orbits and we refer the reader to [**46**]. The Figures 7 and 8, taken from [**45**], describe some of the structure.

From this result again one can derive that there are either two or infinitely many periodic orbits, given the described genericity. We now sketch the proof of this.

If we have only one spanning orbit, then we have a *global disk-like surface of section*, meaning a surface of section, homeomorphic to the open disk, with a well defined return map, preserving a finite volume form. The assertion then follows from Franks' theorem. Assume that we have two spanning orbits. In that case we must have at least one hyperbolic spanning orbit of period T say. One can show quite easily that there exists a heteroclinic chain connecting several hyperbolic spanning orbits. Using our genericity assumption we can use symbolic dynamics to construct infinitely many periodic orbits. The heteroclinic chain follows immediately from the fact that we only have finitely many spanning orbits and that the stable and unstable manifold of a hyperbolic orbit intersect the nearby leaves of the finite energy foliation in loops which have λ-integral equal to the period T. Essentially for area reasons the assertion follows. The reader should take any of the two Figures 7 or 8 and try to carry out the argument.

In recent papers Hryniewicz and Hryniewicz-Salomoa, [**47, 48, 49**] have been able to give a necessary and sufficient condition for when a sphere-like energy surface possesses a global disk-like surface of section.

Recent work by Albers, Frauenfelder, van Koert and Paternain, [**1**] and Cieliebak, Frauenfelder and van Koert, [**17**] make it feasible to use finite energy foliations in the study of the classical restricted circular planar three-body problem.

Using quite different singular foliations in a symplectic setting note recent work of Pelayo and Vũ Ngọc [**68, 69**]. They combined symplectic and dynamical ideas to prove surprising classification results for semitoric integrable systems.

Finally let us note that finite energy foliations have important applications in contact geometry as well, see [**77, 78, 79**].

3. Holomorphic Curves and Disk Maps

We saw in the last section, Theorem 2.10, that finite energy foliations exist for generic Reeb flows on the tight 3-sphere. It turns out that not just one, but many such foliations can be constructed if we replace the 3-sphere with a solid torus and restrict to generic Reeb flows which have no contractible periodic orbits.

Any area preserving diffeomorphism of the disk can be put into this framework, with different iterates giving rise to genuinely different global systems of surfaces of section. The question arises how one can profit, dynamically speaking, from this perspective. One could hope that the holomorphic curves provide a book-keeping tool for tracking the history and future of the orbits of a disk map.

As a first step in this direction, we outline how these can be used to prove the Poincaré-Birkhoff fixed point theorem, and a complementary result "Theorem" 3.20 which says that maps with two periodic orbits have some iterate which either has a "twist" or is the identity map. This latter statement might be new. Combining these one recovers, albeit currently with an additional boundary condition, a celebrated theorem of Franks [**30**] in the smooth category, a more recent result of Franks and Handel [**32**], along with sharp growth estimates on periodic orbits known to follow from results of Le Calvez [**61**].

These applications due to the first author depend on more recent developments [**13, 14**] in which it is shown how to construct finite energy foliations with a prescribed spanning orbit. See "Theorem" 3.13 for a precise statement. This is a surprising novelty, implying that in general there are many more finite energy foliations than might naively be expected. This result is labelled "theorem" as it is not completely written up. Results depending on this one are also labelled in quotation marks.

This section explains the existence statements for finite energy foliations, and outlines proofs of the mentioned dynamical applications.

3.1. Reeb-like Mapping Tori. Let $Z_\infty = \mathbb{R} \times D$ be the infinite tube equipped with coordinates $(\bar{z}, \bar{x}, \bar{y})$. There is a $\mathbb{Z}$-action generated by the "1-shift" automorphism $\tau(\bar{z}, \bar{x}, \bar{y}) = (\bar{z} + 1, \bar{x}, \bar{y})$. Quotienting out by some iterate τ^n gives us a solid torus Z_n of "length" n. We denote by $(z + n\mathbb{Z}, x, y)$ the induced coordinates on Z_n, and write z to mean $z + n\mathbb{Z}$ when the context is clear.

DEFINITION 3.1. For $n \in \mathbb{N}$, a *Reeb-like mapping torus* will refer to a contact form λ_n on Z_n, for which the disk slice $D_0 := \{z = 0\}$ is a global surface of section for the Reeb flow, and which lifts to a contact form λ_∞ on Z_∞ having the following properties: it is invariant under pull-back by the 1-shift automorphism, and has contact structure $\ker\{cdz + xdy - ydx\}$, some $c > 0$.

By a *global* surface of section is meant that the trajectory through any point in the solid torus passes through D_0 in forwards and backwards time, and does so transversely to D_0. In particular, $d\lambda_n$ restricts to an area form on D_0, and the flow induces a first return map.

From a Reeb-like mapping torus λ_n on Z_n one can lift and project to obtain a sequence of mapping tori

$$(Z_1, \lambda_1),\ (Z_2, \lambda_2),\ (Z_3, \lambda_3)\ldots.$$

The first return map of the flow on (Z_1, λ_1) is a diffeomorphism $\psi : D \to D$ preserving the area form $\iota^* d\lambda_1$, where $\iota : D \hookrightarrow D_0 = \{0\} \times D$ is inclusion, and the first return map on (Z_n, λ_n) is then the n-th iterate ψ^n.

In this situation, we will say that (Z_1, λ_1) *generates* the disk map ψ, and that (Z_n, λ_n) generates ψ^n.

LEMMA 3.2. *Let $\psi : D \to D$ be any orientation preserving, C^∞-diffeomorphism preserving $dx \wedge dy$. Then there exists a Reeb-like mapping torus (Z_1, λ_1) having first return map ψ for which $\iota^* d\lambda_1 = dx \wedge dy$, the standard Euclidean volume form.*

A proof of this under the assumption that ψ restricts to a rotation on the boundary of the disk is contained in [**11**].

Identifying Z_n with $\mathbb{R}/n\mathbb{Z}\times D$, the product structure gives us a canonical way to assign linking numbers to pairs of homologous periodic orbits, and, along with the S^1-symmetry of the contact structure, canonically assign Conley-Zehnder indices to individual periodic orbits. We explain this now.

There is a canonical basis of $H_1(\partial Z_n; \mathbb{Z}) \simeq \mathbb{Z} \oplus \mathbb{Z}$ we will denote by $\{L_n, [\partial D]\}$, which are the unique elements represented by closed oriented loops of the form $\mathbb{R}/n\mathbb{Z} \times \{\mathrm{pt}\}$ and $\{\mathrm{pt}\} \times \partial D$ respectively. We call L_n the *canonical longitude* on Z_n, and $[\partial D]$ the *canonical meridian* on Z_n. With a little elementary algebraic topology, one can show that if γ is an immersed closed loop in Z_n, disjoint from ∂Z_n, and homologous to L_n, then

$$\{L_n, [\partial D]\}$$

is also a basis for $H_1(Z_n \backslash \gamma(S^1); \mathbb{Z})$ after applying the inclusion $\partial Z_n \hookrightarrow Z_n \backslash \gamma(S^1)$. (This is not true if we replace homology groups by homotopy groups in case γ is knotted). Thus the following is well defined.

DEFINITION 3.3. Let $\gamma_1, \gamma_2 : S^1 \to Z_n$ be two continously embedded closed loops having degree 1 after projecting onto the S^1-factor of $Z_n = \mathbb{R}/n\mathbb{Z} \times D$. Assume that their images are disjoint from each other and from ∂Z_n. Then γ_1 determines an homology class in the complement of γ_2. Define the *linking number* to be the unique integer $\mathrm{lk}(\gamma_1, \gamma_2)$ such that γ_1 is homologous to

$$L_n + \mathrm{lk}(\gamma_1, \gamma_2)[\partial D] \in H_1(Z_n \backslash \gamma_2(S^1); \mathbb{Z})$$

where $\{L_n, [\partial D]\}$ is the canonical basis, longitude and meridian, that we just defined.

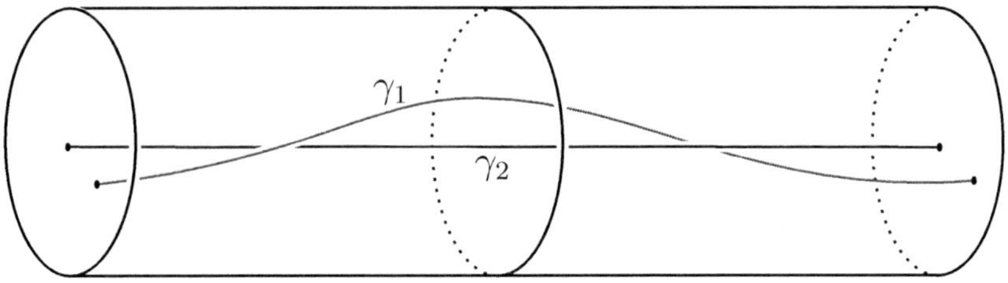

FIGURE 9. γ_1 and γ_2 represent two disjoint continuously embedded closed loops in the solid torus Z_2. Their linking number is $\text{lk}(\gamma_1, \gamma_2) = 1$.

The linking number turns out to be symmetric, that is $\text{lk}(\gamma_1, \gamma_2) = \text{lk}(\gamma_2, \gamma_1)$. Moreover, with these conventions one has linear growth under iterates: if $\gamma_i^k : S^1 \to Z_{kn}$ represents the unique lift of γ_i to the longer mapping torus (Z_{kn}, λ_{kn}), for $i \in \{0, 1\}$, then $\text{lk}(\gamma_1^k, \gamma_2^k) = k \cdot \text{lk}(\gamma_2, \gamma_1)$ if $k \in \mathbb{N}$.

Making use of further symmetry of the contact structure we also obtain canonical Conley-Zehnder indices for periodic orbits of the Reeb flow. Recall that usually, on a general manifold M equipped with a contact form λ, at a periodic orbit γ of the Reeb flow one requires a choice of symplectic trivialization, up to homotopy type, of the contact structure $\xi := \ker \lambda$ along γ to be able to assign a Conley-Zehnder index to γ. In the cases at hand the contact manifold Z_n is covered by $Z_\infty = \mathbb{R} \times D$ on which the contact structure $\xi := \ker \lambda_\infty$ is invariant under the $\mathbb{R}$-action $c \cdot (\bar{z}, \bar{x}, \bar{y}) = (\bar{z} + c, \bar{x}, \bar{y})$ (and hence also globally trivializable). This descends to a circle action $c \cdot (z, x, y) = (z + c, x, y)$ on Z_n. There is a unique homotopy class of trivializations on Z_n admitting a representative which is invariant under this circle action.

DEFINITION 3.4. Let γ be a periodic orbit of the Reeb flow of (Z_n, λ_n). Then we take the *Conley-Zehnder index* of γ to be with respect to the unique homotopy class of global trivializations of the contact structure which admit an S^1-invariant representative. We will denote this by

$$\mu(\gamma) \in \mathbb{Z}.$$

3.2. Notions of Rotation Number. For a more sophisticated approach to rotation numbers of surface maps see [**30, 31, 32, 60**], in which the area preserving property of the map is used to make sense of the rate at which *almost all* orbits rotate about a given fixed point, via the Birkhoff ergodic theorem.

For our discussion it will suffice to talk of the *total* rotation number of a smooth area preserving diffeomorphism of the disk associated to its restriction to the boundary circle and associated to a periodic point, in the latter intuitively describing the infinitesimal rate at which points nearby rotate around the periodic orbit. By total, we mean that the rotation numbers are real valued, as opposed to merely circle valued*.

Recall that any orientation preserving homeomorphism of the circle has a well defined value in $\mathbb{R}/\mathbb{Z}$ called its rotation number. Any choice of lift to a homeomorphism of the real line can be assigned a real valued rotation number, but this depends on the choice of lift. Similarly, any choice of homotopy of the circle map to the identity determines a unique lift and thus allows to assign a real valued rotation number. Indeed, if $F : [0,1] \times S^1 \to S^1$, is a homotopy from $\mathrm{id} = F(0,\cdot)$ to $f = F(1,\cdot)$, there is a unique lift $\bar{F} : [0,1] \times \mathbb{R} \to \mathbb{R}$ of F satisfying $\bar{F}(0,\cdot) = \mathrm{id}_{\mathbb{R}}$, giving us a canonical lift $\bar{F}(1,\cdot)$ of f.

More generally, if one has a diffeomorphism of a surface one can assign a circle valued rotation number to the restriction to any boundary component, and to each fixed point via the differential at the fixed point. But we can obtain the more useful real valued rotation numbers if the map is isotopic to the identity and we choose such an isotopy class. Moreover, this gives a canonical way to assign integer valued linking numbers to pairs of periodic orbits of the same period.

In the framework we are working with here, we are always in the situation of having a diffeomorphism of the disk for which we have chosen a mapping torus generating it as the first return map. This is similar to fixing an isotopy class from the disk map to the identity, and indeed, the choice of a mapping torus allows us to define real valued rotation numbers (and linking numbers as we already saw). Although for all of this it is unnecessary that the map be area preserving, we will nevertheless make use of this to make a short route to a workable definition.

Consider a Reeb-like mapping torus (Z_1, λ_1) generating a disk map $\psi : D \to D$. Let $f : \partial D \to \partial D$ denote the restriction of ψ to the boundary. The choice of mapping torus gives us a canonical lift $\bar{f} : \mathbb{R} \to \mathbb{R}$ of f as follows.

Restricting the Reeb-flow to the boundary of the disk slice $D_0 = \{0\} \times D \subset Z_1$, gives us a smooth map $\phi : \mathbb{R} \times \partial D_0 \to \partial Z_1$. Reparameterising if necessary, we may assume that all points in ∂D_0 have first return time 1. Then ϕ restricts to a map $\phi : [0,1] \times \partial D_0 \to \partial Z_1$ whose projection onto the ∂D factor of $\partial Z_1 = \mathbb{R}/\mathbb{Z} \times \partial D$ gives us a homotopy

$$F : [0,1] \times \partial D \to \partial D$$

from $F(0,\cdot) = \mathrm{id}$ to $F(1,\cdot) = f$. As described above, F has a unique lift to a homotopy of $\mathbb{R}$ starting at the identity and ending at a lift of f, which we take to be $\bar{f}$.

---

*What we refer to here as *total rotation numbers* are perhaps better described as *translation numbers*. The difference between this concept and *rotation number* is explained nicely in Franks [**33**] for example.

DEFINITION 3.5. For each $n \in \mathbb{N}$ define the *total rotation number of* ψ^n *on the boundary* to be the real number

$$\mathrm{Rot}_{\psi^n}(\partial D) := \mathrm{Rot}(\bar{\psi}_n),$$

where $\bar{\psi}_n : \mathbb{R} \to \mathbb{R}$ is the canonical lift, just described, of $\psi^n : \partial D \to \partial D$ induced by the flow on the boundary of Z_n.

It is easy to show that for each $n \in \mathbb{N}$, $\mathrm{Rot}_{\psi^n}(\partial D) = n\,\mathrm{Rot}_{\psi}(\partial D)$.

DEFINITION 3.6. Let $n \in \mathbb{N}$. For $p \in \mathrm{Fix}(\psi^n)$ we define the *infinitesimal rotation number* of ψ^n at p to be

$$\mathrm{Rot}_{\psi^n}(p) := \frac{1}{2} \lim_{k \to \infty} \frac{\mu(\gamma_p^k)}{k}$$

where $\gamma_p : S^1 \to Z_n$ is the periodic orbit passing through the disk slice D_0 at the point p, and $\mu(\gamma)$ denotes the Conley-Zehnder index of γ as described in 3.4.

Again, it is easy to show that for each $k \in \mathbb{N}$, $\mathrm{Rot}_{\psi^{kn}}(p) = k\,\mathrm{Rot}_{\psi^n}(p)$.

DEFINITION 3.7. Let $p \in \mathrm{Fix}(\psi^n)$ be an interior fixed point. Define the *twist interval* of p, as a fixed point of ψ^n, to be the open interval of real numbers

$$\mathrm{twist}_{\psi^n}(p) := \Big(\min\{\mathrm{Rot}_{\psi^n}(p), \mathrm{Rot}_{\psi^n}(\partial D)\}\ ,\ \max\{\mathrm{Rot}_{\psi^n}(\partial D), \mathrm{Rot}_{\psi^n}(p)\} \Big).$$

This interval could be empty.

We emphasize, that all three definitions above are implicitely *with respect to a choice of data* (Z_1, λ_1), that is, a Reeb-like mapping torus generating the disk map ψ.

3.3. Finite Energy Foliations for Mapping Tori. In our discussion here of finite energy foliations associated to Reeb-like mapping tori, as opposed to more general three-manifolds (see definition 2.6), the leaves will come in only two forms; cylinders and half cylinders.

Consider a Reeb-like mapping torus (Z_n, λ_n). Let $\tilde{J}_n$ be an almost complex structure on $\mathbb{R} \times Z_n$ that is compatible with λ_n in the sense described in Section 2.1.

REMARK 3.8. It is possible to do everything that follows under the additional assumption that $\tilde{J}_n$ is the lift of an almost complex structure $\tilde{J}_1$ on $\mathbb{R} \times Z_1$ that is compatible with λ_1. In other words, that the lift $\tilde{J}_\infty$ to $\mathbb{R} \times Z_\infty$ is invariant under the 1-shift automorphism, or deck transformation, $(a, \bar{z}, \bar{x}, \bar{y}) \mapsto (a, \bar{z}+1, \bar{x}, \bar{y})$. This is potentially a very useful symmetry, giving us positivity of intersections between leaves of foliations associated to different iterates. But we do not use this for any of the arguments or applications in this paper except for Theorem 4.1.

A *cylinder* will refer to a finite energy pseudoholomorphic curve of the form

$$[S^2, i, \{0, \infty\}, \tilde{u}],$$

so called because $S^2 \backslash \{0, \infty\}$ can be identified with $\mathbb{R} \times S^1$. In other words, a cylindrical leaf in a finite energy foliaton $\tilde{\mathcal{F}}$ associated to a Reeb-like mapping torus (Z_n, λ_n) with compatible almost complex structure $\tilde{J}_n$, will mean the image of an embedded solution $\tilde{u} : \mathbb{R} \times S^1 \to \mathbb{R} \times Z_n$ to the non-linear Cauchy-Riemann equation, having finite E-energy.

A *half cylinder* leaf in $\tilde{\mathcal{F}}$ will refer to a finite energy pseudoholomorphic curve (with boundary) of the form

$$[D, i, \{0\}, \tilde{v}]$$

where $D = \{z \in \mathbb{C} | \; |z| \leq 1\}$, because $D \backslash \{0\}$ can be holomorphically identified with the half infinite cylinder $[0, \infty) \times S^1$. As this is our first reference to pseudoholomorphic curves with boundary, let us be more precise. Any representative of a half cylinder leaf can be identified with an embedded solution

$$\tilde{v} = (b, v) : [0, \infty) \times S^1 \to \mathbb{R} \times Z_n$$

to the differential equation $T\tilde{v} \circ i = \tilde{J}_n \circ T\tilde{v}$, having finite energy, and satisfying the following boundary condition, that there exists a constant $c \in \mathbb{R}$ such that

$$\tilde{v}(\{0\} \times S^1) \subset \{c\} \times \partial Z_n.$$

The constant c depends on the leaf. Indeed, if $F = \tilde{v}([0, \infty) \times S^1)$ is a leaf with boundary in $\{c\} \times \partial Z_n$, then translation in the $\mathbb{R}$-direction to $F_{c'}$ is a leaf with boundary in $\{c + c'\} \times \partial Z_n$.

For the elliptic theory to work well, a pseudoholomorphic curve with boundary is typically required to have each boundary component lie in a prescribed surface in $\mathbb{R} \times M$ having suitable properties. For example surfaces of the form $\{\text{const}\} \times L$ where λ restricts to a closed form on L, is one possibility. This is the situation we work with here, where $L = \partial Z_n$.

A finite energy foliation $\tilde{\mathcal{F}}$ associated to a Reeb-like mapping torus (Z_n, λ_n), must have a non-empty collection of half cylinder leaves. These determine a unique element in $H_1(\partial Z; \mathbb{Z})$. That is, if $\tilde{v} = (b, v) : [0, \infty) \times S^1 \to \mathbb{R} \times Z_n$ represents a leaf, then the restriction $v(0, \cdot) : S^1 \to \partial Z_n$ is a closed loop representing an homology class that is the same for all leaves in $\tilde{\mathcal{F}}$. This homology class we will refer to as the *boundary condition* of $\tilde{\mathcal{F}}$.

We can visualize the finite energy foliations of a Reeb-like mapping torus as in Figures 10 and 11.

The boundary condition can be described more succinctly in terms of a single integer.

DEFINITION 3.9. Let $\tilde{\mathcal{F}}$ be a finite energy foliation associated to a Reeb-like mapping torus (Z_n, λ_n). We will say that $\tilde{\mathcal{F}}$ has *boundary condition*

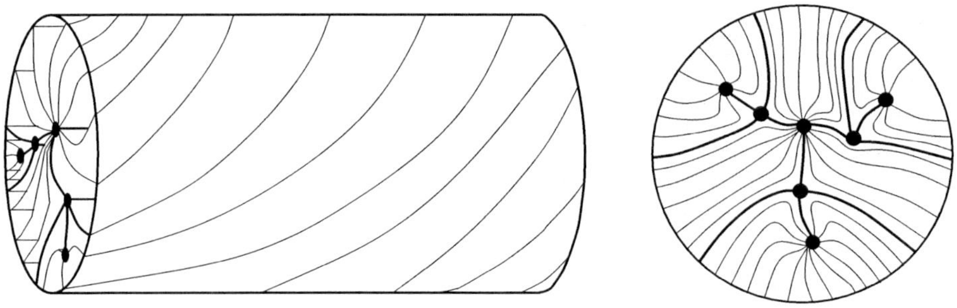

FIGURE 10. A global system of surfaces of section of a Reeb-like mapping torus. On the right a disk-like cross-section, where the spanning orbits are the dots. This can be compared with Figures 7 and 8, although here each spanning orbit corresponds to a single dot rather than a pair.

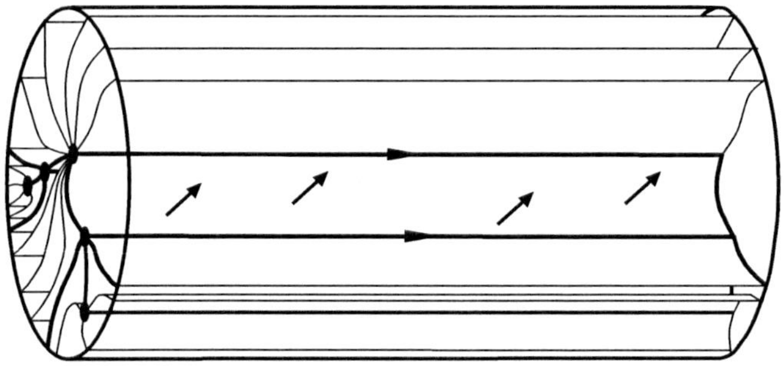

FIGURE 11. An opened out view of a global system of surfaces of section of a mapping torus. Inside, we see that the closure of each leaf is an embedded copy of $S^1 \times [0,1]$. The flow is transversal to the interior of each leaf, and tangent to those components of the boundary which lie in the interior of the solid torus. These are the spanning orbits.

$k \in \mathbb{Z}$ if every half cylinder leaf $F \in \tilde{\mathcal{F}}$ has boundary representing the homology class

$$L_n + k[\partial D] \in H_1(\partial Z_n; \mathbb{Z})$$

where L_n and $[\partial D]$ are the canonical longitude and meridian introduced earlier.

A variety of boundary conditions are illustrated in Figure 12.

REMARK 3.10. The projected leaves from the finite energy foliations always intersect the disk-slice

$$D_0 = \{0\} \times D \in Z_1,$$

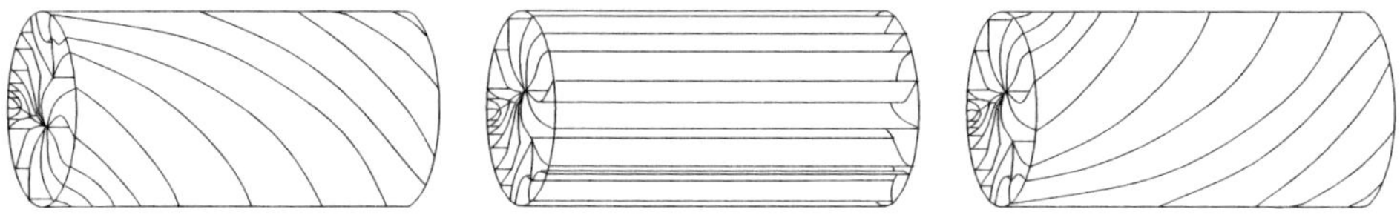

FIGURE 12. From left to right, the boundary condition is represented by the integers -1, 0, and 1 respectively. The winding of the leaves is measured relative to the canonical longitude $L_n \in H_1(\partial Z_n;\mathbb{Z})$.

transversally in the solid torus. Hence the nice cross-sectional pictures in the figures. We justify this now.

It is a simple matter to prescribe the almost complex structure J_1 on a neighborhood of D_0, without losing the necessary genericity. This way one can arrange that there exists an embedded pseudoholomorphic disk $\tilde{v} : D \to \mathbb{R} \times Z_1$ with boundary $\tilde{v}(\partial D) = \{0\} \times \partial D_0 \subset \{0\} \times Z_1 \subset \mathbb{R} \times Z_1$, such that the projection of $\tilde{v}$ to Z_1 is a parameterization of the disk slice D_0. By taking $\mathbb{R}$-translates of $\tilde{v}$ we foliate the whole hypersurface $\mathbb{R} \times D_0 \subset \mathbb{R} \times Z_1$ by a 1-parameter family of embedded holomorphic disks $\mathcal{D}$.

Let $\tilde{F} \in \tilde{\mathcal{F}}$ be a leaf in a finite energy foliation. Let $F = \mathrm{pr}(\tilde{F})$ denote the projection down to Z_1. We claim that F is transverse to D_0. Indeed, $\tilde{F}$ has topological intersection number 1 with each leaf in $\mathcal{D}$, so by positivity of intersections must also intersect each such leaf transversally in $\mathbb{R} \times Z_1$ and at a unique point. The upshot is that $\tilde{F}$ intersects the hypersurface $\mathbb{R} \times D_0$ transversally in $\mathbb{R} \times Z_1$, and so the projection $F = \mathrm{pr}(\tilde{F}) \subset Z_1$ intersects D_0 transversally in Z_1. So $F \cap D_0$ is in each case a connected, compact, non-empty, zero or 1-dimensional embedded submanifold of D_0.

The following was proven in [**11**].

THEOREM 3.11 (Bramham). *Let (Z_1, λ_1) be a Reeb-like mapping torus generating a non-degenerate disk map ψ, which coincides with an irrational rotation on the boundary of D. Let $\beta \in \mathbb{R}$ denote the total rotation number of ψ on the boundary, as determined by the mapping torus. For each $n \in \mathbb{N}$ let $k_n, k_n + 1$ be the two closest integers to $n\beta$. Then there exists an almost complex structure $\tilde{J}_1$ compatible with λ_1, such that for each $n \in \mathbb{N}$, there exist two finite energy foliations, we will denote by $\tilde{\mathcal{F}}_n^{k_n}$ and $\tilde{\mathcal{F}}_n^{k_n+1}$, associated to $(Z_n, \lambda_n, \tilde{J}_n)$, where $\tilde{J}_n$ is the lift of $\tilde{J}_1$, which have the following properties:*

1. ***Spanning orbits:*** *They share a unique spanning orbit. That is, $|\mathcal{P}(\tilde{\mathcal{F}}_n^{k_n}) \cap \mathcal{P}(\tilde{\mathcal{F}}_n^{k_n+1})| = 1$. The shared orbit has odd Conley-Zehnder index $2k_n + 1$.*
2. ***Boundary conditions:*** *The boundary condition for the leaves in $\tilde{\mathcal{F}}_n^j$, for $j \in \{k_n, k_n + 1\}$, is precisely the integer j.*

The following is a basic observation.

LEMMA 3.12. *Any two distinct spanning orbits γ_1, γ_2 of a finite energy foliation $\tilde{\mathcal{F}}$ have linking number $\mathrm{lk}(\gamma_1, \gamma_2) = k$ where k is the integer representing the boundary condition of $\tilde{\mathcal{F}}$.*

The next existence result is far more useful, although it already assumes the existence of a periodic orbit. For the following, note that the orbit γ having odd Conley-Zehnder index, is equivalent to a fixed point of the iterate ψ^n that is either elliptic, or is hyperbolic with unorientable stable and unstable manifolds.

"THEOREM" 3.13 (Bramham). *Let (Z_1, λ_1) be a Reeb-like mapping torus generating a non-degenerate disk map ψ, which coincides with an irrational rotation on the boundary of D*[*]*. Then there exists an almost complex structure $\tilde{J}_1$ compatible with λ_1, such that for each $n \in \mathbb{N}$, the data $(Z_n, \lambda_n, \tilde{J}_n)$ admits "many" finite energy foliations, where $\tilde{J}_n$ is the lift of $\tilde{J}_1$: Let γ be any periodic orbit in (Z_n, λ_n) homologous to the longitude L_n, and having odd parity Conley-Zehnder index. Then for each $k \in \mathbb{Z}$, there exists a finite energy foliation $\tilde{\mathcal{F}}^k(\gamma)$ associated to $(Z_n, \lambda_n, \tilde{J}_n)$ with the following two properties.*

(1) ***Spanning orbits:*** *γ is a spanning orbit. Equivalently $\mathbb{R} \times \gamma$ is a leaf.*
(2) ***Boundary conditions:*** *The boundary condition for the leaves in $\tilde{\mathcal{F}}^k(\gamma)$ is the chosen integer k.*

The strength of this statement is that it allows to choose a spanning orbit. Additional subtleties enter the proof when the boundary condition $k \in \mathbb{Z}$ lies in the twist interval of the pre-selected periodic orbit γ, which will appear in [**14**]. In this case it seems that the only way around certain difficulties is to use a refinement of contact homology developed by Momin [**65, 66**]. The simpler boundary conditions will be covered in [**13**].

Here is a simple way to see that Theorem 3.13 really does produce many different finite energy foliations, even with the same boundary conditions, provided there exist enough periodic orbits. Suppose that γ_1 and γ_2 are two distinct periodic orbits in (Z_n, λ_n) that are homologous to the longitude L_n. They have a linking number $\mathrm{lk}(\gamma_1, \gamma_2) \in \mathbb{Z}$. Now, by Lemma 3.12, for any integer $k \in \mathbb{Z}$ not equal to $\mathrm{lk}(\gamma_1, \gamma_2)$, any finite energy foliation associated to (Z_n, λ_n) that contains γ_1 as a spanning orbit cannot contain γ_2, unless the boundary condition is equal to $\mathrm{lk}(\gamma_1, \gamma_2)$. Thus if $\tilde{\mathcal{F}}^k(\gamma_1)$ and $\tilde{\mathcal{F}}^k(\gamma_2)$ are finite energy foliations with boundary condition $k \in \mathbb{Z}$ where k is not equal to $\mathrm{lk}(\gamma_1, \gamma_2)$, and having γ_1 and γ_2 as spanning orbits respectively, then $\tilde{\mathcal{F}}^k(\gamma_1) \neq \tilde{\mathcal{F}}^k(\gamma_2)$.

The discussion of integrable disk maps in Section 4.1 makes it clear in certain situations what kinds of different finite energy foliations one can

[*]Added in proof: all boundary assumptions in this result, and hence in all the applications here now appear to be unnecessary.

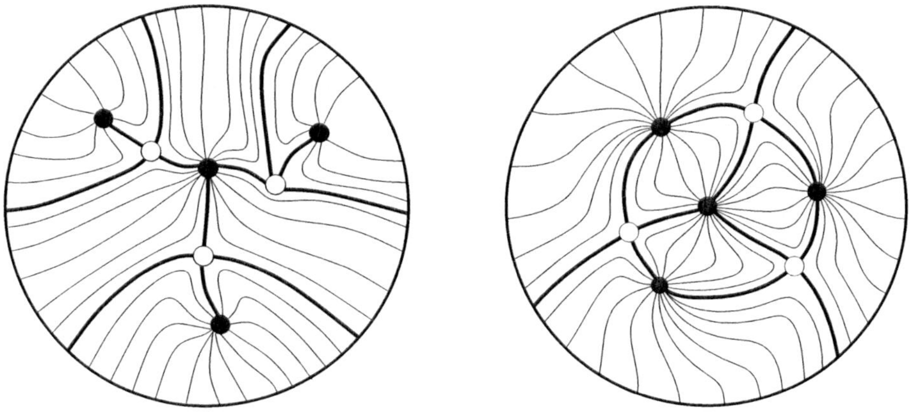

FIGURE 13. On the left a *simple* finite energy foliation. On the right the center spanning orbit is not connected directly to the boundary of the mapping torus by any leaf, so this example is not simple.

expect to find. For example, a finite energy foliation for a given (Z_n, λ_n) is in general not uniquely determined by its boundary condition and a single spanning orbit alone.

For the proofs in the next section it seems useful to distinguish the following feature for finite energy foliations of a Reeb-like mapping torus.

DEFINITION 3.14. Suppose that $\tilde{\mathcal{F}}$ is a finite energy foliation associated to a Reeb-like mapping torus on Z_n. We will say that $\tilde{\mathcal{F}}$ is *simple* if every spanning orbit is connected directly to the boundary of Z_n by a leaf in $\tilde{\mathcal{F}}$.

The following are equivalent characterizations of simple:

- The projection down to Z_n of the rigid leaves intersects the disk slice D_0 in a tree-like graph. Refer also to remark 3.10.
- There are no closed "cycles" of leaves.
- There are no Fredholm index-2 whole cylinders.

Perhaps the simplest example of a situation where one finds a non-simple finite energy foliation is the following.

Figure 14 depicts the flow lines of some autonomous smooth Hamiltonian $H : A \to \mathbb{R}$ on the closed annulus $A = \mathbb{R}/\mathbb{Z} \times [0, 1]$. The time-$T$ map of the flow, for $T > 0$, is an area preserving twist map ϕ in the sense of Poincaré and Birkhoff. In fact for $T > 0$ sufficiently small ϕ is a monotone twist map, meaning that $\phi = (\phi_1, \phi_2) : A \to A$ where $\frac{\partial \phi_1}{\partial x_2} > 0$ in coordinates (x_1, x_2).

Let us now consider the disk map $\psi : D \to D$ which one obtains by shrinking the inner circle to a point, which then corresponds to an elliptic fixed point of ψ at the origin. Let (Z_1, λ_1) be a Reeb-like mapping torus generating ψ, chosen so that the induced total rotation number on the boundary lies in the interval $(0, 1)$. Let us denote by γ_0 the periodic orbit corresponding to the fixed point $0 \in D$. Then a finite energy foliation $\tilde{\mathcal{F}}$ associated

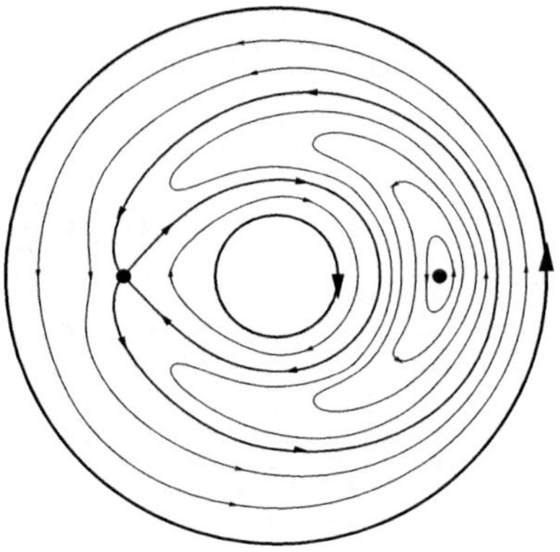

FIGURE 14

to (Z_1, λ_1) that has γ_0 as a spanning orbit, and has boundary condition 0, would, it turns out, have to look as in Figure 15, which is a non-simple foliation.

The property of $\tilde{\mathcal{F}}$ having a chain of rigid leaves surrounding the spanning orbit γ_0 and the "twisting" property going on about the fixed point 0 relative to the boundary behavior, are not coincidental. In the following precise sense there is a twist fixed point if and only if there is a non-simple finite energy foliation.

LEMMA 3.15. *Suppose that $\tilde{\mathcal{F}}$ is a finite energy foliation associated to a Reeb-like mapping torus (Z_n, λ_n). Let $k \in \mathbb{Z}$ be the boundary condition for $\tilde{\mathcal{F}}$. Then:*

(1) *If there exists a spanning orbit that is not connected directly to the boundary of Z_n by a leaf in $\tilde{\mathcal{F}}$ (i.e. if $\tilde{\mathcal{F}}$ is not simple), then there*

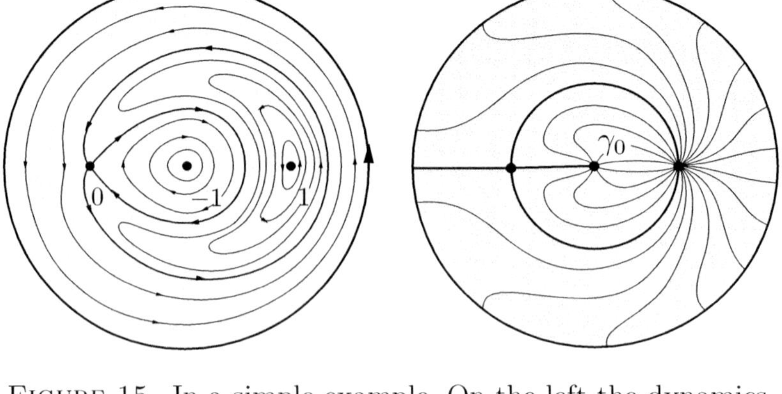

FIGURE 15. In a simple example. On the left the dynamics, with Conley-Zehnder indices, on the right one of the corresponding finite energy foliations.

exists a spanning orbit, not necessarily the same one, having $k \in \mathbb{Z}$ *in its twist interval.*

(2) *Every spanning orbit in* $\tilde{\mathcal{F}}$ *that is connected directly to the boundary of* Z_n *by at least one leaf, corresponds to a fixed point of* ψ^n *that does not have* $k \in \mathbb{Z}$ *in its twist interval. In particular, this implies the converse to (1).*

PROOF. (Of part (2) by pictures) Suppose that there exists a spanning orbit γ which has k in its twist interval. We will argue there cannot be a leaf in $\tilde{\mathcal{F}}$ which connects γ directly to the boundary of Z.

Due to the twist, the infinitesimal rotation number of γ lies on the opposite side of the value k to the rotation number describing the boundary behavior. The behavior of the leaves and the flow near γ and near the boundary are as in the figure.

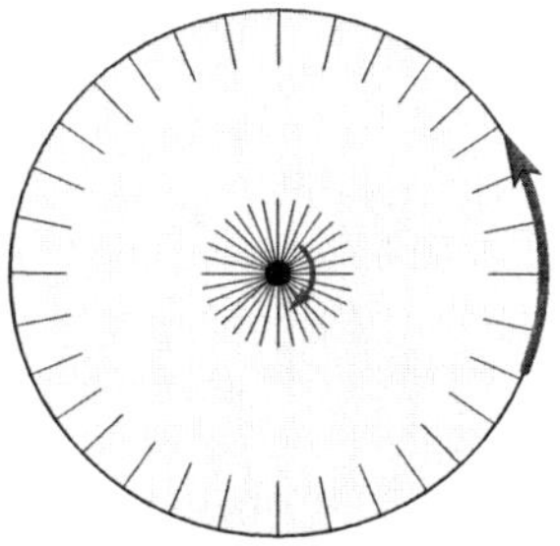

There is no way to complete the picture by connecting a leaf from the spanning orbit γ to the boundary without contradicting the transversality of the flow to the leaf; near the spanning orbit the flow winds around slower than the leaves, while the flow winds faster than the leaves near the boundary. ∎

The example in Figure 15 is just a special case of the integrable maps which are discussed in generality in Section 4.1.

3.4. A Proof of the Poincaré-Birkhoff Fixed Point Theorem. In 1913 Birkhoff in [**8**] proved the following conjecture of Poincaré* known as the Poincaré-Birkhoff fixed point theorem. See also [**9**]. Here A denotes the closed annulus $\mathbb{R}/\mathbb{Z} \times [0,1]$ and $\tilde{A}$ the universal covering $\mathbb{R} \times [0,1]$ with respect to the projection map $\pi(x,y) = ([x], y)$.

THEOREM 3.16 (Birkhoff). *Let* $\psi : A \to A$ *be an area preserving, orientation preserving, homeomorphism of the closed annulus with the following "twist" condition. There exists a lift* $\tilde{\psi} = (\tilde{\psi}_1, \tilde{\psi}_2) : \tilde{A} \to \tilde{A}$ *such that for all* $x \in \mathbb{R}$,

$$\tilde{\psi}_1(x,0) > x \quad \text{and} \quad \tilde{\psi}_1(x,1) < x.$$

Then ψ *has at least two fixed points which also lift to fixed points of* $\tilde{\psi}$.

*Allegedly the effort cost Birkhoff 30 pounds in weight [**5**].

In our framework we consider a smooth disk map $\psi : D \to D$ represented as the first return map of a Reeb-like mapping torus (Z_1, λ_1). We observed in Section 3.2 that the choice of a mapping torus allows us to assign canonical twist intervals to each fixed point of ψ, and an integer to pairs of distinct fixed points, called the linking number. Recall that the twist interval of ψ at a fixed point p was defined as the open interval of real numbers bounded by the (real valued) rotation number of ψ on the boundary of D and the (real valued) infinitesimal rotation number of ψ at p.

We will prove the following reformulation of Theorem 3.16.

THEOREM 3.17. *Suppose that $\psi : D \to D$ is a C^∞-smooth orientation preserving, area preserving, diffeomorphism. Let (Z_1, λ_1) be a Reeb-like mapping torus generating ψ. If ψ has an interior fixed point $p \in D$, for which there exists an integer $k \in \mathbb{Z}$ such that*

$$k \in \mathrm{twist}_\psi(p),$$

then ψ has at least two fixed points x_1, x_2, distinct from p, such that the corresponding periodic orbits $\gamma_1, \gamma_2 : S^1 \to Z_1$ have linking numbers $\mathrm{lk}(\gamma_1, \gamma_p) = \mathrm{lk}(\gamma_2, \gamma_p) = k$. *(In fact one also finds that* $\mathrm{lk}(\gamma_1, \gamma_2) = k$.*)*

The Poincaré-Birkhoff theorem, also in more general formulations than the statement in Theorem 3.16, has a long and beautiful history. We mention just a few references [**16, 19, 20, 21, 28, 29, 37, 38, 52, 62, 67, 70**]. It is perhaps not surprising that a variational approach in the spirit of Floer or Conley and Zehnder should produce a proof in the smooth category. Indeed we recall that this statement led Arnol'd to make his famous conjecture.

Nevertheless, we present now a proof by the first author of Theorem 3.17 using finite energy foliations, which is vaguely reminiscient of the simple argument that applies only to monotone twist maps. The existence of a *second* fixed point arises in a surprising way*, and not by using indices of fixed points.

PROOF. (Of Theorem 3.17) We wish to apply Theorem 3.13 which gives us the existence of certain finite energy foliations. Currently to use this result requires that the behavior of the disk map be a rigid (irrational) rotation on the boundary circle. So we first make an elementary argument to reduce the general case to this one. The reader who wishes to skip this should jump to step 1.

We are given a C^∞-smooth orientation preserving, area preserving, diffeomorphism $\psi : D \to D$ with an interior fixed point p, and an integer k lying in the twist interval of p. Observe that on any ε-neighborhood of the boundary of D we can modify ψ to obtain a new map ψ' with the following

*We later learned that there is some similarity in this part of the argument with that of Cotton-Clay [**20**] in his proof of a generalization of the Poincaré-Birkhoff theorem that uses symplectic Floer homology.

properties. ψ' agrees with ψ outside of the boundary strip, is a rigid rotation on the boundary of D, and is still a C^∞-smooth orientation preserving, area preserving, diffeomorphism. For example by composing ψ with a suitable Hamiltonian diffeomorphism for a Hamiltonian that is constant outside of the ε-neighborhood of ∂D. Moreover, one can do this in such a way that on the open ε-neighborhood ψ' sends points around in one direction further, in an angular sense, than ψ does, and "speeding up" as you approach the boundary circle. Pick this direction so as to enlarge the twist interval. If one does this for $\varepsilon > 0$ sufficiently small, one can arrange that ψ' has the property that any fixed points within the ε-neighborhood of the boundary have linking number with p different from k. To say all this rigorously one should of course work with a lift, on the complement of p, to the universal covering, but the idea is simple. Thus, it suffices to prove the assertion for this modified map ψ', because the two fixed points we find will automatically be fixed points of the unmodified map ψ.

Similarly, to apply the existence Theorem 3.13 it will be convenient to assume that the fixed point p is elliptic. A similar argument to the one just described allows to modify the disk map ψ on an ε-punctured-neighborhood of the fixed point p, this time so as to send points around p faster in the opposite direction, in an angular sense, increasing the twist interval still further. This way we can arrange that for the new map the eigenvalues of the linearization $D\psi'(p)$ lie on the unit circle, so that p is elliptic. The upshot is that without loss of generality we can assume that our disk map is a rigid rotation on the boundary of the disk (with any rotation number, in particular we may take it to be irrational), and that the fixed point p is elliptic. For the rest of the argument we will make these assumptions on ψ.

Pick a Reeb-like mapping torus (Z_1, λ_1) generating the disk map ψ. Let $\gamma_p : S^1 \to Z_1$ be the simply covered periodic orbit corresponding to the fixed point p. Since p is elliptic, γ_p has odd parity Conley-Zehnder index.

Step 1 The non-degenerate case: If the disk map, equivalently the Reeb flow, is non-degenerate, then the existence Theorem 3.13 applies immediately and provides a finite energy foliation $\tilde{\mathcal{F}}$ associated to (Z_1, λ_1) which has the odd index orbit γ_p for a spanning orbit, and has boundary condition the integer k.

By part (2) of Lemma 3.15 γ_p cannot be connected to the boundary of Z_1 by a leaf in $\tilde{\mathcal{F}}$ because $k \in \operatorname{twist}_\psi(p)$. Thus γ_p is "enclosed" by a chain of rigid leaves, as for example in Figure 16.

Note that the chain of rigid leaves has to include at least two spanning orbits, else the enclosed region has the property that points move in only one direction normal to its boundary, which would contradict the area preserving property of ψ.

Any two spanning orbits besides γ_p have linking number k with γ_p because this is the boundary condition for the foliation $\tilde{\mathcal{F}}$. Thus the theorem is proven provided ψ is non-degenerate.

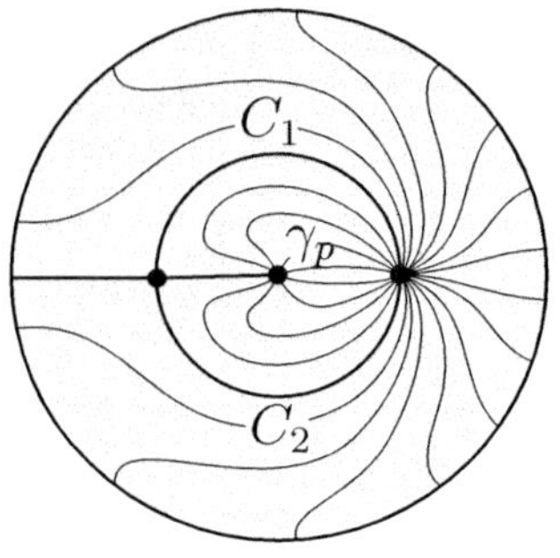

FIGURE 16. A chain of rigid leaves C_1 and C_2 enclosing γ_p.

Step 2 It is not immediately obvious how to complete the proof to the degenerate case while retaining *two* new fixed points. This part really uses compactness of the holomorphic curves, and not merely positivity of intersections.

Suppose ψ has possibly degenerate fixed points. Carrying out the above argument for a sequence of non-degenerate perturbations of ψ, keeping p fixed, we obtain a sequence of finite energy foliations $\{\tilde{\mathcal{F}}_j\}_j \in \mathbb{N}$ having γ_p as a spanning orbit and boundary condition k. By the argument in step 1 each $\tilde{\mathcal{F}}_j$ has at least two spanning orbits besides γ_p. These have uniformly bounded period, so as j goes to infinity they must converge to periodic orbits for the unperturbed mapping torus (Z_1, λ_1) that correspond to fixed points for ψ. The only concern is that all the spanning orbits might collapse onto a single limiting orbit, when the theorem requires two.

We argue as follows. Figure 17 illustrates the key idea. Suppose that indeed all spanning orbits from the sequence $\tilde{\mathcal{F}}_j$, besides γ_p, converge to a single periodic orbit σ in the limit as $j \to \infty$. Note first that the twist condition prevents σ from coinciding with γ_p. Indeed, the sequence of binding orbits which converge to σ have linking number k with γ_p, and this prevents them entering a small neighborhood of γ_p on which all points rotate either much faster or must slower than k depending on the direction of the infinitesimal twisting at p. So $\sigma \neq \gamma_p$.

For each j there exists a closed cycle of rigid leaves in $\tilde{\mathcal{F}}_j$, surrounding γ_p. Let S_j denote the intersection of such a closed cycle with the disk slice D_0. Each S_j is a continuously embedded closed loop in D_0, see remark 3.10. In particular each is non-empty, compact and connected. Taking a subsequence, we may assume that the sequence of sets S_j converges in the Hausdorff metric sense, to a non-empty compact set $S_\infty \subset D$. Each S_j is connected implies that S_∞ is connected. (Although the path connectedness need not pass to the limit.) Moreover, if S_∞ is disjoint from p and from ∂D, it is not hard to see that p and ∂D must both lie in different components of $D \backslash S_\infty$. We will establish that $S_\infty \cap (\{p\} \cup \partial D) = \emptyset$ in a moment.

As $j \to \infty$ the spanning orbits corresponding to fixed points in S_j converge to σ. This means that all their periods converge to that of σ. Which

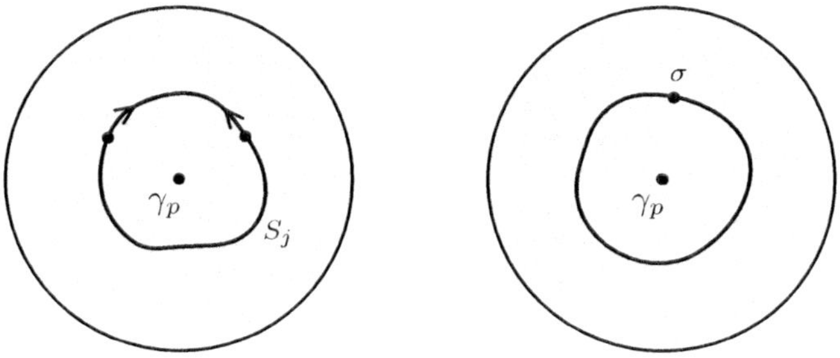

FIGURE 17. Illustrating the worst case scenario when the only two spanning orbits, besides the twist orbit γ_p, collapse onto a single orbit σ when taking a limit of a sequence of non-degenerate perturbations.

means, by Stokes' theorem, that the $d\lambda_j$-energy of the leaves in the S_j cycles converge uniformly to zero as $j \to \infty$. Thus if $F_j \in \tilde{\mathcal{F}}_j$ is any sequence of leaves for which the projection $\mathrm{pr}(F_j)$ meets D_0 in a point in S_j, then any convergent subsequence gives in the limit a pseudoholomorphic whole cylinder C having zero $d\lambda$-energy, and finite E-energy. Such a solution C is either constant or is a spanning cylinder for a periodic orbit (even though the periodic orbits are not necessarily non-degenerate). From the homology class of C we find that it is non-constant, and moreover that its spanning orbit corresponds to a fixed point of ψ.

Applying this idea to various $\mathbb{R}$-translates of such leaves $F_j \in S_j$, we can achieve every point in the set S_∞ as such a limiting fixed point. It follows that every point $q \in S_\infty$ is a fixed point of ψ and satisfies the following two alternatives: either $q = p$, or if $q \neq p$ then it must have linking number k with p. The twist condition implies that on a small punctured neighborhood of p there are no fixed points having this linking number k with p, and so S_∞ is disjoint from some punctured neighborhood of p. Moreover, S_∞ contains at least one point distinct from p because it contains the periodic orbit σ. It follows, since S_∞ is connected, that it is also disjoint from p. Similar arguments show that S_∞ is disjoint from the boundary of D.

Thus, as mentioned above, since p and ∂D are disjoint from S_∞, they lie in different components of the complement $D \backslash S_\infty$. The set S_∞ must therefore have infinitely many points. In particular at least two points. These are fixed points with the desired linking number k. ■

3.5. A Proof of a Theorem of Franks and Handel. We could summarize the last section by saying loosely that if an area preserving disk map has "twisting" in a suitable sense, then one can find periodic orbits and distinguish them by topological means, namely by their linking numbers.

To make this precise, let us say that a diffeomorphism $\psi : D \to D$ has "twisting" if, after making a choice that allows to define real valued rotation numbers, for example an isotopy to the identity, or a generating mapping torus, there exists an interior fixed point $z \in D$ such that

$$\text{Rot}_\psi(z) \neq \text{Rot}_\psi(\partial D), \tag{1}$$

where the left hand side is the total infinitesimal rotation number of ψ at z, and the right hand side is the total rotation number of ψ on the boundary of D. One way to define these is described in Section 3.2, both real valued numbers.

An equivalent way of saying this is that the twist interval of the fixed point z is non-empty. The twist interval grows linearly with iterates, so that for a high enough iterate it contains an integer,

$$\text{twist}_{\psi^n}(z) \cap \mathbb{Z} \neq \emptyset.$$

In this situation the Poincaré-Birkhoff fixed point theorem, as stated in Theorem 3.17, applies to ψ^n and we find that for each integer $k \in \text{twist}_{\psi^n}(z)$, there exists a fixed point $x \in \text{Fix}(\psi^n)$ (actually two fixed points) which has linking number k with z, $\text{lk}_{\psi^n}(x, z) = k$.

It is a simple matter to work with an unbounded sequence of iterates ψ^{n_j}, and pick integers k_j in the twist interval $\text{twist}_{\psi^{n_j}}(z)$ to be primes, and so obtain a sequence of fixed points x_j for which the average linking numbers with z,

$$\frac{\text{lk}_{\psi^{n_j}}(x_j, z)}{n_j} = \frac{k_j}{n_j},$$

are already in lowest form as fractions and are therefore pairwise distinct. Thus the sequence $(x_j)_j \in \mathbb{N}$ is a sequence of periodic points of ψ, lying on pairwise distinct orbits. Moreover, x_j has minimal period n_j, with $n_j \to \infty$ as $j \to \infty$. Thus, a disk map having a twist, or some iterate with a twist, has infinitely many periodic orbits, and moreover periodic orbits of unbounded minimal period.

In [**67**] the argument is pushed as far as possible. If ψ has a fixed point with non-empty twist interval (a, b), $a, b \in \mathbb{R}$, then for each $N \in \mathbb{N}$, the number $\mu_\psi(N)$ of periodic orbits of ψ having minimal period less than or equal to N grows like $|b-a|N^2$. Indeed, for each rational number $p/q \in \mathbb{Q}$ in (a, b) the argument above gives us a periodic point of ψ having average linking number p/q. So $\mu_\psi(N)$ is at least the number of rationals in (a, b) having denominator at most N when written in lowest form. This is equivalent to counting lattice points in a triangle, having relatively prime coordinates,

$$\mu_\psi(N) \geq |\{ (p, q) \in \mathbb{Z} \times \mathbb{Z} \mid Na < p < Nb, \ 1 \leq q \leq N \text{ and } \gcd(p, q) = 1\}|.$$

A little number theory, see [**67, 40**], tells us precisely what the limit of the quotient of the right hand side with N^2 is, leading to the following asymptotic estimate from below

$$\lim_{N\to\infty} \frac{\mu_\psi(N)}{N^2} \geq \frac{3|b-a|}{\pi^2}. \tag{2}$$

To summarize: a disk map with a twist has at least quadratic growth of periodic orbits. Although one might generically expect a far higher growth rate, quadratic is the best possible for general twist maps, see examples in [**67**].

Birkhoff himself applied these ideas to the problem of finding infinitely many closed geodesics on 2-spheres. In many cases he showed that the dynamics can be related to that of an area preserving diffeomorphism of an annulus, possibly without boundary. This holds for example whenever the metric has everywhere positive curvature. When the so called Birkhoff map has a twist he could apply his fixed point theorem as above to complete the argument.

But in the absence of a twist the problem of detecting infinitely many periodic orbits becomes much subtler. Indeed, a twist allows to conclude periodic orbits of unbounded period and to detect them by topological means. But this clearly does not work for example on the identity map of the disk which has infinitely many periodic orbits but which are indistinguishable by linking numbers. It was an open problem for some time, whether every smooth Riemannian metric on S^2 admits infinitely many distinct (prime) closed geodesics.

In 1992 John Franks proved the following celebrated result, [**30**] originally stated for annulus maps.

THEOREM 3.18 (Franks). *Let ψ be any area preserving, orientation preserving, homeomorphism of the open or closed unit disk in the plane. If ψ has 2 fixed points then it has infinitely many interior periodic orbits.*

A remarkable aspect of this statement is that no twisting type of condition is required. Let us point out that by imposing seemingly mild extra assumptions one can unwittingly introduce a twist into the system, thereby allowing much simpler arguments which miss the whole point of this theorem. Two examples which fall into this category, the first of which is a familiar one for symplectic geometers, are the assumption of non-degeneracy of periodic orbits, or that the rotation number on the boundary of the disk is irrational.

Franks' theorem completed the proof of the conjecture regarding closed geodesics on the two-sphere, in the cases where Birkhoff's annulus map is well defined. At the same time Bangert found a clever way to handle the cases where the Birkhoff map is not defined, [**4**]. Let us also note that shortly thereafter Hingston found another route to the closed geodesics question, using equivariant Morse homology [**40**], bypassing the part of the argument that required Theorem 3.18.*

An interesting strengthening of Franks' theorem is the following result due to Franks and Handel [**32**] in 2003. In fact much more generally, they

*Added in proof: recently a proof of the S^2-version of Theorem 3.18, in the smooth category, using Floer homology, has been announced by Collier, Kerman, Reiniger, Turmunkh, and Zimmer [**18**].

proved analogous statements for Hamiltonian diffeomorphisms on general closed oriented surfaces.

We would also like to mention that the Franks-Handel result was extended to the C^0-case by Le Calvez [**59**] using very different techniques. Indeed the approach taken by Le Calvez involved a surprising generalization of the Brouwer plane translation theorem [**57, 58**]. Given a surface homeomorphism isotopic to the identity (not assumed to be area preserving), he constructs a 1-dimensional foliation of the surface that in a certain sense is transverse to the dynamics. It is striking that he achieves this without any recourse to PDE's, or variational principle. Le Calvez' theory is also nicely outlined in the survey [**60**].

THEOREM 3.19 (Franks-Handel). *For a C^∞-smooth Hamiltonian diffeomorphism $\psi : S^2 \to S^2$ having at least three fixed points, either there is no bound on the minimal period of its periodic orbits, or ψ is the identity map.*

In the case when the map is not the identity, they obtain in [**32**] the following lower bound on the growth rate of periodic orbits. There exists a constant $c > 0$, depending on ψ, such that the number of periodic orbits $\mu_\psi(N)$ with period less than or equal to $N \in \mathbb{N}$ satisfies

$$\mu_\psi(N) \geq cN$$

for all $N \in \mathbb{N}$. A significant improvement on this growth rate estimate follows from work by Le Calvez [**61**]. The main result in [**61**] can be combined with the Franks-Handel theorem or [**59**], to conclude at least quadratic growth of periodic orbits. Namely, that there exists $c > 0$ such that $\mu_\psi(N) \geq cN^2$ for all $N \in \mathbb{N}$.

Since this is exactly the growth rate well known for twist maps, described in (2) above, it raises the following question:

> *Suppose that ψ is an area preserving, orientation preserving, diffeomorphism of the disk with at least two fixed points. If ψ is not the identity, is some iterate of it actually a twist map?*

Recall that we said that a disk map ψ is a *twist map* if it has the following property: that there exists an interior fixed point $p \in D \backslash \partial D$ of ψ, such that the annulus map

$$\psi : D\backslash\{p\} \to D\backslash\{p\}$$

has a twist in the following sense. Namely, that it admits a lift to a map on the infinite strip $F = (F_1, F_2) : \mathbb{R} \times (0, 1] \to \mathbb{R} \times (0, 1]$ satisfying the Poincaré-Birkhoff-like twist condition that the component F_1 is uniformly strictly positive on a neighborhood of the line $\mathbb{R} \times \{0\}$, and strictly negative on $\mathbb{R} \times \{1\}$ (or vice versa). Aternatively the twist interval of p, as in definition 3.7, contains an integer. Note that the question only asks for some *iterate* to be a twist map. As far as the authors are aware, the following is new.

"THEOREM" 3.20 (Bramham). *Let $\psi : D \to D$ be a C^∞-smooth, area preserving, orientation preserving, diffeomorphism having at least two fixed points, and coinciding with a rigid rotation on the boundary of the disk*[*]. *Then either ψ is the identity map, or there exists $n \in \mathbb{N}$ for which ψ^n is a twist map in the sense just described above.*

The quotation marks are because it relies on the unpublished "Theorem" 3.13.

Notice that this statement combined with the Poincaré-Birkhoff fixed point theorem gives us the analogous result of Franks and Handel, Theorem 3.19, but for disk maps instead of maps on S^2. By the discussion above it also implies the sharp quadratic growth lower bound on periodic orbits. Moreover, it allows to partition the set of area preserving diffeomorphisms of the disk into the following three disjoint subsets:

(1) **Pseudo-rotations:** those maps with a single periodic orbit.
(2) **Roots of unity:** those maps for which some iterate is the identity.
(3) **Twist maps:** those maps for which some iterate has an interior fixed point on whose complement we have the "twist" behavior just described above.

The following proof will appear in [**12**].

PROOF. (Of Theorem 3.20) Suppose that ψ has two fixed points, and that for all $n \in \mathbb{N}$, ψ^n has no twisting. In other words, every interior fixed point, of every iterate, has empty twist interval. Then we will show that $\psi(z) = z$ for all $z \in D$, by showing that the vertical foliation (definition 2.7) is a finite energy foliation.

Pick a suitable Reeb-like mapping torus (Z_1, λ_1) generating ψ. This gives us for each $n \in \mathbb{N}$ a mapping torus (Z_n, λ_n) generating the iterate ψ^n. With respect to these choices we associate real valued rotation numbers to ψ and its iterates at each fixed point and to the boundary of D. We proceed in three steps.

Step 1: The (real valued) rotation number of ψ on the boundary is an integer.

Suppose not. Then we may assume that ψ has only non-degenerate fixed points (because a degenerate fixed point has integer infinitesimal rotation number), and then the condition of no twisting would imply that the rotation number on the boundary is also an integer and we would be done.

Using that ψ has at least two fixed points one can show, using the foliations for example, or even just the Lefschetz fixed point formula, that there must be a hyperbolic fixed point with orientable stable and unstable manifolds. Since such a hyperbolic fixed point has integer infinitesimal rotation number the lack of twisting implies that the rotation number on the boundary is an integer.

---

[*] Added in proof: it now appears that no assumptions on the boundary are required.

Let us say that a fixed point is *odd* if the eigenvalues of the linearization lie either on the unit circle or the negative real line. Equivalently, the corresponding periodic Reeb orbit is either degenerate or is non-degenerate and has odd parity Conley-Zehnder index.

Step 2: In this step we show that all odd fixed points have the same λ_1-action, where we define the λ_1-action of a fixed point $x \in D$ to be the value

$$\mathcal{A}(x) = \int_{S^1} \gamma^* \lambda_1$$

where $\gamma : S^1 \to Z_1$ is the periodic orbit corresponding to x.

Let $k \in \mathbb{Z}$ be the total rotation number of ψ on the boundary. Being an integer ψ has at least one fixed point on the boundary of D. Let $a_0 > 0$ denote the λ_1-action of any of these boundary fixed points. It suffices to show that every interior odd fixed point of ψ has λ_1-action equal to a_0.

Fix any interior odd fixed point x. Let $\varepsilon > 0$. We will show that

$$|\mathcal{A}(x) - a_0| < \varepsilon.$$

Let $\gamma_x : S^1 \to Z_1$ be the corresponding closed Reeb orbit. Pick $n \in \mathbb{N}$ large enough that $\text{Area}_{d\lambda_1}(D)/n < \varepsilon$. Take a sequence of non-degenerate perturbations λ_1^j of λ_1 for which γ_x is a closed Reeb orbit for each λ_1^j also, with action converging to $\mathcal{A}(x)$. Since x is an odd fixed point of ψ, we may do this in such a way that γ_x has odd Conley-Zehnder index with respect to λ_1^j. Assume that we also perturbed near the boundary of Z_1 so that the first return map ψ_j is a rotation on the boundary of D with rotation number $k + \delta_j$, where for each j, $\delta_j \in (0, 1)$ is irrational.

By assumption ψ^n has no twist fixed points. Thus every twist interval has length zero, and $\text{Rot}_{\psi^n}(p) = \text{Rot}_{\psi^n}(\partial D) = nk$ for all $p \in \text{Fix}(\psi^n)$. Thus for sufficiently small perturbations, that is for j sufficiently large, we may assume that for each fixed point p of ψ_j^n, its twist interval is in a neighborhood of kn. For example we may assume that

$$\text{twist}_{\psi_j^n}(p) \subset \left(kn - \frac{1}{2}, kn + \frac{1}{2}\right).$$

Recall, by part (2) of Lemma 3.15, that any finite energy foliation with boundary condition outside of the twist interval of each of its spanning orbits is simple. Using "Theorem" 3.13 there exists a finite energy foliation $\tilde{\mathcal{F}}_j$ associated to λ_j on Z_n, having γ_x^n as a spanning orbit and any boundary condition we choose. Let us take the boundary condition to be $kn + 1$. As this is outside of every twist interval $\tilde{\mathcal{F}}_j$ must be simple.

It follows that there exists a half cylinder leaf in $\tilde{\mathcal{F}}_j$ which connects γ_x^n to the boundary. Positivity of the $d\lambda_j$-energy of this leaf and Stokes theorem give us the following estimate:

$$\int_{S^1} (\gamma_j^n)^* \lambda_j < \int_{L_n} \lambda_j + (nk + 1) \int_{\partial D} \lambda_j$$

where L_n is any representative of the canonical longitude on ∂Z_n. Dividing by n and letting $j \to \infty$ leads to

$$\mathcal{A}(x) \leq a_0 + \frac{1}{n}\text{Area}_{d\lambda_1}(D) < a_0 + \varepsilon.$$

To complete the argument we need the lower bound $\mathcal{A}(x) > a_0 - \varepsilon$ which can be obtained in the same way by changing the boundary conditions to $nk - 1$.

Step 3: In this step we show that the vertical foliation $\tilde{\mathcal{F}}^{\nu}(Z_1, \lambda_1)$ is also a finite energy foliation!

See definition 2.7 to recall how we define a vertical foliation. Take any sequence of approximating data $\lambda_1^j \to \lambda_1$ with the only requirement that each be non-degenerate (for fixed points at least) and have irrational rotation number on the boundary.

Let $\tilde{\mathcal{F}}_j$ be a finite energy foliation associated to (Z_1, λ_1^j) having boundary condition k. Recall that $k \in \mathbb{Z}$ is the rotation number of ψ on the boundary. There is no need to select a spanning orbit. Then we claim that the $d\lambda$-energies of all leaves converge to zero, that is $E_j \to 0$ where

$$E_j := \sup\left\{ \int_F d\lambda_1^j \mid F \in \tilde{\mathcal{F}}_j \right\}.$$

Roughly this is via the following argument: The value E_j must be achieved by a leaf in $\tilde{\mathcal{F}}_j$ having Fredholm index 2, since all other leaves have index 0 or 1 and lie on the boundary of a family of index-2 leaves, and the $d\lambda_1^j$-integral is lower semi-continuous under C^{∞}_{loc}-convergence. Any leaf having Fredholm index 2 has only odd index asymptotic orbits (or orbit). Any sequence of such leaves has $d\lambda_1^j$-energy decaying to zero because in the limit the asymptotic orbits converge to odd periodic orbits of λ_1, which by step (2) all have the same action.

It follows that the sequence $\tilde{\mathcal{F}}_j$, which has uniformly bounded E-energy, converges to the vertical foliation $\tilde{\mathcal{F}}^{\nu}(Z_1, \lambda_1)$ (one can also show that the limiting leaves are not constants). Thus the vertical foliation is a finite energy foliation, which means that every leaf is a cylinder over a periodic orbit. Thus the mapping torus (Z_1, λ_1) is foliated by periodic orbits. Moreover, each cylinder must be over a periodic orbit homologous to the longitude in Z_1, thus each periodic orbit represents a fixed point of ψ. ■

4. Asymptotic Foliations for Disk Maps

In this final section we begin exploring the following broad question. What is the asymptotic behavior of the foliations associated to arbitrarily high iterates of a disk map? Do the leaves converge in a useful sense, and if so, in what way does the limit reflect anything interesting dynamically? In contrast, the applications in Section 3 only used the framework for arbitrarily high, but finite numbers of iterates.

To develop the observations we present in this section presumably requires a serious study of what one could describe as "locally-finite" energy foliations. The context is as follows.

Let (Z_1, λ_1) be a Reeb-like mapping torus, and J_1 a compatible almost complex structure on $\mathbb{R} \times Z_1$. Let $Z_\infty = \mathbb{R} \times D$ denote the universal covering space of Z_1. The contact form and almost complex structure lift to λ_∞ and J_∞ respectively, which are invariant under the 1-shift automorphism

$$\begin{aligned} \tau : Z_\infty &\to Z_\infty \\ (\bar{z}, \bar{x}, \bar{y}) &\mapsto (\bar{z} + 1, \bar{x}, \bar{y}). \end{aligned} \tag{3}$$

One would like to take a sequence of finite energy foliations $\tilde{\mathcal{F}}_n$ associated to data (Z_n, λ_n, J_n), as n runs over the natural numbers or some subsequence thereof, and extract a limiting object $\tilde{\mathcal{F}}_\infty$ associated to $(Z_\infty, \lambda_\infty, J_\infty)$. One would like to arrange this so that $\tilde{\mathcal{F}}_\infty$ is a foliation by leaves that are the images of properly embedded pseudoholomorphic curves, where each leaf $F \in \tilde{\mathcal{F}}_\infty$ has a parameterization that is the C^∞_{loc}-limit of a sequence of parameterizations of leaves $F_n \in \tilde{\mathcal{F}}_n$. The foliation $\tilde{\mathcal{F}}_\infty$ would be invariant under the $\mathbb{R}$-action $c \cdot (a, m) \mapsto (a + c, m)$ on $\mathbb{R} \times Z_\infty$, but need not be invariant under the $\mathbb{Z}$-action generated by the transformation $\mathrm{id} \times \tau$, or even a finite iterate $\mathrm{id} \times \tau^q$.

When the contact manifold is compact it was discovered in [**41**] that finiteness of the E-energy picks out those holomorphic curves which do not behave too wildly, see Section 2.2. In the non-compact case $(Z_\infty, \lambda_\infty, J_\infty)$ there cannot be any non-constant curves with finite E-energy, since any such curve would pick out a periodic orbit and there are no periodic orbits in $(Z_\infty, \lambda_\infty)$. So presumably another, weaker, condition than global finiteness of the E-energy is required, perhaps some kind of averaging version of the E-energy.

Ignoring these crucial technicalities for the moment, we now outline two situations where nevertheless it is possible to see what happens asymptotically. In the first situation we will describe what happens when the disk map is integrable, the second is when the disk map has only a single periodic orbit, so called irrational pseudo-rotations which seem to have generated renewed interest in recent years [**25, 24, 26, 55, 6, 7**].

4.1. Detecting Invariant Circles. KAM theory guarantees the existence of closed invariant curves near a generic elliptic fixed point, or near to a given invariant circle with certain rotation number and torsion (infinitesimal twisting) conditions. In other words this is a, very successful, perturbation theory. Computer simulations indicate that generically, or often, invariant circles and quasi-periodic behavior should exist in a "global" sense, whatever this means precisely. An interesting source of pictures, numerical observations, and many questions, is MacKay's book [**64**]. More evidence for these global features arises in Aubry-Mather theory, which finds quasi-periodic behavior and invariant circles for monotone twist maps of the annulus. What about without the monotone twist assumption?* Note that if one doesn't view the annulus as a cotangent bundle then the monotone twist condition is not symplectic. An argument against asking this question is of course that the so called family of standard maps $f_\tau : \mathbb{R}/\mathbb{Z} \times \mathbb{R} \to \mathbb{R}/\mathbb{Z} \times \mathbb{R}$, for parameters $\tau \in \mathbb{R}_{>0}$,

$$f_\tau(x, y) = (x + y + \tau \sin(2\pi x), y + \tau \sin(2\pi x)),$$

are all monotone twists, and still there are huge open questions regarding these very explicit examples.

In this section we do not claim any new results, or recover any known ones. We merely describe an intriguing mechanism by which finite energy foliations asymptotically pick out all the invariant circles of an integrable disk map. The circles come filtered through their rotation numbers and the elliptic periodic orbits they enclose. By integrable, we mean the time-1 map of an autonomous Hamiltonian $H : D \to \mathbb{R}$. Of course, since we are in 2-dimensions, the long term behavior of integrable Hamiltonian systems is easily understood without anything so technical as holomorphic curves. But this mechanism uses the integrability in a very weak sense that needs to be better understood, and holomorphic curves have little regard for local versus global issues. One might hope therefore that more sophisticated variations of this approach work as well in very general situations.

Let $H : D \to \mathbb{R}$ be a smooth function, constant on the boundary of D. Viewing the area form $\omega_0 := dx \wedge dy$ as a symplectic form, we have an induced, autonomous, Hamiltonian vector field X_H uniquely solving

$$-dH(x, y) = \omega_0(X_H(x, y), \cdot)$$

for all $(x, y) \in D$, and automatically X_H will vanish on the boundary of D. Let $\varphi_H^t \in \mathrm{Diff}_{\omega_0}(D)$ denote the induced 1-parameter family of diffeomorphisms defined for each $t \in \mathbb{R}$, and let

$$\psi_H : D \to D$$

denote the time-1 map φ_H^1. Then ψ_H is an area preserving and orientation preserving diffeomorphism, and its long term behavior is easily understood because its orbits remain in level curves of H.

Let us imagine that we have picked an H for which the level curves are as in Figure 18, and suppose that λ_1^H is a contact form on $Z_1 = \mathbb{R}/\mathbb{Z} \times D$ giving us a Reeb-like mapping torus for ψ_H.

Let us give names to some of the features in the figure. There are seven isolated fixed points of ψ_H, six of which lie on an "island chain" and are alternately elliptic and hyperbolic $\{h_1, e_1, h_2, e_2, h_3, e_3\}$, which in some sense enclose another elliptic fixed point e_4.

*Added in proof: Le Calvez finds an interesting generalization of Aubry-Mather type phenomena, without any twist-like assumptions, in [**56**].

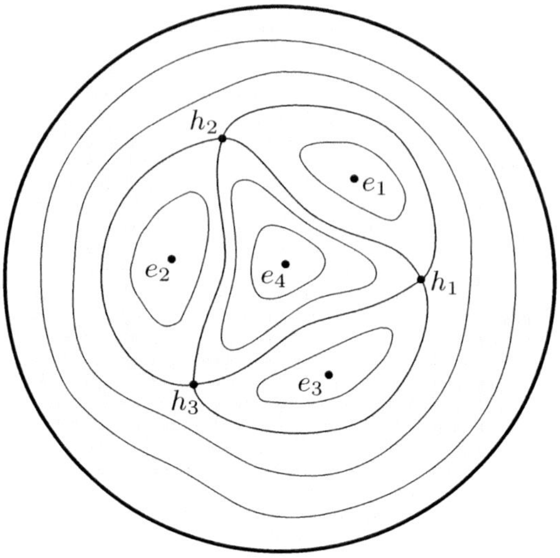

FIGURE 18. A configuration containing six fixed points which encloses an elliptic fixed point.

No matter what the flow on Z_1 is, provided it generates ψ_H as its first return map, the six fixed points on the island chain are "locked together" in the sense that any pair of them must have the same linking number. Let us assume that the flow is arranged so that this linking number is zero. In other words, for any distinct pair $x, y \in \{h_1, e_1, h_2, e_2, h_3, e_3\}$ we have $\mathrm{lk}(x, y) = 0$, if we abbreviate $\mathrm{lk}(x, y)$ to mean the linking number of the corresponding periodic orbits in (Z_1, λ_1^H). Since the island chain lies "outside" of e_4, it follows that for any $x \in \{h_1, e_1, h_2, e_2, h_3, e_3\}$ we have $\mathrm{lk}(x, e_4) = 0$ also. So, since the linking number is a symmetric function of its entries, all seven isolated fixed points have a common linking number which is zero.

Fix any one of the isolated elliptic fixed points, e.g. e_1. Let $\gamma_{e_1} : S^1 \to Z_1$ denote the periodic orbit in the mapping torus corresponding to the fixed point e_1. Let k be any integer. Potentially ψ_H has other non-isolated fixed points if one of the invariant circles has rotation number an integer, in which case Theorem 3.13 only applies to a small perturbation. But to see what is going on imagine that we can nevertheless find a transversal foliation $\tilde{\mathcal{F}}^k(\gamma_{e_1})$ of Z_1, which contains γ_{e_1} as a spanning orbit, and for which the boundary condition is the integer k. Then, by Lemma 3.12, every spanning orbit $\gamma \in \mathcal{P}(\tilde{\mathcal{F}}^k(\gamma_{e_1}))$ distinct from γ_{e_1}, (if such exists) must satisfy

$$\mathrm{lk}(\gamma, \gamma_{e_1}) = k. \tag{4}$$

Let us denote by

$$\mathcal{P}(Z_1, \lambda^H)$$

the collection of periodic Reeb orbits in Z_1 that correspond to fixed points of ψ_H, in other words those which are homologous to the longitude L_1. It turns out that, provided $k \neq 0$, we can characterize the subset of spanning orbits

$\mathcal{P}(\tilde{\mathcal{F}}^k(\gamma_{e_1})) \subset \mathcal{P}(Z_1, \lambda^H)$ as precisely those orbits $\gamma \in \mathcal{P}(Z_1, \lambda^H)$ which are either γ_{e_1} or which satisfy equation (4)! This is not true in general, and uses extremely weak properties of the integrability of ψ_H, and is not true for $k = 0$ without more information about ψ_H. (A more general statement along these lines, without any integrability assumptions, can be made, and may be the subject of a future paper.)

We may carry out this procedure for all iterates. The asymptotic behavior is interesting: First, fix an irrational number $\omega \in \mathbb{R}$. Now, pick any sequence of integers $k_n \in \mathbb{Z}$ such that

$$\lim_{n\to\infty} \frac{k_n}{n} = \omega. \tag{5}$$

Let $(Z_1, \lambda_1^H), (Z_2, \lambda_2^H,), ...$ be the sequence of Reeb-like mapping tori obtained by lifting (Z_1, λ_1^H), where (Z_n, λ_n^H) generates the n-th iterate ψ_H^n. For each $n \in \mathbb{N}$, let

$$\gamma_{e_1}^n : S^1 \to Z_n$$

denote the lift of γ_{e_1}. By Theorem 3.13 we find an almost complex structure J_1 compatible with (Z_1, λ_1^H), such that for each $n \in \mathbb{N}$ there exists a finite energy foliation $\tilde{\mathcal{F}}_n$ associated to (Z_n, λ_n^H, J_n), where J_n denotes the lift of J_1, with the following properties.

- $\gamma_{e_1}^n$ is a spanning orbit for $\tilde{\mathcal{F}}_n$.
- The boundary condition for $\tilde{\mathcal{F}}_n$ is the integer k_n.

Then the sequence of finite energy foliations $\tilde{\mathcal{F}}_n$ converges in a C^∞_{loc}-sense to a locally-finite energy foliation, we will denote by $\tilde{\mathcal{F}}_\infty$, of the symplectization of the universal covering $(Z_\infty, \lambda_\infty^H, J_\infty)$, as described at the beginning of Section 4.

What do the leaves of $\tilde{\mathcal{F}}_\infty$ look like?

Let $C_1, C_2, \ldots, C_m \subset D$ be those invariant circles C of the disk map ψ_H which are characterized by the following properties:

- When restricted to C, ψ_H has rotation number ω.
- C separates e_1 from the boundary of the disk. That is, e_1 and ∂D lie in different components of $D \backslash C$.

We can order these circles so that for each $j \in \{1, \ldots, m-1\}$ the circle C_{j+1} is closer to the boundary than C_j. Figure 19 illustrates a possible scenario. Then it turns out that $\tilde{\mathcal{F}}_\infty$ looks something like in Figure 20.

More precisely, the leaves in $\tilde{\mathcal{F}}_\infty$ which have vanishing $d\lambda_\infty$-energy correspond precisely to these circles and the fixed point e_1 itself. In other words, if $F \in \tilde{\mathcal{F}}_\infty$, has

$$\int_F d\lambda_\infty = 0,$$

then $F = \mathbb{R} \times \phi_z(\mathbb{R})$ is the plane over an orbit $\phi_z : \mathbb{R} \to Z_\infty$ which corresponds to the ψ-orbit of a point z in $\{e_1\} \cup C_1 \cup \ldots \cup C_m$. Conversely, for every point

$$z \in \{e_1\} \cup C_1 \cup \ldots \cup C_m$$

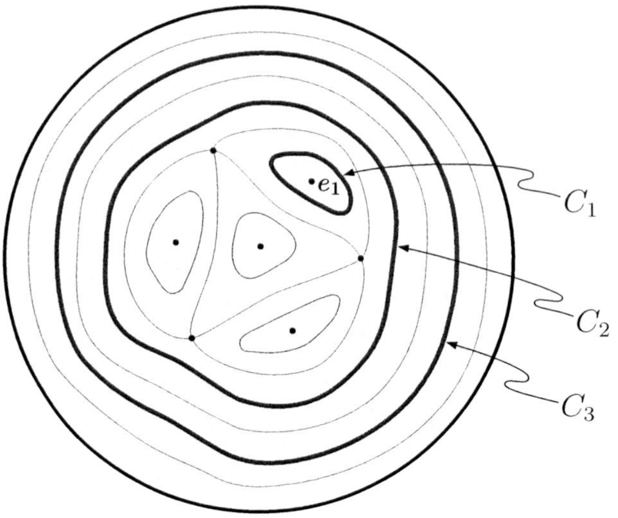

FIGURE 19.

the plane with image $\mathbb{R} \times \phi_z(\mathbb{R})$, where ϕ_z is the Reeb orbit passing through the point $(0, z) \in Z_\infty$, is a leaf in $\tilde{\mathcal{F}}_\infty$.

The remaining leaves can be described schematically as follows. Let $R_1, R_2, \ldots, R_m$ denote the open annular regions in the disk for which R_j has boundary circles C_{j-1}, C_j, for $j = 1, \ldots m$, and R_{m+1} is the remaining region between C_m and ∂D. (So R_{m+1} includes the points in the boundary of D unless $\partial D = C_m$.) For example see Figure 20. Then each leaf in $\tilde{\mathcal{F}}_\infty$ having non-zero $d\lambda_\infty$-energy is a plane which projects down to an infinitely long open strip in the three-manifold Z_∞. With respect to the coordinates $(\bar{z}, \bar{x}, \bar{y})$ on $Z_\infty = \mathbb{R} \times D$, the strip extends to plus and minus infinity in the $\bar{z}$-direction. The two boundary components of the closure of the strip are embedded copies of $\mathbb{R}$. Indeed, each is the image of a Reeb orbit; one corresponding to a point in some C_j or e_1, and the other corresponding to a point in C_{j+1} or C_1. Hence, each leaf with non-zero $d\lambda_\infty$-energy is in some sense asymptotic to a quasi-periodic orbit of ψ_H, with rotation number ω about the chosen fixed point e_1.

To summarize. A number of interesting issues arise when attempting to generalize this approach to finding invariant circles or quasi-periodic behavior. There are compactness issues when taking a limit of a sequence of finite energy foliations because the E-energy must blow up. Before one even takes a limit, there are questions regarding the symmetry of finite energy foliations that also seem important. For example, on the one hand it is much easier to understand symmetry of a foliation with respect to the $\mathbb{Z}$-action described in equation (3) when the map is integrable. On the another hand, the integrability is largely irrelevant.

With the success of KAM results in mind, one knows that Diophantine properties must play a role. It is conceivable that local KAM techniques

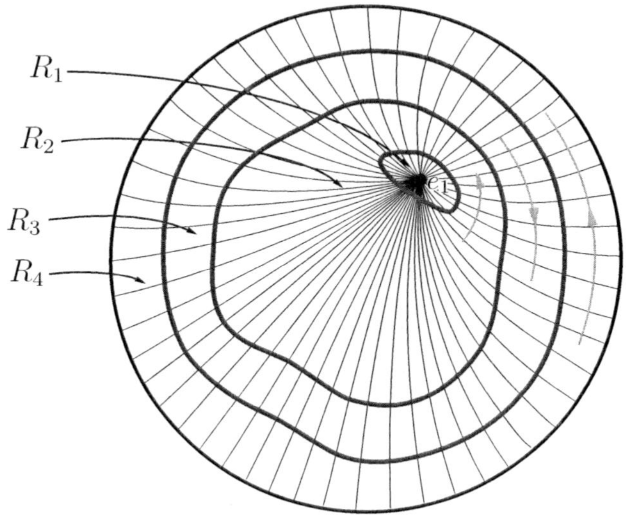

FIGURE 20. A limiting foliation for the integrable disk map ψ_H from Figure 18. The leaves with vanishing $d\lambda$-energy correspond to the quasi-periodic orbits in the invariant circles which have the chosen rotation number ω and which surround the chosen elliptic fixed point e_1. Each leaf with non-zero $d\lambda$-energy is a strip connecting two such orbits on neighboring circles. The flow is transverse to these strips in alternating directions as one passes through a circle, as indicated by the grey arrows.

enter the analysis and become combined with the pseudoholomorphic curves in this framework. In the next section we describe a more concrete situation where this may be a possibility.

4.2. Speculations on Pseudo-Rotations. Pseudo-rotations (also called irrational pseudo-rotations) can be defined as smooth, orientation preserving, area preserving diffeomorphisms of the unit disk which have a single fixed point and no other periodic points. Obvious examples are rigid rotations with irrational rotation number, and smooth conjugacies of these. The question arises whether these are the only examples.

As early as 1970 Anosov and Katok constructed "exotic" examples of pseudo-rotations which are ergodic [**3**] and which therefore cannot be conjugated to a rotation. In all their examples the rotation number on the boundary is a Liouville number - an irrational number well approximated by rationals. Later work of Fayad and Saprykina [**26**] established examples for any Liouville rotation number on the boundary. Their examples are not only ergodic but weak mixing. Meanwhile, unpublished work of Herman precluded such ergodic pseudo-rotations when the rotation number on the boundary is Diophantine - those irrationals which are not Liouvillean.

This seems to have led Herman to raise the following question at his ICM address in '98: is every pseudo-rotation with Diophantine rotation number on the boundary C^∞-smoothly conjugate to a rigid rotation? (That this is even true for circle diffeomorphisms is a deep result of his from '79 [**39**]).

An important step was recently taken in this direction. Fayad and Krikorian [**25**] proved a beautiful result, that in particular answered affirmatively a local version of Herman's question. Their approach, using KAM methods, was apparently to some degree based on ideas of Herman, and they referred to the result as "Herman's Last Geometric Theorem". They showed, that any pseudo-rotation that has Diophantine rotation number α on the boundary is C^∞-smoothly conjugate to the rigid rotation R_α, provided the disk map is already globally sufficiently close to R_α in some C^k-topology, where k is finite and depends on α.

The global question is apparently still open. Holomorphic curves, when one has them, have proven successful at handling global problems in symplectic geometry. Placing pseudo-rotations in the framework of this paper, there are plenty of holomorphic curves available, and we speculate that there might be something new to be gained from this angle. We briefly describe where first observations lead.

Let $\psi : D \to D$ be a pseudo-rotation, which on the boundary is a genuine rotation with rotation number α. Fix a Reeb-like mapping torus (Z_1, λ_1) generating ψ as its first return map. Let $\gamma : S^1 \to Z_1$ be the unique, simply covered, periodic orbit of the Reeb flow. For each $n \in \mathbb{N}$ we have the "longer" mapping torus (Z_n, λ_n) generating ψ^n, and the unique simply covered periodic orbit we denote by γ^n. For each such n Theorem 3.11 provides finite energy foliations $\tilde{\mathcal{F}}_n^-$ and $\tilde{\mathcal{F}}_n^+$ associated to (Z_n, λ_n) which have boundary condition $\lfloor n\alpha \rfloor$ and $\lceil n\alpha \rceil$ respectively. Each has the single spanning orbit, γ^n. A schematic picture of $\tilde{\mathcal{F}}_n^-$ and $\tilde{\mathcal{F}}_n^+$, is shown in Figure 21. In general the arcs connecting the fixed point to the boundary are not perfect radial lines of course, although depicted as such in the figure.

The $\mathbb{R}$-action on the foliations can be translated into an $\mathbb{R}$-action on the projected leaves in the three-manifold Z_n, which on the disk slice D_0 induces something akin to a "radial" coordinate on the disk.

It turns out that each finite energy foliation $\tilde{\mathcal{F}}_n^-$ and $\tilde{\mathcal{F}}_n^+$ gives rise to a smooth diffeomorphism of the disk, not necessarily area preserving,

$$\varphi_n^-, \varphi_n^+ : D \to D,$$

respectively, which fix the fixed point of ψ^n, are each n-th roots of the identity map on D, and the rotation number of φ_n^- on the boundary is $\lfloor n\alpha \rfloor / n$, while the rotation number of φ_n^+ on the boundary is $\lceil n\alpha \rceil / n$. Moreover, one has the following. A similar statement for the φ_n^+ maps holds also.

THEOREM 4.1 (Bramham). *Let $\psi : D \to D$ be a pseudo-rotation with fixed point p, and coinciding with a rigid rotation on the boundary circle. Let $\alpha \in \mathbb{R}/\mathbb{Z}$ be the rotation number on the boundary. Then for each $n \in \mathbb{N}$*

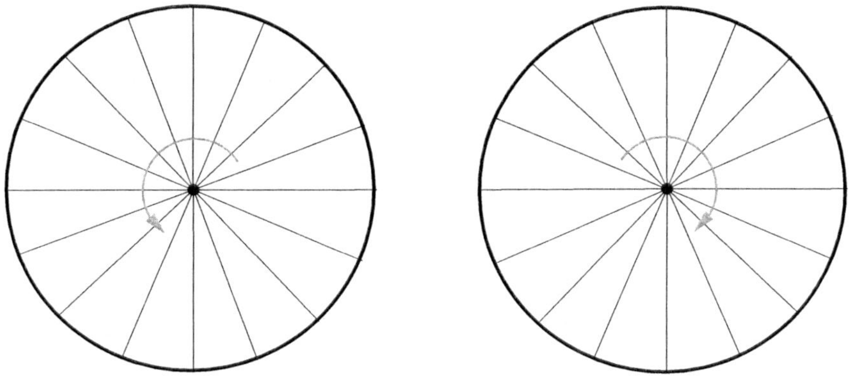

FIGURE 21. Two finite energy foliations associated to an iterate of a pseudo-rotation. On the left the trajectories pass through the leaves positively, on the right they pass through negatively.

there exists a C^∞-diffeomorphism $\varphi_n^- : D \to D$, of finite order, which fixes the point $p \in D$, for which there exists a subsequence $\varphi_{n_j}^-$ which converges in the C^0-topology to ψ. More precisely, $(\varphi_n^-)^n = \mathrm{id}_D$, and the rotation number on the boundary is the projection of $\lfloor n\alpha \rfloor / n$ down to $\mathbb{R}/\mathbb{Z}$.

Thus each pseudo-rotation is the C^0-limit of a sequence of maps which fill the whole disk, minus the fixed point, with smooth invariant circles. Note that no Diophantine conditions are placed on the rotation number α in this statement. Under the assumption that α is Diophantine one would like to control these circles and show that the limiting map, the pseudo-rotation being examined, also fills the disk with invariant circles.

If Herman's question is answered affirmatively, one could also ask the following question about Liouvillean pseudo-rotations. Does every pseudo-rotation with Liouvillean rotation number at least lie in the closure, perhaps in the C^∞-topology, of the set

$$\chi = \{h \circ R_t \circ h^{-1} \,|\, t \in \mathbb{R}/\mathbb{Z},\ h \in \mathrm{Diff}_{\omega_0}(D)\}.$$

Here R_t denotes the rotation through angle $2\pi t$ and $\mathrm{Diff}_{\omega_0}(D)$ refers to the smooth, area preserving, orientation preserving, diffeomorphisms. This is apparently unknown. If this were the case, it would mean that all pseudo-rotations which are conjugate to rotations lie in χ, while all those which are not lie on the boundary of χ. Theorem 4.1 suggests that the finite energy foliations might provide a way to approach this question. Both authors hope to explore these ideas in the future.

4.3. Questions about the Asymptotic Behavior of Finite Energy Foliations and Disk Maps.

QUESTION 4.2. Suppose that the area preserving disk map ψ has an elliptic fixed point $e \in D$ with irrational infinitesimal rotation number ω. See definition 3.6. Suppose moreover that ψ has a unique invariant circle, or quasi-periodic orbit, C surrounding e. If ψ is the time-1 map of an autonomous Hamiltonian then we described above a method by which a suitably chosen sequence of finite energy foliations picks out C. Does the same approach pick out C without the global integrability assumption on ψ?

QUESTION 4.3. Can one, and if so how, recover results of Aubry-Mather theory using the foliations? That is, assume that the disk map ψ is a monotone twist map on the complement of the fixed point $0 \in D$. How do we find quasi-periodic orbits surrounding 0 for all rotation numbers in the twist interval of 0? Can one do this using assumptions of monotone twist maps which are symplectic, that is, which remain true under any symplectic change of coordinates? For example, using only say some inequalities on the Conley-Zehnder indices of period orbits?

Suppose that $\tilde{\mathcal{F}}_j$, over $j \in \mathbb{N}$, is a sequence of finite energy foliations associated to (Z_{n_j}, λ_{n_j}), where (Z_1, λ_1) is a Reeb-like mapping torus generating $\psi : D \to D$, and n_j is a sequence of integers tending to $+\infty$. Let $k_j \in \mathbb{Z}$ be the boundary condition for $\tilde{\mathcal{F}}_j$.

QUESTION 4.4. Suppose that $e \in \text{Fix}(\psi)$ is an interior fixed point, let us say elliptic. Let $\gamma_e : S^1 \to Z_1$ be the corresponding periodic Reeb orbit. Suppose that for each j there exists a closed cycle C_j of rigid leaves in $\tilde{\mathcal{F}}_j$, whose projection down to Z_{n_j} separates $\gamma_e^{n_j}$ from the boundary of Z_{n_j}, and satisfies:

(1) The sequence $E_j := \max\{E_{d\lambda_{n_j}}(F) \,|\, F \in C_j\}$ converges to zero.
(2) The ratio k_j/n_j converges to some irrational number $\omega \in \mathbb{Q}^c$.
(3) Each C_j is invariant under the automorphism $\tau : \mathbb{R}\times Z_{n_j} \to \mathbb{R}\times Z_{n_j}$ where $\tau(a, z, x, y) = (a, z+1, x, y)$, meaning that for each leaf $F \in C_j$, we have $\tau(F) \in C_j$.

Let $S_j \subset D\backslash\{e\}$ be the unparameterized circle given by:

$$S_j := \{z \in D \,|\, \text{there exists } F \in C_j \text{ such that } (0, z) \in \text{pr}(F)\}$$

where $\text{pr} : \mathbb{R} \times Z_{n_j} \to Z_{n_j}$ is the projection.

The question then is: are there further conditions under which the sequence S_j of compact connected subsets of $D\backslash\{e\}$ converges in some sense to a closed ψ-invariant subset of $D\backslash\{e\}$ on which all points have rotation number ω about e?

Recall the deck transformation $\tau : Z_\infty \to Z_\infty$ given by $\tau(\bar{z}, \bar{x}, \bar{y}) = (\bar{z}+1, \bar{x}, \bar{y})$.

DEFINITION 4.5. For an integer $q \in \mathbb{Z}$, let us say that an $\mathbb{R}$-invariant foliation $\tilde{\mathcal{F}}_\infty$ by pseudoholomorphic curves associated to $(Z_\infty, \lambda_\infty, J_\infty)$ is *q-shift invariant* if for every leaf $F \in \tilde{\mathcal{F}}_\infty$ we have $\mathrm{id} \times \tau^q(F)$ is also a leaf.

An easy observation is that if $\tilde{\mathcal{F}}_\infty$ is q-shift invariant, then the subcollection of leaves having vanishing $d\lambda_\infty$-energy picks out a ψ^q-invariant subset of the disk. Indeed, let $\Omega \subset D$ be the set of $z \in D$ such that the plane

$$F = \mathbb{R} \times \phi_z(\mathbb{R}),$$

over the Reeb orbit $\phi_z : \mathbb{R} \to Z_\infty$ characterized by $\phi_z(0) = (0, z)$, is a leaf in $\tilde{\mathcal{F}}_\infty$. Then, since $\tau^q(\mathbb{R} \times \phi_z(\mathbb{R})) = \mathbb{R} \times \phi_{\psi^{-q}(z)}(\mathbb{R})$ is always true, then if $z \in \Omega$, then $\tau^q(F) \in \tilde{\mathcal{F}}_\infty$ implies that $\mathbb{R} \times \phi_{\psi^{-q}(z)}(\mathbb{R})$ is a leaf in $\tilde{\mathcal{F}}_\infty$, and so $\psi^{-q}(z) \in \Omega$. Thus $\psi^{-q}(\Omega) \subset \Omega$. Similarly as q-shift invariance implies $(-q)$-shift invariance, $\psi^q(\Omega) \subset \Omega$, and so $\psi^q(\Omega) = \Omega$.

QUESTION 4.6. Is a statement along the following lines true? Consider a sequence $\tilde{\mathcal{F}}_n$ of finite energy foliations associated to Reeb-like mapping tori $(Z_1, \lambda_1), (Z_2, \lambda_2), \ldots$ where (Z_1, λ_1) generates the area preserving disk map ψ. Suppose that the sequence of foliations $\tilde{\mathcal{F}}_n$ converges in a C^∞_{loc}-sense to a foliation $\tilde{\mathcal{F}}_\infty$ associated to the universal covering $(Z_\infty, \lambda_\infty)$. Suppose that $\tilde{\mathcal{F}}_\infty$ is q-shift invariant, for some $q \in \mathbb{Z}$. Let $\Omega \subset D$ be the set of points $z \in D$ such that the plane

$$F = \mathbb{R} \times \phi_z(\mathbb{R}),$$

over the Reeb orbit $\phi_z : \mathbb{R} \to Z_\infty$ characterized by $\phi_z(0) = (0, z)$, is a leaf in $\tilde{\mathcal{F}}_\infty$. We just saw that ψ^q restricts to a map on Ω. Then the question is, are there interesting general assumptions under which ψ^q must have zero topological entropy on Ω?

Note that in the last question one needs to rule out the vertical foliation $\tilde{\mathcal{F}}^\nu(Z_\infty, \lambda_\infty)$, as Ω for this is the whole disk regardless of the dynamics. Of course the vertical foliation is q-shift invariant for all $q \in \mathbb{Z}$, so this condition does not rule it out. This is another reason to phrase it so that $\tilde{\mathcal{F}}_\infty$ is achieved as a limit of finite energy foliations; then one could perhaps place assumptions on the sequence such as uniform bounds on some average notion of energy, or average of the boundary conditions, to rule out obtaining this trivial foliation.

Acknowledgment

This material is based upon work supported by the National Science Foundation under agreement No. DMS-0635607 (BB) and by NSF DMS-1047602 (HH). Any opinions, findings and conclusions or recommendations expressed in this material are those of the authors and do not necessarily reflect the views of the National Science Foundation.

References

[1] P. Albers, U. Frauenfelder, O. van Koert, G. Paternain, The contact geometry of the restricted 3-body problem, arXiv:1010.2140
[2] P. Albers and H. Hofer, On the Weinstein conjecture in higher dimensions. Comment. Math. Helv. 84 (2009), no. 2, 429–436.
[3] D.V. Anosov, A.B. Katok, New examples in smooth ergodic theory. Ergodic diffeomorphisms. Trans. Moscow Math. Soc. 23 (1970), 1–35.
[4] V. Bangert, On the existence of closed geodesics on two-spheres. Internat. J. Math. 4 (1993), no. 1, 1–10.
[5] J. Barrow-Green, Poincaré and the three body problem. History of Mathematics, 11. American Mathematical Society, Providence, RI; London Mathematical Society, London, 1997.
[6] F. Béguin, S. Crovisier, F. Le Roux, A Patou, Pseudo-rotations of the closed annulus: variation on a theorem of J. Kwapisz. Nonlinearity 17 (2004), 1427–1453.
[7] Béguin, F., Crovisier, S., Le Roux, F., Pseudo-rotations of the open annulus. Bull. Braz. Math. Soc. (N.S.) 37 (2006), no. 2, 275–306.
[8] G.D. Birkhoff, Proof of Poincaré's geometric theorem. Trans. Amer. Math. Soc. 14 (1913), no. 1, 14–22.
[9] G.D. Birkhoff, An extension of Poincaré's last geometric theorem. Acta Math. 47 (1926), no. 4, 297–311.
[10] F. Bourgeois, Y. Eliashberg, H. Hofer, K. Wysocki and E. Zehnder, Compactness Results in Symplectic Field Theory, Geometry and Topology, Vol. 7 (2003), pp. 799–888.
[11] B. Bramham, Pseudoholomorphic Foliations for Area Preserving Disc Maps, Ph.D.-thesis, New York University 2008.
[12] B. Bramham, A dynamical application of finite energy foliations to area preserving disc maps, in preparation.
[13] B. Bramham, Finite energy foliations with chosen binding orbits, in preparation.
[14] B. Bramham, Finite energy foliations with chosen binding orbits II, in preparation.
[15] M. Brown and W. Neumann, Proof of the Poincaré-Birkhoff fixed point theorem. Michigan Math. J. 24 (1977), no. 1, 21–31.
[16] P.H. Carter, An improvement of the Poincaré-Birkhoff fixed point theorem. Trans. Amer. Math. Soc. 269 (1982), no. 1, 285–299.
[17] K. Cieliebak, U. Frauenfelder and O. van Koert, The Cartan geometry of the rotating Kepler problem, preprint.
[18] Collier, Kerman, Reiniger, Turmunkh, Zimmer, A symplectic proof of a theorem of Franks. arXiv:1107.1282.
[19] C. Conley and E. Zehnder, The Birkhoff-Lewis fixed point theorem and a conjecture of V. I. Arnol'd. Invent. Math. 73 (1983), no. 1, 33–49.
[20] A. Cotton-Clay, A sharp bound on fixed points of surface symplectomorphisms in each mapping class, arXiv:1009.0760.
[21] W.-Y. Ding, A generalization of the Poincaré-Birkhoff theorem. Proc. Amer. Math. Soc. 88 (1983), no. 2, 341–346.
[22] Y. Eliashberg, A. Givental, H. Hofer, Introduction to Symplectic Field Theory, Geom. Funct. Anal. (2000), Special Volume, Part II, 560–673.
[23] Y. Eliashberg and H. Hofer, A Hamiltonian characterization of the three-ball. Differential Integral Equations 7 (1994), no. 5–6, 1303–1324.
[24] B. Fayad, A. Katok, Constructions in elliptic dynamics. Ergodic Theory Dynam. Systems 24 (2004), no. 5, 1477–1520.
[25] B. Fayad, R. Krikorian, Herman's last geometric theorem, Ann. Sci. Ecole Norm. Sup. 42 (2009), 193–219.

[26] B. Fayad, M. Saprykina, Weak mixing disc and annulus diffeomorphisms with arbitrary Liouville rotation number on the boundary. Ann. Sci. École Norm. Sup. (4) 38 (2005), 339–364.
[27] A. Floer, H. Hofer, C. Viterbo, The Weinstein conjecture in $P \times C^l$. Math. Z. 203 (1990), no. 3, 469–482.
[28] J. Franks, Generalizations of the Poincaré-Birkhoff theorem. Ann. of Math. (2) 128 (1988), no. 1, 139–151.
[29] J. Franks, A variation on the Poincaré-Birkhoff Theorem, Contemp. Math., 81, Amer. Math. Soc., Providence, RI, 1988, 111–117.
[30] J. Franks, Geodesics on S^2 and periodic points of annulus homeomorphisms. Invent. Math. 108 (1992), no. 2, 403–418.
[31] J. Franks, Rotation vectors and fixed points of area preserving surface diffeomorphisms. Trans. Amer. Math. Soc. 348 (1996), no. 7, 2637–2662.
[32] J. Franks and M. Handel, Periodic points of Hamiltonian surface diffeomorphisms, Geom. Topol. 7 (2003), 713–756.
[33] J. Franks, Rotation and numbers and instability sets. Bull. Amer. Math. Soc. (N.S.) 40 (2003), no. 3, 263–279.
[34] J. Franks and P. Le Calvez, Regions of instability for non-twist maps. Ergodic Theory Dynam. Systems 23 (2003), no. 1, 111–141.
[35] H. Geiges, An Introduction to Contact Topology, Cambridge studies in advanced mathematics, vol. 109.
[36] M. Gromov, Pseudoholomorphic Curves in Symplectic Geometry, Invent. Math. Vol. 82 (1985), 307–347.
[37] L. Guillou, Théorème de translation plane de Brouwer et généralisations du théorème de Poincaré-Birkhoff, Topology, 33 (1994), 331–351.
[38] L. Guillou, A simple proof of P. Carter's theorem, Proc. Amer. Math. Soc., 125 (1997), 1555–1559.
[39] M. Herman, Sur la conjugaison différentiable des difféomorphismes du cercle à des rotations. (French) Inst. Hautes Études Sci. Publ. Math. No. 49 (1979), 5–233.
[40] N. Hingston, On the growth of the number of closed geodesics on the two-sphere, Internat. Math. Res. Notices (1993), no. 9, 253–262.
[41] H. Hofer, Pseudoholomorphic curves in symplectizations with applications to the Weinstein conjecture in dimension three. Invent. Math. 114 (1993), no. 3, 515–563.
[42] H. Hofer, C. Viterbo, The Weinstein conjecture in the presence of holomorphic spheres. Comm. Pure Appl. Math. 45 (1992), no. 5, 583–622.
[43] H. Hofer, K. Wysocki and E. Zehnder, Properties of pseudo-holomorphic curves in symplectisations. II. Embedding controls and algebraic invariants. Geom. Funct. Anal. 5 (1995), no. 2, 270–328.
[44] H. Hofer, K. Wysocki and E. Zehnder, The dynamics on three-dimensional strictly convex energy surfaces. Ann. of Math. (2) 148 (1998), no. 1, 197–289.
[45] H. Hofer, K. Wysocki and E. Zehnder, Pseudoholomorphic curves and dynamics in three dimensions. Handbook of dynamical systems, Vol. 1A, 1129-1188, North-Holland, Amsterdam, 2002.
[46] H. Hofer, K. Wysocki and E. Zehnder, Finite energy foliations of tight three-spheres and Hamiltonian dynamics. Ann. of Math. (2) 157 (2003), no. 1, 125–255.
[47] U. Hryniewicz, Finite Energy Foliations for Convex Sets in $\mathbb{R}^4$, Ph.D.-thesis, New York University 2008.
[48] U. Hryniewcz, Fast finite-energy planes in symplectizations and applications, to appear Transactions of the AMS.
[49] U. Hryniewicz and P. Salomao, On the existence of disk-like global sections for Reeb flows on the tight 3-sphere, arXiv:1006.0049.
[50] A. Katok, Five most resistant problems in dynamics, MSRI-Evans Lecture, Berkeley September 2004.

[51] B. Kerékjártó, Über die periodischen Transformationen der Kreisscheibe und der Kugelfläche, Math. Ann. 80 (1919–1920), 36–38.
[52] B. Kerékjártó, The plane translation theorem of Brouwer and the last geometric theorem of Poincaré, Acta Sci. Math. Szeged, (4) (1928–1929), 86–102.
[53] G. Kuperberg, A volume-preserving counterexample to the Seifert conjecture, Comment. Math. Helv. 71, no. 1:70–97, 1996.
[54] K. Kuperberg, A smooth counter example to the Seifert conjecture in dimension three, Annals of Mathematics, 140:723–732, 1994.
[55] J. Kwapisz, Combinatorics of torus diffeomorphisms. Ergodic Theory Dynam. Systems 23 (2003), 559–586.
[56] P. Le Calvez, Ensembles invariants non enlacs des diffomorphismes du tore et de l'anneau. (Unlinked invariant sets of torus and annulus diffeomorphisms). Invent. Math. 155 (2004), no. 3, 561–603.
[57] P. Le Calvez, Une version feuilletée du théorème de Brouwer. Comment. Math. Helv. 79 (2004), 229–259.
[58] P. Le Calvez, Une version feuilletée équivariante du théorème de translation de Brouwer. Inst. Hautes Ètudes Sci. Publ. Math. 102 (2006), 1–98.
[59] P. Le Calvez, Periodic orbits of Hamiltonian homeomorphisms of surfaces. Duke Math. J. 133 (2006), no. 1, 125–184.
[60] P. Le Calvez, From Brouwer theory to the study of homeomorphisms of surfaces. International Congress of Mathematicians. Vol. III, 77-98, Eur. Math. Soc., Zürich, 2006.
[61] P. Le Calvez, Pourquoi les points periodiques des homeomorphismes du plan tournent-ils autour de certains points fixes?, Ann. Sci. Ec. Norm. Super. (4) 41 (2008), no. 1, 141–176.
[62] P. Le Calvez and J. Wang, Some remarks on the Poincaré-Birkhoff theorem. Proc. Amer. Math. Soc. 138 (2010), no. 2, 703–715.
[63] G. Liu, G. Tian, Weinstein conjecture and GW-invariants. Commun. Contemp. Math. 2 (2000), no. 4, 405–459.
[64] R. S. MacKay, Renormalisation in area-preserving maps. Advanced Series in Nonlinear Dynamics, 6. World Scientific Publishing Co., Inc., River Edge, NJ, 1993.
[65] A. Momin, Cylindrical Contact Homology on Complements of Reeb Orbits, arXiv:1003.3268
[66] A. Momin, Contact Homology of Orbit Complements and Implied Existence, arXiv:1012.1386
[67] W. Neumann, Generalizations of the Poincaré-Birkhoff fixed point theorem. Bull. Austral. Math. Soc. 17 (1977), no. 3, 375–389,
[68] Á Pelayo, S. Vũ Ngọc, Semitoric integrable systems on symplectic 4-manifolds. Invent. Math. 177 (2009) 571–597.
[69] Á Pelayo, S. Vũ Ngọc, Constructing integrable systems of semitoric type. Acta Math. 206 (2011) 93–125.
[70] H. Poincaré, Sur un théorème de géométrie, Rend. Circ. Mat. Palermo, 33 (1912), 375–407.
[71] H. Seifert, Closed integral curves in 3-space and isotopic two-dimensional deformations, Proc. Amer. Math. Soc., 1 (1950), 287–302.
[72] R. Siefring, Intersection theory of punctured pseudoholomorphic curves, arXiv:0907.0470.
[73] C. Taubes, The Seiberg-Witten equations and the Weinstein conjecture. Geom. Topol. 11 (2007), 2117–2202.
[74] C. Viterbo, A proof of Weinstein's conjecture in $\mathbb{R}^{2n}$. Ann. Inst. H. Poincaré Anal. Non Linéaire 4 (1987), no. 4, 337–356.

[75] A. Weinstein, On the hypotheses of Rabinowitz' periodic orbit theorems. J. Differential Equations 33 (1979), no. 3, 353–358.
[76] A. Weinstein, Lectures on symplectic manifolds, Regional Conference Series in Mathematics, No. 29. American Mathematical Society, Providence, R.I., 1977. iv+48 pp.
[77] C. Wendl, Finite Energy Foliations and Surgery on Transverse Links, Ph.D.-thesis, New York University 2005.
[78] C. Wendl, Finite energy foliations on overtwisted contact manifolds. Geom. Topol. 12 (2008), no. 1, 531–616.
[79] C. Wendl, Strongly fillable contact manifolds and J-holomorphic foliations. Duke Math. J. 151 (2010), no. 3, 337–384.

Institute for Advanced Study, Princeton, NJ 08540
E-mail address: bramham@ias.edu
E-mail address: hofer@ias.edu

Rigidity phenomena involving scalar curvature

Simon Brendle

ABSTRACT. We give a survey of various rigidity results involving scalar curvature. Many of these results are inspired by the positive mass theorem in general relativity. In particular, we discuss the recent solution of Min-Oo's Conjecture for the hemisphere (cf. [**13**]). We also analyze the case of equality in Bray's volume comparison theorem.

1. The positive mass theorem and its geometric consequences

In this paper, we discuss various rigidity results involving the scalar curvature. Our starting point is the positive mass theorem in general relativity. Recall that a three-manifold (M, g) is said to be asymptotically flat if there exists a compact set $\Omega \subset M$ such that $M \setminus \Omega$ is diffeomorphic to the region $\{x \in \mathbb{R}^3 : |x| > 1\}$ and the metric satisfies

$$\begin{aligned} g_{ij}(x) &= \delta_{ij} + O(|x|^{-1}), \\ \partial_k g_{ij}(x) &= O(|x|^{-2}), \\ \partial_k \partial_l g_{ij}(x) &= O(|x|^{-3}). \end{aligned}$$

Furthermore, we require that

$$\int_M |R_g| \, d\text{vol}_g < \infty, \tag{1}$$

where R_g denotes the scalar curvature of (M, g). The ADM mass of an asymptotically flat three-manifold (M, g) is defined by

$$m_{ADM} = \lim_{r \to \infty} \frac{1}{16\pi} \int_{\{|x|=r\}} \sum_{i,j} (\partial_j g_{ij}(x) - \partial_i g_{jj}(x)) \, \frac{x^i}{r} \tag{2}$$

The author was supported in part by the National Science Foundation under grant DMS-0905628.

(cf. [**2**], [**3**]). It follows from (1) and the divergence theorem that the limit in (2) exists, and m_{ADM} is well-defined.

THEOREM 1.1 (R. Schoen, S.T. Yau [**44**]; E. Witten [**50**]). *Let (M, g) be an asymptotically flat three-manifold with nonnegative scalar curvature. Then the ADM mass of (M, g) is nonnegative. Moreover, if the ADM mass of (M, g) is zero, then (M, g) is isometric to Euclidean space $\mathbb{R}^3$.*

Theorem 1.1 plays an important role in modern differential geometry. The original proof by Schoen and Yau [**44**] relies on minimal surface techniques. There is an alternative proof due to Witten [**50**], which uses spinors and the Dirac equation (see also [**42**]).

There is an analogous notion of asymptotic flatness for manifolds of dimension greater than three, and the definition of the ADM mass extends naturally to the higher dimensional setting (cf. [**3**]). The minimal surface arguments of Schoen and Yau imply that the positive mass theorem holds for every asymptotically flat manifold of dimension $n < 8$. On the other hand, by generalizing Witten's argument, Bartnik [**3**] was able to extend the positive mass theorem to spin manifolds of arbitrary dimension. It is an interesting question whether the positive mass theorem holds for non-spin manifolds of dimension $n \geq 8$. This question is studied in recent work of Lohkamp [**35**].

We next consider an important special case of the positive mass theorem. Let g be a Riemannian metric on $\mathbb{R}^n$ which agrees with the Euclidean metric outside a compact set. In this case, $(\mathbb{R}^n, g)$ is asymptotically flat and its ADM mass is equal to zero. Using Bartnik's version of the positive mass theorem, one can draw the following conclusion:

THEOREM 1.2. *Let g be a metric on $\mathbb{R}^n$ with nonnegative scalar curvature. Moreover, suppose that g agrees with the Euclidean metric outside a compact set. Then g is flat.*

Similar techniques can be used to show that the n-dimensional torus T^n does not admit a metric of positive scalar curvature.

THEOREM 1.3 (R. Schoen, S.T. Yau [**45**], [**46**]; M. Gromov, H.B. Lawson [**23**], [**24**]). *Let g be a metric on the torus T^n with nonnegative scalar curvature. Then g is flat.*

Theorem 1.3 was first proved for $n = 3$ by Schoen and Yau [**45**]. The proof relies on minimal surface techniques. In [**46**], the result was extended to dimension $n < 8$. The general case was settled by Gromov and Lawson using spinor methods (see [**23**], [**24**]).

We next discuss some rigidity results for bounded domains in $\mathbb{R}^n$. It was observed by Miao [**37**] that the positive mass theorem implies the following rigidity result for metrics on the unit ball:

THEOREM 1.4. *Suppose that g is a smooth metric on the unit ball $B^n \subset \mathbb{R}^n$ with the following properties:*

- *The scalar curvature of g is nonnegative.*
- *The induced metric on the boundary ∂B^n agrees with the standard metric on ∂B^n.*
- *The mean curvature of ∂B^n with respect to g is at least $n-1$.*

Then g is isometric to the standard metric on B^n.

In 2002, Shi and Tam extended Theorem 1.4 to arbitrary convex domains in $\mathbb{R}^n$. The following result is an important special case of Shi and Tam's theorem (cf. [**47**], Theorem 4.1):

THEOREM 1.5 (Y. Shi, L.F. Tam [**47**]). *Let Ω be a strictly convex domain in $\mathbb{R}^n$ with smooth boundary. Moreover, suppose that g is a Riemannian metric on Ω with the following properties:*

- *The scalar curvature of g is nonnegative.*
- *The induced metric on the boundary $\partial\Omega$ agrees with the restriction of the Euclidean metric to $\partial\Omega$.*
- *The mean curvature of $\partial\Omega$ with respect to g is positive.*

Then

$$\int_{\partial\Omega} (H_0 - H_g)\, d\sigma_g \geq 0, \tag{3}$$

where H_g denotes the mean curvature of $\partial\Omega$ with respect to g and H_0 denotes the mean curvature of $\partial\Omega$ with respect to the Euclidean metric. Finally, if equality holds in (3), then g is flat.

In order to prove Theorem 1.5, Shi and Tam glue the metric g to a suitable metric $\tilde{g}$ on the complement $\mathbb{R}^n \setminus \Omega$. The metric $\tilde{g}$ agrees with g along $\partial\Omega$, and the mean curvature of $\partial\Omega$ with respect to $\tilde{g}$ agrees with the mean curvature of $\partial\Omega$ with respect to g. Moreover, the metric $\tilde{g}$ is asymptotically flat, and its scalar curvature is equal to zero. Hence, the positive mass theorem implies that $m_{ADM} \geq 0$. On the other hand, Shi and Tam construct a monotone decreasing function $m(r)$ with the property that

$$m(0) = \int_{\partial\Omega} (H_0 - H_g)\, d\sigma_g$$

and

$$\lim_{r \to \infty} m(r) = c(n)\, m_{ADM},$$

where $c(n)$ is a positive constant (see [**47**], Theorem 2.1 and Lemma 4.2). Putting these facts together gives

$$\int_{\partial\Omega} (H_0 - H_g)\, d\sigma_g \geq c(n)\, m_{ADM} \geq 0,$$

and the last inequality is strict unless g is flat.

Theorems 1.2 and 1.3 show that it is not always possible to deform the metric so that the scalar curvature increases at each point. By contrast, Lohkamp [**34**] proved that for any Riemannian manifold there exist local deformations of the metric which decrease scalar curvature (see also [**33**]).

THEOREM 1.6 (J. Lohkamp [**34**]). *Let (M, g) be a complete Riemannian manifold of dimension $n \geq 3$, and let ψ be a smooth function on M such that $\psi(x) \leq R_g(x)$ for each point $x \in M$. Let $U = \{x \in M : \psi(x) < R_g(x)\}$, and let U_ε denote an ε-neighborhood of the set U. Given any $\varepsilon > 0$, there exists a smooth metric $\hat{g}$ on M such that $\psi(x) - \varepsilon \leq R_{\hat{g}}(x) \leq \psi(x)$ at each point in U_ε and $\hat{g} = g$ outside U_ε.*

Finally, let us mention a result of Fischer and Marsden [**19**] concerning small deformations of the scalar curvature. To that end, we fix a Riemannian manifold (M, g) of dimension $n \geq 3$. Given any symmetric two-tensor h, we define

$$L_g h = \frac{\partial}{\partial t} R_{g+th}\Big|_{t=0}.$$

A straightforward calculation gives

$$L_g h = \sum_{i,j=1}^{n} (D^2_{e_i,e_j} h)(e_i, e_j) - \Delta_g(\mathrm{tr}_g(h)) - \langle \mathrm{Ric}_g, h\rangle_g$$

(see e.g. [**5**], Theorem 1.174). Consequently, the formal adjoint of L_g is given by

$$L_g^* f = D^2 f - (\Delta_g f)\, g - f\, \mathrm{Ric}_g.$$

If the operator L_g^* has trivial kernel, then every function ψ which is sufficiently close to R_g in a suitable sense can be realized as the scalar curvature of a Riemannian metric. A local version of this result was established by Corvino [**17**].

We say that (M, g) is static if the operator L_g^* has non-trivial kernel (cf. [**4**]). Examples of static manifolds include the Euclidean space $\mathbb{R}^n$; the hyperbolic space $\mathbb{H}^n$; and the sphere S^n equipped with its standard metric. In the following sections, we will discuss various rigidity theorems for these model spaces.

2. Rigidity results for hyperbolic space

In this section, we discuss some rigidity theorems for hyperbolic space. The first result of this type was proved by Min-Oo in 1989.

THEOREM 2.1 (M. Min-Oo [**39**]). *Let g be a metric on $\mathbb{H}^n$ with scalar curvature $R_g \geq -n(n-1)$. Moreover, suppose that g agrees with the hyperbolic metric outside a compact set. Then g has constant sectional curvature -1.*

Theorem 2.1 can be viewed as the analogue of Theorem 1.2 in the hyperbolic setting. The proof of Theorem 2.1 uses spinor techniques, and is similar in spirit to Witten's proof of the positive mass theorem. The proof does not actually require the metric g to agree with the hyperbolic metric outside a compact set; it suffices to assume that g satisfies certain asymptotic conditions near infinity (see [**39**] for a precise statement). These asymptotic conditions were later weakened by Andersson and Dahl [**1**].

We note that there is an analogue of the positive mass theorem for asymptotically hyperbolic manifolds with scalar curvature $R_g \geq -n(n-1)$. Results in this direction were obtained by Chruściel and Herzlich [**15**] and Chruściel and Nagy [**16**]. Moreover, the following theorem was established by Wang [**49**]:

THEOREM 2.2 (X. Wang [**49**]). *Let g be a metric on the unit ball B^n with scalar curvature $R_g \geq -n(n-1)$. Moreover, suppose that g satisfies an asymptotic expansion of the form*

$$g = \sinh^{-2}(r) \left(dr^2 + g_0 + \frac{r^n}{n} h + O(r^{n+1}) \right),$$

where g_0 denotes the round metric on the boundary $\partial B^n = S^{n-1}$ and r is a boundary defining function. Then

$$\int_{S^{n-1}} \mathrm{tr}_{g_0}(h)\, d\mathrm{vol}_{g_0} \geq \left| \int_{S^{n-1}} \mathrm{tr}_{g_0}(h)\, x\, d\mathrm{vol}_{g_0} \right|.$$

Moreover, if equality holds, then g is isometric to the hyperbolic metric.

Finally, we point out that Boualem and Herzlich have obtained similar rigidity results for Kähler manifolds that are asymptotic to complex hyperbolic space in a suitable sense (cf. [**6**], [**29**]).

3. Min-Oo's Conjecture for the hemisphere

In this section, we discuss rigidity questions for the hemisphere. For abbreviation, we will denote by $\overline{g}$ the standard metric on S^n. Motivated by the positive mass theorem and its analogue in the asymptotically hyperbolic setting, Min-Oo proposed the following conjecture (cf. [**40**], Theorem 4):

MIN-OO'S CONJECTURE. *Suppose that g is a smooth metric on the hemisphere $S^n_+ = \{x \in S^n : x_{n+1} \geq 0\}$ with the following properties:*

- *The scalar curvature of g is at least $n(n-1)$.*
- *The induced metric on the boundary ∂S^n_+ agrees with the standard metric on ∂S^n_+.*
- *The boundary ∂S^n_+ is totally geodesic with respect to g.*

Then g is isometric to the standard metric on S^n_+.

Min-Oo's Conjecture is very natural given the analogy with the positive mass theorem, and was widely expected to be true; see e.g. [**22**], p. 47, or [**27**], p. 629. Various attempts have been made to prove it (using both spinor and minimal surface techniques), and many partial results have been obtained. In particular, it follows from a classical result of Toponogov that Min-Oo's Conjecture holds in dimension 2.

THEOREM 3.1 (V. Toponogov [**48**]). *Let (M, g) be a compact surface with totally geodesic boundary ∂M. If the Gaussian curvature of (M, g) satisfies $K \geq 1$, then the length of ∂M is at most 2π. Moreover, if equality holds, then (M, g) is isometric to the hemisphere S^2_+ equipped with its standard metric.*

An alternative proof of Theorem 3.1 was given by Hang and Wang (cf. [**27**], Theorem 4). In higher dimensions, Hang and Wang [**27**] showed that Min-Oo's Conjecture holds if the lower bound for the scalar curvature is replaced by a lower bound for the Ricci tensor.

THEOREM 3.2 (F. Hang, X. Wang [**27**]). *Suppose that g is a smooth metric on the hemisphere S^n_+ with the following properties:*

- *The Ricci curvature of g is bounded from below by $\mathrm{Ric}_g \geq (n-1)\, g$.*
- *The induced metric on the boundary ∂S^n_+ agrees with the standard metric on ∂S^n_+.*
- *The second fundamental form of the boundary ∂S^n_+ with respect to g is nonnegative.*

Then g is isometric to the standard metric on S^n_+.

The proof of Theorem 3.2 relies on an interesting application of Reilly's formula. Moreover, Hang and Wang [**26**] were able to verify Min-Oo's Conjecture for metrics conformal to the standard metric. In particular, Min-Oo's Conjecture is true if the metric g is rotationally symmetric.

THEOREM 3.3 (F. Hang, X. Wang [**26**]). *Suppose that $g = e^{2w}\,\overline{g}$ is a metric in the conformal class of $\overline{g}$ with scalar curvature $R_g \geq n(n-1)$. If $g = \overline{g}$ along the boundary ∂S^n_+, then $g = \overline{g}$ at each point in S^n_+.*

Note that Theorem 3.3 does not require any assumptions on the second fundamental form of ∂S^n_+.

In a joint work with F.C. Marques, we have obtained scalar curvature rigidity results for certain geodesic balls in S^n; see [**12**].

THEOREM 3.4 (S. Brendle, F.C. Marques [**12**]). *Fix a real number $c \geq \frac{2}{\sqrt{n+3}}$, and let $\Omega = \{x \in S^n : x_{n+1} \geq c\}$. Moreover, suppose that g is a Riemannian metric on Ω with the following properties:*

- *$R_g \geq n(n-1)$ at each point in Ω.*
- *The metrics g and $\overline{g}$ induce the same metric on $\partial\Omega$.*
- *$H_g \geq H_{\overline{g}}$ at each point on $\partial\Omega$.*

If $g - \overline{g}$ is sufficiently small in the C^2-norm, then $\varphi^*(g) = \overline{g}$ for some diffeomorphism $\varphi : \Omega \to \Omega$ with $\varphi|_{\partial\Omega} = \mathrm{id}$.

The following result is an immediate consequence of Theorem 3.4.

COROLLARY 3.5 (S. Brendle, F.C. Marques [**12**]). *Suppose that g is a Riemannian metric on the hemisphere S^n_+ with the following properties:*

- *$R_g \geq n(n-1)$ at each point in S^n_+.*
- *The metrics g and $\overline{g}$ agree in the region $\{x \in S^n : 0 \leq x_{n+1} \leq \frac{2}{\sqrt{n+3}}\}$.*

If $g - \overline{g}$ is sufficiently small in the C^2-norm, then g is isometric to the standard metric $\overline{g}$.

Eichmair [**18**] has verified Min-Oo's Conjecture for $n = 3$, assuming that the boundary satisfies an isoperimetric condition. The proof of this theorem uses techniques developed by Bray [**7**]. Moreover, Huang and Wu [**30**] showed that Min-Oo's Conjecture holds for graphs in $\mathbb{R}^{n+1}$.

We next describe a non-rigidity theorem for the hemisphere. This result implies that Min-Oo's Conjecture fails in dimension $n \geq 3$. The argument involves two steps. In a first step, we perturb the standard metric on the hemisphere S^n_+ in such a way that the scalar curvature increases at each point and the mean curvature of the boundary becomes positive:

THEOREM 3.6 (S. Brendle, F.C. Marques, A. Neves [**13**]). *Given any integer $n \geq 3$, there exists a smooth metric g on the hemisphere S^n_+ with the following properties:*

- *The scalar curvature of g is strictly greater than $n(n-1)$.*
- *At each point on ∂S^n_+, we have $g - \overline{g} = 0$, where $\overline{g}$ denotes the standard metric on S^n_+.*
- *The mean curvature of ∂S^n_+ with respect to g is strictly positive (i.e. the mean curvature vector points inward).*

The proof of Theorem 3.6 relies on a perturbation analysis. This construction is inspired by the counterexamples to Schoen's Compactness Conjecture for the Yamabe problem (cf. [**10**], [**11**]). We will give an outline of the proof of Theorem 3.6 in Section 4.

In a second step, we construct metrics on the hemisphere S^n_+ which have scalar curvature at least $n(n-1)$ and agree with the standard metric in a neighborhood of the boundary. To that end, we glue the metrics constructed in Theorem 3.6 to a rotationally symmetric model metric. The proof relies on a general gluing theorem, which is of interest in itself.

THEOREM 3.7 (S. Brendle, F.C. Marques, A. Neves [**13**]). *Let M be a compact manifold of dimension n with boundary ∂M, and let g and $\tilde{g}$ be two smooth Riemannian metrics on M such that $g - \tilde{g} = 0$ at each point on ∂M.*

Moreover, we assume that $H_g - H_{\tilde{g}} > 0$ *at each point on* ∂M*. Given any real number* $\varepsilon > 0$ *and any neighborhood* U *of* ∂M*, there exists a smooth metric* $\hat{g}$ *on* M *with the following properties:*

- *We have the pointwise inequality* $R_{\hat{g}}(x) \geq \min\{R_g(x), R_{\tilde{g}}(x)\} - \varepsilon$ *at each point* $x \in M$*.*
- $\hat{g}$ *agrees with* g *outside* U*.*
- $\hat{g}$ *agrees with* $\tilde{g}$ *in a neighborhood of* ∂M*.*

The proof of Theorem 3.7 involves a delicate choice of cut-off functions (see [**13**], Section 4). We will omit the details here.

Combining Theorem 3.6 and Theorem 3.7, we are able to construct counterexamples to Min-Oo's Conjecture in dimension $n \geq 3$.

THEOREM 3.8 (S. Brendle, F.C. Marques, A. Neves [**13**]). *Given any integer* $n \geq 3$*, there exists a smooth metric* $\hat{g}$ *on the hemisphere* S^n_+ *with the following properties:*

- *The scalar curvature of* $\hat{g}$ *is at least* $n(n-1)$ *at each point on* S^n_+*.*
- *The scalar curvature of* $\hat{g}$ *is strictly greater than* $n(n-1)$ *at some point on* S^n_+*.*
- *The metric* $\hat{g}$ *agrees with the standard metric* $\overline{g}$ *in a neighborhood of* ∂S^n_+*.*

Sketch of the proof of Theorem 3.8. Let δ be a small positive number. We first construct a rotationally symmetric metric $\tilde{g}_\delta$ on the hemisphere S^n_+ such that $R_{\tilde{g}_\delta} > n(n-1)$ in the region $\{x \in S^n : \delta < x_{n+1} < 3\delta\}$ and $\tilde{g}_\delta = \overline{g}$ in the region $\{x \in S^n : 0 \leq x_{n+1} \leq \delta\}$.

Let $M_\delta = \{x \in S^n : x_{n+1} \geq 2\delta\}$. Using Theorem 3.7, we can construct a metric g_δ on M_δ with the following properties:

- $R_{g_\delta} > n(n-1)$ at each point in M_δ.
- $g_\delta - \tilde{g}_\delta = 0$ at each point on the boundary ∂M_δ.
- $H_{g_\delta} - H_{\tilde{g}_\delta} > 0$ at each point on ∂M_δ.

Applying Theorem 3.7 to the metrics g_δ and $\tilde{g}_\delta$, we obtain a metric $\hat{g}$ on M_δ with the property that $R_{\hat{g}} > n(n-1)$ at each point in M_δ and $\hat{g} = \tilde{g}_\delta$ in a neighborhood of ∂M_δ. Hence, we may extend the metric $\hat{g}$ to the hemisphere in such a way that the resulting metric has all the required properties. From this, Theorem 3.8 follows.

To conclude this section, we state a corollary of Theorem 3.8. Let $\hat{g}$ be the metric constructed in Theorem 3.8. We may extend $\hat{g}$ to a metric on S^n which is invariant under antipodal reflection. The resulting metric descends to a metric on the real projective space $\mathbb{RP}^n$. Hence, we can draw the following conclusion:

COROLLARY 3.9 (S. Brendle, F.C. Marques, A. Neves [**13**]). *Given any integer* $n \geq 3$*, there exists a smooth metric* g *on the real projective space* $\mathbb{RP}^n$

with the following properties:

- *The scalar curvature of g is at least $n(n-1)$ at each point on $\mathbb{RP}^n$.*
- *The scalar curvature of g is strictly greater than $n(n-1)$ at some point on $\mathbb{RP}^n$.*
- *The metric g agrees with the standard metric in a neighborhood of the equator in $\mathbb{RP}^n$.*

4. Sketch of the proof of Theorem 3.6

In this section, we sketch the main ideas involved in the proof of Theorem 3.6. A complete proof is presented in [**13**]. A crucial issue is that the standard metric $\overline{g}$ on S^n is static. In fact, if we denote by $f : S^n \to \mathbb{R}$ the restriction of the coordinate function x_{n+1} to the unit sphere S^n, then f satisfies the equation

$$\overline{D}^2 f - (\Delta_{\overline{g}} f)\,\overline{g} - f\,\mathrm{Ric}_{\overline{g}} = 0. \tag{4}$$

Consequently, the linearized operator $L_{\overline{g}}$ fails to be surjective. In particular, Corvino's theorem concerning local deformations of the scalar curvature (see [**17**], Theorem 1) does not apply in this situation.

We next consider the upper hemisphere $S^n_+ = \{f \geq 0\}$. To fix notation, let $\Sigma = \{f = 0\}$ denote the equator in S^n and let ν be the outward-pointing unit normal vector field along Σ. Given any Riemannian metric g on S^n_+, we define

$$\mathscr{F}(g) = \int_{S^n_+} R_g\, f\, d\mathrm{vol}_{\overline{g}} + 2\,\mathrm{area}(\Sigma, g). \tag{5}$$

This defines a functional $\mathscr{F}$ on the space of Riemannian metrics on S^n_+. Using the relation (4), one can show that the first variation of $\mathscr{F}$ at $\overline{g}$ vanishes. More precisely, we have the following result:

PROPOSITION 4.1 ([**13**], Proposition 9). *Let $g(t)$ be a smooth one-parameter family of Riemannian metrics on S^n_+ with $g(0) = \overline{g}$. Then $\frac{d}{dt}\mathscr{F}$ $(g(t))\big|_{t=0} = 0$.*

Note that Proposition 4.1 holds for arbitrary variations of the metric, including those that change the induced metric on the boundary. This fact will play a crucial role in the argument.

The strategy is to deform the standard metric $\overline{g}$ in such a way that the scalar curvature is unchanged to first order and the second variation of the functional $\mathscr{F}$ is positive. In order to construct variations with this property, we need the following auxiliary result:

PROPOSITION 4.2 ([**13**], Proposition 10). *Assume that $n \geq 3$. Then there exists a function $\eta : \Sigma \to \mathbb{R}$ such that*

$$\Delta_\Sigma \eta + (n-1)\eta < 0$$

and

$$\int_\Sigma (|\nabla_\Sigma \eta|^2 - (n-1)\eta^2)\, d\sigma_{\overline{g}} > 0.$$

Sketch of the proof of Proposition 4.2. We define a function $\eta : \Sigma \to \mathbb{R}$ by

$$\eta = -c - 1 + \frac{n-1}{2}\, x_n^2 + \frac{(n-1)(n+1)}{24}\, x_n^4 + \frac{(n-1)(n+1)(n+3)}{240}\, x_n^6,$$

where c is a small positive constant. Then

$$\Delta_\Sigma \eta + (n-1)\eta = -(n-1)c - \frac{(n-1)(n+1)(n+3)(n+5)}{48}\, x_n^6 < 0.$$

Moreover, a straightforward calculation shows that

$$\int_\Sigma (|\nabla_\Sigma \eta|^2 - (n-1)\eta^2)\, d\sigma_{\overline{g}} > 0$$

if $c > 0$ is sufficiently small. Hence, the function η has the required properties. This completes the proof of Proposition 4.2.

Note that the proof of Proposition 4.2 fails for $n = 2$. Proposition 4.2 has a natural geometric interpretation: it implies that, for $n \geq 3$, there exist deformations of the equator in S^n which increase area and have positive mean curvature.

COROLLARY 4.3. *Assume that $n \geq 3$. Then there exists a one-parameter family of hypersurfaces Σ_t with the following properties:*

- $\Sigma_0 = \Sigma$.
- Σ_t *has positive mean curvature for each* $t > 0$.
- $\frac{d^2}{dt^2} \text{area}(\Sigma_t, \overline{g})\big|_{t=0} > 0$.

In the remainder of this section, we will always assume that $n \geq 3$. Let $\eta : \Sigma \to \mathbb{R}$ be the function constructed in Proposition 4.2. We can find a smooth vector field X on S^n such that

$$X = \eta\, \nu$$

and

$$\mathscr{L}_X \overline{g} = 0$$

at each point on Σ. The exact choice of X is not important; all that matters is the behavior of X near the equator.

The vector field X generates a one-parameter group of diffeomorphisms, which we denote by $\varphi_t : S^n \to S^n$. For each t, we define two Riemannian metrics $g_0(t)$ and $g_1(t)$ by

$$g_0(t) = \overline{g} + t\, \mathscr{L}_X \overline{g}$$

and

$$g_1(t) = \varphi_t^*(\overline{g}).$$

It follows from our choice of X that the metric $g_0(t)$ agrees with the standard metric $\overline{g}$ at each point on Σ. By contrast, the metric $g_1(t)$ does not agree with $\overline{g}$ along Σ.

PROPOSITION 4.4 ([**13**], Proposition 11). *There exists a smooth function Q on the hemisphere S^n_+ such that*

$$R_{g_0(t)} = n(n-1) + \frac{1}{2}\, t^2\, Q + O(t^3)$$

and

$$\int_{S^n_+} Q\, f\, d\text{vol}_{\overline{g}} > 0.$$

Sketch of the proof of Proposition 4.4. Note that $g_0(t) = g_1(t) + O(t^2)$. Since $R_{g_1(t)} = n(n-1)$, it follows that $R_{g_0(t)} = n(n-1) + O(t^2)$. For abbreviation, let

$$Q = \frac{\partial^2}{\partial t^2} R_{g_0(t)}\Big|_{t=0}.$$

By Proposition 4.1, the first variation of the functional $\mathscr{F}$ at $\overline{g}$ vanishes. This implies

$$\frac{d^2}{dt^2}\mathscr{F}(g_0(t))\Big|_{t=0} = \frac{d^2}{dt^2}\mathscr{F}(g_1(t))\Big|_{t=0}. \tag{6}$$

The left hand side in (6) is given by

$$\frac{d^2}{dt^2}\mathscr{F}(g_0(t))\Big|_{t=0} = \int_{S^n_+} Q\, f\, d\text{vol}_{\overline{g}}.$$

On the other hand, using the identity $R_{g_1(t)} = n(n-1)$, the right hand side in (6) can be rewritten as

$$\begin{aligned}\frac{d^2}{dt^2}\mathscr{F}(g_1(t))\Big|_{t=0} &= 2\,\frac{d^2}{dt^2}\text{area}(\Sigma, g_1(t))\Big|_{t=0} \\ &= 2\,\frac{d^2}{dt^2}\text{area}(\varphi_t(\Sigma), \overline{g})\Big|_{t=0} \\ &= 2\int_\Sigma (|\nabla_\Sigma \eta|^2 - (n-1)\eta^2)\, d\sigma_{\overline{g}}.\end{aligned}$$

Putting these facts together, we obtain

$$\int_{S^n_+} Q\, f\, d\text{vol}_{\overline{g}} = 2\int_\Sigma (|\nabla_\Sigma \eta|^2 - (n-1)\eta^2)\, d\sigma_{\overline{g}} > 0,$$

completing the proof of Proposition 4.4.

We next consider the elliptic equation

$$\Delta_{\overline{g}} u + n u = Q - \frac{\int_{S^n_+} Q\, f\, d\text{vol}_{\overline{g}}}{\int_{S^n_+} f\, d\text{vol}_{\overline{g}}} \tag{7}$$

with Dirichlet boundary condition $u|_\Sigma = 0$. This boundary value problem has a smooth solution $u : S^n_+ \to \mathbb{R}$. We now define

$$g(t) = \overline{g} + t\,\mathscr{L}_X\overline{g} + \frac{1}{2(n-1)}\,t^2\,u\,\overline{g}.$$

The metric $g(t)$ agrees with the standard metric $\overline{g}$ at each point on Σ. Moreover, the scalar curvature of $g(t)$ is given by

$$\begin{aligned} R_{g(t)} &= R_{g_0(t)} - \frac{1}{2}\,t^2\,(\Delta_{\overline{g}}u + nu) + O(t^3) \\ &= n(n-1) + \frac{1}{2}\,t^2\,Q - \frac{1}{2}\,t^2\,(\Delta_{\overline{g}}u + nu) + O(t^3) \\ &= n(n-1) + \frac{1}{2}\,t^2\,\frac{\int_{S^n_+} Q\,f\,d\mathrm{vol}_{\overline{g}}}{\int_{S^n_+} f\,d\mathrm{vol}_{\overline{g}}} + O(t^3). \end{aligned}$$

Since $\int_{S^n_+} Q\,f\,d\mathrm{vol}_{\overline{g}} > 0$, it follows that

$$\inf_{S^n_+} R_{g(t)} > n(n-1)$$

if $t > 0$ is sufficiently small.

It remains to compute the mean curvature of Σ with respect to the metric $g(t)$. Since $g(t) = g_1(t) + O(t^2)$, we have

$$H_{g(t)} = H_{g_1(t)} + O(t^2) = -t\,(\Delta_\Sigma\eta + (n-1)\eta) + O(t^2).$$

By Proposition 4.2, we have $\Delta_\Sigma\eta + (n-1)\eta < 0$ at each point on Σ. Consequently, we have

$$\inf_\Sigma H_{g(t)} > 0$$

if $t > 0$ is sufficiently small.

5. Other rigidity results involving scalar curvature

In this section, we describe other related rigidity results. The following result was established by Llarull [**32**] (see also the survey paper [**22**]).

THEOREM 5.1 (M. Llarull [**32**]). *Let g be a Riemannian metric on S^n with scalar curvature $R_g \geq n(n-1)$. Moreover, suppose that $g \geq \overline{g}$ at each point on S^n. Then $g = \overline{g}$ at each point on S^n.*

In the even-dimensional case, Listing [**31**] was able to generalize Theorem 5.1 as follows:

THEOREM 5.2 (M. Listing [**31**]). *Let $n \geq 4$ be an even integer. Moreover, suppose that g is a Riemannian metric on S^n satisfying $R_g \geq (n-1)\,\mathrm{tr}_g(\overline{g})$ at each point on S^n. Then g is a constant multiple of the standard metric $\overline{g}$.*

We now discuss a rigidity result for Riemannian metrics on $\mathbb{RP}^3$ with scalar curvature at least 6. To fix notation, we denote by $\mathscr{F}$ the set of all embedded surfaces $\Sigma \subset \mathbb{RP}^3$ with the property that Σ is homeomorphic to $\mathbb{RP}^2$.

THEOREM 5.3 (H. Bray, S. Brendle, M. Eichmair, A. Neves [**8**]). *Let g be a Riemannian metric on $\mathbb{RP}^3$ with scalar curvature $R_g \geq 6$. Moreover, suppose that $\Sigma \in \mathscr{F}$ is a surface which has minimal area among all surfaces in $\mathscr{F}$. Then* $\text{area}(\Sigma, g) \leq 2\pi$. *Moreover, if equality holds, then g is isometric to the standard metric on $\mathbb{RP}^3$.*

Sketch of the proof of Theorem 5.3. Given any metric g on $\mathbb{RP}^3$ and any stable minimal surface $\Sigma \in \mathscr{F}$, one can show that

$$\text{area}(\Sigma, g) \inf_{\mathbb{RP}^3} R_g \leq 12\pi \tag{8}$$

(cf. [**8**], Corollary 8). To prove (8), we use special choices of variations in the second variation formula. These variations are constructed by adapting a technique of Hersch [**28**].

We now sketch the proof of the rigidity statement. Suppose that g is a metric on $\mathbb{RP}^3$ with scalar curvature $R_g \geq 6$, and $\Sigma \in \mathscr{F}$ is an area-minizing surface with $\text{area}(\Sigma, g) = 2\pi$. Let $\tilde{g}(t)$, $t \in [0, T)$, denote the unique solution to the Ricci flow with initial metric $\tilde{g}(0) = g$. Fix a real number $\tau \in (0, T)$. By a theorem of Meeks, Simon, and Yau [**36**], we can find a surface $\tilde{\Sigma} \in \mathscr{F}$ which has minimal area with respect to the metric $\tilde{g}(\tau)$. The key idea is to establish a lower bound for the area of $\tilde{\Sigma}$. More precisely, we have

$$\text{area}(\tilde{\Sigma}, \tilde{g}(\tau)) \geq \text{area}(\Sigma, g) - 8\pi\tau = 2\pi\,(1 - 4\tau) \tag{9}$$

(cf. [**8**], Proposition 10). On the other hand, it follows from the maximum principle that

$$\inf_{\mathbb{RP}^3} R_{\tilde{g}(\tau)} \geq \frac{6}{1 - 4\tau}. \tag{10}$$

Moreover, applying (8) to the metric $\tilde{g}(\tau)$ gives

$$\text{area}(\tilde{\Sigma}, \tilde{g}(\tau)) \inf_{\mathbb{RP}^3} R_{\tilde{g}(\tau)} \leq 12\pi.$$

Therefore, the inequalities (9) and (10) hold as equalities. The strict maximum principle then implies that g has constant sectional curvature 1. This completes the proof of Theorem 5.3.

In view of Corollary 3.9, there are non-trivial examples of Riemannian metrics on $\mathbb{RP}^3$ that have scalar curvature at least 6 and agree with the standard metric in a neighborhood of the equator. In this case, the equator is a stable minimal surface of area 2π. Therefore, the rigidity statement in Theorem 5.3 is no longer true if we replace the assumption that Σ is area-minimizing by the weaker condition that Σ is stable.

We next describe an analogous estimate for the area of area-minimizing two-spheres in three-manifolds.

THEOREM 5.4 (H. Bray, S. Brendle, A. Neves [**9**]). *Let (M, g) be a compact three-manifold with scalar curvature $R_g \geq 2$. Moreover, suppose that Σ is an immersed two-sphere which minimizes area in its homotopy class. Then* $\text{area}(\Sigma, g) \leq 4\pi$. *Moreover, if equality holds, then the universal cover of (M, g) is isometric to the cylinder $S^2 \times \mathbb{R}$ equipped with its standard metric.*

Sketch of the proof of Theorem 5.4. The inequality $\text{area}(\Sigma, g) \leq 4\pi$ follows from the stability inequality. We now describe of the rigidity statement. Let Σ be an area-minimizing two-sphere with $\text{area}(\Sigma, g) = 4\pi$, and let ν be a unit normal vector field along Σ. It is easy to see that Σ is totally geodesic and $\text{Ric}(\nu, \nu) = 0$ at each point on Σ. Using the implicit function theorem, we may construct a one-parameter family of stable constant mean curvature surfaces

$$\Sigma_t = \{\exp_x(w(x, t)\, \nu(x)) : x \in \Sigma\},$$

where $w(x, 0) = 0$ and $\frac{\partial}{\partial t} w(x, t)\big|_{t=0} = 1$ for all $x \in \Sigma$. Let ν_t denote the unit normal vector to Σ_t. We assume that ν_t is chosen so that $\nu_0 = \nu$. The mean curvature vector of Σ_t can be written as $-H(t)\, \nu_t$, where $H(t)$ depends only on t. By definition of Σ, we have

$$\text{area}(\Sigma_t, g) \geq \text{area}(\Sigma, g) = 4\pi.$$

Using the inequality $R_g \geq 2$ and the Gauss equations, we obtain

$$\int_{\Sigma_t} (\text{Ric}(\nu_t, \nu_t) + |II_t|^2) \geq 0,$$

where II_t denotes the second fundamental form of Σ_t. Combining this relation with the stability inequality, one can show that $H'(t) \leq 0$ for all $t \in (-\delta, \delta)$ (see [**9**], p. 827). Since $H(0) = 0$, we conclude that

$$\text{area}(\Sigma_t, g) \leq \text{area}(\Sigma, g) = 4\pi$$

for $t \in (-\delta, \delta)$. Consequently, Σ_t must be totally geodesic, and we have $\text{Ric}(\nu_t, \nu_t) = 0$ at each point on Σ_t. From this, we deduce that (M, g) locally splits as a product.

As above, the rigidity statement in Theorem 5.4 fails if we replace the condition that Σ is area-minimizing by the weaker condition that Σ is stable. Moreover, the proof of Theorem 5.4 can be adapted to other settings, see e.g. [**38**], [**41**].

Theorem 5.4 should be compared to the following rigidity theorem due to Cai and Galloway [**14**] (see also [**20**]):

THEOREM 5.5 (M. Cai, G. Galloway [**14**]). *Let (M, g) be a three-manifold with nonnegative scalar curvature, and let Σ be a two-sided minimal torus in (M, g) which is locally area-minimizing. Then g is flat in a neighborhood of Σ.*

Finally, the following important theorem due to Bray gives a sharp upper bound for the volume of a three-manifold with scalar curvature at least 6.

THEOREM 5.6 (H. Bray [7]). *Fix a real number* $\varepsilon \in (0,1)$ *with the property that*

$$\begin{aligned} &\int_0^y \big(36\pi - 27(1-\varepsilon)\, y^{\frac{2}{3}} - 9\varepsilon\, x^{\frac{2}{3}}\big)^{-\frac{1}{2}}\, dx \\ &+ \int_y^{z^{\frac{3}{2}}} \big(36\pi - 18(1-\varepsilon)\, y\, x^{-\frac{1}{3}} - 9\, x^{\frac{2}{3}}\big)^{-\frac{1}{2}}\, dx < \pi^2 \end{aligned} \tag{11}$$

for all pairs $(y,z) \in \mathbb{R} \times \big[\frac{4\pi}{3-2\varepsilon}, 4\pi\big)$ *satisfying* $2(1-\varepsilon)\, y = z^{\frac{1}{2}}\,(4\pi - z)$. *Moreover, let* (M,g) *be a compact three-manifold satisfying* $R_g \geq 6$ *and* $\mathrm{Ric}_g \geq 2\varepsilon\, g$. *If* $\mathrm{vol}(M,g) \geq \mathrm{vol}(S^3,\overline{g})$, *then* (M,g) *is isometric to* $(S^3,\overline{g})$.

Theorem 5.6 is slightly stronger than the result proved in [7]. Theorem 19 in [7] asserts that every three-manifold (M,g) with $R_g \geq 6$ and $\mathrm{Ric}_g \geq 2\varepsilon\, g$ satisfies $\mathrm{vol}(M,g) \leq \mathrm{vol}(S^3,\overline{g})$. However, the case of equality has not been discussed in the literature. For the convenience of the reader, we shall provide a detailed exposition in the following section.

Gursky and Viaclovsky proved that the condition (11) is satisfied for $\varepsilon = \frac{1}{2}$ (see [25], Section 4.1). Hence, we can draw the following conclusion:

COROLLARY 5.7. *Let* (M,g) *be a compact three-manifold satisfying* $R_g \geq 6$ *and* $\mathrm{Ric}_g \geq g$. *If* $\mathrm{vol}(M,g) \geq \mathrm{vol}(S^3,\overline{g})$, *then* (M,g) *is isometric to* $(S^3,\overline{g})$.

6. The case of equality in Bray's volume comparison theorem

In this final section, we describe the proof of Theorem 5.6. We will closely follow the original argument of Bray [7]. Let us fix a real number $\varepsilon \in (0,1)$ satisfying (11). For each $z \in \big(\frac{4\pi}{3-2\varepsilon}, 4\pi\big]$, we define

$$y(z) = \frac{1}{2(1-\varepsilon)}\, z^{\frac{1}{2}}\,(4\pi - z) \in [0, z^{\frac{3}{2}})$$

and

$$\begin{aligned} \gamma(z) = &\int_0^{y(z)} \big(36\pi - 27(1-\varepsilon)\, y(z)^{\frac{2}{3}} - 9\varepsilon\, x^{\frac{2}{3}}\big)^{-\frac{1}{2}}\, dx \\ &+ \int_{y(z)}^{z^{\frac{3}{2}}} \big(36\pi - 18(1-\varepsilon)\, y(z)\, x^{-\frac{1}{3}} - 9\, x^{\frac{2}{3}}\big)^{-\frac{1}{2}}\, dx. \end{aligned}$$

Moreover, for each $z \in \big(0, \frac{4\pi}{3-2\varepsilon}\big]$, we define

$$\gamma(z) = \int_0^{z^{\frac{3}{2}}} \big(9\varepsilon\, z - 9\varepsilon\, x^{\frac{2}{3}}\big)^{-\frac{1}{2}}\, dx.$$

Note that the function $\gamma(z)$ is continuous.

PROPOSITION 6.1. *For each $z \in (0, 4\pi]$, there exists a function $f_z : (-\gamma(z), \gamma(z)) \to (0, z^{\frac{3}{2}}]$ with the following properties:*

- *The function f_z is twice continuously differentiable.*
- *For each $s \in (-\gamma(z), \gamma(z))$, we have*

$$f_z''(s) = \min\Big\{\frac{36\pi - f_z'(s)^2}{6 f_z(s)} - \frac{9}{2}\, f_z(s)^{-\frac{1}{3}}, -3\varepsilon\, f_z(s)^{-\frac{1}{3}}\Big\} \tag{12}$$

- *$f_z(0) = z^{\frac{3}{2}}$ and $f_z'(0) = 0$.*
- *$\lim_{s\to\gamma(z)} f_z(s) = \lim_{s\to-\gamma(z)} f_z(s) = 0$.*

Sketch of the proof of Proposition 6.1. We distinguish two cases:

Case 1: Suppose that $z \in \big(0, \frac{4\pi}{3-2\varepsilon}\big]$. We can find a smooth function $h_z : (0, \gamma(z)) \to (0, z^{\frac{3}{2}})$ such that $\lim_{s\to 0} h_z(s) = z^{\frac{3}{2}}$, $\lim_{s\to\gamma(z)} h_z(s) = 0$, and

$$h_z'(s) = -\big(9\varepsilon\, z - 9\varepsilon\, h_z(s)^{\frac{2}{3}}\big)^{\frac{1}{2}} \tag{13}$$

for all $s \in (0, \gamma(z))$. Clearly, $\lim_{s\to 0} h_z'(s) = 0$. Moreover, the relation (13) implies

$$h_z''(s) = -3\varepsilon\, h_z(s)^{-\frac{1}{3}} \leq \frac{36\pi - h_z'(s)^2}{6 h_z(s)} - \frac{9}{2}\, h_z(s)^{-\frac{1}{3}}$$

for all $s \in (0, \gamma(z))$. Therefore, the function h is a solution of (12). Hence, if we put

$$f_z(s) = \begin{cases} h_z(s) & \text{for } s \in (0, \gamma(z)) \\ h_z(-s) & \text{for } s \in (-\gamma(z), 0) \\ z^{\frac{3}{2}} & \text{for } s = 0, \end{cases}$$

then the function $f_z : (-\gamma(z), \gamma(z)) \to (0, z^{\frac{3}{2}}]$ has all the required properties.

Case 2: Suppose that $z \in \big(\frac{4\pi}{3-2\varepsilon}, 4\pi\big]$. We can find a continuously differentiable function $h_z : (0, \gamma(z)) \to (0, z^{\frac{3}{2}})$ such that $\lim_{s\to 0} h_z(s) = z^{\frac{3}{2}}$, $\lim_{s\to\gamma(z)} h_z(s) = 0$, and

$$h_z'(s) = \begin{cases} -\big(36\pi - 18(1-\varepsilon)\, y(z)\, h_z(s)^{-\frac{1}{3}} - 9\, h_z(s)^{\frac{2}{3}}\big)^{\frac{1}{2}} & \text{if } h_z(s) \geq y(z) \\ -\big(36\pi - 27(1-\varepsilon)\, y(z)^{\frac{2}{3}} - 9\varepsilon\, h_z(s)^{\frac{2}{3}}\big)^{\frac{1}{2}} & \text{if } h_z(s) \leq y(z) \end{cases} \tag{14}$$

for all $s \in (0, \gamma(z))$. It is easy to see that $\lim_{s\to 0} h_z'(s) = 0$. For abbreviation, let $\beta(z) = \sup\{s \in (0, \gamma(z)) : h_z(s) \geq y(z)\}$. Using the relation (14), we obtain

$$h_z''(s) = \frac{36\pi - h_z'(s)^2}{6 h_z(s)} - \frac{9}{2}\, h_z(s)^{-\frac{1}{3}} \leq -3\varepsilon\, h_z(s)^{-\frac{1}{3}}$$

for all $s \in (0, \beta(z))$ and

$$h_z''(s) = -3\varepsilon\, h_z(s)^{-\frac{1}{3}} \leq \frac{36\pi - h_z'(s)^2}{6 h_z(s)} - \frac{9}{2}\, h_z(s)^{-\frac{1}{3}}$$

for all $s \in (\beta(z), \gamma(z))$. Therefore, the function h is twice continuously differentiable, and satisfies the differential equation (12). Hence, if we define

$$f_z(s) = \begin{cases} h_z(s) & \text{for } s \in (0, \gamma(z)) \\ h_z(-s) & \text{for } s \in (-\gamma(z), 0) \\ z^{\frac{3}{2}} & \text{for } s = 0, \end{cases}$$

then the function $f_z : (0, \infty) \to (0, z^{\frac{3}{2}}]$ has all the required properties.

LEMMA 6.2. *We have $\gamma(4\pi) = \pi^2$. Moreover, the function $f_{4\pi}$ satisfies*

$$f_{4\pi}(s - \pi^2) = 6\sqrt{\pi}\, s \left[1 - \frac{3}{10}\left(\frac{3s}{4\pi}\right)^{\frac{2}{3}} - \frac{3}{280}\left(\frac{3s}{4\pi}\right)^{\frac{4}{3}} + O(s^2)\right]$$

if $s > 0$ is small.

Sketch of the proof of Lemma 6.2. Using the substitution $x = (4\pi)^{\frac{3}{2}}$ $\sin^3(r)$, we obtain

$$\gamma(4\pi) = \int_0^{(4\pi)^{\frac{3}{2}}} \left(36\pi - 9\, x^{\frac{2}{3}}\right)^{-\frac{1}{2}} dx = 4\pi \int_0^{\frac{\pi}{2}} \sin^2(r)\, dr = \pi^2.$$

This proves the first statement. We next analyze the asymptotic behavior of the function $f_{4\pi}$. The function $h_{4\pi} : (0, \pi^2) \to \left(0, (4\pi)^{\frac{3}{2}}\right)$ satisfies $\lim_{s\to 0} h_{4\pi}(s) = (4\pi)^{\frac{3}{2}}$, $\lim_{s\to\pi^2} h_{4\pi}(s) = 0$, and

$$h_{4\pi}'(s) = -\left(36\pi - 9\, h_{4\pi}(s)^{\frac{2}{3}}\right)^{\frac{1}{2}}$$

for all $s \in (0, \pi^2)$. This implies

$$h_{4\pi}\left(\pi^2 - 2\pi r + \pi \sin(2r)\right) = (4\pi)^{\frac{3}{2}} \sin^3(r)$$

for each $r \in (0, \frac{\pi}{2})$. Hence, if we put $s = 2\pi r - \pi \sin(2r)$, then we have

$$s = \frac{4\pi}{3}\, r^3 \left(1 - \frac{1}{5}\, r^2 + \frac{2}{105}\, r^4 + O(r^6)\right)$$

and

$$h_{4\pi}(\pi^2 - s) = (4\pi)^{\frac{3}{2}}\, r^3 \left(1 - \frac{1}{2}\, r^2 + \frac{13}{120}\, r^4 + O(r^6)\right).$$

Consequently,

$$h_{4\pi}(\pi^2 - s) = 6\sqrt{\pi}\, s \left[1 - \frac{3}{10}\left(\frac{3s}{4\pi}\right)^{\frac{2}{3}} - \frac{3}{280}\left(\frac{3s}{4\pi}\right)^{\frac{4}{3}} + O(s^2)\right],$$

as claimed.

Lemma 6.3. *We have $\gamma(z) < \pi^2$ for all $z \in (0, 4\pi)$.*

Sketch of the proof of Lemma 6.3. The assumption (11) implies that $\gamma(z) < \pi^2$ for all $z \in \big[\frac{4\pi}{3-2\varepsilon}, 4\pi\big)$. In particular, we have $\gamma\big(\frac{4\pi}{3-2\varepsilon}\big) < \pi^2$. We next consider a real number $z \in (0, \frac{4\pi}{3-2\varepsilon}]$. Using the substitution $x = z^{\frac{3}{2}} \sin^3(r)$, we obtain

$$\gamma(z) = \int_0^{z^{\frac{3}{2}}} \big(9\varepsilon\, z - 9\varepsilon\, x^{\frac{2}{3}}\big)^{-\frac{1}{2}}\, dx = \frac{1}{\sqrt{\varepsilon}}\, z \int_0^{\frac{\pi}{2}} \sin^2(r)\, dr = \frac{\pi}{4\sqrt{\varepsilon}}\, z$$

for $z \in (0, \frac{4\pi}{3-2\varepsilon}]$. Since $\gamma\big(\frac{4\pi}{3-2\varepsilon}\big) < \pi^2$, we conclude that $\gamma(z) < \pi^2$ for all $z \in \big(0, \frac{4\pi}{3-2\varepsilon}\big]$. This completes the proof of Lemma 6.3.

We now assume that (M, g) is a compact three-manifold such that $R_g \geq 6$, $\mathrm{Ric}_g \geq 2\varepsilon\, g$, and

$$\mathrm{vol}(M, g) \geq \mathrm{vol}(S^3, \overline{g}) = 2\pi^2.$$

Let $A : (0, 2\pi^2) \to (0, \infty)$ denote the isoperimetric profile of (M, g). More precisely, for each $s \in (0, 2\pi^2)$, we define

$$A(s) = \inf\big\{\mathrm{area}(\partial\Omega, g) : \mathrm{vol}(\Omega, g) = s\big\}.$$

The following result play a key role in Bray's argument.

Proposition 6.4 (H. Bray [7]). *Given any real number $s_0 \in (0, 2\pi^2)$, there exists a real number $\delta > 0$ and a smooth function $u : (s_0 - \delta, s_0 + \delta) \to (0, \infty)$ with the following properties:*

- $u(s_0) = A(s_0)$.
- $u(s) \geq A(s)$ *for all* $s \in (s_0 - \delta, s_0 + \delta)$.
- $u''(s_0) \leq \min\big\{\frac{4\pi}{u(s_0)^2} - \frac{3u'(s_0)^2}{4u(s_0)} - \frac{3}{u(s_0)}, -\frac{u'(s_0)^2}{2u(s_0)} - \frac{2\varepsilon}{u(s_0)}\big\}$.

Sketch of the proof of Proposition 6.4. We can find a smooth domain $\Omega \subset M$ such that $\mathrm{vol}(\Omega, g) = s_0$ and $\mathrm{area}(\partial\Omega, g) = A(s_0)$. Let ν denote the outward-pointing unit normal vector field along $\partial\Omega$. There exists a smooth one-parameter family of diffeomorphisms $\varphi_s : M \to M$, $s \in (s_0 - \delta, s_0 + \delta)$, with the following properties:

- $\varphi_{s_0}(x) = x$ for each point $x \in M$.
- $\frac{\partial}{\partial s}\varphi_s(x)\big|_{s=s_0} = \frac{1}{A(s_0)}\, \nu(x)$ for each point $x \in \partial\Omega$.
- $\mathrm{vol}(\varphi_s(\Omega), g) = s$ for each $s \in (s_0 - \delta, s_0 + \delta)$.

We define a function $u : (s_0 - \delta, s_0 + \delta) \to (0, \infty)$ by

$$u(s) = \mathrm{area}(\varphi_s(\partial\Omega), g).$$

Clearly, $u(s_0) = A(s_0)$ and $u(s) \geq A(s)$ for each $s \in (s_0 - \delta, s_0 + \delta)$. It follows from the first variation formula that $\partial\Omega$ has constant mean curvature

$H = u'(s_0)$. Using the formula for the second variation of area, we obtain

$$\frac{d^2}{ds^2}\Big(\text{area}(\varphi_s(\partial\Omega), g) - u'(s_0)\,\text{vol}(\varphi_s(\Omega), g)\Big)\Big|_{s=s_0}$$
$$= -\frac{1}{A(s_0)^2}\int_{\partial\Omega}(|II|^2 + \text{Ric}(\nu,\nu)),$$

where II denotes the second fundamental form of the boundary $\partial\Omega$. Since $\text{vol}(\varphi_s(\Omega), g) = s$ for all $s \in (s_0 - \delta, s_0 + \delta)$, we conclude that

$$u(s_0)^2\, u''(s_0) = -\int_{\partial\Omega}(\text{Ric}_g(\nu,\nu) + |II|^2). \tag{15}$$

The inequality $\text{Ric}_g \geq 2\varepsilon\, g$ implies that

$$\int_{\partial\Omega}(\text{Ric}_g(\nu,\nu) + |II|^2) \geq \int_{\partial\Omega}\Big(\frac{1}{2}\,H^2 + 2\varepsilon\Big)$$
$$= \frac{1}{2}\,u(s_0)\,u'(s_0)^2 + 2\varepsilon\,u(s_0). \tag{16}$$

Since (M, g) has positive Ricci curvature, the boundary $\partial\Omega$ is connected (see [**7**], p. 73). Hence, the Gauss-Bonnet theorem implies that $\int_{\partial\Omega} K \leq 4\pi$, where K denotes the Gaussian curvature of $\partial\Omega$. Using the inequality $R_g \geq 6$ and the Gauss equations, we obtain

$$\int_{\partial\Omega}(\text{Ric}_g(\nu,\nu) + |II|^2) = \frac{1}{2}\int_{\partial\Omega}(R_g - 2K + H^2 + |II|^2)$$
$$\geq -4\pi + \int_{\partial\Omega}\Big(\frac{3}{4}\,H^2 + 3\Big)$$
$$= -4\pi + \frac{3}{4}\,u(s_0)\,u'(s_0)^2 + 3\,u(s_0) \tag{17}$$

(cf. [**7**], p. 74). Combining (15), (16), and (17) gives

$$u(s_0)^2\,u''(s_0) \leq \min\Big\{4\pi - \frac{3}{4}\,u(s_0)\,u'(s_0)^2 - 3\,u(s_0),$$
$$-\frac{1}{2}\,u(s_0)\,u'(s_0)^2 - 2\varepsilon\,u(s_0)\Big\},$$

as claimed.

Note that the function $A(s)$ may not be differentiable. For our purposes, it is sufficient that $A(s)$ is continuous.

LEMMA 6.5. *The function $A : (0, 2\pi^2) \to (0, \infty)$ is continuous.*

Sketch of the proof of Lemma 6.5. Let us fix a real number $s_0 \in (0, 2\pi^2)$, and let s_k be a sequence of real numbers with $\lim_{k\to\infty} s_k = s_0$. It follows from Proposition 6.4 that $\limsup_{k\to\infty} A(s_k) \leq A(s_0)$. Hence, it suffices to show that $\liminf_{k\to\infty} A(s_k) \geq A(s_0)$. Suppose this is false. After passing to a subsequence, we may assume that the sequence $A(s_k)$ converges to a real number $\alpha < A(s_0)$. For each k, we can find a smooth domain $\Omega_k \subset M$ such that

$\text{vol}(\Omega_k, g) = s_k$ and $\text{area}(\partial\Omega_k, g) = A(s_k)$. After passing to another subsequence, we may assume that the sequence Ω_k converges to a smooth domain $\Omega_0 \subset M$ (see [**43**], Proposition 2.15). Then $\text{vol}(\Omega_0, g) = s_0$ and $\text{area}(\partial\Omega_0, g) = \alpha < A(s_0)$. This is a contradiction.

Using a comparison argument, one can relate the isoperimetric profile of (M, g) to the function $f_{4\pi}$ constructed above. For abbreviation, we define a function $F : (0, 2\pi^2) \to (0, \infty)$ by $F(s) = A(s)^{\frac{3}{2}}$.

PROPOSITION 6.6. *We have $F(s) \geq f_{4\pi}(s - \pi^2)$ for all $s \in (0, 2\pi^2)$.*

Sketch of the proof of Proposition 6.6. Suppose that $F(s_1) < f_{4\pi}(s_1 - \pi^2)$ for some real number $s_1 \in (0, 2\pi^2)$. By continuity, there exists a real number $z_1 \in (0, 4\pi)$ such that $s_1 \in (\pi^2 - \gamma(z_1), \pi^2 + \gamma(z_1))$ and $F(s_1) < f_{z_1}(s_1 - \pi^2)$. By Lemma 6.3, we can find a real number $\lambda > 1$ such that $\lambda\,\gamma(z) < \pi^2$ for all $z \in (0, z_1]$ and $F(s_1) < \lambda\, f_{z_1}(\frac{s_1 - \pi^2}{\lambda})$.

Let I denote the set of all real numbers $z \in (0, z_1]$ with the property that

$$F(s) \geq \lambda\, f_z\Big(\frac{s - \pi^2}{\lambda}\Big)$$

for all $s \in (\pi^2 - \lambda\,\gamma(z), \pi^2 + \lambda\,\gamma(z))$. It is easy to see that I is closed and $z_1 \notin I$. Moreover, we have $(0, \alpha) \subset I$ if $\alpha > 0$ is sufficiently small. Let $z_0 \in (0, z_1)$ denote the supremum of the set I. Then

$$\inf_{s \in (\pi^2 - \lambda\,\gamma(z_0), \pi^2 + \lambda\,\gamma(z_0))} \Big[F(s) - \lambda\, f_{z_0}\Big(\frac{s - \pi^2}{\lambda}\Big)\Big] = 0.$$

By Lemma 6.5, the function $F : (0, 2\pi^2) \to (0, \infty)$ is continuous. Hence, we can find a real number $s_0 \in (\pi^2 - \lambda\,\gamma(z_0), \pi^2 + \lambda\,\gamma(z_0))$ such that

$$F(s_0) = \lambda\, f_{z_0}\Big(\frac{s_0 - \pi^2}{\lambda}\Big).$$

By Proposition 6.4, we can find a smooth function $u : (s_0 - \delta, s_0 + \delta) \to (0, \infty)$ such that $u(s_0) = A(s_0)$, $u(s) \geq A(s)$ for all $s \in (s_0 - \delta, s_0 + \delta)$, and

$$u''(s_0) \leq \min\Big\{\frac{4\pi}{u(s_0)^2} - \frac{3u'(s_0)^2}{4u(s_0)} - \frac{3}{u(s_0)}, -\frac{u'(s_0)^2}{2u(s_0)} - \frac{2\varepsilon}{u(s_0)}\Big\}. \tag{18}$$

Hence, if we define $v(s) = u(s)^{\frac{3}{2}}$, then we have

$$v(s_0) = F(s_0) = \lambda\, f_{z_0}\Big(\frac{s_0 - \pi^2}{\lambda}\Big)$$

and

$$v(s) \geq F(s) \geq \lambda\, f_{z_0}\Big(\frac{s - \pi^2}{\lambda}\Big)$$

for all $s \in (\pi^2 - \lambda\,\gamma(z_0), \pi^2 + \lambda\,\gamma(z_0)) \cap (s_0 - \delta, s_0 + \delta)$. This implies

$$v'(s_0) = f_{z_0}'\Big(\frac{s_0 - \pi^2}{\lambda}\Big)$$

and

$$v''(s_0) \geq \frac{1}{\lambda} f''_{z_0}\Big(\frac{s_0 - \pi^2}{\lambda}\Big).$$

Using the differential equation (12), we obtain

$$v''(s_0) \geq \min\Big\{\frac{36\pi - v'(s_0)^2}{6v(s_0)} - \frac{9}{2}\lambda^{-\frac{2}{3}} v(s_0)^{-\frac{1}{3}}, -3\varepsilon\,\lambda^{-\frac{2}{3}} v(s_0)^{-\frac{1}{3}}\Big\}.$$

On the other hand, the inequality (18) implies

$$v''(s_0) \leq \min\Big\{\frac{36\pi - v'(s_0)^2}{6v(s_0)} - \frac{9}{2} v(s_0)^{-\frac{1}{3}}, -3\varepsilon\, v(s_0)^{-\frac{1}{3}}\Big\}.$$

However, these inequalities are incompatible since $\lambda > 1$. The proof of Proposition 6.6 is now complete.

Finally, we study the asymptotic behavior of the function $F(s)$ as $s \to 0$. Following an idea of Eichmair [**18**], we consider small geodesic balls in order to obtain an upper bound for the isoperimetric profile of (M, g).

PROPOSITION 6.7. *Fix an arbitrary point* $p \in M$. *If* $s > 0$ *is small, we have*

$$F(s) \leq 6\sqrt{\pi}\, s\,\Big[1 - \frac{3}{2}\,c_1(p)\,\Big(\frac{3s}{4\pi}\Big)^{\frac{2}{3}} - \Big(\frac{35}{24}\,c_1(p)^2 + \frac{5}{2}\,c_2(p)\Big)\Big(\frac{3s}{4\pi}\Big)^{\frac{4}{3}} + O(s^2)\Big].$$

Here, the coefficients $c_1(p)$ *and* $c_2(p)$ *are defined by* $c_1(p) = \frac{1}{30}\,R_g(p)$ *and* $c_2(p) = \frac{1}{6300}\,(\Delta_g R_g(p) + 2\,|\mathrm{Ric}_g(p)|^2 - 4\,R_g(p)^2)$.

Sketch of the proof of Proposition 6.7. It follows from work of Gray and Vanhecke [**21**] that

$$\mathrm{vol}(B(p,r), g) = \frac{4\pi}{3}\,r^3\,(1 - c_1(p)\,r^2 - c_2(p)\,r^4 + O(r^6))$$

and

$$\mathrm{area}(\partial B(p,r), g) = 4\pi\,r^2\,\Big(1 - \frac{5}{3}\,c_1(p)\,r^2 - \frac{7}{3}\,c_2(p)\,r^4 + O(r^6)\Big).$$

Hence, if we put $s = \mathrm{vol}(B(p,r), g)$, then we obtain

$$\begin{aligned} &\mathrm{area}(\partial B(p,r), g)^{\frac{3}{2}} \\ &= 6\sqrt{\pi}\,s\,\Big[1 - \frac{3}{2}\,c_1(p)\,\Big(\frac{3s}{4\pi}\Big)^{\frac{2}{3}} - \Big(\frac{35}{24}\,c_1(p)^2 + \frac{5}{2}\,c_2(p)\Big)\Big(\frac{3s}{4\pi}\Big)^{\frac{4}{3}} + O(s^2)\Big]. \end{aligned}$$

Moreover, we have $F(s) \leq \mathrm{area}(\partial B(p,r), g)^{\frac{3}{2}}$ by definition of $F(s)$. Putting these facts together, the assertion follows.

We now complete the proof of Theorem 5.6. Combining Lemma 6.2, Proposition 6.6, and Proposition 6.7, we obtain

$$\begin{aligned} 0 &\leq \frac{F(s) - f_{4\pi}(s - \pi^2)}{6\sqrt{\pi}\, s} \\ &\leq \Big(\frac{3}{10} - \frac{3}{2}\, c_1(p)\Big) \Big(\frac{3s}{4\pi}\Big)^{\frac{2}{3}} \\ &+ \Big(\frac{3}{280} - \frac{35}{24}\, c_1(p)^2 - \frac{5}{2}\, c_2(p)\Big) \Big(\frac{3s}{4\pi}\Big)^{\frac{4}{3}} + O(s^2) \end{aligned}$$

if $s > 0$ is small enough. From this, we deduce that $\frac{3}{10} - \frac{3}{2}\, c_1(p) \geq 0$, hence $R_g(p) \leq 6$. On the other hand, we have $R_g(p) \geq 6$ by assumption. Since the point $p \in M$ is arbitrary, we conclude that (M, g) has constant scalar curvature 6. We next observe that $\frac{3}{280} - \frac{35}{24}\, c_1(p)^2 - \frac{5}{2}\, c_2(p) \geq 0$. This implies $c_2(p) \leq -\frac{2}{105}$, hence $|\mathrm{Ric}_g(p)|^2 \leq 12$. Since $R_g(p) = 6$, it follows that $\mathrm{Ric}_g = 2g$ for each point $p \in M$. Consequently, the manifold (M, g) has constant sectional curvature 1. Since $\mathrm{vol}(M, g) \geq \mathrm{vol}(S^3, \overline{g})$, we conclude that (M, g) is isometric to $(S^3, \overline{g})$.

References

[1] L. Andersson and M. Dahl, *Scalar curvature rigidity for asymptotically locally hyperbolic manifolds,* Ann. Global Anal. Geom. 16, 1–27 (1998).

[2] R. Arnowitt, S. Deser, and C. Misner, *Coordinate invariance and energy expressions in general relativity,* Phys. Rev. 122, 997–1006 (1961).

[3] R. Bartnik, *The mass of an asymptotically flat manifold,* Comm. Pure Appl. Math. 39, 661–693 (1986).

[4] R. Bartnik, *Energy in general relativity,* Tsing Hua Lectures on Geometry and Analysis (Hsinchu 1990–1991), 5–27, Intl. Press, Cambridge MA (1997).

[5] A. Besse, *Einstein manifolds,* Classics in Mathematics, Springer-Verlag, Berlin (2008).

[6] H. Boualem and M. Herzlich, *Rigidity at infinity for even-dimensional asymptotically complex hyperbolic spaces,* Ann. Scuola Norm. Sup. Pisa (Ser. V), 1, 461–469 (2002).

[7] H. Bray, *The Penrose inequality in general relativity and volume comparison theorems involving scalar curvature,* PhD thesis, Stanford University (1997).

[8] H. Bray, S. Brendle, M. Eichmair, and A. Neves, *Area-minimizing projective planes in three-manifolds,* Comm. Pure Appl. Math. 63, 1237–1247 (2010).

[9] H. Bray, S. Brendle, and A. Neves, *Rigidity of area-minimizing two-spheres in three-manifolds,* Comm. Anal. Geom. 18, 821–830 (2010).

[10] S. Brendle, *Blow-up phenomena for the Yamabe equation,* J. Amer. Math. Soc. 21, 951–979 (2008).

[11] S. Brendle and F.C. Marques, *Blow-up phenomena for the Yamabe equation II,* J. Diff. Geom. 81, 225–250 (2009).

[12] S. Brendle and F.C. Marques, *Scalar curvature rigidity of geodesic balls in S^n,* J. Diff. Geom. 88, 379–394 (2011).

[13] S. Brendle, F.C. Marques, and A. Neves, *Deformations of the hemisphere that increase scalar curvature,* Invent. Math. 185, 175–197 (2011).

[14] M. Cai and G. Galloway, *Rigidity of area minimizing tori in 3-manifolds of nonnegative scalar curvature,* Comm. Anal. Geom. 8, 565–573 (2000).

[15] P.T. Chruściel and M. Herzlich, *The mass of asymptotically hyperbolic Riemannian manifolds,* Pacific J. Math. 212, 231–264 (2003).

[16] P.T. Chruściel and G. Nagy, *The mass of spacelike hypersurfaces in asymptotically anti-de-Sitter space-times,* Adv. Theor. Math. Phys. 5, 697–754 (2001).

[17] J. Corvino, *Scalar curvature deformation and a gluing construction for the Einstein constraint equations,* Comm. Math. Phys. 214, 137–189 (2000).

[18] M. Eichmair, *The size of isoperimetric surfaces in 3-manifolds and a rigidity result for the upper hemisphere,* Proc. Amer. Math. Soc. 137, 2733–2740 (2009).

[19] A.E. Fischer and J.E. Marsden, *Deformations of the scalar curvature,* Duke Math. J. 42, 519–547 (1975).

[20] D. Fischer-Colbrie and R. Schoen, *The structure of complete stable minimal surfaces in 3-manifolds of nonnegative scalar curvature,* Comm. Pure Appl. Math. 33, 199–211 (1980).

[21] A. Gray and L. Vanhecke, *Riemannian geometry as determined by the volumes of small geodesic balls,* Acta Math. 142, 157–198 (1979).

[22] M. Gromov, *Positive curvature, macroscopic dimension, spectral gaps and higher signatures,* Functional analysis on the eve of the 21st century, Vol. II (New Brunswick 1993), 1–213, Progr. Math., 132, Birkhäuser, Boston (1996).

[23] M. Gromov and H.B. Lawson, *Spin and scalar curvature in the presence of a fundamental group,* Ann. of Math. 111, 209–230 (1980).

[24] M. Gromov and H.B. Lawson, *Positive scalar curvature and the Dirac operator on complete Riemannian manifolds,* Publ. Math. IHÉS 58, 83–196 (1983).

[25] M. Gursky and J. Viaclovsky, *Volume comparison and the σ_k-Yamabe problem,* Adv. Math. 187, 447–487 (2004).

[26] F. Hang and X. Wang, *Rigidity and non-rigidity results on the sphere,* Comm. Anal. Geom. 14, 91–106 (2006).

[27] F. Hang and X. Wang, *Rigidity theorems for compact manifolds with boundary and positive Ricci curvature,* J. Geom. Anal. 19, 628–642 (2009).

[28] J. Hersch, *Quatre propriétés isopérimétriques de membranes sphériques homogènes,* C.R. Acad. Sci. Paris 270, 1645–1648 (1970).

[29] M. Herzlich, *Scalar curvature and rigidity of odd-dimensional complex hyperbolic spaces,* Math. Ann. 312, 641–657 (1998).

[30] L. Huang and D. Wu, *Rigidity theorems on hemispheres in non-positive space forms,* Comm. Anal. Geom. 18, 339–363 (2010).

[31] M. Listing, *Scalar curvature on compact symmetric spaces,* arxiv:1007.1832.

[32] M. Llarull, *Sharp estimates and the Dirac operator,* Math. Ann. 310, 55–71 (1998).

[33] J. Lohkamp, *Metrics of negative Ricci curvature,* Ann. of Math. 140, 655–683 (1994).

[34] J. Lohkamp, *Scalar curvature and hammocks,* Math. Ann. 313, 385–407 (1999).

[35] J. Lohkamp, *Positive scalar curvature in dim ≥ 8,* C.R. Acad. Sci. Paris 343, 585–588 (2006).

[36] W. Meeks, L. Simon, and S.T. Yau, *Embedded minimal surfaces, exotic spheres, and manifolds with positive Ricci curvature,* Ann. of Math. 116, 621–659 (1982).

[37] P. Miao, *Positive mass theorem on manifolds admitting corners along a hypersurface,* Adv. Theor. Math. Phys. 6, 1163–1182 (2002).

[38] M. Micallef and V. Moraru, *Splitting of 3-manifolds and rigidity of area-minimizing surfaces,* arxiv:1107.5346.

[39] M. Min-Oo, *Scalar curvature rigidity of asymptotically hyperbolic spin manifolds,* Math. Ann. 285, 527–539 (1989).

[40] M. Min-Oo, *Scalar curvature rigidity of certain symmetric spaces,* Geometry, topology, and dynamics (Montreal, 1995), 127–137, CRM Proc. Lecture Notes vol. 15, Amer. Math. Soc., Providence RI, 1998.

[41] I. Nunes, *Rigidity of area-minimizing hyperbolic surfaces in three-manifolds,* arxiv:1103.4805.

[42] T. Parker and C.H. Taubes, *On Witten's proof of the positive energy theorem,* Comm. Math. Phys. 84, 223–238 (1982).

[43] A. Ros, *The isoperimetric problem,* Global Theory of Minimal Surfaces (Proc. Clay Institute Summer School, 2001), 175–209, Amer. Math. Soc., Providence RI, 2005.

[44] R. Schoen and S.T. Yau, *On the proof of the positive mass conjecture in general relativity,* Comm. Math. Phys. 65, 45–76 (1979).

[45] R. Schoen and S.T. Yau, *Existence of incompressible minimal surfaces and the topology of three dimensional manifolds of non-negative scalar curvature,* Ann. of Math. 110, 127–142 (1979).

[46] R. Schoen and S.T. Yau, *On the structure of manifolds with positive scalar curvature,* Manuscripta Math. 28, 159–183 (1979).

[47] Y. Shi and L.F. Tam, *Positive mass theorem and the boundary behaviors of compact manifolds with nonnegative scalar curvature,* J. Diff. Geom. 62, 79–125 (2002).

[48] V. Toponogov, *Evaluation of the length of a closed geodesic on a convex surface,* Dokl. Akad. Nauk. SSSR 124, 282–284 (1959).

[49] X. Wang, *The mass of asymptotically hyperbolic manifolds,* J. Diff. Geom. 57, 273–299 (2001).

[50] E. Witten, *A new proof of the positive energy theorem,* Comm. Math. Phys. 80, 381–402 (1981).

Department of Mathematics, Stanford University, 450 Serra Mall, Bldg 380, Stanford, CA 94305

Stability, birational transformations and the Kahler-Einstein problem

S. K. Donaldson

1. Introduction

This is the first in a series of papers in which we will discuss the existence problem for Kahler-Einstein metrics on complex projective manifolds.[1] It is well-known that, following the work of Yau in the 1970's, the question is reduced to the "positive" case, so we want to decide when a Fano manifold admits a Kahler-Einstein metric. The present paper is confined to algebro-geometric aspects of the problem. We have three main purposes

- To state some new definitions, of "b-stability" and "$\overline{K}$-stability".
- To indicate the application of these to the differential geometric existence problem.
- To highlight some purely algebro-geometric questions which seem to be important in the existence problem.

In the late 1980's, Yau suggested that the existence of Kahler-Einstein metrics should be related to the algebro-geometric notion of *stability*, whose origins lie in Geometric Invariant Theory [**16**]. This was motivated in part by analogy with the Kobayshi-Hitchin correspondence for Hermitian Yang-Mills connections on holomorphic vector bundles. The conjecture was refined considerably by work of Tian in the 1990's [**14**]. We will indicate this general idea—the *Tian-Yau conjecture*—by the informal slogan

$$\text{KE metric} \iff \text{"algebro} - \text{geometric stability}''.$$

We should emphasise however that, while this slogan summarises the general idea, the exact notion of "stability" needs to be specified, and this specification can be considered as part of the problem. There are a number of established notions in the literature, for example *asymptotic Chow stability* and *K-stability*: a good account of the relation between these is given

[1]An informal document *Discussion of the Kahler-Einstein problem*, giving some more details, is available on the webpage http:/www2.imperial.ac.uk/ skdona/.

in [**9**]. But our purpose here is to introduce two new notions, $\overline{K}$-*stability* and *b-stability*, so that we can formulate a precise version of this conjecture which we hope will be more accessible to proof.

Before going further some general remarks may be helpful. First, there has been a great deal of work in this area, which we will not attempt to summarise in detail here. Notably, Tian showed in [**14**] that his notion of K-stability was a necessary condition for the existence of a Kahler-Einstein metric, and gave an explicit example where this condition goes beyond those known before (see also (4.2) below). Second, the plethora of algebro-geometric notions of stability are all variants of the same basic idea. One expects that, among them, various definitions which are *a priori* different may *a posteriori* turn out to be equivalent. Since we want, ultimately, to establish a necessary and sufficient condition we can to some extent push difficulties from one side to the other: working with a stronger (more restrictive) notion of "algebro-geometric stability" will, as a matter of logic, simplify the proof of stability $\Rightarrow$ KE metric while making the proof of the converse more difficult. We will explain below how this obvious remark applies in the case of $\overline{K}$-stability and b-stability. The third remark is that, in the overall existence problem for Kahler-Einstein metrics, there is some distinction between theory and practice. One may have sufficient criteria which, while not in general necessary, can be applied to give existence statements in certain explicit cases. For example this is true of Tian's theory of the α-invariant. On the other hand one can envisage general, theoretical, solutions which may be difficult or impossible to apply to any specific case. This is probably the situation, as things stand at the moment, with the various notions of stability mentioned above, since they are all exceedingly difficult to verify in examples. We will discuss this further in 4.3 below, but we note here that our notion of "b-stability" comes with an integer parameter m—"b-stability at multiplicity m"—which is related to this issue.

The general shape of all these definitions of stability is that a manifold is stable unless there is a "destabilising object" of an appropriate kind. The notion of b-stability is derived by extending the class of destabilising objects. This extension can be seen as part of a more general trend. In Tian's original definition of K-stability the destabilising objects were projective varieties, smooth or mildly singular, with holomorphic vector fields. In the generalisation of [**4**] the destabilising objects were allowed to be general schemes with $\mathbf{C}^*$-actions. In the new notion of b-stability the destabilising objects are *sequences* of schemes ("webs of descendants"), related by birational transformations (reflected in the prefix "b"). Another point is that we find that we need to work not just with the "test configurations" appearing in standard definitions of stability, but with more general degenerations. Thus in Section 2 we develop background and foundations to do this, review the definition of K-stability and state the definition of $\overline{K}$-stability. Section 4 is intended to explain the motivation for the definition of b-stability. We state, without complete proof, a simple existence theorem and give some general discussion. In Section 5 we collect the proofs of some auxiliary results, not used in an essential way in the core of this paper, and further examples.

The author hopes that the ideas discussed in this paper raise various questions of interest to algebraic geometers. The main question is whether the notion of b-stability is actually needed. The definition is designed to get around a difficulty which it seems (to the author) might in principle occur in taking algebro-geometric limits of Fano manifolds under projective embeddings, defined by higher and higher powers of the anticanonical bundle. But the author does not know of any real example where this occurs (although certainly if one varies the hypotheses slightly it can: see (5.5)). It might be that, with a deeper analysis, one could show that this phenomenon does not occur (or perhaps one could show this under extra hypotheses, such as complex dimension 3). In that case one could forget the definition of b-stability and the existence theory for Kahler-Einstein metrics would be much simpler. On the other hand, it might be that one could produce actual examples of this phenomenon, and such examples should be related to the question of whether Gromov-Haussdorf limits of Kahler-Einstein manifolds have algebraic structures. In another direction, there is a circle of questions related to the distinction between test configurations and general degenerations, which fall into the area of algebraic group actions. We give some simple results and examples in (5.1) and (5.2) but the author would like to understand the general picture better.

The author is very grateful to Alessio Corti, Paolo Cascini, Frances Kirwan, Miles Reid, Julius Ross, Jacopo Stoppa, Gabor Szekelyhidi and Richard Thomas for helpful discussions.

2. Basic definitions

2.1. Algebraic theory for group actions. Let U, V be complex vector spaces of dimensions q, r respectively. We suppose that $G = SL(U)$ acts on V and hence on $\mathbf{P} = \mathbf{P}(V)$. We consider the orbit O in $\mathbf{P}$ of a point x. For simplicity we assume that the stabiliser of x in G is finite. We want to study the closure in $\mathbf{P}$ of O. This is an algebraic variety, so for any point y in the closure which is not actually in O we can find a holomorphic map Γ from the disc $\Delta \subset \mathbf{C}$ to $\mathbf{P}$ with $\Gamma(0) = y$ and $\Gamma(t) \in O$ for $t \neq 0$. We will call such a map an *arc through* y. Really we should work with germs of such maps, allowing us to restrict to a smaller disc, but we will generally ignore this in our notation. We say that two such arcs Γ_1, Γ_2 are equivalent if there is a holomorphic map $h : \Delta \to G$ such that $\Gamma_1(t) = h(t)\Gamma_2(t)$. Any arc Γ can be written in the form $\Gamma(t) = g(t)(x)$ where g is a meromorphic map from Δ to $\mathrm{End}(U)$, which restricts to a holomorphic map from the punctured disc Δ^* to $G \subset \mathrm{End}(U)$. Thus is we choose a vector $\hat{x} \in V$ lying over x we have a meromorphic map $\hat{\Gamma}$ from Δ to V defined by $\gamma(t) = g(t)(\hat{x})$. We define the integer $\nu(\Gamma)$ to be the order of the pole of $\hat{\Gamma}$ at $t = 0$. It is obvious that equivalent arcs give the same value of ν. More explicitly, fixing a basis of V

we represent $\hat{\Gamma}$ by its components $(\gamma_1, \dots, \gamma_r)$ say, where γ_i are meromorphic functions, then $\nu(\Gamma)$ is the maximum over i of the order of pole of γ_i. Thus $\nu(\Gamma) \leq 0$ if and only if $\hat{\Gamma}$ is holomorphic across $t = 0$ and $\nu(\Gamma) < 0$ if and only if $\hat{\Gamma}$ vanishes at $t = 0$. Recall that x is called *stable* if the orbit of $\hat{x}$ is closed and *semistable* if the orbit of $\hat{x}$ does not contain 0 in its closure. It is clear that x is stable if and only if $\nu(\Gamma) > 0$ for all Γ and semistable if and only if $\nu(\Gamma) \geq 0$ for all Γ.

The most familiar case of the above set-up is when $g(t)$ has the form $g(t) = \Lambda(t)R$ where R is a fixed element of G and Λ is a *one parameter subgroup* in G. We call an arc of this type an *equivariant arc*. Replacing x by Rx we can usually reduce to the case when $R = 1$. The *Hilbert-Mumford criterion* asserts that to test stability or semistability it suffices to restrict to equivariant arcs. But it is not true that any point $y \in \overline{O} \setminus O$ is contained in an equivariant arc–there are cases where there are boundary points y which are not "accessible by 1-parameter subgroups", see (5.1). This is related to the stabiliser $\mathrm{Stab}(y)$ of y in G. We have

PROPOSITION 1. *1. If* $\mathrm{Stab}(y)$ *is isomorphic to* $\mathbf{C}^*$ *then any arc through* y *is equivalent to an equivariant arc.*

2. If $\mathrm{Stab}(y)$ *is reductive then there is an equivariant arc through* y.

See (5.2) for the proof.

Recall that a *weighted flag* in U is a chain of subspaces $U_1 \subset \dots U_s = U$ and associated integers $\lambda_1 < \lambda_2 \dots < \lambda_s$. We say that an endomorphism A of U is compatible with the weighted flag if it preserves the subspaces U_i and acts as λ_i on U_i/U_{i-1}. It follows that A is diagonalisable, with eigenvalues λ_i. Conversely, starting with any diagonalisable A, with integer eigenvalues, we can define a weighted flag.

PROPOSITION 2. *Let* g *be a meromorphic function on the disc with values in* $\mathrm{End}(U)$, *holomorphic away from* 0 *and with* $g(t)$ *invertible for* $t \neq 0$. *Then there is a unique weighted flag in* U *with the following property. If* A *is any endomorphism compatible with the weighted flag, then we can write*

$$g(t) = L(t)\Lambda(t)R(t)$$

where L, R *are holomorphic across* 0 *with* $L(0), R(0)$ *invertible and where* $\Lambda(t) = t^A$ *is the* 1*-parameter subgroup generated by* A.

This is a standard result. In different language, we can view g as a meromorphic trivialisation over Δ^* of a bundle (in fact the trivial bundle) over Δ. The statement is that in such a situation we get a *parabolic structure* on the bundle at the point 0. To prove the Proposition we can reduce to the case when g is holomorphic (multiplying by a power of t). So we can think of it as a map of sheaves

$$g : \mathcal{O}^m \to \mathcal{O}^m,$$

and the cokernel of g is a torsion sheaf supported at 0. Then the statement follows from the classification of torsion modules over $\mathbf{C}[t]$.

Now we can define numerical invariants of equivalence classes of arcs Γ from the "eigenvalues" λ_i and their "multiplicities" $\dim U_i/U_{i-1}$ (Of course they are not intrinsically eigenvalues, since the choice of A is not unique, but they are well-defined integers associated to Γ.) For example we can define the " trace"

$$\sum_i \lambda_i \dim U_i/U_{i-1}.$$

This vanishes in the case when g maps into $SL(U)$. We define the "norm" by

$$|\Gamma| = \max_i |\lambda_i|,$$

which is clearly strictly positive. Now define

$$\psi(\Gamma) = \frac{\nu(\Gamma)}{|\Gamma|}. \tag{1}$$

This has the property that it is unchanged if we replace the parameter t by a positive power of t. We can think, informally, of the numbers $\psi(\Gamma)$, for different arcs Γ, as a kind of measure of "how stable" the point x is. Next we define

$$\Psi(y) = \sup_\Gamma \psi(\Gamma), \tag{2}$$

where Γ runs over the arcs through y. In sum, we have attached a numerical invariant $\Psi(y)$ (which *a priori* could be $+\infty$) to each point y in $\overline{O} \setminus O$.

We have

PROPOSITION 3. *The point x is stable if and only if $\Psi(y) > 0$ for all points $y \in \overline{O} \setminus O$.*

This is less obvious than it may appear at first sight. The problem is that we might have a point y such that $\psi(\Gamma) \leq 0$ for some arc through y but not for all. So that while y "destabilises" x we do not have $\Psi(y) \leq 0$. We give the proof in (5.3).

So far our discussion has been entirely algebraic. Now we introduce "metric" geometry. We suppose that we have a hermitian metric on U, and hence a maximal compact subgroup $SU(q) \subset G$, and that we have a norm on the restriction of the tautological line bundle over $\mathbf{P}$ to $\overline{O} \subset \mathbf{P}$. Since $\overline{O}$ is usually singular it is not quite obvious what we mean by a norm and we leave this point for the moment: for example we could think for the time being of the case when this norm extends smoothly to $\mathbf{P}$. We suppose that this bundle norm is invariant under the action of $SU(m) \subset G$. Then we have a moment map

$$M : \overline{O} \to \mathfrak{su}(q)^*.$$

Now suppose we have $y \in \overline{O} \setminus O$ and Γ as above. There is a unique choice of endomorphism A, compatible with the weighted flag, which is self-adjoint with respect to the metric on U. Then iA lies in $\mathfrak{su}(q)$ and we get a real number $\langle M(y), iA\rangle$.

LEMMA 1. *In this situation* $\nu(\Gamma) \leq \langle M(y), iA\rangle$.

In the case of equivariant arcs equality holds, which is a more standard and easy fact. To prove the Lemma, write $g(t) = L(t)t^A R(t)$ as above. Changing the choice of base point x we can reduce to the case when $L(t)$ is identically 1 and $R(0) = 1$. Now write $V = \bigoplus V_\mu$ where A acts with weight $-\mu$ on V_μ. So t^A acts as multiplication by $t^{-\mu}$ on V_μ. Let $\overline{\mu}$ be the largest value of μ such that $\hat{x}$ has a non-zero component in V_μ. Now $R(t)\widehat{(x)}$ has components $x_\mu(t)$ say in V_μ which are (vector valued) holomorphic functions of t and $x_{\overline{\mu}}(0)$ is non-zero by construction. Thus the $V_{\overline{\mu}}$ component of $t^A R(t)\hat{x}$ grows at least as fast as $t^{-\overline{\mu}}$ as $t \to 0$. For $\mu < \overline{\mu}$ the V_μ component of $t^A R(t)\hat{x}$ grows strictly slower than $t^{\overline{\mu}}$ and for $\mu > \mu_0$ it grows strictly slower than $t^{-\mu}$. It follows that $\nu \geq \mu_0$ and the components y_μ of $\hat{y}$ vanish if $\mu < \overline{\mu}$. Let μ_1 be the smallest value of μ such that y_μ is not zero, so we know that $\mu_1 \geq \overline{\mu}$. Then it follows from the above that we must have $\nu \leq \mu_1$ (with strict inequality unless $\mu_1 = \overline{\mu}$).

Now let $z \in \mathbf{P}$ be the limit of $t^{-A}y$ as $t \to 0$, and let $\hat{z} \in V$ be a representative. Then it is clear that $\hat{z}$ is in V_{μ_1}. The definition of the moment map implies that

$$\langle M(z), iA\rangle = \mu_1.$$

On the other hand the definition also implies that

$$\langle M(t^{-A}y, iA\rangle$$

is an decreasing function of t so

$$\langle M(y), iA\rangle \geq \langle M(z), iA\rangle = \mu_1.$$

Combining with the inequality $\nu \leq \mu_1$ from above we have established the Lemma.

In sum, we now have a "differential geometric" way to obtain a bound on the algebro-geometric invariant $\Psi(y)$. With metric structures as above we have

$$\Psi(y) \leq \max_A \frac{\langle M(y), iA\rangle}{\|A\|}, \tag{3}$$

where $\|A\|$ denotes the usual operator norm of A. Of course, it is equivalent to say that

$$\Psi(y) \leq \|M(y)\|_1$$

where $\|\ \|_1$ is the dual "trace-norm".

2.2. Application to Chow varieties. We will apply this theory to the particular case of Chow varieties. Thus we start with a projective space $\mathbf{P}(U)$ and for n, d we consider the set Ch of n-dimensional cycles in $\mathbf{P}(U)$ of degree d. It is a fact that this can be embedded as a projective variety in $\mathbf{P}(V)$ for a certain representation V of $G = SL(U)$ but we do not need to know the details of this embedding. We take a projective manifold $X \subset \mathbf{P}(U)$ which defines a point in the Chow variety and hence a G-orbit. We also have a Hilbert Scheme Hilb which parametrises subschemes with the same Hilbert polynomial as X and a G-equivariant regular map from Hilb to Ch which is a birational isomorphism. (Note that the scheme structure on Hilb—that is to say, infinitesimal deformations which do not extend to actual deformations—will not be relevant, so it would be better to take about the underlying variety.)

Now suppose that Y is a an algebraic cycle in the closure of the orbit of X and choose an arc Γ as above. Then Γ lifts to a map $\tilde{\Gamma}$ to the Hilbert scheme and $\tilde{\Gamma}(0)$ is a scheme W with underlying cycle Y. The lift $\tilde{\Gamma}$ defines a *projective degeneration* of X: a flat family $\pi : \mathcal{X} \to \Delta$, embedded in $\mathbf{P}(U) \times \Delta$, with fibre over t isomorphic to X for $t \neq 0$ and to W for $t = 0$. Conversely, a projective degeneration of X defines an equivalence class of arcs Γ. An equivariant arc corresponds to an equivariant degeneration, with a $\mathbf{C}^*$-action, also called a "test configuration".

Next we go on to the metric theory, so we suppose that U has a Hermitian metric. The basic fact is that there is then a natural induced metric on the restriction of the tautological bundle to Ch $\subset \mathbf{P}(V)$, and in particular to the closure $\overline{O}$ of our orbit, so we are in the situation considered above. This theory is explained well by Phong and Sturm in [**8**]. All we really need to know is the corresponding moment map $M : \text{Ch} \to \mathfrak{su}^*$. Let A be a trace-free self-adjoint endomorphism of U and define a function $H = H_A$ on $\mathbf{P}(U)$ by

$$H = \frac{1}{|x|^2}\langle x, Ax\rangle.$$

This is the Hamiltonian for the action of the 1-parameter group $s \mapsto e^{iAs}$ on $\mathbf{P}(U)$, with respect to the Fubini-Study symplectic form. Now let Z be an n-dimensional algebraic cycle in $\mathbf{P}$. Then the formula which defines the moment map is

$$\langle M(Z), iA\rangle = \int_Z H d\mu$$

Here integration over Z is defined in the obvious way, using the volume form $d\mu$ induced by the Fubini-Study metric (normalised so that the volume of Z is equal to to its degree). It is convenient to extend this definition to general Hermitian A by decreeing that the moment map vanishes on multiples of the identity. It will also be convenient to introduce a factor, so we define the

"Chow number" of Z, A to be

$$\mathrm{Ch}(Z,A) = \frac{1}{\mathrm{Vol}(Z)}\int_Z H\,d\mu - \frac{\mathrm{Tr}A}{\dim U}. \tag{4}$$

Thus $\mathrm{Ch}(Z,A)$ is the difference between the average of H over Z and the average eigenvalue of A. With this explicit formula in place we can go back to the point we skimmed over before, involving precisely what we mean by a hermitian structure on a line bundle over a singular space. There is no need to produce a general definition, since one can directly check that the argument above applies in our setting. The crucial points are

- If Z is preserved by the 1-parameter subgroup generated by A then $\mathrm{Ch}(Z,A)$ is independent of the metric. On the one hand it is equal to the weight of the induced action on the fibre of the line bundle $\mathcal{L}^{-1}$ over Z, and on the on the other hand it has an interpretation in equivariant cohomology (see [**5**]).
- $\mathrm{Ch}(Z,A)$ is monotone if Z moves under the 1-parameter subgroup generated by A (see (5.4)).

To sum up, we attach to each algebraic cycle Y in the closure of the G-orbit of X (but not in the orbit) an invariant $\Psi(Y)$. The projective variety X is *Chow stable* if and only if $\Psi(Y) > 0$ for all such Y. If we choose any Hermitian metric on U we get a bound

$$\Psi(Y) \leq \mathrm{Vol}(Y)\max_A \frac{\mathrm{Ch}(Y,A)}{\|A\|}. \tag{5}$$

Notice that from the form of the definition—using arcs—it does not really matter whether we talk about algebraic cycles or schemes here. For a scheme W which is in the closure of the orbit of X in the Hilbert scheme we can define $\Psi(W)$ by taking the supremum over degenerations with central fibre W of the same quantity we used in defining $\Psi(Y)$, so $\Psi(W) \leq \Psi(Y)$ for the cycle Y underlying W and in particular the inequality (5) gives a bound on $\Psi(W)$.

2.3. K-stability and $\overline{K}$-stability. We will now think slightly more abstractly of a compact complex n-manifold X and positive line bundle L over X. Suppose that L^m is very ample so its sections define an embedding in $\mathbf{P}(U)$ with $U = H^0(X, L^m)^*$. Suppose that we have a degeneration, as considered above. This is a flat family $\mathcal{X} \to \Delta$ with $\mathcal{X} \subset \mathbf{P}(U) \times \Delta$. The central fibre is a scheme $W \subset \mathbf{P}(U)$ and the line bundle $\mathcal{O}(1)$ on $\mathbf{P}(U)$ is isomorphic to L^m on the non-zero fibres. Take a positive integer p such that $\mathcal{O}(p)$ is also very ample on all fibres of $\mathcal{X}$ (this is certainly true if p is sufficiently large). The direct image of $\mathcal{O}(p)$ is a locally free sheaf over Δ which can thus be trivialised. If we fix a trivialisation we get an embedding of $\mathcal{X}$ in $\mathbf{P}(U_p)$ where

$$U_p = H^0(X, L^{pm})^*.$$

Now the fixed degeneration $\mathcal{X}$ has a sequence of projective embeddings. We will write $\mathcal{X}_p$ when we want to emphasise the difference, and W_p for the central fibre. For each one we have a numerical invariant $\nu(\mathcal{X}_p)$ say. The notion of *K-stability* involves the asymptotics of these as $p \to \infty$. The theory developed in the literature deals with the case when $\mathcal{X}$ is an equivariant degeneration (or test configuration), defined by a $\mathbf{C}^*$ action, so we will now restrict to that situation. Then it is known that $p^{-n}\nu(\mathcal{X}_p)$ has a limit as $p \to \infty$ and we define the *Futaki invariant* $F(\mathcal{X})$ to be this limit. We say that (X, L) is K-stable if $F(\mathcal{X}) > 0$ for all degenerations $\mathcal{X}$, and for all m. It is important to realise that, even if L is itself very ample over X, degenerations may occur at some large multiplicity m which cannot be realised by any smaller value, such as $m = 1$.

Let $\tilde{X}$ be the blow-up of X at a point x. For large integers γ the line bundle $\gamma L - [E]$ is ample on $\tilde{X}$, where E is the exceptional divisor.

Definition 1. *The polarised manifold (X, L) is $\overline{K}$-stable if there is a γ_0 such that the blow-up $\tilde{X}$ at any point of X, with the polarisation $\gamma L - [E]$ is K-stable, for $\gamma \geq \gamma_0$.*

There is a slightly different notion in which we use Chow stability of the blow-up rather than K-stability; indeed this may turn out to be more relevant to the existence problem. But for the discussion in the present paper we stick to $\overline{K}$-stability.

3. Birational modifications

3.1. Families over the disc. We begin by considering a flat family $\pi : \mathcal{X} \to \Delta$ over the disc. We write W for the central fibre. Suppose we have an embedding of the family in $\mathbf{P}(U^*) \times \Delta = \mathbf{P} \times \Delta$ such that each fibre $\pi^{-1}(t)$, for nonzero t, maps to a smooth projective variety V_t. We suppose that for all $p \geq 1$ and all $t \in \Delta^*$ the restriction map

$$ev_t : H^0(\mathbf{P}; \mathcal{O}(p)) \to H^0(V_t, \mathcal{O}(p))$$

is surjective. We also suppose that the central fibre W contains a component B which is reduced at its generic point. For each $p \geq 1$ we will define another flat family $\mathcal{X}' \to \Delta$ through a certain birational modification of $\mathcal{X}$.

Shrinking the disc if necessary we can suppose that the kernel of ev_t has a fixed dimension. This family of kernels has a limit as t tends to 0. Choose a fixed subspace $J \subset H^0(\mathbf{P}, \mathcal{O}(p)) = s^p(U)$ which is a complement to this limit. Then the ev_t to J yields an isomorphism for all $t \in \Delta^*$ and when $t = 0$ the restriction of the map $ev_0 : H^0(\mathbf{P}; \mathcal{O}(p)) \to H^0(W, \mathcal{O}(p))$ to J has the same image as ev_0. Then we get an embedding of $\mathcal{X}$ in $\mathbf{P}(J^*) \times \Delta$ which is just the composite of the original embedding, the Veronese map and a linear projection.

Now let s be a nonzero element of J, so $e_0(s)$ is a section of $\mathcal{O}(p)$ over the central fibre. We consider an extension σ of s over $\mathcal{X}$. Thus σ will have the form $\sigma = s + \sum_{i\geq 1} t^i \tau_i$ for τ_i in J. Such an extension has an *order of vanishing* $\nu(\sigma)$ on the component $B \subset \mathcal{W}$.

LEMMA 2. *For each $s \in J \setminus \{0\}$ there is a $\overline{\nu}(s)$ such that $\nu(\sigma) \leq \overline{\nu}(s)$ for all extensions σ of s.*

Given $\mu > 1$ the set of τ_1 such that $s + \tau_1 t$ admits a higher order extension vanishing to order μ along B is an affine subspace $K_\mu \subset J$. Clearly the K_μ decrease as μ increases and hence they are eventually constant. So if there is no such upper bound $\overline{\nu}(s)$ we can find a τ_1 such that $s + \tau_1 t$ admits higher order extensions vanishing to arbitrarily high order. Repeating the argument we can find a sequence τ_i such that for each μ the finite sum $s + \sum_{i=1}^{\mu} \tau_i t^i$ vanishes to order μ along B. Thus we get a formal power series $s + \sum_{i=1}^{\infty} \tau_i t^i$ which vanishes to infinite order. Standard general arguments show that this formal power series can be arranged to be convergent and this contradicts the fact that i_t is an isomorphism for $t \neq 0$.

Of course we specify $\overline{\nu}(s)$ by defining it to be the least possible upper bound and we set $\overline{\nu}(0) = +\infty$. Define

$$J_\mu = \{s \in J : \overline{\nu}(s) \geq \mu\}.$$

Then the J_μ are linear subspaces of J, defining a flag. Choose a corresponding direct sum decomposition

$$J = \bigoplus I_\mu,$$

where

$$J_{\mu'} = \bigoplus_{\mu \geq \mu'} I_\mu.$$

(Here of course the sums run over a finite subset of integers μ.) Let $M : J^* \to J^*$ be the endomorphism which acts as multiplication by μ on I_μ^*. Then for non-zero t, we have an automorphism t^M of J^* and hence of $\mathbf{P}(J^*)$. Now we set

$$V_t' = t^M(V_t).$$

This gives a family $\mathcal{V} \subset \mathbf{P}(J^*) \otimes \Delta^*$ over the punctured disc and by general theory there is a unique way to extend this to a scheme $\mathcal{X}' \subset \mathbf{P}(J^*) \times \Delta$, flat over Δ. We will see presently that $\mathcal{X}'$ is independent of the various choices made in the construction. We write W' for the central fibre of $\mathcal{X}'$.

Notice that we can perform this construction with $p = 1$ and then $W' = W$ if and only if B does not lie in any hyperplane in $\mathbf{P}$.

A simple example

Let $V_t \subset \mathbf{CP}^2$ be a family of cubics degenerating to the union of a conic B and a line R with two intersection points X, Y. Take $p = 2$. Then J is the 6 dimensional space of quadratic polynomials and our decomposition is $J = I_1 \oplus I_0$ where I_1 is the one dimensional subspace spanned by the polynomial P defining the conic B. In the Veronese embedding of the original family $\mathcal{X} \subset \mathbf{P}^5 \times \Delta$ the component B lies in $\mathbf{P}^4 = \mathbf{P}(\mathbf{C}^5)$ and the component R lies in a $\mathbf{P}^2 \subset \mathbf{P}^5$. The modified family $\mathcal{X}' \subset \mathbf{P}^5$ is different. The central fibre has just one component B', which is a rational curve with a double point at $[P]$. The effect of the modification is to collapse the component R and identify the two intersection points X, Y.

Fix a basis s_α of J compatible with the direct sum decomposition so s_α lies in $I_{\mu(\alpha)}$. Let σ_α be extensions vanishing to maximal order. By construction $\underline{\sigma}_\alpha = t^{-\mu(\alpha)}\sigma_\alpha$ extends holomorphically over the generic point in B. Let f_α be the restriction to B. This is meromorphic section of $\mathcal{O}(p)$ over B. More precisely if we lift to the normalisation of B the only poles of f_α will be on the intersection $D \subset B$ of the component B in W and the other components.

LEMMA 3. *The f_α are linearly independent.*

Suppose that there is a linear relation $\sum c_\alpha f_\alpha = 0$, for $c_\alpha \in \mathbf{C}$. Let $\lambda_{\max}$ be the largest value of $\lambda(\alpha)$ for terms with $c_\alpha \neq 0$. Choose ordering so that the relation is

$$c_1 f_1 + \dots c_q f_q + \sum_{\beta > q} c_\beta f_\beta = 0$$

where $\lambda(\alpha) = \lambda_{\max}$ for $\alpha \leq q$ and $\lambda(\alpha) > \lambda_{\max}$ for $\alpha > q$. Then

$$c_1 \sigma_1 + \dots c_q \sigma_q + \sum_{\beta > q} t^{\lambda_{\max} - \lambda(\beta)} c_\beta \sigma_\beta$$

is an extension of $c_1 s_1 + \dots c_q s_q$ which vanishes to order at least $\lambda_{\max} + 1$ along B. But by construction the s_1, s_q are linearly independent elements of the space $I_{\lambda_{\max}}$ and all elements of this space vanish to order exactly $\lambda_{\max}$ so we have our contradiction.

Let R_p^+ denote the vector space of meromorphic sections of $\mathcal{O}(p)$ over $\mathcal{X}$, holomorphic away from the central fibre. Thus elements of R_p^+ can be written as semi-finite Laurent series $\sum \rho_i t^i$ where the sum has only finitely many negative terms and the coefficients ρ_i are in $s^m(U)$. Let $R_p \subset R_p^+$ be the subspace of sections which extend holomorphically over the generic point of $B \subset\subset \mathcal{W}$. We know that the extensions $\underline{\sigma}_\alpha$ lie in R_p. Further we have

LEMMA 4. *The elements of R_p are exactly the sums of the form $\sum_\alpha a_\alpha (t)\underline{\sigma}_\alpha$, for holomorphic functions a_α on Δ.*

It is obvious that such a sum does lie in R_p; we have to establish the converse. Any element σ of R_p^+ can be written in the form

$$\sigma = \sum a_\alpha(t) \underline{\sigma}_\alpha$$

where $a_\alpha(t)$ is meromorphic on Δ, holomorphic away from 0. Suppose that at least one of the $a_\alpha(t)$ has a pole. Let k be the maximal order of pole of the a_α and suppose that $a_1, \dots a_r$ have poles of order k while $a_{r+1}, \dots$ have poles of lower order. Then, along B, the meromorphic section σ is $t^{-k}(b_1 f_1 + \cdots + b_r f_r)$ where b_i is the coefficient of t^{-k} in a_i. By the linear independence of the f_i this section σ cannot be extended over the generic point of B, so does not lie in R_m.

Now suppose we made different choices in our construction (of the subspace J, of the extensions with maximal vanishing order, of the summands I_μ). We get another collection of sections, $\underline{\sigma}^*_\beta$ say, and by the preceding Lemma we can write

$$\underline{\sigma}^*_\beta = \sum g_{\beta\alpha}(t)\underline{\sigma}_\alpha,$$

for a holomorphic matrix-value function $(g_{\alpha\beta}(t))$. Symmetrically the inverse is also holomorphic, so defines an automorphism of $\mathbf{P}(J^*) \times \Delta$, covering the identity on Δ. It follows from the definition that this automorphism takes the family $\mathcal{X}'$, constructed using one set of choices, to that constructed using the other set of choices. So we have

Corollary 1. *The family $\mathcal{X}' \subset \mathbf{P}(J^*) \times \Delta$ is uniquely defined, up to automorphisms of $\mathbf{P}(J^*) \times \Delta$ covering the identity on Δ.*

We record three simple properties of this construction.

Proposition 4. *1. The birational map from $\mathcal{X}$ to $\mathcal{X}'$ maps B to a component $B' \subset W'$ which it is reduced at its generic point.*

2. We have

$$\frac{\deg(B')}{\deg(W')} \geq \frac{\deg(B)}{\deg(W)}.$$

3. Suppose we start with $\mathcal{X}'_p$ and perform the same construction with an integer q, so we obtain a family $(\mathcal{X}'_p)'_q$ say. Then $(W'_p)'_q$ is isomorphic to W'_{pq}.

The proof is sketched in (5.6). (The author is grateful to Julius Ross for pointing out item (3) above.)

To sum up as p varies we get a collection of flat families, say $\mathcal{X}'_p \to \Delta$ with central fibres W'_p, all derived from the original family $\mathcal{X}$. This is very likely a standard construction in algebraic geometry but the authors lack of expertise in that field limits our treatment to the rather pedestrian but self-contained account above. Let us just include two remarks.

1. There is a discussion in the language of graded rings. Let $R_{\mathcal{W}} = \bigoplus_p R_p$ (notice that the definition makes sense for any $p \geq 0$). This is a graded ring with the functions on Δ as the sub-ring R_0. We

can also evaluate elements of R_p on B: this gives a vector space, Q_p, say, of meromorphic functions on B and $Q = \bigoplus Q_p$ is a graded ring. Now we can ask whether either

- The ring R is finitely generated over R_0;
- The ring Q is finitely generated.

It seems reasonable to expect that there properties should be related to the question of the *stabilisation* of $\mathcal{X}_p$ or W_p as p tends to infinity (or perhaps tends to infinity through multiples of some given integer). We do not go into the matter further here, but see (5.5).

2. There is a discussion from the point of view of line bundles over the total space of the family. Consider the component B as a subset of the $\mathcal{X}$. At least in favourable situations this defines a line bundle $L_B \to \mathcal{X}$ whose local sections correspond to functions with a pole along B. Likewise, at least in favourable situations, the remainder R of W defines a line bundle L_R and $L_B \otimes L_R$ is trivial, since $B \cup R$ is the divisor of the holomorphic function t on $\mathcal{X}$. Given p, let $\nu = \nu(p)$ be the maximal order of vanishing as considered above and consider the line bundle $\mathcal{L}' = \mathcal{L}^p \otimes L_B^\nu = \mathcal{L}^p \otimes L_R^{-\nu}$. Our construction of $\mathcal{X}'_p$ amounts to considering the birational image of $\mathcal{X}$ defined by the sections of $\mathcal{L}'$.

3.2. b-stability. Let (X, L) be a polarised manifold and fix m such that L^m is very ample. We can also choose m large enough so that the sections of L^m generate those of L^{pm} for all p. Let $\mathcal{X} \subset \mathbf{P}(U) \times \Delta$ (where $U = H^0(X, L^m)^*$) be a projective degeneration with central fibre W. We say that $\mathcal{X}$ is "admissible"if it contains a component B, as considered above, which is "large" in that the degree (i.e. volume) of B is greater than half the degree of X. Then we call W an *admissible limit at multiplicity m*. For large enough p we can apply the construction of the previous subsection to get a new degeneration $\mathcal{X}'_p$ with central fibre W'_p. We call W'_p a *descendant of W at the power p*. We define a "web of descendants" at multiplicity m to be a sequence of schemes $(W'_1, W'_2, W'_3, \dots)$ such that

1. For each p, the scheme W'_p is an admissible limit at multiplicity pm.
2. For all p, q the scheme W'_{pq} is a descendant at the power q of W'_p.

(REMARK. Note that this definition includes the statement that $W = W'_1$ is a descendant of itself, which just means that $B \subset W$ does not lie in any hyperplane.)

Thus by Proposition 4, any family $\mathcal{X}$, and in particular any test configuration, defines a web of descendants and if $\mathrm{Aut}W = \mathbf{C}^*$ any web of descendants beginning with W is obtained in this way. We define the "birationally modified Futaki invariant" of a web of descendants to be

$$F_b = \limsup_{p \to \infty} p^{1-n} \Psi(W'_p).$$

We define a *destabilising object* for (X, L), at multiplicity m, to be a web of descendants at multiplicity m with $F_b \leq 0$.

DEFINITION 2. *The polarised manifold (X, L) is b-stable at multiplicity m if there are no destabilising objects for (X, L) at multiplicity m. We say that (X, L) is b-stable if it is b-stable at multiplicity m for all large enough m.*

The upshot of all this is that we can formulate a new precise form of Yau's conjecture: that *the existence of a Kahler-Einstein metric should be equivalent to b-stability.* Of course this is in the case of a Fano manifold X with positive line bundle $L = K_X^{-1}$.

Discussion

1. The definition of b-stability may seem a little complicated, but much of the complication arises from the fact that we take account of the possibility of different degenerations with the same central fibre, and hence (perhaps) different descendants of W at the same power p. In the case of a web of descendants derived from a W with $\text{Aut}W = \mathbf{C}^*$ things are much simpler. Then by Proposition 1 we only need to consider a test configuration and its web of descendants. Suppose we are in the case when B is the whole of W. Then our birational modification construction becomes vacuous and the descendant W_p' is isomorphic to W but embedded by the linear system $\mathcal{O}(p)$. The $\mathbf{C}^*$-action on the sections of $\mathcal{O}(p)$ over W has a generator A_p and it is a standard fact that $\|A_p\| = p\|A_1\|$. So

$$\Psi(W_p') = p^{-1}\|A_1\|^{-1}\nu(W_p)$$

and hence $p^{1-n}\Psi(W_p')$ tends to the limit $\|A_1\|^{-1}F(\mathcal{X})$, as $p \to \infty$, where F is the usual Futaki invariant. So, up to the fixed positive factor $\|A_1\|$, our definition reproduces the usual Futaki invariant in this case. It is possible that for any web of descendants $p^{1-n}\Psi(W_p')$ has a limit as p tends to infinity, but because we do not know this we take lim sup instead, in the definition.

2. The definition of b-stability is meant to have a provisional character. Modifying the definition of an "admissible" degeneration allows us to adjust the definition of b-stability by a notch or two, making the notion less stringent or more. For example we could consider imposing a condition that the degree of the complement of B is very small relative to that of B, or that the automorphism group of W is reductive. As we explained in the introduction, such adjustments shift the difficulty from one side of the problem (b-stable $\Longrightarrow$ KE metric) to the other (KE metric $\Longrightarrow$ b-stable).

3.3. Families over a general base. So far, we have considered families over a 1-dimensional base, in fact the disc Δ. Now we want to discuss the general situation of a flat family $\mathcal{X}$ over a general variety N with a base point which we write as $0 \in N$. As usual, we are really working with germs, we can shrink the neighbourhood if necessary. We suppose, first, that the fibre W over 0 contains a component B and, second, that there is a dense open set $N_0 \subset N$ such that the fibres over N_0 are smooth, just as before. It is not possible to directly extend our construction to define a new family $\mathcal{X}'$ over N but this can be dome after blowing up N suitably. Thus given a positive integer p we want to construct a variety $\hat{N}$ with a regular birational isomorphism $q : \hat{N} \to N$ and a family $\mathcal{X}'_p \to \hat{N}$ which is characterised by the following property. Any map $\gamma : \Delta \to N$ with $\gamma(0) = 0$ lifts to $\hat{\gamma} : \Delta \to \hat{N}$ and the pull-back by $\hat{\gamma}$ of $\mathcal{X}'_p$ is isomorphic to the family we have constructed in (3.1), beginning with the pull back $\gamma^* \mathcal{X}$ of $\mathcal{X}$ to Δ.

This construction of $\hat{\mathcal{X}} \to \hat{N}$ is most likely rather routine, as a matter of algebraic geometry, but we will develop it in a way which is adapted to our differential-geometric application in the next section.

We begin with a basic general fact.

LEMMA 5. *Let Ω be a connected complex manifold and let $f_0, \dots f_m$ be holomorphic functions on $\Omega \times N$. For $\underline{y} \in \mathbf{C}^{m+1}$ write $f^{\underline{y}} = \sum y_\alpha f_\alpha$. Given a pre-compact open subset $\Omega_0 \subset \Omega$ we can find a finite set of points $z_1, \dots z_r \in \Omega_0$, a neighbourhood A of 0 in N and $C > 0$ such that for all $z \in \Omega_0$, $\tau \in A$ and $\underline{y} \in \mathbf{C}^{m+1}$ we have*

$$|f^{\underline{y}}(z, \tau)| \leq C \sum_{i=1}^{r} |f^{\underline{y}}(z_i, \tau)|.$$

To begin note that it suffices to prove this for $[\underline{X}]$ in a small neighbourhood in $\mathbf{CP}^m$, using the compactness of projective space. So write $\underline{X} = (1, y_1, \dots, y_m)$ and define a function S on $\mathbf{C}^m \times \Omega \times N$ by

$$S(\underline{y}, z, \tau) = f_0(z, \tau) + \sum y_i f_i(z, \tau).$$

For each fixed $z \in \Omega$ we get a function $S(\underline{y}, z, \tau)$ on $\mathbf{C}^m \times N$, let I be the ideal generated by all of these functions. Now restrict to the local ring of germs of functions about $(0, 0) \in \mathbf{C}^m \times N$. This is Noetherian so the ideal is generated by a finite number of functions $\psi_i(\underline{y}, \tau) = S(\underline{y}, z_i, \tau)$. It is clear that we can choose the z_i to lie in any given open set, in particular in Ω_0. Now let ζ be any fixed point of Ω_0. Let J be the ideal of functions $g(\underline{y}, z, \tau)$ in the local ring at $(0, \zeta, 0)$ such that for each fixed z the function $g(\underline{y}, z, \tau)$ lies in I. Then J is finitely generated and by Nakayama's lemma we can take the generators to be polynomials in z. The co-efficients of these polynomials must lie in I and it follows that any $g \in J$ can be written in the form

$$g(\underline{y}, z, \tau) = \sum \chi_i(\underline{y}, z, \tau) \psi_i(\underline{y}, \tau).$$

In particular S lies in J, by construction so we can write

$$S(\underline{y}, z, \tau) = \sum \chi_i(\underline{y}, z, \tau) S(\underline{y}, z_i, \tau),$$

for points z in a small neighbourhood of ζ and for small $\underline{y}, \tau$. Then for such points the desired inequality holds, with $C = \max |\chi_i(\underline{y}, z, \tau)|$. Now use the precompact hypothesis to cover Ω_0 with a finite number of such small neighbourhoods.

Now we proceed with our construction. Just as before we can fix a subspace $J \subset s^p(U)$ and an embedding of $\mathcal{X}$ in $P(J^*) \times N$. Choose an open set Ω, biholomorphic to a polydisc say, whose closure lies in the smooth part of B. Then the family $\mathcal{X} \to N$ can be trivialised around Ω, in the sense that, after possibly restricting to a smaller neighbourhood of $0 \in N$, there is a holomorphic embedding $\iota : \Omega \times N \to \mathcal{X}$ compatible with the projection $\pi : \mathcal{X} \to N$. We also fix a trivialisation of $\iota^*(\mathcal{O}(p))$. Then any element s of J defines a holomorphic function $s(z, \tau)$ on $\Omega \times N$. We choose a large finite set F in Ω as in the Lemma above, adapted to these functions $s(z, \tau)$. For each $\tau \in N$ we have a linear map $e_\tau : J \to \mathbf{C}^r$ defined by $e_\tau(s) = (s(z_1, \tau), \ldots s(z_r, \tau))$. When τ is in the open dense subset N_0 this map e_τ is injective, because an element of the kernel has to vanish on all of Ω by our choice of F and hence on the whole fibre of $\mathcal{X}$ (since this is irreducible). Thus, taking the images of the e_τ, we get a map from N_0 to the Grassmannian of $m+1$ dimensional subspaces of $\mathbf{C}^r$. This map need not extend to N but taking the closure of the graph, we can find a blow-up $\hat{N}$ containing a copy of N_0, to which the map does extend. Let E be the pull-back to $\hat{N}$ of the tautological bundle over the Grassmannian. From this point of view the maps e_τ give a trivialisation of E over N_0. Thus for $\tau \in N_0$ we can map the fibre $\pi^{-1}(\tau) \subset \mathcal{X}$ into $\mathbf{P}(E^*)_\tau$. So we get a subvariety $\mathcal{X}_0'$ in the projective bundle $\mathbf{P}(E^*)$ over N_0. After perhaps blowing up $\hat{N}$ further, the closure of this defines a flat family $\mathcal{X}'$ over $\hat{N}$.

Suppose that we make this construction in the case when N is the disc Δ, so no blowing up is required and we get a map from Δ to the Grassmannian. Clearly this maps 0 to the subspace of $\mathbf{C}^r$ with basis $f_\alpha(z_i)$, in the notation of (3.1), and one sees that this construction agrees with the previous one. Likewise for the case when we pull back $\mathcal{X}'$ by a map from Δ to $\hat{N}$.

We can make this construction in the case when N is the closure of the orbit of X in the Hilbert scheme, N_0 is the orbit and 0 corresponds to some limiting scheme W. Then *the descendants of W at the power p are exactly the schemes parametrised by $q^{-1}(0) \subset \hat{N}$*. For, in one direction, an arc Γ through 0 has a unique lift to $\hat{\Gamma}$ and we take $\hat{\Gamma}(0)$ which is a point in $q^{-1}(0)$. In the other direction, given a point $\hat{\tau}$ in $q^{-1}(0)$ we can find an arc $\tilde{\Gamma}$ through $\hat{\tau}$ and we obtain the corresponding scheme as a descendant by starting with the arc $q \circ \tilde{\Gamma}$ in N.

REMARK. Suppose that we adjoin an extra point z_{r+1} to F. Then it is clear that the family $\mathcal{X}'$ we construct will be the same, up to isomorphism. It follows easily that the construction does not depend on the choice of the set F, the choice of trivialisations etc.

4. Applications of the definition

In this section we will outline the relevance of the definitions above to the existence problem for Kahler-Einstein metrics. So we consider a Fano manifold X and take $L = K_X^{-1}$. For simplicity we assume that the automorphism group of X is finite.

4.1. KE metric $\Longrightarrow$ b-stability. We have

PROPOSITION 5. *If X has a Kahler-Einstein metric then (X, K_X^{-1}) is $\overline{K}$-stable.*

This is a consequence of the results of the author [**5**], Arezzo-Pacard [**2**] and Stoppa [**10**]. For, according to Arezzo and Pacard, the blow up of X admits a constant scalar curvature metric when the parameter γ is sufficiently large. Then [**5**] shows that the blow-up is at least K-semistable and the refinement of Stoppa shows that it is actually K-stable. (The proof of this refinement will involve blowing up a second time.) In short: KE $\Longrightarrow$ $\overline{K}$-stable. It seems reasonable to hope that the argument of Stoppa in [**10**] can be extended to show that

$$\overline{K} - \text{stability} \quad \Longrightarrow \quad \text{b} - \text{stability},$$

and, assuming this can be done, we get KE $\Longrightarrow$ b-stability.

In fact one might hope ultimately to prove the chain

$$\text{KE metric} \quad \Longrightarrow \quad \overline{K} - \text{stable} \quad \Longrightarrow \quad \text{b} - \text{stable} \quad \Longrightarrow \quad \text{KEmetric},$$

and if this could be done it would be just as good to take a formulation of the main conjecture involving $\overline{K}$-stability in place of b-stability. This means that the result in one direction (KE $\Longrightarrow$ $\overline{K}$-stability) is already in place, but exactly the same work is involved in the extra difficulty of proving the converse ($\overline{K}$-stability $\Longrightarrow$ KE). So it is really a matter of taste which formulation one prefers. The definition of $\overline{K}$-stability is quicker to state but puts into prominence the blow-up, which we prefer to see as a device used in the proofs, rather than something fundamental to the problem.

4.2. b-stability $\Longrightarrow$ KE metric. Here we will discuss a model problem which does not bear immediately on the general existence question but which illustrates ideas which apply in other, more complicated, situations. We suppose that we have a sequence of Kahler-metrics ω_i on X, in the class $c_1(X)$ and that $\mathrm{Ric}(\omega_i) - \omega_i$ tends to zero in C^∞, in the sense that for all $l \geq 0$

$$\max_X |\nabla^l(\mathrm{Ric}(\omega_i - \omega_i)|$$

tends to zero as i tends to infinity. (Here ∇^l denotes the iterated covariant derivative.) We will also assume that X has complex dimension 3, although this is not fundamental. Then we have

THEOREM 1. *If X is b-stable then it admits a Kahler-Einstein metric*

This statement is not intended to be optimal (for example, the arguments could probably be made to work if we restrict to $l = 0$), but it has some content. If X is a suitable small deformation of the Mukai-Umemura manifold X_0, then Tian showed in [**14**] that X does not admit a Kahler-Einstein metric, but since X_0 has a Kahler-Einstein metric (as explained in [**6**]), there is a sequence ω_i on X satisfying the condition above. (See also the related results of Sun and Wang [**11**] in terms of Ricci flow.)

We do not give a complete proof of the Theorem here, but we will give the part of the argument which brings in the b-stability condition, taking as input four "Hypotheses".

Before beginning it may be helpful to emphasise an elementary but important general point. Suppose we have a sequence of projective spaces $\mathbf{P}_i$, all of the same dimension N. Suppose we have varieties $V_i \subset \mathbf{P}_i$, all of the same degree. Does it make sense to take the "limit" of the V_i? Certainly we can choose isomorphisms χ_i from $\mathbf{P}_i$ to the standard model $\mathbf{CP}^N$ and then (at least after passing to a subsequence) we can take a limit of the $\chi_i(V_i)$. But if we change the isomorphisms by automorphisms g_i of $\mathbf{CP}^m$ then the limit of the sequence $g_i\chi_i V_i$ may be completely different. So, as the questions stands, the limit has no intrinsic meaning. Suppose now that the $\mathbf{P}_i$ are "metrized projective spaces" *i.e.* projectivisations of hermitian vector spaces. Then we can choose the isomorphisms χ_i to preserve metrics, the automorphisms are reduced to the compact group $PU(N)$ and the limits we get are isomorphic. So the answer to the question is positive if we work with metrised projective spaces. Essentially, for the purposes of taking limits, we can treat metrized projective spaces as being canonically isomorphic.

Now we begin the proof. For each i and all $k > 0$ we get a standard L^2-norm $\| \|_{i,k}$ on $H^0(X, L^k)$, using the metric ω_i. For large enough k, these sections give a "Tian embedding" $T_{k,i} : X \to \mathbf{P}^{N_k}$ of X in a metrized projective space of dimension N_k. Of course, as above, we can identify this with the standard space $\mathbf{CP}^{N_k}$ with the standard metric. For fixed k and varying i these embeddings differ by the action of $SL(N_k + 1, \mathbf{C})$ so $T_{k,i} = g_{k,i} \circ T_{k,0}$ say.

HYPOTHESIS 1. We can fix a large m so that if the sequence $g_{m,i}$ has a bounded subsequence, then X admits a Kahler-Einstein metric.

For any self-adjoint endomorphism A, with respect to the L^2 metric, A we have a Chow number $\mathrm{Ch}(T_{k,i}X, A)$.

HYPOTHESIS 2. There is a function $\epsilon(k)$ with $\epsilon(k) \to 0$ as $k \to \infty$ such that, for all A and k we have

$$|\mathrm{Ch}(T_{k,i}X, A)| \leq \|A\|\epsilon(k).$$

(We recall that $\|A\|$ denotes the operator norm: the modulus of the largest eigenvalue.)

Now focus attention on the case $k = m$. By compactness of the Hilbert scheme we can suppose (taking a subsequence) that the projective varieties $T_{m,i}X$ have a limit W which is a subscheme of $\mathbf{CP}^{N_m}$. If W is projectively equivalent to $T_{p,0}X$ then the $g_{m,i}$ are bounded and we conclude that X has a Kahler-Einstein metric, by Hypothesis 1. So we suppose the contrary, that W is not equivalent to X and we want to show that X is not b-stable.

HYPOTHESIS 3. The scheme W is an admissible degeneration, that is, it contains a component B of degree greater than one half the degree of W which is reduced at its generic point.

Now consider the universal family $\mathcal{U}$ over a neighbourhood N of 0 (the point corresponding to W) in the closure of the orbit of X in the Hilbert scheme. So the sequence $T_{p,i}X$ yields a sequence $\tau_i \in N$ converging to 0. Fix an open set $\Omega \subset B$ as in (3.3), whose closure lies in the reduced, smooth part of B. Then (after perhaps shrinking N) we can trivialise the universal family $\mathcal{U}$ in a neighbourhood of B and define an analytic embedding $\Omega \times N \to \mathcal{U}$ compatible with the projection $\mathcal{U} \to N$. Thus we have an open set Ω_τ in the fibre of $\mathcal{U}$ over τ and in particular open sets $\Omega_{\tau_i} = \Omega_i$ in $T_{p,i}X$ which, in an obvious sense, tend to Ω as $i \to \infty$. Now we define a norm $\| \ \|_{\Omega,i}$ on sections of L^{pm} by restricting to Ω_i and using the L^2 norm induced by the standard fibre metric and the Fubini-Study volume form.

HYPOTHESIS 4. The L^2 norm defined by ω_i and the norm $\| \ \|_{\Omega,i}$ are uniformly equivalent (i.e. with constants independent of i).

REMARK. Note that this implies that the exact choice of Ω is unimportant: any two choices give equivalent norms. Similarly for the choice of trivialisations etc.

Now we are all set up to state:

PROPOSITION 6. *Assuming Hypotheses (1)–(4) above, if X does not admit a Kahler-Einstein metric then it is not b-stable at multiplicity m.*

This will be a consequence of the following, which is the central result of this paper. For each $p > 0$ we can suppose that the sequence $T_{pm,i}X$ converges to some scheme W''_p.

PROPOSITION 7. *The sequence W_p'' is a web of descendants.*

We give the proof of Proposition 6 assuming Proposition 7. For any self-adjoint endomorphism A we have $\mathrm{Ch}(W_p'', A) \leq \epsilon(pm)|A|$ and this implies by (5) that $p^{1-n}\Psi(W_p'') \leq \epsilon(pm)$. Since $\epsilon(pm)$ tends to 0 as p tends to infinity we see that $F_b \leq 0$, so this web of descendants is a destabilising object.

The proof of Proposition 6 is also easy, given the background we have developed. It suffices to show that each W_p'' is a descendant of W. (For the other conditions for a web of descendants follow by replacing m by a multiple.) The orbit of X yields an open subset N_0 of N. We have a blow up $q : \hat{N} \to N$ and a lift $L : N_0 \to \hat{N}$. We know that points of $q^{-1}(0)$ correspond to descendants of W. Our sequence τ_i is a sequence in N_0 converging to 0 and we thus have a sequence $L\tau_i$ in $\hat{N}$. Taking a subsequence we can suppose that $L\tau_i$ converges in $\hat{N}$ to some descendant W_p'. What we need to show is that this limit is isomorphic to the "differential geometric" limit W_p''.

Now recall that the family $\hat{\mathcal{U}}$ over $\hat{N}$ is defined by taking the closure of a family over $q^{-1}N_0$. For each point τ in N_0 we have a map $e_\tau : J \to \mathbf{C}^r$. The standard Hermitian metric on $\mathbf{C}^r$ restricts to a metric on the image of e_τ. On the other hand we can identify J^* with the space of sections of L^{pm} over the fibre V_τ in $\mathcal{U}$. In particular we can do all this for the points τ_i so that V_{τ_i} is $T_{m,i}X$. Then for each i we have three norms on the space of sections $H^0(X, L^{pm})$.

1. The standard L^2 norm defined by ω_i.
2. The norm $\| \ \|_{\Omega,i}$ defined by the L^2 norm over Ω.
3. The norm defined by the map $e_{\tau_i} : J \to \mathbf{C}^r$, as above.

Unwinding the constructions, to prove that W_p' and W_p'' are isomorphic it suffices to show that the first and third norms are uniformly equivalent, with constants independent of i. Hypothesis 4 states that the first norm is equivalent to the second norm, so it suffices to show that the second and third norms are uniformly equivalent. But this is clear from the way the finite set F was chosen (Lemma 5). In one direction, the third norm dominates the L^∞ norm of sections over a slightly smaller open set (and we have pointed out above that the precise choice of Ω is not important). In the other direction the L^2-norm over Ω dominates the L^∞ norm over an interior set by standard elliptic estimates.

As we mentioned above, the four Hypotheses hold for the sequence of "approximate KE metrics" ω_i, thus proving the Theorem. The proofs will be given elsewhere. Hypotheses 1,3 and 4 are essentially known results, and very similar statements can be found in the recent paper [**15**] of Tian. Thus the main new input is Hypothesis 2. The proof of this (which leads us to make the restriction on the dimension of X for the time being) depends in turn on joint work with X-X Chen.

4.3. Miscellaneous remarks.

1. **Uniform stability**

   A general difficulty which arises in applying the usual definitions of stability is that these definitions state that the Futaki invariant is positive but do not supply any definite lower bound. In the special case of toric manifolds (and in the context of extremal metrics) establishing such a lower bound was one of the main points of [**4**] (Proposition 5.2.2 in [**4**]). In general, this issue was considered by Szekelyhidi [**12**] who introduced a notion of "uniform stability". His definition has the shape

   $$F(\mathcal{X}) \geq c\|A\|,$$

   for all test configurations $\mathcal{X}$ and for an appropriate norm on the generator of the action (and, in a general context, it could be that different norms lead to different notions). Taking such a definition of stability makes it easier to prove the direction "stable" $\Longleftarrow$ KE metric, but it seems hard to establish the converse. Our definition of b-stability adopts this idea to some extent: roughly it asserts uniform stability over a restricted class of test configurations. The fundamental point is that the result of Arezzo and Pacard gives a little extra control of the Futaki invariants (or Chow weights).
2. **Testing stability**

   As we mentioned in the Introduction, the impact of this whole discussion is rather limited unless one has a way of testing "stability" in explicit situations. The difficulty (in the case of K-stability, say) is that the definition requires checking test configurations of arbitrarily high multiplicity–that is to say, degenerations of X embedded in arbitrarily large projective spaces. For a fixed, reasonably small, multiplicity it may be possible to analyse all the test configurations, but this gets more and more complicated as the dimension grows. In this regard, it is relevant that the argument outlined above in the direction b-stability $\Longrightarrow$ KE metric works with b-stability at an explicit multiplicity m, which can be computed in principle from analytical information. If this argument can be refined to produce a multiplicity m which is reasonably small then one could hope to verify b-stability in some explicit cases.

   Again, this issue arises in the toric case. In dimension 2, there is a straightforward test (for K-stability) involving certain "simple" piecewise-linear functions on the polytope of the variety (see [**4**]). But in higher dimensions, even in the toric case, the situation is less clear.
3. **Finite generation and Gromov-Haussdorf limits**

   Recall that we associated a graded ring Q to an admissible degeneration, and we expect that the finite generation of this ring should be related to the stabilisation of the sequence of descendants. It also seems likely that Q can be obtained as the limits of L^2 sections over the Gromov-Hausdorff limit considered by Ding and Tian [**3**]. The finite generation question could then be seen as the problem of endowing this Gromov-Haussdorf limit with an algebraic structure. If Q is finitely generated then the obvious candidate is Proj(Q).

4. **Extremal and constant scalar curvature metrics**

The definition of b-stability is aimed at the Kahler-Einstein case and is not intended to be appropriate, as it stands, in the more general setting of extremal and constant scalar curvature metrics. On the other hand an example of Apostolov *et al* [**1**] shows that K-stability is probably not the correct criterion for the existence of these metrics. It may be the same general idea—involving not just one test configuration but a sequence—will be relevant in that case too.

5. Examples and subsidiary results

5.1. Points not accessible by one-parameter subgroups. We consider the action of $SL(3, \mathbf{C})$ on $s^d(\mathbf{C}^3)$, polynomials in x, y, z, so points of the projectivisation corresponds to plane curves. Suppose that a curve C meets the line $z = 0$ in d points $p_1, \ldots, p_d$. Then the action of the 1-parameter group $x \mapsto tx, y \mapsto ty, z \mapsto t^{-2}z$ deforms C into the union of the d lines Op_i, where O is $x = y = 0$. If $d > 5$ then simple dimension counting shows that a typical singular curve consisting of d lines through a point is not projectively equivalent to one which arises in this way from the fixed curve C. Now consider a curve C of degree 6 defined by the equation

$$z \prod_{i=1}^{5} (x - \lambda_i y) + p(x, y) = 0.$$

Thus there are 5 branches of the curve passing through $x = y = 0$. The inverse of the 1-parameter subgroup above deforms this curve into the union of the 5 lines $x = \lambda_i y$ through O and the line at infinity. Making a projective transformation, fixing lines through O, we can move the line at infinity to a line $x - \mu y = z$ say. Now apply the same 1-parameter subgroup to deform this to the curve C' which is the union of 6 lines through O: the 5 lines $x = \lambda_i y$ and the sixth line $x = \mu y$. Thus C' is in the closure of the orbit of C but there is no reason why this configuration of 6 lines should occur from the intersection of C with a line, so that in general C' will not be accessible by a 1-parameter subgroup.

5.2. Proof of Proposition 1. The result is similar to Luna's slice theorem, but the author has not found this exact statement in the literature.

For $\xi \in \mathfrak{g}$ and z in $\mathbf{P}(V)$ we write ξz for the corresponding tangent vector to $\mathbf{P}(V)$ at z. We suppose that the stabiliser of y is the complexification K^c of a compact group K. (The point of the condition is that representations of K^c decompose as sums of irreducibles.) Then we find an equivariant slice for the action a projective subspace $\mathbf{P}' \subset \mathbf{P}(V)$ with the properties

1. $\mathbf{P}'$ contains y and is preserved by K^c;
2. $\mathbf{P}'$ is transverse to the G-orbit of y at y;
3. for $y' \in \mathbf{P}'$ near y and $\xi \in \mathfrak{g}$ the tangent vector $\xi y'$ lies in the tangent space of $\mathbf{P}'$ if and only if ξ is in $\mathfrak{k}$.

To do this we decompose $\mathfrak{g} = \mathfrak{k}^c \oplus \mathfrak{p}$ as representations of K^c. The derivative of the action at x gives us a decomposition of K^c-representations $V = \mathbf{C}\hat{x} \oplus \mathfrak{p} \oplus S$ say. Then we can take $\mathbf{P}' = \mathbf{P}(\mathbf{C}\hat{x} \oplus S)$. Now the third condition above implies that the identity component of the stabiliser in G of any point in $\mathbf{P}'$ close to y is a subgroup of K^c. Let V_0 be the intersection of the of the G-orbit of O of x with $\mathbf{P}'$. The second condition implies that the G-orbit of any point in $\mathbf{P}(V)$ close to y meets $\mathbf{P}'$ in a point which is also close to y. It follows that y lies in the closure V of V_0. The third condition implies that if y' is close to y and y' is in V_0 then the for small ball B around y' the intersection of B with V_0 is equal to the intersection of B with the K^c-orbit of y'. Since V is an algebraic variety it is clear that, near to y, V_0 is contained in a finite union of K^c orbits. So there is a single K^c orbit $K^c y'$ in V_0 which contains y in its closure. But the action of K^c on $\mathbf{P}'$ is determined by the linear action on S and it follows from the Hilbert-Mumford criterion, applied to the reductive group K^c, shows that there is a 1-parameter subgroup Λ of K^c such that y lies in the closure of the Λ-orbit of y'. Viewing Λ as a 1-parameter subgroup in G we obtain the first statement of the Proposition.

For the second statement, we see from the transversality condition (2) that any arc through $y \in \overline{O}$ is equivalent to an arc mapping into V. Near to y, we know that V_0 is a finite union of $\mathbf{C}^*$ orbits so the arc must map into a single one of these orbits and it follows that the arc is equivalent to an equivariant arc.

Remark. It seems likely that in fact V_0 is, near to y, equal to a single K^c-orbit but the author has not managed to prove this. If this were the case it would follow that when the stabiliser is $\mathbf{C}^*$ there is a unique equivalence class of arcs through y, up to the obvious fact that we can take a covering, replacing t by a power of t.

5.3. Proof of Proposition 3. The argument is related to ideas of Thaddeus [**13**]. In one direction, it is clear that if x is stable then $\Psi(y) > 0$ for all y. In the other direction, suppose that $x \in \mathbf{P}$ is a point which is not stable for the G-action. We choose some other representation of G and hence another projective space Q on which G acts. We can choose Q to contain a point q which *is* stable. We fix metrics so that we have moment maps $\mu_{\mathbf{P}}, \mu_Q$.

Now consider the G-action on the product $\mathbf{P} \times Q$. The notion of stability depends on the choice of a class in $H^2(\mathbf{P} \times Q)$ which we take to be $\omega_{\mathbf{P}} + \eta\omega_Q$ where η is a real parameter, and $\omega_{\mathbf{P}}, \omega_Q$ are the standard generators. When η is irrational we move outside the algebro-geometric framework, but we can still apply the symplectic theory of Kirwan [**7**]. The moment map for the product is $\mu_{\mathbf{P}} + \eta\mu_Q$. When η is very large the point (x, q) is stable and there is a point in its orbit where $\mu_{\mathbf{P}} + \eta\mu_Q$ vanishes. But when $\eta = 0$ there is no point in the orbit where $\mu_{\mathbf{P}}$ vanishes, since x is not stable. It follows that there is some $\eta_0 \geq 0$ and a point (y, r) in the closure of the G-orbit in $\mathbf{P} \times Q$ such that $\mu_{\mathbf{P}}(y) + \eta_0\mu_Q(r) = 0$. We claim that $\Psi(y) \leq 0$. Let $\Gamma_{\mathbf{P}} : \Delta \to \mathbf{P}$ be an arc through y. We can lift this to the G-orbit in $\mathbf{P} \times Q$, so we get a map $(\Gamma_{\mathbf{P}}, \Gamma_Q) : \Delta \to \mathbf{P} \times Q$. The maps $\Gamma_{\mathbf{P}}, \Gamma_Q$ define the same weighted flag and hence the same self-adjoint endomorphism A. We have integers $\nu_{\mathbf{P}}(y), \nu_Q(r)$. The inequality of Lemma 1 gives

$$\nu_{\mathbf{P}}(y) \leq \langle \mu_{\mathbf{P}}(y), iA \rangle, \quad \nu_Q(r) \leq \langle \mu_Q(r), iA \rangle,$$

so $\nu_{\mathbf{P}}(y) + \eta_0\nu_Q(r) \leq 0$. But the condition that q is stable implies that $\nu_Q(r) > 0$ so we see that $\nu_{\mathbf{P}} \leq 0$. This implies that $\Psi(y) \leq 0$, as required.

5.4. Monotonicity of the Chow invariant. Let A be a self-adjoint endomorphism of a Hermitian vector space U and let Z be an algebraic cycle in $\mathbf{P}(U)$. For real s let $Z_s = e^{As}Z$ and consider the function $f(s) = \mathrm{Ch}(Z_s, A)$. We want to show that f is an increasing function of s. This is a well-known fact, first proved by Zhang [Z], and there are a number of proofs in the literature, but we include a short proof here for completeness. Consider for the moment a more general situation of a compact Riemannian manifold P, a real valued function h on P and a submanifold V of P. Let $\phi_s : P \to P$ be the gradient flow of h and let $V_s = \phi_s(V)$. We consider the function

$$g(s) = \int_{V_s} h \, d\mu, \tag{6}$$

where $d\mu$ is the induced Riemannian volume element. Then

$$g'(s) = \int_{V_s} |(\mathrm{grad}_\perp h)|^2 d\mu + \int_{V_s} \langle M, \mathrm{grad}_\perp h \rangle d\mu,$$

where $\mathrm{grad}_\perp h$ is the component of the gradient vector field normal to V_s and M is the mean curvature vector of V_s. Here the second term arises when we differentiate the volume element in (7). The relevance of this is that when P is complex projective space the flow e^{As} is the gradient flow of the function H. Since the mean curvature of a complex subvariety vanishes, the second term drops out and we derive the desired monotonicity. (It is not hard to see that the presence of singularities of Z does not affect the argument.)

5.5. Zariski's example. We explain how a well-known example of Zariski [**18**], fits into the framework of this paper. Let S be the blow-up of the projective plane in 12 points q_i and L be the line bundle defined by the divisor $D = 2(4H - \sum E_i)$ on S. If the points are in general position then L is very ample on S but if the points lie on a cubic $C \subset \mathbf{P}^2$ then the proper transform $\tilde{C}$ has intersection number 0 with D so L is not ample. If q_i are in general position on C then no multiple L^p is trivial on $\tilde{C}$. All sections of $\mathcal{O}(pD)$ vanish on C but, for all p, there are sections which vanish with multiplicity 1. It follows that the ring $\bigoplus H^0(S; L^p)$ is not finitely generated. Now vary q_i in curves $q_i(t)$ for $t \in \Delta$ and hence construct a family $\mathcal{S} \to \Delta$. We suppose that the $q_i(t)$ are in general position for non-zero t but $q_i(0)$ are in general position on a cubic curve, as above. We have a line bundle $\mathcal{L} \to \mathcal{S}$ and the direct image is locally free so we can regard the fibres S_t, for non-zero t as projective varieties in a fixed projective space. Taking the closure we get a family $\mathcal{X} \to \Delta$ where the central fibre has two components. One is S_0, embedded by sections of $L^p \otimes [-\tilde{C}]$, and the other is a $\mathbf{P}^1$ bundle R over $\tilde{C}$. For any $p > 0$ we can apply our construction from (3.1) (taking $B = S_0$) to form $\mathcal{X}'_p$. This is exactly the same as the family we get by replacing L by L^p in the construction of $\mathcal{X}$ above. For all p we get the same central fibre $S_0 \cup_{\tilde{C}} R$ but with a different line bundle over it. The ratio $\deg(R)/\deg(S_0)$ is $O(p^{-1})$ as $p \to \infty$. The graded ring Q defined in (3.1) is isomorphic to $\bigoplus H^0(S_0, L^p)$ and is not finitely generated.

REMARK. In this example there is a natural family, independent of p, in the background–the family $\mathcal{S}$. The problem is that we are trying to embed this using sections of a line bundle which is not positive on the central fibre, and this gives rise to the infinite series of "descendants". Whether an example like this can occur for a degeneration of a Fano manifold and, conversely, whether this is the only mechanism by which we can obtain an infinite sequence of "descendants" are questions beyond the authors knowledge, but which seem to be important.

5.6. Proof of Proposition 4. We can regard W as being embedded in $\mathbf{P}(J^*)$. If $B \subset W$ does not lie a hyperplane in $\mathbf{P}(J^*)$ then $W' = W$ and we are done, so suppose that there $J^* = U_1 \oplus U_2$ and B lies in $\mathbf{P}(U_1)$. Let $B_0 \subset B$ be the complement of the intersection of B with the other components of W and of the support of nilpotents. Then, following through the construction, we see that B_0 maps isomorphically to $B'_0 \subset B'$ by a map of the form $x \mapsto (x, f(x)) \in \mathbf{P}(U_1 \oplus U_2)$ where f is a holomorphic section of $\mathcal{O}(1) \otimes U_2$ over B_0. Thus it is clear that B' is reduced at its generic point. We can choose a linear subspace $\mathbf{P}(R) \subset \mathbf{P}(U_1)$ of complementary dimension such that $B \cap \mathbf{P}(R)$ lies in B_0. Then we get the same number of intersection points of the image of B_0 with $\mathbf{P}(R \oplus U_2)$. Since any other intersection points give a positive contribution to the intersection number we see that $\frac{\deg(B')}{\deg(W')} \geq \frac{\deg(B)}{\deg(W)}$.

To see that $(W'_p)'_q = W'_{pq}$ we can use the approach through the norms on sections of $\mathcal{O}(pq)$ defined by restricting to a large finite set F. Both $(W'_p)'_q$ and W'_{pq} can be obtained as the limits of the nonzero fibres V_t under embeddings in $H^0(V_t, \mathcal{O}(pq))^*$, so, as in Section 4, we have to show that two norms on this space are uniformly equivalent (with constants independent of t). But this is rather clear from the construction using the fact that our local trivialisation of the family $\mathcal{X}$ around $\Omega \subset B$ maps holomorphically to a local trivialisation of $\mathcal{X}'_p$ around the image Ω' by a map of the same form as above.

References

[1] Apostolov, V., Calderbank, D., Gauduchon, P. and Tonnesen-Friedman, C. *Hamiltonian 2-forms in Kahler geometry III: extremal metrics and stability* Inventiones Math. 173 547–60 (2008).

[2] Arezzo, C. and Pacard, F. *Blowing up and desingularising constant scalar curvature Kahler metrics, I* Acta Math. 196 179–228 (2006).

[3] Ding, W. and Tian, G. *Kahler-Einstein metrics and generalised Futaki invariants* Inventiones Math. 110 315–335 (1992).

[4] Donaldson, S. *Scalar curvature and stability of toric varieties* Jour. Differential Geometry 62 289–349 (2002).

[5] Donaldson, S. *Lower bounds on the Calabi functional* Jour. Differential Geometry 70 453–472 (2005).

[6] Donaldson, S. *Kahler geometry on toric manifolds, and some other manifolds with large symmetry* In: handbook of Geometric Analysis, Vol. I Ji et al eds. International Press 29–74 (2008).

[7] Kirwan, F.C. *The cohomology of quotients in symplectic and algebraic geometry* Princeton UP 1983.

[8] Phong, D.H. and Sturm, J. *Scalar curvature, moment maps and the Deligne pairing* Amer. J. Math. 126 693–712 (2004).

[9] Ross, J. and Thomas, R. *A study of the Hilbert-Mumford criterion for the stability of projective varieties* Jour. Algebraic Geometry 10 201–255 (2007).

[10] Stoppa, J. *K-stability of constant scalar curvature Kahler manifolds* Advances Math. 221 1397–1408 (2009).

[11] Sun, S. and Wang, Y. *On the Kahler-Ricci flow near a Kahler-Einstein metric* Arxiv 10004.2018.

[12] Szekelyhidi, G. *Extremal metrics and K-stability* PhD. Thesis 2006 DG/061102.

[13] Thaddeus, M. *Geometric Invariant Theory and flips* Jour. Amer. Math. Soc. 9 691–723 (1996).

[14] Tian, G. *Kahler-Einstein metrics of positive scalar curvature* Inventiones Math. 130 1–57 (1997).

[15] Tian, G. *Existence of Einstein metrics on Fano manifolds* Preprint 2010.

[16] Yau, S-T. *Review of Kahler-Einstein metrics in algebraic geometry* Israel Math. Conference Proc., Bar-Ilan Univ. 9 433–443 (1996).

[17] Zhang, S. *Heights and reductions of semi-stable varieties* Compositio Math. 104 77–105 (1996).

[18] Zariski,O. *The theorem of Riemann-Roch for high multiples of an effective divisor on an algebraic surface* Ann. Math. 76 560–615 (1962).

Lagrangian Floer theory on compact toric manifolds: survey

Kenji Fukaya, Yong-Geun Oh, Hiroshi Ohta, Kaoru Ono

Abstract. This article is a survey of the Lagrangian Floer theory of toric manifolds, which summarizes the results obtained in a series of the present authors' papers [**FOOO3, FOOO4, FOOO5**]. In this survey, we discuss calculations of the Floer cohomology of Lagrangian T^n orbits in compact toric manifolds. Applications to symplectic topology and to mirror symmetry are also discussed.

Contents

Key words and phrases. Floer cohomology, mirror symmetry, toric manifolds, open-closed Gromov-Witten invariant, Saito theory, Landau-Ginzburg model, weakly unobstructed Lagrangian submanifolds, potential function, Jacobian ring, Frobenius structure.

1. Introduction

This is the survey of a series of present authors' papers [**FOOO3, FOOO4, FOOO5**] on the Lagrangian Floer theory of toric manifolds and its applications to mirror symmetry. The main purposes of the present survey are to convey systematic means of computations of Lagrangian Floer cohomology of the T^n-orbits in toric manifolds, to demonstrate various operations introduced in [**FOOO1**] Section 3.8 and apply them to the mirror symmetry between toric A model and Landau-Ginzburg B model and to symplectic topology of toric manifolds.

Let X be a compact toric manifold with complex dimension n and let $L(\mathbf{u})$ be a T^n orbit. (Here $\mathbf{u}$ is an element of the interior of the moment polytope which parametrizes the location of the corresponding T^n orbit. See Section 4 Formula (12).) We show that the cardinality (counted with multiplicity) of the pairs $(L(\mathbf{u}), b)$, for which Floer cohomology $HF((L(\mathbf{u}), b), (L(\mathbf{u}), b); \Lambda)$ is nontrivial, is equal to the Betti number of X. (Theorem 11.9.) Here b is an element of $H^1(L(\mathbf{u}); \Lambda_0)/H^1(L(\mathbf{u}); 2\pi\sqrt{-1}\mathbb{Z})$ which deforms the Floer (pre-)coboundary map and so does the Floer cohomology of L.

The set of such pairs $(L(\mathbf{u}), b)$ one-one corresponds to the set of the critical point of a certain function $\mathfrak{PO}$, called the potential function. Then $\mathbf{u}$, the location of $L(\mathbf{u})$, is given by the valuation of the coordinates of the corresponding critical point. For the given toric manifold X, valuations of the critical points of $\mathfrak{PO}$ can be obtained by solving a system of explicitly calculable algebraic equations in a finite number of times. We illustrate these examples in sections 6 and 10. (We use the result of Cho-Oh [**CO**] for this calculation.)

The above mentioned one-one correspondence is induced by an isomorphism between quantum cohomology $QH(X; \Lambda_0)$ of X and the Jacobian ring Jac($\mathfrak{PO}$) of the potential function $\mathfrak{PO}$. The origin of such a correspondence goes back to Givental [**Gi1, Gi2**] and Batyrev [**B1, B2**] in the case when

X is Fano. We remark that the rank of $QH(X;\Lambda_0)$ is the same as the Betti number of X, and the rank of $\mathrm{Jac}(\mathfrak{PO})$ is the same as the number of critical points of $\mathfrak{PO}$ counted with multiplicity.

The isomorphism $QH(X;\Lambda_0) \cong \mathrm{Jac}(\mathfrak{PO})$ is a ring isomorphism. In case $QH(X;\Lambda)$ is semi-simple, the ring $QH(X;\Lambda)$ splits into the product of copies of the field Λ and each factor of the product corresponds to a critical point of $\mathfrak{PO}$. (Proposition 11.8.)

In this way we associate a non-displaceable Lagrangian submanifold $L(\mathbf{u})$ to each of the direct factors of $QH(X;\Lambda)$. Entov-Polterovich [**E, EP1, EP2, EP3**] and others [**Os, Us**] associated a Calabi quasi-homomorphism to each direct factor of $QH(X;\Lambda)$, and also associates non-displaceable Lagrangian submanifolds $L(\mathbf{u})$ to such Calabi quasi-homomorphisms. The non-displaceable Lagrangian submanifold associated by the Entov-Polterovich's theory coincides with the one associated by the Lagrangian Floer theory, as we prove in [**FOOO7**]. (Our construction and proof are very different from Entov-Polterovich's, though.)

The ring isomorphism $QH(X;\Lambda_0) \cong \mathrm{Jac}(\mathfrak{PO})$ is generalized to the context of big quantum cohomology in the left hand side and the potential function of the Lagrangian Floer theory with bulk deformations in the right hand side. We call the corresponding potential function the potential function with bulk.

Furthermore this ring isomorphism intertwines the pairings, which are the Poincaré duality pairing in the left hand side and (a version of) the residue pairing in the right hand side. This gives rise to an isomorphism of the relevant Frobenius manifold structures present in the two sides: One is the Frobenius manifold structure induced by big quantum cohomology, which is due to Dubrovin [**Dub**], and the other is the one associated to the isolated singularity by Saito [**Sa, MSa**]. This isomorphism can be regarded as a version of mirror symmetry between Toric A model and Landau-Ginzburg B model. It is closely related to the story of Hori-Vafa [**HV**] and also of Givental.

The mirror symmetry between the toric manifold and the singularity theory have been studied by many mathematicians. Besides those already mentioned above, we provide a list of some of them, which is not exhaustive by any means.

In this survey, we focus on the case in which we study the A model (symplectic geometry and pseudo-holomorphic curve) on the toric manifold side and the B model (deformation theory and complex geometry) on singularity theory side. The papers [**Aur1, Aur2, Bar, CLe, CO, Gro1, Gro2, GPS, Iri1, Iri2, Iri3, OT, Ta, W**] also deal with that case.

The other side of the story, namely B model in toric side and A model in singularity theory side, has been more extensively studied than the side of present survey. The papers [**Ab1, AKO, FLTZ, Se2, Ue, UY**] deal with this side.

2. Preliminary

2.1. Notations and terminologies. The *universal Novikov ring* Λ_0 is the set of all formal sums

$$\sum_{i=0}^{\infty} a_i T^{\lambda_i} \tag{1}$$

where $a_i \in \mathbb{C}$ and $\lambda_i \in \mathbb{R}_{\geq 0}$ such that $\lim_{i\to\infty} \lambda_i = \infty$, and T is a formal parameter. We allow $\lambda_i \in \mathbb{R}$ in (1) (namely negative λ_i) to define Λ which we call *universal Novikov field.* It is a field of fraction of Λ_0. We require $\lambda_i > 0$ in (1) to define Λ_+, which is the maximal ideal of Λ_0.

We define a valuation $\mathfrak{v}_T$ on Λ by

$$\mathfrak{v}_T\left(\sum_{i=0}^{\infty} a_i T^{\lambda_i}\right) = \inf\{\lambda_i \mid a_i \neq 0\}. \tag{2}$$

(Here we assume $\lambda_i \neq \lambda_j$ for $i \neq j$.) $\Lambda, \Lambda_0, \Lambda_+$ are complete with respect to $\mathfrak{v}_T$ and (Λ_0, Λ_+) is a valuational ring with valuation $\mathfrak{v}_T$.

REMARK 2.1. In [**FOOO1**] a slightly different Novikov ring $\Lambda_{0,\mathrm{nov}}$ which contains another formal parameter e is used. The role of e is to adjust all the operators appearing in the story so that they have well-defined degree. (e has degree 2.) In [**FOOO3, FOOO4, FOOO5**] and this paper we use Λ_0 since ring theoretical properties of Λ_0 is better than one of $\Lambda_{0,\mathrm{nov}}$. As a drawback only the parities of various operators are well-defined.

Let $Z_1, \ldots, Z_m$ be variables. We define the *strictly convergent power series ring*

$$\Lambda_0\langle\langle Z_1, \ldots, Z_m\rangle\rangle$$

as the set of all formal sums

$$\sum_{k_1=0}^{\infty} \cdots \sum_{k_m=0}^{\infty} C_{k_1\ldots k_m} Z_1^{k_1} \ldots Z_m^{k_m}$$

where $C_{k_1\ldots k_m} \in \Lambda_0$ such that

$$\lim_{k_1+\cdots+k_m\to\infty} \mathfrak{v}_T(C_{k_1\ldots k_m}) = +\infty.$$

We define *strictly convergent Laurent power series ring*

$$\Lambda_0\langle\langle Z_1, Z_1^{-1}, \ldots, Z_m, Z_m^{-1}\rangle\rangle$$

as the set of all formal sums

$$\sum_{k_1\in\mathbb{Z}} \cdots \sum_{k_m\in\mathbb{Z}} C_{k_1\ldots k_m} Z_1^{k_1} \ldots Z_m^{k_m}$$

where $C_{k_1\ldots k_m} \in \Lambda_0$ such that

$$\lim_{|k_1|+\cdots+|k_m|\to\infty} \mathfrak{v}_T(C_{k_1\ldots k_m}) = +\infty.$$

See [**BGR**] about those rings.

We also define

$$\Lambda\langle\langle Z_1, Z_1^{-1}, \dots, Z_m, Z_m^{-1}\rangle\rangle = \Lambda_0\langle\langle Z_1, Z_1^{-1}, \dots, Z_m, Z_m^{-1}\rangle\rangle \otimes_{\Lambda_0} \Lambda.$$

The definition of $\Lambda\langle\langle Z_1, \dots, Z_m\rangle\rangle$ is similar.

Let C be a graded free Λ_0 module. The valuation $\mathfrak{v}_T$ induces a norm on C in an obvious way, by which C is complete. We define its degree shift $C[1]$ by $C[1]^k = C^{k+1}$. The shifted degree $\deg'$ is defined by

$$\deg' x = \deg x - 1.$$

We put

$$B_k C = \underbrace{C \otimes \cdots \otimes C}_{k \text{ times}}. \tag{3}$$

Let $\widehat{B}C = \widehat{\bigoplus}_{k=0}^{\infty} B_k C$ be the completed direct sum of them. Let $\mathfrak{S}_k$ be the symmetric group of order $k!$. It acts on $B_k C$ by

$$\sigma \cdot (x_1 \otimes \cdots \otimes x_k) = (-1)^* x_{\sigma(1)} \otimes \cdots \otimes x_{\sigma(k)} \tag{4}$$

where $* = \sum_{i<j:\sigma(i)>\sigma(j)} \deg x_i \deg x_j$. We define $E_k C$ as the subset of $\mathfrak{S}_k$ invariant elements of $B_k C$ and put $\widehat{E}C = \widehat{\bigoplus}_{k=0}^{\infty} E_k C$ its completed direct sum.

On BC we define a coalgebra structure $\Delta : BC \to (BC)^{\otimes 2}$ by

$$\Delta(x_1 \otimes \cdots \otimes x_k) = \sum_{i=0}^{k} (x_1 \otimes \cdots \otimes x_i) \otimes (x_{i+1} \otimes \cdots \otimes x_k). \tag{5}$$

(Note the summand in the case $i = 0$ is $1 \otimes (x_1 \otimes \cdots \otimes x_k)$.) Δ is coassociative.

We can define $\Delta : EC \to (EC)^{\otimes 2}$ by restriction. It is coassociative and graded cocommutative.

We also consider a map $\Delta^{k-1} : BC \to (BC)^{\otimes k}$

$$\Delta^{k-1} = (\Delta \otimes \underbrace{id \otimes \cdots \otimes id}_{k-2}) \circ (\Delta \otimes \underbrace{id \otimes \cdots \otimes id}_{k-3}) \circ \cdots \circ \Delta.$$

For an indecomposable element $\mathbf{x} \in BC$, it can be expressed as

$$\Delta^{k-1}(\mathbf{x}) = \sum_c \mathbf{x}_c^{k;1} \otimes \cdots \otimes \mathbf{x}_c^{k;k} \tag{6}$$

where c runs over some index set. We use the same notation for EC.

2.2. Moduli spaces of pseudo-holomorphic disks. Lagrangian Floer theory is based on the moduli space of pseudo-holomorphic disks. We recall its definition below. See [**FOOO1**] subsection 2.1.2 for detail.

Let $X = (X, \omega)$ be a symplectic manifold and L its Lagrangian submanifold. We pick a compatible almost complex structure J on X. Let $\beta \in H_2(X, L; \mathbb{Z})$.

The moduli space $\mathcal{M}^{\text{main}}_{k+1;\ell}(\beta)$ is the compactified moduli space of the genus zero bordered holomorphic maps $u : (\Sigma, \partial\Sigma) \to (X, L)$, in class $\beta \in H_2(X, L(u); \mathbb{Z})$ with $k+1$ boundary marked points and ℓ interior marked points. This means the following:

CONDITIONS 2.2. 1) Σ is a connected union of disks and spheres, which we call *(irreducible) components*. We assume the intersection of two different irreducible components is either one point or empty. The intersection of two disk components is if nonempty, a boundary point of both of the components. The intersection of a disk and a sphere component is an interior point of the disk component. We assume that intersection of three different components is empty. We also require Σ to be simply connected. A point which belongs to two different components is called a *singular point*.

2) $u : \Sigma \to X$ is a continuous map which is J-holomorphic on each of the components. $u(\partial\Sigma) \subset L$. Here $\partial\Sigma$ is the union of the boundary of disk components.

3) There are $k+1$ points $z_0, \ldots, z_k$ on $\partial\Sigma$. (We call them *boundary marked points.*) They are mutually distinct. None of them are singular point. We require the order of $k+1$ boundary marked points to respect the counter-clockwise cyclic order of the boundary of Σ.

4) There are ℓ points $z_1^+, \ldots, z_\ell^+$ on $\Sigma \backslash \partial\Sigma$. (We call them *interior marked points.*) They are mutually distinct. None of them are singular point.

5) For each of the components Σ_a of Σ, one of the following conditions hold:
   a) u is not a constant map on Σ_a.
   b) Σ_a is a disk component. We have $2n_{\text{int}} + n_{\text{bdry}} \geq 3$. Here n_{int} is the sum of the numbers of the interior marked points and the interior singular points. n_{bdry} is the sum of the numbers of the boundary marked points and the boundary singular points.
   c) Σ_a is a sphere component. The sum of the numbers of the marked points and the singular points on Σ_a is ≥ 3.

The condition 5) is called the *stability condition*. It is equivalent to the condition that the automorphism group of this element is a finite group.

In case $\ell = 0$ we write $\mathcal{M}^{\text{main}}_{k+1}(\beta)$ in place of $\mathcal{M}^{\text{main}}_{k+1;0}(\beta)$.

We define the evaluation maps

$$\text{ev} : \mathcal{M}^{\text{main}}_{k+1;\ell}(\beta) \to X^\ell \times L^{k+1} \tag{7}$$

where we put

$$\mathrm{ev} = (\mathrm{ev}^+, \mathrm{ev}) = (\mathrm{ev}_1^+, \dots, \mathrm{ev}_\ell^+; \mathrm{ev}_0, \dots, \mathrm{ev}_k),$$

as follows:

$$\mathrm{ev}_i(\Sigma, u) = u(z_i)$$

where z_i is the i-th boundary marked point as in 3).

$$\mathrm{ev}_i^+(\Sigma, u) = u(z_i^+)$$

where z_i^+ is the interior marked point as in 4).

Our moduli spaces $\mathcal{M}_{k+1;\ell}^{\mathrm{main}}(\beta)$ have Kuranishi structure in the sense of [**FO**] section 5 and [**FOOO1**] section A1.

Its boundary is described by using fiber product. For example, in case $\ell = 0$ we have the equality

$$\partial \mathcal{M}_{k+1}^{\mathrm{main}}(\beta) = \bigcup_{k_1+k_2=k+1} \bigcup_{\beta_1+\beta_2=\beta} \bigcup_{i=1}^{k_2} \mathcal{M}_{k_1+1}^{\mathrm{main}}(\beta_1) \ {}_{\mathrm{ev}_0}\times_{\mathrm{ev}_i} \ \mathcal{M}_{k_2+1}^{\mathrm{main}}(\beta_2). \tag{8}$$

as spaces with Kuranishi structures. ([**FOOO1**] subsection 7.1.1.)

3. A quick review of Lagrangian Floer theory

Let $X = (X, \omega)$ be a symplectic manifold and L its Lagrangian submanifold. We assume L is oriented and spin. (Actually relative spinness in the sense of [**FOOO1**] Definition 1.6 is enough.)

In [**FOOO1**] Theorem A, we defined a structure of gapped unital filtered A_∞ algebra $\{\mathfrak{m}_k \mid k = 0, 1, \dots\}$ on the cohomology group $H(L; \Lambda_0)$ of L with Λ_0 coefficient.

Namely there exists a sequence of operators

$$\mathfrak{m}_k : B_k H(L; \Lambda_0)[1] \to H(L; \Lambda_0)[1]$$

of odd degree[1] (for $k \geq 0$).

THEOREM 3.1. 1)

$$\sum_{k_1+k_2=k+1} \sum_{i=1}^{k_2} (-1)^* \mathfrak{m}_{k_2}(x_1, \dots, \mathfrak{m}_{k_1}(x_i, \dots, x_{i+k_1-1}), \dots, x_k) = 0, \tag{9}$$

where $* = \deg' x_1 + \dots + \deg' x_{i-1}$.

2) $\mathfrak{m}_0(1) \equiv 0 \mod \Lambda_+$.

[1] See Remark 2.1. Only the parity of the degree is well-defined in Floer cohomology over Λ_0.

3) (Unitality) $\mathbf{e} = \mathrm{PD}[L] \in H^0(L;\Lambda_0)$ *is the strict unit. (Here* PD : $H_k(L) \to H^{n-k}(L)$ *is the Poincaré duality.) Namely*

$$\mathfrak{m}_{k+1}(x_1,\dots,\mathbf{e},\dots,x_k) = 0 \quad \textit{for } k \geq 2 \textit{ or } k=0.$$

and

$$\mathfrak{m}_2(\mathbf{e},x) = (-1)^{\deg x}\mathfrak{m}_2(x,\mathbf{e}) = x.$$

4) (G-gappedness) *There exists an additive discrete submonoid* $G = \{\lambda_i \mid i = 0,1,2,\dots\}$ $(\lambda_0 = 0 < \lambda_1 < \lambda_2 < \cdots, \lim_{i\to\infty}\lambda_i = \infty)$ *of* $\mathbb{R}_{\geq 0}$ *such that our structure is* G*-gapped. Namely* $\mathfrak{m}_k$ *is written as*

$$\mathfrak{m}_k = \sum_{i=0}^{\infty} T^{\lambda_i}\mathfrak{m}_{k,i}$$

where $\mathfrak{m}_{k,i} : B_kH(L;\mathbb{C})[1] \to H(L;\mathbb{C})[1]$ *is* $\mathbb{C}$*-linear.*

5) $\mathfrak{m}_{2,0}$ *coincides with cup product up to sign.*

The triple $(C, \{\mathfrak{m}_k\}, \mathbf{e})$ that satisfies 1)–4) of Theorem 3.1 (with $H(L;\Lambda_0)$ being replaced by C) is called a *G-gapped unital filtered A_∞ algebra.*

The operator $\mathfrak{m}_k$ is constructed by using the moduli spaces $\mathcal{M}^{\mathrm{main}}_{k+1}(\beta)$ as follows. (Here we use de Rham cohomology, following [**FOOO3, FOOO4, FOOO5, Fu2, Fu3**]. In [**FOOO1**] singular homology is used. Morse homology version is in [**FOOO2**].)

Let $h_1,\dots,h_k$ be differential forms on L. We define a differential form $\mathfrak{m}_{k,\beta}(h_1,\dots,h_k)$ on L as follows:

$$\mathfrak{m}_{k,\beta}(h_1,\dots,h_k) = \mathrm{ev}_{0!}(\mathrm{ev}_1,\dots,\mathrm{ev}_k)^*(h_1\times\cdots\times h_k) \tag{10}$$

for $(k,\beta) \neq (1,0)$. We use evaluation maps (7) in (10). We put

$$\mathfrak{m}_{1,0}(h) = (-1)^{n+\deg h+1}dh,$$

where d is the de Rham differential. (See [**FOOO1**] Remark 3.5.8.)

Here we regard $h_1\times\cdots\times h_k$ as a differential form on L^k. Then the pull back $(\mathrm{ev}_1,\dots,\mathrm{ev}_k)^*$ defines a differential form on $\mathcal{M}^{\mathrm{main}}_{k+1}(\beta)$. The symbol $\mathrm{ev}_{0!}$ denotes the integration along the fiber associated to the map $\mathrm{ev}_0 : \mathcal{M}^{\mathrm{main}}_{k+1}(\beta) \to L$. We remark that $\mathcal{M}^{\mathrm{main}}_{k+1}(\beta)$ itself is not necessarily transversal. So it may have wrong dimension. However we can use general theory of Kuranishi structure to obtain a multisection $\mathfrak{s}$ ([**FO**] section 5, [**FOOO1**] section A1) so that the perturbed moduli space $\mathcal{M}^{\mathrm{main}}_{k+1}(\beta)^{\mathfrak{s}}$ (that is the zero point set of the multisection $\mathfrak{s}$) has a virtual fundamental chain (over $\mathbb{Q}$). However still after perturbation, the map $\mathrm{ev}_0 : \mathcal{M}^{\mathrm{main}}_{k+1}(\beta)^{\mathfrak{s}} \to L$ may not be a submersion on the perturbed moduli space $\mathcal{M}^{\mathrm{main}}_{k+1}(\beta)^{\mathfrak{s}}$. So we take a continuous family of perturbations written as $\{\mathfrak{s}_w\}_{w\in W}$ parametrized by a certain smooth manifold W so that

$$\mathrm{ev}_0^W : \bigcup_{w\in W}\left(\mathcal{M}^{\mathrm{main}}_{k+1}(\beta)^{\mathfrak{s}_w}\times\{w\}\right) \to L$$

is a submersion.[2] So we can justify (10) as

$$\mathfrak{m}_{k,\beta}(h_1,\dots,h_k) = \mathrm{ev}^W_{0!}\left((\mathrm{ev}_1,\dots,\mathrm{ev}_k)^*(h_1\times\cdots\times h_k)\wedge\omega_W\right).$$

Here ω_W is a smooth form of degree $\dim W$ on W that has compact support and satisfies $\int_W \omega_W = 1$. We pull it back to $\bigcup_{w\in W}\left(\mathcal{M}^{\mathrm{main}}_{k+1}(\beta)^{\mathfrak{s}_w}\times\{w\}\right)$ in an obvious way. The fiberwise evaluation map ev^W_0 is ev_0 on $\mathcal{M}^{\mathrm{main}}_{k+1}(\beta)^{\mathfrak{s}_w}\times\{w\}$.

We omit the detail of this construction and refer [**FOOO4**] section 12 or [**Fu2**] section 13. In the toric case, which is the case of our main interest in this article, this construction can be simplified in most of the cases. Namely $\mathrm{ev}_0 : \mathcal{M}^{\mathrm{main}}_{k+1}(\beta)^{\mathfrak{s}} \to L$ itself can be taken to be a submersion (without using continuous family). See Section 5.

We now put

$$\mathfrak{m}_k = \sum_{\beta\in H_2(X,L;\mathbb{Z})} T^{(\beta\cap[\omega])/2\pi}\mathfrak{m}_{k,\beta}.$$

We can use various properties of the moduli space to check Theorem 3.1. In fact, for example, Theorem 3.1 1) is a consequece of Formula (8) and Theorem 3.1 4) is a consequence of Gromov compactness.

Thus we obtain a structure of G-gapped unital filtered A_∞ algebra on de Rham complex of L. Then it induces one on *cohomology* $H(X,L;\Lambda_0)$, by a purely algebraic result. ([**FOOO1**] Theorem 5.4.2.)

The filtered A_∞ algebra $(H(X,L;\Lambda_0),\{\mathfrak{m}_k\mid k=0,1,\dots\})$ is independent of the choices (such as compatible almost complex structures and perturbations etc.) up to an isomorphism of a gapped unital filtered A_∞ algebra, (that is gapped unital filtered A_∞ homomorphism which has an inverse). We omit the precise definition of this notion and refer readers to [**FOOO1**] Definition 3.2.29 and Proposition 5.4.5.

Let $(C,\{\mathfrak{m}_k\mid k=0,1,\dots\},\mathbf{e})$ be a unital filtered A_∞ algebra. We define its *weak Maurer-Cartan scheme* $\widehat{\mathcal{M}}_{\mathrm{weak}}(C)$ as the set of solutions of the equation

$$\sum_{k=0}^{\infty}\mathfrak{m}_k(b,\dots,b)\equiv 0 \mod \Lambda_0\mathbf{e}, \tag{11}$$

for $b\in C^{odd}$, with $b\equiv 0 \mod \Lambda_+$. (Here and hereafter $\mathbf{e}$ denotes the unit.)

For $b\in C^{odd}$, with $b\equiv 0 \mod \Lambda_+$, we define $\mathfrak{m}^b_k$ by

$$\mathfrak{m}^b_k(x_1,\dots,x_k) = \sum_{m_0=0}^{\infty}\cdots\sum_{m_k=0}^{\infty}\mathfrak{m}_k(\underbrace{b,\dots,b}_{m_0},x_1,\underbrace{b,\dots,b}_{m_1},\dots,x_k,\underbrace{b,\dots,b}_{m_k}).$$

The right hand side converges in $\mathfrak{v}_T$ topology. We can show that $(C,\{\mathfrak{m}^b_k\mid k=0,1,\dots\},\mathbf{e})$ is a filtered A_∞ algebra.

In our geometric situation, where $C=H(L;\Lambda_0)$, we can remove the assumption $b\equiv 0 \mod \Lambda_+$ using a trick due to Cho [**Cho3**] and can define

[2]Actually the parameter space W is defined only locally. See [**FOOO4**] section 12.

$\mathfrak{m}_k^b$ for any $b \in H^{odd}(L;\Lambda_0)$. (See [**FOOO3**] section 12 for toric case and [**Fu3**] section 5 for the general case.) Moreover the left hand side of (11) makes sense for any $b \in H^{odd}(L;\Lambda_0)$. In case we need to distinguish it from the case $b \in H^{odd}(L;\Lambda_+)$, we denote the former by $\widehat{\mathcal{M}}_{\text{weak}}(H(L;\Lambda_0);\Lambda_0)$.

It is easy to see that $\mathfrak{m}_0^b(1)$ coincides with the left hand side of (11). Therefore if $b \in \widehat{\mathcal{M}}_{\text{weak}}(C)$ then $\mathfrak{m}_0^b(1) = c\mathbf{e}$ for some $c \in \Lambda_+$. It follows that

$$(\mathfrak{m}_1^b \circ \mathfrak{m}_1^b)(x) = -c\left(\mathfrak{m}_2^b(\mathbf{e},x) + (-1)^{\deg' x}\mathfrak{m}_2^b(x,\mathbf{e})\right) = 0.$$

Here we use Theorem 3.1 1) in the first equality and Theorem 3.1 3) in the second equality. Now we define

DEFINITION 3.2. Let $b \in H^{odd}(L;\Lambda_0)$. We define Floer cohomology by:

$$HF((L,b),(L,b);\Lambda_0) = \frac{\operatorname{Ker}(\mathfrak{m}_1^b)}{\operatorname{Im}(\mathfrak{m}_1^b)}.$$

$HF((L,b),(L,b);\Lambda)$ is defined by taking $\otimes_{\Lambda_0}\Lambda$.

It is proved in [**FOOO1**] Proposition 3.7.75 and the discussion right after that (general case, singular homology version) [**FOOO4**] section 8 (toric case, de Rham homology version) that $HF((L,b),(L,b);\Lambda) \neq 0$ implies that L is Hamiltonian non-displaceable.[3] Namely for any Hamiltonian diffeomorphism $F : X \to X$ we have $F(L) \cap L \neq \emptyset$.

Let $b \in \widehat{\mathcal{M}}_{\text{weak}}(C)$. Then there exists $\mathfrak{PO}(b) \in \Lambda_+$ such that

$$\sum_{k=0}^{\infty} \mathfrak{m}_k(b,\dots,b) = \mathfrak{PO}(b)\mathbf{e}.$$

DEFINITION 3.3. We call $\mathfrak{PO} : \widehat{\mathcal{M}}_{\text{weak}}(C) \to \Lambda_+$, the *potential function.*

In the geometric situation we have $\mathfrak{PO} : \widehat{\mathcal{M}}_{\text{weak}}(H(L;\Lambda_0);\Lambda_0) \to \Lambda_+$.

4. A quick review of toric manifold

In this section we review a very small portion of the theory of toric variety. We explain only the points we use in this article. See for example [**Ful**] for an account of toric variety.

Let (X,ω,J) be a Kähler manifold, where J is its complex structure and ω is its Kähler form. Let n be the complex dimension of X. We assume n dimensional real torus $T^n = (S^1)^n$ acts effectively on X such that J and ω are preserved by the action. We call such (X,ω,J) a *Kähler toric manifold* if the T^n action has a moment map in the sense we describe below. Hereafter we simply say (X,ω,J) (or X) is a toric manifold.

[3] We need to take Λ (not Λ_0) for the coefficient ring for this statement. Actually $HF((L,b),(L,b);\Lambda_0) = 0$ never occurs when Floer cohomology is defined.

Let (X, ω, J) be as above. We say a map $\pi = (\pi_1, \dots, \pi_n) : X \to \mathbb{R}^n$ is a *moment map* if the following holds. We consider the i-th factor S^1_i of T^n. (Here $i = 1, \dots, n$.) Then $\pi_i : X \to \mathbb{R}$ is the moment map of the action of S^1_i. In other words, we have the following identity of π_i

$$d\pi_i(X) = \omega(X, \tilde{\mathfrak{t}}),$$

where $\tilde{\mathfrak{t}}$ is the Killing vector field associated to the action of the circle S^1_i on X.

Let $\mathbf{u} \in \text{Int}P$. Then the inverse image $\pi^{-1}(\mathbf{u})$ is a Lagrangian submanifold which is an orbit of the T^n action. We put

$$L(\mathbf{u}) = \pi^{-1}(\mathbf{u}). \tag{12}$$

This is a Lagrangian torus. The main purpose of this article is to study Lagrangian Floer cohomology for such $L(\mathbf{u})$.

It is well-known that $P = \pi(X)$ is a convex polytope. We can find a finitely many affine functions $\ell_j : \mathbb{R}^n \to \mathbb{R}$ $(j = 1, \dots, m)$ such that

$$P = \{\mathbf{u} \in \mathbb{R}^n \mid \ell_j(\mathbf{u}) \geq 0, \quad \forall j = 1, \dots, m\}. \tag{13}$$

We put $\partial_j P = \{\mathbf{u} \in P \mid \ell_j(\mathbf{u}) = 0\}$ and $D_j = \pi^{-1}(\partial_j P)$. ($\dim_{\mathbb{R}} \partial_j P = n - 1$.) $D_1 \cup \cdots \cup D_m$ is called the *toric divisor*.

Moreover we may choose ℓ_j so that the following holds.

CONDITIONS 4.1. 1) We put

$$d\ell_j = \vec{v}_j = (v_{j,1}, \dots, v_{j,n}) \in \mathbb{R}^n.$$

Then $v_{j,i} \in \mathbb{Z}$.

2) Let p be a vertex of P. Then the number of faces $\partial_j P$ which contain p is n. Let $\partial_{j_1} P, \dots, \partial_{j_n} P$ be those faces. Then $\vec{v}_{j_1}, \dots, \vec{v}_{j_n}$ (which is contained in $\mathbb{Z}^n$ by item 1)) is a basis of $\mathbb{Z}^n$.

The affine function ℓ_j has the following geometric interpretation. Let $\mathbf{u} \in \text{Int}P$. There exists m elements $\beta_j \in H_2(X, L(\mathbf{u}); \mathbb{Z})$ such that

$$\beta_j \cap D_{j'} = \begin{cases} 1 & j = j' \\ 0 & j \neq j'. \end{cases} \tag{14}$$

Then we have

$$2\pi \ell_j(\mathbf{u}) = \int_{\beta_j} \omega. \tag{15}$$

The existence of such ℓ_j and the property above is proved in [**Gu**] Theorem 4.5. (See [**FOOO3**] section 2 also.)

EXAMPLE 4.2. We consider the complex projective space $\mathbb{C}P^n$. Using homogeneous coordinate $[x_0 : x_1 : \cdots : x_n]$ we define T^n action by

$$(t_1, \ldots, t_n) \cdot [x_0 : \cdots : x_n] = [x_0 : e^{2\pi\sqrt{-1}t_1} x_1 : \cdots : e^{2\pi\sqrt{-1}t_n} x_n].$$

(Here we identify $\mathbb{R}/\mathbb{Z} \cong S^1$.) The moment map $\pi = (\pi_1, \ldots, \pi_n)$ is given by

$$\pi_i([x_0 : \cdots : x_n]) = \frac{|x_i|^2}{|x_0|^2 + \cdots + |x_n|^2}.$$

Its moment polytope P_0 is a simplex that is:

$$P_0 = \left\{ (u_1, \ldots, u_n) \mid 0 \le u_i, i = 1, \ldots, n, \ \sum_{i=0}^{n} u_i \le 1 \right\}.$$

We have

$$\ell_i(u_1, \ldots, u_n) = \begin{cases} u_i & i \ne 0 \\ 1 - \sum_{j=0}^{n} u_j & i = 0. \end{cases} \tag{16}$$

EXAMPLE 4.3. We consider $\mathbb{C}P^2$ as above. For $1 > \alpha > 0$, let us consider

$$P(\alpha) = P_0 \setminus \{(u_1, u_2) \in P_0 \mid u_2 > 1 - \alpha\} = \{(u_1, u_2) \in P_0 \mid u_2 \le 1 - \alpha\}.$$

The inverse image $\pi^{-1}(\{(u_1, u_2) \in P_0 \mid u_2 > 1 - \alpha\})$ is a ball of radius $\sqrt{\alpha/2}$ centered at $[0 : 1 : 0]$. The boundary of $\pi^{-1}(P(\alpha))$ has an induced contact form which is identified with the standard contact form of S^3. We identify two points on $\partial\pi^{-1}(P(\alpha))$ if they lie on the same orbit of Reeb flow. After this identification we obtain from $\pi^{-1}(P(\alpha))$ a symplectic manifold which we write $X(\alpha) = \mathbb{C}P^2 \# \overline{\mathbb{C}P}^2(\alpha)$.

It is well-known (see for example [**MS**] section 6.2) and can be proved from the above description that $X(\alpha)$ is a blow up of $\mathbb{C}P^2$ with Kähler form ω such that the symplectic area of the exceptional divisor is α.

The T^2 action on $\mathbb{C}P^2$ induces a T^2 action on $X(\alpha)$ so that it becomes a toric manifold. The moment polytope is $P(\alpha)$.

There are 4 faces of $P(\alpha)$ and 4 affine functions ℓ_i ($i = 0, 1, 2, 3$). Three of them are ℓ_0, ℓ_1, ℓ_2 as in (16). The fourth one is given by

$$\ell_3(u_1, u_2) = 1 - \alpha - u_2. \tag{17}$$

EXAMPLE 4.4. We can blow up again and may regard a two points blow up of $\mathbb{C}P^2$ as a toric manifold. For $\alpha, \alpha' > 0$, with $\alpha + \alpha' < 1$ we consider the polytope

$$P(\alpha, \alpha') = \{(u_1, u_2) \in P_0 \mid u_2 \le 1 - \alpha, \ u_1 + u_2 \ge \alpha'\}.$$

There exists a toric manifold $X(\alpha, \alpha')$ that is a two points blow up of $\mathbb{C}P^2$ and whose moment polytope is $P(\alpha, \alpha')$.

$P(\alpha, \alpha')$ has 5 faces. There are 5 affine functions $\ell_0, \ldots, \ell_4$ associated to each of the faces. ℓ_0, ℓ_1, ℓ_2 are as in (16) and ℓ_3 is as in (17). ℓ_4 is given by

$$\ell_4(u_1, u_2) = u_1 + u_2 - \alpha'. \tag{18}$$

5. Floer cohomology and potential function of the T^n orbits

In this section we give a description of Floer cohomology of the T^n orbit $L(\mathbf{u})$ of the toric manifold X. Here $\mathbf{u} \in \mathrm{Int}P$ and P is the moment polytope of X.

In this toric case the calculation of the Floer cohomology becomes significantly simpler. This is because in this case the calculation of Floer cohomology is reduced to the calculation of the potential function. Moreover the leading order term of the potential function is calculated by the work of Cho-Oh [**CO**]. We will explain those points in this section.

We first fix a basis of $H^1(L(\mathbf{u}); \mathbb{Z})$ as follows. In Section 4 we fix a splitting $T^n = (S^1)^n$ and the associated coordinate $(t_1, \ldots, t_n) \in (\mathbb{R}/\mathbb{Z})^n$. Let $\mathbf{e}_i \in H^1(T^n; \mathbb{Z})$ be the element represented by dt_i in de Rham cohomology, where t_i is the coordinate of the i-th factor of $(S^1)^n$. (Here we identify S^1 with $\mathbb{R}/\mathbb{Z}$.) The elements $\mathbf{e}_i$, $i = 1, \ldots, n$ form a basis of $H^1(T^n; \mathbb{Z}) \cong \mathbb{Z}^n$. Since the T^n action on $L(\mathbf{u})$ is free and transitive, we may identify $H^1(T^n; \mathbb{Z}) = H^1(L(\mathbf{u}); \mathbb{Z})$. Hence we have a basis $\mathbf{e}_i$, $i = 1, \ldots, n$ of $H^1(L(\mathbf{u}); \mathbb{Z})$.

Let $b \in H^1(L(\mathbf{u}); \Lambda_0)$. We can write $b = \sum_{i=1}^n x_i \mathbf{e}_i$. Hence we take $(x_1, \ldots, x_n)$ as a coordinate of $H^1(L(\mathbf{u}); \Lambda_0)$. We also put $y_i^{\mathbf{u}} = e^{x_i}$.

REMARK 5.1. The expression e^{x_i} determines an element of Λ_0 in case $x_i \in \Lambda_0$ as follows. We write $x_i = x_{i,0} + x_{i,+}$ where $x_{i,0} \in \mathbb{C}$ and $x_{i,+} \in \Lambda_+$. Then we put

$$y_i^{\mathbf{u}} = e^{x_i} = e^{x_{i,0}} \sum_{k=0}^{\infty} x_{i,+}^k / k!.$$

Note $e^{x_{i,0}} \in \mathbb{C}$ is defined as usual. The sum $\sum_{k=0}^{\infty} x_{i,+}^k / k!$ converges in $\mathfrak{v}_T$-topology.

Now we consider a toric manifold X with its moment polytope P. We consider affine functions ℓ_j $(j = 1, \ldots, m)$. We define $v_{j,i} \in \mathbb{Z}$ as in Properties 4.1 1). We define

$$z_j = T^{\ell_j(\mathbf{u})} (y_1^{\mathbf{u}})^{v_{j,1}} \ldots (y_n^{\mathbf{u}})^{v_{j,n}}. \tag{19}$$

THEOREM 5.2. 1) *$H^1(L(\mathbf{u}); \Lambda_0)$ is contained in $\widehat{\mathcal{M}}_{\mathrm{weak}}(H(L; \Lambda_0); \Lambda_0)$.*

2) *Let $b = \sum x_i \mathbf{x}_i \in H^1(L(\mathbf{u}); \Lambda_0)$. Then we have*

$$\mathfrak{PO}(b) = z_1 + \cdots + z_m + \sum_{k=1}^{N} T^{\rho_k} P_k(z_1, \ldots, z_m). \tag{20}$$

Here $N \in \mathbb{Z}_{\geq 0}$ or $N = \infty$. The numbers $\rho_k > 0$ are positive and real. In case $N = \infty$, the sequence of numbers ρ_k goes to ∞ as k goes to ∞. $P_k(z_1, \dots, z_m)$ are monomials of $z_1, \dots, z_m$ of degree ≥ 1 with Λ_0 coefficient. We remark that z_j is defined from $y_i^{\mathbf{u}} = e^{x_j}$ by (19).

3) *If X is Fano then P_k are all zero.*
4) *The monomials P_k and the numbers ρ_k are independent of $\mathbf{u}$ and depends only on X.*

Item 1) is [**FOOO3**] Proposition 4.3 plus the last line of [**FOOO3**] section 4.

Item 2) is [**FOOO3**] Theorem 4.6 in the form (slightly) improved in [**FOOO4**] Theorem 3.4. In [**FOOO3, FOOO4**] this formula is written using $y_i^{\mathbf{u}}$ in place of z_j. But it is easy to see that they are the same by the identification (19). We use the result of Cho-Oh [**CO**] to calculate the term $z_1 + \cdots + z_m$ in the right hand side of (20).

Item 3) is [**FOOO3**] Theorem 4.5.

Item 4) follows from [**FOOO3**] Lemma 11.7.

Sketch of the proof. The linear terms z_j in (20) come from the contribution (that is $\mathfrak{m}_{k,\beta_j}(b, \dots, b)$) of $\mathcal{M}_1^{\text{main}}(\beta_j)$ to $\mathfrak{m}_k(b, \dots, b)$, where $\beta_j \in H_2(X, L(\mathbf{u}); \mathbb{Z})$ is as in (14). Its coefficient 1 is the degree of the map

$$\mathrm{ev}_0 : \mathcal{M}_1^{\text{main}}(\beta_j) \to L(\mathbf{u}), \tag{21}$$

which is calculated by [**CO**].

The term $T^{\rho_k} P_k$ is a contribution of $\mathcal{M}_1^{\text{main}}(\beta)$ for some β. We will assume $\beta \neq 0$ in the rest of the argument.

We can use a T^n equivariant multisection to define virtual fundamental chain. To see this we first observe that the T^n action on $\mathcal{M}_1^{\text{main}}(\beta)$ is free. This is because T^n action on $L(\mathbf{u})$ is free and (21) is T^n equivariant. Therefore to find a transversal multisection we can proceed as follows. We first take the quotient with respect to T^n action, next find transversal multisection on the quotient space and then lift it.

Let $\mathfrak{s}$ be a T^n equivariant multisection which is transversal to 0. Then T^n acts freely on its zero set $\mathcal{M}_1^{\text{main}}(\beta)^{\mathfrak{s}}$. Therefore the dimension of $\mathcal{M}_1^{\text{main}}(\beta)^{\mathfrak{s}}$ is not smaller than n if it is nonempty. We can show

$$\dim \mathcal{M}_1^{\text{main}}(\beta)^{\mathfrak{s}} = n + \mu(\beta) - 2$$

where $\mu : H_2(X, L(\mathbf{u}); \mathbb{Z}) \to \mathbb{Z}$ is the Maslov index. It implies that $\mu(\beta) \geq 2$ if $\mathcal{M}_1^{\text{main}}(\beta)^{\mathfrak{s}} \neq \emptyset$.

This is the key point of the proof.

Remark 5.3. In case X is Fano, $\mu(\beta) \geq 2$ automatically holds if $\mathcal{M}_1^{\text{main}}(\beta) \neq \emptyset$. But in non-Fano case this holds only after taking T^n equivariant perturbation.

Moreover T^n equivariance implies that $\mathrm{ev}_0 : \mathcal{M}_1^{\mathrm{main}}(\beta) \to L(\mathbf{u})$ is a submersion if $\mathcal{M}_1^{\mathrm{main}}(\beta) \neq \emptyset$. Therefore we may use this T^n equivariant $\mathfrak{s}$ to define $\mathfrak{m}_{k,\beta}$. Namely we do not need to use a continuous family of multisections in this case.

Now if $\deg b = 1$ then

$$\deg \mathfrak{m}_{k,\beta}(b, \dots, b) = 2 - \mu(\beta) \leq 0.$$

Namely $\mathfrak{m}_{k,\beta}(b, \dots, b)$ is either 0 or is proportional to the unit. This proves item 1).

To study $\mathfrak{m}_{k,\beta}(b, \dots, b)$ for $\beta \neq \beta_j$, we again use the classification of J holomorphic disks in **[CO]** to find that if $\mathcal{M}_1^{\mathrm{main}}(\beta)$ is nonempty the homology class β is decomposed to a sum of β_j's ($j = 1, \dots, m$) and sphere bubbles. Therefore

$$\beta = \beta_{j_1} + \cdots + \beta_{j_e} + \alpha_1 + \cdots + \alpha_f$$

where b_{j_k} is one of b_j's and $\alpha_i \in H_2(X; \mathbb{Z})$ is represented by J-holomorphic sphere. We put

$$c_\beta = \deg[\mathrm{ev}_0 : \mathcal{M}_1^{\mathrm{main}}(\beta)^{\mathfrak{s}} \to L(\mathbf{u})].$$

Here the right hand side is the mapping degree of the map ev_0. It is well-defined since in case $\mu(\beta) = 2$ the boundary of $\mathcal{M}_1^{\mathrm{main}}(\beta)^{\mathfrak{s}}$ is empty. (This is because $\mathcal{M}_1^{\mathrm{main}}(\beta')^{\mathfrak{s}}$ is empty if $\mu(\beta') \leq 0$, $\beta' \neq 0$.)

Then we can show that

$$\sum_{k=0}^{\infty} \mathfrak{m}_{k,\beta}(b, \dots, b) = c_\beta T^{\sum_{i=1}^{f}(\alpha_i \cap \omega)/2\pi} z_{j_1} \cdots z_{j_e}.$$

Item 2) follows from this formula.

Item 3) follows from the fact that in the Fano case, $\mathcal{M}_1^{\mathrm{main}}(\beta) \neq \emptyset$ and $\mu(\beta) = 2$ imply $\beta = \beta_j$ for some j.

Item 4) follows from the fact that c_β is independent of $\mathbf{u}$. □

REMARK 5.4. In the general situation, the filtered A_∞ structure associated to a Lagrangian submanifold is well-defined only up to isomorphism. In particular potential function $\mathfrak{PO}$ is well-defined only up to a coordinate change. (Namely it may depend on the choice of perturbation etc.) However in our toric case we can use a T^n equivariant perturbation $\mathfrak{s}$ and then $\mathfrak{PO}$ is well-defined as a function on $H^1(L(\mathbf{u}); \Lambda_0)$ without ambiguity. This is a consequence of well-definedness of c_β and is **[FOOO3]** Lemma 11.7.

We have the following useful criterion which reduces computation of Floer cohomology to the critical point theory of potential function.

THEOREM 5.5. *Let $b = \sum x_i \mathbf{e}_i \in H^1(L(\mathbf{u}); \Lambda_0)$. Then the following three conditions are equivalent.*

1) *For each of* $i = 1, \ldots, n$ *we have:*

$$\left.\frac{\partial \mathfrak{PO}}{\partial x_i}\right|_b = 0.$$

2)

$$HF((L(\mathbf{u}), b), (L(\mathbf{u}), b); \Lambda_0) \cong H(T^n; \Lambda_0).$$

3)

$$HF((L(\mathbf{u}), b), (L(\mathbf{u}), b); \Lambda) \neq 0.$$

SKETCH OF THE PROOF. By definition

$$\mathfrak{PO}(b)\mathbf{e} = \sum_{k=0}^{\infty} \mathfrak{m}_k(b, \ldots, b). \tag{22}$$

We differentiate (22) by x_i. Then using $\partial b / \partial x_i = \mathbf{e}_i$ we obtain:

$$\left.\frac{\partial \mathfrak{PO}}{\partial x_i}\right|_b \mathbf{e} = \sum_{k_1=0}^{\infty}\sum_{k_2=0}^{\infty} \mathfrak{m}_{k_1+k_2+1}(\underbrace{b, \ldots, b}_{k_1}, \mathbf{e}_i, \underbrace{b, \ldots, b}_{k_2}) = \mathfrak{m}_1^b(\mathbf{e}_i). \tag{23}$$

Here the second equality is the definition of $\mathfrak{m}_1^b$.

Now we assume item 2). Then we have $\mathfrak{m}_1^b(\mathbf{e}_i) = 0$. Therefore (23) implies item 1).

We next assume item 1). Then (23) implies $\mathfrak{m}_1^b(\mathbf{e}_i) = 0$. We use it together with the fact that $\mathbf{e}_i$ generates $H(L(\mathbf{u}); \Lambda_0)$ by cup product, and A_∞ formula to prove that $\mathfrak{m}_1^b = 0$. (See [**FOOO3**] proof of Lemma 13.1.) Item 2) follows.

The equivalence between item 2) and item 3) is proved in [**FOOO3**] Remark 13.9. □

To apply Theorems 5.2 and 5.5 for the calculation of Floer cohomology of T^n, we need some algebraic discussion, which is in order.

Let $y_1, \ldots, y_n$ be n formal variables. We consider the ring $\Lambda[y_1, \ldots, y_n, y_1^{-1}, \ldots, y_n^{-1}]$ of Laurent polynomials of n variables with Λ coefficient. We write it as $\Lambda[y, y^{-1}]$ for simplicity.

Let $\mathbf{u} = (u_1, \ldots, u_n) \in P$. We put

$$y_i^{\mathbf{u}} = T^{-u_i} y_i \in \Lambda[y, y^{-1}]. \tag{24}$$

By an easy computation we have

$$T^{\ell_j(\mathbf{u})}(y_1^{\mathbf{u}})^{v_{j,1}} \cdots (y_n^{\mathbf{u}})^{v_{j,n}} = T^{\ell_j(\mathbf{u}')}(y_1^{\mathbf{u}'})^{v_{j,1}} \cdots (y_n^{\mathbf{u}'})^{v_{j,n}}. \tag{25}$$

for $\mathbf{u}, \mathbf{u}' \in P$. Therefore (19) defines an elements $z_j \in \Lambda[y, y^{-1}]$ in a way independent of $\mathbf{u} \in P$.

We next introduce a family of valuations $\mathfrak{v}_T^{\mathbf{u}}$ on $\Lambda[y, y^{-1}]$ parametrized by $\mathbf{u} \in P$.

Let $F \in \Lambda[y, y^{-1}]$. Then for each $\mathbf{u} \in \operatorname{Int} P$ there exists $F^{\mathbf{u}}_{i_1\dots i_n} \in \Lambda$ for $i_1, \dots, i_n \in \mathbb{Z}^n$ such that

$$F = \sum_{i_1,\dots,i_n \in \mathbb{Z}^n} F^{\mathbf{u}}_{i_1\dots i_n} (y^{\mathbf{u}}_1)^{i_1} \cdots (y^{\mathbf{u}}_n)^{i_n}.$$

Here only finitely many of $F^{\mathbf{u}}_{i_1\dots i_n}$ are nonzero. So the right hand side is actually a finite sum.

DEFINITION 5.6.

$$\mathfrak{v}^{\mathbf{u}}_T(F) = \inf\{\mathfrak{v}_T(F^{\mathbf{u}}_{i_1\dots i_n}) \mid F^{\mathbf{u}}_{i_1\dots i_n} \neq 0\},$$

if $F \neq 0$ and $\mathfrak{v}^{\mathbf{u}}_T(0) = +\infty$.

$\mathfrak{v}^{\mathbf{u}}_T$ defines a valuation on $\Lambda[y, y^{-1}]$.

We denote the completion of $\Lambda[y, y^{-1}]$ with respect to $\mathfrak{v}^{\mathbf{u}}_T$ by $\Lambda\langle\langle y, y^{-1}\rangle\rangle^{\mathbf{u}}$.

By definition we have

$$\mathfrak{v}^{\mathbf{u}}_T(z_j) = \ell_j(\mathbf{u}) \geq 0$$

for $\mathbf{u} \in P$. The following lemma is its immediate consequence.

LEMMA 5.7. *The right hand side of* (20) *converges with respect to* $\mathfrak{v}^{\mathbf{u}}_T$ *for any* $\mathbf{u} \in P$.

We remark that according to the general theory described in section 3, the potential function $\mathfrak{PO}$ associated to a Lagrangian submanifold $L(\mathbf{u})$ is a Λ_+ valued function on $\widehat{\mathcal{M}}_{\text{weak}}(L(\mathbf{u}); \Lambda_0)$. By Theorem 5.2 (1), we have the inclusion $H^1(L(\mathbf{u}); \Lambda_0) \subset \widehat{\mathcal{M}}_{\text{weak}}(L(\mathbf{u}); \Lambda_0)$. Since $x_1, \dots, x_n \in \Lambda_0$ forms a coordinate of $H^1(L(\mathbf{u}); \Lambda_0)$ with respect to the basis $\mathbf{e}_i$, we may regard $\mathfrak{PO}$ restricted to $H^1(L(\mathbf{u}); \Lambda_0)$ as a function on $(x_1, \dots, x_n) \in \Lambda_0^n \cong H^1(L(\mathbf{u}); \Lambda_0)$.

Then by Theorem 5.2 2) we have

$$\mathfrak{PO}(x_1, \dots, x_n) = \mathfrak{PO}(x'_1, \dots, x'_n)$$

if $x_i - x'_i \in 2\pi\sqrt{-1}\mathbb{Z}$ for each i. In other words, we may regard $\mathfrak{PO}$ as a function of $y^{\mathbf{u}}_i = e^{x_i}$. Note $x_i \in \Lambda_0$ implies that $y^{\mathbf{u}}_i - 1 \in \Lambda_+$. We next extend the domain of $\mathfrak{PO}$ by using Theorem 5.2 2).

We put $\lambda_j = \ell_j(\mathbf{0})$. Then it is easy to see from definition that

$$z_j = T^{\lambda_j} y_1^{v_{j,1}} \dots y_n^{v_{j,n}}. \tag{26}$$

LEMMA 5.8. *Let* $(\mathfrak{y}_1, \dots, \mathfrak{y}_n) \in (\Lambda \setminus \{0\})^n$. *We assume*

$$(\mathfrak{v}_T(\mathfrak{y}_1), \dots, \mathfrak{v}_T(\mathfrak{y}_n)) \in P. \tag{27}$$

We put $\mathfrak{z}_j = T^{\lambda_j} \mathfrak{y}_1^{v_{j,1}} \dots \mathfrak{y}_n^{v_{j,n}}$. *Then*

$$\mathfrak{z}_1 + \cdots + \mathfrak{z}_m + \sum_{k=1}^{N} T^{\rho_k} P_k(\mathfrak{z}_1, \dots, \mathfrak{z}_m) \in \Lambda_+$$

converges as $N \to \infty$ *with respect to the valuation* $\mathfrak{v}_T$.

PROOF. (27) implies $\mathfrak{v}_T(\mathfrak{z}_j) = \ell_j(\mathbf{u}) \geq 0$. The lemma then follows easily from $\lim_{k\to\infty} \rho_k = \infty$ in the statement of Theorem 5.2 (20). □

We define

$$\mathfrak{A}(P) = \{(\mathfrak{y}_1, \dots, \mathfrak{y}_n) \in (\Lambda \setminus \{0\})^n \mid (\mathfrak{v}_T(\mathfrak{y}_1), \dots, \mathfrak{v}_T(\mathfrak{y}_n)) \in P\}. \tag{28}$$

By Lemma 5.8 we may regard $\mathfrak{PO}$ as a function

$$\mathfrak{PO} : \mathfrak{A}(P) \to \Lambda_+.$$

We remark that $\mathfrak{A}(P)$ is not a manifold. So we can not define differentiation of $\mathfrak{PO}$ in the sense of usual calculus. Instead we will define it as follows. We remark that z_j and $P_k(z_1, \dots, z_m)$ are Laurent monomials of $y_1, \dots, y_n$ with Λ_0 coefficient. So we can differentiate it by y_i in an obvious way. Moreover

$$y_i \frac{\partial}{\partial y_i} P_k(z_1, \dots, z_m)$$

is again a monomial of $z_1, \dots, z_m$ with Λ_0 coefficient. Therefore for $\mathfrak{y} = (\mathfrak{y}_1, \dots, \mathfrak{y}_n) \in \mathfrak{A}(P)$ the limit

$$\lim_{N\to\infty} \left(\mathfrak{y}_i \frac{\partial z_1}{\partial y_i}(\mathfrak{y}) + \cdots + \mathfrak{y}_i \frac{\partial z_m}{\partial y_i}(\mathfrak{y}) + \sum_{k=1}^{N} T^{\rho_k} \mathfrak{y}_i \frac{\partial P_k}{\partial y_i}(\mathfrak{z}_1, \dots, \mathfrak{z}_m) \right)$$

converges. (Here we put $\mathfrak{z}_j = T^{\lambda_j} \mathfrak{y}_1^{v_{j,1}} \dots \mathfrak{y}_n^{v_{j,n}}$.) We write its limit as

$$\mathfrak{y}_i \frac{\partial \mathfrak{PO}}{\partial y_i}(\mathfrak{y}).$$

Thus we have defined

$$y_i \frac{\partial \mathfrak{PO}}{\partial y_i} : \mathfrak{A}(P) \to \Lambda_+.$$

We now have the following:

THEOREM 5.9. *For $\mathbf{u} \in \operatorname{Int} P$ the following two conditions are equivalent.*

1) *There exists $b \in H^1(L(\mathbf{u}); \Lambda_0)$ such that*

$$HF((L(\mathbf{u}), b), (L(\mathbf{u}), b); \Lambda_0) \cong H(T^n; \Lambda_0).$$

2) *There exists $\mathfrak{y} = (\mathfrak{y}_1, \dots, \mathfrak{y}_n) \in \mathfrak{A}(P)$ such that*

$$\mathfrak{y}_i \frac{\partial \mathfrak{PO}}{\partial y_i}(\mathfrak{y}) = 0 \tag{29}$$

for $i = 1, \dots, n$ and that

$$(\mathfrak{v}_T(\mathfrak{y}_1), \dots, \mathfrak{v}_T(\mathfrak{y}_n)) = \mathbf{u}.$$

DEFINITION 5.10. We say that $L(\mathbf{u})$ is a *strongly balanced* if the Condition 1) (= Condition 2)) in Theorem 5.9 is satisfied.

PROOF. 2) $\Longrightarrow$ 1): Let $\mathfrak{y}$ be as in 2). We put $y_i^{\mathbf{u}} = T^{-u_i}\mathfrak{y}_i$. Then $\mathfrak{v}_T(y_i^{\mathbf{u}}) = 0$. Therefore there exist $y_{i,0}^{\mathbf{u}} \in \mathbb{C}$ and $y_{i,+}^{\mathbf{u}} \in \Lambda_+$ such that $y_i^{\mathbf{u}} = y_{i,0}^{\mathbf{u}} + y_{i,+}^{\mathbf{u}}$. We put $x_{i,0} = \log(y_{i,0}^{\mathbf{u}})$ and

$$x_{i,+} = \log(1 + (y_{i,0}^{\mathbf{u}})^{-1} y_{i,+}^{\mathbf{u}})).$$

Note $(y_{i,0}^{\mathbf{u}})^{-1} y_{i,+}^{\mathbf{u}} \in \Lambda_+$. Therefore we can define the right hand side by the Taylor expansion of $\log(1 + z)$.

We put $x_i = x_{i,0} + x_{i,+}$ and $b = \sum_{i=1}^m x_i \mathbf{e}_1$. Then using Theorem 5.5 it is easy to see that 1) is satisfied.

1) $\Longrightarrow$ 2): Let $b = \sum x_i \mathbf{e}_i$ be as in 1). We put $\mathfrak{y}_i = T^{u_i} e^{x_i}$. It is easy to see that $\mathfrak{y} = (\mathfrak{y}_1, \ldots, \mathfrak{y}_n)$ satisfies $\mathfrak{y}_i \frac{\partial \mathfrak{PO}}{\partial y_i}(\mathfrak{y}) = 0$. □

REMARK 5.11. It is easy to see that $y_i^{\mathbf{0}} = y_i$, where $\mathbf{0} \in \mathbb{R}^n$ is the origin. Note that the moment polytope P is well-defined only up to parallel translation. Namely we can replace it by $P + \mathbf{u}$ for any $\mathbf{u} \in \mathbb{R}^n$, then $P + \mathbf{u}$ corresponds to the same toric manifold as P.

Thus the choice $y_i^{\mathbf{0}} = y_i$ is quite ad-hoc, and we may take any $y_i^{\mathbf{u}}$ in place of y_i in our story. In fact the ring $\Lambda[y, y^{-1}]$ can be canonically identified with the Laurent polynomial rings over $y_i^{\mathbf{u}}$ $(i = 1, \ldots, n)$ using $y_i^{\mathbf{u}} \in \Lambda[y, y^{-1}]$.

On the other hand, the valuation $\mathfrak{v}_T^{\mathbf{u}}$ and the completion $\Lambda\langle\langle y, y^{-1}\rangle\rangle^{\mathbf{u}}$ is canonically associated to the Lagrangian submanifold $L(\mathbf{u})$.

The variables $y_i^{\mathbf{u}}$ also is defined in a way independent of the choice of the origin of the affine space in which P is embedded.

In some reference such as [**Aur1, HV**] 'renormalization' is discussed. It seems that this process depends on the choice of the origin in the affine space $\mathbb{R}^n$. Namely it is related to the homothetic transformation $y_i \mapsto Cy_i$ where $C \to \infty$.

As we mentioned above the choice of $\mathbf{0}$ is not intrinsic. More canonical way seems to be as follows. We consider each of $\mathbf{u}_0$ such that $HF((L(\mathbf{u}_0), b), (L(\mathbf{u}_0), b); \Lambda) \neq 0$ for some b. We then replace P by $P - \mathbf{u}_0$, so this orbit $L(\mathbf{u}_0)$ becomes $L(\mathbf{0})$. We now use $y_i \mapsto Cy_i$ to 'renormalize'.

Thus there exists a 'renormalization' for each such $\mathbf{u}_0$. This process of 'renormalization' seems to be related to the study of leading term equation, which we discuss in section 9.

6. Examples 1

EXAMPLE 6.1. We first consider the case of $\mathbb{C}P^n$. We use (16) and Theorem 5.2 2), 3) to obtain

$$\mathfrak{PO} = z_1 + \cdots + z_n + z_0 = y_1 + \cdots + y_n + T(y_1 \cdots y_n)^{-1}.$$

Therefore the equation (29) becomes

$$0 = y_i \frac{\partial \mathfrak{PO}}{\partial y_i} = y_i - T(y_1 \cdots y_n)^{-1}.$$

The solutions are

$$y_1 = \cdots = y_n = T^{1/(n+1)} \exp(2\pi\sqrt{-1}k/(n+1))$$

where $k = 0, 1, \ldots, n$. The valuation of y_i are $1/(n+1)$. Thus $\mathbf{u}_0 = (1/(n+1), \ldots, 1/(n+1))$ is the unique strongly balanced fiber.

EXAMPLE 6.2. We next consider $X(\alpha)$, one point blow up of $\mathbb{C}P^2$ as in Example 4.3. Using the discussion in Example 4.3 and Theorem 5.2 2), 3) we obtain

$$\mathfrak{PO} = y_1 + y_2 + T(y_1 y_2)^{-1} + T^{1-\alpha} y_2^{-1}.$$

The equation (29) becomes

$$1 - T y_1^{-2} y_2^{-1} = 0, \quad 1 - T y_1^{-1} y_2^{-2} - T^{1-\alpha} y_2^{-2} = 0.$$

By eliminating $y_2 = T y_1^{-2}$ we obtain

$$y_1^4 + T^{\alpha} y_1^3 - T^{\alpha+1} = 0. \tag{30}$$

We put $u_1 = \mathfrak{v}_T(y_1)$.

(Case 1) $u_1 < \alpha$.

We take $\mathfrak{v}_T$ of (30) and obtain $4u_1 = \alpha + 1$. Namely $u_1 = (\alpha + 1)/4$. $u_1 > \alpha$ then implies $\alpha > 1/3$.

Conversely if $\alpha > 1/3$ and $u_1 = (\alpha + 1)/4$ we put $y_1 = T^{u_1} y$ then (30) becomes

$$y^4 + T^{(3\alpha-1)/4} y^3 - 1 = 0.$$

Since $(3\alpha - 1)/4 > 0$, this equation has 4 simple roots y which are congruent to $\pm 1, \pm\sqrt{-1}$ modulo Λ_+, respectively.

(Case 2) $u_1 > \alpha$.

We take $\mathfrak{v}_T$ of (30) and have $3u_1 + \alpha = \alpha + 1$. Namely $u_1 = 1/3$. $u_1 > \alpha$ then implies $\alpha < 1/3$.

Conversely if $\alpha < 1/3$ and $u_1 = 1/3$ we put $y_1 = T^{1/3} y$ then (30) becomes

$$T^{1/3-\alpha} y^4 + y^3 - 1 = 0.$$

This equation has 3 simple roots y which are congruent to $1, e^{2\pi\sqrt{-1}/3}$, $e^{4\pi\sqrt{-1}/3}$ modulo Λ_+, respectively.

(Case 3) $u_1 = \alpha$.

We put $y_1 = T^{u_1} y$. Then $\mathfrak{v}_T(y) = 0$ and we have

$$y^3(1 + y) - T^{1-3\alpha} = 0. \tag{31}$$

(Case 3-1) $\alpha = 1/3$.

In this case there exist exactly 4 roots $y \in \mathbb{C}$ of (31).

(Case 3-2) $\alpha \neq 1/3$.

By (31) $\alpha < 1/3$. Then $\mathfrak{v}_T(1+y) = 1 - 3\alpha$. We put $y = T^{1-3\alpha} w - 1$. Then $\mathfrak{v}_T(w) = 0$. Then (31) becomes

$$(1 - T^{1-3\alpha} w)^3 w + 1 = 0.$$

There is one root of this equation with $w \equiv -1$ modulo Λ_+. Three other roots do not satisfy $\mathfrak{v}_T(w) = 0$. Thus there exists one solution in this case such that $u_1 = \mathfrak{v}_T(y_1) = \alpha$.

In sum we have the following.

If $\alpha < 1/3$ there exists one solution with $u_1 = \mathfrak{v}_T(y_1) = \alpha$ and three solutions with $u_1 = 1/3$. Note $u_2 = \mathfrak{v}_T(y_2) = 1 - 2u_1$. Therefore $L(\alpha, 1-2\alpha)$ and $L(1/3, 1/3)$ are the strongly balanced fibers.

If $\alpha \geq 1/3$ we have 4 solutions with $u_1 = (\alpha+1)/4$, $u_2 = (1-\alpha)/2$. Namely there is exactly one strong balanced fiber $L((\alpha+1)/4, (1-\alpha)/2)$.

In this section we discuss the Fano case only, where we can explicitly calculate $\mathfrak{PO}$. The non-Fano case will be discussed in section 10.

In the case of Example 6.1 and 6.2, McDuff [**Mc**] proved that all the T^n orbits where Floer cohomology vanish for all choices of b, are displaceable by Hamiltonian diffeomorphism.

However there is an example of toric surface and its T^2 orbit, such that one can not displace it from itself by the method of [**Mc**] but all the known versions of Floer cohomology over Λ vanish for this T^2 orbit. (See [**Mc**] Lemma 4.4.) We do not know whether they are displaceable or not.

7. Open-closed Gromov-Witten theory and operators $\mathfrak{q}$

In this section, we discuss the operator $\mathfrak{q}$ introduced in [**FOOO1**] section 3.8. Let (X, ω) be a symplectic manifold and L its Lagrangian submanifold as in section 3. Let $h_1, \dots, h_k$ be differential forms on L and $g_1, \dots, g_\ell$ differential forms on X. Let $\beta \in H_2(X, L; \mathbb{Z})$. We define

(32)

$$\begin{aligned}&\mathfrak{q}_{\ell,k,\beta}(g_1, \dots, g_\ell; h_1, \dots, h_k)\\ &\quad = \frac{1}{\ell!}\mathrm{ev}_{0!}\left((\mathrm{ev}_1^+, \dots, \mathrm{ev}_\ell^+, \mathrm{ev}_1, \dots, \mathrm{ev}_k)^* (g_1 \times \cdots \times g_\ell \times h_1 \times \cdots \times h_k)\right).\end{aligned}$$

We also put

$$\mathfrak{q}_{0;1;0}(h) = (-1)^n dh.$$

We remark that $g_1 \times \cdots \times g_\ell \times h_1 \times \cdots \times h_k$ is a differential form on $X^\ell \times L^k$ and its pull back is a differential form on $\mathcal{M}^{\mathrm{main}}_{k+1;\ell}(\beta)$. The map $\mathrm{ev}_{0!}$ is integration along fiber by the map $\mathrm{ev}_0 : \mathcal{M}^{\mathrm{main}}_{k+1;\ell}(\beta) \to L$. More precisely we use a continuous family of perturbations in the same way as we defined $\mathfrak{m}_k$ in section 3.

We then put

$$\mathfrak{q}_{\ell,k} = \sum_{\beta \in H_2(X,L;\mathbb{Z})} T^{(\beta \cap \omega)/2\pi} \mathfrak{q}_{\ell,k,\beta}.$$

It defines a map

$$\mathfrak{q}_{\ell;k} : E_\ell(\Omega(X)[2] \otimes \Lambda_0) \otimes B_k(\Omega(L)[1] \otimes \Lambda_0) \to \Omega(L)[1] \otimes \Lambda_0.$$

This operator has the following properties. We omit the suffix ℓ, k in $\mathfrak{q}_{\ell;k}$ and write $\mathfrak{q}$ in the formula below. We use the convention (6) introduced at the end of subsection 2.1.

THEOREM 7.1. 1) *Let* $\mathbf{x} \in B_k(\Omega(L)[1] \otimes \Lambda_0)$, $\mathbf{y} \in E_\ell(\Omega(X)[2] \otimes \Lambda_0)$. *Suppose* $\mathbf{y}$ *is a linear combination of the elements of the form* $y_1 \otimes \cdots \otimes y_\ell$ *where each of* y_i *are closed forms. We then have the following:*

$$0 = \sum_{c_1, c_2} (-1)^* \mathfrak{q}(\mathbf{y}_{c_1}^{2;1}; \mathbf{x}_{c_2}^{3;1} \otimes \mathfrak{q}(\mathbf{y}_{c_1}^{2;2}; \mathbf{x}_{c_2}^{3;2}) \otimes \mathbf{x}_{c_2}^{3;3}) \tag{33}$$

where $* = \deg' \mathbf{x}_{c_2}^{3;1} + \deg' \mathbf{x}_{c_2}^{3;1} \deg \mathbf{y}_{c_1}^{2;2} + \deg \mathbf{y}_{c_1}^{2;1}$.

2) *If* $\mathbf{y} = 1 \in E_0(\Omega(X)[2] \otimes \Lambda_0) = \Lambda_0$ *then*

$$\mathfrak{q}_{0,k}(1, \mathbf{x}) = \mathfrak{m}_k(\mathbf{x}). \tag{34}$$

3) *Let* $\mathbf{e} = \mathrm{PD}([L])$ *be the Poincaré dual to the fundamental class of* L. *Let* $\mathbf{x}_i \in B(\Omega(L)[1] \otimes \Lambda_0)$ *and we put* $\mathbf{x} = \mathbf{x}_1 \otimes \mathbf{e} \otimes \mathbf{x}_2 \in B(\Omega(L)[1] \otimes \Lambda_0)$. *Then*

$$\mathfrak{q}(\mathbf{y}; \mathbf{x}) = 0 \tag{35a}$$

except the following case:

$$\mathfrak{q}(1; \mathbf{e} \otimes x) = (-1)^{\deg x} \mathfrak{q}(1; x \otimes \mathbf{e}) = x, \tag{35b}$$

where $x \in \Omega(L)[1] \otimes \Lambda_0 = B_1(\Omega(L)[1] \otimes \Lambda_0)$.

4) *There exists a discrete submonoid* $G = \{\lambda_i \mid i = 0, 1, 2, \dots\}$ *such that*

$$\mathfrak{q}_{\ell,k} = \sum_{i=1}^{\infty} T^{\lambda_i} \mathfrak{q}_{\ell,k,i}$$

where $\mathfrak{q}_{\ell,k,i} : E_\ell(\Omega(X)[2]) \otimes B_k(\Omega(L)[1]) \to \Omega(L)[1]$.

5) *Let* $i : L \to X$ *be the inclusion and* $y \in \Omega(X) \otimes \Lambda_0$. *Then*

$$\mathfrak{q}_{1,0}(y, 1) \equiv i^*(y) \mod \Omega(L) \otimes \Lambda_+.$$

REMARK 7.2. Formula (33) above implies that the operator $\mathfrak{q}$ (after modifying the sign appropriately) define a homomorphism $E\mathcal{A}[2] \to HH^*(L; \Lambda)$ to the Hochschild cohomology of de Rham cohomology ring of L. See [**FOOO1**] Section 7.4.

This is de Rham version of [**FOOO1**] Theorem 3.8.32. Namely item 1) is [**FOOO1**] (3.8.33), Item 2) is [**FOOO1**] Theorem 3.8.32 (3). Item 3) is [**FOOO1**] (3.8.34.2). Item 4) follows immediately from definition. Item 5) follows from [**FOOO1**] (3.8.34).

Let $\mathfrak{b} \in \Omega^{even}(X) \otimes \Lambda_+$ and $b \in \Omega^{odd}(L) \otimes \Lambda_+$. Suppose $d\mathfrak{b} = 0$. We put $\mathbf{b} = (\mathfrak{b}, b)$ and define

$$\mathfrak{m}_k^{\mathbf{b}} : B_k(\Omega(L)[1] \otimes \Lambda_0) \to \Omega(L)[1] \otimes \Lambda_0$$

by

$$\mathfrak{m}_k^{\mathbf{b}}(x_1,\dots,x_k) = \sum_{\ell=0}^{\infty}\sum_{m_0=0}^{\infty}\cdots\sum_{m_k=0}^{\infty} \mathfrak{q}_{\ell,k}(\mathfrak{b}^{\ell}; \underbrace{b,\dots,b}_{m_0}, x_1, \underbrace{b,\dots,b}_{m_1},\dots,x_k,\underbrace{b,\dots,b}_{m_k}). \tag{36}$$

It is easy to see that $\{\mathfrak{m}_k^{\mathbf{b}} \mid k=0,1,2,\dots\}$ defines a unital and gapped filtered A_∞ structure.

We define $\widehat{\mathcal{M}}_{\text{def,weak}}(L)$ as the set of all $\mathbf{b}=(\mathfrak{b},b)$ such that

$$\mathfrak{m}_0^{\mathbf{b}}(1) = c\mathbf{e}. \tag{37}$$

Here $\mathbf{e} = 1 \in \Omega^0(L)$.

If $\mathbf{b}\in\widehat{\mathcal{M}}_{\text{def,weak}}(L)$ then we have

$$\mathfrak{m}_1^{\mathbf{b}} \circ \mathfrak{m}_1^{\mathbf{b}} = 0.$$

Definition 7.3. For $\mathbf{b}\in\widehat{\mathcal{M}}_{\text{def,weak}}(L)$, we define *Floer cohomology with bulk deformation* by

$$HF((L,\mathbf{b}),(L,\mathbf{b});\Lambda_0) \cong \frac{\text{Ker }\mathfrak{m}_1^{\mathbf{b}}}{\text{Im }\mathfrak{m}_1^{\mathbf{b}}}. \tag{38}$$

$HF((L,\mathbf{b}),(L,\mathbf{b});\Lambda)$ is defined by taking $\otimes_{\Lambda_0}\Lambda$.

We define the potential function $\mathfrak{PO} : \widehat{\mathcal{M}}_{\text{def,weak}}(L) \to \Lambda_+$ by the equation

$$\mathfrak{PO}\mathbf{e} = \mathfrak{m}_0^{\mathbf{b}}(1). \tag{39}$$

We also put $\mathfrak{PO}^{\mathfrak{b}}(b) = \mathfrak{PO}(\mathfrak{b},b)$.

If $HF((L,\mathbf{b}),(L,\mathbf{b});\Lambda)\neq 0$ then L is non-displaceable. This is [**FOOO4**] Proposition 3.15 which is proved in [**FOOO4**] section 8.

8. Floer cohomology with bulk deformation in the toric case

Now we apply the construction explained in the last section to the case of toric manifolds. In this section we use cycles (submanifolds) rather than differential forms to represent the (co)homology classes of ambient manifold X, by a reason we will mention in Remark 8.3.

Let $D_1,\dots,D_m$ be the irreducible components of toric divisors. Let $J = \{j_1,\dots,j_k\}\subseteq\{1,\dots,m\}$. If $D_J = D_{j_1}\cap\cdots\cap D_{j_k}$ is non-empty, it is a (real) codimension $2k$ submanifold of X. We include the case $J=\emptyset$. In that case $D_J = X$. We denote by $\mathcal{A}$ the free abelian group generated by D_J. We put cohomology degree on it. Namley $\deg D_J = 2k$ if codimension of D_J is $2k$. We define $\mathcal{A}(\Lambda_0) = \mathcal{A}\otimes\Lambda_0$.

There is an obvious homomorphism

$$\mathcal{A} \to H^*(X;\mathbb{Z}) \tag{40}$$

which is surjective but not injective. We write the generator of $\mathcal{A}$ as $\mathbf{p}_i$, $(i=0,\dots,B)$, where $\mathbf{p}_0 = X$ and $\mathbf{p}_i = D_i$ for $i=1,\dots,m$ are degree 2 classes. For $I=(i_1,\dots,i_\ell)\in\{1,\dots,B\}^\ell$ we put

$$\mathbf{p}_I = \mathbf{p}_{i_1}\otimes\cdots\otimes\mathbf{p}_{i_\ell},\quad [\mathbf{p}_I] = \frac{1}{\ell!}\sum_{\sigma\in\mathfrak{S}_\ell}\mathbf{p}_{i_{\sigma(1)}}\otimes\cdots\otimes\mathbf{p}_{i_{\sigma(\ell)}}\in E_\ell\mathcal{A}[2].$$

Here $\mathfrak{S}_\ell$ is the symmetric group of order $\ell!$.

Let $\mathbf{u}\in \operatorname{Int} P$, $\beta\in H_2(X,L(\mathbf{u});\mathbb{Z})$ and $I\in\{1,\dots,B\}^\ell$. We define:

$$\mathcal{M}^{\text{main}}_{k+1,\ell}(\beta,\mathbf{p}_I) = \mathcal{M}^{\text{main}}_{k+1,\ell}(\beta)_{(\text{ev}^+_1,\dots,\text{ev}^+_\ell)}\times_{X^\ell}\mathbf{p}_I. \tag{41}$$

Note ev^+_i are evaluation maps at interior marked points. We then still have evaluation maps at boundary marked points:

$$\text{ev} = (\text{ev}_0,\dots,\text{ev}_k) : \mathcal{M}^{\text{main}}_{k+1,\ell}(\beta,\mathbf{p}_I)\to L^{k+1}. \tag{42}$$

We use it to define an operator

$$\mathfrak{q}_{\ell,k;\beta} : E_\ell\mathcal{A}[2]\otimes B_kH(L(\mathbf{u});\mathbb{C})[1]\to H(L(\mathbf{u});\mathbb{C})[1]$$

as follows. We remark that there is a transitive and free action of T^n on $L(\mathbf{u})$. We put a T^n invariant metric on $L(\mathbf{u})$. Harmonic forms with respect to this metric are nothing but the T^n invariant differential forms. We identify the cohomology group $H(L(\mathbf{u});\mathbb{C})$ with the set of the T^n invariant forms on $L(\mathbf{u})$ from now on.

Let $h_1,\dots,h_k\in H(L(\mathbf{u});\mathbb{C})$. The pull-back

$$(\text{ev}_1,\dots,\text{ev}_k)^*(h_1\times\cdots\times h_k)$$

is a differential form on $\mathcal{M}^{\text{main}}_{k+1,\ell}(\beta,\mathbf{p}_I)$. We use integration along fiber of the evaluation map $\text{ev}_0 : \mathcal{M}^{\text{main}}_{k+1,\ell}(\beta,\mathbf{p}_I)\to L$ and define:

$$\mathfrak{q}_{\ell,k;\beta}([\mathbf{p}_I];h_1\times\cdots\times h_k) = \text{ev}_{0!}(\text{ev}_1,\dots,\text{ev}_k)^*(h_1\times\cdots\times h_k). \tag{43}$$

We can perform all the constructions in a T^n equivariant way. So the right hand side is a T^n invariant differential form, which we identify with an element of cohomology group.

REMARK 8.1. To define integration along the fiber, we need the map $\text{ev}_0 : \mathcal{M}^{\text{main}}_{k+1,\ell}(\beta,\mathbf{p}_I)\to L$ to be a submersion. We also need the moduli space to be transversal after taking an appropriate perturbation.

We can do so by using multisection in the same way as section 5 as follows. We remark that the fiber product moduli space $\mathcal{M}^{\text{main}}_{k+1,\ell}(\beta,\mathbf{p}_I)$ has a Kuranishi structure. The group T^n acts on it. Moreover the T^n action is free. (This is because ev_0 is T^n equivariant and the T^n action on $L(\mathbf{u})$ is free.) Thus by the same argument as we explained during the proof of Theorem 5.2, we can take multisection $\mathfrak{s}$ which is T^n equivariant and transversal to 0. Then $\text{ev}_0 : \mathcal{M}^{\text{main}}_{k+1,\ell}(\beta,\mathbf{p}_I)^{\mathfrak{s}}\to L(\mathbf{u})$ automatically becomes a submersion if $\mathcal{M}^{\text{main}}_{k+1,\ell}(\beta,\mathbf{p}_I)^{\mathfrak{s}}$ is nonempty.

We can also choose our perturbation so that it is invariant under the permutation of the interior marked points so descents to $E_\ell\mathcal{A}[2]$. Therefore the right hand side of (43) depends only on $[\mathbf{p}_I]$ rather than on $\mathbf{p}_I$.

We now define

$$\mathfrak{q}_{\ell,k} : E_\ell\mathcal{A}(\Lambda_0)[2] \otimes B_k H(L(\mathbf{u});\Lambda_0)[1] \to H(L(\mathbf{u});\Lambda_0)[1]$$

by

$$\mathfrak{q}_{\ell,k} = \sum_{\beta\in H_2(X;L(\mathbf{u});\mathbb{Z})} T^{(\omega\cap\beta)/2\pi}\mathfrak{q}_{\ell,k;\beta}.$$

In case we consider elements of $E_\ell\mathcal{A}(\Lambda_0)[2]$ which contain $\mathbf{p}_0$, the Poincaré dual to $[X]$, we define $\mathfrak{q}_{\ell,k}$ as follows:

$$\mathfrak{q}_{1,0}(\mathbf{p}_0;1) = \mathbf{e}. \tag{44}$$

In all the other cases, $\mathfrak{q}_{\ell,k}$ is zero if the first factor $E_\ell\mathcal{A}(\Lambda_0)[2]$ contains $\mathbf{p}_0$.

Then our $\mathfrak{q}_{\ell,k}$ satisfies the conclusion of Theorem 7.1.

For $\mathbf{b} = (\mathfrak{b}, b) \in \mathcal{A}(\Lambda_+)\times H^{odd}(L(\mathbf{u});\Lambda_+)$, we define $\mathfrak{m}_k^{\mathbf{b}}$ by (36). It defines a unital gapped filtered A_∞ structure on $H(L(\mathbf{u});\Lambda_0)$.

Now we define

$$\widehat{\mathcal{M}}_{\text{def,weak}}(L(\mathbf{u});\Lambda_+) \subset \mathcal{A}(\Lambda_+) \times H^{odd}(L(\mathbf{u});\Lambda_+)$$

as the set of all $\mathbf{b} = (\mathfrak{b}, b) \in \mathcal{A}(\Lambda_+) \times H^{odd}(L(\mathbf{u});\Lambda_+)$ such that $\mathfrak{m}_0^{\mathbf{b}}(1) \equiv 0 \mod \Lambda_+\mathbf{e}$. In other words it is the set of $(\mathfrak{b}, b)$ such that

$$\sum_{\ell=0}^{\infty}\sum_{k=0}^{\infty} \mathfrak{q}_{\ell;k}(\mathfrak{b}^\ell; b^k) \equiv 0 \mod \Lambda_+\mathbf{e}. \tag{45}$$

We define the potential function $\mathfrak{PO} : \widehat{\mathcal{M}}_{\text{def,weak}}(L(\mathbf{u});\Lambda_+) \to \Lambda_+$ by

$$\sum_{\ell=0}^{\infty}\sum_{k=0}^{\infty} \mathfrak{q}_{\ell;k}(\mathfrak{b}^\ell; b^k) = \mathfrak{PO}(\mathfrak{b}; b)\mathbf{e}. \tag{46}$$

Using a similar trick as the one used in section 5 we can extend the story to the cohomology groups with Λ_0 coefficient. Namely we obtain a Maurer-Cartan scheme

$$\widehat{\mathcal{M}}_{\text{def,weak}}(L(\mathbf{u});\Lambda_0) \subset \mathcal{A}(\Lambda_0) \times H^{odd}(L(\mathbf{u});\Lambda_0)$$

and Floer cohomology parametrized thereover. We also have a potential function

$$\mathfrak{PO} : \widehat{\mathcal{M}}_{\text{def,weak}}(L(\mathbf{u});\Lambda_0) \to \Lambda_+.$$

Most of the stories in section 5 can be generalized to the current situation.

THEOREM 8.2. *1)* $\mathcal{A}(\Lambda_0) \times H^1(L(\mathbf{u});\Lambda_0)$ *is contained in* $\widehat{\mathcal{M}}_{\text{def,weak}}(H(L;\Lambda_0);\Lambda_0)$.

2) *Let* $b = \sum x_i \mathbf{x}_i \in H^1(L(\mathbf{u}); \Lambda_0)$ *and* $\mathfrak{b} \in \mathcal{A}(\Lambda_+)$. *Then we have*

$$\mathfrak{PO}(\mathfrak{b}, b) = z_1 + \cdots + z_m + \sum_{k=1}^{N} T^{\rho_k} P_k(\mathfrak{b}; z_1, \ldots, z_m). \tag{47}$$

Here $N \in \mathbb{Z}_{\geq 0}$ *or* $N = \infty$. *The numbers* ρ_k *are all positive and real. In case* $N = \infty$, *the sequence of numbers* ρ_k *goes to* ∞ *as* k *goes to* ∞. $P_k(\mathfrak{b}; z_1, \ldots, z_m)$ *are monomials of* $z_1, \ldots, z_m$ *of degree* ≥ 1 *with* Λ_0 *coefficient. (Here degree means that of monomials of* z_i.) *We remark that* z_j *is defined from* $y_i^{\mathbf{u}} = e^{x_j}$ *by* (19).

3) *Let* $b = \sum x_i \mathbf{x}_i \in H^1(L(\mathbf{u}); \Lambda_0)$ *and* $\mathfrak{b} \in \mathcal{A}(\Lambda_0)$.

$$\begin{aligned} \mathfrak{PO}(\mathfrak{b}, b) &= \mathfrak{c}_1 z_1 + \cdots + \mathfrak{c}_m z_m + P_0(\mathfrak{b}; z_1, \ldots, z_m) \\ &\quad + \sum_{k=1}^{N} T^{\rho_k} P_k(\mathfrak{b}; z_1, \ldots, z_m). \end{aligned} \tag{48}$$

$P_0(\mathfrak{b}; z_1, \ldots, z_m)$ *is a formal power series of* $z_1, z_2, \ldots, z_m$ *with* Λ_0 *coefficient such that each term has degree* ≥ 2. *The numbers* $\mathfrak{c}_j$ *are defined as follows. Let* $\mathfrak{b} = \sum w_j \mathbf{p}_j$. *We put* $w_j \equiv \overline{w}_j \mod \Lambda_+$ *and* $\overline{w}_j \in \mathbb{C}$. *Then* $\mathfrak{c}_j = e^{\overline{w}_j} \in \mathbb{C} \setminus \{0\}$. *Other notations are the same as in* (47).

4) *The monomials* P_k *and the numbers* ρ_k *are independent of* $\mathbf{u}$ *and depends only on* X *and* $\mathfrak{b}$.

Item 1) is [**FOOO4**] Proposition 3.1. (In [**FOOO4**] Proposition 3.1 it is assumed that $\mathfrak{b} \in \mathcal{A}(\Lambda_+)$. It holds also for $\mathfrak{b} \in \mathcal{A}(\Lambda_0)$. See [**FOOO4**] section 11.)

Item 2) is [**FOOO4**] Theorem 3.4.

Item 3) follows from [**FOOO4**] sections 8 and 11. (Formulas (9.3), (11.1) etc.)

Item 4) follows from [**FOOO4**] Lemma 6.8.

The proof of Theorem 8.2 is similar to that of Theorem 5.2. We here mention only a few points. Let $I = (i_1, \ldots, i_\ell) \in \{1, \ldots, B\}^\ell$. We put $\mathbf{p}_I = \mathbf{p}_{i_1} \otimes \cdots \otimes \mathbf{p}_{i_\ell}$. We have

$$\dim \mathcal{M}_{1,\ell}^{\text{main}}(\beta; \mathbf{p}_I) = n - 2 + \mu(\beta) - \sum_{i=1}^{\ell} (\deg \mathbf{p}_i - 2). \tag{49}$$

Here dim is the virtual dimension that is the dimension in the sense of Kuranishi structure. As we explained in Remark 8.1 the perturbed moduli space $\mathcal{M}_{1,\ell}^{\text{main}}(\beta; \mathbf{p}_I)^{\mathfrak{s}}$ is empty if (49) $< n$.

REMARK 8.3. This is the reason why we use cycles $\mathbf{p}_i$ rather than differential forms on X to represent cohomology classes of X. This point is crucial to prove item 1) in Theorem 8.2.

In the case (49) $= n$ we define

$$c(\beta; I) = \deg[\mathrm{ev}_0 : \mathcal{M}_{1,\ell}^{\mathrm{main}}(\beta; \mathbf{p}_I)^{\mathfrak{s}} \to L(\mathbf{u})] \in \mathbb{Q}. \tag{50}$$

Here and hereafter $\mathcal{M}_{1,\ell}^{\mathrm{main}}(\beta; \mathbf{p}_I)^{\mathfrak{s}}$ denotes the perturbation of the moduli space $\mathcal{M}_{1,\ell}^{\mathrm{main}}(\beta; \mathbf{p}_I)$. Namely it is the zero set of the multisection $\mathfrak{s}$. This zero set has a triangulation and each simplex of maximal degree comes with a weight $\in \mathbb{Q}$. Thus it has a virtual fundamental cycle. See [**FOOO1**] Section A1.

The number (50) is well-defined. Namely it is independent of the perturbation $\mathfrak{s}$ as far as it is T^n equivariant. It is also independent of $\mathbf{u}$. ([**FOOO4**] Lemma 6.8.) The potential function is calculated by using $c(\beta; I)$ as follows. Let $\vec{\ell} = (\ell_1, \dots, \ell_B) \in \mathbb{Z}_{\geq 0}^B$. We put

$$I(\vec{\ell}) = (\underbrace{1, \dots, 1}_{\ell_1}, \dots, \underbrace{B, \dots, B}_{\ell_B}) \in \{1, \dots, B\}^{\sum_{i=1}^B \ell_i},$$

and

$$c(\beta; \vec{\ell}) = c(\beta; I(\vec{\ell})).$$

Let $\mathfrak{b} = \sum_{i=0}^B w_i \mathbf{p}_i$.

We define $\partial_i(\beta) \in \mathbb{Z}$ by

$$\partial_i(\beta) = \langle \partial\beta, \mathbf{e}_i^* \rangle,$$

and put

$$(y^{\mathbf{u}})^{\partial\beta} = (y_1^{\mathbf{u}})^{\partial_1\beta} \cdots (y_n^{\mathbf{u}})^{\partial_n\beta} = T^{-\langle \partial\beta, \mathbf{u} \rangle} y_1^{\partial_1\beta} \cdots y_n^{\partial_n\beta}.$$

Now we have

$$\begin{aligned} \mathfrak{PO}(\mathfrak{b}, b) = w_0 + & \sum_{\beta \in H_2(X, L(\mathbf{u}); \mathbb{Z})} \sum_{\ell_1=0}^{\infty} \cdots \sum_{\ell_B=0}^{\infty} \\ & \times \frac{T^{(\beta \cap \omega)/2\pi}}{\ell_1! \cdots \ell_B!} c(\beta; \vec{\ell}) w_1^{\ell_1} \cdots w_B^{\ell_B} (y^{\mathbf{u}})^{\partial\beta}. \end{aligned} \tag{51}$$

For the proof of (51) see [**FOOO4**] section 9.

(47) follows from (51) and $c(\beta_j; (0, \dots, 0)) = 1$. This follows from [**CO**]. (See [**FOOO4**] section 7.)

Theorem 5.5 is generalized to our situation without change. Namely we have the following theorem. Hereafter we put $\mathfrak{PO}^{\mathfrak{b}}(b) = \mathfrak{PO}(\mathfrak{b}, b)$.

THEOREM 8.4. *Let $b = \sum x_i \mathbf{e}_i \in H^1(L(\mathbf{u}); \Lambda_0)$ and $\mathfrak{b} \in \mathcal{A}(\Lambda_0)$. Then the following three conditions are equivalent.*

1) *For each of $i = 1, \dots, n$ we have:*

$$\left. \frac{\partial \mathfrak{PO}^{\mathfrak{b}}}{\partial x_i} \right|_b = 0$$

2)

$$HF((L(\mathbf{u}),(\mathfrak{b},b)),(L(\mathbf{u}),(\mathfrak{b},b));\Lambda_0)\cong H(T^n;\Lambda_0).$$

3)

$$HF((L(\mathbf{u}),(\mathfrak{b},b)),(L(\mathbf{u}),(\mathfrak{b},b));\Lambda)\neq 0.$$

The proof is the same as the proof of Theorem 5.5 except some technical points, which we omit and refer [**FOOO4**].

The discussion in section 5 on the domain of the function $\mathfrak{PO}$ as a function of y_i is also generalized.

We put:

$$\mathfrak{A}(\overset{\circ}{P})=\{(\mathfrak{y}_1,\dots,\mathfrak{y}_n)\in\Lambda^n \mid (\mathfrak{v}_T(\mathfrak{y}_1),\dots,\mathfrak{v}_T(\mathfrak{y}_n))\in \operatorname{Int} P\}. \tag{52}$$

We remark that by Theorem 8.2 $\mathfrak{PO}^{\mathfrak{b}}$ may be regarded as a function of $y_1,\dots,y_n$.

LEMMA 8.5. *Let* $(\mathfrak{y}_1,\dots,\mathfrak{y}_n)\in\mathfrak{A}(\overset{\circ}{P})$. *We put* $\mathfrak{z}_j=T^{\lambda_j}\mathfrak{y}_1^{v_{j,1}}\cdots\mathfrak{y}_n^{v_{j,n}}$. *Then*

$$\mathfrak{z}_1+\cdots+\mathfrak{z}_m+P_0(\mathfrak{b};\mathfrak{z}_1,\dots,\mathfrak{z}_m)+\sum_{k=1}^{N}T^{\rho_k}P_k(\mathfrak{b};\mathfrak{z}_1,\dots,\mathfrak{z}_m)\in\Lambda_+$$

converges as $N\to\infty$ *with respect to the valutaion* $\mathfrak{v}_T$.

In case $\mathfrak{b}\in\mathcal{A}(\Lambda_+)$ *where the term* $P_0(\mathfrak{b};\mathfrak{z}_1,\dots,\mathfrak{z}_m)$ *is absent, we may relax the assumption to* $(\mathfrak{y}_1,\dots,\mathfrak{y}_n)\in\mathfrak{A}(P)$.

Thus we may regard $\mathfrak{PO}^{\mathfrak{b}}$ as a function either $:\mathfrak{A}(\overset{\circ}{P})\to\Lambda_+$ or $:\mathfrak{A}(P)\to\Lambda_0$.

We can define

$$\mathfrak{y}_i\frac{\partial\mathfrak{PO}^{\mathfrak{b}}}{\partial y_i}$$

in the same way as section 5. It defines either a function $:\mathfrak{A}(\overset{\circ}{P})\to\Lambda_+$ or $:\mathfrak{A}(P)\to\Lambda_0$. Theorem 5.9 can be generalized as follows:

THEOREM 8.6. *For* $\mathbf{u}\in\operatorname{Int}P$, $\mathfrak{b}\in\mathcal{A}(\Lambda_0)$, *the following two conditions are equivalent.*

1) *There exists* $b\in H^1(L(\mathbf{u});\Lambda_0)$ *such that*

$$HF((L(\mathbf{u}),(\mathfrak{b},b)),(L(\mathbf{u}),(\mathfrak{b},b));\Lambda_0)\cong H(T^n;\Lambda_0).$$

2) *There exists* $\mathfrak{y}=(\mathfrak{y}_1,\dots,\mathfrak{y}_n)\in\mathfrak{A}(P)$ *such that*

$$\mathfrak{y}_i\frac{\partial\mathfrak{PO}^{\mathfrak{b}}}{\partial y_i}(\mathfrak{y})=0 \tag{53}$$

for $i = 1, \ldots, n$ *and that*

$$(\mathfrak{v}_T(\mathfrak{y}_1), \ldots, \mathfrak{v}_T(\mathfrak{y}_n)) = \mathbf{u}.$$

This is [**FOOO4**] Theorem 3.12.

9. Leading term equation

Theorem 8.6 provides a means of determining the Floer cohomology in terms of the potential function. The main obstacle to directly apply the theorem in practice is that we do not know how to calculate the extra terms $P_k(\mathfrak{b}; z_1, \cdots, z_n)$ unless X is Fano and $\mathfrak{b}$ has degree 2. (There has been some computation carried out in this direction for the nef case. See e.g. [**CLa**].)

Fortunately to determine all the T^n orbits $L(\mathbf{u})$ for which some Floer cohomology with bulk does not vanish, we do not need to calculate those terms. We will explain it in this section.

In this and the next sections we fix $\mathfrak{b}$ and $\mathbf{u}$ and consider $\mathfrak{PO}^{\mathfrak{b}}$ as a function of variables $y_i^{\mathbf{u}}$. In this section we write $\overline{y}_i$ instead of $y_i^{\mathbf{u}}$. We remark that $\mathfrak{v}_T^{\mathbf{u}}(\overline{y}_i) = 0$ and

$$z_j = T^{\ell_j(\mathbf{u})} \overline{y}_1^{v_{j,1}} \cdots \overline{y}_1^{v_{j,n}}.$$

Definition 9.1. We denote the sum of linear terms z_j's in $\mathfrak{PO}^{\mathfrak{b}}$ by

$$\mathfrak{PO}_0^{\mathfrak{b}} = \mathfrak{c}_1 z_1 + \cdots + \mathfrak{c}_m z_m = \sum_{j=1}^{m} T^{\ell_j(\mathbf{u})} \mathfrak{c}_j \overline{y}_1^{v_{j,1}} \cdots \overline{y}_n^{v_{j,n}}$$

and call it the *leading order potential function.* Here $\mathfrak{c}_j$ is defined as in Theorem 8.2 3).

Note this function appears frequently in the literature (see [**Gi1, HV, Iri1**]), is denoted as W, and is called the (Landau-Ginzburg) superpotential.

Remark 9.2. Note in our situation of toric manifold, superpotential in physics literature is basically the same as our potential function. However in other situation they may be different. For example in the case of Calabi-Yau 3 fold X and its special Lagrangian submanifold L, our potential function is identically 0. (In other words, if b is a weak bounding chain then it is a bounding chain automatically.) On the other hand, the physisists' superpotential coincides with the invariant introduced in [**Fu4**].

We remark that the leading order potential function $\mathfrak{PO}_0^{\mathfrak{b}}$ is explicitly read off from the moment polytope P and $\mathbf{u}$. The leading term equation we will define below depends only on leading order potential function and so is also explicitly calculable.

We renumber the values $\ell_i(\mathbf{u})$ according to its order. Namely we take $j(l, r) \in \{1, \ldots, m\}$ for $l = 1, \ldots, K_0$, $r = 1, \ldots, a(l')$ with the following conditions.

CONDITIONS 9.3. 1) $\{j(l,r) \mid l = 1,\dots,K_0, r = 1,\dots,a(l)\} = \{1,\dots,m\}$.
2) $a(1) + \cdots + a(K_0) = m$.
3) $\ell_{j(l,r)}(\mathbf{u}) = \ell_{j(l,r')}(\mathbf{u})$ for $1 \le r, r' \le a(l)$.
4) $\ell_{j(l,r)}(\mathbf{u}) < \ell_{j(l',r')}(\mathbf{u})$ if $l < l'$.

We put

$$S_l = \ell_{j(l,r)}(\mathbf{u}). \tag{54}$$

This is independent of r. Set

$$\vec{v}_{l,r} = \vec{v}_{j(l,r)} = (v_{j(l,r),1},\dots,v_{j(l,r),n}) \in \mathbb{Z}^n. \tag{55}$$

It is an element of the dual vector space of $\mathbb{A}(\mathbb{Q}) = \mathbb{Q}^n$, which we denote by $\mathbb{A}(\mathbb{Q})^*$. Here $\mathbb{A}(\mathbb{R}) = \mathbb{A}(\mathbb{Q}) \otimes \mathbb{R}$ is the $\mathbb{R}$ vector space associated to the affine space which contains the moment polytope P. Let $\mathbb{A}_l^\perp$ is a vector subspace of $\mathbb{A}(\mathbb{Q})^*$ generated by $\{\vec{v}_{l',r} \mid l' \le l, r = 1,\dots,a(l')\}$. We denote by $K \le K_0$ the smallest integer such that $\mathbb{A}_K^\perp = \mathbb{A}(\mathbb{Q})^*$. We have a filtration

$$0 \subset \mathbb{A}_1^\perp \subset \mathbb{A}_2^\perp \subset \cdots \subset \mathbb{A}_K^\perp = \mathbb{A}(\mathbb{Q})^*. \tag{56}$$

We put

$$d(l) = \dim \mathbb{A}_l^\perp - \dim \mathbb{A}_{l-1}^\perp. \tag{57}$$

We have

$$d(1) + \cdots + d(K) = n = \dim \mathbb{A}(\mathbb{Q})^*. \tag{58}$$

Note $\mathbb{A} \cong \mathbb{Z}^n \subset \mathbb{A}(\mathbb{Q}) = \mathbb{Q}^n$. So $\mathbb{Z}^n \subset \mathbb{A}(\mathbb{Q})^*$ is determined canonically. (We remark that $\mathbb{Z}^n \subset \mathbb{A}(\mathbb{Q})^*$ is generated by $\vec{v}_j$, $j = 1,\dots,m$.) Let $\{\mathbf{e}_i^* \mid i = 1,\dots,n\}$ be the standard basis of $\mathbb{Z}^n \subset \mathbb{A}(\mathbb{Q})^*$. We take $\mathbf{e}_{l,s}^*$ for $l = 1,\dots,K$, $s = 1,\dots,d(l)$ satisfying the following conditions.

CONDITIONS 9.4. 1) $\{\mathbf{e}_{l',s}^* \mid l' \le l,\ s = 1,\dots,d(l')\}$ is a $\mathbb{Q}$ basis of $\mathbb{A}_l^\perp$.
2) $\vec{v}_{l,r}$ is contained in the $\mathbb{Z}$ module generated by $\{\mathbf{e}_{l',s}^* \mid l' \le l,\ s = 1,\dots,d(l')\}$.

We define $b_{l',s;i} \in \mathbb{Q}$ by

$$\mathbf{e}_{l',s}^* = \sum_{i=1}^n b_{l',s;i}\mathbf{e}_i^*$$

and put

$$\overline{y}_{l',s} = \prod_{i=1}^n \overline{y}_i^{\,b_{l',s;i}} = \exp\left(\sum_{i=1}^n b_{l',s;i}x_i\right).$$

(Note $\overline{y}_i = e^{x_i}$.) Since $b_{l',s;i}$ may not be an integer, $\overline{y}_{l',s}$ may not be contained in the Laurent polynomial ring $\Lambda_0[\overline{y}, \overline{y}^{-1}]$ of the variables $\overline{y}_i$ $(i = 1, \dots, n)$. But it is contained in the finite extention of it.

By Condition 9.4 2),

$$z_{j(l,r)} = T^{S_l} \overline{y}_1^{v_{j(l,r),1}} \cdots \overline{y}_n^{v_{j(l,r),n}}$$

is contained in $\Lambda_0[\overline{y}, \overline{y}^{-1}]$. Moreover it is contained in the Laurent polynomial ring of the variables $\overline{y}_{l',s}$, $l' = 1, \dots, l$, $s = 1, \dots, d(l')$.

We define $c_{l,r;l',s} \in \mathbb{Z}$ by

$$z_{j(l,r)} = T^{S_l} \prod_{l' \le l} \prod_{s \le d(l')} \overline{y}_{l',s}^{c_{l,r;l',s}}. \tag{59}$$

In other words

$$\vec{v}_{l,r} = \sum_{l' \le l} \sum_{s \le d(l')} c_{l,r;l',s} e^*_{l',s}.$$

We put

$$\left(\mathfrak{PO}_0^{\mathfrak{b}}\right)_l = T^{-S_\ell} \sum_{r=1}^{a(l)} \mathfrak{c}_{j(l,r)} z_{j(l,r)} = \sum_{r=1}^{a(l)} \mathfrak{c}_{j(l,r)} \prod_{l' \le l} \prod_{s \le d(l')} \overline{y}_{l',s}^{c_{l,r;l',s}}. \tag{60}$$

The numbers $\mathfrak{c}_{j(l,r)} \in \{c \in \Lambda_0 \mid \mathfrak{v}_T(c) = 0\}$ are defined in Definition 9.1.

We remark $(\mathfrak{PO}_0^{\mathfrak{b}})_l$ is a Laurent polynomial of variables $\overline{y}_{l',s}$, $l' \le l$, $s = 1, \dots, d(l')$ with coefficients in a field of complex numbers.

DEFINITION 9.5. The *leading term equation* is a system of n equations of n variables $\overline{y}_{l,s}$ with complex number coefficients. We define it by

$$\begin{cases} \overline{y}_{1,s} \dfrac{\partial (\mathfrak{PO}_0^{\mathfrak{b}})_1}{\partial \overline{y}_{1,s}} = 0 & s = 1, \dots, d(1), \\ \overline{y}_{2,s} \dfrac{\partial (\mathfrak{PO}_0^{\mathfrak{b}})_2}{\partial \overline{y}_{2,s}} = 0 & s = 1, \dots, d(2), \\ \qquad \cdots & \\ \overline{y}_{l,s} \dfrac{\partial (\mathfrak{PO}_0^{\mathfrak{b}})_l}{\partial \overline{y}_{l,s}} = 0 & s = 1, \dots, d(l), \\ \qquad \cdots & \\ \overline{y}_{K,s} \dfrac{\partial (\mathfrak{PO}_0^{\mathfrak{b}})_K}{\partial \overline{y}_{K,s}} = 0 & s = 1, \dots, d(K). \end{cases} \tag{61}$$

Note the first equation in (61) contains $\overline{y}_{1,s}$ $s = 1, \dots, d(1)$, the second equation in (61) contains $\overline{y}_{1,s}$ $s = 1, \dots, d(1)$ and $\overline{y}_{2,s}$ $s = 1, \dots, d(2)$ etc.

If $\mathfrak{b} - \mathfrak{b}' \in \mathcal{A}^2(\Lambda_+) \oplus \bigoplus_{k \ne 2} \mathcal{A}^k(\Lambda_0)$ then $(\mathfrak{PO}_0^{\mathfrak{b}})_l = \left(\mathfrak{PO}_0^{\mathfrak{b}'}\right)_l$. So the leading term equation is the same for such $\mathfrak{b}$ and $\mathfrak{b}'$.

One of the main results of [**FOOO4**] is as follows.

THEOREM 9.6. *Let* $\mathbf{u} \in \mathrm{Int}\, P$ *and* $\mathfrak{b} \in \mathcal{A}(\Lambda_0)$. *Then the following two conditions are equivalent.*

1) *The leading term equation* (61) *has a solution* $\overline{y}_{l,s} \in \mathbb{C} \setminus \{0\}$.
2) *There exists* $b \in H^1(L(\mathbf{u}); \Lambda_0)$ *and* $\mathfrak{b}' \in \mathcal{A}(\Lambda_0)$ *with* $\mathfrak{b} - \mathfrak{b}' \in \mathcal{A}^2(\Lambda_+)$ *such that*

$$HF((L(\mathbf{u}), (\mathfrak{b}, b)), (L(\mathbf{u}), (\mathfrak{b}, b)); \Lambda_0) \cong H(T^n; \Lambda_0).$$

This is [**FOOO4**] Theorem 4.5 and Proposition 11.3. We omit the proof and refer [**FOOO4**].

DEFINITION 9.7. We say that $L(\mathbf{u})$ is *strongly bulk balanced* if there exists $\mathfrak{b} \in \mathcal{A}(\Lambda_0)$ and $b \in H^1(L(\mathbf{u}); \Lambda_0)$ such that

$$HF((L(\mathbf{u}), (\mathfrak{b}, b)), (L(\mathbf{u}), (\mathfrak{b}, b)); \Lambda_0) \cong H(T^n; \Lambda_0).$$

See [**FOOO4**] Definition 3.13 for a related definition.

Theorem 9.6 gives a way to locate strongly bulk balanced $L(\mathbf{u})$ in terms of the leading term equation.

10. Examples 2

EXAMPLE 10.1. We consider Hirzebruch surface F_n, $n \geq 2$. We take its Kähler form so that the moment polytope is

$$P = \{(u_1, u_2) \mid 0 \leq u_1, u_2,\ u_1 + nu_2 \leq n,\ u_2 \leq 1 - \alpha\},$$

$0 < \alpha < 1$. The leading order potential function is

$$\mathfrak{PO}_0 = y_1 + y_2 + T^n y_1^{-1} y_2^{-n} + T^{1-\alpha} y_2^{-1}.$$

We put

$$\ell_1(u_1, u_2) = u_1, \quad \ell_2(u_1, u_2) = u_2,$$
$$\ell_3(u_1, u_2) = n - u_1 - nu_2, \quad \ell_4(u_1, u_2) = 1 - \alpha - u_2.$$

We put $S_1(u_1, u_2) = \inf\{\ell_j(u_1, u_2) \mid j = 1, 2, 3, 4\}$.

Suppose the first of the leading term equation (61) has a nonzero solution. Then it is easy to see that $d(1) \geq 2$. Namely

$$\#\{j \mid S_1(u_1, u_2) = \ell_j(u_1, u_2)\} \geq 2.$$

This is satisfied on the 5 line segUments $l_1, \ldots, l_5$, where

$$l_1 : u_1 = u_2 \leq (1 - \alpha)/2, \quad l_2 : u_1 = 1 - \alpha - u_2 \leq (1 - \alpha)/2,$$
$$l_3 : u_1 = n - (n + 1)u_2 \geq n - (n + 1)(1 - \alpha)/2$$
$$l_4 : u_1 = n - 1 + \alpha - (n - 1)u_2 \geq n - (n - 1)(1 - \alpha)/2,$$
$$l_5 : u_2 = (1 - \alpha)/2, (1 - \alpha)/2 \leq u_1 \leq n - (n - 1)(1 - \alpha)/2.$$

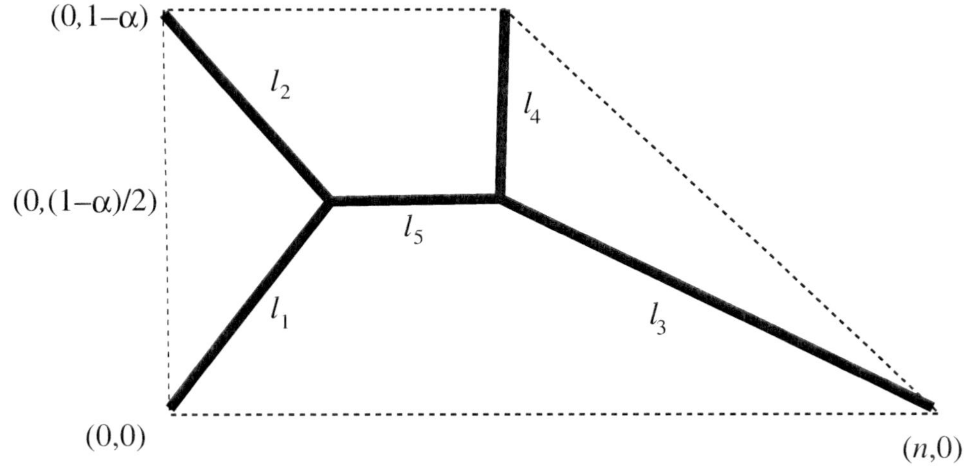

FIGURE 10.1

Note

$$v_1 = (1, 0),\ v_2 = (0, 1),\ v_3 = (-1, -n),\ v_4 = (0, -1).$$

Let $\mathbf{u} = (u_1, u_2) \in l_5$. Then $\mathbb{A}_1^\perp$ is $\mathbb{Q} \cdot (0, 1)$ and

$$(\mathfrak{PO}_0^{\mathbf{u}})_1 = \overline{y}_2 + \overline{y}_2^{-1} \tag{62}$$

(Here $\mathfrak{b} = 0$ and so we do not write $\mathfrak{b}$ in the above notation. We put $\overline{y}_i = y_i^{\mathbf{u}}$.)

We also have

$$(\mathfrak{PO}_0^{\mathbf{u}})_2 = \begin{cases} \overline{y}_1 & \text{If } u_1 < (1+\alpha)n/4 \\ \overline{y}_1^{-1}\overline{y}_2^{-n} & \text{If } u_1 > (1+\alpha)n/4 \\ \overline{y}_1 + \overline{y}_1^{-1}\overline{y}_2^{-n} & \text{If } u_1 = (1+\alpha)n/4. \end{cases} \tag{63}$$

(62) gives the first leading term equation $1 - \overline{y}_2^{-2} = 0$ whose solutions are $\overline{y}_2 = \pm 1$.

Then (63) gives the second of the leading term equation which are $1 = 0$, $-(\pm 1)^{-n}\overline{y}_1^{-2} = 0$, $1 - (\pm 1)^{-n}\overline{y}_1^{-2} = 0$, where $u_1 < (1+\alpha)n/4$, $u_1 > (1+\alpha)n/4$ and $u_1 = (1+\alpha)n/4$, respectively.

The solution $\overline{y}_1 \neq 0$ exists only in the case $u_1 = (1+\alpha)n/4$. In that case the solutions of leading term equations are $(1, \pm 1)$ and $(-1, \pm(-1)^{n/2})$. Thus $L((1+\alpha)n/4, (1-\alpha)/2)$ is strongly bulk balanced.

We can check that there are no other strongly bulk balanced T^2 orbits. (This follows from Theorem 11.9 also.)

See [**FOOO3**] Example 8.2 where the same conclusion is proved by basically the same but a slightly different calculation.

REMARK 10.2. For the case of Example 10.1 we can actually prove that $L((1+\alpha)n/4, (1-\alpha)/2)$ is strongly balanced. Namely some Floer cohomology without bulk deformation is non-zero. This follows from [**FOOO3**] Theorem 10.4.

EXAMPLE 10.3. ([**FOOO4**] section 5, [**FOOO3**] Example 10.17.) We consider two points blow up $X(\alpha, \alpha')$ of $\mathbb{C}P^2$. (Example 4.4.) We consider the case $\alpha > 1/3$, $\alpha' = (1-\alpha)/2$. The moment polytope is

$$P = \{(u_1, u_2) \mid 0 \le u_1 \le 1,\ 0 \le u_2 \le 1-\alpha,\ (1-\alpha)/2 \le u_1 + u_2 \le 1\}.$$

We consider

$$\mathbf{u}(t) = (t, (1-\alpha)/2), \quad t \in ((1-\alpha)/2, (1+\alpha)/4). \tag{64}$$

We have

$$\mathfrak{PO} = T^{(1-\alpha)/2}(\overline{y}_2 + \overline{y}_2^{-1}) + T^t(\overline{y}_1 + \overline{y}_1\overline{y}_2) + T^{(1+\alpha)/2+t}(\overline{y}_1\overline{y}_2)^{-1}$$

where

$$(1-\alpha)/2 < t < (1+\alpha)/2 + t.$$

Therefore

$$(\mathfrak{PO})_1 = \overline{y}_2 + \overline{y}_2^{-1}, \quad (\mathfrak{PO})_2 = \overline{y}_1 + \overline{y}_1\overline{y}_2.$$

Thus the leading term equation is

$$1 - \overline{y}_2^{-2} = 0, \quad 1 + \overline{y}_2 = 0.$$

This has a solution $\overline{y}_2 = -1$ ($\overline{y}_1$ is any number $\in \mathbb{C} \setminus \{0\}$.)

Theorem 9.6 implies that all of $L(\mathbf{u}(t))$ as in (64) are strongly bulk balanced. In particular they are non-displaceable.

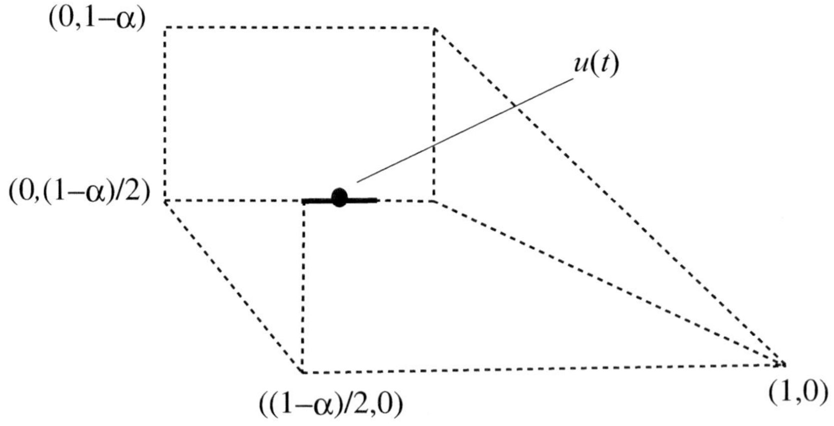

FIGURE 10.2

REMARK 10.4. In the toric case, for each given $\mathfrak{b}$, the number of $L(\mathbf{u})$ with nontrivial Floer cohomology for a pair $(\mathfrak{b}, b)$ for some $b \in H^1(L(\mathbf{u}); \Lambda_0)$ is finite. (It is not greater than the Betti number of X by Theorem 11.9.) So to obtain infinitely many $L(\mathbf{u})$ with nontrivial Floer cohomology we need to include bulk deformations.

In the examples we discussed in this section, we do not need to change the variables from y_j to $y_{l,s}$. An example where we need this change of variables is given in [**FOOO3**] Example 10.10.

In Example 10.3 we obtain a continuum of non-displaceable Lagrangian torus in certain two points blow up of $\mathbb{C}P^2$. ([**FOOO4**]). We can also use bulk deformation to obtain a continuum of Lagrangian tori in $S^2 \times S^2$. They are not of the type of T^2 orbit but is obtained from the T^2 orbit of singular Hirzebruch surface $F_2(0)$ by deforming the singularity, that is of orbifold of A_1-type. ([**FOOO6**].) Closely related construction is in [**NNU1, NNU2**]

11. Quantum cohomology and Jacobian ring

11.1. Jacobian ring over Novikov ring. In this section we discuss the isomorphism between the Jacobian ring of $\mathfrak{PO}^{\mathfrak{b}}$ and the quantum cohomology ring of X deformed by $\mathfrak{b}$. We start with defining Jacobian ring precisely.

Usually Jacobian ring is studied in the case of (Laurent) polynomial or holomorphic function germ. Our function $\mathfrak{PO}^{\mathfrak{b}}$ is neither a Laurent polynomial and nor a holomorphic function. So we first define a function space in which $\mathfrak{PO}^{\mathfrak{b}}$ is contained.

We consider the Laurent polynomial ring $\Lambda[y, y^{-1}]$ of n variables with Λ coefficients. We defined a valuation $\mathfrak{v}_T^{\mathbf{u}}$ for each $\mathbf{u} \in \mathbb{R}^n$ in section 5 Definition 5.6. Let P be a compact subset of $\mathbb{R}^n$. (We use the case when P is a convex polytope only in this article.)

DEFINITION 11.1. For $F \in \Lambda[y, y^{-1}]$ we define

$$\mathfrak{v}_T^P(F) = \inf\{\mathfrak{v}_T^{\mathbf{u}}(F) \mid \mathbf{u} \in P\}.$$

This is not a valuation but is a norm. Therefore it defines a metric on $\Lambda[y, y^{-1}]$ by $d_P(F_1, F_2) = e^{-\mathfrak{v}_T^P(F_1 - F_2)}$. We denote the completion of $\Lambda[y, y^{-1}]$ with respect to d_P by $\Lambda\langle\langle y, y^{-1}\rangle\rangle^P$. It is a normed ring.

We define $\Lambda\langle\langle y, y^{-1}\rangle\rangle_0^P$ as the set of all $F \in \Lambda\langle\langle y, y^{-1}\rangle\rangle^P$ such that $\mathfrak{v}_T^P(F) \geq 0$.

Let P be a moment polytope of our toric manifold X. We take ℓ_j ($j = 1, \ldots, m$) as in Condition 4.1 and put

$$P_\epsilon = \{\mathbf{u} \in \mathbb{R}^n \mid \ell_j(\mathbf{u}) \geq \epsilon, \ \ j = 1, \ldots, m\}$$

for $\epsilon > 0$. We define $\mathfrak{v}_T^{P_\epsilon}$, $\Lambda\langle\langle y, y^{-1}\rangle\rangle^{P_\epsilon}$ and $\Lambda\langle\langle y, y^{-1}\rangle\rangle_0^{P_\epsilon}$ in a similar way.

DEFINITION 11.2. We define a metric $d_{\overset{\circ}{P}}$ on $\Lambda[y, y^{-1}]$ by

$$d_{\overset{\circ}{P}}(F_1, F_2) = \sum_{k=1}^{\infty} 2^{-k} \min(d_{P_{1/k}}(F_1, F_2), 1).$$

Let $\Lambda\langle\langle y, y^{-1}\rangle\rangle^{\overset{\circ}{P}}$ be the completion of $\Lambda[y, y^{-1}]$ with respect to $d_{\overset{\circ}{P}}$ and define $\Lambda\langle\langle y, y^{-1}\rangle\rangle_0^{\overset{\circ}{P}} = \bigcap_{k=1}^{\infty} \Lambda\langle\langle y, y^{-1}\rangle\rangle_0^{P_{1/k}}$.

It is easy to see that an element of $\Lambda\langle\langle y,y^{-1}\rangle\rangle^P$ may be regarded as a function : $\mathfrak{A}(P)\to\Lambda$ and an element of $\Lambda\langle\langle y,y^{-1}\rangle\rangle^{\overset{\circ}{P}}$ may be regarded as a function : $\mathfrak{A}(\overset{\circ}{P})\to\Lambda$.

LEMMA 11.3. *If* $\mathfrak{b}\in\mathcal{A}(\Lambda_0)$ *then*

$$\mathfrak{PO}^{\mathfrak{b}}\in\Lambda\langle\langle y,y^{-1}\rangle\rangle_0^{\overset{\circ}{P}},\quad y_i\frac{\partial\mathfrak{PO}^{\mathfrak{b}}}{\partial y_i}\in\Lambda\langle\langle y,y^{-1}\rangle\rangle_0^{\overset{\circ}{P}}. \tag{65}$$

If $\mathfrak{b}\in\mathcal{A}(\Lambda_+)$ *then*

$$\mathfrak{PO}^{\mathfrak{b}}\in\Lambda\langle\langle y,y^{-1}\rangle\rangle_0^{P},\quad y_i\frac{\partial\mathfrak{PO}^{\mathfrak{b}}}{\partial y_i}\in\Lambda\langle\langle y,y^{-1}\rangle\rangle_0^{P}. \tag{66}$$

We omit the proof, which follows from Theorem 8.2. See [**FOOO5**] Lemma 2.6. Now we define

DEFINITION 11.4.

$$\mathrm{Jac}(\mathfrak{PO}^{\mathfrak{b}})=\frac{\Lambda\langle\langle y,y^{-1}\rangle\rangle_0^{\overset{\circ}{P}}}{\mathrm{Clos}_{d_{\overset{\circ}{P}}}\left(y_i\frac{\partial\mathfrak{PO}^{\mathfrak{b}}}{\partial y_i}:i=1,\dots,n\right)}.$$

(We may replace $\Lambda\langle\langle y,y^{-1}\rangle\rangle_0^{\overset{\circ}{P}}$ by $\Lambda\langle\langle y,y^{-1}\rangle\rangle_0^{P}$ in the above formula in case $\mathfrak{b}\in\mathcal{A}(\Lambda_+)$.)

Here the denominator is the closure of the ideal generated by $y_i\frac{\partial\mathfrak{PO}^{\mathfrak{b}}}{\partial y_i}$: $i=1,\dots,n$. The closure is taken with respect to the metric $d_{\overset{\circ}{P}}$.

11.2. Big quantum cohomology: a quick review. We next review briefly the well established story of deformed quantum cup product. Let (X,ω) be a symplectic manifold. For $\alpha\in H_2(X;\mathbb{Z})$ let $\mathcal{M}_\ell(\alpha)$ be the moduli space of stable maps from genus zero semi-stable curves with ℓ marked points and of homology class α. There exists an evaluation map

$$\mathrm{ev}:\mathcal{M}_\ell(\alpha)\to X^\ell.$$

$\mathcal{M}_\ell(\alpha)$ has a virtual fundamental cycle and hence defines a class

$$\mathrm{ev}_*[\mathcal{M}_\ell(\alpha)]\in H_*(X^\ell;\mathbb{Q}).$$

(See [**FO**].) Here $*=2n+2c_1(X)\cap\alpha+2\ell-6$. Let $Q_1,\dots,Q_\ell$ be cycles such that

$$\sum\mathrm{codim}\,Q_i=2n+2c_1(X)\cap\alpha+2\ell-6. \tag{67}$$

We define Gromov-Witten invariant by

$$GW_\ell(\alpha:Q_1,\dots,Q_\ell)=\mathrm{ev}_*[\mathcal{M}_\ell(\alpha)]\cap(Q_1\times\cdots\times Q_\ell)\in\mathbb{Q}.$$

We put $GW_\ell(\alpha:Q_1,\dots,Q_\ell)=0$ when (67) is not satisfied.

We now define

$$GW_\ell(Q_1,\dots,Q_\ell) = \sum_\alpha T^{(\alpha\cap\omega)/2\pi} GW_\ell(\alpha : Q_1,\dots,Q_\ell). \tag{68}$$

The formula (68) extends to a Λ_0 module homomorphism

$$GW_\ell : H(X;\Lambda_0)^{\otimes \ell} \to \Lambda_0.$$

DEFINITION 11.5. Let $\mathfrak{b} \in H(X;\Lambda_0)$ be given. For each given pair $\mathfrak{c}, \mathfrak{d} \in H(X;\Lambda_0)$, we define a product $\mathfrak{c} \cup^{\mathfrak{b}} \mathfrak{d} \in H(X;\Lambda_0)$ by the following formula

$$\langle \mathfrak{c} \cup^{\mathfrak{b}} \mathfrak{d}, \mathfrak{e}\rangle_{\mathrm{PD}_X} = \sum_{\ell=0}^{\infty} \frac{1}{\ell!} GW_{\ell+3}(\mathfrak{c},\mathfrak{d},\mathfrak{e},\mathfrak{b},\dots,\mathfrak{b}). \tag{69}$$

Here $\langle\cdot,\cdot\rangle_{\mathrm{PD}_X}$ denotes the Poincaré duality pairing. The right hand side converges if $\mathfrak{b} \in H^2(X;\Lambda_+) \oplus \bigoplus_{k>2} H^k(X;\Lambda_0)$. We can extend it to arbitrary $\mathfrak{b} \in H^*(X;\Lambda_0)$. (This is well-known. See for example [**FOOO5**] section 2.)

$\cup^{\mathfrak{b}}$ defines a graded commutative and associative ring structure on $H(X;\Lambda_0)$. We call $\cup^{\mathfrak{b}}$ the *deformed quantum cup product.*

11.3. The isomorphism 'Jacobian ring = quantum cohomology' and its applications.

THEOREM 11.6. *There exists a ring isomorphism*

$$(H(X;\Lambda_0), \cup^{\mathfrak{b}}) \cong \mathrm{Jac}(\mathfrak{PO}^{\mathfrak{b}}).$$

This is [**FOOO5**] Theorem 1.1.1 (1). We explain some parts of the proof later in this section. We first discuss some applications.

DEFINITION 11.7. Let $\mathrm{Crit}(\mathfrak{PO}^{\mathfrak{b}})$ be the set of all $\mathfrak{y} \in \mathfrak{A}(\mathring{P})$ such that

$$\frac{\partial \mathfrak{PO}^{\mathfrak{b}}}{\partial y_i}(\mathfrak{y}) = 0$$

for $i = 1,\dots,n$. An element of $\mathrm{Crit}(\mathfrak{PO}^{\mathfrak{b}})$ is said to be a *critical point* of $\mathfrak{PO}^{\mathfrak{b}}$.

A critical point $\mathfrak{y}$ of $\mathfrak{PO}^{\mathfrak{b}}$ is said to be *non-degenerate* if the matrix

$$\left[\mathfrak{y}_i\mathfrak{y}_j \frac{\partial^2 \mathfrak{PO}^{\mathfrak{b}}}{\partial y_i \partial y_j}(\mathfrak{y})\right]_{i,j=1}^{i,j=n}$$

is invertible, as an $n \times n$ matrix with Λ entries.

The function $\mathfrak{PO}^{\mathfrak{b}}$ is said to be a *Morse function* if all of its critical points are non-degenerate.

We put

$$\mathfrak{M}(X;\mathfrak{b}) = \left\{ (\mathbf{u}, b) \,\middle|\, \begin{array}{l} \mathbf{u} \in \operatorname{Int} P, b \in H^1(L(\mathbf{u});\Lambda_0)/H^1(L(\mathbf{u});2\pi\sqrt{-1}\mathbb{Z}), \\ HF((L(\mathbf{u}),(\mathfrak{b},b)),(L(\mathbf{u}),(\mathfrak{b},b));\Lambda_0) \cong H(T^n;\Lambda_0) \end{array} \right\}.$$

Theorem 8.6 implies the following.

$$\#\mathfrak{M}(X;\mathfrak{b}) = \#\mathrm{Crit}(\mathfrak{PO}^{\mathfrak{b}}). \tag{70}$$

PROPOSITION 11.8. *There exists a direct product decomposition*

$$\mathrm{Jac}(\mathfrak{PO}^{\mathfrak{b}}) \otimes_{\Lambda_0} \Lambda = \prod_{\mathfrak{y}\in \mathrm{Crit}(\mathfrak{PO}^{\mathfrak{b}})} \mathrm{Jac}(\mathfrak{PO}^{\mathfrak{b}};\mathfrak{y}), \tag{71}$$

as a ring.

The factor $\mathrm{Jac}(\mathfrak{PO}^{\mathfrak{b}};\mathfrak{y})$ *in the right hand side is a local ring.*

The ring $\mathrm{Jac}(\mathfrak{PO}^{\mathfrak{b}};\mathfrak{y})$ *is one dimensional if and only if* $\mathfrak{y}$ *is non-degenerate.*

This is a standard result in the case, for example, when the function ($\mathfrak{PO}^{\mathfrak{b}}$ in our case) is a polynomial or a holomorphic function. We can prove Proposition 11.8 in a similar way to those cases. It is proved in [**FOOO5**] section 2.2.

Theorem 11.6 and Proposition 11.8 imply that $(H(X;\Lambda),\cup^{\mathfrak{b}})$ is semi-simple if and only if $\mathfrak{PO}^{\mathfrak{b}}$ is a Morse function.

Theorem 11.6 together with Proposition 11.8 and Formula (70) imply the following:

THEOREM 11.9. 1) *If* $\mathfrak{PO}^{\mathfrak{b}}$ *is a Morse function then*

$$\operatorname{rank} H(X;\mathbb{Q}) = \#\mathfrak{M}(X;\mathfrak{b}).$$

2) *If* $\mathfrak{PO}^{\mathfrak{b}}$ *is not a Morse function then*

$$0 < \#\mathfrak{M}(X;\mathfrak{b}) < \operatorname{rank} H(X;\mathbb{Q}).$$

This is [**FOOO5**] Theorem 1.1.3. Some of the earlier partial results is given in [**FOOO3**] Theorems 1.9 and 1.12.

REMARK 11.10. Theorem 11.9 in particular implies that there exists at least one non-displaceable T^n orbit. This fact also follows from an earlier result by Entov-Polterovich [**EP2, EP3**].

Another application is the following:

THEOREM 11.11. ([**FOOO5**] Theorem 1.1.4.) *Assume* $\mathfrak{b} \in H^2(X;\Lambda_0)$. *The set of eigenvalues of the map* $x \mapsto c_1(X)\cup^{\mathfrak{b}} x : H(X;\Lambda) \to H(X;\Lambda)$ *coincides with the set of critical values of* $\mathfrak{PO}^{\mathfrak{b}}$, *with multiplicities counted.*

REMARK 11.12. Theorem 11.11 was conjectured by M. Kontsevich. See also [**Aur1**].

PROOF. The proof uses the following:

LEMMA 11.13. *Let us consider the situation of Theorem 11.11. Then, by the isomorphism in Theorem 11.6, the first Chern class* $c_1(X) \in H^2(X;\mathbb{C})$ *is sent to the equivalence class of* $\mathfrak{PO}^{\mathfrak{b}}$ *in* $\mathrm{Jac}(\mathfrak{PO}^{\mathfrak{b}})$.

This is [**FOOO5**] Proposition 2.12.1.

Now we consider $x \mapsto c_1(X) \cup^{\mathfrak{b}} x$. We use Thoerem 11.6 and Proposition 11.8 then it is identified to the direct sum of maps

$$[F] \mapsto [\mathfrak{PO}^{\mathfrak{b}} F], \quad \mathrm{Jac}(\mathfrak{PO}^{\mathfrak{b}};\mathfrak{y}) \to \mathrm{Jac}(\mathfrak{PO}^{\mathfrak{b}};\mathfrak{y}).$$

The eigenvalue of this map is $\mathfrak{PO}^{\mathfrak{b}}(\mathfrak{y})$. This implies Theorem 11.11. □

11.4. Construction of the homomorphism $\mathfrak{ks}_{\mathfrak{b}}$. In various applications of Thoerem 11.6 it is also important to know the way how the isomorphism is defined, which we describe in this subsection.

Let $\mathbf{p}_i$ be the basis of $\mathcal{A}$ as in section 8. We write an element $\mathfrak{b} \in \mathcal{A}(\Lambda_0)$ as

$$\mathfrak{b} = \sum_{i=0}^{B} w_i \mathbf{p}_i.$$

We put $\mathfrak{w}_i = e^{w_i}$ for $i = 1,\dots,m$. (Note $\mathbf{p}_i$, $i = 1,\dots,m$ are degree 2 classes.) We define $P_{j_0\cdots j_B}(y)$ by

$$\mathfrak{PO}(\mathfrak{b};y) = \sum_{j_0=0}^{\infty}\cdots\sum_{j_B=0}^{\infty} P_{j_0\cdots j_B}(y) w_0^{j_0}\mathfrak{w}_1^{j_1}\cdots\mathfrak{w}_m^{j_m} w_{m+1}^{j_{m+1}}\cdots w_B^{j_B}. \tag{72}$$

We can show that

$$P_{j_0\cdots j_B}(y) \in T^{\rho_{j_0\cdots j_B}} \Lambda\langle\langle y, y^{-1}\rangle\rangle_0^{\overset{\circ}{P}}$$

with

$$\lim_{j_0+\cdots+j_B\to\infty} \rho_{j_0\cdots j_B} = \infty.$$

Therefore the right hand side of

$$\frac{\partial}{\partial w_i}\mathfrak{PO}(\mathfrak{b};y)$$
$$= \begin{cases} \displaystyle\sum_{j_0=0}^{\infty}\cdots\sum_{j_B=0}^{\infty} j_i P_{j_0\cdots j_B}(y) w_0^{j_0}\cdots w_i^{j_i-1}\cdots w_B^{j_B} & i \ne 1,\dots,m \\ \displaystyle\sum_{j_0=0}^{\infty}\cdots\sum_{j_B=0}^{\infty} j_i P_{j_0\cdots j_B}(y) w_0^{j_0}\cdots \mathfrak{w}_i^{j_i}\cdots w_B^{j_B} & i = 1,\dots,m \end{cases} \tag{73}$$

makes sense and is contained in $\Lambda\langle\langle y, y^{-1}\rangle\rangle^{\overset{\circ}{P}}$ for each $\mathfrak{b} \in \mathcal{A}(\Lambda_0)$.

We define the map

$$\widetilde{\mathfrak{ks}}_{\mathfrak{b}_0} : \mathcal{A}(\Lambda_0) \to \Lambda\langle\langle y, y^{-1}\rangle\rangle_0^{\overset{\circ}{P}}$$

by

$$\widetilde{\mathfrak{ks}}_{\mathfrak{b}_0}(\mathbf{p}_i) = \frac{\partial}{\partial w_i}\mathfrak{PO}(\mathfrak{b}; y)\Big|_{\mathfrak{b}=\mathfrak{b}_0}. \tag{74}$$

THEOREM 11.14. *There exists a Λ_0 module homomorphism $\mathfrak{ks}_{\mathfrak{b}}$ such that the following diagram commutes:*

$$\begin{array}{ccc} \mathcal{A}(\Lambda_0) & \xrightarrow{\widetilde{\mathfrak{ks}}_{\mathfrak{b}}} & \Lambda\langle\langle y, y^{-1}\rangle\rangle_0^{\overset{\circ}{P}} \\ \downarrow & & \downarrow \\ H(X;\Lambda_0) & \xrightarrow{\mathfrak{ks}_{\mathfrak{b}}} & \mathrm{Jac}(\mathfrak{PO}^{\mathfrak{b}}). \end{array} \tag{75}$$

The map $\mathfrak{ks}_{\mathfrak{b}}$ is the isomorphism mentioned in Theorem 11.6.
Theorem 11.14 is [**FOOO5**] Theorem 2.4.1.

SKETCH OF THE PROOF. By definition, we have

$$\mathfrak{PO}(\mathfrak{b}; y) = \sum_{k=0}^{\infty}\sum_{\ell=0}^{\infty}\int_{L(\mathbf{u})} \mathfrak{q}_{\ell;k}(\mathfrak{b}^{\ell}, b^k). \tag{76}$$

Here $b = \sum_{i=1}^n x_i \mathbf{e}_i$ and $y_i = e^{x_i}$. Using $\partial\mathfrak{b}/\partial w_i = \mathbf{p}_i$ we have

$$\frac{\partial \mathfrak{PO}(\mathfrak{b}; y)}{\partial w_i} = \sum_{k=0}^{\infty}\sum_{\ell_1=0}^{\infty}\sum_{\ell_2=0}^{\infty}\int_{L(\mathbf{u})} \mathfrak{q}_{\ell;k}(\mathfrak{b}^{\ell_1}\mathbf{p}_i\mathfrak{b}^{\ell_2}, b^k). \tag{77}$$

The homomorphism

$$\mathbf{p}_i \mapsto \sum_{k=0}^{\infty}\sum_{\ell_1=0}^{\infty}\sum_{\ell_2=0}^{\infty} \mathfrak{q}_{\ell;k}(\mathfrak{b}^{\ell_1}\mathbf{p}_i\mathfrak{b}^{\ell_2}, b^k) \tag{78}$$

induces a homomorphism

$$H(X;\Lambda_0) \to HF((L(\mathbf{u}),(\mathfrak{b}, b)),(L(\mathbf{u}),(\mathfrak{b}, b));\Lambda_0). \tag{79}$$

This fact was proved in [**FOOO1**] Theorem 3.8.62 for arbitrary $L \subset X$.

Note that to define (79) by (78) we fix $\mathfrak{b}$, b and regard the right hand side of (78) as an element of $H(L(\mathbf{u}),\Lambda_0)$. When we define $\widetilde{\mathfrak{ks}}_{\mathfrak{b}}$, we regard $b = \sum_{i=1}^n x_i\mathbf{e}_i$, as a $H(L(\mathbf{u}),\Lambda_0)$ valued function of x_i. So the right hand side of (77) is a function of $y_i = e^{x_i}$.

In other words we need to study the 'family version' of the well-definedness of (79).

We consider the boundary operator

$$a \in H(L(\mathbf{u}),\Lambda_0) \mapsto \mathfrak{m}_1^{\mathfrak{b},b}(a) = \sum_{k_1=0}^{\infty}\sum_{k_2=0}^{\infty}\sum_{\ell=0}^{\infty} \mathfrak{q}_{\ell;k}(\mathfrak{b}^{\ell}, b^{k_1} a b^{k_2}).$$

The well-definedness of (79) means the following Claim 11.15. Let $i^*_{\mathrm{qm},(\mathfrak{b},b)}$ $(\mathbf{p}_i)$ be the right hand side of (78).

CLAIM 11.15. *If* $\sum_{i=0}^{B} c_i\mathbf{p}_i$ *is zero in* $H(X;\Lambda_0)$*, then* $\sum_{i=0}^{B} c_i i^*_{\mathrm{qm},(\mathfrak{b},b)}(\mathbf{p}_i)$ *lies in the image of* $\mathfrak{m}_1^{\mathfrak{b},b}$.

We can prove the same claim when we regard b as a function of x_i. By the proof of Theorem 5.5 (especially by Formula (23)), the image of $\mathfrak{m}_1^{\mathfrak{b},b}$ (where b is regarded as a function of x_i) is in the Jacobian ideal (the ideal generated by $y_i\partial\mathfrak{PO}^{\mathfrak{b}}/\partial y_i$).

Thus the kernel of $\mathcal{A}(\Lambda_0)\to H(X;\Lambda_0)$ is mapped to the Jacobian ideal by $\widetilde{\mathfrak{ks}}_{\mathfrak{b}}$. This implies the theorem. □

Before closing this subsection, we state Theorem 11.17 which is a non-linear version of Theorem 11.14.

The potential function with bulk $\mathfrak{PO}^{\mathfrak{b}}$ is parametrized by $\mathfrak{b}\in\mathcal{A}(\Lambda_0)$. Theorem 11.17 says that it depends only on the cohomology class $\mathfrak{b}$ up to appropriate change of variables. $\Lambda\langle\langle y,y^{-1}\rangle\rangle_+^{\overset{\circ}{P}}$ denotes the set of elements R of $\Lambda\langle\langle y,y^{-1}\rangle\rangle_0^{\overset{\circ}{P}}$ such that $T^{-\epsilon}R\in\Lambda\langle\langle y,y^{-1}\rangle\rangle_0^{\overset{\circ}{P}}$ for some $\epsilon>0$.

DEFINITION 11.16. We consider n elements $y_i'\in\Lambda\langle\langle y,y^{-1}\rangle\rangle^{\overset{\circ}{P}}$ $(i=1,\dots,n)$.

1) We say that $y'=(y_1',\dots,y_n')$ is a *coordinate change converging on* $\mathrm{Int}\,P$ (or a *coordinate change on* $\mathrm{Int}\,P$) if

$$y_i'\equiv c_iy_i \mod y_i\Lambda\langle\langle y,y^{-1}\rangle\rangle_+^{\overset{\circ}{P}} \tag{80}$$

$c_i\in\mathbb{C}\setminus\{0\}$.

2) We say that the coordinate change is *strict* if $c_i=1$ for all i.
3) We say that the *coordinate change converges on* P if $y_i'\in\Lambda\langle\langle y,y^{-1}\rangle\rangle^{P}$ $(i=1,\dots,n)$ in addition. Its strictness is defined in the same way. We also say that y' is a *coordinate change on* P.

The set of all coordinate changes forms a group. It is regarded as a kind of group of self automorphisms of the filtered A_∞ algebra associated to $L(\mathbf{u})$. (The domain of convergence assumed in Definition 11.16 requires that it converges not only by the norm $\mathfrak{v}_T^{\mathbf{u}}$ but also by $\mathfrak{v}_T^{\mathbf{u}'}$ with any $\mathbf{u}'$. This is the reason we write "a kind of" in the above sentence.) A closely related group appears in [**KS2**] and [**GPS**].

THEOREM 11.17. *Let* $\mathfrak{b},\mathfrak{b}'\in\mathcal{A}(\Lambda_0)$. *We assume that* $[\mathfrak{b}]=[\mathfrak{b}']\in H(X;\Lambda_0)$.

Then there exists a coordinate change y' *on* $\mathrm{Int}\,P$*, such that*

$$\mathfrak{PO}^{\mathfrak{b}}(y')=\mathfrak{PO}^{\mathfrak{b}'}(y). \tag{81}$$

If $\mathfrak{b}-\mathfrak{b}'\in\mathcal{A}(\Lambda_+)$*, then* y' *can be taken to be strict.*

If both $\mathfrak{b},\mathfrak{b}'\in\mathcal{A}(\Lambda_+)$*, then* y' *can be taken to be a strict coordinate change on* P.

This is [**FOOO5**] Theorem 8.7.

We remark that $\Lambda\langle\langle y, y^{-1}\rangle\rangle_0^{\overset{\circ}{P}}$ parametrizes the deformation of the potential function. Then the Jacobian ideal corresponds to the part induced by the coordinate change. Thus Theorem 11.17 follows from Theorem 11.14 by some 'integration' (that is solving appropriate ordinary differential equation.) See [**FOOO5**] section 8.

11.5. The homomorphism $\mathfrak{ks}_{\mathfrak{b}}$ is an isomorphism. The main geometric input to the proof of Theorem 11.6 is the following:

THEOREM 11.18. *The map* $\mathfrak{ks}_{\mathfrak{b}} : (H(X;\Lambda_0), \cup^{\mathfrak{b}}) \to \mathrm{Jac}(\mathfrak{PO}^{\mathfrak{b}})$ *is a ring homomorphism.*

Theorem 11.18 is [**FOOO5**] Theorem 2.6.1.

Note this theorem is a version of a result which holds in greater generality. Namely there exists a ring homomorphism

$$QH(X;\Lambda_0) \to HH(Fuk(X,\omega)), \tag{82}$$

where the right hand side is the Hochschild cohomology of the Fukaya category (see [**Fu1, FOOO8**] for its definition.) The existence of such homomorphism was first suggested by [**Ko**] and conjectured explicitly by [**Se3**] etc. See [**FOOO5**] section 4.5 and the reference therein for some of the related works.

We remark that $HH(Fuk(X,\omega))$ parametrizes the deformation of the Lagrangian Floer theory on X. The Jacobian ring $\mathrm{Jac}(\mathfrak{PO}^{\mathfrak{b}})$ parametrizes the deformation of a part of the structures, that is the part described by $\mathfrak{m}_0^b(1)$. So there is a natural ring homomorphism $HH(Fuk(X,\omega)) \to \mathrm{Jac}(\mathfrak{PO}^{\mathfrak{b}})$ in the toric case. Combining them we obtain the ring homomorphism in Theorem 11.18.

More precise and down-to-earth proof of Theorem 11.18 is given as follows.

We recall that the map $\mathfrak{ks}_{\mathfrak{b}} : (H(X;\Lambda_0), \cup^{\mathfrak{b}}) \to \mathrm{Jac}(\mathfrak{PO}^{\mathfrak{b}})$ is induced from the map

$$\mathbf{p}_i \mapsto \sum_{k=0}^{\infty}\sum_{\ell_1=0}^{\infty}\sum_{\ell_2=0}^{\infty}\int_{L(\mathbf{u})} \mathfrak{q}_{\ell_1+\ell_2+1;k}(\mathfrak{b}^{\ell_1}\mathbf{p}_i\mathfrak{b}^{\ell_2}, b^k) : \mathcal{A} \to \Lambda\langle\langle y, y^{-1}\rangle\rangle_0^P. \tag{83}$$

(See (78).) Note $b = \sum x_i \mathbf{e}_i$ and the right hand side is a function of x_i. It then turns out to be a function of $y_i^{\mathbf{u}} = e^{x_i}$. Moreover by changing the variables to y_i by the formula $y_i = T^{u_i} y_i^{\mathbf{u}}$, the right hand side becomes a function of y_i and is an element of $\Lambda\langle\langle y, y^{-1}\rangle\rangle_0^P$.

We consider the case $\mathfrak{b} = \mathbf{0}$ for simplicity.

We consider the moduli space $\mathcal{M}_{k+1;2}(\beta)$ of J-holomorphic disks with $k+1$ boundary and ℓ interior marked points, (See subsection 2.2.) and take a fiber product

$$\mathcal{M}_{k+1;2}^{\mathrm{main}}(\beta)_{(\mathrm{ev}_1^+,\mathrm{ev}_2^+)} \times (\mathbf{p} \times \mathbf{p}')$$

where $\mathbf{p}, \mathbf{p}' \in \mathcal{A}$. We denote this fiber product by

$$\mathcal{M}^{\text{main}}_{k+1;2}(\beta; \mathbf{p}, \mathbf{p}').$$

Let $\mathcal{M}_{1;2}$ be the moduli space of bordered Riemann surface of genus 0 with two interior and one boundary marked points. This moduli space is a two dimensional disk. We consider two points $[\Sigma_1], [\Sigma_2] \in \mathcal{M}_{1;2}$ as in the figure below.

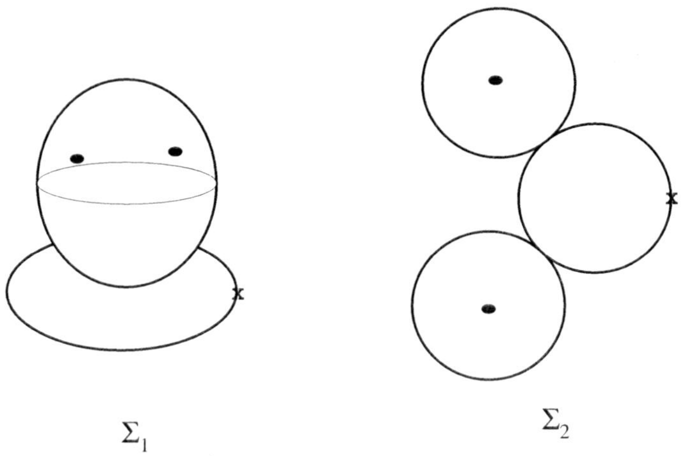

FIGURE 11.1

We have a forgetful map

$$\mathfrak{forget} : \mathcal{M}^{\text{main}}_{k+1;2}(\beta) \to \mathcal{M}_{1;2}. \tag{84}$$

Namely we put

$$\mathfrak{forget}([\Sigma; z_0, \dots, z_k, z_1^+, z_2^+, u]) = [\Sigma; z_0; z_1^+, z_2^+].$$

It induces a map

$$\mathfrak{forget} : \mathcal{M}^{\text{main}}_{k+1;2}(\beta; \mathbf{p}, \mathbf{p}') \to \mathcal{M}_{1;2}.$$

For $i = 1, 2$, we denote by

$$\mathcal{M}^{\text{main}}_{k+1;2}(\beta; \mathbf{p}, \mathbf{p}'; \Sigma_i)$$

the inverse image of $\{[\Sigma_i]\}$ in $\mathcal{M}^{\text{main}}_{k+1;2}(\beta; \mathbf{p}, \mathbf{p}')$.

Let $h_j \in H^1(L(\mathbf{u}); \mathbb{C})$ $(j = 1, \dots, k)$. (Note we identify the cohomology group with the set of T^n invariant forms.) We pull back $h_1 \times \cdots \times h_k$ to $\mathcal{M}^{\text{main}}_{k+1;2}(\beta; \mathbf{p}, \mathbf{p}'; \Sigma_i)$ by $(\text{ev}_1, \dots, \text{ev}_k)$ and consider the integration along fiber by ev_0. We denote it by

$$\text{Corr}(h_1 \times \cdots \times h_k; \mathcal{M}^{\text{main}}_{k+1;2}(\beta; \mathbf{p}, \mathbf{p}'; \Sigma_i)).$$

More precisely we take a T^n invariant multisection $\mathfrak{s}$ so that the zero set $\mathcal{M}^{\text{main}}_{k+1;2}(\beta; \mathbf{p}, \mathbf{p}'; \Sigma_i)^{\mathfrak{s}}$ is transversal to zero. Then integration along the fiber is well-defined. This is because ev_0 on $\mathcal{M}^{\mathrm{main}}_{k+1;2}(\beta;\mathbf{p},\mathbf{p}';\Sigma_i)^{\mathfrak{s}}$ must become a submersion by the T^n equivariance.

We put

$$\begin{aligned}&\mathrm{Corr}(h_1\times\cdots\times h_k;\mathcal{M}^{\mathrm{main}}_{k+1;2}(\mathbf{p},\mathbf{p}';\Sigma_i))\\&=\sum_{\beta}T^{(\beta\cap\omega)/2\pi}\mathrm{Corr}(h_1\times\cdots\times h_k;\mathcal{M}^{\mathrm{main}}_{k+1;2}(\beta;\mathbf{p},\mathbf{p}';\Sigma_i))\end{aligned}$$

and extend $\mathrm{Corr}(\cdots;\mathcal{M}^{\mathrm{main}}_{k+1;2}(\mathbf{p},\mathbf{p}';\Sigma_i))$ to

$$H^1(L(\mathbf{u});\Lambda_0)^{\otimes k}\to\Lambda_0.$$

We then can prove the following two formulas:

$$\mathrm{Corr}(\underbrace{b,\dots,b}_{k};\mathcal{M}^{\mathrm{main}}_{k+1;2}(\mathbf{p},\mathbf{p}';\Sigma_1))=\mathfrak{q}_{1;k}(\mathbf{p}\cup^{Q}\mathbf{p}';b^k).\tag{85}$$

$$\mathrm{Corr}(\underbrace{b,\dots,b}_{k};\mathcal{M}^{\mathrm{main}}_{k+1;2}(\mathbf{p},\mathbf{p}';\Sigma_2))=\sum_{k_1+k_2=k}\mathfrak{q}_{1;k_1}(\mathbf{p};b^{k_1})\mathfrak{q}_{1;k_2}(\mathbf{p}';b^{k_2}).\tag{86}$$

Note the sum over k of the right hand sides of (85) and (86) are

$$\mathfrak{ks}_{\mathbf{0}}(\mathbf{p}\cup^{Q}\mathbf{p}')\quad\text{and}\quad\mathfrak{ks}_{\mathbf{0}}(\mathbf{p})\mathfrak{ks}_{\mathbf{0}}(\mathbf{p}')$$

respectively. (Note we are studying the case $\mathfrak{b}=\mathbf{0}$.)

We finally use cobordism argument to show that the left hand side of (85) coincides with the left hand side of (86) modulo elements in the Jacobian ideal. This is an outline of the proof of Theorem 11.18. See [**FOOO5**] section 2.6 for detail.

The outline of the rest of the proof of Theorem 11.6 is as follows.

We first prove the surjectivity of $\mathfrak{ks}_{\mathfrak{b}}$. For this purpose we consider the map obtained by reducing the coefficient to $\mathbb{C}=\Lambda_0/\Lambda_+$. Then the quantum cohomology of the domain becomes ordinary cohomology. We can calculate the $\mathbb{C}=\Lambda_0/\Lambda_+$ reduction of the Jacobian ring using Cho-Oh's result (namely by studying the leading order term $z_1+\cdots+z_m$. See Theorem 8.2.) Then the $\mathbb{C}$-reduction of $\mathfrak{ks}_{\mathfrak{b}}$ is an isomorphism by a classical result of Stanley which calculates the cohomology ring of toric manifold. (See for example [**Ful**].) It implies that $\mathfrak{ks}_{\mathfrak{b}}$ itself is surjective.

We remark that the fact that $\mathbb{C}$-reduction of $\mathfrak{ks}_{\mathfrak{b}}$ is an isomorphism does not imply that $\mathfrak{ks}_{\mathfrak{b}}$ is isomorphism. In fact we need to eliminate the possibility that $\mathrm{Jac}(\mathfrak{PO}^{\mathfrak{b}})$ has a component such as $\Lambda_0/(T^{\lambda})$. Note that the (quantum) cohomology $H(X;\Lambda_0)$ is a free Λ_0 module. Therefore to prove the injectivity of $\mathfrak{ks}_{\mathfrak{b}}$ and complete the proof of Theorem 11.6 it suffices to prove the following inequality.

$$\mathrm{rank}_{\Lambda}(\mathrm{Jac}(\mathfrak{PO}^{\mathfrak{b}})\otimes_{\Lambda_0}\Lambda)\geq\mathrm{rank}_{\mathbb{Q}}H(X;\mathbb{Q}).\tag{87}$$

We remark that in many explicit examples we can prove the equality (87) directly by finding critical points of $\mathfrak{PO}^{\mathfrak{b}}$, for example by solving leading term equation. However the proof of (87) is in general more involved, which we briefly describe now. We consider the case $\mathfrak{b} = \mathbf{0}$, for simplicity.

We prove (87) in two steps. We first use a result of McDuff-Tolman [**MT**] (which is based on Seidel's work [**Se1**]), to find elements $z'_1, \dots, z'_m \in QH(X; \Lambda_0)$ with the following properties.

1) $z'_1, \dots, z'_m$ satisfies quantum Stanley-Reisner relation.
2) There exists $\mathfrak{P}_i(Z_1, \dots, Z_m) = \sum_{j=1}^m v_{j,i} Z_i + \sum_{k=1}^\infty T^{\rho_k} P_{i,k}(Z_1, \dots, Z_m)$ such that

$$\mathfrak{P}_i(z'_1, \dots, z'_m) = 0 \tag{88}$$

   and $\rho_k \to \infty$, $\rho_k > 0$, $P_{i,k} \in \mathbb{C}[Z_1, \dots, Z_m]$. (We recall $d\ell_j = (v_{j,1}, \dots, v_{j,n}) \in \mathbb{Z}^n$.)
3) The relations in the above (1),(2) are all the relations among z'_i. Moreover z'_i generates $QH(X; \Lambda_0)$.

Let us explain the above statement briefly. By putting $Z_i = T^{\lambda_i} y_1^{v_{i,1}} \cdots y_n^{v_{i,n}}$ we obtain a surjective ring homomorphism

$$\Lambda[Z_1, \dots, Z_m] \to \Lambda[y_1, y_1^{-1}, \dots, y_n, y_n^{-1}].$$

The quantum Stanley-Reisner relations are the generators of the kernel of this homomorphism. (See [**FOOO3**] Definition 6.4.) The quantum Stanley-Reisner relation appeared in the Batyrev's work on quantum cohomology of toric manifold and is given explicitly by using moment polytope P.

We put $z_i = T^{\lambda_i} y_1^{v_{i,1}} \cdots y_n^{v_{i,n}} \in \mathrm{Jac}(\mathfrak{PO}^{\mathbf{0}})$. Then (47) implies that it satisfies the formula

$$\sum_{j=1}^m v_{j,i} z_i + \sum_{k=1}^\infty T^{\rho_k} \frac{\partial P_k}{\partial x_i}(z_1, \dots, z_m) = 0. \tag{89}$$

(Note we put $y_i = e^{x_i}$ so (89) is $\frac{\partial P_k}{\partial x_i} = y_i \frac{\partial P_k}{\partial y_i}$.) We remark that the first term of the left hand side of (89) is

$$\frac{\partial \mathfrak{PO}^{\mathbf{0}}}{\partial x_i}(z_1, \dots, z_m).$$

We also remark that the left hand side of (88) is similar to (89). Namely their leading order terms coincide.

The element z'_i is the invariant of [**Se1**] associated to the Hamiltonian S^1 action. Here S^1 is the component of T^n which fixes D_i. The fact that they satisfy the quantum Stanley-Reisner relation is proved in [**MT**] using the relation between those S^1 actions and basic properties of Seidel invariant. The property (2) can be proved using the fact $z'_i \equiv [D_i] \mod \Lambda_+$.

Let $(QSR) \subset \Lambda_0\langle\langle Z_1, \ldots, Z_m\rangle\rangle$ be the ideal generated by the quantum Stanley-Reisner relations. Then (1), (2) above imply the existence of homomorphism

$$\frac{\Lambda_0\langle\langle Z_1, \ldots, Z_m\rangle\rangle}{\mathrm{Clos}((QSR) \cup \{\mathfrak{P}_i : i = 1, \ldots, m\})} \to QH(X; \Lambda_0). \tag{90}$$

Here Clos means a closure with respect to an appropriate topology. By reduction to $\mathbb{C} = \Lambda_0/\Lambda_+$ we can show that (90) is an isomorphism. (We use the fact that $QH(X; \Lambda_0)$ is torsion free here.)

Now the proof of (87) goes as follows. For $\mathfrak{s} \in \Lambda$ we put

$$\mathfrak{P}_i^{\mathfrak{s}} = \mathfrak{s}\frac{\partial \mathfrak{PO}^{\mathbf{0}}}{\partial x_i} + (1 - \mathfrak{s})\mathfrak{P}_i.$$

We remark $\mathfrak{P}_i^{\mathfrak{s}}$ has the form

$$\mathfrak{P}_i^{\mathfrak{s}}(Z_1, \ldots, Z_m) = \sum_{j=1}^{m} v_{j,i} Z_i + \sum_{k=1}^{\infty} T^{\rho_k} P_{i,k}^{\mathfrak{s}}(Z_1, \ldots, Z_m).$$

We define the ring $\mathfrak{R}_{\mathfrak{s}}$ by

$$\mathfrak{R}_{\mathfrak{s}} = \frac{\Lambda_0\langle\langle Z_1, \ldots, Z_m\rangle\rangle}{\mathrm{Clos}((QSR) \cup \{\mathfrak{P}_i^{\mathfrak{s}} : i = 1, \ldots, m\})} \otimes_{\Lambda_0} \Lambda.$$

We have

$$\mathfrak{R}_0 \cong QH(X; \Lambda)$$

since (90) is an isomorphism. On the other hand

$$\mathfrak{R}_1 \cong \mathrm{Jac}(\mathfrak{PO}^{\mathbf{0}}) \otimes_{\Lambda_0} \Lambda.$$

Thus it suffices to show that $\dim_\Lambda \mathfrak{R}_{\mathfrak{s}}$ is independent of $\mathfrak{s}$. We regard $\cup_{\mathfrak{s}\in\Lambda}$ $Spec(\mathfrak{R}_{\mathfrak{s}})$ as a family of affine schemes parametrized by $\mathfrak{s} \in \Lambda$. If we can prove that this family is flat and proper then the independence of $\dim_\Lambda \mathfrak{R}_{\mathfrak{s}}$ is a standard result of algebraic geometry.

We prove the properness using the fact that the valuation of the solution of the equation $\mathfrak{P}_1^{\mathfrak{s}} = \cdots = \mathfrak{P}_m^{\mathfrak{s}} = 0$ can not escape from moment polytope. The flatness is a consequence of the fact that our scheme is a local complete intersection and also of standard facts about the regular sequence of Cohen-Macauley ring.

In general $\mathfrak{P}_i^{\mathfrak{s}}$ is an infinite series rather than a polynomial. So we first need to change the coordinate y_i so that $\mathfrak{P}_i^{\mathfrak{s}}$ becomes a polynomial. Such a process is known in algebraic geometry as a algebraization of singularity. See [**FOOO5**] section 2.9.

This is an outline of the proof of (87). See [**FOOO5**] especially its section 2.11 for details. □

REMARK 11.19. We regard

$$\bigcup_{\mathfrak{b}\in H(X;\Lambda_0)} Spec(\mathfrak{PO}^{\mathfrak{b}}) \tag{91}$$

as a $H(X;\Lambda_0)$ parametrized 'family of schemes'[4]

The same argument to show the flatness and properness of the family $\bigcup_{\mathfrak{s}\in\Lambda} Spec(\mathfrak{R}_{\mathfrak{s}})$ seems to be applicable to show that the family (91) is also flat and proper.

In the study of K. Saito theory of Laurent polynomials (such as one described in [**Sab**]), the properness of the family of the critical point sets is an important issue. When one works over $\mathbb{C}$ the properness is not necessarily satisfied. When we work over a Novikov ring in place of $\mathbb{C}$, properness of the family of the critical point sets (that is the geometric points of $Spec(\mathfrak{PO}^{\mathfrak{b}})$) is always satisfied at least for the potential function appearing as the mirror of a toric manifold. The authors believe that this is an important advantage of working with Novikov ring over working with $\mathbb{C}$.

REMARK 11.20. Let us consider the family (91). For the $H^2(X;\Lambda_0)$ part of $\mathfrak{b}$ it is natural to replace the coordinate w_i by its exponential $\mathfrak{w}_i = e^{w_i}$. Then we may extend the domain $\{\mathfrak{w}_i \mid \mathfrak{v}_T(\mathfrak{w}_i) = 0\}$ to $\mathfrak{w}_i \in \Lambda$. Note in $\mathfrak{PO}^{\mathfrak{b}}$ the leading order term is $\sum \mathfrak{w}_i z_i$. So if we extend $\mathfrak{w}_i$ and allow for example $\mathfrak{w}_i = T^c$, we have a term $T^c z_i$. We may regard this insertion $\mathfrak{w}_i = T^c$ as changing the moment polytope. Namely appearance of the term $T^c z_i$ is equivalent to moving $\partial_i P = \{\mathbf{u} \mid \ell_i(\mathbf{u}) = 0\}$ to $\{\mathbf{u} \mid \ell_i(\mathbf{u}) = -c\}$[5].

Thus for this extended family the flatness and properness still hold as far as the corresponding moment polytope is combinatorially equivalent to the original one.

There is some flavor of this kind of arguments in [**FOOO5**] subsection 2.11.2.

12. Poincaré duality and Residue pairing

In this section we explain that the isomorphism in Theorem 11.6 can be enhanced to give an isomorphism between two Frobenius manifold structures.

12.1. Big quantum cohomology and Frobenius manifold.

DEFINITION 12.1. A *Frobenius manifold structure* on a manifold M is a quintet $(\langle\cdot\rangle, \nabla, \circ, e, \Phi)$ with the following properties.

---

[4]It is proved in [**FOOO5**] that each of $\mathfrak{PO}^{\mathfrak{b}}$ can be transformed to a Laurent polynomial by change of variables. So we can define its *Spec*. It is not verified that the whole family can be regarded to be a scheme. So we put quotation mark.

[5]In other words the parameter $\mathfrak{v}_T(\mathfrak{w}_i)$ corresponds to the Kähler cone of our toric manifold X. This is similar to the fact that the valuation of $\mathfrak{y}_i$ corresponds to the parameter $\mathbf{u}$ of the Lagrangian submanifold $L(\mathbf{u})$

1) $\langle \cdot \rangle$ is a non-degenerate inner product on the tangent bundle TM.
2) ∇ is a connection of TM.
3) ∇ is a metric connection. Namely :

$$X\langle Y, Z\rangle = \langle \nabla_X Y, Z\rangle + \langle Y, \nabla_X Z\rangle.$$

4) ∇ is flat and torsion free. Namely :

$$\begin{aligned} \nabla_X \nabla_Y - \nabla_Y \nabla_X - \nabla_{[X,Y]} &= 0, \\ \nabla_X Y - \nabla_Y X - [X, Y] &= 0. \end{aligned}$$

5) $\circ$ defines a ring structure on T_pM which depends smoothly on p and satisfies

$$\langle X \circ Y, Z\rangle = \langle X, Y \circ Z\rangle. \tag{92}$$

An associative algebra with unit which satisfies (92) is called a *Frobenius algebra.*

6) e is a section of TM such that $e(p)$ is the unit of $(T_pM, \circ, +)$ for each p. Moreover

$$\nabla e = 0.$$

7) Φ is a function on M such that

$$\left\langle \frac{\partial}{\partial x_i} \circ \frac{\partial}{\partial x_j}, \frac{\partial}{\partial x_k} \right\rangle = \frac{\partial^3 \Phi}{\partial x_i \partial x_j \partial x_k}. \tag{93}$$

Here x_i $(i = 1, \ldots, \dim M)$ is a local coordinate of M such that $\nabla_{\frac{\partial}{\partial x_i}} (\frac{\partial}{\partial x_j}) = 0$. We call Φ the *potential.*

In some case we have a vector field $\mathfrak{E}$ on M that satisfies the following

$$\begin{aligned} \mathfrak{E}\langle X, Y\rangle - \langle [\mathfrak{E}, X], Y\rangle - \langle X, [\mathfrak{E}, Y]\rangle &= d_1\langle X, Y\rangle, \\ [\mathfrak{E}, X \circ Y] - [\mathfrak{E}, X] \circ Y - X \circ [\mathfrak{E}, Y] &= d_2 X \circ Y, \\ [\mathfrak{E}, e] &= d_3 e, \end{aligned} \tag{94}$$

where $d_1, d_2, d_3 \in \mathbb{Q}$. We call $\mathfrak{E}$ the *Euler vector field.*

REMARK 12.2. In various situations where a Frobenius manifold arises the tangent space T_pM appears as either a $\mathbb{C}$ vector space or a Λ vector space. In that case the inner product $\langle \cdot \rangle$ is bilinear over $\mathbb{C}$ or Λ. (In this case $\langle \cdot \rangle$ is required to be complex symmetric *not hermitian.*) Moreover Φ is a $\mathbb{C}$ or Λ valued function.

We do not try to define what connection, function, coordinate etc. mean in case TM is a Λ vector space. At the present stage of development, we do not meet the situation where we need to seriously study it. In the main example of our consideration, M is a Λ_0 affine space, hence we can easily make sense out of them.

This structure first appeared in K. Saito's work [**Sa**] (see the next subsection). Dubrovin [**Dub**] discovered this structure in Gromov-Witten theory, which we recall below.

Let X be a symplectic manifold. We take $M = H^{evev}(X;\Lambda_0)$ the even degree cohomology group of X with Λ_0 coefficients. (We may include odd degree part by regarding X as a supermanifold. Since in the case of our main interest (toric manifold), there is no cohomology class of odd degree, we do not discuss odd degree part.)

In subsection 11.2 we associate a deformed quantum cup product $\cup^{\mathfrak b}$ on $H(X;\Lambda)$ for each $\mathfrak b \in H^{even}(X;\Lambda_0)$. We regard $T_{\mathfrak b}M = H(X;\Lambda)$ and put $\circ = \cup^{\mathfrak b}$ there. It is associative.

REMARK 12.3. Note $H^{even}(X;\Lambda_0)$ is not an open set of $H^{even}(X;\Lambda)$. So $T_{\mathfrak b}H^{even}(X;\Lambda_0) = H(X;\Lambda)$ does not make sense in a usual sense of manifold. This is regarded only as a convention here.

We have Poincaré duality pairing

$$H^d(X;\Lambda) \otimes_\Lambda H^{2n-d}(X;\Lambda) \to \Lambda.$$

The inner product $\langle\cdot\rangle$ is the Poincaré duality pairing. We remark that then the Levi-Civita connection, that is the connection which is a torsion free metric connection of the metric $\langle\cdot\rangle$, is the standard affine connection of the vector space $H^{even}(X;\Lambda_0)$. It is obviously flat.

(92) follows from

$$\langle \mathfrak c \cup^{\mathfrak b} \mathfrak d, \mathfrak e\rangle_{\mathrm{PD}_X} = \sum_{\ell=0}^{\infty} \frac{1}{\ell!} GW_{\ell+3}(\mathfrak c, \mathfrak d, \mathfrak e, \mathfrak b, \dots, \mathfrak b).$$

(See (69).) and the fact that $GW_\ell(Q_1,\dots,Q_\ell)$ is independent of the permutation of Q_i.

The element e is the unit of the cohomology group that is the Poincaré dual to the fundamental homology class $[X]$.

The potential Φ is defined by

$$\Phi(\mathfrak b) = \sum_{\ell=0}^{\infty} \frac{1}{\ell!} GW_\ell(\mathfrak b,\dots,\mathfrak b) \tag{95}$$

for which the formula (93) can be easily checked. The potential Φ in (95) is called the *Gromov-Witten potential.*

The Euler vector field $\mathfrak E$ is defined by the vector field:

$$\mathfrak E = \frac{\partial}{\partial w_0} + \sum_{i=1}^{m} r_i \frac{\partial}{\partial \mathfrak w_i} + \sum_{i=m+1}^{B} \left(1 - \frac{\deg \mathbf p_i}{2}\right) w_i \frac{\partial}{\partial w_i}, \tag{96}$$

where $c_1(X) = \sum_{i=1}^m r_i \mathbf p_i$. We remark that $\mathbf p_i$, $i = 0,\dots,B$ are basis of H^{even} $(X;\mathbb Q)$ such that $\deg \mathbf p_0 = 0$, $\deg \mathbf p_i = 2$ for $i = 1,\dots,m$ and $\deg \mathbf p_i > 2$ for $i > m$.

By using the dimension formula

$$\dim_{\mathbb{C}} \mathcal{M}_\ell(\alpha) = n + \ell - 3 + c_1(X) \cap \alpha$$

of the moduli space $\mathcal{M}_\ell(\alpha)$ of pseudo-holomorphic spheres with ℓ interior marked points and of homology class α, we can prove (94), where $d_1 = 2-n$, $d_2 = 1$, $d_3 = 0$. Thus we have:

THEOREM 12.4. (Dubrovin) $(\langle\cdot\rangle, \nabla, \cup^{\mathfrak{b}}, \Phi, e)$ *is a structure of Frobenius manifold on* $H(X;\Lambda_0)$. (96) *is its Euler vector field.*

12.2. A fragment of K. Saito theory. Let

$$F(x_1,\dots,x_n; w_0, w_1, \dots, w_B) : U \times V \to \mathbb{C} \tag{97}$$

be a holomorphic function on $U \times V \subset \mathbb{C}^n \times \mathbb{C}^{B+1}$. Here U and V are small neighborhoods of origin in $\mathbb{C}^n$ and $\mathbb{C}^{B+1}$, respectively.

We assume F is of the form

$$F(x_1,\dots,x_n; w_0, w_1, \dots, w_B) = w_0 + F(x_1,\dots,x_n; 0, w_1, \dots, w_B).$$

We put

$$F^{\vec{w}}(x_1,\dots,x_n) = F(x_1,\dots,x_n; w_0, w_1, \dots, w_B),$$

for $\vec{w} = (w_0,\dots,w_B)$. We assume that $F^{\vec{0}}(x_1,\dots,x_n)$ has $\vec{x} = \vec{0}$ as an isolated critical point. Namely $(dF^{\vec{0}})(0,\dots,0) = 0$, and $(dF^{\vec{0}})(\vec{x}) \neq 0$ for $\vec{x} \in U \backslash \{\vec{0}\}$.

DEFINITION 12.5. We define the *Jacobian ring* $\mathrm{Jac}(F^{\vec{w}})$ by

$$\mathrm{Jac}(F^{\vec{w}}) = \frac{\mathcal{O}(U)}{\left(\frac{\partial F^{\vec{w}}}{\partial x_i}; i = 1,\dots,n\right)}. \tag{98}$$

Here $\mathcal{O}(U)$ is the ring of holomorphic functions on U and the denominator is its ideal generated by $\frac{\partial F^{\vec{w}}}{\partial x_i}$, $i = 1,\dots,n$.

We define the *Kodaira-Spencer map* $\mathfrak{ks}_{\vec{w}} : T_{\vec{w}}V \to \mathrm{Jac}(F^{\vec{w}})$ by

$$\mathfrak{ks}_{\vec{w}}\left(\frac{\partial}{\partial w_i}\right) \equiv \frac{\partial F}{\partial w_i}(x_1,\dots,x_n;\vec{w}) \in \mathrm{Jac}(F^{\vec{w}}). \tag{99}$$

F is called a *universal unfolding* of $F^{\vec{0}}$ if $\mathfrak{ks}_{\vec{0}} : T_{\vec{0}}V \to \mathrm{Jac}(F^{\vec{0}})$ is an isomorphism.

We remark that if F is a universal unfolding of $F^{\vec{0}}$ then by shrinking V if necessary we may assume that $\mathfrak{ks}_{\vec{w}}$ is an isomorphism for any $\vec{w} \in V$. We assume it in the rest of this subsection.

We remark that $\mathrm{Jac}(F^{\vec{w}})$ is a ring. On the other hand $T_{\vec{w}}V$ does not have a ring structure a priori. We *define*

$$X \circ Y = (\mathfrak{ks}_{\vec{w}})^{-1}(\mathfrak{ks}_{\vec{w}}(X)\mathfrak{ks}_{\vec{w}}(Y)), \tag{100}$$

for $X, Y \in T_{\vec{w}}V$. Thus $(T_{\vec{w}}V, \circ, +)$ forms a ring. Note $\partial/\partial w_0 \in T_{\vec{w}}V$ is sent to $[1] \in \mathrm{Jac}(F^{\vec{w}})$. Therefore

$$e(\vec{w}) = \partial/\partial w_0 \in T_{\vec{w}}V$$

is a unit.

THEOREM 12.6. (K.Saito-M.Saito) *There exists a $\mathbb{C}$ valued metric $\langle\cdot\rangle$ on TV, its Levi-Civita connection ∇ and a holomorphic function $\Phi : V \to \mathbb{C}$ such that $(\langle\cdot\rangle, \circ, e, \nabla, \Phi)$ is a Frobenius manifold.*

K. Saito [**Sa**] constructed a Frobenius manifold structure assuming the existence of a primitive form. We do not explain the notion of primitive form here. (See [**SaTa**] for its description in a way closely related to the discussion here.) Existence of primitive form for a universal unfolding of a germ of isolated singularity is established in [**MSa**]. We remark that Theorem 12.6 had been proved before Gromov-Witten theory started.

The metric $\langle\cdot\rangle$ is called a *residue paring.* Since ∇ is flat there exists a local coordinate $t_0, t_1, \dots, t_B$ of V so that $\nabla_{\partial/\partial t_i}(\partial/\partial t_j) = 0$. Such a coordinate $(t_0, t_1, \dots, t_B)$ is called a *flat coordinate.* $(t_0 = w_0.)$

For some F associated to an ADE singularity, the primitive form takes a simple form $dx_1 \wedge dx_2 \wedge dx_3$. In such a case we have the following description of the residue pairing.

We put

$$\mathrm{Crit}(F^{\vec{w}}) = \{\mathfrak{y} \in U \mid dF^{\vec{w}}(\mathfrak{y}) = 0\}.$$

Let $\mathcal{O}_{\mathfrak{y}}$ be the ring of germs of holomorphic functions at $\mathfrak{y} \in U$. We put

$$\mathrm{Jac}(F^{\vec{w}}; \mathfrak{y}) = \frac{\mathcal{O}_{\mathfrak{y}}}{\left(\frac{\partial F^{\vec{w}}}{\partial x_i}; i = 1, \dots, n\right)}. \tag{101}$$

The following fact is standard:

PROPOSITION 12.7. *We have*

$$\mathrm{Jac}(F^{\vec{w}}) \cong \prod_{\mathfrak{y} \in \mathrm{Crit}(F^{\vec{w}})} \mathrm{Jac}(F^{\vec{w}}; \mathfrak{y}).$$

$\mathrm{Jac}(F^{\vec{w}}; \mathfrak{y})$ is one dimensional if and only if the critical point $\mathfrak{y}$ is non-degenerate.

Let $\vec{w}$ be a vector such that $F^{\vec{w}}$ is a Morse function. Let $1_{\mathfrak{y}} \in \mathrm{Jac}(F^{\vec{w}}; \mathfrak{y})$ be the unit. Then Proposition 12.7 implies that $\{1_{\mathfrak{y}} \mid \mathfrak{y} \in \mathrm{Crit}(F^{\vec{w}})\}$ forms a $\mathbb{C}$ basis of the vector space $\mathrm{Jac}(F^{\vec{w}})$. If $\mathfrak{y} \neq \mathfrak{y}'$ we obtain

$$\langle 1_{\mathfrak{y}}, 1_{\mathfrak{y}'} \rangle = \langle 1_{\mathfrak{y}}, 1_{\mathfrak{y}'} \circ 1 \rangle = \langle 1_{\mathfrak{y}} \circ 1_{\mathfrak{y}'}, 1 \rangle = 0,$$

from the equation $1_{\mathfrak{y}} \circ 1_{\mathfrak{y}'} = 0$ and (92). Namely $\{1_{\mathfrak{y}} \mid \mathfrak{y} \in \mathrm{Crit}(F^{\vec{w}})\}$ is an orthogonal basis with respect to the residue pairing.

LEMMA 12.8. *If the primitive form is* $dx_1 \wedge \cdots \wedge dx_n$ *and* $F^{\vec{w}}$ *is a Morse function then we have*

$$\langle 1_{\mathfrak{y}}, 1_{\mathfrak{y}} \rangle = \left(\det \left[\frac{\partial^2 F^{\vec{w}}}{\partial x_i \partial x_j} \right]_{i=1,j=1}^{i=n,j=n} (\mathfrak{y}) \right)^{-1}.$$

This lemma follows from the definition. We remark that in general the primitive form is not necessarily equal to $dx_1 \wedge \cdots \wedge dx_n$.

12.3. Residue pairing on $\mathrm{Jac}(\mathfrak{PO}^{\mathfrak{b}})$**.** We now consider the case $F(x_1, \ldots, x_n, \vec{w}) = \mathfrak{PO}^{\mathfrak{b}}(y_1, \ldots, y_n)$ where $\mathfrak{b} = \sum w_i \mathbf{p}_i$ and $e^{x_i} = y_i$.

We however remark that our situation is different from that of subsection 12.2 in the following two points.

1) The tangent space $T_{\mathfrak{b}}(H(X; \Lambda_0))$ is a Λ vector space and is not a $\mathbb{C}$ vector space.
2) The 'open set' on which $\mathfrak{PO}^{\mathfrak{b}}$ is defined is the set $\mathfrak{A}(\overset{\circ}{P})$ which is not a 'small' neighborhood of a point.

However, many parts of the story are directly translated to the case $\mathfrak{PO}^{\mathfrak{b}}$. (See however Remark 12.26.) Note V in subsection 12.2 corresponds to $H(X; \Lambda_0)$.

In this subsection we describe a pairing on $\mathrm{Jac}(\mathfrak{PO}^{\mathfrak{b}})$ which we expect to be the version of residue pairing in our situation.

DEFINITION 12.9. Let C be a $\mathbb{Z}_2$ graded finitely generated free Λ module. A structure of *unital Frobenius algebra* of dimension n is $\langle \cdot, \cdot \rangle : C^k \otimes C^{n-k} \to \Lambda$, $\cup : C^k \otimes C^\ell \to C^{k+\ell}$, $1 \in C^0$, such that:

1) $\langle \cdot, \cdot \rangle$ is a graded symmetric bilinear form which induces an isomorphism $x \mapsto (y \mapsto \langle x, y \rangle)$, $C^k \to Hom_\Lambda(C^{n-k}, \Lambda)$.
2) $\cup$ is an associative product on C. 1 is its unit.
3) $\langle x \cup y, z \rangle = \langle x, y \cup z \rangle$.

The cohomology group of an oriented closed manifold becomes a unital Frobenius algebra in an obvious way.

DEFINITION 12.10. *Let* $(C, \langle \cdot, \cdot \rangle, \cup, 1)$ *be a unital Frobenius algebra. We take a basis* $\mathbf{e}_I$, $I \in \mathfrak{I}$ *of* C *such that* $\mathbf{e}_0$ *is the unit. Let* $g_{IJ} = \langle \mathbf{e}_I, \mathbf{e}_J \rangle$ *and let* g^{IJ} *be its inverse matrix. We define an invariant of* C *by*

$$\begin{aligned} Z(C) = \sum_{I_1, I_2, I_3 \in \mathfrak{I}} \sum_{J_1, J_2, J_3 \in \mathfrak{I}} & (-1)^* g^{I_1 J_1} g^{I_2 J_2} g^{I_3 0} g^{J_3 0} \\ & \times \langle \mathbf{e}_{I_1} \cup \mathbf{e}_{I_2}, \mathbf{e}_{I_3} \rangle \langle \mathbf{e}_{J_1} \cup \mathbf{e}_{J_2}, \mathbf{e}_{J_3} \rangle \end{aligned} \tag{102}$$

where $* = \deg \mathbf{e}_{I_1} \deg \mathbf{e}_{J_2} + \frac{n(n-1)}{2}$*. We call* $Z(C)$ *the* trace *of unital Frobenius algebra* C*.*

It is straightforward to check that $Z(C)$ is independent of the choice of the basis. This invariant is an example of 1-loop partition function and can be described by the following Feynman diagram.

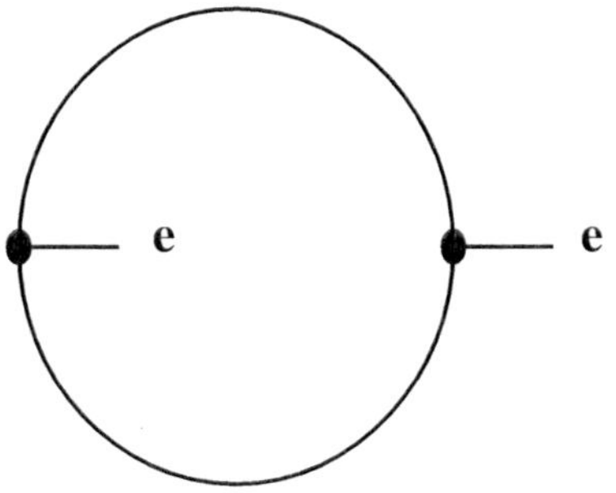

FIGURE 12.1

Let us consier $\mathbf{u} \in \operatorname{Int} P$ and $b \in H^1(L(\mathbf{u});\Lambda_0)$ such that the Floer cohomology $HF((L(\mathbf{u}),(\mathfrak{b},b)),(L(\mathbf{u}),(\mathfrak{b},b));\Lambda)$ is isomorphic to $H(T^n;\Lambda)$.

We have a binary operator $\mathfrak{m}_2^{\mathfrak{c},\mathfrak{b},b}$ on it. The Poincaré duality induces a Λ valued non-degenerate inner product $\langle\cdot\rangle_{\mathrm{PD}_{L(\mathbf{u})}}$ of it.

We define

$$x \cup^{\mathfrak{c},\mathfrak{b},b} y = (-1)^{\deg x(\deg y+1)}\mathfrak{m}_2^{\mathfrak{c},\mathfrak{b},b}(x,y), \tag{103}$$

$$\langle x,y\rangle_{\mathrm{cyc}} = (-1)^{\deg x(\deg y+1)}\langle x,y\rangle_{\mathrm{PD}_{L(u)}}. \tag{104}$$

Then $(H(L(\mathbf{u});\Lambda),\langle\cdot,\cdot\rangle_{\mathrm{cyc}},\cup^{\mathfrak{c},\mathfrak{b},b},\mathrm{PD}[L(\mathbf{u})])$ becomes a unital Frobenius algebra.

REMARK 12.11. We remark that the operation $\mathfrak{m}_2^{\mathfrak{c},\mathfrak{b},b}$ is slightly different from the operation $\mathfrak{m}_2^{\mathfrak{b},b}$ which is obtained from the operation $\mathfrak{q}_{\ell,k}$ by (36). In fact $\mathfrak{q}_{\ell,k}$ may not satisfy the cyclic symmetry:

$$\begin{aligned}&\langle\mathfrak{q}_{\ell;k}(\mathbf{y};h_1,\dots,h_k),h_0\rangle_{\mathrm{cyc}}\\&=(-1)^{\deg' h_0(\deg' h_1+\cdots+\deg' h_k)}\langle\mathfrak{q}_{\ell;k}(\mathbf{y};h_0,h_1,\dots,h_{k-1}),h_k\rangle_{\mathrm{cyc}}.\end{aligned} \tag{105}$$

This is because the way how we perturb the moduli space $\mathcal{M}^{\mathrm{main}}_{k+1;\ell}(\beta)$, which we described in sections 3 and 7, breaks cyclic symmetry.

However we can modify the construction of $\mathfrak{q}_{\ell;k}$ to obtain $\mathfrak{q}^{\mathfrak{c}}_{\ell;k}$ for which (105) is satisfied. Using it in place of $\mathfrak{q}_{\ell;k}$ we define $\mathfrak{m}_2^{\mathfrak{c},\mathfrak{b},b}$, which appears in (103). Then Definition 12.9 3) is satisfied for $\cup^{\mathfrak{b},b}$.

This point is quite technical and delicate. So we do not discuss its detail in this survey and refer readers to [**FOOO5**] sections 3.2–3.3. However it is inevitable and essential, especially in the non-Fano case. It might be related to the fact that primitive form may be different from $dx_1\wedge\cdots\wedge dx_n$ in general.

We put

$$Z(\mathfrak{b}, b) = Z((H(L(u); \Lambda), \langle \cdot, \cdot \rangle_{\text{cyc}}, \cup^{\mathfrak{c},\mathfrak{b},b}, \text{PD}([L(u)])). \tag{106}$$

DEFINITION 12.12. Assume that $\mathfrak{PO}^{\mathfrak{b}}$ is a Morse function. We then define a *residue pairing*

$$\langle \cdot, \cdot \rangle_{\text{res}} : (\text{Jac}(\mathfrak{PO}^{\mathfrak{b}}) \otimes_{\Lambda_0} \Lambda) \otimes (\text{Jac}(\mathfrak{PO}^{\mathfrak{b}}) \otimes_{\Lambda_0} \Lambda) \to \Lambda$$

by

$$\langle 1_{\mathfrak{y}}, 1_{\mathfrak{y}'} \rangle_{\text{res}} = \begin{cases} 0 & \text{if } \mathfrak{y} \neq \mathfrak{y}', \\ (Z(\mathfrak{b}, b))^{-1} & \text{if } \mathfrak{y} = \mathfrak{y}'. \end{cases} \tag{107}$$

We remark that we use the decomposition in Proposition 11.8 and $1_{\mathfrak{y}}$ is the unit of $\text{Jac}(\mathfrak{PO}^{\mathfrak{b}}; \mathfrak{y})$. $\mathbf{u} = (u_1, \dots, u_n)$ is defined by the valuation of $\mathfrak{y} = (\mathfrak{y}_1, \dots, \mathfrak{y}_n)$. Namely $u_i = \mathfrak{v}_T(\mathfrak{y}_i)$. $b \in H^1(L(\mathbf{u}); \Lambda_0)$ is defined from $\mathfrak{y}_i$ by $b = \sum_{i=1}^n \mathfrak{x}_i \mathbf{e}_i$, $T^{u_i} e^{\mathfrak{x}_i} = \mathfrak{y}_i$.

The name 'residue pairing' is justified by the following Theorem 12.13 and Lemma 12.8.

THEOREM 12.13. 1) *Assume that $\mathfrak{y}$ is a nondegenerate critical point of $\mathfrak{PO}^{\mathfrak{b}}$. Suppose $b = \sum_{i=1}^n \mathfrak{x}_i \mathbf{e}_i$, $T^{u_i} e^{\mathfrak{x}_i} = \mathfrak{y}_i$ as above. Then*

$$Z(\mathfrak{b}, b) \equiv \det \left[y_i y_j \frac{\partial^2 \mathfrak{PO}^{\mathfrak{b}}}{\partial y_i \partial y_j} \right]_{i,j=1}^{i,j=n} (\mathfrak{y}) \mod T^{\lambda} \Lambda_+. \tag{108}$$

Here $\lambda = v_T(Z(\mathfrak{b}, b))$ and $\mathfrak{y} = (T^{u_1} e^{\mathfrak{x}_1}, \dots, T^{u_n} e^{\mathfrak{x}_n})$.

2) *If $\dim_{\mathbb{C}} X = 2$, then we have*

$$Z(\mathfrak{b}, b) = \det \left[y_i y_j \frac{\partial^2 \mathfrak{PO}^{\mathfrak{c},\mathfrak{b}}}{\partial y_i \partial y_j} \right]_{i,j=1}^{i,j=n} (\mathfrak{y}). \tag{109}$$

3) *If X is nef and $\deg \mathfrak{b} = 2$, then we have*

$$Z(\mathfrak{b}, b) = \det \left[y_i y_j \frac{\partial^2 \mathfrak{PO}^{\mathfrak{b}}}{\partial y_i \partial y_j} \right]_{i,j=1}^{i,j=n} (\mathfrak{y}). \tag{110}$$

REMARK 12.14. We use $\mathfrak{m}_k^{\mathfrak{c},\mathfrak{b}}$ in place of $\mathfrak{m}_k^{\mathfrak{b}}$ to define $\mathfrak{PO}^{\mathfrak{c},\mathfrak{b}}$ by

$$\mathfrak{PO}^{\mathfrak{c},\mathfrak{b}}(b) = \sum_{k=0}^{\infty} \int_{L(\mathbf{u})} \mathfrak{m}_k^{\mathfrak{c},\mathfrak{b}}(\underbrace{b, \dots, b}_{k}).$$

$\mathfrak{PO}^{\mathfrak{c},\mathfrak{b}}$ appears in (109).

Theorem 12.13 is Theorem 1.2.25 [**FOOO5**].

SKETCH OF THE PROOF. We discuss only the case X is nef and $\mathfrak{b} = \mathbf{0}$. We will prove that the algebra $(H(L(u);\Lambda), \cup^b)$ is a Clifford algebra, modifying the proof of a related result by Cho [**Cho2**]. More precisely we prove the following Proposition 12.15.

Let $\mathbf{e}'_1, \dots, \mathbf{e}'_n$ be formal variables and $d_i \in \Lambda \setminus \{0\}$ $(i = 1, \dots, n)$. We consider relations

$$\begin{cases} \mathbf{e}'_i\mathbf{e}'_j + \mathbf{e}'_j\mathbf{e}'_i = 0, & i \neq j \\ \mathbf{e}'_i\mathbf{e}'_i = d_i 1. \end{cases} \tag{111}$$

We take a free (non-commutative) Λ algebra generated by $\mathbf{e}'_1, \dots, \mathbf{e}'_n$ and divide it by the two-sided ideal generated by (111). We denote it by Cliff_Λ $(n;\vec{d})$, where we set $\vec{d} = (d_1, \dots, d_n)$.

Let $I = (i_1, \dots, i_k)$, $1 \le i_1 < \cdots < i_k \le n$. We write the set of such I's by $2^{\{1,\dots,n\}}$. We put

$$\mathbf{e}'_I = \mathbf{e}'_{i_1}\mathbf{e}'_{i_2} \cdots \mathbf{e}'_{i_{k-1}}\mathbf{e}'_{i_k} \in \mathrm{Cliff}_\Lambda(n;\vec{d}).$$

It is well-known and can be easily checked that $\{\mathbf{e}'_I \mid I \in 2^{\{1,\dots,n\}}\}$ forms a basis of $\mathrm{Cliff}_\Lambda(n;\vec{d})$ as a Λ vector space.

Assume moreover that there exists a Λ valued non-degenerate inner product $\langle\cdot\rangle$ on $\mathrm{Cliff}_\Lambda(n;\vec{d})$ such that $\mathrm{Cliff}_\Lambda(n;\vec{d})$ becomes a Frobenius algebra. We say that $\mathbf{e}'_i$ forms a *cyclic Clifford basis* if

$$\langle \mathbf{e}'_I, \mathbf{e}'_J\rangle = \begin{cases} (-1)^{*(I)} & J = I^c, \\ 0 & \text{otherwise}. \end{cases} \tag{112}$$

Here $I^c = \{1, \dots, n\} \setminus I$ and $*(I) = \#\{(i,j) \mid i \in I, j \in I^c, j < i\}$.

PROPOSITION 12.15. *Suppose X is nef and* $\deg \mathfrak{b} = 2$. *We also assume that $L(\mathbf{u})$ and $b \in H^1(L(\mathbf{u});\Lambda_0)$ satisfy $HF((L(\mathbf{u}),(\mathfrak{b},b)),(L(\mathbf{u}),(\mathfrak{b},b));\Lambda) \cong H(T^n;\Lambda)$.*

Then there exists a basis $(\mathbf{e}'_1, \dots, \mathbf{e}'_n)$ of $H^1(L(\mathbf{u});\Lambda)$ such that the algebra $((H(L(\mathbf{u});\Lambda), \cup^{\mathfrak{c},\mathfrak{b},b})$ is isomorphic to the Clifford algebra $\mathrm{Cliff}_\Lambda(n;\vec{d})$ *where $(d_1, \dots, d_n)$ satisfies*

$$^tA\left[y_iy_j\frac{\partial^2\mathfrak{PO}_{\mathfrak{b}}}{\partial y_i\partial y_j}(\mathfrak{y})\right]_{i,j=1}^{i,j=n} A = 2\begin{bmatrix} d_1 & 0 & \dots & 0 \\ 0 & d_2 & \dots & 0 \\ \vdots & \vdots & \ddots & \vdots \\ 0 & 0 & \dots & d_n \end{bmatrix}.$$

for a matrix A with $\det A = 1$.

Moreover $(\mathbf{e}'_1, \dots, \mathbf{e}'_n)$ is a cyclic Clifford basis.

Furthermore

$$\int_{L(\mathbf{u})} \mathbf{e}'_1 \cup^{\mathfrak{b},b} \cdots \cup^{\mathfrak{b},b} \mathbf{e}'_n = 1.$$

This is [**FOOO5**] Theorem 3.6.2. Once Proposition 12.15 is established we can prove Theorem 12.13 by a direct calculation. (See [**FOOO5**] section 3.7) □

SKETCH OF THE PROOF OF PROPOSITION 3.4.8. Note

$$\mathfrak{PO}^{\mathfrak{b}}(b) = \sum_{k=0}^{\infty} \int_{L(\mathbf{u})} \mathfrak{m}_k^{\mathfrak{b}}(\underbrace{b, \dots, b}_{k}).$$

Its first derivative at $\mathfrak{y}$ is zero since $\mathfrak{y}$ is a critical point. We calculate its second derivative $\partial^2 \mathfrak{PO}^{\mathfrak{b}}/\partial x_i \partial x_j = y_i y_j \partial^2 \mathfrak{PO}^{\mathfrak{b}}/\partial y_i \partial y_j$. Then we have

$$\mathfrak{m}_2^{\mathfrak{b},b}(\mathbf{e}_i, \mathbf{e}_j) + \mathfrak{m}_2^{\mathfrak{b},b}(\mathbf{e}_j, \mathbf{e}_i) = \left(\left(y_i y_j \frac{\partial^2 \mathfrak{PO}^{\mathfrak{b}}}{\partial y_i \partial y_j} \right) (\mathfrak{y}) \right) 1 \tag{113}$$

Here $1 \in H^0(L(\mathbf{u}); \mathbb{Q})$ is the unit and $\{\mathbf{e}_i\}$ is the basis of $H^1(L(\mathbf{u}); \mathbb{Q})$ which we fixed before. (Note $b = \sum x_i \mathbf{e}_i$.)

We take basis $(\mathbf{e}'_1, \dots, \mathbf{e}'_n)$ of $H^1(L(\mathbf{u}); \Lambda)$ so that the quadratic form associated to the Hessian matrix becomes diagonal and $\int_{L(\mathbf{u})} \mathbf{e}'_1 \cup \dots \cup \mathbf{e}'_n = 1$. Then (113) implies that $(\mathbf{e}'_1, \dots, \mathbf{e}'_n)$ satisfies the Clifford relation (111). Using this fact we can prove that $((H(L(\mathbf{u}); \Lambda), \cup^{\mathfrak{c},\mathfrak{b},b})$ is a Clifford algebra. (We do not use the assumption X is nef and $\mathfrak{b}$ is degree two, up to this point.)

The proof of (112) is as follow. We use the assumption that X is nef and $\mathfrak{b}$ is degree two to show

$$\mathfrak{a} \cup^{\mathfrak{c},\mathfrak{b},b} \mathfrak{a}' - \mathfrak{a} \cup \mathfrak{a}' \in \bigoplus_{k < \deg \mathfrak{a} + \deg \mathfrak{a}'} H^k(T^n; \Lambda) \tag{114}$$

for $\mathfrak{a}, \mathfrak{a}' \in HF((L(\mathbf{u}), (\mathfrak{b}, b)), (L(\mathbf{u}), (\mathfrak{b}, b)); \Lambda) \cong H(T^n; \Lambda)$. Here the second term is the usual cup product. We use cyclic symmetry to show

$$\langle \mathbf{e}'_I, \mathbf{e}'_J \rangle_{\mathrm{PD}_{L(\mathbf{u})}} = \langle \mathbf{e}'_I \cup^{\mathfrak{c},\mathfrak{b},b} \mathbf{e}'_J, 1 \rangle_{\mathrm{PD}_{L(\mathbf{u})}} = \int_{L(\mathbf{u})} \mathbf{e}'_I \cup^{\mathfrak{c},\mathfrak{b},b} \mathbf{e}'_J.$$

Using (114) and the Clifford relation, we can see that $\mathbf{e}'_I \cup^{\mathfrak{c},\mathfrak{b},b} \mathbf{e}'_J$ has no $H^n(L(\mathbf{u}); \Lambda)$ component unless $I^c = J$. This implies Proposition 12.15. □

12.4. Residue pairing is Poincaré duality.

THEOREM 12.16. *Let X be a compact toric manifold and $\mathfrak{b} \in \mathcal{A}(\Lambda_0)$. Suppose $\mathfrak{PO}^{\mathfrak{b}}$ is a Morse function. Then for each $\mathfrak{a}_1, \mathfrak{a}_2 \in H(X; \Lambda)$ we have*

$$\langle \mathfrak{a}_1, \mathfrak{a}_2 \rangle_{\mathrm{PD}_X} = \langle \mathfrak{ks}_{\mathfrak{b}} \mathfrak{a}_1, \mathfrak{ks}_{\mathfrak{b}} \mathfrak{a}_2 \rangle_{\mathrm{res}}. \tag{115}$$

Here the pairing in the right hand side is defined in Definition 12.12 and the map $\mathfrak{ks}_{\mathfrak{b}}$ is the isomorphism in Theorem 11.14. The pairing in the left hand side is the Poincaré duality.

Theorem 12.16 is [**FOOO5**] Theorem 1.1.1 (2) and is proved in [**FOOO5**] Chapter 3. Before explaining an outline of its proof, we mention some of its consequences.

COROLLARY 12.17. 1) *The inner product* $\langle\cdot\rangle_{\rm res}$, *whose definition was given only in case* $\mathfrak{PO}^{\mathfrak b}$ *is a Morse function (in Definition 12.12), extends to arbitrary* $\mathfrak b$*'s.*

2) *The Levi-Civita connection* ∇ *of this extended* $\langle\cdot\rangle_{\rm res}$ *is flat.*

3) $(H(X;\Lambda_0), \langle\cdot\rangle_{\rm res}, \nabla, \circ, \Phi, 1)$ *is a Frobenius manifold.*

4) *The Frobenius manifold structure of Item* 3) *above is equal to one in Theorem 12.4.*

PROOF. 1) is an immediate consequence of Theorem 12.16 and the fact that the Poincaré duality pairing is independent of $\mathfrak b$ and is obviously extended.

The Levi-Civita connection of the Poincaré duality pairing is the canonical affine connection of $H(X;\Lambda_0)$ and hence is flat. 2) follows.

3) then follows from Theorem 12.4.

4) is obvious. □

REMARK 12.18. The Frobenius manifold in Corollary 12.17 3) has an Euler vector field (96) with $r_i = 1$. We also have

$$\mathfrak{E}(\mathfrak{PO}) = \mathfrak{PO}, \tag{116}$$

here $\mathfrak{PO}$ is a function of $\mathfrak b = \sum w_i \mathbf{p}_i$ and y_i. The formula (116) is proved in [**FOOO4**] Theorem 10.2.

REMARK 12.19. Corollary 12.17 first appeared as a conjecture in [**Ta**], where the case of $\mathbb{C}P^1$ was checked. It was further studied in [**Bar**]. See the papers mentioned at the end of the introduction for some of the other related works.

The above proof of the coincidense of the two Frobenius manifold structures is not so satisfactory since the proof of Items 1), 2) uses the isomorphism of Item 4). It is preferable that we construct Frobenius manifold structure on $H(X;\Lambda_0)$ using the family of functions $\mathfrak{PO}^{\mathfrak b}$ and without going to the quantum cohomology theory side, and then prove Item 4) for that Frobenius manifold structure.

PROBLEM 12.20. *Develop an analogue of K. Saito theory for our family of* Λ *valued functions* $\mathfrak{PO}^{\mathfrak b}$.

Define the notion of primitive form for it and prove its existence.

Construct the Frobenius manifold structure on $H(X;\Lambda_0)$ *using primitive form and prove that it is isomorphic to one obtained in Theorem 12.4.*

Another corollary of Theorem 12.16 is the following. Let $\mathrm{Crit}(\mathfrak{PO}^{\mathfrak{b}})$ be the critical point set of $\mathfrak{PO}^{\mathfrak{b}}$. For $\mathfrak{y} = (\mathfrak{y}_1, \dots, \mathfrak{y}_n) \in \mathfrak{A}(P)$ we put

$$\mathfrak{y}_i = T^{u_i} e^{\mathfrak{x}_i}, \quad b = \sum_{i=1}^{n} \mathfrak{x}_i \mathbf{e}_i \in H^1(L(\mathbf{u}), \Lambda_0). \tag{117}$$

Here $\mathbf{u} = (u_1, \dots, u_n) \in P$ and $\mathfrak{x}_i \in \Lambda_0$. Note $u_i = \mathfrak{v}_T(\mathfrak{y}_i)$. In this way we may regard $\mathrm{Crit}(\mathfrak{PO}^{\mathfrak{b}})$ as a set of pairs $(\mathbf{u}_c, b_c)$, $c = 1, \dots, B'$. Here we put $B' = \#\mathrm{Crit}(\mathfrak{PO}^{\mathfrak{b}})$.

COROLLARY 12.21. *Suppose $\mathfrak{PO}^{\mathfrak{b}}$ is a Morse function. Then we have*

$$0 = \sum_{c=1}^{B'} \frac{1}{Z(\mathfrak{b}, b_c)}. \tag{118}$$

PROOF. Let $1_X \in H^0(X;\Lambda)$ be the unit. Then $\langle 1_X, 1_X \rangle_{\mathrm{PD}_X} = 0$. By Proposition 11.8 we have $1_X = \sum_{\mathfrak{y} \in \mathfrak{M}(X,\mathfrak{b})} 1_{\mathfrak{y}}$ where $1_{\mathfrak{y}}$ is the unit of the Jacobian ring $\mathrm{Jac}(\mathfrak{PO}_{\mathfrak{b}}; \mathfrak{y})$. Corollary 12.21 now follows from (107) and Theorem 12.16. □

12.5. Operator $\mathfrak{p}$ and the Poincaré dual to $\mathfrak{ks}_{\mathfrak{b}}$. In this and the next subsections we sketch a proof of Theorem 12.16. We assume $\mathfrak{PO}^{\mathfrak{b}}$ is a Morse function in this and next subsections. Let $\mathfrak{y} \in \mathrm{Crit}(\mathfrak{PO}^{\mathfrak{b}})$. It defines $\mathbf{u}, b$ by (117). We define a homomorphism

$$i^*_{\mathrm{qm},(\mathfrak{b},b,\mathbf{u})} : H(X;\Lambda_0) \to HF((L(\mathbf{u}),(\mathfrak{b},b));(L(\mathbf{u}),(\mathfrak{b},b));\Lambda_0) \tag{119}$$

by

$$i^*_{\mathrm{qm},(\mathfrak{b},b,\mathbf{u})}(Q) = \sum_{k=0}^{\infty} \sum_{\ell_1=0}^{\infty} \sum_{\ell_2=0}^{\infty} \mathfrak{q}^{\mathfrak{c}}_{\ell_1+\ell_2;k}(\mathfrak{b}^{\ell_1} Q \mathfrak{b}^{\ell_2}, b^k). \tag{120}$$

(See (78) and [**FOOO1**] Theorem 3.8.62.)

Here $\mathfrak{q}^{\mathfrak{c}}_{\ell;k}$ is a cyclically symmetric version of the operator $\mathfrak{q}_{\ell;k}$. (See Remark 12.14.)

We define

$$i_{*,\mathrm{qm},(\mathfrak{b},b,\mathbf{u})} : HF((L(\mathbf{u}),\mathfrak{b},b);(L(\mathbf{u}),\mathfrak{b},b);\Lambda) \to H(X;\Lambda) \tag{121}$$

by

$$\langle i^*_{\mathrm{qm},(\mathfrak{b},b,\mathbf{u})}(Q), P \rangle_{\mathrm{PD}_{L(\mathbf{u})}} = \langle Q, i_{*,\mathrm{qm},(\mathfrak{b},b,\mathbf{u})}(P) \rangle_{\mathrm{PD}_X}. \tag{122}$$

The main part of the proof of Theorem 12.16 is the proof of Theorem 12.22 below. Let $\mathrm{vol}_{L(\mathbf{u})} \in H^n(X;\mathbb{Q})$ be the degree n cohomology class such that $\int_{L(\mathbf{u})} \mathrm{vol}_{L(\mathbf{u})} = 1$. Let $\{\mathbf{e}_I \mid I \in 2^n\}$ be a basis of

$$H^n(L(\mathbf{u});\Lambda) \cong HF((L(\mathbf{u}),(\mathfrak{b},b));(L(\mathbf{u}),(\mathfrak{b},b));\Lambda).$$

We put $g_{IJ} = \langle \mathbf{e}_I, \mathbf{e}_J \rangle_{PD_X}$. Let g^{IJ} be the inverse matrix of g_{IJ}.

THEOREM 12.22. *We have:*

$$\begin{aligned}&\langle i_{*,\mathrm{qm},(\mathfrak{b},b,\mathbf{u})}(\mathrm{vol}_{L(\mathbf{u})}), i_{*,\mathrm{qm},(\mathfrak{b},b,\mathbf{u})}(\mathrm{vol}_{L(\mathbf{u})})\rangle_{\mathrm{PD}_X}\\ &= \sum_{I,J\in 2^{\{1,\dots,n\}}} (-1)^{\frac{n(n-1)}{2}} g^{IJ} \langle \mathfrak{m}_2^{\mathfrak{c},\mathfrak{b},b}(\mathbf{e}_I, \mathrm{vol}_{L(\mathbf{u})}), \mathfrak{m}_2^{\mathfrak{c},\mathfrak{b},b}(\mathbf{e}_J, \mathrm{vol}_{L(\mathbf{u})})\rangle_{\mathrm{PD}_{L(u)}}.\end{aligned}$$

This is [**FOOO5**] Theorem 3.4.1.

THEOREM 12.22 $\Rightarrow$ THEOREM 12.16. Let $Q_{\mathfrak{y}} \in H(X;\Lambda)$ be an element such that $\mathfrak{ks}_{\mathfrak{b}}(Q_{\mathfrak{y}}) = 1_{\mathfrak{y}}$, where $1_{\mathfrak{y}}$ is the unit of the factor $\mathrm{Jac}(\mathfrak{PO}^{\mathfrak{b}};\mathfrak{y})$ of $\mathrm{Jac}(\mathfrak{PO}^{\mathfrak{b}})$. Let $b, \mathbf{u}$ corresponds to $\mathfrak{y}$ by (117).

Then we have

$$i^*_{\mathrm{qm},(\mathfrak{b},b,\mathbf{u})}(Q_{\mathfrak{y}'}) = \begin{cases} 1 & \text{if } \mathfrak{y}' = \mathfrak{y} \\ 0 & \text{if } \mathfrak{y}' \neq \mathfrak{y}. \end{cases}$$

Here $1 \in H^0(L(\mathbf{u});\Lambda)$ is the unit. This is a consequence of the definition of $\mathfrak{ks}_{\mathfrak{b}}$. Therefore

$$\langle Q_{\mathfrak{y}}, i_{*,\mathrm{qm},(\mathfrak{b},b,\mathbf{u})}(\mathrm{vol}_{L(\mathbf{u})})\rangle_{\mathrm{PD}_X} = 1. \tag{123}$$

We remark $\langle Q_{\mathfrak{y}}, Q_{\mathfrak{y}'}\rangle = \langle Q_{\mathfrak{y}} \cup^{\mathfrak{b}} Q_{\mathfrak{y}'}, 1\rangle = 0$ if $\mathfrak{y} \neq \mathfrak{y}'$. Therefore

$$i_{*,\mathrm{qm},(\mathfrak{b},b,\mathbf{u})}(\mathrm{vol}_{L(u)}) = \frac{1}{\langle Q_{\mathfrak{y}}, Q_{\mathfrak{y}}\rangle_{\mathrm{PD}_X}} Q_{\mathfrak{y}}. \tag{124}$$

Theorem 12.22 implies

$$\langle i_{*,\mathrm{qm},(\mathfrak{b},b,\mathbf{u})}(\mathrm{vol}_{L(\mathbf{u})}), i_{*,\mathrm{qm},(\mathfrak{b},b,\mathbf{u})}(\mathrm{vol}_{L(\mathbf{u})})\rangle_{\mathrm{PD}_X} = Z(\mathfrak{b}, b). \tag{125}$$

(See [**FOOO5**] subsection 3.10.2 for sign.) Theorem 12.16 follows from (124) and (125). □

To prove Theorem 12.22 we need a geometric description of the homomorphism $i_{*,\mathrm{qm},(\mathfrak{b},b,\mathbf{u})}$. We use the operator $\mathfrak{p}$ introduced in [**FOOO1**] section 3.8, for this purpose. To simplify the notation we consider only the case $\mathfrak{b} = \mathbf{0}$. Let C be a filtered A_∞ algebra and define an automorphism $\mathrm{cyc} : B_kC[1] \to B_kC[1]$ by

$$\mathrm{cyc}(x_1 \otimes \cdots \otimes x_k) = (-1)^{\deg' x_k \times (\sum_{i=1}^{k-1} \deg' x_i)} x_k \otimes x_1 \otimes \cdots \otimes x_{k-1}.$$

It induces a $\mathbb{Z}_k$ action on $B_kC[1]$. Let $B_k^{\mathrm{cyc}}C[1]$ be the invariant set of the $\mathbb{Z}_k$ action and $B^{\mathrm{cyc}}C[1] = \widehat{\bigoplus}_k B_k^{\mathrm{cyc}}C[1]$ the completed direct sum of them. We call $B_k^{\mathrm{cyc}}C[1]$ the *cyclic bar complex.*

THEOREM 12.23. *For a relatively spin Lagrangian submanifold L there exists a sequence of operators*

$$\mathfrak{p}_k : B_k^{\mathrm{cyc}} H(L;\Lambda_0)[1] \longrightarrow H(X;\Lambda_0)$$

$(k = 0, 1, 2, \dots)$ *of degree $n+1$ with the following properties.*

Let $\mathfrak{p} : B^{\rm cyc}H(L;\Lambda_0)[1] \longrightarrow H(X;\Lambda_0)$ *be the operator whose restriction on* $B^{\rm cyc}_k H(L;\Lambda_0)[1]$ *is* $\mathfrak{p}_k$. *We denote by* $\mathfrak{m}^{\mathfrak{c}}_k$ *the cyclically symmetric version of* $\mathfrak{m}_k$ *and write* $\mathfrak{m}^{\mathfrak{c}}$ *instead of* $\mathfrak{m}^{\mathfrak{c}}_k$.

1)

$$\mathfrak{p}_1 \equiv i_! \mod \Lambda_+.$$

Here $i_! = H^k(L;\Lambda_0) \to H^{k+n}(X;\Lambda)$ *is the Gysin homomorphism.*

2)

$$\sum_c \mathfrak{p}(\mathbf{x}^{3;1}_c \otimes \mathfrak{m}^{\mathfrak{c}}(\mathbf{x}^{3;2}_c) \otimes \mathbf{x}^{3;3}_c) = 0 \tag{126}$$

for $\mathbf{x} \in B^{\rm cyc}_k H(L;\Lambda_0)[1]$, $k > 0$. *We use the notation* (6).

3)

$$(\mathfrak{p}_1 \circ \mathfrak{m}^{\mathfrak{c}}_0)(1) + GW_1(L) = 0.$$

Here the second term is defined by $\langle GW_1(L), Q\rangle_{{\rm PD}_X} = GW_2(L,Q)$, *where the right hand side is as in* (68).

This is [**FOOO1**] Theorem 3.8.9. (Here we use cohomology group instead of appropriate chain complex. The latter is used in [**FOOO1**] Theorem 3.8.9. We also omit the statement on the unit in [**FOOO1**] Theorem 3.8.9.) See also [**FOOO5**] section 3.1–3.3.

The operator $\mathfrak{p}_k$ is constructed as follows. We consider the moduli space $\mathcal{M}^{\rm main}_{k;1}(\beta)$ described in section 2.2. Note the number of interior marked point is 1 and the number of exterior marked points is k. We have an evaluation map

$$({\rm ev}_1,\dots,{\rm ev}_k,{\rm ev}^+) = ({\rm ev},{\rm ev}^+) : \mathcal{M}^{\rm main}_{k;1}(\beta) \to L^k \times X.$$

Let $h_1,\dots,h_k$ be differential forms on L. We consider the pull back ${\rm ev}^*(h_1 \times \cdots \times h_k)$, which is a differential form on $\mathcal{M}^{\rm main}_{k;1}(\beta)$. We use integration along fiber by the map ${\rm ev}^+$ to obtain a differential form on X, which we put $\mathfrak{p}_{k,\beta}(h_1,\dots,h_k)$. Namely

$$\mathfrak{p}_{k,\beta}(h_1,\dots,h_k) = {\rm ev}^+_!({\rm ev}^*(h_1 \times \cdots \times h_k)).$$

This is a map between differential forms. By an algebraic argument it induces a map between tensor products of the de Rham cohomology groups of L and of X. Thus obtain the operator

$$\mathfrak{p}_k = \sum_{\beta \in H_2(X,L)} T^{(\beta\cap\omega)/2\pi}\mathfrak{p}_{k,\beta}.$$

We can prove (126) by studying the stable map compactification of $\mathcal{M}^{\rm main}_{k;1}(\beta)$. In case $k = 0$ the compactification of $\mathcal{M}_{0;1}(\beta)$ is slightly different from the case of $k > 0$. The second term of Item 3) appears by this reason. In our case of toric manifold and T^n orbit L, this term drops since L is homologous to 0 in X. So we do not discuss it here but refer to [**FOOO1**] subsections 3.8.3 and 7.4.1 for more detail.

Now we go back to the case where X is a toric manifold and $L = L(\mathbf{u})$ is a T^n orbit. Let $b \in H^1(L(\mathbf{u});\Lambda_0)$. For $P \in H(L(\mathbf{u});\Lambda_0)$ we put

$$[Pe^b] = \sum_{k_1=0}^{\infty}\sum_{k_2=0}^{\infty} \underbrace{b\otimes\cdots\otimes b}_{k_1}\otimes P\otimes \underbrace{b\otimes\cdots\otimes b}_{k_2}.$$

Suppose $H(L(\mathbf{u});\Lambda) \cong HF((L(\mathbf{u}),(\mathbf{0},b));(L(\mathbf{u}),(\mathbf{0},b));\Lambda)$.

PROPOSITION 12.24. *Let $P \in H(L(\mathbf{u});\Lambda_0)$. Then we have:*

$$i_{*,\mathrm{qm},(\mathbf{0},b,\mathbf{u})}(P) = \mathfrak{p}([Pe^b]). \tag{127}$$

REMARK 12.25. We remark that $[Pe^b]$ is an element of $B^{\mathrm{cyc}}H(L(\mathbf{u});\Lambda_0)$ if $b \equiv 0 \mod \Lambda_+$. So $\mathfrak{p}([Pe^b])$ is defined in that case. Otherwise we write $b = b_0 + b_+$ such that $b_0 \in H^1(L(\mathbf{u});\mathbb{C})$ and $b_+ \in H^1(L(\mathbf{u});\Lambda_+)$, and define

$$\mathfrak{p}([Pe^b]) = \sum_{\beta\in H_2(X,L:\mathbb{Z})} T^{(\beta\cap\omega)/2\pi}\exp(b_0\cap\partial\beta)\mathfrak{p}_\beta([Pe^{b_+}]).$$

We omit the discussion of this point. See [**FOOO4**] section 9 and [**FOOO5**] section 3.3.

SKETCH OF THE PROOF. Let $Q \in H(X;\Lambda_0)$. We remark that $i_{*,\mathrm{qm},(0,b,\mathbf{u})}(P)$ is defined by (122). Therefore it suffices to prove

$$\sum_{k=0}^{\infty}\langle \mathfrak{q}^{\mathfrak{c}}_{1,k}(Q;b^k),P\rangle_{\mathrm{PD}_{L(\mathbf{u})}} = \langle Q,\mathfrak{p}([Pe^b])\rangle_{\mathrm{PD}_X}. \tag{128}$$

This is [**FOOO5**] Theorem 19.8. Let us sketch its proof for the case $b = 0$. In case $b = 0$, Formula (128) reduced to

$$\langle \mathfrak{q}^{\mathfrak{c}}_{1,0}(Q;1),P\rangle_{\mathrm{PD}_{L(\mathbf{u})}} = \langle Q,\mathfrak{p}_1(P)\rangle_{\mathrm{PD}_X}. \tag{129}$$

We take ρ and h which are closed forms on X and $L(\mathbf{u})$, representing the cohomology class Q and P, respectively. Then it is easy to see that the left and the right hand sides of (129) both become

$$\sum_{\beta\in H_2(X,L(\mathbf{u});\mathbb{Z})} T^{(\beta\cap\omega)/2\pi}\int_{\mathcal{M}_{1;1}(\beta)} (\mathrm{ev}^+)^*\rho\wedge \mathrm{ev}^* h. \tag{130}$$

Here $(\mathrm{ev},\mathrm{ev}^+): \mathcal{M}_{1;1}(\beta) \to L(\mathbf{u})\times X$ is evaluation maps at marked points. (129) follows. □

REMARK 12.26. In fact, we need to perturb $\mathcal{M}_{1;1}(\beta)$ appropriately so that the integration in (130) makes sense. It is a nontrivial thing to prove that after perturbation (129) still holds. Actually we need to consider cyclically symmetric version of the operator $\mathfrak{q}$ for this purpose. (See [**FOOO5**] Remark 3.3.12.) We omit the discussion about perturbation and refer the reader to [**FOOO5**] section 3.3.

12.6. Annulus argument. We continue the sketch of the proof of Theorem 12.22. We assume $\mathfrak{b} = \mathbf{0}$ in this subsection for simplicity. We consider the class $\mathrm{vol}_{L(\mathbf{u})}$. (It is the Poincaré dual to the point class.) Then the left hand side is

(131)
$$\sum_{\substack{\beta_1,\beta_2 \in H_2(X,L(\mathbf{u});\mathbb{Z}),\\ \beta=\beta_1+\beta_2}} T^{((\beta_1+\beta_2)\cap\omega)/2\pi} \left\langle \mathfrak{p}_{\beta_1}([\mathrm{vol}_{L(\mathbf{u})} e^b]), \mathfrak{p}_{\beta_2}([\mathrm{vol}_{L(\mathbf{u})} e^b]) \right\rangle_{\mathrm{PD}_X}.$$

We show that (131) can be regarded as an appropriate integration of the differential form $\mathrm{vol}_{L(\mathbf{u})} \times \mathrm{vol}_{L(\mathbf{u})}$ on a moduli space of pseudo-holomorphic annuli, as follows. For simplicity we assume $b = 0$.

We consider a pair $((\Sigma; z_1, z_2), u)$ with the following properties.

1) Σ is a bordered curve of genus zero such that $\partial\Sigma$ is a disjoint union of two circles, which we denote by $\partial_1\Sigma$, $\partial_2\Sigma$.
2) The singularity of Σ is at worst the double point.
3) $z_i \in \partial_i\Sigma$ for $i = 1, 2$.
4) $u : \Sigma \to X$ is a pseudo-holomorphic map. $u(\partial\Sigma) \subset L(\mathbf{u})$.
5) $u_*([\Sigma]) = \beta \in H_2(X, L(\mathbf{u}); \mathbb{Z})$.
6) The set of maps $v : \Sigma \to \Sigma$ which is biholomorphic, $v(z_i) = z_i$ for $i = 1, 2$, and $u \circ v = u$ is finite.

We denote by $\mathcal{M}_{(1,1);0}(\beta)$ the totality of such $((\Sigma; z_1^+, z_2^+), u)$. There exists an evaluation map

$$\mathrm{ev} = (\mathrm{ev}_1, \mathrm{ev}_2) : \mathcal{M}_{(1,1);0}(\beta) \to L(\mathbf{u})^2,$$

which is defined by

$$\mathrm{ev}((\Sigma; z_1, z_2), u) = (u(z_1), u(z_2)).$$

We consider the set of all $(\Sigma; z_1, z_2)$ which satisfies 1), 2), 3) above and

7) The set of all biholomorphic maps $v : \Sigma \to \Sigma$ with $v(z_i) = z_i$ for $i = 1, 2$ is finite.

We denote it by $\mathcal{M}_{(1,1);0}$. There is a forgetful map

(132)
$$\mathfrak{forget} : \mathcal{M}_{(1,1);0}(\beta) \to \mathcal{M}_{(1,1);0},$$

which is obtained by forgetting the map u.

We can show that $\mathcal{M}_{(1,1);0}$ is homeomorphic to a disk and so is connected. We take two points $(\Sigma^{(j)}; z_1^{(j)}, z_2^{(j)}) \in \mathcal{M}_{(1,1);0}$ ($j = 1, 2$) which we show in the figure below.

We denote by $\mathcal{M}_{(1,1);0}(\beta; \Sigma^{(j)})$ the inverse image of $\{(\Sigma^{(j)}; z_1^{(j)}, z_2^{(j)})\}$ by the map (132).

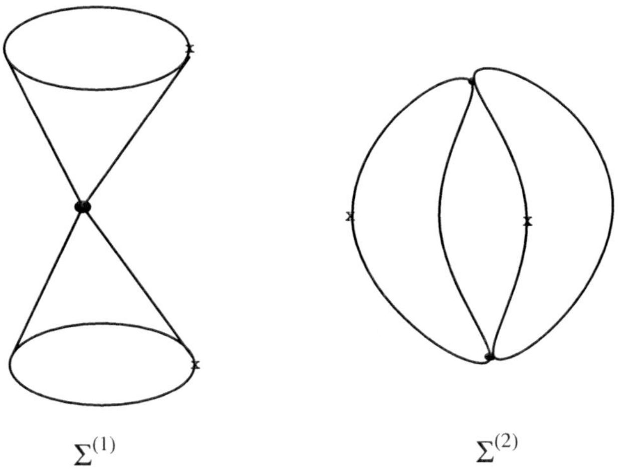

FIGURE 12.2

LEMMA 12.27.

$$\sum_{\substack{\beta_1,\beta_2\in H_2(X,L(\mathbf{u});\mathbb{Z}),\\ \beta=\beta_1+\beta_2}} \langle \mathfrak{p}_{1,\beta_1}(\mathrm{vol}_{L(\mathbf{u})}), \mathfrak{p}_{1,\beta_2}(\mathrm{vol}_{L(\mathbf{u})})\rangle_{\mathrm{PD}_X}$$
$$= \int_{\mathcal{M}_{(1,1);0}(\beta;\Sigma^{(1)})} \mathrm{ev}_1^*\mathrm{vol}_{L(\mathbf{u})} \wedge \mathrm{ev}_2^*\mathrm{vol}_{L(\mathbf{u})}.$$

Geometric origin of this lemma is clear from Figure 12.2. To prove the lemma rigorously we need to work out the way to perturb our moduli space $\mathcal{M}_{(1,1);0}(\beta;\Sigma^{(1)})$ so that the integration of the right hand side makes sense and the lemma holds. The detail is given in [**FOOO5**] section 3.4 as the proof of Proposition 3.4.8.

LEMMA 12.28.

$$\sum_{I,J\in 2^{\{1,\dots,n\}}} (-1)^{\frac{n(n-1)}{2}} g^{IJ}\langle \mathfrak{m}_2^{\mathfrak{c},0,0}(\mathbf{e}_I, \mathrm{vol}_{L(\mathbf{u})}), \mathfrak{m}_2^{\mathfrak{c},0,0}(\mathbf{e}_J, \mathrm{vol}_{L(\mathbf{u})})\rangle_{\mathrm{PD}_{L(u)}}$$
$$= \int_{\mathcal{M}_{(1,1);0}(\beta;\Sigma^{(2)})} \mathrm{ev}_1^*\mathrm{vol}_{L(\mathbf{u})} \wedge \mathrm{ev}_2^*\mathrm{vol}_{L(\mathbf{u})}.$$

Geometric origin of this lemma is also clear from Figure 12.2 and the equality

$$[\{(x,x) \mid x \in L(\mathbf{u})\}] = \sum_{I,J} (-1)^{\deg \mathbf{e}_I \deg \mathbf{e}_J} g^{IJ} \mathbf{e}_I \times \mathbf{e}_J \in H_n(L(\mathbf{u}) \times L(\mathbf{u});\mathbb{Z}). \tag{133}$$

The detail is given in [**FOOO5**] section 3.4 as the proof of Proposition 3.4.10. (The sign in (133) is proved in [**FOOO5**] Lemma 3.10.7.)

Now we can use the fact that $\mathcal{M}_{(1,1);0}$ is connected to find a cobordism between $\mathcal{M}_{(1,1);0}(\beta;\Sigma^{(1)})$ and $\mathcal{M}_{(1,1);0}(\beta;\Sigma^{(2)})$. The differential form $\mathrm{ev}_1^*\mathrm{vol}_{L(\mathbf{u})} \wedge \mathrm{ev}_2^*\mathrm{vol}_{L(\mathbf{u})}$ extends to this cobordism. Therefore Lemmas 12.27 and 12.28 imply Theorem 12.22 in case $\mathfrak{b} = b = 0$. The general case is similar. □

REMARK 12.29. According to E. Getzler, the fact $\mathcal{M}_{(1,1);0}(\beta;\Sigma^{(1)})$ is cobordant to $\mathcal{M}_{(1,1);0}(\beta;\Sigma^{(2)})$ is called the Cardy relation.

REMARK 12.30. A similar trick using the annulus is used in [**Ab2, BC**] for a similar but a slightly different purpose.

13. Examples 3

EXAMPLE 13.1. We consider the case of $\mathbb{C}P^n$ and $\mathfrak{b} = \mathbf{0}$. The moment polytope P is a simplex $\{(u_1,\dots,u_n) \mid 0 \le u_i, \sum u_i \le 1\}$ and the potential function is

$$\mathfrak{PO}^{\mathbf{0}} = \sum_{i=1}^{n} y_i + T(y_1\cdots y_n)^{-1}.$$

The critical points are $\mathfrak{y}^{(k)} = (T^{\frac{1}{n+1}}e^{\frac{2\pi\sqrt{-1}k}{n+1}},\dots,T^{\frac{1}{n+1}}e^{\frac{2\pi\sqrt{-1}k}{n+1}})$, $k = 0,\dots,n$ which are all non-degenerate. The isomorphism $\mathrm{Jac}(\mathfrak{PO}^{\mathbf{0}}) \otimes_{\Lambda_0} \Lambda \cong \prod_{k=0}^{n} \Lambda 1_{\mathfrak{y}^{(k)}}$. is induced by

$$P \mapsto \sum_{k=0}^{n} P(\mathfrak{y}^{(k)}) 1_{\mathfrak{y}^{(k)}}.$$

We put $\mathbf{f}_k = \pi^{-1}(\{(u_1,\dots,u_n) \in P \mid u_i = 0,\ i = n-k+1,\dots,n\})$ and $\mathbf{p}_k = \mathrm{PD}(\mathbf{f}_k)$ the Poincaré dual to $\mathbf{f}_k$. We derive

$$\mathfrak{PO}^{w\mathbf{p}_1}(y) = \mathfrak{PO}^{\mathbf{0}}(y) + (e^w - 1)y_n$$

from Proposition 4.9 [**FOOO4**] and hence

$$\mathfrak{ks}_{\mathbf{0}}(\mathbf{p}_1) = [y_n] = T^{\frac{1}{n+1}} \sum_{k=0}^{n} e^{\frac{2\pi\sqrt{-1}k}{n+1}} 1_{\mathfrak{y}^{(k)}} \tag{134}$$

by definition of $\mathfrak{ks}_{\mathbf{0}}$. Using the fact that $\mathfrak{ks}_{\mathbf{0}}$ is a ring homomorphism, we have

$$\mathfrak{ks}_{\mathbf{0}}(\mathbf{p}_\ell) = T^{\frac{\ell}{n+1}} \sum_{k=0}^{n} e^{\frac{2\pi\sqrt{-1}k\ell}{n+1}} 1_{\mathfrak{y}^{(k)}}. \tag{135}$$

Note this holds for $\ell = 0$ also since $\mathbf{f}_0$ is a unit and $\mathfrak{ks}_{\mathbf{0}}$ is unital.

The Hessian of $\mathfrak{PO}^{\mathbf{0}}$ is given by

$$\mathrm{Hess}_{\mathfrak{x}^{(k)}}\mathfrak{PO}^{\mathbf{0}} = \left[T^{\frac{1}{n+1}} \frac{\partial^2}{\partial x_i \partial x_j}\left(e^{x_1} + \cdots + e^{x_n} + e^{-(x_1+\cdots+x_n)}\right)\right]_{i,j=1}^{i,j=n} (\mathfrak{x}^{(k)})$$

with $\mathfrak{x}^{(k)} = \frac{2\pi\sqrt{-1}k}{n+1}$. Therefore

$$\mathrm{Hess}_{\mathfrak{x}^{(k)}}\mathfrak{PO}^{\mathbf{0}} = T^{\frac{1}{n+1}} e^{\frac{2\pi\sqrt{-1}k}{n+1}} \left[\delta_{ij}+1\right]_{i,j=1}^{i,j=n}.$$

It is easy to see that the determinant of the matrix $[\delta_{ij}+1]_{i,j=1}^{i,j=n}$ is $n+1$. Therefore the residue pairing is given by

$$\langle 1_{\mathfrak{y}^{(k)}}, 1_{\mathfrak{y}^{(k')}}\rangle_{\mathrm{res}} = T^{-\frac{n}{n+1}} e^{-\frac{2\pi\sqrt{-1}kn}{n+1}} \frac{\delta_{kk'}}{1+n}. \tag{136}$$

Combining (135) and (136), we obtain

$$\langle \mathfrak{ks}_{\mathbf{0}}(\mathbf{p}_\ell), \mathfrak{ks}_{\mathbf{0}}(\mathbf{p}_{\ell'})\rangle_{\mathrm{res}} = \frac{1}{n+1} T^{-\frac{n}{n+1}} \sum_{k=0}^{n} e^{-\frac{2\pi\sqrt{-1}kn}{n+1}} T^{\frac{\ell+\ell'}{n+1}} e^{\frac{2\pi\sqrt{-1}(\ell+\ell')k}{n+1}}. \tag{137}$$

It follows that (137) is 0 unless $\ell+\ell' = n$ and

$$\langle \mathfrak{ks}_{\mathbf{0}}(\mathbf{p}_\ell), \mathfrak{ks}_{\mathbf{0}}(\mathbf{p}_{n-\ell})\rangle_{\mathrm{res}} = 1 = \langle \mathbf{p}_\ell, \mathbf{p}_{n-\ell}\rangle_{\mathrm{PD}_{\mathbb{C}P^n}}.$$

Thus Theorem 12.16 holds in this case.

REMARK 13.2. There are various works in the case of $\mathbb{C}P^n$. See [**Ta, Bar, Gro1**].

EXAMPLE 13.3. We consider the Hirzebruch surface $F_2(\alpha)$. We use the notation of Example 10.1. In this case the full potential function for $\mathfrak{b} = \mathbf{0}$ is calculated in [**Aur2**], [**FOOO5**] section 2.13 and [**FOOO6**] section 5 as follows.

$$\mathfrak{PO}^{\mathbf{0}} = y_1 + y_2 + T^2 y_1^{-1} y_2^{-2} + T^{1-\alpha}(1+T^{2\alpha}) y_2^{-1}. \tag{138}$$

The valuation of the critical points are

$$(\mathfrak{v}_T(\mathfrak{y}_1), \mathfrak{v}_T(\mathfrak{y}_2)) = ((1-\alpha)/2, (1+\alpha)/2) = \mathbf{u}.$$

It is the same for 4 critical points. Then using the variables $\overline{y}_i = y_i^{\mathbf{u}}$ we have

$$\mathfrak{PO}^{\mathbf{0}} = T^{(1-\alpha)/2}(\overline{y}_2 + (1+T^{2\alpha})\overline{y}_2^{-1}) + T^{(1+\alpha)/2}(\overline{y}_1 + \overline{y}_1^{-1}\overline{y}_2^{-2}). \tag{139}$$

(See Example 10.1.) (We remark $\mathfrak{v}_T(\overline{y}_i) = 0$.) The critical point equation is

$$0 = 1 - \overline{y}_1^{-2}\overline{y}_2^{-2}. \tag{140}$$

$$0 = 1 - 2T^{\alpha}\overline{y}_1^{-1}\overline{y}_2^{-3} - (1+T^{2\alpha})\overline{y}_2^{-2}. \tag{141}$$

This has 4 solutions.

The Hessian matrix of (139) is

$$\begin{bmatrix} T^{(1+\alpha)/2}(\overline{y}_1 + \overline{y}_1^{-1}\overline{y}_2^{-2}) & 2T^{(1+\alpha)/2}\overline{y}_1^{-1}\overline{y}_2^{-2} \\ 2T^{(1+\alpha)/2}\overline{y}_1^{-1}\overline{y}_2^{-2} & \begin{matrix} T^{(1-\alpha)/2}(\overline{y}_2 + (1+T^{2\alpha})(\overline{y}_2^{-1}) \\ +4T^{(1+\alpha)/2}\overline{y}_1^{-1}\overline{y}_2^{-2} \end{matrix} \end{bmatrix}.$$

We can easily calculate the determinants of this matrix at the four solutions of (140), (141). The determinants are $4T, 4T, -4T, -4T$. (See [**FOOO5**] section 2.13 for the detail of the calculation.)

The Hirzebruch suface $F_2(\alpha)$ is symplectomorphic to $S^2(1-\alpha)\times S^2(1-\alpha)$, where $S^2(1-\alpha)$ is the sphere S^2 with total area $1-\alpha$. This fact is proved in [**FOOO6**] Proposition 5.1.

The quantum cohomology of $S^2(1-\alpha)\times S^2(1-\alpha)$ is generated by x, y that correspond to the fundamental class of the factors $S^2(1-\alpha)$ and $S^2(1+\alpha)$ respectively. The fundamental relations among them are

$$x^2 = T^{1-\alpha}1, \quad y^2 = T^{1+\alpha}1, \quad xy = yx.$$

We put

$$e_\pm = \frac{1}{2}T^{-(1-\alpha)/2}(T^{(1-\alpha)/2} \pm x), \quad f_\pm = \frac{1}{2}T^{-(1+\alpha)/2}(T^{(1+\alpha)/2} \pm y).$$

Then $e_-f_-, e_-f_+, e_+f_-, e_+f_+$ are the units of the 4 direct product factors of $QH(S^2(1-\alpha)\times S^2(1+\alpha);\Lambda)$. We have

$$\int_{S^2(1-\alpha)\times S^2(1+\alpha)} e_-f_-e_-f_- = \frac{1}{4T}$$

Hence

$$\langle e_-f_-, e_-f_-\rangle_{\mathrm{PD}_{S^2(1-\alpha)\times S^2(1+\alpha)}} = \frac{1}{4T}.$$

We obtain $-1/4T, -1/4T, 1/4T$ from e_-f_+, e_+f_-, e_+f_+ in the same way. Thus, Theorem 12.16 holds in this case also.

EXAMPLE 13.4. We take the monotone toric blow up of $\mathbb{C}P^2$ at one point, whose moment polytope is $\{(u_1,u_2)\mid 0\ge u_1,u_2, u_1+u_2\le 1, u_1\le 2/3\}$. Its unique monotone fiber is $\mathbf{u} = (1/3,1/3)$. We put $\overline{y}_1 = y_1^{\mathbf{u}}$, $\overline{y}_2 = y_2^{\mathbf{u}}$. Then the potential function (for $\mathfrak{b} = \mathbf{0}$) is:

$$\mathfrak{PO}^{\mathbf{0}} = T^{1/3}(\overline{y}_1 + \overline{y}_2 + (\overline{y}_1\overline{y}_2)^{-1} + \overline{y}_1^{-1}). \tag{142}$$

The condition for $(\overline{y}_1,\overline{y}_2)$ to be critical gives rise to the equation:

$$1-\overline{y}_1^{-2}\overline{y}_2^{-1} - \overline{y}_1^{-2} = 0, \quad 1-\overline{y}_1\overline{y}_2^2 = 0. \tag{143}$$

We put $\overline{y}_2 = z$. Then $\overline{y}_1 = 1/z$ and

$$z^4 + z^3 - 1 = 0. \tag{144}$$

By Theorem 12.13 (3) we have

$$\begin{aligned} Z(0,(\overline{y}_1,\overline{y}_2)) &= T^{2/3}\det\begin{bmatrix} \overline{y}_1 + (\overline{y}_1\overline{y}_2)^{-1} + \overline{y}_1^{-1} & (\overline{y}_1\overline{y}_2)^{-1} \\ (\overline{y}_1\overline{y}_2)^{-1} & \overline{y}_2 + (\overline{y}_1\overline{y}_2)^{-1}\end{bmatrix} \\ &= T^{2/3}\frac{4-z^3}{z}.\end{aligned}$$

Let z_i $(i = 1, 2, 3, 4)$ be the 4 solutions of (144). Then the left hand side of (118) becomes:

$$T^{-2/3}\sum_{i=1}^{4}\frac{z_i}{4-z_i^3}. \tag{145}$$

We can directly check that (145)$= 0$. (See **[FOOO5]** Example 1.2.36.) Thus we checked that Corollary 12.21 holds in this case.

Acknowledgements

KF is supported partially by JSPS Grant-in-Aid for Scientific Research No. 18104001 and Global COE Program G08, YO by US NSF grant # 0904197, HO by JSPS Grant-in-Aid for Scientific Research No. 19340017 and 23340015, KO by JSPS Grant-in-Aid for Scientific Research No. 21244002.

References

[Ab1] M. Abouzaid, *Morse homology, tropical geometry, and homological mirror symmetry for toric varieties*, Geom. Topol. 10 (2006), 1097–1157.

[Ab2] M. Abouzaid, *A geometric criterion for generating the Fukaya category*, Publ. Math. Inst. Hautes Études Sci. No. 112 (2010), 191–240, arXiv:1001.4593.

[AKO] D. Auroux, L. Katzarkov and D. Orlov, *Mirror symmetry for weighted projective planes and their noncommutative deformations*, Ann. of Math. (2) 167 (2008), no. 3, 867–943, mathAG/0404281.

[Aur1] D. Auroux, *Mirror symmetry and T-duality in the complement of an anticanonical divisor*, J. Gökova Geom. Topol. 1 (2007), 51–91, arXiv:0706.3207v2.

[Aur2] D. Auroux, *Special Lagrangian fibrations, wall-crossing, and mirror symmetry*, Surveys in Differential Geometry 13 (2009) 1–47, arXiv:0902.1595.

[Bar] S. Barannikov, *Semi-infinite Hodge structures and mirror symmetry for projective spaces*, preprint, arXiv:math/0010157.

[B1] V. Batyrev, *Quamtum cohomology rings of toric manifolds,* Journées de Géométrie Algébrique d'Orsay, 1992, Astérisque 218, 9–34.

[B2] V. Batyrev, *Dual polyhedra and mirror symmetry for Calabi-Yau hypersurfaces in toric varieties,* J. Alge. Geom. 3 (1994), no. 3, 493–535.

[BC] P. Biran and O. Cornea, *Quantum structures for Lagrangian submanifolds*, preprint, arXiv:0708.4221.

[BGR] S. Bosch, U. Güntzer and R. Remmert, Non-Archimedean Analysis. Asystematic Approach to Rigid Analytic Geometry, Grundlehren der Mathematischen Wissenschaften 261, Springer-Verlag, Berline, 1984.

[CLa] K. Chan, and S.-C. Lau *Open Gromov-Witten invariants and superpotentials for semi-Fano toric surfaces*, preprint 2010, arXiv:1010.5287.

[CLe] K. Chan and N.C. Leung, *Mirror symmetry for toric Fano manifolds via SYZ transformations*, Adv. Math. 223 (2010), no. 3, 797–839, arXiv:0801.2830.

[Cho1] C.-H. Cho, *Holomorphic discs, spin structures and the Floer cohomology of the Clifford torus*, Internat. Math. Res. Notices 35 (2004), 1803–1843.

[Cho2] C.-H. Cho, *Products of Floer cohomology of torus fibers in toric Fano manifolds*, Commun. Math. Phys. 260 (2005) 613–640, math.SG/0412414.

[Cho3] C.-H. Cho, *Non-displaceable Lagrangian submanifolds and Floer cohomology with non-unitary line bundle*, J. Geom. Phys. 58 (2008), 213–226.

[CO] C.-H. Cho and Y.-G. Oh, *Floer cohomology and disc instantons of Lagrangian torus fibers in Fano toric manifolds*, Asian J. Math. 10 (2006), 773–814.

[Dub] B. Dubrovin, *Geometry of 2D topological field theories*, Lecture note in Math. 1620 (1996), 120–348.

[E] M. Entov, *K-area, Hofer metric and geometry of conjugacy classes in Lie groups,* Invent. Math. 146 (2001), 93–141.

[EP1] M. Entov and L. Polterovich, *Calabi quasimorphism and quantum homology,* Int. Math. Res. Not. 2003, no. 30, 1635–1676.

[EP2] M. Entov and L. Polterovich, *Quasi-states and symplectic intersections*, Comment. Math. Helv. 81 (2006), 75–99.

[EP3] M. Entov and L. Polterovich, *Rigid subsets of symplectic manifolds*, Compositio Math. 145 (2009), 773–826.

[FLTZ] B. Fang, C.-C. M. Liu, D. Treumann, and E. Zaslow *A categorification of Morelli's theorem and homological mirror symmetry for toric varieties*, Invent math (2011) 186:79–114, arXiv:0811.1228v3.

[Fl1] A. Floer, *Morse theory for Lagrangian intersections,* J. Differ. Geom. 28 (1988), 513–547.

[Fl2] A. Floer, *Cup length estimates on Lagrangian intersections.* Comm. Pure Appl. Math. 42 (1989), no. 4, 335–356.

[Fu1] K. Fukaya, *Floer homology and mirror symmetry II*, Minimal Surfaces, Geometric Analysis and Symplectic Geometry (Baltimore, MD, 1999) Adv. Stud. Pure Math. 34, 31–127, 2002.

[Fu2] K. Fukaya, *Differentiable operad, Kuranishi correspondence, and Foundation of topological field theories based on pseudo-holomorphic curve*, in Arithmetic and Geometry Around Quantization, Ceyhan et. ed. Birkhäuser Prog. Math. 279 (2010) 123–200.

[Fu3] K. Fukaya, *Cyclic symmetry and adic convergence in Lagrangian Floer theory,* Kyoto J. Math. 50 (2010) 521–590, arXiv:0907.4219.

[Fu4] K. Fukaya, *Counting pseudo-holomorphic discs in Calabi-Yau 3 fold*, to appear in Tohoku Math. J., arXiv:0908.0148.

[FOOO1] K. Fukaya, Y.-G. Oh, H. Ohta and K. Ono, Lagrangian Intersection Floer theory – Anomaly and Obstructions –, Part I & II AMS/IP Studies in Advanced Mathematics, vol 46-1, vol 46-2, 2009, Amer. Math. Soc./International Press, MR2553465, MR2548482.

[FOOO2] K. Fukaya, Y.-G. Oh, H. Ohta and K. Ono, *Canonical models of filtered A_∞-algebras and Morse complexes*, In: New Perspectives and Challenges in Symplectic Field Theory, CRM Proceedings and Lecture Notes 49, 201–227, Amer. Math. Soc. 2009.

[FOOO3] K. Fukaya, Y.-G. Oh, H. Ohta and K. Ono, *Lagrangian Floer theory on compact toric manifolds I,* Duke. Math. J. 151 (2010), 23–174.

[FOOO4] K. Fukaya, Y.-G. Oh, H. Ohta and K. Ono, *Lagrangian Floer theory on compact toric manifolds II: bulk deformations,* Selecta Math. New Series, 17 no. 3, (2011), 609–711. arXiv:0810.5654.

[FOOO5] K. Fukaya, Y.-G. Oh, H. Ohta and K. Ono, *Lagrangian Floer theory and mirror symmetry on compact toric manifolds,* preprint, arXiv:1009.1648.

[FOOO6] K. Fukaya, Y.-G. Oh, H. Ohta and K. Ono, *Toric degeneration and nondisplaceable Lagrangian tori in $S^2 \times S^2$,* to appear in Int. Math. Res. Not., doi:10.1093/imrn/rnr128, arXiv:1002.1666.

[FOOO7] K. Fukaya, Y.-G. Oh, H. Ohta and K. Ono, *Spectral invariants with bulk, quasimorphisms and Lagrangian Floer theory,* submitted, arXiv:1105. 5123.

[FOOO8] K. Fukaya, Y.-G. Oh, H. Ohta and K. Ono, *Anchored Lagrangian submanifolds and their Floer theory,* Contemp. Math. 527 (2010) 15–54.

[FO] K. Fukaya and K. Ono, *Arnold conjecture and Gromov-Witten invariant*, Topology, 38 (1999), 933–1048.

[Ful] W. Fulton, Introduction to Toric Varieties, Annals of Math. Studies, 131, Princeton University Press, Princeton, 1993.

[Gi1] A. Givental, *Homological geometry and mirror symmetry*, in Proceeding of the ICM-1994, Zürich, Basel, 1995, 472–480.

[Gi2] A. Givental, *A mirror theorem for toric complete intersections*, Topological field theory, primitive forms and related topics (Kyoto, 1996), 141–175, Progr. Math., 160, Birkhäuser Boston, MA, 1998.

[GrPa] T. Graber and R. Pandharipande, *Localization of virtual classes*, Invent. Math. 135 (1999), 487–518.

[Gro1] M. Gross, *Mirror symmetry for* $\mathbb{P}^2$ *and tropical geometry*, Adv. Math. 224 (2010), no. 1, 169–245, arXiv:0903.1378.

[Gro2] M. Gross, *Toric geometry and mirror symmetry*, CBMS Regional Conference Series in Mathematics, 114. Published for the Conference Board of the Mathematical Sciences, Washington, DC; by the American Mathematical Society, Providence, RI, 2011.

[GPS] M. Gross, R. Pandharipande, and B. Siebert, *The tropical vertex*, Duke Math. J. 153 (2010), no. 2, 297–362, arXiv:0902.0779.

[Gu] V. Guillemin, *Kähler structures on toric varieties*, J. Differ. Geom. 43 (1994), 285–309.

[Ho] K. Hori, *Linear models in supersymmetric D-branes,* Proceedings of the KIAS conference - Mirror Symmetry and Symplectic Geometry- (Seoul, 2000), eds by K. Fukaya, Y.-G. Oh, K. Ono and G. Tian, World Sci., River Edge, New Jersey, 2001.

[HV] K. Hori and C. Vafa, *Mirror symmetry,* preprint, 2000, hep-th/0002222.

[MIRROR] K. Hori, S. Katz, A. Klemm, R. Pandharipande, R. Thomas, C. Vafa, R. Vakil and E. Zaslow, Mirror Symmetry, Clay Mathematics Monographs, Amer. Math. Soc., 2003

[Iri1] H. Iritani, *Quantum D-modules and generalized mirror transformations*, Topology 47 (2008), 225–276.

[Iri2] H. Iritani, *An integral structure in quantum cohomology and mirror symmetry for toric orbifolds*, Adv. Math. 222 (2009), 1016–1079.

[Iri3] H. Iritani, *Real and integral structures in quantum cohomology I: Toric orbifolds*, preprint, arXiv:0712.2204.

[Ko] M. Kontsevich, *Homological algebra of mirror symmetry,* ICM-1994 Proceedings, Zürich, Birkhäuser, 1995.

[KS1] M. Kontsevich and Y. Soibelman, *Homological mirror symmetry and torus fibration,* in: Symplectic Geometry and Mirror Symmetry, (Seoul, 2000), eds by K. Fukaya, Y.-G. Oh, K. Ono and G. Tian, pp 203–263, World Sci. River Edge, 2001.

[KS2] M. Kontsevich and Y. Soibelman, *Affine structures and non-archimedean analytic spaces*, in: The Unity of Mathematics (P. Etingof, V. Retakh, I. M. Singer, eds), pp 321–385, Progr. Math. 244, Birkhäuser 2006.

[Mc] D. McDuff, *Displacing Lagrangian toric fibers via probes*, preprint, arXiv:0904.1686.

[MS] D. McDuff and D. Salamon, Introduction to symplectic topology. Second edition. Oxford Mathematical Monographs. The Clarendon Press, Oxford University Press, New York, 1998.

[MT] D. McDuff and S. Tolman, *Topological properties of Hamiltonian circle actions*, Int. Math. Res. Pap. 2006, 72826, 1–77.

[NNU1] T. Nishinou, Y. Nohara, and K. Ueda, *Toric degenerations of Gelfand-Cetlin systems and potential functions*, Adv. Math. 224 (2010), 648–706., arXiv:0810.3470.

[NNU2] T. Nishinou, Y. Nohara, and K. Ueda, *Potential functions via toric degenerations*, preprint, arXiv:0812.0066.

[Os] Y. Ostrover, *Calabi quasi-morphisms for some non-monotone symplectic manifolds*, Algebr. Geom. Topol. 6 (2006), 405–434.

[OT] Y. Ostrover and I. Tyomkin, *On the quantum homology algebra of toric Fano manifolds,* Selecta Math. 15 (2009), 121–149, arXiv:0804.0270.

[Sab] C. Sabbah, *Universal unfolding of Laurent polynomials and tt^* structures*, in "From Hodge Theory to Integrabiliy and TQFT tt^* geometry", Proceeding Symposia in Pure Mathematics 78 (2008) 1–31.

[Sa] K. Saito, *Period mapping associated to a primitive form*, Publ. R.I.M.S. 19 (1983), 1231–1261.

[SaTa] K. Saito and A. Takahashi, *From primitive forms to Frobenius manifolds*, in "From Hodge Theory to Integrabiliy and TQFT tt^* geometry", Proceeding Symposia in Pure Mathematics 78 (2008) 31–48.

[MSa] M. Saito, *On the structure of Brieskorn lattices*, Ann. Institute Fourier (Grenoble) 39 (1989) 27–72.

[Se1] P. Seidel, π_1 *of symplectic automorphism groups and invertibles in quantum homology rings*, Geom. Funct. Anal. 7 (1997), no. 6, 1046–1095.

[Se2] P. Seidel, *More about vanishing cycles and mutation,* Symplectic geometry and mirror symmetry (Seoul, 2000), 429–465, World Sci. Publ., River Edge, NJ, 2001.

[Se3] P. Seidel, *Symplectic homology as Hochschild homology.* Algebraic Geometry—Seattle 2005. Part 1, 415–434, Proc. Sympos. Pure Math., 80, Part 1, Amer. Math. Soc., Providence, RI, 2009.

[SYG] A. Strominger, S. Yau and E. Zaslow *Mirror symmetry is T-duality.* Nucl. Phys. B 479 (1996) 243–259.

[Ta] A. Takahashi, *Primitive forms, topological LG model coupled with gravity, and mirror symmetry*, preprint, arXiv:9802059.

[Us] M. Usher, *Spectral numbers in Floer theories*, Compos. Math. 144 (2008), 1581–1592.

[Ue] K. Ueda, *Homological mirror symmetry for toric del Pezzo surfaces*, Comm. Math. Phys. 264 (2006), 71–85.

[UY] K. Ueda and M. Yamazaki, *Homological mirror symmetry for toric orbifolds of toric del Pezzo surfaces*, preprint, arXiv:math/0703267.

[W] C. Woodward, *Gauged Floer theory of toric moment fibers,* Geom. Funct. Anal. 21 (2011), no. 3, 680–749, arXiv:1004.2841.

Department of Mathematics, Kyoto University, Kyoto, Japan
E-mail address: fukaya@math.kyoto-u.ac.jp

Department of Mathematics, University of Wisconsin, Madison, WI, USA, & Department of Mathematics, POSTECH, Pohang, Korea
E-mail address: oh@math.wisc.edu

Graduate School of Mathematics, Nagoya University, Nagoya, Japan, & Korea Institute for Advanced Study, Seoul, Korea
E-mail address: ohta@math.nagoya-u.ac.jp

Research Institute for Mathematical Sciences, Kyoto Univerisity, Kyoto, Japan, & Korea Institute for Advanced Study, Seoul, Korea
E-mail address: ono@kurims.kyoto-u.ac.jp

An introduction to C^∞-schemes and C^∞-algebraic geometry

Dominic Joyce

1. Introduction

If X is a manifold then the $\mathbb{R}$-algebra $C^\infty(X)$ of smooth functions $c : X \to \mathbb{R}$ is a C^∞*-ring*. That is, for each smooth function $f : \mathbb{R}^n \to \mathbb{R}$ there is an n-fold operation $\Phi_f : C^\infty(X)^n \to C^\infty(X)$ acting by $\Phi_f : c_1, \ldots, c_n \mapsto f(c_1, \ldots, c_n)$, and these operations Φ_f satisfy many natural identities. Thus, $C^\infty(X)$ actually has a far richer structure than the obvious $\mathbb{R}$-algebra structure.

In [**7**] the author set out the foundations of a version of algebraic geometry in which rings or algebras are replaced by C^∞-rings, focussing on C^∞*-schemes*, a category of geometric objects which generalize manifolds, and whose morphisms generalize smooth maps, *quasicoherent* and *coherent sheaves* on C^∞-schemes, and C^∞*-stacks*, in particular *Deligne–Mumford* C^∞*-stacks*, a 2-category of geometric objects which generalize orbifolds. This paper is a survey of [**7**].

C^∞-rings and C^∞-schemes were first introduced in *synthetic differential geometry*, see for instance Dubuc [**3**], Moerdijk and Reyes [**14**] and Kock [**9**]. Following Dubuc's discussion of 'models of synthetic differential geometry' [**2**] and oversimplifying a bit, symplectic differential geometers are interested in C^∞-schemes as they provide a category $\mathbf{C^\infty Sch}$ of geometric objects which includes smooth manifolds and certain 'infinitesimal' objects, and all fibre products exist in $\mathbf{C^\infty Sch}$, and $\mathbf{C^\infty Sch}$ has some other nice properties to do with open covers, and exponentials of infinitesimals.

Synthetic differential geometry concerns proving theorems about manifolds using synthetic reasoning involving 'infinitesimals'. But one needs to check these methods of synthetic reasoning are valid. To do this you need a 'model', some category of geometric spaces including manifolds and infinitesimals, in which you can think of your synthetic arguments as happening. Once you know there exists at least one model with the properties you want, then as far as synthetic differential geometry is concerned the job is done. For this reason C^∞-schemes were not developed very far in synthetic differential geometry.

Recently, C^∞-rings and C^∞-ringed spaces appeared in a very different context, as part of Spivak's definition of *derived manifolds* [**17**], which are an extension to differential geometry of Lurie's 'derived algebraic geometry' programme [**11**]. The author [**8**] is developing an alternative theory of derived differential geometry which simplifies, and goes beyond, Spivak's derived manifolds. Our notion of derived manifolds are called *d-manifolds*. We also study *d-manifolds with boundary*, and *d-manifolds with corners*, and orbifold versions of all these, *d-orbifolds*. To define d-manifolds and d-orbifolds we need theories of C^∞-schemes, C^∞-stacks, and quasicoherent sheaves upon them, much of which had not been done, so the author set up the foundations of these in [**7**].

D-manifolds and d-orbifolds will have important applications in symplectic geometry, and elsewhere. Many areas of symplectic geometry involve moduli spaces $\overline{\mathcal{M}}_{g,m}(J,\beta)$ of stable J-holomorphic curves in a symplectic manifold (M,ω). The original motivation for [**8**] was to find a good geometric description for the geometric structure on such moduli spaces $\overline{\mathcal{M}}_{g,m}(J,\beta)$. In the Lagrangian Floer cohomology theory of Fukaya, Oh, Ohta and Ono [**5**], moduli spaces $\overline{\mathcal{M}}_{g,m}(J,\beta)$ are given the structure of *Kuranishi spaces*. The notion of Kuranishi space seemed to the author to be unsatisfactory. In trying improve it, using ideas from Spivak [**17**], the author arrived at the theory of [**8**]. The author believes the 'correct' definition of Kuranishi space in the work of Fukaya et al. [**5**] should be that a Kuranishi space is a d-orbifold with corners.

Section 2 explains C^∞-rings and their modules, §3 introduces C^∞-schemes, and quasicoherent and coherent sheaves upon them, and §4 discusses C^∞-stacks, particularly Deligne–Mumford C^∞-stacks, their relation to orbifolds, and quasicoherent and coherent sheaves on Deligne–Mumford C^∞-stacks.

2. C^∞-rings

We begin by explaining the basic objects out of which our theories are built, *C^∞-rings*, or *smooth rings*, following [**7**, §2, §3 & §5]. With the exception of the material on *good* C^∞-rings in §2.2, almost everything in §2.1–§2.2 was already known in synthetic differential geometry, and can be found in Moerdijk and Reyes [**14**, Ch. I], Dubuc [**2**–**4**] or Kock [**9**, §III].

2.1. Two definitions of C^∞-ring.

DEFINITION 2.1. A *C^∞-ring* is a set $\mathfrak{C}$ together with operations $\Phi_f : \mathfrak{C}^n \to \mathfrak{C}$ for all $n \geqslant 0$ and smooth maps $f : \mathbb{R}^n \to \mathbb{R}$, where by convention when $n = 0$ we define $\mathfrak{C}^0$ to be the single point $\{\emptyset\}$. These operations must satisfy the following relations: suppose $m, n \geqslant 0$, and $f_i : \mathbb{R}^n \to \mathbb{R}$ for

$i=1,\dots,m$ and $g:\mathbb{R}^m\to\mathbb{R}$ are smooth functions. Define a smooth function $h:\mathbb{R}^n\to\mathbb{R}$ by

$$h(x_1,\dots,x_n)=g\big(f_1(x_1,\dots,x_n),\dots,f_m(x_1\dots,x_n)\big),$$

for all $(x_1,\dots,x_n)\in\mathbb{R}^n$. Then for all $(c_1,\dots,c_n)\in\mathfrak{C}^n$ we have

$$\Phi_h(c_1,\dots,c_n)=\Phi_g\big(\Phi_{f_1}(c_1,\dots,c_n),\dots,\Phi_{f_m}(c_1,\dots,c_n)\big).$$

We also require that for all $1\leqslant j\leqslant n$, defining $\pi_j:\mathbb{R}^n\to\mathbb{R}$ by $\pi_j:(x_1,\dots,x_n)\mapsto x_j$, we have $\Phi_{\pi_j}(c_1,\dots,c_n)=c_j$ for all $(c_1,\dots,c_n)\in\mathfrak{C}^n$.

Usually we refer to $\mathfrak{C}$ as the C^∞-ring, leaving the operations Φ_f implicit.

A *morphism* between C^∞-rings $\big(\mathfrak{C},(\Phi_f)_{f:\mathbb{R}^n\to\mathbb{R}\ C^\infty}\big)$, $\big(\mathfrak{D},(\Psi_f)_{f:\mathbb{R}^n\to\mathbb{R}\ C^\infty}\big)$ is a map $\phi:\mathfrak{C}\to\mathfrak{D}$ such that $\Psi_f\big(\phi(c_1),\dots,\phi(c_n)\big)=\phi\circ\Phi_f(c_1,\dots,c_n)$ for all smooth $f:\mathbb{R}^n\to\mathbb{R}$ and $c_1,\dots,c_n\in\mathfrak{C}$. We will write $\mathbf{C^\infty Rings}$ for the category of C^∞-rings.

Here is the motivating example:

EXAMPLE 2.2. Let X be a manifold, and write $C^\infty(X)$ for the set of smooth functions $c:X\to\mathbb{R}$. For $n\geqslant 0$ and $f:\mathbb{R}^n\to\mathbb{R}$ smooth, define $\Phi_f: C^\infty(X)^n\to C^\infty(X)$ by

$$\big(\Phi_f(c_1,\dots,c_n)\big)(x)=f\big(c_1(x),\dots,c_n(x)\big), \tag{1}$$

for all $c_1,\dots,c_n\in C^\infty(X)$ and $x\in X$. It is easy to see that $C^\infty(X)$ and the operations Φ_f form a C^∞-ring.

Now let $f:X\to Y$ be a smooth map of manifolds. Then pullback $f^*: C^\infty(Y)\to C^\infty(X)$ mapping $f^*:c\mapsto c\circ f$ is a morphism of C^∞-rings. Furthermore, every C^∞-ring morphism $\phi:C^\infty(Y)\to C^\infty(X)$ is of the form $\phi=f^*$ for a unique smooth map $f:X\to Y$.

Write $\mathbf{C^\infty Rings}^{\rm op}$ for the opposite category of $\mathbf{C^\infty Rings}$, with directions of morphisms reversed, and $\mathbf{Man}$ for the category of manifolds without boundary. Then we have a full and faithful functor $F_{\mathbf{Man}}^{\mathbf{C^\infty Rings}}:\mathbf{Man}\to\mathbf{C^\infty Rings}^{\rm op}$ acting by $F_{\mathbf{Man}}^{\mathbf{C^\infty Rings}}(X)=C^\infty(X)$ on objects and $F_{\mathbf{Man}}^{\mathbf{C^\infty Rings}}(f)=f^*$ on morphisms. This embeds $\mathbf{Man}$ as a full subcategory of $\mathbf{C^\infty Rings}^{\rm op}$.

Note that C^∞-rings are far more general than those coming from manifolds. For example, if X is any topological space we could define a C^∞-ring $C^0(X)$ to be the set of *continuous* $c:X\to\mathbb{R}$, with operations Φ_f defined as in (1). For X a manifold with $\dim X>0$, the C^∞-rings $C^\infty(X)$ and $C^0(X)$ are different.

There is a more succinct definition of C^∞-rings using category theory:

DEFINITION 2.3. Write $\mathbf{Euc}$ for the full subcategory of $\mathbf{Man}$ spanned by the Euclidean spaces $\mathbb{R}^n$. That is, the objects of $\mathbf{Euc}$ are the manifolds $\mathbb{R}^n$ for $n=0,1,2,\dots$, and the morphisms in $\mathbf{Euc}$ are smooth maps $f:\mathbb{R}^n\to\mathbb{R}^m$. Write $\mathbf{Sets}$ for the category of sets. In both $\mathbf{Euc}$ and $\mathbf{Sets}$ we have notions of (finite) products of objects (that is, $\mathbb{R}^{n+m} = \mathbb{R}^n \times \mathbb{R}^m$, and products $S \times T$ of sets S, T), and products of morphisms. Define a *(category-theoretic) C^∞-ring* to be a product-preserving functor $F : \mathbf{Euc} \to \mathbf{Sets}$.

Here is how this relates to Definition 2.1. Suppose $F : \mathbf{Euc} \to \mathbf{Sets}$ is a product-preserving functor. Define $\mathfrak{C} = F(\mathbb{R})$. Then $\mathfrak{C}$ is an object in **Sets**, that is, a set. Suppose $n \geqslant 0$ and $f : \mathbb{R}^n \to \mathbb{R}$ is smooth. Then f is a morphism in **Euc**, so $F(f) : F(\mathbb{R}^n) \to F(\mathbb{R}) = \mathfrak{C}$ is a morphism in **Sets**. Since F preserves products $F(\mathbb{R}^n) = F(\mathbb{R}) \times \cdots \times F(\mathbb{R}) = \mathfrak{C}^n$, so $F(f)$ maps $\mathfrak{C}^n \to \mathfrak{C}$. We define $\Phi_f : \mathfrak{C}^n \to \mathfrak{C}$ by $\Phi_f = F(f)$. Then $\bigl(\mathfrak{C}, (\Phi_f)_{f:\mathbb{R}^n \to \mathbb{R}\ C^\infty}\bigr)$ is a C^∞ ring.

Since all small colimits exist in **Sets**, regarding C^∞-rings as functors $F : \mathbf{Euc} \to \mathbf{Sets}$ as in Definition 2.3, to take small colimits in the category of C^∞-rings we can take colimits in **Sets** object-wise in **Euc**, so as in [**7**, Prop. 2.5], [**14**, p. 21-22] we have:

PROPOSITION 2.4. *In the category* $\mathbf{C^\infty Rings}$ *of C^∞-rings, all small colimits exist, and so in particular pushouts and all finite colimits exist.*

DEFINITION 2.5. Let $\mathfrak{C}$ be a C^∞-ring. Then we may give $\mathfrak{C}$ the structure of a *commutative $\mathbb{R}$-algebra*. Define addition '$+$' on $\mathfrak{C}$ by $c+c' = \Phi_f(c, c')$ for $c, c' \in \mathfrak{C}$, where $f : \mathbb{R}^2 \to \mathbb{R}$ is $f(x, y) = x+y$. Define multiplication '$\cdot$' on $\mathfrak{C}$ by $c \cdot c' = \Phi_g(c, c')$, where $g : \mathbb{R}^2 \to \mathbb{R}$ is $f(x, y) = xy$. Define scalar multiplication by $\lambda \in \mathbb{R}$ by $\lambda c = \Phi_{\lambda'}(c)$, where $\lambda' : \mathbb{R} \to \mathbb{R}$ is $\lambda'(x) = \lambda x$. Define elements 0 and 1 in $\mathfrak{C}$ by $0 = \Phi_{0'}(\emptyset)$ and $1 = \Phi_{1'}(\emptyset)$, where $0' : \mathbb{R}^0 \to \mathbb{R}$ and $1' : \mathbb{R}^0 \to \mathbb{R}$ are the maps $0' : \emptyset \mapsto 0$ and $1' : \emptyset \mapsto 1$. One can then show using the relations on the Φ_f that all the axioms of a commutative $\mathbb{R}$-algebra are satisfied. In Example 2.2, this yields the obvious $\mathbb{R}$-algebra structure on the smooth functions $c : X \to \mathbb{R}$.

An *ideal* I in $\mathfrak{C}$ is an ideal $I \subset \mathfrak{C}$ in $\mathfrak{C}$ regarded as a commutative $\mathbb{R}$-algebra. Then we make the quotient $\mathfrak{C}/I$ into a C^∞-ring as follows. If $f : \mathbb{R}^n \to \mathbb{R}$ is smooth, define $\Phi_f^I : (\mathfrak{C}/I)^n \to \mathfrak{C}/I$ by

$$\bigl(\Phi_f^I(c_1 + I, \ldots, c_n + I)\bigr)(x) = f\bigl(c_1(x), \ldots, c_n(x)\bigr) + I.$$

To show this is well-defined, we must show it is independent of the choice of representatives $c_1, \ldots, c_n$ in $\mathfrak{C}$ for $c_1 + I, \ldots, c_n + I$ in $\mathfrak{C}/I$. By Hadamard's Lemma there exist smooth functions $g_i : \mathbb{R}^{2n} \to \mathbb{R}$ for $i = 1, \ldots, n$ with

$$f(y_1, \ldots, y_n) - f(x_1, \ldots, x_n) = \textstyle\sum_{i=1}^n (y_i - x_i) g_i(x_1, \ldots, x_n, y_1, \ldots, y_n)$$

for all $x_1, \ldots, x_n, y_1, \ldots, y_n \in \mathbb{R}$. If $c_1', \ldots, c_n'$ are alternative choices for $c_1, \ldots, c_n$, so that $c_i' + I = c_i + I$ for $i = 1, \ldots, n$ and $c_i' - c_i \in I$, we have

$$\begin{aligned} &f\bigl(c_1'(x), \ldots, c_n'(x)\bigr) - f\bigl(c_1(x), \ldots, c_n(x)\bigr) \\ &\quad = \textstyle\sum_{i=1}^n (c_i' - c_i) g_i\bigl(c_1'(x), \ldots, c_n'(x), c_1(x), \ldots, c_n(x)\bigr). \end{aligned}$$

The second line lies in I as $c_i'-c_i\in I$ and I is an ideal, so Φ_f^I is well-defined, and clearly $\bigl(\mathfrak{C}/I,(\Phi_f^I)_{f:\mathbb{R}^n\to\mathbb{R}\ C^\infty}\bigr)$ is a C^∞-ring.

We will use the notation $(f_a : a\in A)$ to denote the ideal in a C^∞-ring $\mathfrak{C}$ generated by a collection of elements $f_a\in\mathfrak{C}$, $a\in A$. That is,

$$(f_a : a\in A)=\bigl\{\textstyle\sum_{i=1}^n f_{a_i}\cdot c_i : n\geqslant 0,\ a_1,\ldots,a_n\in A,\ c_1,\ldots,c_n\in\mathfrak{C}\bigr\}.$$

2.2. Special classes of C^∞-ring. We define *finitely generated, finitely presented, local, fair,* and *good* C^∞-rings.

DEFINITION 2.6. A C^∞-ring $\mathfrak{C}$ is called *finitely generated* if there exist $c_1,\ldots,c_n$ in $\mathfrak{C}$ which generate $\mathfrak{C}$ over all C^∞-operations. That is, for each $c\in\mathfrak{C}$ there exists smooth $f:\mathbb{R}^n\to\mathbb{R}$ with $c=\Phi_f(c_1,\ldots,c_n)$. Given such $\mathfrak{C},c_1,\ldots,c_n$, define $\phi: C^\infty(\mathbb{R}^n)\to\mathfrak{C}$ by $\phi(f)=\Phi_f(c_1,\ldots,c_n)$ for smooth $f:\mathbb{R}^n\to\mathbb{R}$, where $C^\infty(\mathbb{R}^n)$ is as in Example 2.2 with $X=\mathbb{R}^n$. Then ϕ is a surjective morphism of C^∞-rings, so $I=\operatorname{Ker}\phi$ is an ideal in $C^\infty(\mathbb{R}^n)$, and $\mathfrak{C}\cong C^\infty(\mathbb{R}^n)/I$ as a C^∞-ring. Thus, $\mathfrak{C}$ is finitely generated if and only if $\mathfrak{C}\cong C^\infty(\mathbb{R}^n)/I$ for some $n\geqslant 0$ and ideal I in $C^\infty(\mathbb{R}^n)$.

An ideal I in $C^\infty(\mathbb{R}^n)$ is called *finitely generated* if $I=(f_1,\ldots,f_k)$ for some $f_1,\ldots,f_k\in C^\infty(\mathbb{R}^n)$. A C^∞-ring $\mathfrak{C}$ is called *finitely presented* if $\mathfrak{C}\cong C^\infty(\mathbb{R}^n)/I$ for some $n\geqslant 0$, where I is a finitely generated ideal in $C^\infty(\mathbb{R}^n)$.

A difference with conventional algebraic geometry is that $C^\infty(\mathbb{R}^n)$ is not noetherian, so ideals in $C^\infty(\mathbb{R}^n)$ may not be finitely generated, and $\mathfrak{C}$ finitely generated does not imply $\mathfrak{C}$ finitely presented.

DEFINITION 2.7. A C^∞-ring $\mathfrak{C}$ is called a C^∞-*local ring* if regarded as an $\mathbb{R}$-algebra, as in Definition 2.5, $\mathfrak{C}$ is a local $\mathbb{R}$-algebra with residue field $\mathbb{R}$. That is, $\mathfrak{C}$ has a unique maximal ideal $\mathfrak{m}_{\mathfrak{C}}$ with $\mathfrak{C}/\mathfrak{m}_{\mathfrak{C}}\cong\mathbb{R}$.

If $\mathfrak{C},\mathfrak{D}$ are C^∞-local rings with maximal ideals $\mathfrak{m}_{\mathfrak{C}},\mathfrak{m}_{\mathfrak{D}}$, and $\phi:\mathfrak{C}\to\mathfrak{D}$ is a morphism of C^∞ rings, then using the fact that $\mathfrak{C}/\mathfrak{m}_{\mathfrak{C}}\cong\mathbb{R}\cong\mathfrak{D}/\mathfrak{m}_{\mathfrak{D}}$ we see that $\phi^{-1}(\mathfrak{m}_{\mathfrak{D}})=\mathfrak{m}_{\mathfrak{C}}$, that is, ϕ is a *local* morphism of C^∞-local rings. Thus, there is no difference between morphisms and local morphisms.

EXAMPLE 2.8. For $n\geqslant 0$ and $p\in\mathbb{R}^n$, define $C_p^\infty(\mathbb{R}^n)$ to be the set of germs of smooth functions $c:\mathbb{R}^n\to\mathbb{R}$ at $p\in\mathbb{R}^n$. That is, $C_p^\infty(\mathbb{R}^n)$ is the quotient of the set of pairs (U,c) with $p\in U\subset\mathbb{R}^n$ open and $c:U\to\mathbb{R}$ smooth by the equivalence relation $(U,c)\sim(U',c')$ if there exists $p\in V\subseteq U\cap U'$ open with $c|_V\equiv c'|_V$. Define operations $\Phi_f:(C_p^\infty(\mathbb{R}^n))^m\to C_p^\infty(\mathbb{R}^n)$ for $f:\mathbb{R}^m\to\mathbb{R}$ smooth by (1). Then $C_p^\infty(\mathbb{R}^n)$ is a C^∞-local ring, with maximal ideal $\mathfrak{m}=\bigl\{[(U,c)] : c(p)=0\bigr\}$.

DEFINITION 2.9. An ideal I in $C^\infty(\mathbb{R}^n)$ is called *fair* if for each $f\in C^\infty(\mathbb{R}^n)$, f lies in I if and only if $\pi_p(f)$ lies in $\pi_p(I)\subseteq C_p^\infty(\mathbb{R}^n)$ for all $p\in\mathbb{R}^n$, where $C_p^\infty(\mathbb{R}^n)$ is as in Example 2.8 and $\pi_p: C^\infty(\mathbb{R}^n)\to C_p^\infty(\mathbb{R}^n)$ is the natural projection $\pi_p: c\mapsto[(\mathbb{R}^n,c)]$. A C^∞-ring $\mathfrak{C}$ is called *fair* if it is isomorphic to $C^\infty(\mathbb{R}^n)/I$, where I is a fair ideal.

Let X be a closed subset of $\mathbb{R}^n$. Define $\mathfrak{m}_X^\infty$ to be the ideal of all $g \in C^\infty(\mathbb{R}^n)$ such that $\partial^k g|_X \equiv 0$ for all $k \geqslant 0$, that is, g and all its derivatives vanish at each $x \in X$. An ideal I in $C^\infty(\mathbb{R}^n)$ is called *good* if it is of the form $I = (f_1, \ldots, f_k, \mathfrak{m}_X^\infty)$ for some $f_1, \ldots, f_k \in C^\infty(\mathbb{R}^n)$ and closed $X \subseteq \mathbb{R}^n$. A C^∞-ring $\mathfrak{C}$ is called *good* if $\mathfrak{C} \cong C^\infty(\mathbb{R}^n)/I$ for $n \geqslant 0$, where I is a good ideal.

Our term 'fair' was introduced in [**7**] for brevity, but the idea was already well-known. They were introduced by Dubuc [**3**, Def. 11] under the name 'C^∞-rings of finite type presented by an ideal of local character', and in more recent work would be called 'finitely generated and germ-determined C^∞-rings'.

As in [**7**, §2], if $C^\infty(\mathbb{R}^m)/I \cong C^\infty(\mathbb{R}^n)/J$ then I is finitely generated, or fair, or good, if and only if J is. Thus, to decide whether a C^∞-ring $\mathfrak{C}$ is finitely presented, or fair, or good, it is enough to test one presentation $\mathfrak{C} \cong C^\infty(\mathbb{R}^n)/I$. Also, $\mathfrak{C}$ finitely presented implies $\mathfrak{C}$ good implies $\mathfrak{C}$ fair implies $\mathfrak{C}$ finitely generated. Write $\mathbf{C^\infty Rings^{fp}}, \mathbf{C^\infty Rings^{go}}, \mathbf{C^\infty Rings^{fa}}$ and $\mathbf{C^\infty Rings^{fg}}$ for the full subcategories of finitely presented, good, fair, and finitely generated C^∞-rings in $\mathbf{C^\infty Rings}$, respectively. Then

$$\mathbf{C^\infty Rings^{fp}} \subset \mathbf{C^\infty Rings^{go}} \subset \mathbf{C^\infty Rings^{fa}} \subset \mathbf{C^\infty Rings^{fg}} \subset \mathbf{C^\infty Rings}.$$

From [**7**, Prop.s 2.28 & 2.30] we have:

PROPOSITION 2.10. *The subcategories* $\mathbf{C^\infty Rings^{fg}}, \mathbf{C^\infty Rings^{fp}}, \mathbf{C^\infty Rings^{go}}$ *are closed under pushouts and all finite colimits in* $\mathbf{C^\infty Rings}$, *but* $\mathbf{C^\infty Rings^{fa}}$ *is not. Nonetheless, pushouts and finite colimits exist in* $\mathbf{C^\infty Rings^{fa}}$, *though they may not coincide with pushouts and finite colimits in* $\mathbf{C^\infty Rings}$.

Given morphisms $\phi : \mathfrak{C} \to \mathfrak{D}$ and $\psi : \mathfrak{C} \to \mathfrak{E}$ in $\mathbf{C^\infty Rings}$, the pushout $\mathfrak{D} \amalg_{\phi,\mathfrak{C},\psi} \mathfrak{E}$ in $\mathbf{C^\infty Rings}$ should be thought of as a *completed tensor product* $\mathfrak{D} \hat{\otimes}_{\mathfrak{C}} \mathfrak{E}$. The tensor product $\mathfrak{D} \otimes_{\mathfrak{C}} \mathfrak{E}$ is an $\mathbb{R}$-algebra, but in general not a C^∞-ring, and to get a C^∞-ring we must take a completion $\mathfrak{D} \hat{\otimes}_{\mathfrak{C}} \mathfrak{E}$. When $\mathfrak{C} = \mathbb{R}$, the trivial C^∞-ring, the pushout $\mathfrak{D} \amalg_{\mathbb{R}} \mathfrak{E}$ is the coproduct $\mathfrak{D} \amalg \mathfrak{E} = \mathfrak{D} \hat{\otimes}_{\mathbb{R}} \mathfrak{E}$. For example, one can show that $C^\infty(\mathbb{R}^m) \hat{\otimes}_{\mathbb{R}} C^\infty(\mathbb{R}^n) \cong C^\infty(\mathbb{R}^{m+n})$.

Here is [**7**, Prop. 3.2]. Part (b) is one reason for introducing *good* C^∞-rings.

PROPOSITION 2.11. **(a)** *If X is a manifold without boundary then the C^∞-ring $C^\infty(X)$ of Example* 2.2 *is finitely presented.*

(b) *If X is a manifold with boundary, or with corners, and $\partial X \neq \emptyset$, then the C^∞-ring $C^\infty(X)$ of Example* 2.2 *is good, but is not finitely presented.*

To save space we will say no more about manifolds with boundary or corners and C^∞-geometry in this paper. More information can be found in [**7**, **8**].

EXAMPLE 2.12. A *Weil algebra* [**2**, Def. 1.4] is a finite-dimensional commutative $\mathbb{R}$-algebra W which has a maximal ideal $\mathfrak{m}$ with $W/\mathfrak{m} \cong \mathbb{R}$ and $\mathfrak{m}^n = 0$ for some $n > 0$. Then by Dubuc [**2**, Prop. 1.5] or Kock [**9**, Th. III.5.3], there is a unique way to make W into a C^∞-ring compatible with the given underlying commutative $\mathbb{R}$-algebra. This C^∞-ring is finitely presented [**9**, Prop. III.5.11]. C^∞-rings from Weil algebras are important in synthetic differential geometry, in arguments involving infinitesimals.

2.3. Modules over C^∞-rings, and cotangent modules. In [**7**, §5] we discuss modules over C^∞-rings.

DEFINITION 2.13. Let $\mathfrak{C}$ be a C^∞-ring. A $\mathfrak{C}$-*module* M is a module over $\mathfrak{C}$ regarded as a commutative $\mathbb{R}$-algebra as in Definition 2.5. $\mathfrak{C}$-modules form an abelian category, which we write as $\mathfrak{C}$-mod. For example, $\mathfrak{C}$ is a $\mathfrak{C}$-module, and more generally $\mathfrak{C} \otimes_{\mathbb{R}} V$ is a $\mathfrak{C}$-module for any real vector space V.

A $\mathfrak{C}$-module M is called *finitely presented* if there exists an exact sequence $\mathfrak{C} \otimes_{\mathbb{R}} \mathbb{R}^m \to \mathfrak{C} \otimes_{\mathbb{R}} \mathbb{R}^n \to M \to 0$ in $\mathfrak{C}$-mod for some $m, n \geqslant 0$. We write $\mathfrak{C}$-mod^{fp} for the full subcategory of finitely presented $\mathfrak{C}$-modules in $\mathfrak{C}$-mod. Then $\mathfrak{C}$-mod^{fp} is closed under cokernels and extensions in $\mathfrak{C}$-mod. But it may not be closed under kernels, so $\mathfrak{C}$-mod^{fp} may not be an abelian category.

Let $\phi : \mathfrak{C} \to \mathfrak{D}$ be a morphism of C^∞-rings. If M is a $\mathfrak{C}$-module then $\phi_*(M) = M \otimes_{\mathfrak{C}} \mathfrak{D}$ is a $\mathfrak{D}$-module. This induces a functor $\phi_* : \mathfrak{C}\text{-mod} \to \mathfrak{D}\text{-mod}$, which maps $\mathfrak{C}\text{-mod}^{\mathrm{fp}} \to \mathfrak{D}\text{-mod}^{\mathrm{fp}}$.

EXAMPLE 2.14. Let X be a manifold, and $E \to X$ be a vector bundle. Write $C^\infty(E)$ for the vector space of smooth sections e of E. Then $C^\infty(X)$ acts on $C^\infty(E)$ by $(c, e) \mapsto c \cdot e$ for $c \in C^\infty(X)$ and $e \in C^\infty(E)$, so $C^\infty(E)$ is a $C^\infty(X)$-module, which is finitely presented.

Now let X, Y be manifolds and $f : X \to Y$ a smooth map. Then $f^* : C^\infty(Y) \to C^\infty(X)$ is a morphism of C^∞-rings. If E is a vector bundle over Y, then $f^*(E)$ is a vector bundle over X. Under the functor $(f^*)_* : C^\infty(Y)\text{-mod} \to C^\infty(X)\text{-mod}$ of Definition 2.13, we see that $(f^*)_*\bigl(C^\infty(E)\bigr) = C^\infty(E) \otimes_{C^\infty(Y)} C^\infty(X)$ is isomorphic as a $C^\infty(X)$-module to $C^\infty\bigl(f^*(E)\bigr)$.

Every commutative algebra A has a natural module Ω_A called the *module of Kähler differentials*, which is a kind of analogue for A of the cotangent bundle T^*X of a manifold X. In [**7**, §5.3] we define the *cotangent module* $\Omega_{\mathfrak{C}}$ of a C^∞-ring $\mathfrak{C}$, which is the C^∞-version of the module of Kähler differentials.

DEFINITION 2.15. Let $\mathfrak{C}$ be a C^∞-ring, and M a $\mathfrak{C}$-module. A C^∞-*derivation* is an $\mathbb{R}$-linear map $\mathrm{d} : \mathfrak{C} \to M$ such that whenever $f : \mathbb{R}^n \to \mathbb{R}$ is a smooth map and $c_1, \ldots, c_n \in \mathfrak{C}$, we have

$$\mathrm{d}\Phi_f(c_1, \ldots, c_n) = \textstyle\sum_{i=1}^n \Phi_{\frac{\partial f}{\partial x_i}}(c_1, \ldots, c_n) \cdot \mathrm{d}c_i.$$

Note that d is *not* a morphism of $\mathfrak{C}$-modules. We call such a pair M, d a *cotangent module* for $\mathfrak{C}$ if it has the universal property that for any $\mathfrak{C}$-module M' and C^∞-derivation $\mathrm{d}' : \mathfrak{C} \to M'$, there exists a unique morphism of $\mathfrak{C}$-modules $\phi : M \to M'$ with $\mathrm{d}' = \phi \circ \mathrm{d}$.

Define $\Omega_{\mathfrak{C}}$ to be the quotient of the free $\mathfrak{C}$-module with basis of symbols $\mathrm{d}c$ for $c \in \mathfrak{C}$ by the $\mathfrak{C}$-submodule spanned by all expressions of the form $\mathrm{d}\Phi_f(c_1, \ldots, c_n) - \sum_{i=1}^n \Phi_{\frac{\partial f}{\partial x_i}}(c_1, \ldots, c_n) \cdot \mathrm{d}c_i$ for $f : \mathbb{R}^n \to \mathbb{R}$ smooth and $c_1, \ldots, c_n \in \mathfrak{C}$, and define $\mathrm{d}_{\mathfrak{C}} : \mathfrak{C} \to \Omega_{\mathfrak{C}}$ by $\mathrm{d}_{\mathfrak{C}} : c \mapsto \mathrm{d}c$. Then $\Omega_{\mathfrak{C}}, \mathrm{d}_{\mathfrak{C}}$ is a cotangent module for $\mathfrak{C}$. Thus cotangent modules always exist, and are unique up to unique isomorphism.

Let $\mathfrak{C}, \mathfrak{D}$ be C^∞-rings with cotangent modules $\Omega_{\mathfrak{C}}, \mathrm{d}_{\mathfrak{C}}$, $\Omega_{\mathfrak{D}}, \mathrm{d}_{\mathfrak{D}}$, and $\phi : \mathfrak{C} \to \mathfrak{D}$ be a morphism of C^∞-rings. Then ϕ makes $\Omega_{\mathfrak{D}}$ into a $\mathfrak{C}$-module, and there is a unique morphism $\Omega_\phi : \Omega_{\mathfrak{C}} \to \Omega_{\mathfrak{D}}$ in $\mathfrak{C}$-mod with $\mathrm{d}_{\mathfrak{D}} \circ \phi = \Omega_\phi \circ \mathrm{d}_{\mathfrak{C}}$. This induces a morphism $(\Omega_\phi)_* : \Omega_{\mathfrak{C}} \otimes_{\mathfrak{C}} \mathfrak{D} \to \Omega_{\mathfrak{D}}$ in $\mathfrak{D}$-mod with $(\Omega_\phi)_* \circ (\mathrm{d}_{\mathfrak{C}} \otimes \mathrm{id}_{\mathfrak{D}}) = \mathrm{d}_{\mathfrak{D}}$. If $\phi : \mathfrak{C} \to \mathfrak{D}$, $\psi : \mathfrak{D} \to \mathfrak{E}$ are morphisms of C^∞-rings then $\Omega_{\psi \circ \phi} = \Omega_\psi \circ \Omega_\phi$.

EXAMPLE 2.16. Let X be a manifold. Then the cotangent bundle T^*X is a vector bundle over X, so as in Example 2.14 it yields a $C^\infty(X)$-module $C^\infty(T^*X)$. The exterior derivative $\mathrm{d} : C^\infty(X) \to C^\infty(T^*X)$ is a C^∞-derivation. These $C^\infty(T^*X), \mathrm{d}$ have the universal property in Definition 2.15, and so form a *cotangent module* for $C^\infty(X)$.

Now let X, Y be manifolds, and $f : X \to Y$ be smooth. Then $f^*(TY)$, TX are vector bundles over X, and the derivative of f is a vector bundle morphism $\mathrm{d}f : TX \to f^*(TY)$. The dual of this morphism is $(\mathrm{d}f)^* : f^*(T^*Y) \to T^*X$. This induces a morphism of $C^\infty(X)$-modules $((\mathrm{d}f)^*)_* : C^\infty(f^*(T^*Y)) \to C^\infty(T^*X)$. This $((\mathrm{d}f)^*)_*$ is identified with $(\Omega_{f^*})_*$ in Definition 2.15 under the natural isomorphism $C^\infty(f^*(T^*Y)) \cong C^\infty(T^*Y) \otimes_{C^\infty(Y)} C^\infty(X)$.

Definition 2.15 abstracts the notion of cotangent bundle of a manifold in a way that makes sense for any C^∞-ring. From [**7**, Th.s 5.13 & 5.16] we have:

THEOREM 2.17. **(a)** *Suppose $\mathfrak{C}$ is a finitely presented or good C^∞-ring. Then $\Omega_{\mathfrak{C}}$ is a finitely presented $\mathfrak{C}$-module.*

(b) *Suppose we are given a pushout diagram of finitely generated C^∞-rings:*

$$\begin{array}{ccc} \mathfrak{C} & \xrightarrow{\beta} & \mathfrak{E} \\ \downarrow \alpha & & \delta \downarrow \\ \mathfrak{D} & \xrightarrow{\gamma} & \mathfrak{F}, \end{array}$$

so that $\mathfrak{F} = \mathfrak{D} \amalg_{\mathfrak{C}} \mathfrak{E}$. Then the following sequence of $\mathfrak{F}$-modules is exact:

$$\Omega_{\mathfrak{C}} \otimes_{\mu_{\mathfrak{C}}, \mathfrak{C}, \gamma \circ \alpha} \mathfrak{F} \xrightarrow{\substack{(\Omega_\alpha)_* \oplus \\ -(\Omega_\beta)_*}} \begin{array}{c} \Omega_{\mathfrak{D}} \otimes_{\mu_{\mathfrak{D}}, \mathfrak{D}, \gamma} \mathfrak{F} \oplus \\ \Omega_{\mathfrak{E}} \otimes_{\mu_{\mathfrak{E}}, \mathfrak{E}, \delta} \mathfrak{F} \end{array} \xrightarrow{(\Omega_\gamma)_* \oplus (\Omega_\delta)_*} \Omega_{\mathfrak{F}} \longrightarrow 0.$$

Here $(\Omega_\alpha)_* : \Omega_{\mathfrak{C}} \otimes_{\mu_{\mathfrak{C}}, \mathfrak{C}, \gamma\circ\alpha} \mathfrak{F} \to \Omega_{\mathfrak{D}} \otimes_{\mu_{\mathfrak{D}}, \mathfrak{D}, \gamma} \mathfrak{F}$ *is induced by* $\Omega_\alpha : \Omega_{\mathfrak{C}} \to \Omega_{\mathfrak{D}}$, *and so on.*

3. C^∞-schemes

We now summarize material in [**7**, §4] on C^∞-schemes, and in [**7**, §6] on coherent and quasicoherent sheaves on C^∞-schemes. Much of §3.1 goes back to Dubuc [**3**].

3.1. The definition of C^∞-schemes.

The basic definitions are modelled on the definitions of schemes in Hartshorne [**6**, §II.2], but replacing rings by C^∞-rings throughout.

DEFINITION 3.1. A *C^∞-ringed space* $\underline{X} = (X, \mathcal{O}_X)$ is a topological space X with a sheaf $\mathcal{O}_X$ of C^∞-rings on X. That is, for each open set $U \subseteq X$ we are given a C^∞ ring $\mathcal{O}_X(U)$, and for each inclusion of open sets $V \subseteq U \subseteq X$ we are given a morphism of C^∞-rings $\rho_{UV} : \mathcal{O}_X(U) \to \mathcal{O}_X(V)$, called the *restriction maps*, and all this data satisfies the usual sheaf axioms [**6**, §II.1].

A *morphism* $\underline{f} = (f, f^\sharp) : (X, \mathcal{O}_X) \to (Y, \mathcal{O}_Y)$ of C^∞ ringed spaces is a continuous map $f : X \to Y$ and a morphism $f^\sharp : \mathcal{O}_Y \to f_*(\mathcal{O}_X)$ of sheaves of C^∞-rings on Y. That is, for each open $U \subset Y$ we are given a morphism of C^∞-rings $f^\sharp(U) : \mathcal{O}_Y(U) \to \mathcal{O}_X(f^{-1}(U))$ satisfying the obvious compatibilities with the restriction maps ρ_{UV} in $\mathcal{O}_X$ and $\mathcal{O}_Y$.

A *local C^∞-ringed space* $\underline{X} = (X, \mathcal{O}_X)$ is a C^∞-ringed space for which the stalks $\mathcal{O}_{X,x}$ of $\mathcal{O}_X$ at x are C^∞-local rings for all $x \in X$. Since morphisms of C^∞-local rings are automatically local morphisms, morphisms of local C^∞-ringed spaces $(X, \mathcal{O}_X), (Y, \mathcal{O}_Y)$ are just morphisms of C^∞-ringed spaces, without any additional locality condition. Write $\mathbf{C^\infty RS}$ for the category of C^∞-ringed spaces, and $\mathbf{LC^\infty RS}$ for the full subcategory of local C^∞-ringed spaces.

For brevity, we will use the notation that underlined upper case letters $\underline{X}, \underline{Y}, \underline{Z}, \ldots$ represent C^∞-ringed spaces $(X, \mathcal{O}_X), (Y, \mathcal{O}_Y), (Z, \mathcal{O}_Z), \ldots$, and underlined lower case letters $\underline{f}, \underline{g}, \ldots$ represent morphisms of C^∞-ringed spaces $(f, f^\sharp), (g, g^\sharp), \ldots$. When we write '$x \in \underline{X}$' we mean that $\underline{X} = (X, \mathcal{O}_X)$ and $x \in X$. When we write '*$\underline{U}$ is open in $\underline{X}$*' we mean that $\underline{U} = (U, \mathcal{O}_U)$ and $\underline{X} = (X, \mathcal{O}_X)$ with $U \subseteq X$ an open set and $\mathcal{O}_U = \mathcal{O}_X|_U$.

DEFINITION 3.2. Write $\mathbf{C^\infty Rings}^{\mathrm{op}}$ for the opposite category of $\mathbf{C^\infty Rings}$. The *global sections functor* $\Gamma : \mathbf{LC^\infty RS} \to \mathbf{C^\infty Rings}^{\mathrm{op}}$ acts on objects $(X, \mathcal{O}_X)$ in $\mathbf{LC^\infty RS}$ by $\Gamma : (X, \mathcal{O}_X) \mapsto \mathcal{O}_X(X)$ and on morphisms $(f, f^\sharp) : (X, \mathcal{O}_X) \to (Y, \mathcal{O}_X)$ by $\Gamma : (f, f^\sharp) \mapsto f^\sharp(X)$. As in [**3**, Th. 8] there is a *spectrum functor* $\mathrm{Spec} : \mathbf{C^\infty Rings}^{\mathrm{op}} \to \mathbf{LC^\infty RS}$, defined explicitly in [**7**, Def. 4.5], which is a right adjoint to Γ, that is, for all $\mathfrak{C} \in \mathbf{C^\infty Rings}$ and $\underline{X} \in \mathbf{LC^\infty RS}$ there are functorial isomorphisms

$$\mathrm{Hom}_{\mathbf{C^\infty Rings}}(\mathfrak{C}, \Gamma(\underline{X})) \cong \mathrm{Hom}_{\mathbf{LC^\infty RS}}(\underline{X}, \mathrm{Spec}\,\mathfrak{C}). \tag{2}$$

For any C^∞-ring $\mathfrak{C}$ there is a natural morphism of C^∞-rings $\Phi_{\mathfrak{C}} : \mathfrak{C} \to \Gamma(\operatorname{Spec}\mathfrak{C})$ corresponding to $\underline{\mathrm{id}}_{\underline{X}}$ in (2) with $\underline{X} = \operatorname{Spec}\mathfrak{C}$. By [**3**, Th. 13], the restriction of Spec to $(\mathbf{C^\infty Rings^{fa}})^{\mathrm{op}}$ is full and faithful.

A local C^∞-ringed space $\underline{X}$ is called an *affine C^∞-scheme* if it is isomorphic in $\mathbf{LC^\infty RS}$ to $\operatorname{Spec}\mathfrak{C}$ for some C^∞-ring $\mathfrak{C}$. We call $\underline{X}$ a *finitely presented*, or *good*, or *fair*, affine C^∞-scheme if $X \cong \operatorname{Spec}\mathfrak{C}$ for $\mathfrak{C}$ that kind of C^∞-ring.

Let $\underline{X} = (X, \mathcal{O}_X)$ be a local C^∞-ringed space. We call $\underline{X}$ a *C^∞-scheme* if X can be covered by open sets $U \subseteq X$ such that $(U, \mathcal{O}_X|_U)$ is an affine C^∞-scheme. We call a C^∞-scheme $\underline{X}$ *locally fair*, or *locally good*, or *locally finitely presented*, if X can be covered by open $U \subseteq X$ with $(U, \mathcal{O}_X|_U)$ a fair, or good, or finitely presented, affine C^∞-scheme, respectively.

Write $\mathbf{C^\infty Sch^{lf}}, \mathbf{C^\infty Sch^{lg}}, \mathbf{C^\infty Sch^{lfp}}, \mathbf{C^\infty Sch}$ for the full subcategories of locally fair, and locally good, and locally finitely presented, and all, C^∞-schemes in $\mathbf{LC^\infty RS}$, respectively. We call a C^∞-scheme $\underline{X}$ *separated*, or *paracompact*, if the underlying topological space X is Hausdorff, or paracompact.

Example 3.3. Let X be a manifold. Define a C^∞-ringed space $\underline{X} = (X, \mathcal{O}_X)$ to have topological space X and $\mathcal{O}_X(U) = C^\infty(U)$ for each open $U \subseteq X$, where $C^\infty(U)$ is the C^∞-ring of smooth maps $c : U \to \mathbb{R}$, and if $V \subseteq U \subseteq X$ are open define $\rho_{UV} : C^\infty(U) \to C^\infty(V)$ by $\rho_{UV} : c \mapsto c|_V$. Then $\underline{X} = (X, \mathcal{O}_X)$ is a local C^∞-ringed space. It is canonically isomorphic to $\operatorname{Spec} C^\infty(X)$, and so is an affine C^∞-scheme. It is locally finitely presented.

Define a functor $F_{\mathbf{Man}}^{\mathbf{C^\infty Sch}} : \mathbf{Man} \to \mathbf{C^\infty Sch^{lfp}} \subset \mathbf{C^\infty Sch}$ by $F_{\mathbf{Man}}^{\mathbf{C^\infty Sch}} = \operatorname{Spec} \circ F_{\mathbf{Man}}^{\mathbf{C^\infty Rings}}$. Then $F_{\mathbf{Man}}^{\mathbf{C^\infty Sch}}$ is full and faithful, and embeds $\mathbf{Man}$ as a full subcategory of $\mathbf{C^\infty Sch}$.

By [**7**, Prop.s 4.3, 4.4, 4.18, 4.25, Cor.s 4.11, 4.14 & Th. 4.26] we have:

Theorem 3.4. **(a)** *All finite limits exist in the category* $\mathbf{C^\infty RS}$.

(b) *The full subcategories* $\mathbf{C^\infty Sch^{lfp}}, \mathbf{C^\infty Sch^{lg}}, \mathbf{C^\infty Sch^{lf}}, \mathbf{C^\infty Sch}$, $\mathbf{LC^\infty RS}$ *in* $\mathbf{C^\infty RS}$ *are closed under all finite limits in* $\mathbf{C^\infty RS}$. *Hence, fibre products and all finite limits exist in each of these subcategories.*

(c) *If* $\mathfrak{C}$ *is a finitely generated C^∞-ring then* $\operatorname{Spec}\mathfrak{C}$ *is a fair affine C^∞-scheme.*

(d) *Let $(X, \mathcal{O}_X)$ be a finitely presented, or good, or fair, affine C^∞-scheme, and $U \subseteq X$ be an open subset. Then $(U, \mathcal{O}_X|_U)$ is also a finitely presented, or good, or fair, affine C^∞-scheme, respectively. However, this does not hold for general affine C^∞-schemes.*

(e) *Let $(X, \mathcal{O}_X)$ be a locally finitely presented, locally good, locally fair, or general, C^∞-scheme, and $U \subseteq X$ be open. Then $(U, \mathcal{O}_X|_U)$ is also a locally finitely presented, or locally good, or locally fair, or general, C^∞-scheme, respectively.*

(f) *The functor* $F_{\mathbf{Man}}^{\mathbf{C^\infty Sch}}$ *takes transverse fibre products in* $\mathbf{Man}$ *to fibre products in* $\mathbf{C^\infty Sch}$.

In [**7**, Def. 4.27 & Prop. 4.28] we discuss *partitions of unity* on C^∞-schemes, building on ideas of Dubuc [**4**].

DEFINITION 3.5. Let $\underline{X}=(X,\mathcal{O}_X)$ be a C^∞-scheme. Consider a formal sum $\sum_{a\in A} c_a$, where A is an indexing set and $c_a\in\mathcal{O}_X(X)$ for $a\in A$. We say $\sum_{a\in A} c_a$ is a *locally finite sum on* $\underline{X}$ if X can be covered by open $U\subseteq X$ such that for all but finitely many $a\in A$ we have $\rho_{XU}(c_a)=0$ in $\mathcal{O}_X(U)$.

By the sheaf axioms for $\mathcal{O}_X$, if $\sum_{a\in A} c_a$ is a locally finite sum there exists a unique $c\in\mathcal{O}_X(X)$ such that for all open $U\subseteq X$ such that $\rho_{XU}(c_a)=0$ in $\mathcal{O}_X(U)$ for all but finitely many $a\in A$, we have $\rho_{XU}(c)=\sum_{a\in A}\rho_{XU}(c_a)$ in $\mathcal{O}_X(U)$, where the sum makes sense as there are only finitely many nonzero terms. We call c the *limit* of $\sum_{a\in A} c_a$, written $\sum_{a\in A} c_a=c$.

Let $c\in\mathcal{O}_X(X)$. Suppose $V_i\subseteq X$ is open and $\rho_{XV_i}(c)=0\in\mathcal{O}_X(V_i)$ for $i\in I$, and let $V=\bigcup_{i\in I}V_i$. Then $V\subseteq X$ is open, and $\rho_{XV}(c)=0\in\mathcal{O}_X(V)$ as $\mathcal{O}_X$ is a sheaf. Thus taking the union of all open $V\subseteq X$ with $\rho_{XV}(c)=0$ gives a unique maximal open set $V_c\subseteq X$ such that $\rho_{XV_c}(c)=0\in\mathcal{O}_X(V_c)$. Define the *support* $\operatorname{supp} c$ of c to be $X\setminus V_c$, so that $\operatorname{supp} c$ is closed in X. If $U\subseteq X$ is open, we say that *c is supported in U* if $\operatorname{supp} c\subseteq U$.

Let $\{U_a:a\in A\}$ be an open cover of X. A *partition of unity on* $\underline{X}$ *subordinate to* $\{U_a:a\in A\}$ is $\{\eta_a:a\in A\}$ with $\eta_a\in\mathcal{O}_X(X)$ supported on U_a for $a\in A$, such that $\sum_{a\in A}\eta_a$ is a locally finite sum on $\underline{X}$ with $\sum_{a\in A}\eta_a=1$.

PROPOSITION 3.6. *Suppose $\underline{X}$ is a separated, paracompact, locally fair C^∞-scheme, and $\{\underline{U}_a:a\in A\}$ an open cover of $\underline{X}$. Then there exists a partition of unity $\{\eta_a:a\in A\}$ on $\underline{X}$ subordinate to $\{\underline{U}_a:a\in A\}$.*

Here are some differences between ordinary schemes and C^∞-schemes:

REMARK 3.7. **(i)** If A is a ring or algebra, then points of the corresponding scheme $\operatorname{Spec} A$ are prime ideals in A. However, if $\mathfrak{C}$ is a C^∞-ring then (by definition) points of $\operatorname{Spec}\mathfrak{C}$ are maximal ideals in $\mathfrak{C}$ with residue field $\mathbb{R}$, or equivalently, $\mathbb{R}$-algebra morphisms $x:\mathfrak{C}\to\mathbb{R}$. This has the effect that if X is a manifold then points of $\operatorname{Spec} C^\infty(X)$ are just points of X.

(ii) In conventional algebraic geometry, affine schemes are a restrictive class. Central examples such as $\mathbb{CP}^n$ are not affine, and affine schemes are not closed under open subsets, so that $\mathbb{C}^2$ is affine but $\mathbb{C}^2\setminus\{0\}$ is not. In contrast, affine C^∞-schemes are already general enough for many purposes. For example:

- All manifolds are affine C^∞-schemes.
- Open C^∞-subschemes of fair affine C^∞-schemes are fair and affine.
- If $\underline{X}$ is a separated, paracompact, locally fair C^∞-scheme then $\underline{X}$ is affine.

Affine C^∞-schemes are always separated (Hausdorff), so we need general C^∞-schemes to include non-Hausdorff behaviour.

(iii) In conventional algebraic geometry the Zariski topology is too coarse for many purposes, so one has to introduce the étale topology. In C^∞-algebraic geometry there is no need for this, as affine C^∞-schemes are Hausdorff.

(iv) Even very basic C^∞-rings such as $C^\infty(\mathbb{R}^n)$ for $n>0$ are not noetherian as $\mathbb{R}$-algebras. So C^∞-schemes should be compared to non-noetherian schemes in conventional algebraic geometry.

3.2. Quasicoherent and coherent sheaves on C^∞-schemes. In [**7**, §6] we discuss sheaves of modules on C^∞-schemes.

DEFINITION 3.8. Let $\underline{X}=(X,\mathcal{O}_X)$ be a C^∞-scheme. An *$\mathcal{O}_X$-module* $\mathcal{E}$ on $\underline{X}$ assigns a module $\mathcal{E}(U)$ over $\mathcal{O}_X(U)$ for each open set $U\subseteq X$, with $\mathcal{O}_X(U)$-action $\mu_U:\mathcal{O}_X(U)\times\mathcal{E}(U)\rightarrow\mathcal{E}(U)$, and a linear map $\mathcal{E}_{UV}:\mathcal{E}(U)\rightarrow\mathcal{E}(V)$ for each inclusion of open sets $V\subseteq U\subseteq X$, such that the following commutes:

$$\begin{array}{ccc} \mathcal{O}_X(U)\times\mathcal{E}(U) & \xrightarrow{\mu_U} & \mathcal{E}(U) \\ \downarrow{\scriptstyle \rho_{UV}\times\mathcal{E}_{UV}} & & {\scriptstyle \mathcal{E}_{UV}}\downarrow \\ \mathcal{O}_X(V)\times\mathcal{E}(V) & \xrightarrow{\mu_V} & \mathcal{E}(V), \end{array}$$

and all this data $\mathcal{E}(U),\mathcal{E}_{UV}$ satisfies the usual sheaf axioms [**6**, §II.1].

A *morphism of $\mathcal{O}_X$-modules* $\phi:\mathcal{E}\rightarrow\mathcal{F}$ assigns a morphism of $\mathcal{O}_X(U)$-modules $\phi(U):\mathcal{E}(U)\rightarrow\mathcal{F}(U)$ for each open set $U\subseteq X$, such that $\phi(V)\circ\mathcal{E}_{UV}=\mathcal{F}_{UV}\circ\phi(U)$ for each inclusion of open sets $V\subseteq U\subseteq X$. Then $\mathcal{O}_X$-modules form an *abelian category*, which we write as $\mathcal{O}_X$-mod.

As in [**7**, §6.2], the spectrum functor $\mathrm{Spec}:\mathbf{C^\infty Rings}^{\mathrm{op}}\rightarrow\mathbf{C^\infty Sch}$ has a counterpart for modules: if $\mathfrak{C}$ is a C^∞-ring and $(X,\mathcal{O}_X)=\mathrm{Spec}\,\mathfrak{C}$ we can define a functor $\mathrm{MSpec}:\mathfrak{C}\text{-mod}\rightarrow\mathcal{O}_X\text{-mod}$. Let $\underline{X}=(X,\mathcal{O}_X)$ be a C^∞-scheme, and $\mathcal{E}$ an $\mathcal{O}_X$-module. We call $\mathcal{E}$ *quasicoherent* if $\underline{X}$ can be covered by open $\underline{U}$ with $\underline{U}\cong\mathrm{Spec}\,\mathfrak{C}$ for some C^∞-ring $\mathfrak{C}$, and under this identification $\mathcal{E}|_{\underline{U}}\cong\mathrm{MSpec}\,M$ for some $\mathfrak{C}$-module M. We call $\mathcal{E}$ *coherent* if furthermore we can take these $\mathfrak{C}$-modules M to be finitely presented. We call $\mathcal{E}$ a *vector bundle of rank* $n\geqslant 0$ if $\underline{X}$ may be covered by open $\underline{U}$ such that $\mathcal{E}|_{\underline{U}}\cong\mathcal{O}_U\otimes_{\mathbb{R}}\mathbb{R}^n$. Write $\mathrm{qcoh}(\underline{X}),\mathrm{coh}(\underline{X})$, and $\mathrm{vect}(\underline{X})$ for the full subcategories of quasicoherent sheaves, coherent sheaves, and vector bundles in $\mathcal{O}_X$-mod, respectively.

DEFINITION 3.9. Let $\underline{f}=(f,f^\sharp):(X,\mathcal{O}_X)\rightarrow(Y,\mathcal{O}_Y)$ be a morphism of C^∞-schemes, and $\mathcal{E}$ be an $\mathcal{O}_Y$-module. Following Hartshorne [**6**, p. 65, p. 110], define the *pullback* $\underline{f}^*(\mathcal{E})$ to be the sheaf of $\mathcal{O}_X$-modules on $(X,\mathcal{O}_X)$ associated to the presheaf $U\mapsto\lim_{V\supseteq f(U)}\mathcal{E}(V)\otimes_{\mathcal{O}_Y(V)}\mathcal{O}_X(U)$, where U is an open set in X, and the limit is over open sets V in Y containing $f(U)$.

If $\phi:\mathcal{E}\rightarrow\mathcal{F}$ is a morphism in $\mathcal{O}_Y$-mod we have an induced morphism $\underline{f}^*(\phi)=f^{-1}(\phi)\otimes\mathrm{id}_{\mathcal{O}_X}:\underline{f}^*(\mathcal{E})\rightarrow\underline{f}^*(\mathcal{F})$ in $\mathcal{O}_X$-mod. Then $\underline{f}^*:\mathcal{O}_Y\text{-mod}\rightarrow$

$\mathcal{O}_X$-mod is a right exact functor between abelian categories, which restricts to a right exact functor $\underline{f}^*:\mathrm{qcoh}(\underline{Y})\to\mathrm{qcoh}(\underline{X})$.

Pullbacks $\underline{f}^*(\mathcal{E})$ are a kind of fibre product, and may be characterized by a universal property. So they should be regarded as being *unique up to canonical isomorphism*, rather than unique. We use the Axiom of Choice to choose $\underline{f}^*(\mathcal{E})$ for all $\underline{f},\mathcal{E}$, and so speak of 'the' pullback $\underline{f}^*(\mathcal{E})$. However, it may not be possible to make these choices functorial in $\underline{f}$. That is, if $\underline{f}:\underline{X}\to\underline{Y}$, $\underline{g}:\underline{Y}\to\underline{Z}$ are morphisms and $\mathcal{E}\in\mathcal{O}_Z$-mod then $(\underline{g}\circ\underline{f})^*(\mathcal{E})$ and $\underline{f}^*(\underline{g}^*(\mathcal{E}))$ are canonically isomorphic in $\mathcal{O}_X$-mod, but may not be equal. We will write $I_{\underline{f},\underline{g}}(\mathcal{E}):(\underline{g}\circ\underline{f})^*(\mathcal{E})\to\underline{f}^*(\underline{g}^*(\mathcal{E}))$ for these canonical isomorphisms. Then $I_{\underline{f},\underline{g}}:(\underline{g}\circ\underline{f})^*\Rightarrow\underline{f}^*\circ\underline{g}^*$ is a natural isomorphism of functors.

Similarly, when $\underline{f}$ is the identity $\underline{\mathrm{id}}_X:\underline{X}\to\underline{X}$ and $\mathcal{E}\in\mathcal{O}_X$-mod we may not have $\underline{\mathrm{id}}_X^*(\mathcal{E})=\mathcal{E}$, but there is a canonical isomorphism $\delta_{\underline{X}}(\mathcal{E}):\underline{\mathrm{id}}_X^*(\mathcal{E})\to\mathcal{E}$, and $\delta_{\underline{X}}:\underline{\mathrm{id}}_X^*\Rightarrow\mathrm{id}_{\mathcal{O}_X\text{-mod}}$ is a natural isomorphism of functors.

EXAMPLE 3.10. Let X be a manifold, and $\underline{X}$ the associated C^∞-scheme from Example 3.3, so that $\mathcal{O}_X(U)=C^\infty(U)$ for all open $U\subseteq X$. Let $E\to X$ be a vector bundle. Define an $\mathcal{O}_X$-module $\mathcal{E}$ on $\underline{X}$ by $\mathcal{E}(U)=C^\infty(E|_U)$, the smooth sections of the vector bundle $E|_U\to U$, and for open $V\subseteq U\subseteq X$ define $\mathcal{E}_{UV}:\mathcal{E}(U)\to\mathcal{E}(V)$ by $\mathcal{E}_{UV}:e_U\mapsto e_U|_V$. Then $\mathcal{E}\in\mathrm{vect}(\underline{X})$ is a vector bundle on $\underline{X}$, which we think of as a lift of E from manifolds to C^∞-schemes.

Suppose $f:X\to Y$ is a smooth map of manifolds, and $\underline{f}:\underline{X}\to\underline{Y}$ is the corresponding morphism of C^∞-schemes. Let $F\to Y$ be a vector bundle over Y, so that $f^*(F)\to X$ is a vector bundle over X. Let $\mathcal{F}\in\mathrm{vect}(\underline{Y})$ be the vector bundle over $\underline{Y}$ lifting F. Then $\underline{f}^*(\mathcal{F})$ is canonically isomorphic to the vector bundle over $\underline{X}$ lifting $f^*(F)$.

The next theorem comes from [**7**, Cor. 6.11 & Prop. 6.12]. In part (a), the reason $\mathrm{coh}(\underline{X})$ is not closed under kernels is that the C^∞-rings we are interested in are generally *not noetherian* as commutative $\mathbb{R}$-algebras, and this causes problems with coherence; in conventional algebraic geometry, one usually only considers coherent sheaves over noetherian schemes.

THEOREM 3.11. **(a)** *Let $\underline{X}$ be a C^∞-scheme. Then* $\mathrm{qcoh}(\underline{X})$ *is closed under kernels, cokernels and extensions in* $\mathcal{O}_X$-mod, *so it is an abelian category. Also* $\mathrm{coh}(\underline{X})$ *is closed under cokernels and extensions in* $\mathcal{O}_X$-mod, *but may not be closed under kernels in* $\mathcal{O}_X$-mod, *so* $\mathrm{coh}(\underline{X})$ *may not be an abelian category.*

(b) *Suppose $\underline{f}:\underline{X}\to\underline{Y}$ is a morphism of C^∞-schemes. Then pullback $\underline{f}^*$:* $\mathcal{O}_Y$-mod $\to\mathcal{O}_X$-mod *maps* $\mathrm{qcoh}(\underline{Y})\to\mathrm{qcoh}(\underline{X})$ *and* $\mathrm{coh}(\underline{Y})\to\mathrm{coh}(\underline{X})$ *and* $\mathrm{vect}(\underline{Y})\to\mathrm{vect}(\underline{X})$. *Also* $\underline{f}^*:\mathrm{qcoh}(\underline{Y})\to\mathrm{qcoh}(\underline{X})$ *is a right exact functor.*

(c) *Let $\underline{X}$ be a locally fair C^∞-scheme. Then every $\mathcal{O}_X$-module $\mathcal{E}$ on $\underline{X}$ is quasicoherent, that is,* $\mathrm{qcoh}(\underline{X})=\mathcal{O}_X$-mod.

Let $\underline{X}$ be a separated, paracompact, locally fair C^∞-scheme. Then partitions of unity exist on $\underline{X}$ subordinate to any open cover by Proposition 3.6. As in [**7**, §6.3], this shows that quasicoherent sheaves $\mathcal{E}$ on $\underline{X}$ are *fine*, which implies that their cohomology groups $H^i(\mathcal{E})$ are zero for all $i>0$. In [**7**, Prop. 6.13] we deduce an exactness property for sections of quasicoherent sheaves on $\underline{X}$:

PROPOSITION 3.12. *Suppose* $\underline{X}=(X,\mathcal{O}_X)$ *is a separated, paracompact, locally fair* C^∞*-scheme, and* $\cdots\rightarrow\mathcal{E}^i\xrightarrow{\phi^i}\mathcal{E}^{i+1}\xrightarrow{\phi^{i+1}}\mathcal{E}^{i+2}\rightarrow\cdots$ *an exact sequence in* $\mathrm{qcoh}(\underline{X})$*. Then* $\cdots\rightarrow\mathcal{E}^i(U)\xrightarrow{\phi^i(U)}\mathcal{E}^{i+1}(U)\xrightarrow{\phi^{i+1}(U)}\mathcal{E}^{i+2}(U)\rightarrow\cdots$ *is an exact sequence of* $\mathcal{O}_X(U)$*-modules for each open* $U\subseteq X$*.*

We define *cotangent sheaves*, the sheaf version of cotangent modules in §2.3.

DEFINITION 3.13. Let $\underline{X}$ be a C^∞-scheme. Define $\mathcal{P}T^*\underline{X}$ to associate to each open $U\subseteq X$ the cotangent module $\Omega_{\mathcal{O}_X(U)}$, and to each inclusion of open sets $V\subseteq U\subseteq X$ the morphism of $\mathcal{O}_X(U)$-modules $\Omega_{\rho_{UV}}:\Omega_{\mathcal{O}_X(U)}\rightarrow\Omega_{\mathcal{O}_X(V)}$ associated to the morphism of C^∞-rings $\rho_{UV}:\mathcal{O}_X(U)\rightarrow\mathcal{O}_X(V)$. Then $\mathcal{P}T^*\underline{X}$ is a *presheaf of* $\mathcal{O}_X$*-modules on* $\underline{X}$. Define the *cotangent sheaf* $T^*\underline{X}$ *of* $\underline{X}$ to be the sheafification of $\mathcal{P}T^*\underline{X}$, as an $\mathcal{O}_X$-module.

If $\underline{f}:\underline{X}\rightarrow\underline{Y}$ is a morphism of C^∞-schemes, then $\underline{f}^*\bigl(T^*\underline{Y}\bigr)$ is the sheafification of the presheaf $\underline{f}^*(\mathcal{P}T^*\underline{Y})$ acting by

$$\begin{aligned}U\longmapsto \underline{f}^*(\mathcal{P}T^*\underline{Y})(U)&=\lim_{V\supseteq f(U)}\mathcal{P}T^*\underline{Y}(V)\otimes_{\mathcal{O}_Y(V)}\mathcal{O}_X(U)\\&=\lim_{V\supseteq f(U)}\Omega_{\mathcal{O}_Y(V)}\otimes_{\mathcal{O}_Y(V)}\mathcal{O}_X(U).\end{aligned}$$

Define a morphism of presheaves $\mathcal{P}\Omega_{\underline{f}}:\underline{f}^*(\mathcal{P}T^*\underline{Y})\rightarrow\mathcal{P}T^*\underline{X}$ on X by

$$(\mathcal{P}\Omega_{\underline{f}})(U)=\lim_{V\supseteq f(U)}(\Omega_{\rho_{f^{-1}(V)\,U}\circ f^\sharp(V)})_*,$$

where $(\Omega_{\rho_{f^{-1}(V)\,U}\circ f^\sharp(V)})_*:\Omega_{\mathcal{O}_Y(V)}\otimes_{\mathcal{O}_Y(V)}\mathcal{O}_X(U)\rightarrow\Omega_{\mathcal{O}_X(U)}=(\mathcal{P}T^*\underline{X})(U)$ is constructed as in Definition 2.15 from the C^∞-ring morphisms $f^\sharp(V):\mathcal{O}_Y(V)\rightarrow\mathcal{O}_X(f^{-1}(V))$ in $\underline{f}$ and $\rho_{f^{-1}(V)U}:\mathcal{O}_X(f^{-1}(V))\rightarrow\mathcal{O}_X(U)$ in $\mathcal{O}_X$. Define $\Omega_{\underline{f}}:\underline{f}^*\bigl(T^*\underline{Y}\bigr)\rightarrow T^*\underline{X}$ to be the induced morphism of the associated sheaves.

EXAMPLE 3.14. Let X be a manifold, and $\underline{X}$ the associated C^∞-scheme. Then $T^*\underline{X}$ is a vector bundle on $\underline{X}$, and is canonically isomorphic to the lift to C^∞-schemes from Example 3.10 of the cotangent vector bundle T^*X of X.

Here [**7**, Th.s 6.16 & 6.17] are some properties of cotangent sheaves.

THEOREM 3.15. **(a)** *Let* $\underline{X}$ *be a locally good* C^∞*-scheme. Then* $T^*\underline{X}$ *is a coherent sheaf.*

(b) *Let* $\underline{f}:\underline{X}\rightarrow\underline{Y}$ *and* $\underline{g}:\underline{Y}\rightarrow\underline{Z}$ *be morphisms of* C^∞*-schemes. Then*

$$\Omega_{\underline{g}\circ\underline{f}}=\Omega_{\underline{f}}\circ\underline{f}^*(\Omega_{\underline{g}})\circ I_{\underline{f},\underline{g}}(T^*\underline{Z})$$

as morphisms $(\underline{g}\circ\underline{f})^*(T^*\underline{Z})\rightarrow T^*\underline{X}$ *in* $\mathcal{O}_X$-mod. *Here* $\Omega_{\underline{g}}:\underline{g}^*(T^*\underline{Z})\rightarrow T^*\underline{Y}$ *in* $\mathcal{O}_Y$-mod, *so applying* $\underline{f}^*$ *gives* $\underline{f}^*(\Omega_{\underline{g}}):\underline{f}^*(\underline{g}^*(T^*\underline{Z}))\rightarrow\underline{f}^*(T^*\underline{Y})$ *in* $\mathcal{O}_X$-mod, *and* $I_{\underline{f},\underline{g}}(T^*\underline{Z}):(\underline{g}\circ\underline{f})^*(T^*\underline{Z})\rightarrow\underline{f}^*(\underline{g}^*(T^*\underline{Z}))$ *is as in Definition* 3.9.

(c) *Suppose* $\underline{W},\underline{X},\underline{Y},\underline{Z}$ *are locally fair* C^∞*-schemes with a Cartesian square*

$$\begin{array}{ccc}\underline{W} & \xrightarrow{\underline{f}} & \underline{Y}\\ \downarrow{\scriptstyle \underline{e}} & & {\scriptstyle \underline{h}}\downarrow\\ \underline{X} & \xrightarrow{\underline{g}} & \underline{Z}\end{array}$$

in $\mathbf{C^\infty Sch^{lf}}$, *so that* $\underline{W}=\underline{X}\times_{\underline{Z}}\underline{Y}$. *Then the following is exact in* $\mathrm{qcoh}(\underline{W})$:

$$(\underline{g}\circ\underline{e})^*(T^*\underline{Z})\xrightarrow{\substack{\underline{e}^*(\Omega_{\underline{g}})\circ I_{\underline{e},\underline{g}}(T^*\underline{Z})\oplus\\ -\underline{f}^*(\Omega_{\underline{h}})\circ I_{\underline{f},\underline{h}}(T^*\underline{Z})}}\underline{e}^*(T^*\underline{X})\oplus\underline{f}^*(T^*\underline{Y})\xrightarrow{\Omega_{\underline{e}}\oplus\Omega_{\underline{f}}}T^*\underline{W}\longrightarrow 0.$$

4. C^∞-stacks

In [**7**, §7–§8] we discuss C^∞*-stacks*, which are related to C^∞-schemes in the same way that Artin stacks and Deligne–Mumford stacks in algebraic geometry are related to schemes. Stacks are a rather technical subject which take a lot of work and many pages to set up properly, so to keep this section short we will give less detail than in §2 and §3.

We are most interested in a subclass of C^∞-stacks called *Deligne–Mumford* C^∞*-stacks*. Here are some of their important properties:

- Deligne–Mumford C^∞-stacks are geometric objects locally modelled on quotients $\underline{U}/G$, for $\underline{U}$ an affine C^∞-scheme and G a finite group.
- Deligne–Mumford C^∞-stacks are related to C^∞-schemes in exactly the same way that orbifolds are related to manifolds.
- Any C^∞-scheme yields an example of a Deligne–Mumford C^∞-stack.
- Deligne–Mumford C^∞-stacks form a 2-*category* $\mathbf{DMC^\infty Sta}$. That is, we have *objects* $\mathcal{X},\mathcal{Y}$, 1-*morphisms* $f,g:\mathcal{X}\rightarrow\mathcal{Y}$, and 2-*morphisms* $\eta:f\Rightarrow g$. All 2-morphisms are invertible, that is, they are 2-*isomorphisms.*

  The geometric meaning of 1- and 2-morphisms is not obvious; to get a feel for it, it helps to consider the case when $\mathcal{X},\mathcal{Y}$ are quotients $[\underline{X}/G],[\underline{Y}/H]$ for C^∞-schemes $\underline{X},\underline{Y}$ and finite groups G,H acting on $\underline{X},\underline{Y}$. Oversimplifying somewhat, a 1-morphism $f:[\underline{X}/G]\rightarrow[\underline{Y}/H]$ is roughly a pair $(\underline{f},\rho)$ where $\rho:G\rightarrow H$ is a group morphism and $\underline{f}:\underline{X}\rightarrow\underline{Y}$ is a morphism of C^∞-schemes with

$\underline{f}\circ\gamma=\rho(\gamma)\circ\underline{f}$ for all $\gamma\in G$. If $f=(\underline{f},\rho)$ and $g=(\underline{g},\sigma)$ are two such 1-morphisms, then a 2-morphism $\eta:f\Rightarrow g$ is roughly an element $\delta\in H$ such that $\sigma(\gamma)=\delta\,\rho(\gamma)\delta^{-1}$ for all $\gamma\in G$, and $\underline{g}=\delta\circ\underline{f}$.
- There is a good notion of *fibre product* in a 2-category. All fibre products of Deligne–Mumford C^∞-stacks exist, as Deligne–Mumford C^∞-stacks.

4.1. The definition of C^∞-stacks. The next few definitions assume a lot of standard material from stack theory, which is summarized in [**7**, §7].

DEFINITION 4.1. Define a *Grothendieck topology* $\mathcal{J}$ on the category $\mathbf{C^\infty Sch}$ of C^∞-schemes to have coverings $\{\underline{i}_a:\underline{U}_a\to\underline{U}\}_{a\in A}$ where $V_a=i_a(U_a)$ is open in U with $\underline{i}_a:\underline{U}_a\to(V_a,\mathcal{O}_U|_{V_a})$ an isomorphism for all $a\in A$, and $U=\bigcup_{a\in A}V_a$. Up to isomorphisms of the $\underline{U}_a$, the coverings $\{\underline{i}_a:\underline{U}_a\to\underline{U}\}_{a\in A}$ of $\underline{U}$ correspond exactly to open covers $\{V_a:a\in A\}$ of U. Then $(\mathbf{C^\infty Sch},\mathcal{J})$ is a *site*.

The *stacks on* $(\mathbf{C^\infty Sch},\mathcal{J})$ form a 2-category $\mathbf{Sta}_{(\mathbf{C^\infty Sch},\mathcal{J})}$. The site $(\mathbf{C^\infty Sch},\mathcal{J})$ is *subcanonical*. Thus, if $\underline{X}$ is any C^∞-scheme we have an associated stack on $(\mathbf{C^\infty Sch},\mathcal{J})$ which we write as $\bar{\underline{X}}$. A C^∞*-stack* is a stack $\mathcal{X}$ on $(\mathbf{C^\infty Sch},\mathcal{J})$ such that the diagonal 1-morphism $\Delta_{\mathcal{X}}:\mathcal{X}\to\mathcal{X}\times\mathcal{X}$ is representable, and there exists a surjective 1-morphism $\Pi:\bar{\underline{U}}\to\mathcal{X}$ called an *atlas* for some C^∞-scheme $\underline{U}$. C^∞-stacks form a 2-category $\mathbf{C^\infty Sta}$. All 2-morphisms in $\mathbf{C^\infty Sta}$ are invertible, that is, they are 2-isomorphisms.

REMARK 4.2. So far as the author knows, [**7**] is the first paper to consider stacks on the site $(\mathbf{C^\infty Sch},\mathcal{J})$. Note that Behrend and Xu [**1**, Def. 2.15] use the term 'C^∞-stack' to mean something different, a geometric stack over the site $(\mathbf{Man},\mathcal{J}_{\mathbf{Man}})$ of manifolds without boundary with Grothendieck topology $\mathcal{J}_{\mathbf{Man}}$ given by open covers. These are called 'smooth stacks' by Metzler [**12**].

Write $\mathbf{ManSta}$ for the 2-category of geometric stacks on $(\mathbf{Man},\mathcal{J}_{\mathbf{Man}})$, as in [**1, 10, 12**]. The functor $F_{\mathbf{Man}}^{\mathbf{C^\infty Sch}}$ of Example 3.3 embeds the site $(\mathbf{Man},\mathcal{J}_{\mathbf{Man}})$ into $(\mathbf{C^\infty Sch},\mathcal{J})$. Thus, restricting from $(\mathbf{C^\infty Sch},\mathcal{J})$ to $(\mathbf{Man},\mathcal{J}_{\mathbf{Man}})$ defines a natural truncation 2-functor $F_{\mathbf{C^\infty Sta}}^{\mathbf{ManSta}}:\mathbf{C^\infty Sta}\to\mathbf{ManSta}$.

A C^∞-stack $\mathcal{X}$ encodes all morphisms $F:\underline{U}\to\mathcal{X}$ for C^∞-schemes $\underline{U}$, whereas its image $F_{\mathbf{C^\infty Sta}}^{\mathbf{ManSta}}(\mathcal{X})$ remembers only morphisms $F:U\to\mathcal{X}$ for manifolds U. Thus the truncation functor $F_{\mathbf{C^\infty Sta}}^{\mathbf{ManSta}}$ *loses information*, as it forgets morphisms from C^∞-schemes which are not manifolds. This includes any information about *nonreduced* C^∞-schemes. For our applications in [**8**] this nonreduced information will be essential, so we must consider stacks on $(\mathbf{C^\infty Sch},\mathcal{J})$ rather than on $(\mathbf{Man},\mathcal{J}_{\mathbf{Man}})$.

DEFINITION 4.3. A *groupoid object* $(\underline{U},\underline{V},\underline{s},\underline{t},\underline{u},\underline{i},\underline{m})$ in $\mathbf{C^\infty Sch}$, or simply *groupoid* in $\mathbf{C^\infty Sch}$, consists of objects $\underline{U},\underline{V}$ in $\mathbf{C^\infty Sch}$ and morphisms

$\underline{s},\underline{t}:\underline{V}\to\underline{U}$, $\underline{u}:\underline{U}\to\underline{V}$, $\underline{i}:\underline{V}\to\underline{V}$ and $\underline{m}:\underline{V}\times_{\underline{s},\underline{U},\underline{t}}\underline{V}\to\underline{V}$ satisfying the identities

$$\underline{s}\circ\underline{u}=\underline{t}\circ\underline{u}=\mathrm{id}_{\underline{U}},\ \ \underline{s}\circ\underline{i}=\underline{t},\ \ \underline{t}\circ\underline{i}=\underline{s},\ \ \underline{s}\circ\underline{m}=\underline{s}\circ\underline{\pi}_2,\ \ \underline{t}\circ\underline{m}=\underline{t}\circ\underline{\pi}_1,$$
$$\underline{m}\circ(\underline{i}\times\underline{\mathrm{id}}_{\underline{V}})=\underline{u}\circ\underline{s},\ \ \underline{m}\circ(\underline{\mathrm{id}}_{\underline{V}}\times\underline{i})=\underline{u}\circ\underline{t},$$
$$\underline{m}\circ(\underline{m}\times\underline{\mathrm{id}}_{\underline{V}})=\underline{m}\circ(\underline{\mathrm{id}}_{\underline{V}}\times\underline{m}):\underline{V}\times_{\underline{U}}\underline{V}\times_{\underline{U}}\underline{V}\longrightarrow\underline{V},$$
$$\underline{m}\circ(\underline{\mathrm{id}}_{\underline{V}}\times\underline{u})=\underline{m}\circ(\underline{u}\times\underline{\mathrm{id}}_{\underline{V}}):\underline{V}=\underline{V}\times_{\underline{U}}\underline{U}\longrightarrow\underline{V}.$$

We write groupoids in $\mathbf{C^\infty Sch}$ as $\underline{V}\rightrightarrows\underline{U}$ for short, to emphasize the morphisms $\underline{s},\underline{t}:\underline{V}\to\underline{U}$. To any such groupoid we can associate a *groupoid stack* $[\underline{V}\rightrightarrows\underline{U}]$, which is a C^∞-stack. Conversely, if $\mathcal{X}$ is a C^∞-stack and $\Pi:\bar{\underline{U}}\to\mathcal{X}$ is an atlas one can construct a groupoid $\underline{V}\rightrightarrows\underline{U}$ in $\mathbf{C^\infty Sch}$, and $\mathcal{X}$ is equivalent (in the 2-category sense) to $[\underline{V}\rightrightarrows\underline{U}]$. Thus, every C^∞-stack is equivalent to a groupoid stack.

Suppose $\underline{U}$ is a C^∞-scheme and G is a finite group which acts on the left on $\underline{U}$ by automorphisms, with action $\underline{\mu}:G\times\underline{U}\to\underline{U}$. Then

(3) $\big(\underline{U},G\times\underline{U},\underline{\pi}_{\underline{U}},\underline{\mu},1\times\underline{\mathrm{id}}_{\underline{U}},(i\circ\underline{\pi}_G)\times\underline{\mu},(m\circ((\underline{\pi}_G\circ\underline{\pi}_1)\times(\underline{\pi}_G\circ\underline{\pi}_2)))\times(\underline{\pi}_{\underline{U}}\circ\underline{\pi}_2)\big)$

is a groupoid object in $\mathbf{C^\infty Sch}$, where $1\in G$ is the identity, $i:G\to G$ is the inverse map, $m:G\times G\to G$ is group multiplication, and in the final morphism $\underline{\pi}_1,\underline{\pi}_2$ are the projections from $(G\times\underline{U})\times_{\underline{\pi}_{\underline{U}},\underline{U},\underline{\mu}}(G\times\underline{U})$ to the first and second factors $G\times\underline{U}$. Write $[\underline{U}/G]$ for the groupoid stack associated to (3). It is a C^∞-stack, which we call a *quotient stack*.

We define some classes of morphisms of C^∞-schemes.

DEFINITION 4.4. Let $\underline{f}=(f,f^\sharp):\underline{X}=(X,\mathcal{O}_X)\to\underline{Y}=(Y,\mathcal{O}_Y)$ be a morphism in $\mathbf{C^\infty Sch}$. Then:

- We call $\underline{f}$ an *open embedding* if $V=f(X)$ is an open subset in Y and $(f,f^\sharp):(X,\mathcal{O}_X)\to(U,\mathcal{O}_Y|_V)$ is an isomorphism.
- We call $\underline{f}$ *étale* if each $x\in X$ has an open neighbourhood U in X such that $V=f(U)$ is open in Y and $(f|_U,f^\sharp|_U):(U,\mathcal{O}_X|_U)\to(V,\mathcal{O}_Y|_V)$ is an isomorphism. That is, $\underline{f}$ is a local isomorphism.
- We call $\underline{f}$ *proper* if $f:X\to Y$ is a proper map of topological spaces, that is, if $S\subseteq Y$ is compact then $f^{-1}(S)\subseteq X$ is compact.
- We call $\underline{f}$ *separated* if $f:X\to Y$ is a separated map of topological spaces, that is, $\Delta_X=\{(x,x):x\in X\}$ is a closed subset of the topological fibre product $X\times_{f,Y,f}X=\{(x,x')\in X\times X:f(x)=f(x')\}$.
- We call $\underline{f}$ *universally closed* if whenever $\underline{g}:\underline{W}\to\underline{Y}$ is a morphism then $\pi_W:X\times_{f,Y,g}W\to W$ is a closed map of topological spaces, that is, it takes closed sets to closed sets.

Each one is invariant under base change and local in the target in $(\mathbf{C^\infty Sch},\mathcal{J})$. Thus, they are also defined for representable 1-morphisms of C^∞-stacks.

Definition 4.5. Let $\mathcal{X}$ be a C^∞-stack. We say that $\mathcal{X}$ is *separated* if the diagonal 1-morphism $\Delta_{\mathcal{X}}:\mathcal{X}\to\mathcal{X}\times\mathcal{X}$ is universally closed. If $\mathcal{X}=\bar{\underline{X}}$ for some C^∞-scheme $\underline{X}=(X,\mathcal{O}_X)$ then $\mathcal{X}$ is separated if and only if $\Delta_X:X\to X\times X$ is closed, that is, if and only if X is Hausdorff, so $\underline{X}$ is separated.

Definition 4.6. Let $\mathcal{X}$ be a C^∞-stack. A *C^∞-substack* $\mathcal{Y}$ in $\mathcal{X}$ is a substack of $\mathcal{X}$ which is also a C^∞-stack. It has a natural inclusion 1-morphism $i_{\mathcal{Y}}:\mathcal{Y}\hookrightarrow\mathcal{X}$. We call $\mathcal{Y}$ an *open C^∞-substack* of $\mathcal{X}$ if $i_{\mathcal{Y}}$ is a representable open embedding. An *open cover* $\{\mathcal{Y}_a:a\in A\}$ of $\mathcal{X}$ is a family of open C^∞-substacks $\mathcal{Y}_a$ in $\mathcal{X}$ with $\coprod_{a\in A}i_{\mathcal{Y}_a}:\coprod_{a\in A}\mathcal{Y}_a\to\mathcal{X}$ surjective.

Deligne–Mumford stacks in algebraic geometry are locally modelled on quotient stacks $[X/G]$ for X an affine scheme and G a finite group. This motivates:

Definition 4.7. A *Deligne–Mumford C^∞-stack* is a C^∞-stack $\mathcal{X}$ which admits an open cover $\{\mathcal{Y}_a:a\in A\}$ with each $\mathcal{Y}_a$ equivalent to a quotient stack $[\underline{U}_a/G_a]$ for $\underline{U}_a$ an affine C^∞-scheme and G_a a finite group. We call $\mathcal{X}$ a *locally fair*, or *locally good*, or *locally finitely presented*, Deligne–Mumford C^∞-stack if it has such an open cover with each $\underline{U}_a$ a fair, good, or finitely presented, affine C^∞-scheme, respectively. Write $\mathbf{DMC^\infty Sta^{lf}}$, $\mathbf{DMC^\infty Sta^{lg}}$, $\mathbf{DMC^\infty Sta^{lfp}}$ and $\mathbf{DMC^\infty Sta}$ for the full 2-subcategories of locally fair, locally good, locally finitely presented, and all, Deligne–Mumford C^∞-stacks in $\mathbf{C^\infty Sta}$.

From [**7**, Th.s 8.5, 8.21, 8.26 & Prop. 8.17] we have:

Theorem 4.8. **(a)** *All fibre products exist in the 2-category* $\mathbf{C^\infty Sta}$.

(b) $\mathbf{DMC^\infty Sta},\mathbf{DMC^\infty Sta^{lf}},\mathbf{DMC^\infty Sta^{lg}}$ *and* $\mathbf{DMC^\infty Sta^{lfp}}$ *are closed under fibre products in* $\mathbf{C^\infty Sta}$.

(c) $\mathbf{DMC^\infty Sta},\mathbf{DMC^\infty Sta^{lf}},\mathbf{DMC^\infty Sta^{lg}}$ *and* $\mathbf{DMC^\infty Sta^{lfp}}$ *are closed under taking open C^∞-substacks in* $\mathbf{C^\infty Sta}$.

(d) *A C^∞-stack $\mathcal{X}$ is separated and Deligne–Mumford if and only if it is equivalent to a groupoid stack $[\underline{V}\rightrightarrows\underline{U}]$ where $\underline{U},\underline{V}$ are separated C^∞-schemes, $\underline{s}:\underline{V}\to\underline{U}$ is étale, and $\underline{s}\times\underline{t}:\underline{V}\to\underline{U}\times\underline{U}$ is universally closed.*

(e) *A C^∞-stack $\mathcal{X}$ is separated, Deligne–Mumford and locally fair (or locally good, or locally finitely presented) if and only if it is equivalent to some $[\underline{V}\rightrightarrows\underline{U}]$ with $\underline{U},\underline{V}$ separated, locally fair (or locally good, or locally finitely presented) C^∞-schemes, $\underline{s}:\underline{V}\to\underline{U}$ étale, and $\underline{s}\times\underline{t}:\underline{V}\to\underline{U}\times\underline{U}$ proper.*

A C^∞-stack $\mathcal{X}$ has an *underlying topological space* $\mathcal{X}_{\rm top}$.

Definition 4.9. Let $\mathcal{X}$ be a C^∞-stack. Write $\underline{*}$ for the point $\operatorname{Spec}\mathbb{R}$ in $\mathbf{C^\infty Sch}$, and $\bar{\underline{*}}$ for the associated point in $\mathbf{C^\infty Sta}$. Define $\mathcal{X}_{\rm top}$ to be the set of 2-isomorphism classes $[x]$ of 1-morphisms $x:\bar{\underline{*}}\rightarrow\mathcal{X}$. When $i_{\mathcal{U}}:\mathcal{U}\rightarrow\mathcal{X}$ is an open C^∞-substack in $\mathcal{X}$, write

$$\mathcal{U}_{\mathcal{X},\mathrm{top}}=\bigl\{[u\circ i_{\mathcal{U}}]\in\mathcal{X}_{\mathrm{top}}:u:\bar{\underline{*}}\rightarrow\mathcal{U}\text{ is a 1-morphism}\bigr\}\subseteq\mathcal{X}_{\mathrm{top}}.$$

Define $\mathcal{T}_{\mathcal{X}_{\mathrm{top}}}=\bigl\{\mathcal{U}_{\mathcal{X},\mathrm{top}}:i_{\mathcal{U}}:\mathcal{U}\rightarrow\mathcal{X}$ is an open C^∞-substack in $\mathcal{X}\bigr\}$. Then $(\mathcal{X}_{\mathrm{top}},\mathcal{T}_{\mathcal{X}_{\mathrm{top}}})$ is a topological space, which we call the *underlying topological space* of $\mathcal{X}$, and usually write as $\mathcal{X}_{\mathrm{top}}$.

If $f:\mathcal{X}\rightarrow\mathcal{Y}$ is a 1-morphism of C^∞-stacks then there is a natural continuous map $f_{\mathrm{top}}:\mathcal{X}_{\mathrm{top}}\rightarrow\mathcal{Y}_{\mathrm{top}}$ defined by $f_{\mathrm{top}}([x])=[f\circ x]$. If $f,g:\mathcal{X}\rightarrow\mathcal{Y}$ are 1-morphisms and $\eta:f\Rightarrow g$ is a 2-isomorphism then $f_{\mathrm{top}}=g_{\mathrm{top}}$. Mapping $\mathcal{X}\mapsto\mathcal{X}_{\mathrm{top}}$, $f\mapsto f_{\mathrm{top}}$ and 2-morphisms to identities defines a 2-functor $F_{\mathbf{C^\infty Sta}}^{\mathbf{Top}}:\mathbf{C^\infty Sta}\rightarrow\mathbf{Top}$, where the category of topological spaces $\mathbf{Top}$ is regarded as a 2-category with only identity 2-morphisms.

If $\underline{X}=(X,\mathcal{O}_X)$ is a C^∞-scheme, so that $\bar{\underline{X}}$ is a C^∞-stack, then $\bar{\underline{X}}_{\mathrm{top}}$ is naturally homeomorphic to X, and we will identify $\bar{\underline{X}}_{\mathrm{top}}$ with X. If $\underline{f}=(f,f^\sharp):\underline{X}=(X,\mathcal{O}_X)\rightarrow\underline{Y}=(Y,\mathcal{O}_Y)$ is a morphism of C^∞-schemes, so that $\bar{\underline{f}}:\bar{\underline{X}}\rightarrow\bar{\underline{Y}}$ is a 1-morphism of C^∞-stacks, then $\bar{\underline{f}}_{\mathrm{top}}:\bar{\underline{X}}_{\mathrm{top}}\rightarrow\bar{\underline{Y}}_{\mathrm{top}}$ is $f:X\rightarrow Y$.

We call a Deligne–Mumford C^∞-stack $\mathcal{X}$ *paracompact* if the underlying topological space $\mathcal{X}_{\mathrm{top}}$ is paracompact.

DEFINITION 4.10. Let $\mathcal{X}$ be a C^∞-stack, and $[x]\in\mathcal{X}_{\mathrm{top}}$. Pick a representative x for $[x]$, so that $x:\bar{\underline{*}}\rightarrow\mathcal{X}$ is a 1-morphism. Let G be the group of 2-morphisms $\eta:x\Rightarrow x$. There is a natural C^∞-scheme $\underline{G}=(G,\mathcal{O}_G)$ with $\bar{\underline{G}}\cong\bar{\underline{*}}\times_{x,\mathcal{X},x}\bar{\underline{*}}$, which makes $\underline{G}$ into a *C^∞-group* (a group object in $\mathbf{C^\infty Sch}$, just as a Lie group is a group object in $\mathbf{Man}$). With $[x]$ fixed, this C^∞-group $\underline{G}$ is independent of choices up to noncanonical isomorphism; roughly, $\underline{G}$ is canonical up to conjugation in $\underline{G}$. We define the *stabilizer group* (or *isotropy group*, or *orbifold group*) $\mathrm{Iso}([x])$ of $[x]$ to be this C^∞-group $\underline{G}$, regarded as a C^∞-group up to noncanonical isomorphism.

If $f:\mathcal{X}\rightarrow\mathcal{Y}$ is a 1-morphism of C^∞-stacks and $[x]\in\mathcal{X}_{\mathrm{top}}$ with $f_{\mathrm{top}}([x])=[y]\in\mathcal{Y}_{\mathrm{top}}$, for $y=f\circ x$, then we define $f_*:\mathrm{Iso}([x])\rightarrow\mathrm{Iso}([y])$ by $f_*(\eta)=\mathrm{id}_f*\eta$. Then f_* is a group morphism, and extends to a C^∞-group morphism. It is independent of choices of $x\in[x]$, $y\in[y]$ up to conjugation in $\mathrm{Iso}([x]),\mathrm{Iso}([y])$.

If $\mathcal{X}$ is a Deligne–Mumford C^∞-stack then $\mathrm{Iso}([x])$ is a finite group for all $[x]$ in $\mathcal{X}_{\mathrm{top}}$, which is discrete as a C^∞-group. Here are [**7**, Prop. 8.31 & Th. 8.32].

PROPOSITION 4.11. *Let $\mathcal{X}$ be a Deligne–Mumford C^∞-stack and $[x]\in\mathcal{X}_{\mathrm{top}}$, so that $\mathrm{Iso}([x])\cong H$ for some finite group H. Then there exists an open C^∞-substack $\mathcal{U}$ in $\mathcal{X}$ with $[x]\in\mathcal{U}_{\mathcal{X},\mathrm{top}}\subseteq\mathcal{X}_{\mathrm{top}}$ and an equivalence $\mathcal{U}\simeq[\underline{Y}/H]$, where $\underline{Y}=(Y,\mathcal{O}_Y)$ is an affine C^∞-scheme with an action of H, and $[x]\in\mathcal{U}_{\mathcal{X},\mathrm{top}}\cong Y/H$ corresponds to a fixed point y of H in Y.*

THEOREM 4.12. *Suppose $\mathcal{X}$ is a Deligne–Mumford C^∞-stack with $\mathrm{Iso}([x]) \cong \{1\}$ for all $[x] \in \mathcal{X}_{\mathrm{top}}$. Then $\mathcal{X}$ is equivalent to $\bar{\underline{X}}$ for some C^∞-scheme $\underline{X}$.*

In conventional algebraic geometry, a stack with all stabilizer groups trivial is (equivalent to) an *algebraic space*, but may not be a scheme, so the category of algebraic spaces is larger than the category of schemes. Here algebraic spaces are spaces which are locally isomorphic to schemes in the étale topology, but not necessarily locally isomorphic to schemes in the Zariski topology.

In contrast, as Theorem 4.12 shows, in C^∞-algebraic geometry there is no difference between C^∞-schemes and C^∞-algebraic spaces. This is because in C^∞-geometry the Zariski topology is already fine enough, as in Remark 3.7(iii), so we gain no extra generality by passing to the étale topology.

4.2. Orbifolds as Deligne–Mumford C^∞-stacks. Example 3.3 defined a functor $F_{\mathbf{Man}}^{\mathbf{C^\infty Sch}} : \mathbf{Man} \to \mathbf{C^\infty Sch}$ embedding manifolds as a full subcategory of C^∞-schemes. Similarly, one might expect to define a (2)-functor $F_{\mathbf{Orb}}^{\mathbf{DMC^\infty Sta}} : \mathbf{Orb} \to \mathbf{DMC^\infty Sta}$ embedding the (2-)category of orbifolds as a full (2-)subcategory of Deligne–Mumford C^∞-stacks. In fact, in [**7**, §8.8] we took a slightly different approach: we defined a full 2-subcategory **Orb** in $\mathbf{DMC^\infty Sta}$, and then showed this is equivalent to other definitions of the (2-)category of orbifolds. The reason for this is that there is not one definition of orbifolds, but several, and our new definition of orbifolds as examples of C^∞-stacks may be as useful as some of the other definitions.

Orbifolds (without boundary) are spaces locally modelled on $\mathbb{R}^n/G$ for G a finite group acting linearly on $\mathbb{R}^n$, just as manifolds are spaces locally modelled on $\mathbb{R}^n$. They were introduced by Satake [**16**], who called them V-manifolds. Moerdijk [**13**] defines orbifolds as *proper étale Lie groupoids* in **Man**. Both [**13**, **16**] regard orbifolds as a 1-category (an ordinary category), if the issue arises at all. However, for issues such as fibre products or pullbacks of vector bundles this 1-category structure is badly behaved, and it becomes clear that orbifolds should really be a 2-category, as for stacks in algebraic geometry.

There are two main routes in the literature to defining a 2-category of orbifolds **Orb**. The first, as in Pronk [**15**] and Lerman [**10**, §3.3], is to define a 2-category **Gpoid** of proper étale Lie groupoids, and then to define **Orb** as a (weak) 2-category localization of **Gpoid** at a suitable class of 1-morphisms. The second, as in Behrend and Xu [**1**, §2], Lerman [**10**, §4] and Metzler [**12**, §3.5], is to define orbifolds as a class of Deligne–Mumford stacks on the site $(\mathbf{Man}, \mathcal{J}_{\mathbf{Man}})$ of manifolds with Grothendieck topology $\mathcal{J}_{\mathbf{Man}}$ coming from open covers. Our approach is similar to the second route, but defines orbifolds as a class of C^∞-stacks, that is, as stacks on the site $(\mathbf{C^\infty Sch}, \mathcal{J})$ rather than on $(\mathbf{Man}, \mathcal{J}_{\mathbf{Man}})$.

DEFINITION 4.13. A C^∞-stack $\mathcal{X}$ is called an *orbifold* if it is equivalent to a groupoid stack $[\underline{V} \rightrightarrows \underline{U}]$ for some groupoid $(\underline{U}, \underline{V}, \underline{s}, \underline{t}, \underline{u}, \underline{i}, \underline{m})$ in $\mathbf{C^\infty Sch}$ which is the image under $F_{\mathbf{Man}}^{\mathbf{C^\infty Sch}}$ of a groupoid (U, V, s, t, u, i, m) in $\mathbf{Man}$, where $s : V \to U$ is an étale smooth map, and $s \times t : V \to U \times U$ is a proper smooth map. That is, $\mathcal{X}$ is the C^∞-stack associated to a *proper étale Lie groupoid* in $\mathbf{Man}$. As a C^∞-stack, every orbifold $\mathcal{X}$ is a separable, paracompact, locally finitely presented Deligne–Mumford C^∞-stack. Write $\mathbf{Orb}$ for the full 2-subcategory of orbifolds in $\mathbf{C^\infty Sta}$.

Here is [**7**, Th. 8.39 & Cor. 8.40]. Since equivalent (2-)categories are considered to be 'the same', the moral of Theorem 4.14 is that our orbifolds are essentially the same objects as those considered by other recent authors.

THEOREM 4.14. *The* 2*-category* $\mathbf{Orb}$ *of orbifolds defined above is equivalent to the* 2*-categories of orbifolds considered as stacks on* $\mathbf{Man}$ *defined in Metzler* [**12**, §3.4] *and Lerman* [**10**, §4], *and also equivalent as a weak* 2*-category to the weak* 2*-categories of orbifolds regarded as proper étale Lie groupoids defined in Pronk* [**15**] *and Lerman* [**10**, §3.3].

Furthermore, the homotopy 1*-category* $\mathbf{Orb^{ho}}$ *of* $\mathbf{Orb}$ *(that is, the category whose objects are objects in* $\mathbf{Orb}$*, and whose morphisms are* 2*-isomorphism classes of* 1*-morphisms in* $\mathbf{Orb}$*) is equivalent to the* 1*-category of orbifolds regarded as proper étale Lie groupoids defined in Moerdijk* [**13**]. *Transverse fibre products in* $\mathbf{Orb}$ *agree with the corresponding fibre products in* $\mathbf{C^\infty Sta}$.

4.3. Quasicoherent and coherent sheaves on C^∞-stacks. In [**7**, §9] the author studied sheaves on Deligne–Mumford C^∞-stacks.

DEFINITION 4.15. Let $\mathcal{X}$ be a Deligne–Mumford C^∞-stack. Define a category $\mathcal{C}_\mathcal{X}$ to have objects pairs $(\underline{U}, u)$ where $\underline{U}$ is a C^∞-scheme and $u : \underline{U} \to \mathcal{X}$ is an étale 1-morphism, and morphisms $(\underline{f}, \eta) : (\underline{U}, u) \to (\underline{V}, v)$ where $\underline{f} : \underline{U} \to \underline{V}$ is an étale morphism of C^∞-schemes, and $\eta : u \Rightarrow v \circ \underline{f}$ is a 2-isomorphism. If $(\underline{f}, \eta) : (\underline{U}, u) \to (\underline{V}, v)$ and $(\underline{g}, \zeta) : (\underline{V}, v) \to (\underline{W}, w)$ are morphisms in $\mathcal{C}_\mathcal{X}$ then we define the composition $(\underline{g}, \zeta) \circ (\underline{f}, \eta)$ to be $(\underline{g} \circ \underline{f}, \theta) : (\underline{U}, u) \to (\underline{W}, w)$, where θ is the composition of 2-morphisms across the diagram:

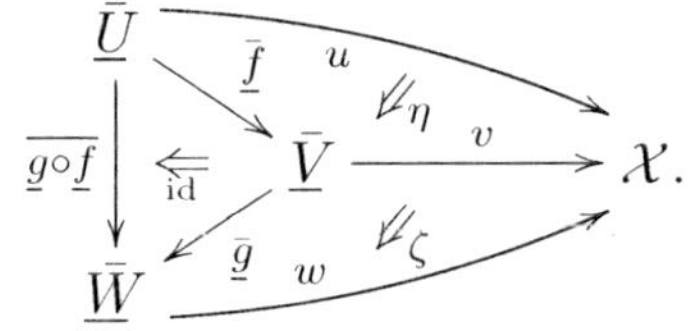

Define an $\mathcal{O}_\mathcal{X}$*-module* $\mathcal{E}$ to assign an $\mathcal{O}_U$-module $\mathcal{E}(\underline{U}, u)$ on $\underline{U} = (U, \mathcal{O}_U)$ for all objects $(\underline{U}, u)$ in $\mathcal{C}_\mathcal{X}$, and an isomorphism $\mathcal{E}_{(\underline{f},\eta)} : \underline{f}^*(\mathcal{E}(\underline{V}, v)) \to \mathcal{E}(\underline{U}, u)$ for all morphisms $(\underline{f}, \eta) : (\underline{U}, u) \to (\underline{V}, v)$ in $\mathcal{C}_\mathcal{X}$, such that for all

$(\underline{f},\eta),(\underline{g},\zeta),(\underline{g}\circ\underline{f},\theta)$ as above the following diagram of isomorphisms of $\mathcal{O}_U$-modules commutes:

$$\begin{array}{ccccc} (\underline{g}\circ\underline{f})^*(\mathcal{E}(\underline{W},w)) & & \xrightarrow{\quad\mathcal{E}_{(\underline{g}\circ\underline{f},\theta)}\quad} & & \mathcal{E}(\underline{U},u), \\ & \searrow^{I_{\underline{f},\underline{g}}(\mathcal{E}(\underline{W},w))} & & \nearrow^{\mathcal{E}_{(\underline{f},\eta)}} & \\ & & \underline{f}^*(\underline{g}^*(\mathcal{E}(\underline{W},w))) \xrightarrow{\underline{f}^*(\mathcal{E}_{(\underline{g},\zeta)})} \underline{f}^*(\mathcal{E}(\underline{V},v)) & & \end{array} \tag{4}$$

for $I_{\underline{f},\underline{g}}(\mathcal{E})$ as in Definition 3.9.

A *morphism of $\mathcal{O}_{\mathcal{X}}$-modules* $\phi:\mathcal{E}\rightarrow\mathcal{F}$ assigns a morphism of $\mathcal{O}_U$-modules $\phi(\underline{U},u):\mathcal{E}(\underline{U},u)\rightarrow\mathcal{F}(\underline{U},u)$ for each object $(\underline{U},u)$ in $\mathcal{C}_{\mathcal{X}}$, such that for all morphisms $(\underline{f},\eta):(\underline{U},u)\rightarrow(\underline{V},v)$ in $\mathcal{C}_{\mathcal{X}}$ the following commutes:

$$\begin{array}{ccc} \underline{f}^*(\mathcal{E}(\underline{V},v)) & \xrightarrow{\mathcal{E}_{(\underline{f},\eta)}} & \mathcal{E}(\underline{U},u) \\ \downarrow{\scriptstyle \underline{f}^*(\phi(\underline{V},v))} & & \downarrow{\scriptstyle \phi(\underline{U},u)} \\ \underline{f}^*(\mathcal{F}(\underline{V},v)) & \xrightarrow{\mathcal{F}_{(\underline{f},\eta)}} & \mathcal{F}(\underline{U},u). \end{array}$$

We call $\mathcal{E}$ *quasicoherent*, or *coherent*, or a *vector bundle of rank n*, if $\mathcal{E}(\underline{U},u)$ is quasicoherent, or coherent, or a vector bundle of rank n, respectively, for all $(\underline{U},u)\in\mathcal{C}_{\mathcal{X}}$. Write $\mathcal{O}_{\mathcal{X}}$-mod for the category of $\mathcal{O}_{\mathcal{X}}$-modules, and $\mathrm{qcoh}(\mathcal{X})$, $\mathrm{coh}(\mathcal{X})$, $\mathrm{vect}(\mathcal{X})$ for the full subcategories of quasicoherent sheaves, coherent sheaves, and vector bundles, respectively.

Here are [**7**, Prop. 9.3 & Ex. 9.4].

PROPOSITION 4.16. *Let $\mathcal{X}$ be a Deligne–Mumford C^∞-stack. Then* $\mathcal{O}_{\mathcal{X}}$-mod *is an abelian category, and* $\mathrm{qcoh}(\mathcal{X})$ *is closed under kernels, cokernels and extensions in* $\mathcal{O}_{\mathcal{X}}$-mod, *so it is also an abelian category. Also* $\mathrm{coh}(\mathcal{X})$ *is closed under cokernels and extensions in* $\mathcal{O}_{\mathcal{X}}$-mod, *but it may not be closed under kernels in* $\mathcal{O}_{\mathcal{X}}$-mod, *so may not be abelian. If $\mathcal{X}$ is locally fair then* $\mathrm{qcoh}(\mathcal{X})=\mathcal{O}_{\mathcal{X}}$-mod.

EXAMPLE 4.17. Let $\underline{X}$ be a C^∞-scheme. Then $\mathcal{X}=\bar{\underline{X}}$ is a Deligne–Mumford C^∞-stack. We will define an *inclusion functor* $\mathcal{I}_{\underline{X}}:\mathcal{O}_X\text{-mod}\rightarrow\mathcal{O}_{\mathcal{X}}\text{-mod}$. Let $\mathcal{E}$ be an object in $\mathcal{O}_X$-mod. If $(\underline{U},u)$ is an object in $\mathcal{C}_{\mathcal{X}}$ then $u:\bar{\underline{U}}\rightarrow\mathcal{X}=\bar{\underline{X}}$ is 1-isomorphic to $\bar{\underline{u}}:\bar{\underline{U}}\rightarrow\bar{\underline{X}}$ for some unique morphism $\underline{u}:\underline{U}\rightarrow\underline{X}$. Define $\mathcal{E}'(\underline{U},u)=\underline{u}^*(\mathcal{E})$. If $(\underline{f},\eta):(\underline{U},u)\rightarrow(\underline{V},v)$ is a morphism in $\mathcal{C}_{\mathcal{X}}$ and $\underline{u},\underline{v}$ are associated to u,v as above, so that $\underline{u}=\underline{v}\circ\underline{f}$, then define

$$\mathcal{E}'_{(\underline{f},\eta)}=I_{\underline{f},\underline{v}}(\mathcal{E})^{-1}:\underline{f}^*(\mathcal{E}'(\underline{V},v))=\underline{f}^*(\underline{v}^*(\mathcal{E}))\longrightarrow(\underline{v}\circ\underline{f})^*(\mathcal{E})=\mathcal{E}'(\underline{U},u).$$

Then (4) commutes for all $(\underline{f},\eta),(\underline{g},\zeta)$, so $\mathcal{E}'$ is an $\mathcal{O}_{\mathcal{X}}$-module.

If $\phi:\mathcal{E}\rightarrow\mathcal{F}$ is a morphism of $\mathcal{O}_X$-modules then we define a morphism $\phi':\mathcal{E}'\rightarrow\mathcal{F}'$ in $\mathcal{O}_{\mathcal{X}}$-mod by $\phi'(\underline{U},u)=\underline{u}^*(\phi)$ for $\underline{u}$ associated to u as above. Then defining $\mathcal{I}_{\underline{X}}:\mathcal{E}\mapsto\mathcal{E}'$, $\mathcal{I}_{\underline{X}}:\phi\mapsto\phi'$ gives a functor $\mathcal{O}_X\text{-mod}\rightarrow\mathcal{O}_{\mathcal{X}}\text{-mod}$, which induces equivalences between the categories $\mathcal{O}_X$-mod, $\mathrm{qcoh}(\underline{X})$, $\mathrm{coh}(\underline{X})$ defined in §3.2 and $\mathcal{O}_{\mathcal{X}}$-mod, $\mathrm{qcoh}(\mathcal{X})$, $\mathrm{coh}(\mathcal{X})$ above.

In [7, §9.2] we explain how to describe sheaves on a Deligne–Mumford C^∞-stack $\mathcal{X}$ in terms of sheaves on $\underline{U}$ for an étale atlas $\Pi : \bar{\underline{U}} \to \mathcal{X}$ for $\mathcal{X}$. Here are [7, Def. 9.5 & Th. 9.6].

Definition 4.18. Let $\mathcal{X}$ be a Deligne–Mumford C^∞-stack. Then $\mathcal{X}$ admits an étale atlas $\Pi : \bar{\underline{U}} \to \mathcal{X}$, and as in Definition 4.3 from Π we can construct a groupoid $(\underline{U}, \underline{V}, \underline{s}, \underline{t}, \underline{u}, \underline{i}, \underline{m})$ in $\mathbf{C^\infty Sch}$, with $\underline{s}, \underline{t} : \underline{V} \to \underline{U}$ étale, such that $\mathcal{X}$ is equivalent to the groupoid stack $[\underline{V} \rightrightarrows \underline{U}]$. Define a $(\underline{V} \rightrightarrows \underline{U})$-*module* to be a pair (E, Φ) where E is an $\mathcal{O}_U$-module and $\Phi : \underline{s}^*(E) \to \underline{t}^*(E)$ is an isomorphism of $\mathcal{O}_V$-modules, such that

$$
\begin{aligned}
(5)\qquad I_{\underline{m},\underline{t}}(E)^{-1} \circ \underline{m}^*(\Phi) \circ I_{\underline{m},\underline{s}}(E) = \bigl(&I_{\underline{\pi}_1,\underline{t}}(E)^{-1} \circ \underline{\pi}_1^*(\Phi) \circ I_{\underline{\pi}_1,\underline{s}}(E)\bigr) \\
&\circ \bigl(I_{\underline{\pi}_2,\underline{t}}(E)^{-1} \circ \underline{\pi}_2^*(\Phi) \circ I_{\underline{\pi}_2,\underline{s}}(E)\bigr)
\end{aligned}
$$

in morphisms of $\mathcal{O}_W$-modules $(\underline{s} \circ \underline{m})^*(E) \to (\underline{t} \circ \underline{m})^*(E)$, where $\underline{W} = \underline{V} \times_{\underline{s},\underline{U},\underline{t}} \underline{V}$ and $\underline{\pi}_1, \underline{\pi}_2 : \underline{W} \to \underline{V}$ are the projections. Define a *morphism of* $(\underline{V} \rightrightarrows \underline{U})$-*modules* $\phi : (E, \Phi) \to (F, \Psi)$ to be a morphism of $\mathcal{O}_U$-modules $\phi : E \to F$ such that $\Psi \circ \underline{s}^*(\phi) = \underline{t}^*(\phi) \circ \Phi : \underline{s}^*(E) \to \underline{t}^*(F)$. Then $(\underline{V} \rightrightarrows \underline{U})$-modules form an *abelian category* $(\underline{V} \rightrightarrows \underline{U})$-mod. Write $\mathrm{qcoh}(\underline{V} \rightrightarrows \underline{U})$ and $\mathrm{coh}(\underline{V} \rightrightarrows \underline{U})$ for the full subcategories of (E, Φ) in $(\underline{V} \rightrightarrows \underline{U})$-mod with E quasicoherent, or coherent, respectively. Then $\mathrm{qcoh}(\underline{V} \rightrightarrows \underline{U})$ is abelian. Define a functor F_Π : $\mathcal{O}_\mathcal{X}$-mod $\to (\underline{V} \rightrightarrows \underline{U})$-mod by $F_\Pi : \mathcal{E} \mapsto \bigl(\mathcal{E}(\underline{U}, \Pi), \mathcal{E}_{(\underline{t},\eta)}^{-1} \circ \mathcal{E}_{(\underline{s},\mathrm{id}_{\Pi \circ \underline{s}})}\bigr)$ and F_Π : $\phi \mapsto \phi(\underline{U}, \Pi)$. As in [7, §9.2], $F_\Pi(\mathcal{E})$ does satisfy (5) and so lies in $(\underline{V} \rightrightarrows \underline{U})$-mod, and it also maps $\mathrm{qcoh}, \mathrm{coh}(\mathcal{X})$ to $\mathrm{qcoh}, \mathrm{coh}(\underline{V} \rightrightarrows \underline{U})$.

Theorem 4.19. *The functor* F_Π *above induces equivalences between* $\mathcal{O}_\mathcal{X}$-mod, $\mathrm{qcoh}(\mathcal{X})$, $\mathrm{coh}(\mathcal{X})$ *and* $(\underline{V} \rightrightarrows \underline{U})$-mod, $\mathrm{qcoh}(\underline{V} \rightrightarrows \underline{U})$, $\mathrm{coh}(\underline{V} \rightrightarrows \underline{U})$, *respectively.*

For example, if $\mathcal{X} = [\underline{Y}/G]$ for $\underline{Y}$ a C^∞-scheme acted on by a finite group G, then Theorem 4.19 shows that $\mathrm{qcoh}(\mathcal{X})$ is equivalent to the abelian category $\mathrm{qcoh}^G(\underline{Y})$ of G-equivariant quasicoherent sheaves on $\underline{Y}$.

In §3.2, for a morphism of C^∞-schemes $\underline{f} : \underline{X} \to \underline{Y}$ we defined a right exact *pullback functor* $\underline{f}^* : \mathcal{O}_Y$-mod $\to \mathcal{O}_X$-mod. Pullbacks may not be strictly functorial in $\underline{f}$, that is, we do not have $\underline{f}^*(\underline{g}^*(\mathcal{E})) = (\underline{g} \circ \underline{f})^*(\mathcal{E})$ for all $\underline{f} : \underline{X} \to \underline{Y}$, $\underline{g} : \underline{Y} \to \underline{Z}$ and $\mathcal{E} \in \mathcal{O}_Z$-mod, but instead we have canonical isomorphisms $I_{\underline{f},\underline{g}}(\mathcal{E}) : (\underline{g} \circ \underline{f})^*(\mathcal{E}) \to \underline{f}^*(\underline{g}^*(\mathcal{E}))$. We now generalize this to sheaves on Deligne–Mumford C^∞-stacks. We must interpret pullback for 2-morphisms as well as 1-morphisms.

Definition 4.20. Let $f : \mathcal{X} \to \mathcal{Y}$ be a 1-morphism of Deligne–Mumford C^∞-stacks, and $\mathcal{F}$ be an $\mathcal{O}_\mathcal{Y}$-module. A *pullback* of $\mathcal{F}$ to $\mathcal{X}$ is an $\mathcal{O}_\mathcal{X}$-module $\mathcal{E}$, together with the following data: if $\underline{U}, \underline{V}$ are C^∞-schemes and $u : \bar{\underline{U}} \to \mathcal{X}$ and $v : \bar{\underline{V}} \to \mathcal{Y}$ are étale 1-morphisms, then there is a C^∞-scheme $\underline{W}$ and morphisms $\underline{\pi}_{\underline{U}} : \underline{W} \to \underline{U}$, $\underline{\pi}_{\underline{V}} : \underline{W} \to \underline{V}$ giving a 2-Cartesian diagram:

$$
\begin{array}{ccc}
\underline{W} & \xrightarrow{\underline{\pi}_{\underline{V}}} & \underline{V} \\
\downarrow{\scriptstyle \underline{\pi}_{\underline{U}}} & \zeta \Uparrow & \downarrow{\scriptstyle v} \\
\underline{U} & \xrightarrow{f \circ u} & \mathcal{Y}.
\end{array} \tag{6}
$$

Then an isomorphism $i(\mathcal{F}, f, u, v, \zeta) : \underline{\pi}_{\underline{U}}^*\big(\mathcal{E}(\underline{U}, u)\big) \to \underline{\pi}_{\underline{V}}^*\big(\mathcal{F}(\underline{V}, v)\big)$ of $\mathcal{O}_W$-modules should be given, which is functorial in $(\underline{U}, u)$ in $\mathcal{C}_{\mathcal{X}}$ and $(\underline{V}, v)$ in $\mathcal{C}_{\mathcal{Y}}$ and the 2-isomorphism ζ in (6). We usually write pullbacks $\mathcal{E}$ as $f^*(\mathcal{F})$. By [7, Prop. 9.9], pullbacks $f^*(\mathcal{F})$ exist, and are unique up to unique isomorphism. Using the Axiom of Choice, we choose a pullback $f^*(\mathcal{F})$ for all such $f : \mathcal{X} \to \mathcal{Y}$ and $\mathcal{F}$.

Let $f : \mathcal{X} \to \mathcal{Y}$ be a 1-morphism, and $\phi : \mathcal{E} \to \mathcal{F}$ be a morphism in $\mathcal{O}_{\mathcal{Y}}$-mod. Then $f^*(\mathcal{E}), f^*(\mathcal{F}) \in \mathcal{O}_{\mathcal{X}}$-mod. Define the *pullback morphism* $f^*(\phi) : f^*(\mathcal{E}) \to f^*(\mathcal{F})$ to be the unique morphism in $\mathcal{O}_{\mathcal{X}}$-mod such that whenever $u : \underline{U} \to \mathcal{X}$, $v : \underline{V} \to \mathcal{Y}$, $\underline{W}, \underline{\pi}_{\underline{U}}, \underline{\pi}_{\underline{V}}$ are as above, the following diagram of morphisms of $\mathcal{O}_W$-modules commutes:

$$
\begin{array}{ccc}
\underline{\pi}_{\underline{U}}^*\big(f^*(\mathcal{E})(\underline{U}, u)\big) & \xrightarrow[i(\mathcal{E},f,u,v,\zeta)]{} & \underline{\pi}_{\underline{V}}^*\big(\mathcal{E}(\underline{V}, v)\big) \\
\downarrow{\scriptstyle \underline{\pi}_{\underline{U}}^*(f^*(\phi)(\underline{U},u))} & & \downarrow{\scriptstyle \underline{\pi}_{\underline{V}}{}^*(\phi(\underline{V},v))} \\
\underline{\pi}_{\underline{U}}^*\big(f^*(\mathcal{F})(\underline{U}, u)\big) & \xrightarrow{i(\mathcal{F},f,u,v,\zeta)} & \underline{\pi}_{\underline{V}}^*\big(\mathcal{F}(\underline{V}, v)\big).
\end{array}
$$

This defines a functor $f^* : \mathcal{O}_{\mathcal{Y}}$-mod $\to \mathcal{O}_{\mathcal{X}}$-mod, which also maps $\mathrm{qcoh}(\mathcal{Y}) \to \mathrm{qcoh}(\mathcal{X})$ and $\mathrm{coh}(\mathcal{Y}) \to \mathrm{coh}(\mathcal{X})$. It is right exact by [7, Prop. 9.12].

Let $f : \mathcal{X} \to \mathcal{Y}$ and $g : \mathcal{Y} \to \mathcal{Z}$ be 1-morphisms of Deligne–Mumford C^∞-stacks, and $\mathcal{E} \in \mathcal{O}_{\mathcal{Z}}$-mod. Then $(g \circ f)^*(\mathcal{E})$ and $f^*(g^*(\mathcal{E}))$ both lie in $\mathcal{O}_{\mathcal{X}}$-mod. One can show that $f^*(g^*(\mathcal{E}))$ is a possible pullback of $\mathcal{E}$ by $g \circ f$. Thus as in Definition 3.9, we have a canonical isomorphism $I_{f,g}(\mathcal{E}) : (g \circ f)^*(\mathcal{E}) \to f^*(g^*(\mathcal{E}))$. This defines a natural isomorphism of functors $I_{f,g} : (g \circ f)^* \Rightarrow f^* \circ g^*$.

Let $f, g : \mathcal{X} \to \mathcal{Y}$ be 1-morphisms of Deligne–Mumford C^∞-stacks, $\eta : f \Rightarrow g$ a 2-morphism, and $\mathcal{E} \in \mathcal{O}_{\mathcal{Y}}$-mod. Then we have $\mathcal{O}_{\mathcal{X}}$-modules $f^*(\mathcal{E})$, $g^*(\mathcal{E})$. Define $\eta^*(\mathcal{E}) : f^*(\mathcal{E}) \to g^*(\mathcal{E})$ to be the unique isomorphism such that whenever $\underline{U}, \underline{V}, \underline{W}, u, v, \underline{\pi}_{\underline{U}}, \underline{\pi}_{\underline{V}}$ are as above, so that we have 2-Cartesian diagrams

$$
\begin{array}{ccc}
\underline{W} & \xrightarrow{\underline{\pi}_{\underline{V}}} & \underline{V} \\
\downarrow{\scriptstyle \underline{\pi}_{\underline{U}}} & \zeta \odot (\eta * \mathrm{id}_{u \circ \underline{\pi}_{\underline{U}}}) \Uparrow & \downarrow{\scriptstyle v} \\
\underline{U} & \xrightarrow{f \circ u} & \mathcal{Y},
\end{array}
\qquad
\begin{array}{ccc}
\underline{W} & \xrightarrow{\underline{\pi}_{\underline{V}}} & \underline{V} \\
\downarrow{\scriptstyle \underline{\pi}_{\underline{U}}} & \zeta \Uparrow & \downarrow{\scriptstyle v} \\
\underline{U} & \xrightarrow{g \circ u} & \mathcal{Y},
\end{array}
$$

as in (6), where in $\zeta\odot(\eta*\mathrm{id}_{u\circ\underline{\pi}_U})$ '$*$' is horizontal and '$\odot$' vertical composition of 2-morphisms, then we have commuting isomorphisms of $\mathcal{O}_W$-modules:

$$\begin{array}{ccc}\underline{\pi}_U^*\big(f^*(\mathcal{E})(\underline{U},u)\big) & \xrightarrow{i(\mathcal{E},f,u,v,\zeta\odot(\eta*\mathrm{id}_{u\circ\underline{\pi}_U}))} & \\ \Big\downarrow{\scriptstyle \underline{\pi}_U^*((\eta^*(\mathcal{E}))(\underline{U},u))} & & \underline{\pi}_V^*\big(\mathcal{E}(\underline{V},v)\big). \\ \underline{\pi}_U^*\big(g^*(\mathcal{E})(\underline{U},u)\big) & \xrightarrow{i(\mathcal{E},g,u,v,\zeta)} & \end{array}$$

This defines a natural isomorphism $\eta^*:f^*\Rightarrow g^*$.

If $\mathcal{X}$ is a Deligne–Mumford C^∞-stack with identity 1-morphism $\mathrm{id}_{\mathcal{X}}:\mathcal{X}\to\mathcal{X}$ then for each $\mathcal{E}\in\mathcal{O}_{\mathcal{X}}$-mod, $\mathcal{E}$ is a possible pullback $\mathrm{id}_{\mathcal{X}}^*(\mathcal{E})$, so we have a canonical isomorphism $\delta_{\mathcal{X}}(\mathcal{E}):\mathrm{id}_{\mathcal{X}}^*(\mathcal{E})\to\mathcal{E}$. These define a natural isomorphism $\delta_{\mathcal{X}}:\mathrm{id}_{\mathcal{X}}^*\Rightarrow\mathrm{id}_{\mathcal{O}_{\mathcal{X}}\text{-mod}}$.

Here is [**7**, Th. 9.11]:

THEOREM 4.21. *Mapping $\mathcal{X}$ to $\mathcal{O}_{\mathcal{X}}$-mod for objects $\mathcal{X}$ in* **DMC$^\infty$Sta**, *and mapping 1-morphisms $f:\mathcal{X}\to\mathcal{Y}$ to $f^*:\mathcal{O}_{\mathcal{Y}}\text{-mod}\to\mathcal{O}_{\mathcal{X}}\text{-mod}$, and mapping 2-morphisms $\eta:f\Rightarrow g$ to $\eta^*:f^*\Rightarrow g^*$ for 1-morphisms $f,g:\mathcal{X}\to\mathcal{Y}$, and the natural isomorphisms $I_{f,g}:(g\circ f)^*\Rightarrow f^*\circ g^*$ for all 1-morphisms $f:\mathcal{X}\to\mathcal{Y}$ and $g:\mathcal{Y}\to\mathcal{Z}$ in* **DMC$^\infty$Sta**, *and $\delta_{\mathcal{X}}$ for all $\mathcal{X}\in$* **DMC$^\infty$Sta**, *together make up a* ***pseudofunctor*** (**DMC$^\infty$Sta**)$^{\mathrm{op}}\to$ **AbCat**, *where* **AbCat** *is the 2-category of abelian categories. That is, they satisfy the conditions:*

(a) *If $f:\mathcal{W}\to\mathcal{X}$, $g:\mathcal{X}\to\mathcal{Y}$, $h:\mathcal{Y}\to\mathcal{Z}$ are 1-morphisms in* **DMC$^\infty$Sta** *and $\mathcal{E}\in\mathcal{O}_{\mathcal{Z}}$-mod then the following diagram commutes in $\mathcal{O}_{\mathcal{X}}$-mod:*

$$\begin{array}{ccc}(h\circ g\circ f)^*(\mathcal{E}) & \xrightarrow[I_{f,h\circ g}(\mathcal{E})]{} & f^*\big((h\circ g)^*(\mathcal{E})\big) \\ \Big\downarrow{\scriptstyle I_{g\circ f,h}(\mathcal{E})} & & \Big\downarrow{\scriptstyle f^*(I_{g,h}(\mathcal{E}))} \\ (g\circ f)^*\big(h^*(\mathcal{E})\big) & \xrightarrow{I_{f,g}(h^*(\mathcal{E}))} & f^*\big(g^*(h^*(\mathcal{E}))\big).\end{array}$$

(b) *If $f:\mathcal{X}\to\mathcal{Y}$ is a 1-morphism in* **DMC$^\infty$Sta** *and $\mathcal{E}\in\mathcal{O}_{\mathcal{Y}}$-mod then the following pairs of morphisms in $\mathcal{O}_{\mathcal{X}}$-mod are inverse:*

$$\begin{array}{c}f^*(\mathcal{E})=\\(f\circ\mathrm{id}_{\mathcal{X}})^*(\mathcal{E})\end{array}\underset{\delta_{\mathcal{X}}(f^*(\mathcal{E}))}{\overset{I_{\mathrm{id}_{\mathcal{X}},f}(\mathcal{E})}{\rightleftarrows}}\mathrm{id}_{\mathcal{X}}^*(f^*(\mathcal{E})),\quad \begin{array}{c}f^*(\mathcal{E})=\\(\mathrm{id}_{\mathcal{Y}}\circ f)^*(\mathcal{E})\end{array}\underset{f^*(\delta_{\mathcal{Y}}(\mathcal{E}))}{\overset{I_{f,\mathrm{id}_{\mathcal{Y}}}(\mathcal{E})}{\rightleftarrows}}f^*(\mathrm{id}_{\mathcal{Y}}^*(\mathcal{E})).$$

Also $(\mathrm{id}_f)^*(\mathrm{id}_{\mathcal{E}})=\mathrm{id}_{f^*(\mathcal{E})}:f^*(\mathcal{E})\to f^*(\mathcal{E})$.

(c) *If $f,g,h:\mathcal{X}\to\mathcal{Y}$ are 1-morphisms and $\eta:f\Rightarrow g$, $\zeta:g\Rightarrow h$ are 2-morphisms in* **DMC$^\infty$Sta**, *so that $\zeta\odot\eta:f\Rightarrow h$ is the vertical composition, and $\mathcal{E}\in\mathcal{O}_{\mathcal{Y}}$-mod, then*

$$\zeta^*(\mathcal{F})\circ\eta^*(\mathcal{E})=(\zeta\odot\eta)^*(\mathcal{E}):f^*(\mathcal{E})\to h^*(\mathcal{E})\quad \textit{in } \mathcal{O}_{\mathcal{X}}\text{-mod}.$$

(d) *If* $f,\tilde f:\mathcal{X}\to\mathcal{Y}$, $g,\tilde g:\mathcal{Y}\to\mathcal{Z}$ *are* 1*-morphisms and* $\eta:f\Rightarrow f'$, $\zeta:g\Rightarrow g'$ 2*-morphisms in* $\mathbf{DMC^\infty Sta}$, *so that* $\zeta*\eta:g\circ f\Rightarrow\tilde g\circ\tilde f$ *is the horizontal composition, and* $\mathcal{E}\in\mathcal{O}_{\mathcal{Z}}$-mod, *then the following commutes in* $\mathcal{O}_{\mathcal{X}}$-mod :

$$\begin{array}{ccccc}
(g\circ f)^*(\mathcal{E}) & & \xrightarrow{\quad(\zeta*\eta)^*(\mathcal{E})\quad} & & (\tilde g\circ\tilde f)^*(\mathcal{E}) \\
\downarrow{\scriptstyle I_{f,g}(\mathcal{E})} & & & & \downarrow{\scriptstyle I_{\tilde f,\tilde g}(\mathcal{E})} \\
f^*(g^*(\mathcal{E})) & \xrightarrow{f^*(\zeta^*(\mathcal{E}))} & f^*(\tilde g^*(\mathcal{E})) & \xrightarrow{\eta^*(\tilde g^*(\mathcal{E}))} & \tilde f^*(\tilde g^*(\mathcal{E})).
\end{array}$$

Definition 4.22. Let $\mathcal{X}$ be a Deligne–Mumford C^∞-stack. Define an $\mathcal{O}_{\mathcal{X}}$-module $T^*\mathcal{X}$ called the *cotangent sheaf* of $\mathcal{X}$ by $(T^*\mathcal{X})(\underline{U},u)=T^*\underline{U}$ for all objects $(\underline{U},u)$ in $\mathcal{C}_{\mathcal{X}}$ and $(T^*\mathcal{X})_{(\underline{f},\eta)}=\Omega_{\underline{f}}:\underline{f}^*(T^*\underline{V})\to T^*\underline{U}$ for all morphisms $(\underline{f},\eta):(\underline{U},u)\to(\underline{V},v)$ in $\mathcal{C}_{\mathcal{X}}$, where $T^*\underline{U}$ and $\Omega_{\underline{f}}$ are as in §3.2.

Let $f:\mathcal{X}\to\mathcal{Y}$ be a 1-morphism of Deligne–Mumford C^∞-stacks. Then $f^*(T^*\mathcal{Y}),T^*\mathcal{X}$ are $\mathcal{O}_{\mathcal{X}}$-modules. Define $\Omega_f:f^*(T^*\mathcal{Y})\to T^*\mathcal{X}$ to be the unique morphism characterized as follows. Let $u:\bar{\underline{U}}\to\mathcal{X}$, $v:\bar{\underline{V}}\to\mathcal{Y}$, $\underline{W},\underline{\pi}_U$, $\underline{\pi}_V$ be as in Definition 4.20, with (6) Cartesian. Then the following diagram of morphisms of $\mathcal{O}_{\underline{W}}$-modules commutes:

$$\begin{array}{ccccc}
\underline{\pi}_U^*(f^*(T^*\mathcal{Y})(\underline{U},u)) & \xrightarrow[i(T^*\mathcal{Y},f,u,v,\zeta)]{} & \underline{\pi}_V^*((T^*\mathcal{Y})(\underline{V},v)) & = & \underline{\pi}_V^*(T^*\underline{V}) \\
\downarrow{\scriptstyle \underline{\pi}_U^*(\Omega_f(\underline{U},u))} & & & & \downarrow{\scriptstyle \Omega_{\underline{\pi}_V}} \\
\underline{\pi}_U^*((T^*\mathcal{X})(\underline{U},u)) & \xrightarrow{(T^*\mathcal{X})_{(\underline{\pi}_U,\mathrm{id}_{u\circ\underline{\pi}_U})}} & (T^*\mathcal{X})(\underline{W},u\circ\underline{\pi}_U) & = & T^*\underline{W}.
\end{array}$$

If $\Pi:\bar{\underline{U}}\to\mathcal{X}$, $(\underline{U},\underline{V},\underline{s},\underline{t},\underline{u},\underline{i},\underline{m})$ and the functor $F_\Pi:\mathcal{O}_{\mathcal{X}}\text{-mod}\to(\underline{V}\rightrightarrows\underline{U})\text{-mod}$ are as in Definition 4.18 then by definition $F_\Pi(T^*\mathcal{X})=(T^*\underline{U},\Omega_{\underline{t}}^{-1}\circ\Omega_{\underline{s}})$, and so we write $T^*(\underline{V}\rightrightarrows\underline{U})=(T^*\underline{U},\Omega_{\underline{t}}^{-1}\circ\Omega_{\underline{s}})$ in $(\underline{V}\rightrightarrows\underline{U})$-mod.

Here [**7**, Prop. 9.14 & Th. 9.15] is the analogue of Theorem 3.15.

Theorem 4.23. **(a)** *Suppose* $\mathcal{X}$ *is an* n*-orbifold. Then* $T^*\mathcal{X}$ *is a rank* n *vector bundle on* $\mathcal{X}$.

(b) *Let* $\mathcal{X}$ *be a locally good Deligne–Mumford* C^∞*-stack. Then* $T^*\mathcal{X}$ *is coherent.*

(c) *Let* $f:\mathcal{X}\to\mathcal{Y}$ *and* $g:\mathcal{Y}\to\mathcal{Z}$ *be* 1*-morphisms of Deligne–Mumford* C^∞*-stacks. Then*

$$\Omega_{g\circ f}=\Omega_f\circ f^*(\Omega_g)\circ I_{f,g}(T^*\mathcal{Z})$$

as morphisms $(g\circ f)^*(T^*\mathcal{Z})\to T^*\mathcal{X}$ *in* $\mathcal{O}_{\mathcal{X}}$-mod.

(d) *Let* $f,g:\mathcal{X}\to\mathcal{Y}$ *be* 1*-morphisms of Deligne–Mumford* C^∞*-stacks and* $\eta:f\Rightarrow g$ *a* 2*-morphism. Then* $\Omega_f=\Omega_g\circ\eta^*(T^*\mathcal{Y}):f^*(T^*\mathcal{Y})\to T^*\mathcal{X}$.

(e) *Suppose* $\mathcal{W},\mathcal{X},\mathcal{Y},\mathcal{Z}$ *are locally fair Deligne–Mumford* C^∞*-stacks with a* 2*-Cartesian square*

$$\begin{array}{ccc}
\mathcal{W} & \xrightarrow{\quad f\quad} & \mathcal{Y} \\
\downarrow{\scriptstyle e} & \eta\Uparrow & \downarrow{\scriptstyle h} \\
\mathcal{X} & \xrightarrow{\quad g\quad} & \mathcal{Z}
\end{array}$$

in $\mathbf{DMC^\infty Sta^{lf}}$, *so that* $\mathcal{W}=\mathcal{X}\times_{\mathcal{Z}}\mathcal{Y}$. *Then the following is exact in* $\mathrm{qcoh}(\mathcal{W})$:

$$(g\circ e)^*(T^*\mathcal{Z})\xrightarrow{\substack{e^*(\Omega_g)\circ I_{e,g}(T^*\mathcal{Z})\oplus \\ -f^*(\Omega_h)\circ I_{f,h}(T^*\mathcal{Z})\circ\eta^*(T^*\mathcal{Z})}}\begin{array}{c}e^*(T^*\mathcal{X})\oplus\\ f^*(T^*\mathcal{Y})\end{array}\xrightarrow{\Omega_e\oplus\Omega_f}T^*\mathcal{W}\longrightarrow 0.$$

Acknowledgements

I would like to thank Eduardo Dubuc and Jacob Lurie for helpful conversations.

References

[1] K. Behrend and P. Xu, *Differentiable stacks and gerbes*, J. Symplectic Geom. 9 (2011), 285–341. math.DG/0605694.

[2] E.J. Dubuc, *Sur les modèles de la géométrie différentielle synthétique*, Cah. Topol. Géom. Différ. Catég. 20 (1979), 231–279.

[3] E.J. Dubuc, *C^∞-schemes*, Amer. J. Math. 103 (1981), 683–690.

[4] E.J. Dubuc, *Open covers and infinitary operations in C^∞-rings*, Cah. Topol. Géom. Différ. Catég. 22 (1981), 287–300.

[5] K. Fukaya, Y.-G. Oh, H. Ohta and K. Ono, *Lagrangian intersection Floer theory – anomaly and obstruction*, Parts I & II. AMS/IP Studies in Advanced Mathematics, 46.1 & 46.2, A.M.S./International Press, 2009.

[6] R. Hartshorne, *Algebraic Geometry*, Graduate Texts in Math. 52, Springer, New York, 1977.

[7] D. Joyce, *Algebraic Geometry over C^∞-rings*, arXiv:1001.0023, 2010.

[8] D. Joyce, *D-manifolds and d-orbifolds: a theory of derived differential geometry*, book in preparation, 2012. Preliminary version available at `http://people.maths.ox.ac.uk/~joyce/`.

[9] A. Kock, *Synthetic differential geometry*, second edition, L.M.S. Lecture Notes 333, Cambridge University Press, 2006.

[10] E. Lerman, *Orbifolds as stacks?*, Enseign. Math. 56 (2010), 315–363. arXiv:0806.4160.

[11] J. Lurie, *Derived Algebraic Geometry I: Stable ∞-categories*, math.CT/0608228, 2006.

[12] D.S. Metzler, *Topological and smooth stacks*, math.DG/0306176, 2003.

[13] I. Moerdijk, *Orbifolds as groupoids: an introduction*, pages 205–222 in A. Adem, J. Morava and Y. Ruan, editors, *Orbifolds in Mathematics and Physics*, Cont. Math. 310, A.M.S., Providence, RI, 2002. math.DG/0203100.

[14] I. Moerdijk and G.E. Reyes, *Models for smooth infinitesimal analysis*, Springer-Verlag, New York, 1991.

[15] D. Pronk, *Etendues and stacks as bicategories of fractions*, Compositio Math. 102 (1996), 243–303.

[16] I. Satake, *On a generalization of the notion of manifold*, Proc. Nat. Acad. Sci. U.S.A. 42 (1956), 359–363.

[17] D.I. Spivak, *Derived smooth manifolds*, Duke Mathematical Journal 153 (2010), 55–128. arXiv:0810.5174.

The Mathematical Institute, 24-29 St. Giles, Oxford, OX1 3LB, U.K.
E-mail address: `joyce@maths.ox.ac.uk`

Complex Monge-Ampère Equations[1]

D.H. Phong, Jian Song, and Jacob Sturm

Abstract. This is a survey of some of the recent developments in the theory of complex Monge-Ampère equations. The topics discussed include refinements and simplifications of classical a priori estimates, methods from pluripotential theory, variational methods for big cohomology classes, semiclassical constructions of solutions of homogeneous equations, and envelopes.

Contents

[1]Contribution to the proceedings of the Journal of Differential Geometry Conference in honor of Professor C.C. Hsiung, Lehigh University, May 2010. Work supported in part by National Science Foundation grants DMS-07-57372, DMS-09-05873, and DMS-08-47524.

1. Introduction

Monge-Ampère equations are second-order partial differential equations whose leading term is the determinant of the Hessian of a real unknown function φ. As such, they are arguably the most basic of fully non-linear equations. The Hessian is required to be positive or at least non-negative, so the equations are elliptic or degenerate elliptic. Monge-Ampère equations can be divided into real or complex, depending on whether φ is defined on a real or complex manifold. In the real case, the Hessian is $\nabla_j \nabla_k \varphi$, so the positivity of the Hessian is a convexity condition. In the complex case, the Hessian is $\partial_j \partial_{\bar{k}} \varphi$, and its positivity is rather a plurisubharmonicity condition. Unlike convex functions, plurisubharmonic functions can have singularities, and this accounts for many significant differences between the theories of real and complex Monge-Ampère equations. In these lectures, we shall concentrate on the complex case.

The foundations of an existence and regularity theory for complex Monge-Ampère equations in the elliptic case, with smooth data, were laid by Yau [**Y78**] and Caffarelli, Kohn, Nirenberg, and Spruck [**CNS, CKNS**]. In [**Y78**], a complete solution was given for the Calabi conjecture, which asserts the existence of a smooth solution to the equation

$$\left(\omega_0 + \frac{i}{2}\partial\bar{\partial}\varphi\right)^n = e^{f(z)}\,\omega_0^n, \tag{1.1}$$

on a compact n-dimensional Kähler manifold (X, ω_0) without boundary, where $f(z)$ is a given smooth function satisfying the necessary condition $\int_X e^f \omega_0^n = \int_X \omega_0^n$. The solution was by the method of continuity, and the key estimates for the C^0 norms of φ, $\Delta\varphi$, and $\nabla_j \nabla_{\bar{k}} \nabla_l \varphi$ were formulated and derived there. In [**CKNS**], a complete solution was given for the Dirichlet problem

$$\det(\partial_j \partial_{\bar{k}} \varphi) = F(z, \varphi) \text{ on } D, \quad \varphi = \varphi_b \text{ on } \partial D, \tag{1.2}$$

where D is a smooth, bounded, strongly pseudoconvex domain in $\mathbf{C}^n$, $F \in C^\infty(\bar{D} \times \mathbf{R})$, $F(z, \varphi) > 0$, $F_\varphi(z, \varphi) \geq 0$, and $\varphi_b \in C^\infty(\partial D)$. A crucial ingredient of the existence and regularity developed there is the C^0 boundary estimates for the second order derivatives and their modulus of continuity.

In his paper [**Y78**], Yau also began an existence and regularity theory for singular complex Monge-Ampère equations on Kähler manifolds. Here the term "singular" should be interpreted in a broad sense. It encompasses situations where the right hand side may be degenerate or have singularities [**Y78**], or where the manifold X may not be compact or have singularities [**CY80, MY83, CY86, TY86**], or where the boundary condition may be infinite [**CY80**]. Such extensions were required by geometric applications, and many important results were obtained, of which the references we just gave are just a small sample (see e.g. [**TY90, TY91, K83, W08, LYZ**], and especially [**Y93, Y94, Y96**] and references therein).

The last fifteen years or so have witnessed remarkable progresses in the theory of singular Monge-Ampère equations. A particularly strong impetus was provided by related problems from the minimal model program in algebraic geometry ([**EGZ, ST08, TZ, BEGZ**]) and from the problem of finding metrics of constant scalar curvature in a given Kähler class (see [**Y93, T97, D02**] and [**PS08**] for a survey). The solutions in these problems are often inherently singular, and thus they must be understood in a generalized sense. The foundations of a theory of generalized solutions for the complex Monge-Ampère equation - or pluripotential theory - had been laid out by Bedford and Taylor in [**BT76, BT82**]. There they constructed Monge-Ampère measures for bounded potentials and capacities, established monotonicity theorems for their convergence, and obtained generalized solutions of the Dirichlet problem for degenerate right hand sides by the Perron method. A key catalyst for the several of the recent progresses is the theorem of Kolodziej [**K98**], based on pluripotential theory, which provided C^0 estimates for Monge-Ampère equations with right hand sides in L^p for any $p > 1$. Other important ingredients have been the extensions of pluripotential theory to unbounded potentials ([**GZ, BBGZ, Ceg, B06, CG**] and references therein), the Tian-Yau-Zelditch theorem [**Y93, T90a, Z, Cat, L**] on approximations of smooth metrics by Fubini-Study metrics [**PS06, PS07, PS09b, SZ07, SZ10, RZ08, RZ10a, RZ10b**], and refinements and extensions [**Gb, B09b, PS09a, PS09c, GL, Gp, Ch, TW1, TW2, DK**] of the classic estimates in [**Y78, CKNS**].

The main goal of this paper is to survey some of the recent progresses. There have been many of them, and the theory is still in full flux. While definitive answers may not yet be available to many questions, we thought it would be useful to gather here in one place, for the convenience of students and newcomers to the field, some of what is known. It was not possible to be comprehensive, and our selection of material necessarily reflects our own limitations. At the same time, we hope that the survey would be useful to a broad audience of people with relatively little familiarity with complex Monge-Ampère equations, and we have provided reasonably complete derivations in places, when the topics are of particular importance or the literature not easily accessible. Each of us has lectured on parts of this paper at our home institutions, and at various workshops. In particular, the first-named author spoke at the 2010 conference at Lehigh University in honor of Professor C.C. Hsiung, one of the founders of the Journal of Differential Geometry. We would like to contribute this paper to the volume in his honor.

2. Some General Perspective

Let (X, ω_0) be a compact Kähler manifold. We consider complex Monge-Ampère equations of the form

$$\left(\omega_0 + \frac{i}{2}\partial\bar\partial\varphi\right)^n = F(z,\varphi)\,\omega_0^n \tag{2.1}$$

where $F(z,\varphi)$ is a non-negative function. The solution φ is required to be ω_0-plurisubharmonic, that is, $\varphi \in PSH(X,\omega_0)$, with

$$PSH(X,\omega_0) = \Big\{\varphi : X \to [-\infty,\infty); \varphi \text{ is upper semicontinuous}, \omega_\varphi \equiv \omega_0 + \frac{i}{2}\partial\bar\partial\varphi \geq 0\Big\}.$$

We shall consider both the case of X compact without boundary, and the case of $\bar X$ compact with smooth boundary ∂X, in which case we also impose a Dirichlet condition $\varphi = \varphi_b$, where $\varphi_b \in C^\infty(\partial X)$ is a given function.

2.1. Geometric interpretation. Equations of the form (2.1) are fundamentally geometric in nature. The form ω_φ can be viewed as a form in the same cohomology class as ω_0. It defines a regular Kähler metric when it is >0, or a Kähler metric with degeneracies when it does have zeroes. It is well-known that the Ricci curvature form $Ricci(\omega_\varphi)$ of a Kähler form ω_φ is given by

$$Ricci(\omega_\varphi) = -\frac{i}{2}\partial\bar{\partial}\log\omega_\varphi^n. \tag{2.2}$$

Thus the equation (2.1) is just an equation for a possibly degenerate metric ω_φ in the same Kähler class as ω_0, satisfying a given constraint on its volume form ω_φ^n or, equivalently upon differentiation, a given constraint on its Ricci curvature $Ricci(\omega_\varphi)$.

The modern theory of complex Monge-Ampère equations began with the following two fundamental theorems, due respectively to Yau [**Y78**] and to Yau [**Y78**] and Aubin [**A**].

THEOREM 1. *Let (X, ω_0) be a compact Kähler manifold without boundary, and let $F(z) = e^{f(z)}$, where $f(z)$ is a smooth function satisfying the condition*

$$\int_X e^f \omega_0^n = \int_X \omega_0^n. \tag{2.3}$$

Then the equation (2.1) admits a smooth solution $\varphi \in PSH(X, \omega_0)$, unique up to an additive constant.

THEOREM 2. *Let (X, ω_0) be a compact Kähler manifold without boundary, and let $F(z, \varphi) = e^{f+\varphi}$ where $f(z)$ is a smooth function. Then the equation (2.1) admits a unique smooth solution $\varphi \in PSH(X, \omega_0)$.*

Geometrically, Theorem 1 provides a solution of the Calabi conjecture, which asserts that, on a compact Kähler manifold X with $c_1(X) = 0$, there is a unique metric ω_φ with $Ricci(\omega_\varphi) = 0$ in any Kähler class $[\omega_0]$. Indeed, the formula (2.2) shows that the Ricci form of any Kähler metric must be in $c_1(X)$. The assumption that $c_1(X) = 0$ implies that $Ricci(\omega_0) = \frac{i}{2}\partial\bar{\partial}f$ for some smooth function $f(z)$. It is now readily verified, by taking $F(z) = e^{f(z)}$ in the equation (2.1) and taking $\frac{i}{2}\partial\bar{\partial}$ of both sides, that the solution of (2.1) satisfies the condition

$$Ricci(\omega_\varphi) = 0. \tag{2.4}$$

Similarly, Theorem 2 implies the existence of a Kähler-Einstein metric with negative curvature on any compact Kähler manifold X with $c_1(X) < 0$. In this case, since $c_1(X) < 0$, we can choose a Kähler form ω_0 in the cohomology class $-c_1(X)$. But the Ricci curvature form $Ricci(\omega_0)$ is still in $c_1(X)$, and thus there is a smooth function $f(z)$ with $Ricci(\omega_0) + \omega_0 = \frac{i}{2}\partial\bar{\partial}f$. Taking $F(z, \varphi) = e^{f+\varphi}$ in the equation (2.1) and taking again $i\partial\bar{\partial}$ of both sides, we see that the solution of (2.1) satisfies now the Kähler-Einstein condition

$$Ricci(\omega_\varphi) = -\omega_\varphi. \tag{2.5}$$

We note that the Kähler-Einstein problem for compact Kähler manifolds X with $c_1(X) > 0$ is still open at this time, despite a lot of progress [**TY87, Si, N, T90b, T97, D10, D11a, D11b, CDa, CDb**]. A well-known conjecture of Yau [**Y93**] asserts the equivalence between the existence of such a metric on X and the stability of X in geometric invariant theory. This can be reduced, just as above for the cases $c_1(X) = 0$ and $c_1(X) < 0$, to a complex Monge-Ampère equation of the form (2.1), but with $F(z, \varphi) = e^{f(z)-\varphi}$. Thus the conjecture of Yau asserts the equivalence between the solvability of a complex Monge-Ampère equation and a global, algebraic-geometric, condition. Clearly, bringing the algebraic-geometric conditions into play in the solution of a non-linear partial differential equation is an important and challenging problem. The two major successes in this direction are the theorem of Donaldson-Uhlenbeck-Yau [**D87, UY**], on the equivalence between the existence of a Hermitian-Einstein metric on a holomorphic vector bundle $E \to (X, \omega)$ and the Mumford-Takemoto stability of E, and the recent results of Donaldson [**D08**] on the equivalence between the existence of metrics of constant scalar curvature on toric 2-folds and their K-stability. However, there are still many unanswered questions in this direction.

2.2. The method of continuity. The original proof of Theorems 1 and 2 is by the method of continuity, and this has remained a prime method for solving complex Monge-Ampère equations to this day. In this method, the equation to be solved is deformed continuously to an equation which we know how to solve. For example, one introduces for Theorem 1 the deformation,

$$\left(\omega_0 + \frac{i}{2}\partial\bar{\partial}\varphi\right)^n = \frac{\int_X \omega_0^n}{\int_X e^{tf(z)}\omega_0^n} e^{tf(z)}\omega_0^n, \quad 0 \leq t \leq 1 \tag{2.6}$$

and for Theorem 2 the deformation,

$$\left(\omega_0 + \frac{i}{2}\partial\bar{\partial}\varphi\right)^n = e^{tf(z)+\varphi}\omega_0^n, \quad 0 \leq t \leq 1. \tag{2.7}$$

The equations admit trivially the smooth solution $\varphi = 0$ at $t = 0$. It is not difficult to show, by the implicit function theorem, that the set of parameters t for which the equation is solvable is open. So to show that this set is the full interval $[0, 1]$ reduces to show that it is closed. This in turn reduces to the proof of a priori estimates for the solutions φ, assuming that they already exist and are smooth.

3. A Priori Estimates: C^0 Estimates

We begin by discussing C^0 estimates for the most basic complex Monge-Ampère equation. Let (X, ω_0) be a compact Kähler manifold without boundary, and consider the equation

$$\left(\omega_0 + \frac{i}{2}\partial\bar{\partial}\varphi\right)^n = F(z)\,\omega_0^n \tag{3.1}$$

for a smooth function φ satisfying the condition $\varphi \in PSH(X,\omega_0)$, with $F(z)$ a smooth strictly positive function. Since the equation is invariant under shifts of φ by constants, we may assume that $\sup_X \varphi = 0$. It is well-known that all functions in $PSH(X,\omega_0)$ satisfy an exponential integrability condition, and hence their L^p norms are all uniformly bounded by constants depending only on the Kähler class $[\omega_0]$ and on p, for any $1 \leq p < \infty$ (see e.g. Appendix A). But the L^∞, or C^0 estimate, is fundamentally different. In this section, we discuss several methods for obtaining C^0 estimates.

3.1. Yau's original method. Yau's original method was by Moser iteration. Set $\psi = \sup_X \varphi - \varphi + 1 \geq 1$ and let $\alpha \geq 0$. Since $(F-1)\omega_0^n = (\omega_0 + \frac{i}{2}\partial\bar\partial\varphi)^n - \omega_0^n = \frac{i}{2}\partial\bar\partial\varphi \sum_{j=0}^{n-1}(\omega_0 + \frac{i}{2}\partial\bar\partial\varphi)^{n-1-j}\omega_0^j$, we find, after multiplying by $\psi^{\alpha+1}$ and integrating by parts.

$$\int_X \psi^{\alpha+1}(F-1)\omega_0^n = (\alpha+1)\sum_{j=0}^{n-1}\int_X \psi^\alpha\, i\partial\psi \wedge \bar\partial\psi \left(\omega_0 + \frac{i}{2}\partial\bar\partial\varphi\right)^{n-1-j} \omega_0^j. \tag{3.2}$$

All the integrals on the right hand side are positive. Keeping only the contribution with $j = n-1$, we obtain

$$\begin{aligned}\left|\int_X \psi^{\alpha+1}(F-1)\omega_0^n\right| &\geq (\alpha+1)\int_X \psi^\alpha\, i\partial\psi \wedge \bar\partial\psi\, \omega_0^{n-1} \\ &= \frac{(\alpha+1)}{2(\frac{\alpha}{2}+1)^2}\int_X i\partial(\psi^{\frac{\alpha}{2}+1}) \wedge \bar\partial(\psi^{\frac{\alpha}{2}+1}) \wedge \omega_0^{n-1}. \end{aligned} \tag{3.3}$$

and hence, with C_1 depending only on $\|F\|_{L^\infty}$, and all norms and covariant derivatives with respect to the metric ω_0,

$$\|\nabla(\psi^{\frac{\alpha}{2}+1})\|^2 \leq C_1 \frac{n(\frac{\alpha}{2}+1)^2}{\alpha+1}\int_X \psi^{\alpha+1}\omega_0^n. \tag{3.4}$$

On the other hand, the Sobolev inequality asserts that

$$\|u\|^2_{L^{\frac{2n}{n-1}}} \leq C_2(\|\nabla u\|^2_{L^2} + \|u\|^2_{L^2}) \tag{3.5}$$

with C_2 the Sobolev constant of (X,ω_0). Applied to $u = \psi^{\frac{p}{2}}$, it can be expressed as

$$\|\psi\|^p_{L^{p\beta}} \leq C_2(\|\nabla(\psi^{\frac{p}{2}})\|^2_{L^2} + \|\psi\|^p_{L^p}), \tag{3.6}$$

with $\beta = \frac{n}{n-1} > 1$. Setting $p = \alpha + 2$, and applying the inequality (3.4), we find

$$\|\psi\|_{L^{p\beta}} \leq (C_3 p)^{\frac{1}{p}}\|\psi\|_{L^p}, \quad p \geq 2, \tag{3.7}$$

with a constant C_3 depending only on n, $\|F\|_{L^\infty}$, and the Sobolev constant of (X, ω_0). We can iterate $p \to p\beta \to \cdots \to p\beta^k$ and get

$$\log \|\psi\|_{L^\infty} \le \sum_{k=0}^{\infty} \frac{\log(C_3 p\beta^k)}{p\beta^k} + \log \|\psi\|_{L^p} = C_{4,p} + \log \|\psi\|_{L^p}. \tag{3.8}$$

An a priori bound for $\|\psi\|_{L^p}$ for any fixed finite p can be obtained from the exponential estimate for plurisubharmonic functions in Appendix A. Alternatively, we can obtain an a priori bound for $\|\psi\|_{L^2}$ from (3.4) by taking $\alpha = 0$, and applying the Poincaré inequality to the left hand side. Either way gives

THEOREM 3. *Let φ be a smooth solution of the equation (3.1) on a compact Kähler manifold (X, ω_0) without boundary, $F > 0$, and $\varphi \in PSH(X, \omega_0)$. Then $\|\psi\|_{L^\infty(X)}$ is bounded by a constant depending only on n, an upper bound for $\|F\|_{L^\infty(X)}$, and the Kähler form ω_0. The dependence on the Kähler form ω_0 can be stated more precisely as a dependence on the Sobolev constant and the Poincaré constant of ω_0, or on the exponential bound for ω_0.*

The Moser iteration method is now widely used in the study of Monge-Ampère and other non-linear equations. An important variant has been introduced by Weinkove [**W**], where the Moser iteration is applied to e^φ instead of φ. Applications of this variant are in [**SW, ST06, TWY**].

3.2. Reduction to Alexandrov-Bakelman-Pucci estimates. It was suggested early on by Cheng and Yau that the Alexandrov-Bakelman-Pucci estimate can be applied to the complex Monge-Ampère equation. They did not publish their work, but a detailed account was subsequently provided by Bedford [**B**] and Cegrell and Persson [**CP**]. Using the Alexandrov-Bakelman-Pucci estimate, Blocki [**B11a**] gives the following proof of the C^0 estimate. This proof is of particular interest as it is almost a local argument. We follow closely Blocki's presentation.

Let D be any bounded domain in $\mathbf{C}^n$, $u \in C^2(\bar{D})$, $u_{\bar{k}j} \ge 0$, and $u = 0$ on ∂D. Then

$$\|u\|_{C^0} \le C \, \|\det u_{\bar{k}j}\|_{L^2}^{\frac{1}{n}} \tag{3.9}$$

where $C = C(n, \operatorname{diam} D)$ depends only on the diameter of D and the dimension n. To see this, we apply the ABP estimate (from [**GT**], Lemma 9.2) to get

$$\|u\|_{C^0} \le c_n (\operatorname{diam} D) \left(\int_\Gamma \det D^2 u \right)^{\frac{1}{2n}} \tag{3.10}$$

where Γ is the contact set, defined by

$$\Gamma = \{ z \in D ; u(w) \ge u(z) + \langle Du(z), w - z \rangle, \text{ for all } w \in D \}. \tag{3.11}$$

On the contact set Γ, the function u satisfies $D^2u \geq 0$, and for such functions, we have the following inequality between the determinants of the real and complex Hessians,

$$\det u_{\bar{k}j} \geq 2^{-n}(\det D^2 u)^{\frac{1}{2}}. \tag{3.12}$$

This proves the estimate (3.9).

Let now $z \in D$, $h > 0$, and define the sublevel set $S(z,h)$ by

$$S(z,h) = \{w \in D; u(w) < u(z) + h\} \tag{3.13}$$

If $S(z,h) \subset\subset D$, then applying the previous inequality to $S(z,h)$ instead of D gives

$$\begin{aligned} \|u - u(z) - h\|_{C^0(S(z,h))} &\leq C(n, \operatorname{diam} D)\|\det u_{\bar{k}j}\|^{\frac{1}{n}}_{L^2(S(z,h))} \\ &\leq C(n, \operatorname{diam} D)|S(z,h)|^{\frac{1}{2nq}}\|\det u_{\bar{k}j}\|^{\frac{1}{n}}_{L^{2p}} \end{aligned} \tag{3.14}$$

for any $p > 1$, $\frac{1}{p} + \frac{1}{q} = 1$. In particular, we obtain the following lower bound for $|S(z,h)|$

$$h \leq C(n, \operatorname{diam} D)|S(z,h)|^{\frac{1}{2qn}}\|\det u_{\bar{k}j}\|^{\frac{1}{n}}_{L^{2p}}. \tag{3.15}$$

On the other hand, we have the following easy upper bound for $|S(z,h)|$,

$$|S(z,h)|(-u(z) - h) \leq \int_{S(z,h)} (-u) \leq \|u\|_{L^1(D)} \tag{3.16}$$

If we choose z to be the minimum point for u, and eliminate $|S(z,h)|$ between the two inequalities, we obtain a lower bound for u in terms of h and h^{-1}.

This can be applied to the C^0 estimate for the Monge-Ampère equation $(\omega_0 + \frac{i}{2}\partial\bar{\partial}\varphi)^n = F(z)\,\omega_0^n$ on a compact Kähler manifold (X, ω_0). Let z be the minimum point for φ on X, and let $K(w, \bar{w})$ be a Kähler potential for ω_0 in a neighborhood of z. By adding a negative constant and shifting $K(w, \bar{w})$ by the real part of a second order polynomial in w if necessary, we can assume that $K(w, \bar{w}) \leq 0$ in a ball $B(z, 2r)$ around z, $K(w, \bar{w}) \geq K(0) + h$ for $r \leq |w| \leq 2r$, and $K(w, \bar{w})$ attains its minimum in $B(z, 2r)$ at 0. The constant $h > 0$ depends only on the Kähler form ω_0. Then the function $u = K + \varphi$ attains its minimum in $B(z, 2r)$ at z, and the corresponding set $S(z,h) \subset B(z, 2r) \setminus B(z, r)$ has compact closure. By the preceding inequalities, we obtain a lower bound for $u(z)$, depending only on ω_0 and the L^{2p} norm of F, for any $p > 1$. Thus

THEOREM 4. *Let the setting be the same as in Theorem 3. Then for any $p > 1$, $\|\varphi\|_{L^\infty(X)}$ can be bounded by a constant depending only on n, an upper bound for $\|F\|_{L^{2p}(X)}$, and the Kähler form ω_0.*

3.3. Methods of pluripotential theory. A third method for C^0 estimates was introduced by Kolodziej [**K98**]. This method combines the classic approach of De Giorgi with modern techniques of pluripotential theory. It produces C^0 bounds even when the right hand side F is only in $L^p(X)$ for some $p>1$. As shown by Eyssidieux, Guedj, and Zeriahi [**EGZ, EGZ08**] and Demailly and Pali [**DP**], it can also be extended to a family setting, where the background Kähler form ω_t is allowed to degenerate to a closed form χ which is just non-negative. Other family versions of Kolodziej's C^0 estimates are in [**KT, DZ, TZ**]. As we shall see later, such family versions are important for the study of singular Kähler-Einstein metrics and Monge-Ampère equations on complex manifolds with singularities.

Let (X, ω_0) be a compact Kähler manifold. Let $\chi \geq 0$ be a C^∞ closed semi-positive $(1,1)$-form which is not identically 0. Set

$$\omega_t = \chi + (1-t)\omega_0, \quad t \in (0,1) \tag{3.17}$$

and let $[\omega_t^n] = \int_X \omega_t^n$. Consider the equation

$$\left(\omega_t + \frac{i}{2}\partial\bar{\partial}\varphi_t\right)^n = F_t \omega_t^n \tag{3.18}$$

for some strictly positive function F_t and $\varphi_t \in PSH(X, \omega_t) \cap L^\infty(X)$. Then we have the following family version of the C^0 estimates of Kolodziej [**K98**], due to Eyssidieux-Guedj-Zeriahi [**EGZ, EGZ08**] and Demailly-Pali [**DP**]:

THEOREM 5. *Let $A>0$ and and $p>1$. Assume that $\chi \leq A\omega_0$ and $\frac{1}{[\omega_t^n]}\frac{\omega_t^n}{\omega_0^n} \leq A$. Assume also that the functions F_t are in $L^p(X, \omega_t^n)$ and that*

$$\frac{1}{[\omega_t^n]}\int_X F_t^p \omega_t^n \leq A^p < \infty \tag{3.19}$$

for all $t \in (0,1)$. Normalize φ_t so that $\sup_X \varphi_t = 0$. Then there exists a constant $C>0$, depending only on n, ω_0 and A, so that

$$\sup_{t\in[0,1)} \|\varphi_t\|_{L^\infty(X)} \leq C. \tag{3.20}$$

PROOF. Recall the notion of capacity of a Borel set E with respect to a Kähler form ω,

$$\mathrm{Cap}_\omega(E) = \sup\left\{\int_E \left(\omega + \frac{i}{2}\partial\bar{\partial}u\right)^n ;\ u \in PSH(X,\omega),\ 0 \leq u \leq 1\right\}. \tag{3.21}$$

Set

$$f_t(s) = \left(\frac{\mathrm{Cap}_{\omega_t}(\varphi_t < -s)}{[\omega_t^n]}\right)^{\frac{1}{n}}. \tag{3.22}$$

It suffices to show that there exists $s_\infty < \infty$ independent of t so that

$$f_t(s) = 0 \quad \text{for } s > s_\infty. \tag{3.23}$$

Since $f_t(s)^n \geq \frac{1}{[\omega_t^n]} \int_{\varphi<-s} \omega_t^n$, it would follow that $\varphi_t \geq -s_\infty$ a.e. with respect to the measure ω_t^n, and since φ_t is upper semi-continuous, that $\varphi_t \geq -s_\infty$ everywhere. The following classic lemma of De Giorgi provides sufficient conditions for the existence of s_∞:

LEMMA 1. *Let $f : \mathbf{R}^+ \to \mathbf{R}^+$ satisfy the following conditions:*

(a) *f is right-continuous;*
(b) *f decreases to 0;*
(c) *There exist positive constants α, A_α so that for all $s \geq 0$ and all $0 \leq r \leq 1$, we have*

$$r\, f(s+r) \leq A_\alpha f(s)^{1+\alpha}. \tag{3.24}$$

Then there exists s_∞, depending only on α, A_α and the smallest value s_0 for which we have $f(s_0)^\alpha \leq (2A_\alpha)^{-1}$ so that $f(s) = 0$ for $s > s_\infty$. In fact, we can take $s_\infty = s_0 + 2A_\alpha(1 - 2^{-\alpha})^{-1} f(s_0)^\alpha$. □

We shall show that the above functions $f_t(s)$ satisfy the conditions of Lemma 1. The right-continuity (a) of the function $f_t(s)$ is a consequence of the fact that, for any Kähler form ω, and any sequence of increasing sequence of Borel sets $E_j \subset E_{j+1}$, we have $\mathrm{Cap}_\omega(\cup_{j=1}^\infty E_j) = \lim_{j\to\infty} \mathrm{Cap}_\omega(E_j)$. Clearly $f_t(s)$ decreases as s increases. In fact, it does so uniformly to 0 in t as is shown by the following lemma:

LEMMA 2. *There exists a constant C depending only on ω_0 and an upper bound A for χ so that*

$$f_t(s)^n \leq C\, s^{-1}. \tag{3.25}$$

PROOF OF LEMMA 2. Let $u \in PSH(X, \omega_t)$. Then

$$\begin{aligned}
\int_{\varphi_t<-s} &\left(\omega_t + \frac{i}{2}\partial\bar\partial u\right)^n \\
&\leq \frac{1}{s}\int_X (-\varphi_t)\left(\omega_t + \frac{i}{2}\partial\bar\partial u\right)^n \\
&= \frac{1}{s}\int_X (-\varphi_t)\omega_t^n + \frac{1}{s}\int_X (-\varphi_t)\frac{i}{2}\partial\bar\partial u \sum_{j=0}^{n-1} \omega_t^j \left(\omega_t + \frac{i}{2}\partial\bar\partial u\right)^{n-1-j}
\end{aligned} \tag{3.26}$$

Writing $\omega_t^n \leq A[\omega_t^n]\omega_0^n$, and noting that $PSH(X, \omega_t) \subset PSH(X, (A+1)\omega_0)$, we can bound the first integral on the right hand side by $C[\omega_t^n]$, in view of Theorem A.1 on exponential estimates for plurisubharmonic functions.

The other integrals can be re-expressed as

$$\int_X \varphi_t \frac{i}{2}\partial\bar\partial u\, \omega_t^j \left(\omega_t + \frac{i}{2}\partial\bar\partial u\right)^{n-1-j}$$
$$= -\int_X \frac{i}{2}\partial\bar\partial\varphi_t\, u\, \omega_t^j \left(\omega_t + \frac{i}{2}\partial\bar\partial u\right)^{n-1-j}$$
$$= -\int_X u \left(\omega_t + \frac{i}{2}\partial\bar\partial\varphi_t\right) \omega_t^j \left(\omega_t + \frac{i}{2}\partial\bar\partial u\right)^{n-1-j}$$
$$+ \int_X u\, \omega_t^{j+1} \left(\omega_t + \frac{i}{2}\partial\bar\partial u\right)^{n-1-j}.$$

For $0 \le u \le 1$, we can write

$$\left| \int_X u \left(\omega_t + \frac{i}{2}\partial\bar\partial\varphi_t\right) \omega_t^j \left(\omega_t + \frac{i}{2}\partial\bar\partial u\right)^{n-1-j} \right|$$
$$\le \int_X \left(\omega_t + \frac{i}{2}\partial\bar\partial\varphi_t\right) \omega_t^j \left(\omega_t + \frac{i}{2}\partial\bar\partial u\right)^{n-1-j} = [\omega_t^n]$$

and similarly for the other integral. Thus we obtain an upper bound $C[\omega_t^n]$, and taking the supremum in u establishes the desired inequality. Q.E.D.

It remains to establish the property (c) for $f_t(s)$. For this, we need the following two lemmas:

LEMMA 3. *Let $\varphi \in PSH(X,\omega) \cap L^\infty(X)$. Then for all $s > 0$, $0 \le r \le 1$,*

$$r^n \mathrm{Cap}_\omega(\varphi < -s - r) \le \int_{\varphi < -s} \left(\omega + \frac{i}{2}\partial\bar\partial\varphi\right)^n. \tag{3.27}$$

LEMMA 4. *There exist constants $\delta, C > 0$ so that for any open set $E \subset X$, and any $t \in [0,1]$, we have*

$$\frac{1}{[\omega_t^n]} \int_E \omega_t^n \le C \exp\left[-\delta \left(\frac{[\omega_t^n]}{\mathrm{Cap}_{\omega_t}(E)}\right)^{1/n}\right]. \tag{3.28}$$

Assuming these two lemmas for the moment, we can readily establish the inequality (c) in Lemma 1. For $\alpha > 0$ we have

$$[r\, f_t(s+r)]^n = r^n \frac{\mathrm{Cap}_\omega(\varphi_t < -s - r)}{[\omega_t^n]} \le \frac{1}{[\omega_t^n]} \int_{\varphi_t < -s} \left(\omega + \frac{i}{2}\partial\bar\partial\varphi_t\right)^n$$
$$= \frac{1}{[\omega_t^n]} \int_{\varphi_t < -s} F_t \omega_t^n \le \left(\frac{1}{[\omega_t^n]} \int_{\varphi_t < -s} F_t^p \omega_t^n\right)^{\frac{1}{p}} \left(\frac{1}{[\omega_t^n]} \int_{\varphi_t < -s} \omega_t^n\right)^{\frac{1}{q}}$$
$$\le A \exp\left(-\frac{\delta}{q}\left[\frac{[\omega_t^n]}{\mathrm{Cap}_\omega(\varphi_t < -s)}\right]^{\frac{1}{n}}\right) \le A_\alpha\, f^{(1+\alpha)n}. \tag{3.29}$$

It remains to prove the two lemmas. Let $u \in PSH(X,\omega)$ with $0 \le u \le 1$, and write

$$r^n \int_{\varphi<-s-r} \left(\omega + \frac{i}{2}\partial\bar\partial u\right)^n = \int_{\varphi<-s-r} \left(r\omega + \frac{i}{2}\partial\bar\partial ru\right)^n$$

$$\le \int_{\varphi<-s-r+ru} \left(\omega + \frac{i}{2}\partial\bar\partial(ru-s-r)\right)^n$$

$$\le \int_{\varphi<-s-r+ru} \left(\omega + \frac{i}{2}\partial\bar\partial\varphi\right)^n \tag{3.30}$$

where we have applied the comparison principle. Since $-r+ru$ is negative, this last integral is bounded by the integral over the larger region $\{\varphi < -s\}$, and Lemma 3 is proved. The next lemma requires some properties of global extremal functions [**GZ, Ze**]:

LEMMA 5. *Let $E \subset X$ be an open set, and define its global extremal function $\psi_{E,\omega}$ as the upper semi-continuous envelope of the following function $\tilde\psi_{E,\omega}$,*

$$\tilde\psi_{E,\omega} = \sup\{u \in PSH(X,\omega); u = 0 \text{ on } E\}. \tag{3.31}$$

Then

(a) $\psi_{E,\omega} \in PSH(X,\omega) \cap L^\infty(X)$
(b) $\psi_{E,\omega} = 0$ *on* E
(c) $(\omega + \frac{i}{2}\partial\bar\partial\psi_{E,\omega})^n = 0$ *on* $X \setminus \bar E$.

We can now prove Lemma 4. Let $E' \subset E$ be any relatively compact open subset. Then

$$\frac{1}{[\omega_t^n]}\int_{E'} \omega_t^n = \frac{1}{[\omega_t^n]} e^{-\delta \sup_X \psi_{E',\omega_t}} \int_{E'} e^{-\delta(\psi_{E'\omega_t} - \sup_X \psi_{E',\omega_t})}\omega_t^n$$

$$\le e^{-\delta \sup_X \psi_{E',\omega_t^n}} A \int_X e^{-\delta(\psi_{E',\omega_t} - \sup_X \psi_{E',\omega_t})}\omega_0^n \tag{3.32}$$

where A is an upper bound for $\frac{1}{[\omega_t^n]}\frac{\omega_t^n}{\omega_0^n}$. Since $\chi \le A\omega_0$ also by assumption, $PSH(X,\omega_t) \subset PSH(X,(A+1)\omega_0)$, and Theorem A.1 implies that the integral on the right hand side is bounded by a constant independent of t and E'. We can now complete the proof of Lemma 4. First, observe that if $\sup_X \psi_{E',\omega_t} \le 1$, then

$$[\omega_t^n] = \int_{\bar E'} \left(\omega_t + \frac{i}{2}\partial\bar\partial\psi_{E',\omega_t}\right) \le \mathrm{Cap}_{\omega_t}(\bar E') \le \mathrm{Cap}_{\omega_t}(E) \le \mathrm{Cap}_{\omega_t}(X) = [\omega_t^n]. \tag{3.33}$$

Thus $\frac{[\omega_t^n]}{\mathrm{Cap}_{\omega_t}(E)} = 1$, and a constant C_δ can clearly be chosen so that the desired inequality holds. Next, assume that $\sup_X \psi^n_{E',\omega_t} > 1$. We can write

$$(\sup_X \psi_{E',\omega_t})^{-n} = (\sup_X \psi_{E',\omega_t})^{-n} \frac{\int_X \left(\omega_t + \frac{i}{2}\partial\bar\partial \psi_{E',\omega_t}\right)^n}{[\omega_t^n]}$$

$$= (\sup_X \psi_{E',\omega_t})^{-n} \frac{\int_{\bar E'} \left(\omega_t + \frac{i}{2}\partial\bar\partial \psi_{E',\omega_t}\right)^n}{[\omega_t^n]}$$

$$\leq \frac{\int_{\bar E'} \left(\omega_t + \frac{i}{2}\partial\bar\partial \left(\frac{\psi_{E',\omega_t}}{\sup_X \psi_{E',\omega_t}}\right)\right)^n}{[\omega_t^n]}. \tag{3.34}$$

This last term is bounded by $[\omega_t^n]^{-1}\mathrm{Cap}_{\omega_t}(\bar E') \leq [\omega_t^n]^{-1}\mathrm{Cap}_{\omega_t}(E)$. Thus we obtain

$$\frac{1}{[\omega_t^n]} \int_{E'} \omega_t^n \leq \exp\left(-\delta \left(\frac{[\omega_t^n]}{\mathrm{Cap}_{\omega_t}(E)}\right)^{\frac{1}{n}}\right). \tag{3.35}$$

Taking limits as E' increases to E establishes Lemma 4. The proof of the theorem is complete.

We note that all three proofs of C^0 estimates can be extended to the equation (1.1) on Hermitian manifolds. For the Moser iteration method, this is carried out in [**Ch, TW1, TW2**]. It is interesting that, in the Kähler case, only one term in the right hand side of (3.2) was needed, while the other terms are also needed in the Hermitian case. The extension of the pluripotential method to the Hermitian case is in [**DK**], while the extension of the Alexandrov-Bakelman-Pucci method is in [**B11a**].

4. Stability Estimates

In this section, we shall establish the stability and uniqueness of the continuous solutions of the complex Monge-Ampère equations due to Kolodiej [**K03**]. We shall closely follow the arguments in [**K05**]. Let (X,ω) be an n-dimensional compact Kähler manifold and let

$$\mathcal{F}_{p,A} = \left\{ F \in L^p(X) | F \geq 0, \int_X F^p \omega^n \leq A, \int_X F\omega^n = \int_X \omega^n \right\} \tag{4.1}$$

for $p > 1$ and $A > 0$.

THEOREM 6. *For any two $F, G \in \mathcal{F}_{p,A}$, let φ and $\psi \in PSH(X,\omega) \cap C(X)$ be solutions of the following Monge-Ampère equations*

$$\left(\omega + \frac{i}{2}\partial\bar\partial\varphi\right)^n = F(z)\omega^n, \quad \left(\omega + \frac{i}{2}\partial\bar\partial\psi\right)^n = G(z)\omega^n \tag{4.2}$$

normalized by

$$\sup_X \varphi = \sup_X \psi = 0. \tag{4.3}$$

Then for any $\varepsilon > 0$, there exists $C > 0$ depending only on ε, p, A and (X, ω), so that

$$\|\varphi - \psi\|_{L^\infty(X)} \leq C\|F - G\|_{L^1(X)}^{\frac{1}{n+3+\varepsilon}}. \tag{4.4}$$

Theorem 6 has been generalized in [**DZ**] to nonnegative, big and smooth closed $(1,1)$-form ω if ω is chosen appropriately. The uniqueness of the solutions $\varphi \in PSH(X, \omega) \cap C^0(X)$, $\sup_X \varphi = 0$, to the Monge-Ampère equation (3.1) for $F \in \mathcal{F}_{p,A}$ for some $p > 1$ and $A > 0$ follows immediately from the stability theorem.

As before, we write $\omega_\varphi = \omega + \frac{i}{2}\partial\bar{\partial}\varphi$ for any $\varphi \in PSH(X, \omega)$. The following lemma is a generalization of Lemma 3 and a detailed proof can be found in [**K05**].

LEMMA 6. *Let φ and $\psi \in PSH(X, \omega) \cap C(X)$ with $0 \leq \varphi \leq C$. Then for any $s > 0$,*

$$\mathrm{Cap}_\omega(\{\psi + 2s < \varphi\}) \leq \left(\frac{C+1}{s}\right)^n \int_{\psi+s<\varphi} \omega_\psi^n. \tag{4.5}$$

The following lemma is well-known for smooth plurisubharmonic functions. The general result can also be found in [**K05**].

LEMMA 7. *Let $\mathcal{B}$ be a Euclidean ball in $\mathbf{C}^n$ with the standard Euclidean volume form Ω. For any $u \in PSH(\mathcal{B}) \cap C(\overline{\mathcal{B}})$ and any nonnegative $F \in L^1(\overline{\mathcal{B}})$ with*

$$\left(\frac{i}{2}\partial\bar{\partial}u\right)^n \geq F\,\Omega, \quad \left(\frac{i}{2}\partial\bar{\partial}v\right)^n \geq F\,\Omega, \tag{4.6}$$

we have

$$\left(\frac{i}{2}\partial\bar{\partial}u\right)^k \wedge \left(\frac{i}{2}\partial\bar{\partial}v\right)^{n-k} \geq F\,\Omega \tag{4.7}$$

for all $k = 0, \ldots, n$.

By Theorem 5, there exists $a > 0$ depending on $3^p A$, p and (X, ω) such that for any $F \in \mathcal{F}_{p,3^pA}$, the solution $u \in PSH(X, \omega) \cap L^\infty(X)$ of $(\omega + \frac{i}{2}\partial\bar{\partial}u)^n = F\omega^n$ satisfies

$$\sup_X u - \inf_X u \leq a. \tag{4.8}$$

Without loss of generality, we can assume that $\int_X F\omega^n = \int_X G\omega^n = \int_X \omega^n = 1$ and

$$\int_{\psi<\varphi} (F + G)\omega^n \leq 1. \tag{4.9}$$

Furthermore, we let $0 < t_0 < (q-1)/2$, where $q = (3/2)^{1/n}$ and we further assume $t_0^{n+3+\varepsilon} < 1/3$. For any $0 < t < t_0$, we define for all nonnegative integers k,

$$E_k = \{\psi < \varphi - kat\} \tag{4.10}$$

for a fixed $\varepsilon > 0$. From now on we assume that for $0 < t < t_0$,

$$\|F - G\|_{L^1(X)} = t^{n+3+\varepsilon}. \tag{4.11}$$

We have immediately

$$\int_{E_0} G\omega^n = \frac{1}{2}\int_{E_0} ((F+G) + (G-F))\omega^n \le \frac{1}{2}\left(1 + \frac{1}{3}\right) = \frac{2}{3}. \tag{4.12}$$

Now we define a new function H such that $H = 3G/2$ on E_0 and $H = c_0$ on the complement of E_0 so that $\int_X H\omega^n = 1$. Obviously, $c_0 > 0$ and $H \in \mathcal{F}_{p,(3/2)^p A}$. Then there exists a unique $\rho \in PSH(X,\omega) \cap C(X)$ such that

$$\omega_\rho^n = H\omega^n, \quad \sup_X \rho = 0. \tag{4.13}$$

Furthermore,

$$-a \le \rho \le 0. \tag{4.14}$$

We now define

$$E = \{\psi < (1-t)\varphi + t\rho - at\}, \quad S = \{F < (1-t^2)G\}. \tag{4.15}$$

The following lemma can be easily verified.

Lemma 8.

$$E_2 \subset E \subset E_0. \tag{4.16}$$

Lemma 9. *On $E_0 \setminus S$, for $k = 0, \ldots, n$, we have*

$$\omega_\varphi^k \wedge \omega_\rho^{n-k} \ge q^{n-k}(1-t^2)^{k/n} G\omega^n. \tag{4.17}$$

Proof. On $E_0 \setminus S$, we have

$$((1-t^2)^{-1/n}\omega_\varphi)^n \ge G\omega^n, \quad (q^{-1}\omega_\rho)^n = G\omega^n. \tag{4.18}$$

The lemma then follows from Lemma 7. Q.E.D.

Lemma 10. *Let $B = \int_{E_2} G\omega^n$. Then*

$$B \le \frac{3}{q-1} t^{n+\varepsilon}. \tag{4.19}$$

PROOF. On $E_0 \setminus S$, we have

$$\begin{aligned}\omega^n_{(1-t)\varphi+t\rho} &= \sum_{k=0}^{n} \binom{n}{k} (1-t)^k t^{n-k} \omega_\varphi^k \wedge \omega_\rho^{n-k}\\ &\geq \sum_{k=0}^{n} \binom{n}{k} q^{n-k}(1-t^2)^{k/n}(1-t)^k t^{n-k} G\omega^n. \qquad (4.20)\end{aligned}$$

The right hand side can in turn be estimated by

$$\begin{aligned}&\sum_{k=0}^{n} \binom{n}{k} q^{n-k}(1-t^2)\frac{k}{n}(1-t)^k t^{n-k} G\omega^n\\ &\quad = (qt+(1-t)(1-t^2)^{1/n})^n G \geq ((1-t)(1-t^2)+qt)^n G\omega^n\\ &\quad \geq (1+(q-1)t-t^2)^n G\omega^n \geq (1+(q-1)t/2)G\omega^n, \qquad (4.21)\end{aligned}$$

where we make use of the additional assumption that $t < t_0 < (q-1)/2$. On the other hand, since $\int_S F\omega^n \leq (1-t^2)\int_S G\omega^n$ by the definition of S, we have

$$t^2 \int_S G\omega^n \leq \int_S (G-F)\omega^n \leq t^{n+3+\varepsilon} \qquad (4.22)$$

and so

$$\int_S G\omega^n \leq t^{n+1+\varepsilon}. \qquad (4.23)$$

The above inequality implies that

$$\int_{E\setminus S} G\omega^n \leq \frac{2}{q-1} t^{n+\varepsilon}. \qquad (4.24)$$

Thus by (4.21) and (4.23),

$$B \leq \int_E G\omega^n \leq \int_{E\setminus S} G\omega^n + \int_S G\omega^n \leq \frac{3}{q-1} t^{n+\varepsilon}. \qquad (4.25)$$

The lemma is then proved. Q.E.D.

LEMMA 11.

$$\mathrm{Cap}_\omega(E_4) \leq \frac{3}{q-1}(2a)^{-n}(a+1)^n t^\epsilon. \qquad (4.26)$$

PROOF. By Lemma 6, we have

$$\mathrm{Cap}_\omega(E_4) \leq \frac{(a+1)^n}{(2at)^n} \int_{E_2} G\omega^n = \frac{(a+1)^n}{(2at)^n} B \leq \frac{(a+1)^n}{(2a)^n} \frac{3}{q-1} t^\varepsilon. \qquad \square$$

The following lemma is used in Kolodziej's original proof of the L^∞ estimates. We refer the readers to the detailed proof in [**K05**].

LEMMA 12. *Let $\varphi, \psi \in PSH(X,\omega) \cap C(X)$ with $0 \le \varphi \le C$. Let*

$$U(s) = \{\psi - s < \varphi\}, \quad \alpha(s) = \mathrm{Cap}_\omega(U(s)). \tag{4.27}$$

Assume that

(1) $\{\psi - S < \varphi\} \neq \emptyset$ *for some* S,
(2) *For any Borel set* K,

$$\int_K \omega_\psi^n \le f(\mathrm{Cap}_\omega(K)), \tag{4.28}$$

where $f(x) = \frac{x}{h(x^{-1/n})}$ *and* $h(x) : \mathbf{R}^+ \to (0,\infty)$ *is a continuous strictly increasing function satisfying* $\int_1^\infty \frac{1}{t h^{1/n}(t)} dt < \infty$.

Then for any $D < 1$, *we have*

$$D \le \kappa(\alpha(S+D)), \tag{4.29}$$

where

$$\kappa(s) = c(n)(1+C)\left(\int_{s^{-1/n}}^\infty \frac{dx}{x h^{1/n}(x)} + \frac{1}{h^{1/n}(s^{-1/n})}\right) \tag{4.30}$$

for some constant $c(n)$ *depending only on* n.

PROOF OF THEOREM 6. By Lemma 4, for any $\delta > 0$ and open set K, there exists $C_\delta > 0$,

$$\int_K \omega^n \le C_1 e^{-(C_2 \mathrm{Cap}_\omega(K))^{-1}} \le C_\delta (\mathrm{Cap}_\omega(K))^{1/\delta}. \tag{4.31}$$

Then $0 \le \varphi + a \le a$. We can easily check that we can choose $h(x) = x^{1/\delta}$ and there exists $C'_\delta > 0$ such that

$$\kappa(s) = C'_\delta s^{1/(\delta n^2)}. \tag{4.32}$$

Now we can prove the theorem by contradiction. Suppose that

$$\{\psi < \varphi - (4a+1)t\} = \{\psi + a < \varphi + a - (4a+1)t\} \neq \emptyset. \tag{4.33}$$

Then by applying Lemma 12 with $\psi + a$, $\varphi + a$, $S = -(4a+1)t$ and $D = t$,

$$t \le \kappa(\mathrm{Cap}_\omega(E_4)) \le \kappa\left(\frac{3}{q-1}(2a)^{-n}(a+1)^n t^\epsilon\right)$$
$$= C'_\delta \left(\frac{3}{q-1}(2a)^{-n}(a+1)^n t^\epsilon\right)^{1/(\delta n^2)}.$$

This is a contradiction if we choose $\delta > 0$ sufficiently small and then $t > 0$ sufficiently small. Therefore $\{\psi < \varphi - (4a+1)t\} = \emptyset$ and so

$$\sup_X(\varphi - \psi) = \sup_X(\psi - \varphi) \le (4a+1)t = (4a+1)(\|F-G\|_{L^1(X)})^{1/(n+3+\varepsilon)} \tag{4.34}$$

if we choose t_0 sufficiently small. The theorem is proved. Q.E.D.

5. A Priori Estimates: C^1 Estimates

In Yau's original solution of the Calabi conjecture [**Y78**], the C^2 estimates were shown to follow directly from the C^0 estimates. The C^1 estimates follow from the C^0 and C^2 estimates by general linear elliptic theory. However, for more general Monge-Ampère equations where the right hand side may be an expression $F(z,\varphi)$ depending on the unknown φ as well as for the Dirichlet problem, the C^1 estimates cannot be bypassed. In this section, we describe the sharpest C^1 estimates available at this time. They are due to [**PS09a, PS09c**], and they exploit a key differential inequality discovered by Blocki [**B09a**].

Let (X,ω_0) be a compact Kähler manifold with smooth boundary ∂X (which may be empty) and complex dimension n. We consider the Monge-Ampère equation on $\bar{X}$

$$\left(\omega_0 + \frac{i}{2}\partial\bar{\partial}\varphi\right)^n = F(z,\varphi)\,\omega_0^n. \tag{5.1}$$

Here $F(z,\varphi)$ is a C^2 function on $\bar{X}\times\mathbf{R}$ which is assumed to be strictly positive on the set $\bar{X}\times[\inf\varphi,\infty)$. The gradient estimates allow φ to be singular along a subset $Z\subset X$, possibly empty, which does not intersect ∂X. All covariant derivatives and curvatures listed below are with respect to the metric ω_0. Then [**PS09c**]

THEOREM 7. *Let (X,ω_0) be a compact Kähler manifold, with smooth boundary ∂X (possibly empty). Assume that $\varphi\in C^4(\bar{X}\setminus Z)$ is a solution of the equation (5.1) on $\bar{X}\setminus Z$. If Z is not empty, assume further that Z does not intersect ∂X, and that there exists a constant $B>0$ so that*

$$\begin{aligned} &\varphi(z)\to+\infty \quad \text{as } z\to Z,\\ &\log|\nabla\varphi(z)|^2 - B\,\varphi(z)\to-\infty \quad \text{as } z\to Z. \end{aligned} \tag{5.2}$$

Then we have the a priori estimate

$$|\nabla\varphi(z)|^2 \le C_1\exp(A_1\,\varphi(z)), \quad z\in\bar{X}\setminus Z, \tag{5.3}$$

where C_1 and A_1 are constants that depend only on upper bounds for $\inf_X\varphi$, $\sup_{X\times[\inf\varphi,\infty)}F$, $\sup_{X\times[\inf\varphi,\infty)}(|\nabla F^{\frac{1}{n}}|+|\partial_\varphi F^{\frac{1}{n}}|)$, $\sup_{\partial X}|\varphi|$, $\sup_{\partial X}|\nabla\varphi|$, and the following constant,

$$\Lambda = -\inf_X\inf_{M>0}\frac{M^j{}_k R^k{}_j{}^p{}_q (M^{-1})^q{}_p}{\operatorname{Tr}M\,\operatorname{Tr}M^{-1}}, \tag{5.4}$$

where $M=(M^q{}_p)$ runs over all self-adjoint and positive definite endomorphisms.

When there is no boundary, and the function $F(z,\varphi)$ is a function $F(z)$ of z alone, the equation (5.1) is unchanged under shifts of φ by an additive constant. Thus the infimum of $\varphi(z)$ can be normalized to be 0 by replacing $\varphi(z) \to \varphi(z) - \inf_X \varphi$, so we obtain the estimate

$$(5.5) \qquad |\nabla\varphi(z)|^2 \le C_1 \exp(A_1(\varphi(z) - \inf_X \varphi)) \quad z \in \bar{X}$$

where the constant C_1 does not depend on $\inf \varphi$, but depends only on the other quantities listed above. We shall see that the Laplacian $\Delta\varphi$ satisfies the same pointwise estimate.

Not surprisingly, the constants $\sup_{M\times[\inf\varphi,\infty]} F$ and $\sup_{M\times[\inf\varphi,\infty]} |\nabla F^{\frac{1}{n}}| + |\partial_\varphi F^{\frac{1}{n}}|$ in (5.3) can be replaced by $\sup_{M\times[\inf\varphi,\sup\varphi]} F$ and $\sup_{M\times[\inf\varphi,\sup\varphi]} |\nabla F^{\frac{1}{n}}| + |\partial_\varphi F^{\frac{1}{n}}|$ respectively. Thus, when $\|\varphi\|_{C^0}$ is bounded, we obtain gradient bounds for φ for completely general smooth and strictly positive functions $F(z,\varphi)$. We have however stated them in the above form since we are particularly interested in the cases when there is no upper bound for $\sup\varphi$. This is crucial for certain applications [**PS09a, PS09b**]. If a dependence on $\|\varphi\|_{C^0}$ is allowed, then there are many earlier direct approaches. The first appears to be due to Hanani [**Ha**]. More recently, Blocki [**B09a**] gave a different proof, and our approach builds directly on his. The method of P. Guan [**Gp**] can be extended to Hessian equations, while the method of B. Guan-Q. Li [**GL**] allows a general Hermitian metric ω as well as a more general right hand side $F(z)\chi^n$, where χ is a Kähler form.

The proof is an application of the maximum principle. Let $g_{\bar{k}j}$ and $g'_{\bar{k}j}$ be the two metrics defined by the Kähler forms ω_0 and $\omega_0 + \frac{i}{2}\partial\bar{\partial}\varphi$. The covariant derivatives and Laplacians with respect to $g_{\bar{k}j}$ and $g'_{\bar{k}j}$ are denoted by ∇, Δ, and ∇', Δ' respectively. A subindex g or g' will denote the metric with respect to which a norm is taken. It is convenient to introduce the endomorphisms

$$(5.6) \qquad h^j{}_k = g^{j\bar{p}} g'_{\bar{p}k}, \quad (h^{-1})^j{}_k = (g')^{j\bar{p}} g_{\bar{p}k}.$$

Their traces are $\mathrm{Tr}\, h = n + \Delta\varphi$ and $\mathrm{Tr}\, h^{-1} = n + \Delta'\varphi$.

As a preliminary, we calculate $\Delta' \log |\nabla\varphi|_g^2$, $|\nabla\varphi|_g^2$ being the expression of interest, and Δ' being the natural Laplacian to use, as it arises from differentiating the Monge-Ampère equation. We have

$$(5.7) \qquad \Delta' \log |\nabla\varphi|_g^2 = \frac{\Delta'|\nabla\varphi|_g^2}{|\nabla\varphi|_g^2} - \frac{|\nabla|\nabla\varphi|_g^2|_{g'}^2}{|\nabla\varphi|_g^4}.$$

If we express Δ' on scalars as $\Delta' = (g')^{p\bar{q}}\nabla_p\nabla_{\bar{q}}$, then we can write

$$(5.8) \quad \Delta'|\nabla\varphi|_g^2 = \Delta'(\nabla_m\varphi)\nabla^m\varphi + \nabla_m\varphi\Delta'(\nabla^m\varphi) + |\nabla\nabla\varphi|_{gg'}^2 + |\bar{\nabla}\nabla\varphi|_{gg'}^2.$$

However, making use of the Monge-Ampère equation, we obtain

$$\Delta'(\nabla_m\varphi) = (g')^{p\bar{q}}\nabla_p\nabla_{\bar{q}}\nabla_m\varphi = (g')^{p\bar{q}}\nabla_m(\nabla_p\nabla_{\bar{q}}\varphi)$$

$$(5.9) \qquad = \partial_m \log\frac{(\omega')^n}{\omega^n} = \partial_m \log F,$$

while

$$\Delta'(\nabla^m\varphi) = (g')^{p\bar q}\nabla^m\nabla_{\bar q}\nabla_p\varphi + (g')^{p\bar q}R_{\bar q p}{}^m{}_\ell\nabla^\ell\varphi$$
$$= \nabla^m \log F + (h^{-1})^p{}_r R^r{}_p{}^m{}_\ell \nabla^\ell\varphi. \tag{5.10}$$

Thus

$$\Delta' \log |\nabla\varphi|_g^2 \geq \frac{2\mathrm{Re}\nabla_m \log F\nabla^m\varphi}{|\nabla\varphi|_g^2} - \Lambda\,\mathrm{Tr}\,h^{-1}$$
$$+ \frac{|\nabla\nabla\varphi|^2_{gg'} + |\bar\nabla\nabla\varphi|^2_{gg'}}{|\nabla\varphi|_g^2} - \frac{|\nabla|\nabla\varphi|_g^2|^2_{g'}}{|\nabla\varphi|_g^4} \tag{5.11}$$

The first term on the right is easily bounded: first write,

$$\frac{2\mathrm{Re}\nabla_m \log F\nabla^m\varphi}{|\nabla\varphi|_g^2} \geq -2|\nabla \log F|_g \frac{1}{|\nabla\varphi|_g} = -2nF^{-\frac{1}{n}}|\nabla F^{\frac{1}{n}}|_g \frac{1}{|\nabla\varphi|_g}, \tag{5.12}$$

and note that

$$|\nabla F^{\frac{1}{n}}|_g \leq \sup\nolimits_{X\times[0,\infty)}|\partial_z F(z,\varphi)^{\frac{1}{n}}| + |\nabla\varphi|_g \sup\nolimits_{X\times[0,\infty)}|\partial_\varphi F(z,\varphi)^{\frac{1}{n}}|_g$$
$$\equiv F_1'' + |\nabla\varphi|_g F_1',$$

while, using the Monge-Ampère equation and the arithmetic-geometric mean inequality,

$$nF^{-\frac{1}{n}} \leq \mathrm{Tr}\,h^{-1}. \tag{5.13}$$

Thus we find

$$\Delta' \log |\nabla\varphi|_g^2 \geq -\left(\Lambda + 2F_1' + 2\frac{F_1''}{|\nabla\varphi|_g}\right)\mathrm{Tr}\,h^{-1}$$
$$+ \frac{|\nabla\nabla\varphi|^2_{gg'} + |\bar\nabla\nabla\varphi|^2_{gg'}}{|\nabla\varphi|_g^2} - \frac{|\nabla|\nabla\varphi|_g^2|^2_{g'}}{|\nabla\varphi|_g^4}. \tag{5.14}$$

The only troublesome term is the negative last term to the right. The key to handling it is a partial cancellation with the two squares preceding it. This cancellation is rather general, and we formalize in the following lemma:

LEMMA 13. *Let X be a Kähler manifold and $g_{\bar kj}, g'_{\bar kj}$ a pair of Kähler metrics on X (not necessarily in the same Kähler class). Let $\varphi \in C^\infty(X)$ and define*

$$S = \langle \nabla\nabla\varphi, \nabla\varphi\rangle_g, \quad T = \langle \nabla\varphi, \bar\nabla\nabla\varphi\rangle_g, \tag{5.15}$$

Then we have

$$\frac{|\nabla\nabla\varphi|^2_{gg'} + |\nabla\bar\nabla\varphi|^2_{gg'}}{|\nabla\varphi|_g^2} \geq \frac{|\nabla|\nabla\varphi|_g^2|^2_{g'}}{|\nabla\varphi|_g^4} - 2\mathrm{Re}\left\langle \frac{\nabla|\nabla\varphi|_g^2}{|\nabla\varphi|_g^4}, T\right\rangle_{g'} + 2\frac{|T|^2_{g'}}{|\nabla\varphi|_g^4}. \tag{5.16}$$

PROOF OF LEMMA 13. First, we observe that for all tensors A_{pi} and B_j on X,

$$|\langle A, B\rangle_g|_{g'} = |A_{pi}g^{i\bar{j}}\overline{B_j}|_{g'} \leq |A|_{gg'}|B|_g. \tag{5.17}$$

Now $\nabla|\nabla\varphi|_g^2 = S + T$, and applying (5.17) to S and T gives:

$$\begin{aligned} |\nabla\varphi|_g^2 \cdot (|\nabla\nabla\varphi|_{gg'}^2 + |\nabla\bar{\nabla}\varphi|_{gg'}^2) &\geq |S|_{g'}^2 + |T|_{g'}^2 = |\nabla|\nabla\varphi|_g^2 - T|_{g'}^2 + |T|_{g'}^2 \\ &= |\nabla|\nabla\varphi|_g^2|_{g'}^2 - 2\mathrm{Re}\langle\nabla|\nabla\varphi|_g^2, T\rangle_{g'} + 2|T|_{g'}^2 \end{aligned} \tag{5.18}$$

This proves the inequality (5.16). □

Returning to the problem of C^1 estimates, we can now formulate and prove an important inequality due to Blocki at interior critical points of an expression of the form

$$\log\, |\nabla\varphi|_g^2 - \gamma(\varphi) \tag{5.19}$$

where γ is an arbitrary function of a real variable. We apply Lemma 13 as follows: on the right side of (5.16), we drop the third term $2|T|_{g'}^2/|\nabla\varphi|_g^4$. In the second term, the tensor T simplifies upon replacing $\bar{\nabla}\nabla\varphi$ by $g' - g$, so that T becomes $T_j = (\nabla_i\varphi)g^{i\bar{k}}g'_{\bar{k}j} - \nabla_j\varphi$. We obtain

$$\begin{aligned} &\frac{|\nabla\nabla\varphi|_{gg'}^2 + |\nabla\bar{\nabla}\varphi|_{gg'}^2}{|\nabla\varphi|_g^2} - \frac{|\nabla|\nabla\varphi|_g^2|_{g'}^2}{|\nabla\varphi|_g^4} \\ &\geq 2\mathrm{Re}\left\langle\frac{\nabla|\nabla\varphi|_g^2}{|\nabla\varphi|_g^2}, \frac{\nabla\varphi}{|\nabla\varphi|_g^2}\right\rangle_{g'} - 2\mathrm{Re}\left\langle\frac{\nabla|\nabla\varphi|_g^2}{|\nabla\varphi|_g^2}, \frac{\nabla\varphi}{|\nabla\varphi|_g^2}\right\rangle_g \\ &= 2\gamma'(\varphi)\frac{|\nabla\varphi|_{g'}^2}{|\nabla\varphi|_g^2} - 2\gamma'(\varphi) \end{aligned} \tag{5.20}$$

In the last line, we made use of the fact that $\nabla\log|\nabla\varphi|_g^2 = \gamma'(\varphi)\nabla\varphi$ at an interior critical point of the function $\log\, |\nabla\varphi|_g^2 - \gamma(\varphi)$.

On the other hand,

(5.21)

$$-\Delta'\gamma(\varphi) = -\gamma'(\varphi)\Delta'\varphi - \gamma''(\varphi)|\nabla\varphi|_{g'}^2 = \gamma'(\varphi)\,\mathrm{Tr}\,h^{-1} - n\gamma'(\varphi) - \gamma''(\varphi)|\nabla\varphi|_{g'}^2.$$

Combining this with the preceding inequality, we obtain Blocki's inequality [**B09a**],

$$\begin{aligned} \Delta'(\log|\nabla\varphi|_g^2 - \gamma(\varphi)) \geq &\left[\gamma'(\varphi) - \Lambda - 2F_1' - 2\frac{F_1''}{|\nabla\varphi|_g}\right]\mathrm{Tr}\,h^{-1} \\ &- (n+2)\gamma'(\varphi) - \gamma''(\varphi)|\nabla\varphi|_{g'}^2 + 2\gamma'(\varphi)\frac{|\nabla\varphi|_{g'}^2}{|\nabla\varphi|_g^2}. \end{aligned} \tag{5.22}$$

The key to the desired estimate is the following choice of $\gamma(\varphi)$ [**PS09a, PS09c**]

$$\gamma(\varphi) = A\varphi - \frac{1}{\varphi + C_1} \tag{5.23}$$

where C_1 is chosen to be $C_1 = -\inf_X \varphi + 1$, and A is a large positive constant. Then

$$A\varphi - 1 \leq \gamma(\varphi) \leq A\varphi, \quad A \leq \gamma'(\varphi) \leq A+1, \quad \gamma''(\varphi) = -\frac{2}{(\varphi + C_1)^3} < 0 \tag{5.24}$$

and we obtain

$$\begin{aligned} \Delta'(\log|\nabla\varphi|_g^2 - \gamma(\varphi)) \geq & \left[A - \Lambda - 2F_1' - 2\frac{F_1''}{|\nabla\varphi|_g}\right] \operatorname{Tr} h^{-1} \\ & + \frac{2}{(\varphi + C_1)^3}|\nabla\varphi|_{g'}^2 - C_2. \end{aligned} \tag{5.25}$$

It suffices to show that, at an interior maximum point p, the function $\log|\nabla\varphi|_g^2 - \gamma(\varphi)$ is bounded by an admissible constant. We can assume that $|\nabla\varphi(p)|_g^2 \geq 1$, otherwise the statement follows trivially from the fact that $\gamma(\varphi) \geq A\varphi - 1$, and φ is bounded from below. Choose $A = \Lambda + 2F_1' + 2F_1'' + 1$. Then the preceding inequality simplifies further to

$$\Delta'(\log|\nabla\varphi|_g^2 - \gamma(\varphi)) \geq \operatorname{Tr} h^{-1} + \frac{2}{(\varphi + C_1)^3}|\nabla\varphi|_{g'}^2 - C_2. \tag{5.26}$$

At an interior minimum point p, the left hand side is non-positive. This implies that $\operatorname{Tr} h^{-1}(p)$ is bounded above, and hence the eigenvalues of $h(p)$ are bounded below by a priori constants. In view of the Monge-Ampère equation, they are then bounded above and below by a priori constants, since these constants are allowed to depend on $\sup_X F$. This implies that $|\nabla\varphi|_{g'}^2 \geq C_3|\nabla\varphi|_g^2$, and we obtain

$$|\nabla\varphi|_g^2 \leq C_4(\varphi + C_1)^3. \tag{5.27}$$

But we can assume that $\log|\nabla\varphi(p)|_g^2 - \gamma(\varphi(p)) \geq 0$, otherwise there is nothing to prove. Thus $\gamma(\varphi(p)) \leq \log|\nabla\varphi(p)|_g^2$, and hence

$$A\varphi(p) \leq \gamma(\varphi(p)) + 1 \leq \log|\nabla\varphi(p)|_g^2 + 1. \tag{5.28}$$

Substituting this in the previous inequality, we find

$$|\nabla\varphi(p)|_g^2 \leq C_4(\log|\nabla\varphi(p)|_g^2 + C_5)^3. \tag{5.29}$$

This implies that $|\nabla\varphi(p)|_g^2$ is bounded by an a priori constant. The proof of the C^1 estimates is complete.

We note that Lemma 13 has other uses. For example, the inequality (5.16) also implies, by completing the square,

$$\frac{|\nabla\nabla\varphi|^2_{gg'} + |\nabla\bar{\nabla}\varphi|^2_{gg'}}{|\nabla\varphi|^2_g} \geq \frac{1}{2}\frac{|\nabla|\nabla\varphi|^2_g|^2_{g'}}{|\nabla\varphi|^4_g}. \tag{5.30}$$

This inequality can be used to simplify several estimates in the Kähler-Ricci flow, including the one on the gradient of the Ricci potential.

6. A Priori Estimates: C^2 Estimates

The C^2 estimates for the complex Monge-Ampère equation (5.1) are due to Yau [**Y78**] and Aubin [**A**]. The precise statement is,

THEOREM 8. *Let φ be a C^4 solution of the equation (2.1) on a compact Kähler manifold X, with smooth boundary ∂X (possibly empty). Then*

$$0 \leq n + \Delta\varphi(z) \leq C \exp\left(A_2\left(\varphi(z) - \inf_X \varphi\right)\right) \tag{6.1}$$

where the constant C depends only on an upper bound for F, for $\sup_{X\times[\inf\varphi,\infty)}|(\log F)_\varphi(z,\varphi)|$, for the scalar curvature R, and a lower bound for $\Delta_z \log F$, for $(\log F)_{\varphi\varphi}(z,\varphi)|\nabla\varphi|^2$, and for the lower bound Λ introduced in (5.4) for the bisectional curvature of $g_{\bar{k}j}$.

When ∂X is not empty, the constant also depends on the boundary value φ_b of φ, and on $\|\Delta\varphi\|_{C^0(\partial X)}$.

The conclusion still holds for $z \in X \setminus Z$, if the equation (5.1) holds on $X \setminus Z$, Z is a subset of X not intersecting ∂X, and $\varphi(z) \to +\infty$ as $z \to Z$.

The derivation of the C^2 estimates is particularly transparent if we use the formalism of the relative endomorphisms h of (5.6), as in [**PSS**] and [**PS09a**], which we follow here. As in the proof of the C^1 estimates, we would like to estimate $\mathrm{Tr}\, h$ by the maximum principle. As a preliminary, we calculate

$$\begin{aligned}\Delta'\mathrm{Tr}\, h &= (g')^{p\bar{q}}\partial_{\bar{q}}\partial_p \mathrm{Tr}\, h = (g')^{p\bar{q}}\mathrm{Tr}(\nabla'_{\bar{q}}((\nabla'_p h h^{-1})h) \\ &= (g')^{p\bar{q}}\mathrm{Tr}(\nabla'_{\bar{q}}(\nabla'_p h h^{-1})h) + (g')^{p\bar{q}}\mathrm{Tr}(\nabla'_p h h^{-1}\nabla'_{\bar{q}} h).\end{aligned} \tag{6.2}$$

But $\nabla'_{\bar{q}}(\nabla_p h h^{-1}) = -Rm'_{\bar{q}p} + Rm_{\bar{q}p}$, as a special case of the general formula comparing the curvatures of two Hermitian metrics on the same holomorphic vector bundle. Here the full curvature tensors $Rm_{\bar{q}p}$ and $Rm'_{\bar{q}p}$ are viewed as endomorphisms on the holomorphic tangent bundle. Thus

$$\begin{aligned}(g')^{p\bar{q}}\mathrm{Tr}(\nabla'_{\bar{q}}(\nabla'_p h h^{-1})h) &= -(g')^{p\bar{q}}R'_{\bar{q}p}{}^j{}_k h^k_j + (g')^{p\bar{q}}R_{\bar{q}p}{}^j{}_k h^k_j \\ &= -R'_{\bar{m}k}g^{k\bar{m}} + (h^{-1})^p{}_m R^m{}_p{}^j{}_k h^k{}_j.\end{aligned} \tag{6.3}$$

But the Ricci curvature $R'_{\bar{m}k}$ can be obtained from the Monge-Ampère equation

$$R'_{\bar{m}k} = R_{\bar{m}k} - \partial_k\partial_{\bar{m}} \log F(z,\varphi). \tag{6.4}$$

Thus we obtain

(6.5)
$$\Delta'\mathrm{Tr}\, h = -R + \Delta \log\, F(z,\varphi) + (h^{-1})^p{}_m R^m{}_p{}^j{}_k h^k{}_j + (g')^{p\bar q}\mathrm{Tr}(\nabla'_p h h^{-1}\nabla'_{\bar q} h)$$

and hence

$$\Delta' \log \mathrm{Tr}\, h = \frac{-R + \Delta \log\, F(z,\varphi) + (h^{-1})^p{}_m R^m{}_p{}^j{}_k h^k{}_j}{\mathrm{Tr}\, h}$$

(6.6)
$$+ \left\{ \frac{(g')^{p\bar q}\mathrm{Tr}(\nabla'_p h h^{-1}\nabla'_{\bar q} h)}{\mathrm{Tr}\, h} - \frac{|\nabla' \mathrm{Tr}\, h|^2}{(\mathrm{Tr}\, h)^2} \right\}.$$

A fundamental inequality due to Yau and Aubin is that the expression between brackets is non-negative, as a consequence of the Cauchy-Schwarz inequality. Also

(6.7)
$$\Delta(\log\, F(z,\varphi)) = (\Delta_z \log F)(z,\varphi) + (\log F)_\varphi(\mathrm{Tr}\, h - n) + (\log F)_{\varphi\varphi}|\nabla\varphi|^2.$$

Thus

$$\text{(6.8)} \quad \Delta' \log \mathrm{Tr}\, h \geq -C_1(\mathrm{Tr}\, h)^{-1} - \Lambda \mathrm{Tr}\, h^{-1} - C_2 \geq -(C_1+\Lambda)\mathrm{Tr}\, h^{-1} - C_2$$

since $(\mathrm{Tr}\, h)^{-1} \leq \mathrm{Tr}\, h^{-1}$. Here C_1, C_2 depend only on an upper bound for the scalar curvature R, a lower bound for $(\Delta_z \log\, F)(z,\varphi)$, a lower bound for $(\log F)_{\varphi\varphi}|\nabla\varphi|^2$, and an upper bound for $|(\log F)_\varphi|$. We can write now

$$\Delta'(\log \mathrm{Tr}\, h - A_2\varphi) \geq A_2(\mathrm{Tr}\, h^{-1} - n) - (C_1+\Lambda)\mathrm{Tr}\, h^{-1} - C_2$$

(6.9)
$$\geq \frac{1}{2} A_2\, \mathrm{Tr}\, h^{-1} - C_3$$

for $A_2 \geq 2(C_1+\Lambda)$ and $C_3 = nA_2$. At a maximum point z_0 for $\log \mathrm{Tr}\, h - A_2\varphi$, the eigenvalues of h^{-1} are then bounded from above by absolute constants. Equivalently, the eigenvalues λ_i of h are bounded from below by absolute constants, and hence, in view of the Monge-Ampère equation $\prod_{i=1}^n \lambda_i = F$, they are also bounded from above by constants depending also on $\sup_X F$. Thus for any $z \in X$,

$$\text{(6.10)} \qquad \log \mathrm{Tr}\, h(z) \leq \log \mathrm{Tr}\, h(z_0) + A_2(\varphi(z) - \varphi(z_0)).$$

This establishes the desired C^2 estimates. We note that the C^0 bounds for $\Delta\varphi$ imply similar C^0 bounds for $\partial_j \partial_{\bar k}\varphi$ by plurisubharmonicity, but not for $\nabla_j \nabla_k \varphi$.

7. A Priori Estimates: The Calabi Identity

To obtain estimates for derivatives of order higher than 2, we need the equation to be non-degenerate. Thus we allow constants to depend now on a lower bound for F. In particular, the C^2 estimates imply that the metrics $g_{\bar k j}$ and $g'_{\bar k j}$ are equivalent, up to such constants. We restrict ourselves to the equation (3.1), although the arguments can be extended to certain classes of more general $F(z,\varphi)$, for example $F(z,\varphi)=e^{f(z)\pm\varphi}$.

In Yau's solution of the Calabi conjecture [**Y78**], uniform bounds for the third order derivatives $\nabla_j\nabla_{\bar k}\nabla_m\varphi$ were derived from a generalization to the complex case of an identity due to Calabi [**Ca2**]. We present here a simplified proof of this identity which appeared in [**PSS**], and which is again based on the formalism of the relative endomorphism $h^j{}_k = g^{j\bar p}g'_{\bar pk}$. Norms and lowering and raising of indices are with respect to $g'_{\bar kj}$. Covariant derivatives with respect to $g_{\bar kj}$ and $g'_{\bar kj}$ are denoted by ∇ and ∇' respectively.

Define $S=|\nabla\bar\nabla\nabla\varphi|^2$ as in [**Y78**]. In terms of $h^\alpha{}_\beta$, we have $(g')^{\alpha\bar k}\nabla_j\varphi_{\bar k\beta} = (\nabla'_j h\,h^{-1})^\alpha{}_\beta$, and thus [**PSS**]

$$S=|\nabla' h\,h^{-1}|^2. \tag{7.1}$$

The point is that the Laplacian of S can now be evaluated directly in terms of metrics and curvatures, instead of Kähler potentials. We readily find

$$\begin{aligned}\Delta' S &= (g')^{m\bar\gamma}(\Delta'(\nabla'_m h\,h^{-1})^\beta{}_l\overline{(\nabla'_\gamma h\,h^{-1})_{\bar\beta}{}^{\bar\ell}}\\ &\quad + (\nabla'_m h\,h^{-1})^\beta{}_\ell\overline{\bar\Delta'(\nabla'_\gamma h\,h^{-1})_{\bar\beta}{}^{\bar\ell}}) + |\bar\nabla'(\nabla' h\,h^{-1})|^2 + |\nabla'(\nabla' h\,h^{-1})|^2\end{aligned} \tag{7.2}$$

where $|\bar\nabla'(\nabla' h\,h^{-1})|^2 \equiv (g')^{q\bar p}\nabla'_{\bar p}(\nabla'_j h\,h^{-1})^\alpha{}_\beta\overline{\nabla'_{\bar q}(\nabla'_m h\,h^{-1})_{\bar\alpha}{}^{\bar\beta}}$, and $\Delta' = (g')^{q\bar p}\nabla'_q\nabla'_{\bar p}$, $\bar\Delta' = (g')^{q\bar p}\nabla'_{\bar p}\nabla'_q$. Commuting the ∇'_q and the $\nabla'_{\bar p}$ derivatives gives,

$$\begin{aligned}(\bar\Delta'(\nabla'_j h\,h^{-1}))^\gamma{}_\alpha &= (\Delta'(\nabla'_j h\,h^{-1}))^\gamma{}_\alpha - (R')^\gamma{}_\mu(\nabla'_\gamma h\,h^{-1})^\mu{}_\alpha\\ &\quad + (R')^\mu{}_\alpha(\nabla'_j h\,h^{-1})^\gamma{}_\mu + (R')^\mu{}_j(\nabla'_\mu h\,h^{-1})^\gamma{}_\alpha\end{aligned} \tag{7.3}$$

while, in view of the Bianchi identity,

$$\begin{aligned}\Delta'(\nabla'_j h\,h^{-1})^l{}_m &= (\nabla')^{\bar p}\partial_{\bar p}(\nabla'_j h\,h^{-1}) = -(\nabla')^{\bar p}R'_{\bar pj}{}^l{}_m + (\nabla')^{\bar p}R_{\bar pj}{}^l{}_m\\ &= -\nabla'_j(R')^l{}_m + (\nabla')^{\bar p}R_{\bar pj}{}^l{}_m.\end{aligned}$$

with $R_{\bar pj}{}^l{}_m = -\partial_{\bar p}(g^{l\bar q}\partial_j g_{\bar qm})$. Thus we obtain the exact formula

$$\begin{aligned}\Delta' S &= |\bar\nabla'(\nabla' h\,h^{-1})|^2 + |\nabla'(\nabla' h\,h^{-1})|^2\\ &\quad - ((\nabla')^{\bar\gamma}R'_{\bar\beta\alpha}\overline{(\nabla'_\gamma h\,h^{-1})^{\beta\bar\alpha}} + ((\nabla')^{\bar\gamma}h\,h^{-1})^{\beta\bar\alpha}\overline{\nabla'_\gamma R'_{\bar\beta\ell}})\\ &\quad + (\nabla'_m h\,h^{-1})^\beta{}_l((R')^{l\bar\rho}\overline{((\nabla')^{\bar m}h\,h^{-1})_{\bar\beta\rho}} - R'_{\bar\rho\beta}\overline{((\nabla')^{\bar m}h\,h^{-1})^{\rho\bar l}}\\ &\quad + (R')^{m\bar\rho}\overline{(\nabla'_\rho h\,h^{-1})_{\bar\beta}{}^{\bar l}}) + (\nabla')^{\bar p}R_{\bar pm}{}^\beta{}_l\overline{(\nabla')^{\bar m}h\,h^{-1})_{\bar\beta}{}^{\bar l}}\\ &\quad + ((\nabla')^{\bar\gamma}h\,h^{-1})_{\bar\mu}{}^{\bar\alpha}\overline{(\nabla')^{\bar p}R_{\bar p\gamma}{}^\mu{}_\alpha}.\end{aligned} \tag{7.4}$$

Since $R'_{\bar pm} = R_{\bar pm} - \partial_m\partial_{\bar p}\log F$, it can be viewed as known. As already noted, the metrics $g_{\bar kj}$ and $g'_{\bar kj}$ are uniformly equivalent. Since the connection $\nabla' hh^{-1}$ is of order $O(S^{\frac12})$, the above identity implies

$$\Delta' S \geq -C_1 S - C_2 \tag{7.5}$$

where the constants C_i depend on an upper bound for $\Delta\varphi$, a lower bound for F, the C^3 norm of F, and the C^1 norms of the curvature $R_{\bar{k}j}{}^l{}_m$ of $g_{\bar{k}j}$. Using the expression (6.5) for $\Delta'\mathrm{Tr}\, h$, and the fact that

$$(7.6)\quad (g')^{p\bar{q}}\mathrm{Tr}(\nabla'_p h h^{-1}\nabla'_{\bar{q}}h) = (g')^{p\bar{q}}(g')^{j\bar{k}}g^{m\bar{r}}\nabla_p\nabla_{\bar{m}}\nabla_j\varphi\overline{\nabla_q\nabla_{\bar{m}}\nabla_k\varphi} \geq C_3 S$$

since the metrics $g_{\bar{k}j}$ and $g'_{\bar{k}j}$ are equivalent, we readily see that

$$(7.7)\qquad \Delta'(S + A\,\mathrm{Tr}\, h) \geq C_4 S - C_5$$

for A sufficiently large. We can now apply the maximum principle and obtain the following:

THEOREM 9. *Let φ be a C^5 solution of the equation (5.1) on a compact Kähler manifold (X,ω_0) with smooth boundary ∂X (possibly empty). Then that S is uniformly bounded by constants depending only on the C^0 norms of $\partial_j\partial_{\bar{k}}\log F$ and $\nabla_j\nabla_{\bar{k}}\nabla_m\log F$, a lower bound for F, the C^1 norm of $R_{\bar{k}j}{}^l{}_m$, $\|\mathrm{Tr}h\|_{C^0(X)}$ and, when ∂X is not empty, on $\|S\|_{C^0(\partial X)}$.*

We conclude this section by noting that most of the a priori estimates discussed here have counterparts for parabolic Monge-Ampère equations. They were instrumental in Cao's proof of the all-time existence for the Kähler-Ricci flow on manifolds of definite Chern classes [**Cao**]. They also apply in many situations to manifolds of general type [**Ts, EGZ, ST09**]. For the modified Kähler-Ricci flow, the C^3 estimates and the Calabi identity require an additional argument [**PSSW2**], as well as a full use of the square terms in (7.4) which were dropped in the proof of Theorem 9. Extensions to flows on Hermitian manifolds can be found in [**Gm, ZZ**].

8. Boundary Regularity

In this section, we discuss a priori estimates for the Dirichlet problem for the complex Monge-Ampère equation on a Kähler manifold (X,ω_0) with smooth boundary ∂X.

8.1. C^0 estimates. Let φ_b be a smooth function on ∂X, and consider the Dirichlet problem

$$(8.1)\qquad \left(\omega_0 + \frac{i}{2}\partial\bar{\partial}\varphi\right)^n = F(z,\varphi,\nabla\varphi)\omega_0^n \text{ on } X, \quad \varphi = \varphi_b \text{ on } \partial X,$$

where $n = \dim X$, F is a smooth strictly positive function, and $\varphi \in PSH(X,\omega_0)\cap C^\infty(X)$.

The fact that $\varphi \in PSH(X,\omega_0)$ implies that $n + \Delta\varphi \geq 0$. If h is the solution of the Dirichlet problem $\Delta h = -n$ on X, $h = \varphi_b$ on ∂X, then the comparison principle implies

$$(8.2)\qquad \varphi \leq h.$$

Thus, to obtain C^0 estimates, we need only a lower bound for φ. As shown by Caffarelli, Kohn, Nirenberg, and Spruck [**CNS, CKNS**], this can be effectively obtained if we assume the existence of a smooth subsolution $\underline{\varphi}$ of the Dirichlet problem (9.1), that is, a smooth function $\underline{\varphi}$ satisfying

$$\left(\omega_0 + \frac{i}{2}\partial\bar{\partial}\underline{\varphi}\right)^n > F(z, \underline{\varphi}, \nabla\underline{\varphi})\omega_0^n \text{ on } X, \quad \underline{\varphi} = \varphi_b \text{ on } \partial X. \tag{8.3}$$

Indeed, in the method of continuity, the problem reduces to a priori estimates for the equation

$$\begin{gathered}\left(\omega_0 + \frac{i}{2}\partial\bar{\partial}\varphi\right)^n = tF(z, \varphi, \nabla\varphi)\omega_0^n + (1-t)\left(\omega_0 + \frac{i}{2}\partial\bar{\partial}\underline{\varphi}\right)^n \quad \text{on } X,\\ \varphi = \varphi_b \text{ on } \partial X,\end{gathered} \tag{8.4}$$

for $0 \le t \le 1$. Let $\varphi = \underline{\varphi}$ for $t = 0$. We claim that, if a smooth solution exists in an interval $0 \le t < T$, then

$$\underline{\varphi} < \varphi \text{ in } X, \tag{8.5}$$

for all $t < T$. To see this, note that the derivative in t of $\frac{(\omega_0 + \frac{i}{2}\partial\bar{\partial}\varphi)^n}{(\omega_0 + \frac{i}{2}\partial\bar{\partial}\underline{\varphi})^n}$ is strictly negative at $t = 0$. Thus $(\omega_0 + \frac{i}{2}\partial\bar{\partial}\varphi)^n < (\omega_0 + \frac{i}{2}\partial\bar{\partial}\underline{\varphi})^n$, and $\underline{\varphi} < \varphi$ for t strictly positive and small, by the comparison principle. If there exists t_0, $0 < t_0 < T$, with $\underline{\varphi}(z_0) = \varphi(z_0)$ for some $z_0 \in X$, let t_0 be the first such time. By continuity, $\underline{\varphi}(z) \le \varphi(z)$ for all $z \in X$ and $t = t_0$, so z_0 is a maximum of the function $\underline{\varphi} - \varphi$ at t_0. In particular, at t_0 and z_0, we have $\nabla\underline{\varphi} = \nabla\varphi$ and

$$\left(\omega_0 + \frac{i}{2}\partial\bar{\partial}\underline{\varphi}\right)^n \le \left(\omega_0 + \frac{i}{2}\partial\bar{\partial}\varphi\right)^n. \tag{8.6}$$

But the equation (8.4) implies, again at t_0 and z_0,

$$\left(\omega_0 + \frac{i}{2}\partial\bar{\partial}\varphi\right)^n = tF(z, \underline{\varphi}, \nabla\underline{\varphi}) + (1-t)\left(\omega_0 + \frac{i}{2}\partial\bar{\partial}\underline{\varphi}\right)^n < \left(\omega_0 + \frac{i}{2}\partial\bar{\partial}\underline{\varphi}\right)^n, \tag{8.7}$$

which is a contradiction.

8.2. C^1 boundary estimates. The C^1 estimates at the boundary ∂X follow from the bounds $\underline{\varphi} \le u \le h$, and the fact that all three functions have the same boundary values. When the right hand side $F(z, \varphi, \nabla\varphi)$ does not depend on $\nabla\varphi$, the estimates established earlier in Section 4 show that the interior C^1 estimates can be reduced to the boundary C^1 estimates.

8.3. C^2 boundary estimates of Caffarelli-Kohn-Nirenberg-Spruck and B. Guan. The barrier constructions of Caffarelli, Kohn, Nirenberg, Spruck [**CKNS**] and B. Guan [**Gb**] provide $C^0(\partial X)$ bounds for $\Delta\varphi$, in terms of $C^0(X)$ bounds for φ and for $\nabla\varphi$. The following slightly more precise formulation of their estimates can be found in [**PS09a**], under the simplifying assumption that the boundary ∂X is holomorphically flat[2]:

THEOREM 10. *Assume that ∂X is holomorphically flat, and that φ is a C^3 solution of the equation (8.1), with $F(z)$ on the right-hand side. Then we have*

$$\sup_{\partial X}(n+\Delta\varphi) \leq C \sup_{\partial X}(1+|\nabla\varphi|^2)\sup_X(1+|\nabla\varphi|^2), \tag{8.8}$$

for a constant C depending only on the boundary ∂X, ω_0, and upper bounds for $\sup_X F$, and $\sup_X(\nabla \log F)$, and $\|\varphi\|_{C^0(X)}$.

By the interior estimates of Yau and Aubin in §6, the uniform bound for $\Delta\varphi$ in the whole of X can be reduced to its estimate on ∂X. Thus the above bound implies that $\|\Delta\varphi\|_{C^0(X)}$ is bounded in terms of the constants indicated. By plurisubharmonicity, it follows that all the mixed partials $\|\partial_j\partial_{\bar{k}}\varphi\|_{C^0(X)}$ are bounded as well.

It is an interesting question whether bounds for the un-mixed partials $\|\nabla_j\nabla_k\varphi\|_{C^0}$ can be obtained as well without additional assumptions. Such bounds have been obtained by Blocki [**B09b**] under the additional assumption that the background form ω_0 has non-negative bisectional curvature.

If we allow bounds to depend on a lower bound for F, then the equation (8.3) can be viewed as uniformly elliptic, since the eigenvalues of the relative endomorphism $h^j{}_k = g^{j\bar{p}}g'_{\bar{p}k}$ are already known to be bounded from above, and using the lower bound for F, they are also bounded from below. The Monge-Ampère equation is concave, so we can then apply to the following general theorem of the Evans-Krylov and Krylov theory, which we quote from Chen-Wu [**CW**] (see also Gilbarg-Trudinger [**GT**] p. 482 and Q. Han [**H**]). The statement is local, and can be formulated for domains with smooth boundary in $\mathbf{R}^n$:

THEOREM 11. *Assume that $\Omega \subset \mathbf{R}^n$ has smooth boundary, and the boundary data is smooth. Assume that $F(x, u, Du, D^2u)$ is smooth in all variables (x, u, p, A), uniformly elliptic and concave (or convex) in D^2u, and assume that $\|u\|_{C^{1,\gamma}(\bar{\Omega})}$ is bounded for some $0<\gamma<1$. Then there are constants $0<\alpha<\gamma$ and C so that, for any $0<\beta<\alpha$, we have*

$$\|u\|_{C^{2,\beta}(\bar{\Omega})} \leq C. \tag{8.9}$$

We note that, while both the local [**Ca**] and the global [**TrWa**] $C^{2,\alpha}$ regularity is known for real Monge-Ampère equations when the right hand

[2]A hypersurface ∂X is holomorphically flat if, locally, there exist holomorphic coordinates $(z_1, \ldots, z_n)$ so that ∂X is given by $\mathrm{Re}\, z_n = 0$.

side F is in C^α, the corresponding question is still not completely resolved in the complex case. For some recent progress on this issue, see [**DZZ**], and particularly [**W2**], where it is shown that the solution φ is of class $C^{2,\alpha}$ if the right hand side $F(z)$ is strictly positive, $F^{\frac{1}{n}} \in C^\alpha$, and $\Delta\varphi$ is bounded.

9. The Dirichlet Problem for the Monge-Ampère Equation

The preceding a priori estimates imply the following classic existence theorem due to Caffarelli, Kohn, Nirenberg, and Spruck [**CKNS**] and B. Guan [**Gb**]:

THEOREM 12. *Let $(\bar{X}, \omega_0)$ be a compact Kähler manifold of dimension n, with smooth boundary ∂X. Let $F(z,\varphi)$ be a smooth, strictly positive function of the variables z and φ, and let φ_b be a smooth function on ∂X. Consider the Dirichlet problem*

$$\left(\omega_0 + \frac{i}{2}\partial\bar{\partial}\varphi\right)^n = F(z,\varphi)\,\omega_0^n, \quad \varphi = \varphi_b \text{ on } \partial X. \tag{9.1}$$

If $F_\varphi(z,\varphi) \geq 0$ and the problem admits a smooth subsolution, that is, a smooth function $\underline{\varphi}$ satisfying

$$\left(\omega_0 + \frac{i}{2}\partial\bar{\partial}\underline{\varphi}\right)^n > F(z,\underline{\varphi})\,\omega_0^n, \quad \underline{\varphi} = \varphi_b \text{ on } \partial X, \tag{9.2}$$

then the Dirichlet problem (9.1) admits a unique solution φ, and $\varphi \in C^\infty(\bar{X})$.

Indeed, the C^0 estimates of §8.1, the C^2 estimates of §6 and §8.2, the Evans-Krylov theory for higher derivatives of §8.2 can be applied to show that the equation (8.4) admits a solution for $0 \leq t \leq 1$.

Similar results for the real Monge-Ampère equation can be found in [**GS**] and [**Gb98**]. An extension to Hermitian manifolds can be found in [**GL**].

The a priori estimates show more than just the existence of a solution φ for the equation (9.1): the upper bound for $\Delta\varphi$ does not depend on a lower bound for $F(z,\varphi)$. This allows an immediate application to the existence of solutions to the Dirichlet problem for the completely degenerate, or homogeneous, complex Monge-Ampère equation. For this, we apply Theorem 12 to the Dirichlet problem

$$\left(\omega_0 + \frac{i}{2}\partial\bar{\partial}\varphi_\varepsilon\right)^n = \varepsilon\,\omega_0^n, \quad \varphi_\varepsilon = 0 \text{ on } \partial X, \tag{9.3}$$

where ε is a constant satisfying $0 < \varepsilon < 1$. The function $\underline{\varphi}_\varepsilon = 0$ is a subsolution, and hence Theorem 12 implies the existence of a smooth solution φ_s with $\Delta\varphi_s$ bounded uniformly in ε. Thus a subsequence of the functions φ_ε converges in $C^{1,\alpha}$ to a $C^{1,\alpha}$ solution of the equation (9.4) for all $0 < \alpha < 1$. We obtain in this manner the following theorem, whose present formulation is due to Blocki [**B09b**] and which generalizes the theorem of Chen [**C00**] stated further below as Theorem 24:

THEOREM 13. *Let* (X, ω_0) *be a compact Kähler manifold with smooth boundary* ∂X*. Then the Dirichlet problem*

$$\left(\omega_0 + \frac{i}{2}\partial\bar\partial\varphi\right)^n = 0 \text{ on } \partial X, \quad \varphi = 0 \text{ on } \partial X \tag{9.4}$$

admits a unique solution, which is of class $C^{1,\alpha}(\bar X)$ *for each* $0 < \alpha < 1$.

In some applications, as in the problem of geodesics in the space of Kähler potentials described below in Section §13, it is actually necessary to consider equations of the form (9.4), but with the Kähler form ω_0 replaced by a smooth background $(1,1)$-form ω which is closed, non-negative, but not strictly positive. We discuss a specific situation where the existence and regularity of solutions can still be established by the a priori estimates that we described in sections §5, §6, §8.2, and §8.3 (in fact, some of the C^1 estimates given in §5 were designed for that purpose).

Assume that ω is a smooth, closed, and non-negative $(1,1)$-form, and that there exists an effective divisor E, not intersecting ∂X, with the line bundle $O(E)$ admitting a metric K satisfying

$$\omega_K \equiv \omega + \delta\frac{i}{2}\partial\bar\partial \log K > 0. \tag{9.5}$$

for some strictly positive constant δ. Then we have the following theorem [**PS09a**]:

THEOREM 14. *Let* X *be a compact complex manifold with smooth boundary* ∂X*. Assume that* ω *is a smooth non-negative* $(1,1)$*-form,* E *is an effective divisor not intersecting* ∂X*,* K *is a metric on* $O(E)$*, with* ω_K *satisfying the Kähler condition (9.5). Then the Dirichlet problem*

$$\left(\omega + \frac{i}{2}\partial\bar\partial\varphi\right)^n = 0 \text{ on } X, \quad \varphi = 0 \text{ on } \partial X \tag{9.6}$$

admits a unique bounded solution. The solution is $C^\alpha(\bar X \setminus E)$ *for any* $0 < \alpha < 1$*. If* ∂X *is holomorphically flat (in the sense that there exists holomorphic coordinates* z_i *with* $\partial X = \{\mathrm{Re}\, z_n = 0\}$ *locally), then the solution is* $C^{1,\alpha}(\bar X \setminus E)$ *for any* $0 < \alpha < 1$.

We sketch the proof. Let

$$\omega_s = (1-s)\omega_0 + s\omega_K. \tag{9.7}$$

For $0 < s < 1$, ω_s is strictly positive definite. Consider the equation

$$\left(\omega_s + \frac{i}{2}\partial\bar\partial\varphi_s\right)^n = F_s(z)\omega_s^n \text{ on } X, \quad \varphi_s = 0 \text{ on } \partial X, \tag{9.8}$$

for some smooth functions $F_s > 0$ satisfying $\sup_X F_s < 1$, to be specified more completely later. By Theorem 12, this equation admits a smooth solution in $PSH(X, \omega_s)$ for each $s > 0$. Since the eigenvalues of ω_s are bounded from above with respect to the Kähler form ω_K, the ω_s-plurisubharmonicity of φ_s implies that $\Delta_{\omega_K}\varphi_s \geq -C$ for a uniform constant C. The arguments for C^0 estimates in §8.1 imply that the norms $\|\varphi_s\|_{C^0(X)}$ are uniformly bounded in s.

To obtain C^1 estimates on compact subsets of $X \setminus E$, we choose F_s as follows. First, define

$$\hat{\omega}_s = \omega_K + s\frac{i}{2}\partial\bar{\partial}\log K^{\delta}. \tag{9.9}$$

Then $\hat{\omega}_s$ is uniformly bounded from below for all s sufficiently small. In particular, its curvature tensor is uniformly bounded together with all its derivatives. On the other hand, since $\hat{\omega}_s$ can also be expressed as

$$\hat{\omega}_s = \omega_s + \frac{i}{2}\partial\bar{\partial}\log K^{\delta}. \tag{9.10}$$

the equation (9.8) can be rewritten as

$$\left(\hat{\omega}_s + \frac{i}{2}\partial\bar{\partial}(\varphi_s - \delta\log\|\psi\|_K^2)\right)^n = \hat{F}_s\hat{\omega}_s^n, \tag{9.11}$$

with ψ a holomorphic section of $O(E)$, $\|\psi\|_K^2 = \psi\bar{\psi}K$, and $\hat{F}_s\hat{\omega}_s^n = F_s\omega_s^n$. Choose F_s to be constants tending so fast to 0 that $\limsup_{s\to 0}\|F_s\|_{C^0(X)} = 0$. The desired uniform bounds for $\nabla\varphi_s$ on compact subsets of $X \setminus E$ follow from the C^1 estimates of §5. With these estimates, it is then easy to show the existence of a subsequence of φ_s converging in C^α on compact subsets of $X \setminus E$ to a solution of (9.6).

10. Singular Monge-Ampère Equations

In the seminal paper [**Y78**], Yau not only solved the Calabi conjecture, but he also started the study of complex Monge-Ampère equations in more general settings. These include settings when the right hand side may have zeroes or poles, or when the manifold X is not compact and one looks for a complete Kähler-Einstein metric, or when X is quasi-projective. We recall briefly some of these classical results below, before discussing some more recent developments. In these more recent developments, the underlying manifold may have singularities, and/or the background form ω_0 in the Monge-Ampère equation may be degenerate.

10.1. Classic works. The classical literature on singular Monge-Ampère equations and singular Kähler-Einstein metrics is particularly rich, as different equations are required by different geometric situations. We shall restrict ourselves to describing three results.

First, we consider the case when the right hand side of the equation has zeroes and/or poles. Let (X, ω_0) be a compact Kähler manifold. Let $\{L_i\}_{i=1}^I$

be a family of holomorphic line bundles over X. For each i, let s_i be a holomorphic section of L_i and h_i a smooth hermitian metric on L_i. Let

$$(10.1)\quad u_k = \sum_{i=1}^{I} a_{k,i}|s_i|_{h_i}^{2\alpha_{k,i}},\ k = 1, \ldots, K, \quad v_l = \sum_{i=1}^{I} b_{l,i}|s_i|_{h_i}^{2\beta_{l,i}}\ l = 1, \ldots, L,$$

where $a_{k,i}$ $b_{l,i}$, $\alpha_{k,i}$ and $\beta_{l,i}$ are nonnegative numbers, and consider the following Monge-Ampère equation

$$(10.2)\qquad \left(\omega_0 + \frac{i}{2}\partial\bar\partial\varphi\right)^n = \frac{u_1 u_2 \ldots u_K}{v_1 v_2 \ldots v_L} e^{f(z)}\omega_0^n,$$

where $f = f(z)$ is a smooth function on X,. The following theorem is due to Yau [**Y78**]:

THEOREM 15. *Assume the following two conditions:*

(1) $(u_1 u_2 \ldots u_K)(v_1 v_2 \ldots v_L)^{-1} e^f \in L^n(X)$ *and*

$$\int_X \frac{u_1 u_2 \ldots u_K}{v_1 v_2 \ldots v_L} e^f\, \omega_0^n = \int_X \omega_0^n,$$

(2) *there exists* $\epsilon > 0$ *such that*

$$(v_1 v_2 \ldots v_L)^{-\epsilon}|\Delta \log (v_1 v_2 \ldots v_L)|^{(n-1)/n} \in L^1(X \setminus D),$$

where Δ *is the Laplacian with respect to* ω_0 *and* D *is the union of the zeros of* v_l, $l = 1, \ldots, L$.

Then there exists a bounded ω_0*-psh function* φ *solving the equation (10.2). Furthermore,* φ *is smooth outside the zeros of* u_k *and* v_l, *for* $k = 1, \ldots, K$ *and* $l = 1, \ldots, L$, *and* φ *is unique up to a constant.*

In particular, if for each k and l, $u_k = a_{i_k}|s_{i_k}|_{h_{i_k}}^{2\alpha_{i_k}}$ and $v_l = b_{i_l}|s_{i_l}|_{h_{i_l}}^{2\beta_{i_l}}$ for some $1 \leq i_k \leq K$ and $1 \leq i_l \leq I$, the second assumption in Theorem 15 holds automatically. The first assumption for Theorem 15 is that the right hand side of the equation (10.2) is in $L^n(X)$. Thus the theorem on C^0 estimates of Kolodziej can also be applied here, and we can obtain in this manner a new proof of Theorem 15.

The next important geometric situation is that of open complex manifolds. There one is interested in complete Kähler-Einstein metrics of negative curvature. In [**CY80**], Cheng and Yau gave effective criteria for the existence of such metrics. In particular, they proved the existence of a complete Kähler-Einstein metric of negative scalar curvature on bounded, smooth, strictly pseudoconvex domains in $\mathbf{C}^n$. This corresponds to solving Monge-Ampère equations of the form (2.1), with the solution tending to ∞ at the boundary. This also allowed Cheng and Yau to obtain essentially sharp boundary regularity results for the Dirichlet problem for the closely related equation $J(u) = 1$ of Fefferman [**F76**]. It was subsequently shown by Mok and Yau [**MY83**] that any bounded domain of holomorphy admits a complete Kähler-Einstein metric.

A third important class of non-compact manifolds is the class of quasi-projective manifolds. Let $M = \overline{M} \setminus D$ be a quasi-projective manifold, where $\overline{M}$ is a projective manifold and D is a smooth ample divisor on M. The Calabi conjecture for quasi-projective manifolds asserts that, for any smooth real valued $(1,1)$-form $\eta \in c_1(K_{\overline{M}}^{-1} \otimes [D]^{-1})$, there exists a complete Kähler metric on M with its Ricci curvature equal to $\eta|_M$. This was proved by Tian and Yau in [**TY86, TY90, TY91**].

These works required at that time many new technical tools which remain useful to this day. They include the notion of bounded geometry, the Cheng-Yau Hölder spaces with weights, and particularly the observation repeatedly stressed in these works that the arguments are almost local in nature, and that the manifold can be allowed singularities, as long as the metric admits a non-singular resolution by a local holomorphic map.

10.2. Monge-Ampère equations on normal projective varieties. The original theorems of Yau [**Y78**] and Yau [**Y78**] and Aubin [**A**] establish the existence of Kähler-Einstein metrics on a Kähler manifold X when K_X has zero or positive first Chern class. We discuss now one of the new developments in the theory of complex Monge-Ampère equations, namely an extension of these results to normal projective manifolds. Normal projective manifolds are a very specific class of manifolds with singularities. For the convenience of the reader, we summarize here some of their basic definitions and properties.

Let X be a subvariety of $\mathbf{CP}^N$. A function on a neighborhood of a point $z_0 \in X$ is holomorphic if it extends to a holomorphic function on a neighborhood of $z_0 \in \mathbf{CP}^N$. Let X_{sing} be the smallest subset of X with $X \setminus X_{sing}$ a complex manifold. Then X is said to be normal if for any $z_0 \in X_{sing}$, there is a neighborhood U so that any bounded holomorphic function on $U \setminus X_{sing}$ extends to a holomorphic function on U.

A plurisubharmonic function on $U \subset X$ is by definition the restriction to X of a plurisubharmonic function in a neighborhood $\hat{U}$ of U in $\mathbf{CP}^N$. By a theorem of Fornaess and Narasimhan [**FN**], if X be a normal projective variety, and a function φ is plurisubharmonic on $U \setminus X_{sing}$ and is bounded, then φ is plurisubharmonic on U.

A line bundle L on X is an ample $\mathbf{Q}$-line bundle if mL is the restriction to X of $O(1)$ for some $m \in \mathbf{Z}^+$. More generally, a line bundle $\tilde{L} \to \tilde{X}$ is an ample $\mathbf{Q}$-line bundle if there is an imbedding of $\tilde{X}$ into projective space, with the pull-back of $O(1)$ equal to $m\tilde{L}$ for some $m \in \mathbf{Z}^+$.

We can define now the notion of Monge-Ampère measure on a normal projective variety X. Let $\dim X = n$, and let $\pi : \tilde{X} \to X$ be a smooth resolution of singularities of X. Let $\tilde{L} = \pi^* L$ for any ample $\mathbf{Q}$-line bundle $L \to X$.

By definition, $mL = O(1)$ for some $m \in \mathbf{Z}^+$ and so

$$m\tilde{L} = \pi^* O(1).$$

Let $m\omega$ be the restriction of the Fubini-Study metric on $\mathbf{CP}^M$ to X and let $\tilde{\omega} = \pi^*\omega$. For any bounded ω-plurisubharmonic function φ on X, we let $\tilde{\varphi} = \pi^*\varphi$. The measure $(\tilde{\omega} + \frac{i}{2}\partial\bar{\partial}\tilde{\varphi})^n$ is a well-defined Monge-Ampère measure on $\tilde{X}$. Since $\tilde{\varphi}$ is bounded and $\tilde{\omega}$-plurisubharmonic, by the work of Bedford and Taylor [**BT76**], it puts no mass on the exceptional locus $\pi^{-1}(X_{sing})$. Furthermore,

$$\int_{X\backslash X_{sing}} \left(\omega + \frac{i}{2}\partial\bar{\partial}\varphi\right)^n = \int_{\tilde{X}} \left(\tilde{\omega} + \frac{i}{2}\partial\bar{\partial}\tilde{\varphi}\right)^n = \int_{\tilde{X}} \tilde{\omega}^n < \infty. \tag{10.3}$$

Therefore, $(\tilde{\omega} + \frac{i}{2}\partial\bar{\partial}\tilde{\varphi})^n$ can be pushed forth to a measure on X and it coincides with the trivial extension of $(\omega + \frac{i}{2}\partial\bar{\partial}\varphi)^n$ from $X\backslash X_{sing}$ to X. In particular, the Monge-Ampère measure $(\omega + \frac{i}{2}\partial\bar{\partial}\varphi)^n$ on X does not depend on the resolution of singularities.

Let Ω be a smooth real valued semi-positive (n, n)-form on $X\backslash X_{sing}$ and let $\tilde{\Omega} = \pi^*\Omega$. We consider the following Monge-Ampère equation on X

$$\left(\omega + \frac{i}{2}\partial\bar{\partial}\varphi\right)^n = e^{\alpha\varphi + F}\Omega, \tag{10.4}$$

where $\alpha = -1, 0, 1$ and F is a real valued function on X. This Monge-Ampère equation can be lifted to a Monge-Ampère equation on $\tilde{X}$ if the solution φ is bounded and ω-plurisubharmonic, i.e., $\omega + \frac{i}{2}\partial\bar{\partial}\varphi \geq 0$. More precisely, let $\tilde{\varphi} = \pi^*\varphi$, and consider the Monge-Ampère equation on $\tilde{X}$

$$\left(\tilde{\omega} + \frac{i}{2}\partial\bar{\partial}\tilde{\varphi}\right)^n = e^{\alpha\tilde{\varphi} + \tilde{F}}\tilde{\Omega}, \tag{10.5}$$

where $\tilde{F} = \pi^*F$. If φ is a bounded ω-plurisubharmonic solution of equation (10.4), then $\tilde{\varphi} = \pi^*\varphi$ is bounded and $\tilde{\omega}$-plurisubharmonic, and it solves equation (10.5). Now let us assume that $\tilde{\varphi}$ is a bounded $\tilde{\omega}$-plurisubharmonic solution of the equation (10.5). Any fibre of π over a singular point of X is connected by Zariski's connectedness theorem, and $\tilde{\omega} = 0$ when restricted to the fibre. Therefore $\tilde{\varphi}$ is constant along the fibre because it is plurisubharmonic and bounded. Hence $\tilde{\varphi}$ descends to a function φ on X. The function φ is bounded and ω-plurisubharmonic on $X\backslash X_{sing}$, and so it is ω-plurisubharmonic function on X because X is normal. We have thus shown

LEMMA 14. *The equation (10.4) admits a bounded ω-plurisubharmonic solution φ if and only if the equation (10.5) admits a bounded $\tilde{\omega}$-plurisubharmonic solution $\tilde{\varphi}$.*

Therefore, we can solve equation (10.5) on a smooth manifold $\tilde{X}$ instead of solving equation (10.4) on a singular variety X. Furthermore the construction is resolution independent because given any two resolutions, we can move the measures to the same resolution and apply the uniqueness property of Monge-Ampère equations there. So we obtain the following lemma, which follows immediately from Theorem 5:

LEMMA 15. *Let Θ be a smooth volume form on $\tilde{X}$. Then the equation (10.4) admits a bounded and ω-plurisubharmonic solution for $\alpha=0$, if $\frac{e^{\tilde{F}}\tilde{\Omega}}{\Theta}\in L^p(\tilde{X})$ for some $p>1$.*

In fact, Lemma 15 also holds for $\alpha=1$ by [**EGZ**]. We can apply it now to solving Kähler-Einstein equations on singular varieties. Recall some basic definitions for canonical models of general type:

DEFINITION 1. (a) *A projective variety X is said to be a canonical model of general type if X is a normal and the canonical divisor K_X is an ample* **Q***-line bundle.*

(b) *Let X be a canonical model of general type. A form Ω is said to be a smooth volume form on X if for any point $z\in X$, there exists an open neighborhood U of z such that*

$$\Omega=f_U(\eta\wedge\overline{\eta})^{\frac{1}{m}},$$

where f_U is a smooth positive function on U and η is a local generator of mK_X on U. In particular, any smooth volume Ω induces a smooth hermitian metric $h=\Omega^{-1}$ on K_X.

(c) *X is said to be a canonical model of general type with canonical singularities if for any resolution of singularities $\pi:\tilde{X}\to X$ and any smooth volume form Ω on X,*

$$\tilde{\Omega}=\pi^*\Omega \tag{10.6}$$

is a smooth real valued (n,n)-form on $\tilde{X}$.

We can now describe some recent results of Eyssidieux, Guedj, and Zeriahi [**EGZ**] on the existence of Kähler-Einstein metrics of zero or negative curvature on manifolds with canonical singularities. As above, let $\pi:\tilde{X}\to X$ be a resolution of singularities, let $m\omega$ be the restriction of the Fubini-Study metric of $\mathbf{CP}^M$ on X, and let Ω be a smooth volume form Ω on X such that

$$\frac{i}{2}\partial\bar{\partial}\log\Omega=\omega. \tag{10.7}$$

The following theorem on Kähler-Einstein metrics with negative curvature was proved in [**EGZ**], using the C^0 estimates of Theorem 5:

THEOREM 16. *Let X be a canonical model of general type with canonical singularities. Then there exists a unique bounded and ω-plurisubharmonic function φ solving the following Monge-Ampère equation on X*

$$\left(\omega+\frac{i}{2}\partial\bar{\partial}\varphi\right)^n=e^{\varphi}\Omega. \tag{10.8}$$

In particular, on $X \backslash X_{sing}$, $\omega_{KE} = \omega + \frac{i}{2}\partial\bar{\partial}\varphi$ *is smooth and*

$$Ricci(\omega_{KE}) = -\omega_{KE}.$$

Next we discuss the case of zero curvature. Recall that X is said to be a Calabi-Yau variety if X is a projective normal variety and mK_X is a trivial line bundle on X for some $m \in \mathbf{Z}^+$. Since mK_X is a trivial line bundle on X, there exists a constant global section η of mK_X. Let $\Omega_{CY} = (\eta \wedge \eta)^{\frac{1}{m}}$. Then Ω is a smooth volume form on X. Furthermore, $\frac{i}{2}\partial\bar{\partial}\log\Omega_{CY} = 0$.

DEFINITION 2. *A Calabi-Yau variety* X *is said to be a Calabi-Yau variety with canonical singularities if for any resolution of singularities* $\pi : \tilde{X} \to X$,

$$\tilde{\Omega}_{CY} = \pi^*\Omega_{CY}$$

is a smooth real valued (n, n)*-form on* $\tilde{X}$.

Let X be a Calabi-Yau variety with canonical singularities. We choose the smooth Kähler form $\omega_L \in c_1(L)$ induced from the Fubini-Study metric on $\mathbf{CP}^N$ in the same way as in the earlier discussion. Then we have the following theorem due to [**EGZ**]

THEOREM 17. *Let* X *be a Calabi-Yau variety with canonical singularities. Then for any ample* $\mathbf{Q}$*-line bundle* L, *there exists a unique bounded and* ω_L*-plurisubharmonic function* φ *solving the following Monge-Ampère equation on* X

$$\left(\omega_L + \frac{i}{2}\partial\bar{\partial}\varphi\right)^n = c_L\Omega_{CY}, \tag{10.9}$$

where $c_L \int_X \Omega_{CY} = \int_X \omega_L^n$. *In particular, on* $X \backslash X_{sing}$, $\omega_{CY} = \omega_L + \frac{i}{2}\partial\bar{\partial}\varphi$ *is smooth and*

$$Ricci(\omega_{CY}) = 0. \tag{10.10}$$

10.3. Positivity notions for cohomology classes. Another extension of the theory is the existence of Kähler-Einstein metrics, which are then necessarily singular, on manifolds X whose first Chern class $c_1(K_X)$ is neither zero nor positive definite.

To discuss the classes which are allowed, we recall briefly the definitions of some basic cones in the space of cohomology classes. They were introduced by Demailly [**D1**] and play an important role in his differential geometric approach to positivity problems in algebraic geometry. Let X be a compact Kähler manifold and $\alpha \in H^{1,1}(X, \mathbf{R})$ be a cohomology class. Then

$$\alpha \in \frac{\{\theta : \text{closed } (1,1) \text{ forms}\}}{\{\theta : \text{exact } (1,1) \text{ forms}\}} = \frac{\{T : \text{closed } (1,1) \text{ currents}\}}{\{T : \text{exact } (1,1) \text{ currents}\}} \tag{10.11}$$

We say α is pseudo-effective (psef) if there is a closed $(1,1)$ current $T \in \alpha$ such that $T \geq 0$. We say α is big if there exists $T \in \alpha$ with $T \geq \varepsilon\omega$ for some $\varepsilon > 0$ and some Kähler form ω.

Let PSEF(X) be the set of psef classes and BIG(X) the set of big classes. Then we clearly have BIG$(X) \subseteq$ PSEF(X). Moreover, PSEF(X) is a closed convex cone in the vector space $H^{1,1}(X, \mathbf{R})$ and BIG(X) is an open convex cone. If T is psef then $T + \varepsilon\omega$ is big for all $\varepsilon > 0$. This shows that BIG(X) is precisely the interior of PSEF(X).

Let KAH(X) be the set of Kähler classes in $H^{1,1}(X, \mathbf{R})$. Thus $\alpha \in$ KAH(X) if and only if there exists a Kähler form $\omega \in \alpha$. Thus KAH(X) is an open cone and we clearly have KAH$(X) \subseteq$ BIG(X). This inclusion may be proper. Let NEF(X) be the closure of KAH(X). An element of $\alpha \in$ NEF(X) is called a nef class. In summary,

$$\begin{array}{ccc} \mathrm{BIG}(X) & \subseteq & \mathrm{PSEF}(X) \\ \cup & & \cup \\ \mathrm{KAH}(X) & \subseteq & \mathrm{NEF}(X) \end{array} \tag{10.12}$$

The cones on the left are open and those on the right are their closures.

Let

$$\mathcal{T}(X, \alpha) = \{T \in \alpha : T \text{ a closed } (1,1) \text{ current, } T \geq 0\} \tag{10.13}$$

Then α is pseudo-effective (psef) iff $\mathcal{T}(X, \alpha) \neq \emptyset$. We endow $\mathcal{T}(X, \alpha)$ with the weak topology, so that $T_j \rightharpoonup T$ iff $\int_X T_j \wedge \eta \to \int_X T \wedge \eta$ for all smooth $(n-1, n-1)$ forms η. The space $\mathcal{T}(X, \alpha)$ is compact in the weak topology.

Fix a smooth volume form dV on X and let $L^1(X) = L^1(X, dV)$. For $\theta \in \alpha$ smooth, let $PSH^1(X, \theta) = \{\varphi \in L^1(X) : \theta + \frac{i}{2}\partial\bar\partial\varphi \geq 0\}$ endowed with the $L^1(X)$ topology. The map $\varphi \mapsto \theta + \frac{i}{2}\partial\bar\partial\varphi$ defines $PSH^1(X, \theta)/\mathbf{R} \to \mathcal{T}(X, [\theta])$ a homeomorphism of compact topological spaces. The map $\sup : PSH^1(X, \theta) \to \mathbf{R}$ is continuous (this is Hartogs' lemma). Thus we have a homeomorphism

$$\{\varphi \in PSH^1(X, \theta) : \sup \varphi = 0\} \to \mathcal{T}(X, [\theta]). \tag{10.14}$$

Let X be a compact Kähler manifold and $\alpha \in H^{1,1}(X, \mathbf{R})$ a big class. Fix $\theta \in \alpha$, a closed smooth $(1,1)$ form. Then, by definition, there exists $\varphi \in PSH^1(X, \theta)$ such that $T = \theta + \frac{i}{2}\partial\bar\partial\varphi \geq \varepsilon\omega$ for some Kähler metric ω and some $\varepsilon > 0$. Demailly's theorem says that we may choose φ such that φ has analytic singularities. This means that locally on X,

$$\varphi = c \log \left(\sum_{j=1}^{N} |f_j|^2 \right) + \psi \tag{10.15}$$

where $c > 0$, f_j are holomorphic and ψ is smooth. In particular, the set $Amp(\theta)$ where T is smooth is a Zariski open subset of X.

Let X and θ be as above. Thus θ is big, but it general it will not be positive. We define $V_\theta \in PSH(X, \theta)$, the extremal function of θ (the analogue of the "convex hull") by

$$V_\theta(x) = \sup\{\varphi(x) : \varphi \in PSH^1(X,\theta),\ \sup_X \varphi \leq 0\} \tag{10.16}$$

Thus $V_\theta = 0$ if θ is a Kähler metric.

The extremal function V_θ has a number of nice properties. To describe them, and because V_θ is not bounded in general, we need to extend the definition of Monge-Ampère measures. We shall use the definition that does not charge pluripolar sets, and which can be described as follows.

Let $T_1, \ldots, T_p$ be closed positive $(1,1)$ currents. For any $z \in M$ there there exists an open set U containing z, and pluri-subharmonic functions $u_1, \ldots, u_j$, for which $T_j = \frac{i}{2}\partial\bar\partial u_j$. Let $U_k = \cap_j\{u_j > -k\} \subseteq U$. Then the non-pluripolar product $\langle T_1 \wedge \cdots \wedge T_p\rangle$ is the closed (p,p)-current defined by [**BEGZ**]

$$T_1 \wedge \cdots \wedge T_p|_U = \lim_{k\to\infty} \mathbf{1}_{U_k} \bigwedge_{j=1}^{p} \frac{i}{2}\partial\bar\partial \max(u_j, -k) \tag{10.17}$$

where $\mathbf{1}_{U_k}$ denotes the characteristic function of U_k. The non-pluripolar product coincides with the Bedford-Taylor definition of $T_1 \wedge \cdots \wedge T_p$ if the potentials are all bounded. We still denote $T \wedge \cdots \wedge T$ by T^n. We can now describe the properties of extremal functions:

THEOREM 18. *Let X be a compact Kähler manifold and θ a big $(1,1)$ form. Let V_θ be the extremal function of θ. Then*

1. $V_\theta \in PSH^1(X,\theta)$
2. *V_θ has minimal singularities: if $\varphi \in PSH^1(X,\theta)$ then $\varphi \leq V_\theta + C$ for some $C \geq 0$.*
3. *On the set $Amp(\theta)$ V_θ is continuous, and $\frac{i}{2}\partial\bar\partial V_\theta$ is locally bounded.*
4. *$(\theta + \frac{i}{2}\partial\bar\partial V_\theta)^n$ has L^∞ density with respect to dV. In particular, if $\varphi \in PSH^1(X,\theta)$ then $\int_X |\varphi|\,(\theta + \frac{i}{2}\partial\bar\partial V_\theta)^n < \infty$.*
5. *V_θ is maximal with respect to θ, that is*

$$\left(\theta + \frac{i}{2}\partial\bar\partial V_\theta\right)^n = \mathbf{1}_{\{V_\theta=0\}}\theta^n \tag{10.18}$$

10.4. Prescribing the Monge-Ampère measure. If $[\theta]$ is a big class and $T_1, \ldots, T_n \in \mathcal{T}(X,[\theta])$ then $\int_X\langle T_1 \wedge \cdots \wedge T_n\rangle \leq \int_X [\theta]^n$. If we have $T_1 = \cdots T_n = T = \theta + \frac{i}{2}\partial\bar\partial\varphi$ and if equality holds, we say T has full mass. Define

$$\mathcal{T}^0(X,\theta) = \{T \in T(X,\theta) : T \text{ has full mass}\}$$

$$\mathcal{T}^1(X,\theta) = \left\{T = \theta + \frac{i}{2}\partial\bar\partial\varphi \in \mathcal{T}^1(X,\omega) : \int_X |\varphi| \left(\theta + \frac{i}{2}\partial\bar\partial\varphi\right)^n < \infty\right\} \tag{10.19}$$

Let $\mathcal{M}_X$ be the space of probability measures on X, let $\mathcal{M}_X^0$ consist of those measures which take no mass on pluri-polar sets, and let $\mathcal{M}_X^1$ consist of those measures of "finite energy" (to be defined below). Then it has been shown by Guedj and Zeriahi [**GuZe**] and by Berman, Boucksom, Guedj, and Zeriahi [**BBGZ**] respectively that the map $T \mapsto T^n$ defines bijections:

$$\mathcal{T}^0(X,\omega) \to \mathcal{M}_X^0, \quad \mathcal{T}^1(X,\omega) \to \mathcal{M}_X^1. \tag{10.20}$$

10.5. Singular KE metrics on manifolds of general type. Before stating the results, it is convenient to recast the Kähler-Einstein equation for negative curvature in a slightly different form from usual. Let X be a compact Kähler manifold, and let K_X be its canonical bundle. If $c_1(K_X)$ is a Kähler class, then the Aubin-Yau theorem says that for any Kähler metric ω there is a unique smooth $\psi \in PSH(X, -Ricci(\omega))$ such that

$$\left(-Ricci(\omega) + \frac{i}{2}\partial\bar{\partial}\psi\right)^n = e^{\psi}\omega^n \tag{10.21}$$

To put this in the usual form, let $\eta = -Ricci(\omega) + \frac{i}{2}\partial\bar{\partial}\psi$. Then $\eta > 0$ and

$$Ricci(\omega) - Ricci(\eta) = \frac{i}{2}\partial\bar{\partial}\psi = Ricci(\omega) + \eta \tag{10.22}$$

which implies $Ricci(\eta) = -\eta$.

Now assume that $c_1(K)$ is big and nef. Then Tsuji [**Ts**] proved that there is a subvariety $Z \subseteq X$ and a smooth function $\psi \in C^\infty(X\backslash Z)$ such that $\eta = -Ricci(\omega) + \frac{i}{2}\partial\bar{\partial}\psi > 0$ and such that (10.21) holds on $X\backslash Z$. Thus $Ricci(\eta) = -\eta$ on $X\backslash Z$.

Tian-Zhang [**TZ**] proved that ψ extends to a locally bounded $\psi \in PSH$ $(X, -Ricci(\omega))$ satisfying $\int(-Ricci(\omega) + \frac{i}{2}\partial\bar{\partial}\psi)^n = \int(-Ricci(\omega))^n$ (i.e., ψ has full MA measure) and that (10.21) holds on all of X.

Now assume that $c_1(X)$ is big. Then [EGZ] showed that there is a unique $\psi \in PSH^1(-Ricci(\omega))$ of full Monge-Ampère measure such that (10.21) holds. The [EGZ] proof uses the existence of canonical models. Tsuji described an interesting approach to proving the existence of a singular Kähler-Einstein metric without resorting to the existence of a canonical model in [**Ts**]. Then Song-Tian [**ST09**] gave an independent proof, via the Kähler-Ricci flow. A new proof was also given by [BEGZ] which used a generalized comparison principle. More recently, a proof using variational methods was given in [BBGZ].

11. Variational Methods for Big Cohomology Classes

A basic property of the Monge-Ampère determinant is that it can be interpreted as the variational derivative of a concave energy functional. In fact, if ω is a smooth Kähler form on a compact complex manifold X, and we set

$$(11.1)\qquad E_\omega(\varphi) = \frac{1}{n+1}\sum_{j=0}^{n}\int_X \varphi\left(\omega + \frac{i}{2}\partial\bar{\partial}\varphi\right)^j \wedge \omega^{n-j}$$

for $\omega_\varphi \equiv \omega + \frac{i}{2}\partial\bar{\partial}\varphi > 0$. Then for smooth and small variations $\delta\varphi$, we have

$$(11.2)\qquad \delta E = \int_X \delta\varphi\, \omega_\varphi^n \quad \text{and} \quad \delta^2 E = -n\int_X d\delta(\varphi)\wedge d^c\delta\varphi \wedge \omega_\varphi^{n-1}$$

This shows that $\frac{\delta E}{\delta\varphi} = \omega_\varphi^n$ and that E is concave.

The functional $E(\varphi)$ is actually equal to $E(\varphi) = -(J(\varphi) - \int_X \varphi\omega^n)$, where

$$(11.3)\qquad J(\varphi) = \sum_{j=0}^{n-1}\frac{n-j}{n+1}\int_X i\partial\varphi\wedge\bar{\partial}\varphi\wedge\omega_\varphi^{n-1-j}\wedge\omega^j.$$

This relation will play an important role below. The functional E is sometimes denoted by $-(n+1)VF^0$ in the literature, where F^0 is the Aubin-Yau energy functional.

The goal of this section is to describe some recent work of Berman, Boucksom, Guedj, and Zeriahi [**BBGZ**] taking this variational viewpoint further. Their approach, which works in the generality of big cohomology classes, allows one to use direct methods of the calculus of variations to obtain solutions to a variety of Monge-Ampère equations.

11.1. Finite dimensional motivation. We start with a finite dimensional model for the method introduced in [**BBGZ**]. Let $P \subseteq \mathbf{R}^n$ be a convex domain and $E : P \to \mathbf{R}$ be a strictly concave smooth function. Then $\nabla E : TP \to \mathbf{R}$ where $TP = P \times \mathbf{R}^n$ is the tangent bundle of P. In other words, $\nabla E : P \to (\mathbf{R}^n)^*$ where $(\mathbf{R}^n)^*$ is the dual of $\mathbf{R}^n$. Since E is strictly concave, $\nabla^2 E(\varphi) : \mathbf{R}^n \to (\mathbf{R}^n)^*$ is strictly negative definite for all $\varphi \in P$ and ∇E is a diffeomorphism of P onto a convex domain $\mathcal{M} \subseteq (\mathbf{R}^n)^*$.

More generally, suppose E is concave (but not strictly concave) and that there is a flat direction, that is, an element $\varphi_0 \in P$ such that $E(\varphi + t\varphi_0) = E(\varphi) + t$ for all $t \in \mathbf{R}$ and let $\mathcal{T}^1 = P/(\mathbf{R}\cdot\varphi_0)$ which we view as a domain in the vector space $W = \mathbf{R}^n/(\mathbf{R}\cdot\varphi_0)$. Then $\nabla E : P \to (\mathbf{R}^n)^*$ is invariant under $\mathbf{R}\cdot\varphi_0$ so

$$(11.4)\qquad \nabla E : \mathcal{T}^1 \to \mathcal{M}^0$$

where $\mathcal{M}^0 \subseteq (\mathbf{R}^n)^*$ is the $n-1$ dimensional affine space

$$(11.5)\quad \mathcal{M}^0 = \{\mu \in \mathbf{R}^n : \langle\varphi_0, \mu\rangle = 1\} = \text{ the elements of } (\mathbf{R}^n)^* \text{ of } \textit{full mass}$$

Now we want to impose the following condition: E is strictly concave in every direction other than $\mathbf{R}\cdot\varphi_0$. There are several equivalent ways of making this condition precise:

LEMMA 16. *Let $E : P \to \mathbf{R}$ be a concave function and assume that $E(\varphi + t\varphi_0) = E(\varphi) + t$ for all $t \in \mathbf{R}$. Then the following conditions are equivalent*

(1) *The map $t \mapsto E((1-t)\psi_0 + t\psi_1)$ is strictly concave whenever $\psi_1 - \psi_0 \notin \mathbf{R} \cdot \varphi_0$.*
(2) *The negative definite map $\nabla^2 E(\varphi) : W \to W^*$ is strictly negative definite if $\varphi \in \mathcal{T}$.*
(3) *The map $\nabla E : \mathcal{T}^1 \to \mathcal{M}$ is a diffeomorphism where $\mathcal{M} \subseteq \mathcal{M}^0$ is the image of ∇E.*
(4) *Fix $\psi_0 \in \mathcal{T}^1$, let $\mu_0 = \nabla E(\psi_0)$, $L_0(\varphi) = \varphi \cdot \mu_0$ and define $J_0 : \mathcal{T}^1 \to [0, \infty)$ by*

$$J_{\psi_0}(\varphi) = E(\psi_0) + \nabla E(\varphi_0) \cdot (\varphi - \psi_0) - E(\varphi) = L_0(\varphi) - E(\varphi) + C \tag{11.6}$$

Then $J_0 = J_{\psi_0} : \mathcal{T}^1 \to [0, \infty)$ is strictly convex.

The proof of the lemma is easy and will be omitted. Henceforth, we shall assume that E satisfies any one of the equivalent conditions enumerated in Lemma 1.

The function J_0 is proper so the sets $\{J_0 \leq C\} \subseteq \mathcal{T}^1$ are compact and exhaust $\mathcal{T}^1$. In the case $\mathcal{T}^1$ is bounded, we can define $\mathcal{T}$ to be the closure of $\mathcal{T}^1$ and extend $J_0 : \mathcal{T} \to [0, \infty]$ as a continuous map between compact spaces.

For $\mu \in \mathcal{M}^0$ we let

$$E^*(\mu) = \sup_{\varphi \in \mathcal{T}} (E(\varphi) - \varphi \cdot \mu) = \sup_{\varphi \in \mathcal{T}} F_\mu(\varphi) \ \in (-\infty, \infty] \tag{11.7}$$

We let $\mathcal{M}^1 \subseteq \mathcal{M}^0$ be the elements with *finite energy*:

$$\mathcal{M}^1 = \{\mu \in \mathcal{M}^0 : E^*(\mu) < \infty\}$$

If $\mu \in \mathcal{M}$ then, by definition, there exists $\varphi \in \mathcal{T}^1$ such that $\nabla E(\varphi) = \mu$ and hence φ is a critical point of F_μ. Since F_μ is strictly concave we conclude that F_μ achieves its maximum at φ and thus

$$\mathcal{M} \subseteq \mathcal{M}^1 \subseteq \mathcal{M}^0 \tag{11.8}$$

Conversely, if $\mu \in \mathcal{M}^1$ then $\sup F_\mu < \infty$ and $\mu \in \mathcal{M}$ if and only if F_μ *achieves* its sup (i.e. there exists $\varphi \in \mathcal{T}^1$ such that $F_\mu(\varphi) = E^*(\mu)$).

The inclusion $\mathcal{M} \subseteq \mathcal{M}^1$ may be strict. We give two simple examples:

Example A. Suppose $\mathcal{T} = \mathbf{R}$ and $E : \mathbf{R} \to \mathbf{R}$ is smooth and concave, and $E(x) = \log x - x$ for x large. Then $\mathcal{M} = \mathcal{M}^1 = (-1, 1)$. So for example A,

$$\mathcal{M} = \mathcal{M}^1 \tag{11.9}$$

Example B. Suppose $\mathcal{T} = \mathbf{R}$ and $E : \mathbf{R} \to \mathbf{R}$ is smooth and convex, and $E(x) = -x - \frac{1}{x}$ for x large. Then $\mathcal{M} = (-1, 1)$ but $\mathcal{M}^1 = [-1, 1]$. Thus in example B property (11.9) fails.

There is a simple criterion for guaranteeing that (11.9) holds. Suppose $J : \mathcal{T} \to [0, \infty)$ is a proper function and let $\mu \in \mathcal{M}^1$. We say that F_μ is coercive with respect to J if there exists $\varepsilon, C > 0$, depending on μ, such that

$$L_\mu - E = -F_\mu \geq \varepsilon J - C = \varepsilon(L_{\mu_0} - E) - C' \tag{11.10}$$

We say F is J-proper if $-F_\mu \to \infty$ as $J \to \infty$. Clearly J-coercive implies J-proper. Note that (11.10) holds trivially if $\varepsilon = 0$: If $\mu_0 = \nabla E(\psi_0)$ then $J_0(\varphi) = F_{\mu_0}(\psi_0) - F_{\mu_0}(\varphi) \geq 0$.

LEMMA 17. *Let $\mu \in \mathcal{M}^1$ and assume that F_μ is J-proper for some J. Then $\mu \in \mathcal{M}$.*

To see this, we fix $\varphi_0 \in \mathcal{T}$ and let $A = J(\varphi_0) + 1$. Then the set $\mathcal{T}_A = \{J \leq A\}$ is compact so there exists $\hat{\varphi}$ such that $F_\mu(\hat{\varphi}) \geq F_\mu(\varphi)$ for all $\varphi \in \mathcal{T}_A$ and hence for all $\varphi \in \mathcal{T}$. Thus F_μ achieves its sup at $\hat{\varphi}$ which implies $\nabla F_\mu(\hat{\varphi}) = 0$, that is, $\nabla E(\hat{\varphi}) = \mu$. This shows $\mu \in \mathcal{M}$.

In order to apply Lemma 17 we need to find an appropriate J. It turns out that if F_μ is J-coercive for some J, then it is coercive for the function J_0 constructed in (11.6):

LEMMA 18. *Let $\mu \in \mathcal{M}^1$. Then the following are equivalent*

(1) $\mu \in \mathcal{M}$
(2) *F_μ is J_0-coercive.*
(3) *F_μ is J-coercive for some exhaustion function J.*
(4) *F_μ is J-proper for some exhaustion function J.*

PROOF. We need only show that 1) implies 2). Let $\mu \in \mathcal{M}$. We must show that for some $\varepsilon, C > 0$ the following holds.

$$-F_\mu(\varphi) \geq \varepsilon(F_{\mu_0}(\psi_0) - F_{\mu_0}(\varphi)) - C \tag{11.11}$$

which we rewrite as

$$\varphi \cdot \mu_\varepsilon - E(\varphi) = \varphi \cdot \frac{(\mu - \varepsilon\mu_0)}{1 - \varepsilon} - E(\varphi) \geq \frac{\varepsilon F_{\mu_0}(\psi_0) - C}{1 - \varepsilon} \tag{11.12}$$

But $\mathcal{M}$ is an open set so for ε sufficiently small, $\mu_\varepsilon \in \mathcal{M}$ which implies $\varphi \cdot \mu_\varepsilon - E(\varphi)$ is bounded below. □

Now fix $\psi_0 \in \mathcal{T}$ and define J_0 as in (11.6).

LEMMA 19. *$\mathcal{M} = \mathcal{M}^1$ if and only if F_μ is J_0-coercive for all $\mu \in \mathcal{M}^1$.*

Now we record a condition on J_0 that guarantees F_μ is coercive for all μ. This key condition will hold in the infinite dimensional setting and is used in the proof of coerciveness.

LEMMA 20. *Suppose that J_0 grows quadratically or, more generally, suppose that*

$$|J_0(t\varphi)| \le C(t^p|J(\varphi)| + 1) \quad \text{if } t \le 1 \tag{11.13}$$

for some $p > 1$. Then F_μ is J_0-coercive for all $\mu \in \mathcal{M}^1$.

Observe that we can rewrite the coercive condition (11.10) as follows:

$$L_{\mu_0} - L_\mu \le (1-\varepsilon)J_0 + C_1 \tag{11.14}$$

Now (11.13) implies that for arbitrary $\varphi \in \mathcal{T}$, we have $\varphi_p = \frac{\varphi}{|J(\varphi)|^{1/p}} \in \{J_0 \le C_2\}$ which is a compact set. This implies

$$\frac{|L_{\mu_0} - L_\mu|(\varphi)}{|J_0(\varphi)|^{1/p}} = |L_{\mu_0} - L_\mu|(\varphi_p) \le C_3 \tag{11.15}$$

which clearly implies (11.14).

11.2. The infinite dimensional setting. We describe the work of [**BBGZ**]. Let θ be a big $(1,1)$ form on a compact Kähler manifold X. Let $P = PSH^1(X,\theta)$ and $\mathcal{T} = \mathcal{T}(X,\theta)$ and $\mathcal{M}'$ the space of positive metrics on X. Our goal is to define a concave function $E : PSH^1(X,\theta) \to \mathbf{R}$ with the property: $\nabla E : P \to \mathcal{M}'$ is the map $\varphi \mapsto MA(\varphi)$, where $MA(\varphi)$ is the Monge-Ampère measure, defined by the non-pluripolar product. Moreover, we will show that if $\varphi_0 = 1$, the constant function, then φ_0 is a flat direction for E and $\mathcal{T}$ is a complement to $\mathbf{R}\varphi_0$ in P. Then, as in the finite dimensional case, we obtain a map $\nabla E : \mathcal{T} \to \mathcal{M}'$. Also, in analogy with the finite dimensional case, we define the function $F_\mu(\varphi) = E(\varphi) - \int_X \varphi\, d\mu$ for $\mu \in \mathcal{M}'$. Our goal is to prove (11.9) and this will be done by establishing (11.10) for a suitably chosen J.

The functional $E(\varphi)$ has been defined in (11.1) for $\theta = \omega$ a Kähler form, and φ a smooth form in $PSH^1(X,\theta)$. To extend it to θ a big form and $\varphi \in PSH^1(X,\theta)$, we proceed as follows: Define $E(\varphi) = \inf\{E(\psi) : \psi \in PSH^1(X,\omega) \cap C^\infty(X) \text{ and } \psi \ge \varphi\}$.

Now let θ be a big $(1,1)$ form (not necessarily Kähler). Let $\varphi \in PSH^1(X,\theta)$ and assume φ has minimal singularities (i.e. that $\varphi - V_\theta$ is bounded). Define

$$E(\varphi) = \frac{1}{n+1}\sum_{j=0}^{n}\int_X (\varphi - V_\theta)\left(\theta + \frac{i}{2}\partial\bar\partial\varphi\right)^j \left(\theta + \frac{i}{2}\partial\bar\partial V_\theta\right)^{n-j} \tag{11.16}$$

Of course this coincides with (11.1) in the case where $\theta = \omega$ is Kähler (since in that case, $V_\theta = 0$). If φ is arbitrary, then again extend using the monotonicity property of E as before: $E(\varphi) = \inf\{E(\psi) : \psi \ge \varphi \text{ and } \psi \text{ has minimal singularities}\}$.

We wish to implement the finite dimensional program in the infinite dimensional setting. Since E may assume the value $-\infty$, we must, at the outset, restrict E to the set $\mathcal{E}^1 = \{\varphi \in PSH^1(X,\theta) : |E(\varphi)| < \infty\}$ so that

$E:\mathcal{E}^1 \to \mathbf{R}$ is concave. If $\varphi_0 = 1$, then φ_0 is a flat direction so, as in the finite dimensional program, we let $\mathcal{T}^1 = \mathcal{E}^1/\mathbf{R}$ Again, we let $F_\mu(\varphi) = E(\varphi) - \int_X (\varphi - V_\theta)\, d\mu = E(\varphi) - L_\mu(\varphi)$ and let $\mathcal{M}^1 \subseteq \mathcal{M}^0$ be those measures with finite energy.

11.3. Statement of theorems and sketch of proofs.

THEOREM 19. *([**BBGZ**]) Let θ be a big $(1,1)$ form on a compact complex manifold X. If $\varphi \in \mathcal{T}^1(X,\theta)$ then $\nabla E(\varphi) = MA(\varphi)$ has full mass and finite energy, that is, $MA(\varphi) \in \mathcal{M}^1$. Conversely, if $\mu \in \mathcal{M}^1$ then there exists a unique $\varphi \in \mathcal{T}^1$ such that $MA(\varphi) = \mu$. Moreover, the solution φ satisfies the following bound:*

$$\frac{1}{n}E^*(\mu) \leq J_0(\varphi) \leq nE^*(\mu) \tag{11.17}$$

where $J_0 = J_{\psi_0}$ is defined in (11.6).

We sketch the proof of the converse, using the finite dimensional program as our guide. Recall that

$$\begin{aligned} J_\psi(\varphi) &= E(\psi) - E(\varphi) + \int_X (\varphi - \psi) MA(\psi) \\ &= \sum_{j=0}^{n-1} \frac{j+1}{n+1} \int_X \frac{i}{2} \partial(\varphi - \psi) \wedge \bar{\partial}(\varphi - \psi) \wedge \theta_\psi^j \wedge \theta_\varphi^{n-1-j} \end{aligned} \tag{11.18}$$

The first equality is a definition and the second follows via integration by parts. In particular, we see

$$n^{-1} J_\psi(\varphi) \leq J_\varphi(\psi) \leq n J_\psi(\varphi) \tag{11.19}$$

Now we can establish the bound Now we can establish the bound (11.17): Assume $\varphi \in \mathcal{E}^1$ $MA(\varphi) = T^n = \mu$, where $T = \theta + \frac{i}{2}\partial\bar{\partial}\varphi$. Then (11.18) and (11.19) imply

$$\frac{1}{n} J_0(\varphi) \leq J_\varphi(V_0) = E(\varphi) + \int_X (V_0 - \varphi) MA(\varphi) = E(\varphi) - \int_X (\varphi - V_0)\, d\mu = F_\mu(\varphi)$$

Since $MA(\varphi) = \mu$ one shows, in analogy with the finite dimensional picture, that $E^*(\mu) = \sup_{\psi \in \mathcal{E}^1} F_\mu(\psi) = F_\mu(\varphi)$. This proves (11.17).

Now let us fix $\mu \in \mathcal{M}^1$. We wish to prove the existence of $T \in \mathcal{T}^1$ such that $T^n = \mu$. The first step is to prove that F_μ is J_0 coercive, where J_0 is defined with respect to the potential $\psi_0 = V_0$, as in (11.6). To do this, we wish to use Lemma 20, which means that we must prove that E grows quadratically. But this follows easily from the definitions (here we assume

$V_\theta = 0$ for simplicity): If $\sup_X \varphi = 0$ then

$$|E(t\varphi)| = \frac{1}{n+1}\left|\sum_{j=0}^{n}\int (t\varphi)\left(t\left(\omega + \frac{i}{2}\partial\bar\partial\varphi\right) + (1-t)\omega\right)^j \omega^{n-j}\right|$$
$$\leq c(n)[t^2 E(\varphi) + 1] \tag{11.20}$$

For $C > 0$ let $\mathcal{E}_C$ be the compact set $\mathcal{E}_C = \{\varphi \in \mathcal{E}^1 : E(\varphi) \geq C\}$. Since F_μ is coercive, there exists $C > 0$ such that $\sup_{\varphi\in\mathcal{E}^1} F_\mu = \sup_{\varphi\in\mathcal{E}_C} F_\mu$.

The first main difficulty in implementing the finite dimensional program is proving that F_μ is upper semi continuous on $\mathcal{E}_C$. To do this, [**BBGZ**] restrict first to the case where $\mu \in \mathcal{C} = \{\mu \in \mathcal{M}^1 : \mu \leq A{\cdot}Cap$ for some $A > 0\}$ (here Cap is the pluri-subharmonic capacity).

Fix $\mu \in C$. Since E is easily seen to be usc, it suffices to show that L_μ is continuous. Let $T : K \to L^1(X, d\mu)$ be the map $T(\varphi) = \varphi - V_\theta$ (here $K \subseteq \mathcal{E}^1$ is any compact convex subset). It's not hard to show that $T(K)$ is closed and that it has a closed graph. If we could prove $T(K)$ is compact, then we would be done. To see this, let $\varphi_j \to \varphi$ in K. If $T(K)$ is compact then $T(\varphi_j) \to f$ (after passing to a subsequence). But the closed graph property implies $f = \varphi$. Thus $T(\varphi_j) \to T(\varphi)$ so T is continuous which implies L_μ is continuous.

Instead of proving that $T(K)$ is compact, we prove something weaker, namely that $T(K) \subseteq L^1(X, d\mu)$ is "convex combination compact". Recall that if B is a Banach space and if $T \subseteq B$ is closed and convex, then T is convex combination compact if for every sequence $\tau_1, \tau_2, \ldots \in T$ there exists $\tau_1', \tau_2', \ldots \in T$ such that τ_j' is a finite convex combination of $\tau_j, \tau_{j+1}, \ldots$ and such that τ_j' converges. Observe that showing $T(K)$ is convex combination compact suffices for our purposes: Let $\varphi_j \to \varphi \in K$. Then one shows $L_\mu(\varphi_j)$ is bounded and hence $L_\mu(\varphi_j) = \int T(\varphi_j) \to \ell \in \mathbf{R}$ (after passing to a subsequence). On the other hand, the convex combination compactness of T implies there exist $\psi_1, \psi_2, \ldots \in K$ such that ψ_j is a convex combination of $\varphi_j, \varphi_{j+1} \ldots$ and $T(\psi_j) \to f$ for some $f \in T(K)$. Since $\psi_j \to \varphi$ we see that $f = T(\varphi)$ by the closed graph property. Thus

$$\int T(\psi_j)\, d\mu \to \int f\, d\mu = \int T(\varphi)\, d\mu \tag{11.21}$$

On the other hand, $\lim_j \int T(\psi_j)\, d\mu = \lim_j \int T(\varphi_j)\, d\mu$ so $\int T(\varphi_j) \to \int T(\varphi)$. This shows L_μ is continuous on K.

To show that $T(K)$ is convex combination compact, it suffices to prove $T(K)$ is weakly compact (by the Hahn-Banach Theorem). On the other hand, the Dunford-Pettis theorem says that to show that $T(K)$ is weakly compact, it suffices to show that $T(K)$ is uniformly integrable, that is, there exists $\varepsilon_k \downarrow 0$ such that

$$\left| \int_{T(\varphi)<-k} T(\varphi)\, d\mu \right| \leq \varepsilon_k \quad \text{for all } \varphi \in K \tag{11.22}$$

But $\mu \in \mathcal{C}$ implies

$$\int_0^\infty t\mu\{t < V - \varphi\}\, dt \leq A \int_0^\infty tCap\{t < V - \varphi\}\, dt \leq C_1 \tag{11.23}$$

Thus $T(K)$ is a bounded subset of $L^2(X, d\mu)$ and hence, by Hölder's inequality, $T(K)$ is uniformly integrable.

The second main difficulty is to prove the Euler-Lagrange equation is satisfied at the point where F_μ reaches its maximum. The problem is that the minimum φ may only satisfy $\omega + \frac{i}{2}\partial\bar{\partial}\varphi \geq 0$, and a variation $\varphi_t = \varphi + \delta\varphi$ may no longer satisfy this condition even if $\delta\varphi$ is smooth and small. This is overcome using a technique of [**BB**] which we now describe. Define for each u an upper semi-continuous function on X, its θ-psh envelope,

$$P(u) = \sup\{\varphi \in PSH^1(X, \theta) : \varphi \leq u \text{ on } X\} \tag{11.24}$$

Then $P(u) \in PSH^1(X, \theta)$ so the function $g(t) = E(P(\varphi + t\delta\varphi)) - L_\mu(\varphi + t\delta\varphi)$ is now well-defined for $t \in \mathbf{R}$, and has a maximum at $t = 0$. It is shown in [**BBGZ**] that we still have $0 = g'(0) = \int_X \delta\varphi(MA(\varphi) - d\mu)$. Thus the equation $MA(\varphi) = \mu$ has a solution in the case where $\mu \in \mathcal{C}$.

We now remove the assumption $\mu \in \mathcal{C}$. Thus we let μ be a non-pluripolar probability measure with finite energy. Our goal is to prove that there exists $\varphi \in \mathcal{T}^1$ such that $MA(\varphi) = \mu$. To do this, we make use of the following lemma of Cegrell [**Ceg**]:

LEMMA 21. *Let μ be a non-pluripolar probability measure. Then there exists $\nu \in \mathcal{C}$ and $f \in L^1(X, \nu)$ such that $\mu = f\nu$.*

Write $\mu = f\nu$ as in the lemma. For $k > 0$ choose $\varepsilon_k \geq 0$ so that $\mu_k = (1 + \varepsilon_k)\min(f, k)\nu$ is a probability measure. Then $\mu_k \leq 2k\, Cap$ so $\mu \in \mathcal{C}$. Thus, by what has been proved thus far, $\mu_k = T_k^n$ for some $T_k \in \mathcal{T}^1$. Next we observe that $\mu_k \leq 2\mu$. Using the fact that $E^*(\mu) < \infty$, one shows that $E^*(\mu_k) \leq C < \infty$ for some $C > 0$. The bound (11.17) then implies that the T_k all lie in the compact set $\{J_0 \leq nC\}$. Thus, after passing to a subsequence, we conclude $T_k \to T$ for some $T \in \mathcal{T}^1$. Since $T_k^n = \mu_k$ we can take the limit as $k \to \infty$ to conclude $T^n = \mu$ (this follows by Fatou's lemma in the case $n = 1$, and a generalization of Fatou's lemma, due to [**BEGZ**], in the case $n > 1$).

We remark that if μ is a non-pluripolar probability measure (not necessarily of finite energy), then one can still apply Lemma 21 to conclude $\mu = f\nu$ and one can still construct $\mu_k = (1 + \varepsilon_k)\min(f, k)\nu$ as above. Then Theorem 19 implies the existence of $T_k \in \mathcal{T}^1$ such that $T_k^n = \mu_k$. Since μ is not assumed to have finite energy, we cannot conclude that the T_k lie in a compact subset of $\mathcal{T}^1$. On the other hand, we have $T_k \subseteq \mathcal{T}^1 \subseteq \mathcal{T}$ and $\mathcal{T}$

is compact. Thus, after passing to a subsequence, $T_k \to T$ for some $T \in \mathcal{T}$. Then [**BBGZ**] show that one can again take the limit as $k \to \infty$ to conclude $T^n = \mu$. This gives a variational proof of the following theorem of [**BEGZ**]:

THEOREM 20. *Let μ be a non-pluripolar probability measure on X. Then there exists $T \in \mathcal{T}(X,\theta)$ such that $T^n = \mu$.*

We close by describing one more application of the variational method in [**BBGZ**].

THEOREM 21. *Let (X,ω) be a manifold of general type and let $\theta = -Ricci(\omega)$. Then there exists $\psi \in PSH^1(X,\theta)$ such that*

$$\left(\theta + \frac{i}{2}\partial\bar{\partial}\psi\right)^n = e^{\psi}\omega^n \tag{11.25}$$

To prove this theorem, we consider the functional $F = E - L$ where $L(\varphi) = \frac{1}{2}\log\int_X e^{\varphi}\omega^n$ and follow the same three steps as in the proof of Theorem 19. The first step is to show that F is J_0 coercive. The second step is to show that L is continuous (which implies that F is upper semi-continuous). And the third step is to prove that the critical points for F are solutions to (11.25).

Steps one and three proceed exactly as in the proof of Theorem 19. Thus we restrict ourselves to step two, which is much easier than the corresponding step in Theorem 19. Indeed, if $\varphi_j \to \varphi$ is a convergent sequence in $PSH^1(X,\theta)$ then after passing to a subsequence, $\varphi_j \to \varphi$ almost everywhere. On the other hand, Hartogs' lemma implies that φ_j is bounded above. Thus $\int_X e^{\varphi_j}\omega^n \to \int_X e^{\varphi}\omega^n$ and hence L is continuous.

We observe that Theorem 21 follows from Theorem 16. To see this, let $\pi : X \to X_{\rm can}$ be the canonical model of X and choose $\varphi_{\rm can}$ to be the solution of (10.8)

$$\left(\omega + \frac{i}{2}\partial\bar{\partial}\varphi_{\rm can}\right)^n = e^{\varphi_{\rm can}}\Omega. \tag{11.26}$$

where $\frac{i}{2}\partial\bar{\partial}\log\Omega = \omega$. Choose Θ such that $\frac{i}{2}\partial\bar{\partial}\log\Theta = \theta$. Let $\tilde{\Omega} = \pi^*\Omega$ and $\tilde{\varphi}_{\rm can} = \varphi_{\rm can}\circ\pi$. Define ψ by the equation $e^{\tilde{\varphi}_{\rm can}}\tilde{\Omega} = e^{\psi}\Theta$. Applying π^* to both sides of (11.26) we obtain (11.25).

The variational method establishes the existence of generalized solutions to the complex Monge-Ampère equation. It is then important to determine when the generalized solution is actually smooth. One such result is the theorem of Szekelyhidi-Tosatti [**ST**] which asserts the smoothness of the generalized solution when it is known to be bounded and the right hand side is smooth.

It may also be noteworthy that the above solutions to the Monge-Ampère equation can be obtained as limits of the critical points of certain naturally-defined finite-dimensional analogues of the infinite-dimensional functionals.

In the Fano case, the proof makes use of the Moser-Trudinger inequality proved in [**PSSW**]. Finite-dimensional approximations also play a major role in the construction of solutions to the homogeneous Monge-Ampère equation in § 13.

12. Uniqueness of Solutions

It is a remarkable fact that the map $\mathcal{T}^0(X,\theta) \to \mathcal{M}_X^0$, defined by $T \mapsto T^n$, is bijective. The surjectivity is part of Theorem 20. The injectivity was proved in [**BEGZ**], by adapting the proof of Dinew [**Di09**], who proved the injectivity in the Kähler case. We give here a slightly streamlined version of the proof of Dinew.

THEOREM 22. *Let (X,ω) be a compact Kähler manifold and $\mathcal{E}(X,\omega) \subseteq PSH^1(X,\omega)$ denote the potentials φ such that $(\omega + \frac{i}{2}\partial\bar\partial\varphi)^n$ has full mass. Let $\varphi,\psi \in \mathcal{E}(X,\omega)$. Assume $(\omega + \frac{i}{2}\partial\bar\partial\varphi)^n = (\omega + \frac{i}{2}\partial\bar\partial\psi)^n$. Then $\varphi - \psi$ is constant.*

To prove the theorem, we wish to make use of the comparison principle. If we apply it directly to $V = \{\varphi < \psi\}$ we get $\int_V \omega_\psi^n \le \int_V \omega_\varphi^n$, which is not useful (in fact, the inequality is an equality since $\omega_\varphi^n = \omega_\psi^n$). Instead, we shall apply the comparison principle to the set $V_\varepsilon(\theta,\rho) = \{(1-\varepsilon)\varphi + \varepsilon\theta < (1-\varepsilon)\psi + \varepsilon\rho\}$ where θ, ρ are potentials to be chosen later. We obtain, for T a positive closed current, and $k \ge 1$,

$$\int_{V_\varepsilon} T \wedge ((1-\varepsilon)\omega_\psi + \varepsilon\omega_\rho)^k \le \int_{V_\varepsilon} T \wedge ((1-\varepsilon)\omega_\varphi + \varepsilon\omega_\theta)^k \tag{12.1}$$

Assume $\int_{V_\varepsilon} T \wedge \omega_\psi^k = \int_{V_\varepsilon} T \wedge \omega_\varphi^k$. Then the leading terms cancel so

$$\int_{V_\varepsilon} T \wedge \omega_\psi^{k-1} \wedge \omega_\rho \le \int_{V_\varepsilon} T \wedge \omega_\varphi^{k-1} \wedge \omega_\theta + O(\varepsilon) \tag{12.2}$$

We shall also need the following generalization of Lemma 7 proved in Dinew [**Di09a**]:

THEOREM 23. *Let $\varphi,\psi \in \mathcal{E}(X,\omega)$ and μ a positive non-pluripolar measure. Assume that $\omega_\varphi^n \ge f d\mu$ and $\omega_\psi^n \ge g d\mu$ for some non-negative $f, g \in L^1(d\mu)$. Then for $0 \le k \le n$*

$$\omega_\varphi^k \wedge \omega_\psi^{n-k} \ge f^{\frac{k}{n}} g^{\frac{n-k}{n}} d\mu \tag{12.3}$$

In particular, if $\omega_\varphi^n = \omega_\psi^n$, then $\omega_\varphi^n = \omega_{t\varphi+(1-t)\psi}^n$ for all $t \in [0,1]$.

We return to the proof of Theorem 22 and follow the argument in [**Di09**]. Let $\mu = \omega_\varphi^n$ and define $f : \mathbf{R} \to [0,1]$ by $f(t) = \mu(\{\varphi < \psi + t\}$. Then f is left continuous. Moreover, since f is increasing, there is a countable set Σ such that f is continuous on $\mathbf{R}\backslash\Sigma$.

The key step in the proof is to show that Image$(f) = \{0,1\}$: Assume not. Then there exists $\alpha \in \mathbf{R}\backslash\Sigma$ such that $0 < f(\alpha) < 1$. To see this, let $\beta \in \mathbf{R}$ satisfy $0 < f(\beta) < 1$. Now choose an increasing sequence $t_j \in \mathbf{R}\backslash\Sigma$ such that $t_j \to \beta$. Then we can take $\alpha = t_j$ for any sufficiently large j. After replacing φ by $\varphi - \alpha$ we may assume $\alpha = 0$.

Since $0 < f(0) < 1$ we may choose $0 < q < 1$ such that $1 - q < f(0) < q$. With this choice of q we see that $\mu(\{\varphi < \psi\}) < q$ and $\mu(\{\psi < \varphi\}) < q$ and $\mu(\{\varphi = \psi\}) = 0$ (the last equality follows from the fact that $\alpha \in \mathbf{R}\backslash\Sigma$). To get a contradiction, we consider the probability measure $\hat{\mu} = g\mu$ where $g = \frac{1}{q}$ on $\{\varphi < \psi\}$ and $g = c\mu$ on $\{\varphi \geq \psi\}$ for an appropriately chosen $c > 0$. The theorem of Guedj-Zeriahi [**GuZe**] implies there exists $\rho \in PSH^1(X,\omega)$ with $\sup \rho = 0$ and $\omega_\rho^n = \hat{\mu}$. Then, setting $a = (\frac{1}{q})^{1/n}$, and $\theta = 0$, and $k = n$, Theorem 23 implies

$$\omega_\psi^{n-1} \wedge \omega_\rho \geq a\, \omega_\psi^n \quad \text{on the set } V_\varepsilon(\theta,\rho) \subseteq \{\varphi < \psi\} \tag{12.4}$$

Substituting in (12.2), and taking the limit, $\varepsilon \downarrow 0$, we get $V_\varepsilon \uparrow \{\varphi < \psi\}$ so

$$a \int_{\varphi<\psi} \omega_\varphi^n \leq \int_{\varphi<\psi} \omega_\varphi^{n-1} \wedge \omega \tag{12.5}$$

If instead we take $k = 1$ and $T = \omega_\psi^{n-1}$ we obtain

$$a \int_{\varphi<\psi} \omega_\psi^n \leq \int_{\varphi<\psi} \omega_\psi^{n-1} \wedge \omega$$

Interchanging φ and ψ in the second estimate, we get $a \int_{\varphi\geq\psi} \omega_\varphi^n \leq \int_{\varphi\geq\psi} \omega_\varphi^{n-1} \wedge \omega$, where we make use of the fact that $\mu\{\varphi = \psi\} = 0$. Adding this to the previous inequality we conclude that $a \leq 1$, a contradiction.

The next step is to show

$$\int_{\varphi<\psi} \omega_\varphi^j \omega_\psi^{k-1} \omega^{l+1} \leq \int_{\varphi<\psi} \omega_\varphi^j \omega_\psi^k \omega^l = 0 \tag{12.6}$$

for all j,k,l such that $j + k + l = n$. To see this, we use induction on l. Let $\rho = 0$ and $\theta = \varphi$ and $T = \omega_\varphi^j \omega^l$. Then applying (12.2) we obtain, for every $\delta > 0$,

$$\int_{\varphi<(1-\varepsilon)\psi-\delta} \omega_\varphi^j \omega_\psi^{k-1} \omega^{l+1} \leq \int_{\varphi<(1-\varepsilon)\psi-\delta} \omega_\varphi^j \omega_\psi^k \omega^l + O(\varepsilon)$$

Taking the limit as $\varepsilon \downarrow 0$ we obtain (12.6) but with $\{\varphi < \psi\}$ replaced by $\{\varphi \leq \psi - \delta\}$. Now take the limit $\delta \downarrow 0$ to obtain (12.6).

Taking $l = n$ we obtain $\int_{\varphi<\psi} \omega^n = 0$. Similarly $\int_{\varphi>\psi} \omega^n = 0$. Since these sets are plurifine open, they must be empty. This proves the theorem.

13. Semiclassical Solutions of Monge-Ampère Equations

In this section we discuss a particular method for solving Monge-Ampère equations, namely by semiclassical limits of Bergman kernels. The equation accessible this way is the homogeneous complex Monge-Ampère equation. It is a very specific equation, but one which is of great interest in the problem of finding Kähler metrics of constant scalar curvature, and which is intimately linked with the notion of stability in geometric invariant theory (see [**D99**] and [**PS08**] for a survey). We shall see how semiclassical limits lead to generalized solutions of this equation. This may be noteworthy in itself from the viewpoint of PDE theory, as generalized solutions of partial differential equations usually arise rather from either variational or Perron methods.

13.1. Geodesics in the space of Kähler potentials. We begin with some geometric motivation. Let (X, ω_0) be a compact Kähler manifold without boundary of dimension n. Then the space $\mathcal{K}$ of Kähler potentials

$$\mathcal{K} = \left\{ \varphi \in C^\infty(X);\ \omega_\varphi \equiv \omega_0 + \frac{i}{2}\partial\bar{\partial}\varphi > 0 \right\} \tag{13.1}$$

is formally an infinite-dimensional Riemannian manifold with tangent space $T_\varphi(\mathcal{K}) = \{\delta\varphi \in C^\infty(X)\}$ and metric

$$\|\delta\varphi\|^2 = \int_X |\delta\varphi|^2 \omega_\varphi^n. \tag{13.2}$$

The geodesic equation for $\mathcal{K}$ is the Euler-Lagrange equation for the energy functional

$$\mathcal{E} = \int_0^T \int_X \dot{\varphi}^2 \omega_\varphi^n dt \tag{13.3}$$

for paths $[0,T) \ni t \to \varphi(\cdot, t) \in \mathcal{K}$. Under a variation $\delta\varphi$ of this path, we have

$$\delta\mathcal{E} = \int_0^T \int_X (2\dot{\varphi}\,\delta\dot{\varphi} + \dot{\varphi}^2 \Delta_\varphi \delta\varphi)\omega_\varphi^n dt, \tag{13.4}$$

where Δ_φ is the Laplacian with respect to the Kähler form ω_φ. For variations $\delta\varphi$ fixing the end points, integrating by parts gives

$$\delta\mathcal{E} = -2\int_0^T \int_X \delta\varphi(\ddot{\varphi} - |\nabla\dot{\varphi}|^2_{\omega_\varphi})\omega_\varphi^n dt. \tag{13.5}$$

and thus the geodesic equation for paths $[0,T) \ni t \to \varphi(\cdot, t) \in \mathcal{K}$ is

$$\ddot{\varphi} - |\nabla\dot{\varphi}|^2_{\omega_\varphi} = 0. \tag{13.6}$$

A key observation due to Donaldson [**D99**] and Semmes [**Se**] is that this geodesic equation is equivalent to a homogeneous complex Monge-Ampère equation

$$\left(\pi^*\omega_0 + \frac{i}{2}\partial\bar\partial\Phi\right)^{n+1} = 0. \tag{13.7}$$

for the function

$$\Phi(z,w) \equiv \varphi(z, \log|w|) \tag{13.8}$$

on $M \equiv X \times A$, with $A = \{w \in \mathbf{C}; e^{-T} < |w| < 1\}$, Here $\partial\bar\partial\Phi$ is taken with respect to all $(n+1)$ variables (z,w), and $\pi^*\omega_0$ is the pull-back of ω_0 to M. Note that $\pi^*\omega_0$ is not strictly positive, viewed as a $(1,1)$-form on M.

We shall consider both *geodesic segments*, joining two points φ_0 and φ_1 of $\mathcal{K}$, and *geodesic rays*, extending from a point $\varphi_0 \in \mathcal{K}$ to infinity. There is no loss of generality in taking $\varphi_0 = 0$.

The equation for geodesic segments is then a standard Dirichlet problem on the manifold $M = X \times A$, $\partial A = \{w \in \mathbf{C}; |w| = 1 \text{ or } |w| = e^{-T}\}$ with T finite, and boundary value Φ_b defined by

$$\begin{aligned} \Phi_b(z,w) &= 0 \quad \text{for } |w| = 1 \\ \Phi_b(z,w) &= \varphi_1 \quad \text{for } |w| = e^{-T}. \end{aligned} \tag{13.9}$$

The equation for geodesic rays is more unusual: here $T = \infty$, so the annulus A reduces to the punctured disk $D^\times = \{w \in \mathbf{C}; 0 < |w| < 1\}$, and $M = X \times D^\times$. A boundary value Φ_b is assigned only on the component $|w| = 1$ of the boundary of M,

$$\Phi_b(z,w) = 0 \quad \text{for } |w| = 1 \tag{13.10}$$

but there is no condition near $w = 0$. In practice, and with motivation from geometric invariant theory, we shall restrict to the case where $\omega_0 = c_1(L)$, where $L \to X$ is a positive line bundle, and consider the geodesic rays associated to a *test configuration* of $L \to X$. A test configuration (see the precise definition in §13.2 below) is a one-parameter subgroup (1PS) degeneration of the line bundle $L \to X$. It produces a limiting singular line bundle, or more precisely a polarized scheme, $L_0 \to X_0$, which can be viewed as an implicit boundary value for the Dirichlet problem for the homogeneous complex Monge-Ampère equation at $w = 0$. The net result is that we shall associate a canonical geodesic ray to each test configuration, starting from an arbitrary point $\varphi_0 \in \mathcal{K}$. Thus a test configuration can also be viewed as providing a direction where the Cauchy problem admits a generalized solution for infinite time, and hence as a generalized vector field on the space $\mathcal{K}$ of Kähler metrics.

The motivation for test configurations and geodesic rays is the following. It has been shown by Donaldson [**D99**] and Mabuchi [**M87**] that $\mathcal{K}$ is a symmetric space with non-positive curvature. The geodesic rays in $\mathcal{K}$ are then just a generalization of the one-parameter subgroups of finite-dimensional symmetric spaces of negative curvature. In Donaldson's program for the problem of constant scalar curvature metrics [**D99**], a numerical invariant for geodesic rays can be defined as the limiting value as $t \to \infty$ of the rate of change of the Mabuchi K-energy along geodesic rays. An infinite-dimensional notion of stability in geometric invariant theory (GIT) can then be defined as the negativity of this numerical invariant of geodesic rays, and we obtain in this way an infinite-dimensional version of the Yau-Tian-Donaldson conjecture, which asserts the equivalence between the existence of a metric of constant scalar curvature in the Kähler class $c_1(L)$ and the K-stability of $L \to X$ in the sense of GIT.

Very recently, Lempert and Vivas [**LV**] have produced examples of Kähler manifolds (X, ω_0), specifically tori with certain symmetries, where there are no C^3 geodesic connecting two given potentials $\varphi_1, \varphi_2 \in \mathcal{K}$. A C^∞ geodesic would correspond to a function $\Phi(z, w)$ which is C^∞ in both z and w, a solution of the completely degenerate Monge-Ampère equation, and which is strictly plurisubharmonic with respect to ω_0 for each $w \in A$. On the other hand, generalized solutions of (13.7) in the sense of pluripotential theory, where $\Phi(z, w)$ is only known to be plurisubharmonic in all variables (z, w), and where Φ is only of class $C^{1,\alpha}$, have been known for some time. Their existence can be deduced from the general theory of boundary value problems of §9 [**C00, B09b, PS09a**], as well as from explicit semi-classical constructions [**PS06, PS07, SZ07, SZ10, RZ08**]. Other constructions are due to Arezzo-Tian [**AT**] (by the Cauchy-Kowalevska method), and to Chen [**C06**], Chen-Sun [**CS**], Chen-Tang [**CT**], Ross-Witt Nystrom [**RW**] and other authors under various types of assumptions. A partial regularity theory for the homogeneous complex Monge-Ampère equation has been proposed by Chen and Tian [**CT**]. We discuss some of these developments in this section.

13.2. Geodesics from a priori estimates. We show how generalized geodesics can be obtained from the existence theorems for the Dirichlet problem of Section §8. In the present context, (X, ω_0) is a given compact Kähler manifold without boundary, and the manifolds with boundary of the theorems in Section §8 are now given by $M = X \times A$, with A an annulus, or $M = X \times D^\times$, with $D^\times$ a punctured disk.

First, we consider the case of geodesic segments, linking $\varphi_0 = 0$ to $\varphi_1 \in \mathcal{K}$. Here the set-up of the geodesic equation is exactly the same as for Theorem 13, with the difference that the form $\pi^*\omega_0$ in (13.7) is not strictly positive. However, it is easy to bring ourselves back to the case of a strictly positive form, by constructing a smooth function $\underline{\Phi}$ satisfying

$$\pi^*\omega_0 + \frac{i}{2}\partial\bar\partial\underline{\Phi} > 0 \text{ on } M, \quad \underline{\Phi} = \Phi_b \text{ on } \partial M. \tag{13.11}$$

Setting then $\Omega_0 = \pi^*\omega_0 + \frac{i}{2}\partial\bar\partial\underline{\Phi}$ and $\Phi = \Psi + \underline{\Phi}$, the equation $(\pi^*\omega_0 + \frac{i}{2}\partial\bar\partial\Phi)^{n+1} = 0$ is equivalent to the equation $(\Omega_0 + \frac{i}{2}\partial\bar\partial\Psi)^{n+1} = 0$, with Ω_0 now a Kähler form so that Theorem 13 applies at once. The function $\underline{\Phi}$ is obtained by the following elementary construction (see e.g. Lemma 14 of

[**PS08**]): $\Phi = t\varphi_1 + f(w)$, $t =$ log $|w|$, with $f(w)$ the solution of the Dirichlet problem $\Delta f(w) = C$ on A, $f = 0$ on ∂A, and C is a large positive constant. Thus we have established the following theorem, first proved by Chen [**C00**]:

THEOREM 24. *Let φ_0 and φ_1 be two points in $\mathcal{K}$. Then there exists a unique generalized geodesic of class $C^{1,\alpha}$, for any $0 < \alpha < 1$, joining φ_0 and φ_1.*

We turn next to geodesic rays. The notion of test configuration alluded to in §13.1 can be defined precisely as follows (see Donaldson [**D02**]):

DEFINITION 3. *Let $L \to X$ be a positive line bundle over a compact complex manifold X. A test A test configuration $\mathcal{T}$ for $L \to X$ consists of*

(1) *a scheme $\mathcal{X}$ with a $\mathbf{C}^\times$ action ρ;*
(2) *a $\mathbf{C}^\times$ equivariant line bundle $\mathcal{L} \to \mathcal{X}$, ample on all fibers;*
(3) *and a flat $\mathbf{C}^\times$ equivariant map $\pi : \mathcal{X} \to \mathbf{C}$, where $\mathbf{C}^\times$ acts on $\mathbf{C}$ by multiplication, with the property that $(\pi^{-1}(1), \mathcal{L}|_{\pi^{-1}(1)})$ is isomorphic to (X, L^r) for some $r > 0$.*

We shall also denote $\mathcal{T}$ by

$$\mathcal{T} = \big(\rho : \mathbf{C}^\times \to \mathrm{Aut}(\mathcal{L} \to \mathcal{X} \to \mathbf{C})\big) \tag{13.12}$$

where $\mathrm{Aut}(\mathcal{L} \to \mathcal{X} \to \mathbf{C})$ is the space the space of automorphisms of the fibrations $\mathcal{L} \to \mathcal{X} \to \mathbf{C}$. A test configuration configuration $\mathcal{L}$ is said to be trivial if $\mathcal{L} = L \times \mathbf{C}$, with the action $\rho(\tau)(\ell, w) = (\ell, \tau w)$, for $(\ell, w) \in L \times \mathbf{C}$, $\tau \in \mathbf{C}^\times$.

It is convenient to introduce the notation $X_w = \pi^{-1}(w)$, $L_w = \mathcal{L}|_{\pi^{-1}(w)}$, and to view $\mathcal{X}$ as $\mathcal{X} = \overline{\cup_{w\in\mathbf{C}} X_w}$. A typical example of a test configuration would be $\mathcal{X} = \cup_\tau \sigma_\tau(X)$ where X is a submanifold of $\mathbf{CP}^N$, and $\sigma_\tau = e^{\tau B}$, is a one-parameter subgroup of $GL(N+1)$.

Note that all fibers (X_w, L_w) are biholomorphic to (X_1, L_1) for $w \neq 0$. However, the "central fiber" (X_0, L_0) will usually have singularities. It can be viewed as the limit in the sense of schemes of (X_w, L_w) as $w \to 0$, and it is invariant under the action of ρ.

For the construction of geodesic rays, we need the following geometric properties of a test configuration. Let

$$p : \tilde{\mathcal{X}} \to \mathcal{X} \to \mathbf{C} \tag{13.13}$$

a resolution of singularities, which can be chosen to be equivariant, in the sense that the homomorphism ρ lifts to a homomorphism $\tilde{\rho} : \mathbf{C}^\times \to \mathrm{Aut}(p_*\mathcal{L} \to \mathcal{X} \to \mathbf{C})$, and that all diagrams commute. Let $\mathcal{X}_D = \pi^{-1}(D)$, $\mathcal{X}_D^\times = \pi^{-1}(D^\times)$, and $\mathcal{L}_D \to \mathcal{X}_D = \pi^{-1}(D)$ and $\mathcal{L}_D^\times \to \mathcal{X}_D^\times$ be the fibrations above D and $D^\times$, with similar definitions for $\tilde{\mathcal{X}}_D$, $\tilde{\mathcal{X}}_D^\times$, and $\tilde{\mathcal{L}}_D$ and $\tilde{\mathcal{L}}_D^\times$. Let the Kähler form ω_0 on X be the curvature of a metric h_0 on L. Then an element φ of the space of Kähler potentials $\mathcal{K}$ can be identified with a metric $h = h_0 e^{-\varphi}$ on L. We can now state the properties of test configurations which we need:

- There is a biholomorphism

$$
\begin{aligned}
M = X \times D^\times &\to \tilde{\mathcal{X}}^\times \\
L \times D^\times &\to \tilde{\mathcal{L}}^\times
\end{aligned}
\tag{13.14}
$$

defined by $(\zeta, w) \to \rho(w)(\zeta) \in L_w$, for $\zeta \in L = L_1$. If we extend the metric h_0 on L trivially as a metric on $L \times D^\times$, then it can be identified through the above biholomorphism with a metric H_0 on $\tilde{\mathcal{L}}^\times$. The curvature of H_0 is (the push-forth of) Ω_0. If H is any other metric on $\tilde{\mathcal{L}}^\times$ and Ω its curvature, then $H = H_0 e^{-\Psi}$, and the equation $(\Omega_0 + \frac{i}{2}\partial\bar{\partial}\Phi)^m = 0$ on M is equivalent to the equation

$$
\left(\Omega + \frac{i}{2}\partial\bar{\partial}\tilde{\Phi}\right)^{n+1} = 0 \quad \text{on } \tilde{\mathcal{X}}_D^\times
\tag{13.15}
$$

with $\tilde{\Phi} = \Phi - \Psi$.

- Recall that h_0 is any fixed metric on L with positive curvature ω_0. It is shown in [**PS07a**], [**PS09b**] that there is a metric H on $\tilde{\mathcal{L}}$ (in particular over the whole of $\tilde{\mathcal{X}}_D$, including the central fiber), which restricts to h_0 on X_1, and which has curvature $\Omega \geq 0$ over $\tilde{\mathcal{X}}$, and $\Omega > 0$ on $\tilde{\mathcal{X}}_D^\times$.
- Furthermore, there exists an effective divisor E supported only in the central fiber of $\tilde{\mathcal{X}}$ and a smooth metric K on $O(E)$ so that

$$
\Omega_K \equiv \Omega + \varepsilon \frac{i}{2}\partial\bar{\partial}\log K
\tag{13.16}
$$

is smooth and strictly positive definite everywhere on $\tilde{\mathcal{X}}_D$, for all small, strictly positive ε.

We return now to the problem of constructing generalized geodesic rays. Given a test configuration, the above constructions show that a generalized geodesic ray is a solution of the equation (13.15) on $\tilde{\mathcal{X}}_D^\times$. The above geometric properties of test configurations mean precisely that the hypotheses of Theorem 13.15 are satisfied with $M = \tilde{\mathcal{X}}_D$. Note that in this case, the boundary of the manifold is clearly holomorphically flat. Thus we obtain the following theorem [**PS09a**]:

THEOREM 25. *Let $L \to X$ be a positive line bundle over a compact complex manifold X. Let $\mathcal{T}$ be any test configuration of $L \to X$, and let p be an equivariant resolution as in (13.13). Then for any metric h_0 on L with positive curvature, there is a $C^{1,\alpha}$ generalized geodesic, with bounded potential, starting from h_0. More precisely, there is a bounded solution $\tilde{\Phi}$ of the equation $(\Omega + \frac{i}{2}\partial\bar{\partial}\tilde{\Phi})^{n+1} = 0$ on $\tilde{\mathcal{X}}_D$, which is of class $C^{1,\alpha}$ on $\tilde{\mathcal{X}}_D^\times$.*

13.3. Algebraic approximations: the Tian-Yau-Zelditch theorem. The conjecture of Yau on Kähler-Einstein metrics [**Y93**], and the Yau-Tian-Donaldson conjecture on the existence of metrics of constant scalar curvature in a Kähler class $c_1(L)$ if L is K-stable [**D02**], are conjectures relating the solvability of a geometric partial differential geometric to a global algebraic condition. One strategy, advocated by Yau and implemented particularly successfully by Donaldson [**D01, D10**] in the proof of the necessity of stability, has been to approximate transcendental objects such as metrics by their algebraic counterparts, such as Bergman kernels and Fubini-Study metrics. A precise example is the Tian-Yau-Zelditch theorem [**Y93, T90a, Z**] (also proved independently by Catlin [**Cat**] and refined by Lu [**L**]), which can be stated as follows.

Let $L \rightarrow X$ be a positive line bundle over a compact complex manifold X. If $\underline{s} = \{s_\alpha\}_0^{N_k}$ is a basis for the space $H^0(X, L^k)$ of holomorphic sections of L^k, $\dim H^0(X, L^k) = 1 + N_k$, then the Kodaira map $\iota_{\underline{s}}$ is defined by

$$\iota_{\underline{s}}: \ X \ni z \rightarrow [s_0(z) : \cdots : s_{N_k}(z)] \in \mathbf{CP}^{N_k}. \tag{13.17}$$

The Kodaira imbedding theorem says that $\iota_{\underline{s}}$ is an imbedding for k sufficiently large. The hyperplane bundle $O(1)$ over $\mathbf{CP}^{N_k}$ pulls back to L^k. The Fubini-Study metric $h_{FS} = \frac{1}{\sum_{\alpha=0}^{N_k} |s_\alpha|^2}$ on $O(1)$ and $\omega_{FS} = -\frac{i}{2}\partial\bar\partial \log h_{FS}$ on $\mathbf{CP}^{N_k}$ pull back then to the metrics

$$\iota_{\underline{s}}^*(h_{FS}) = \frac{1}{\sum_{\alpha=0}^{N_k} |s_\alpha(z)|^2}, \quad \iota_{\underline{s}}^*(\omega_{FS}) = -\frac{i}{2}\partial\bar\partial \log \sum_{\alpha=0}^{N_k} |s_\alpha(z)|^2. \tag{13.18}$$

on L^k and X respectively. Note that $h(k) \equiv (\iota_{\underline{s}}^*(h_{FS}))^{\frac{1}{k}}$ is then a metric on L, and its curvature $\omega(k) \equiv \frac{1}{k}\iota_{\underline{s}}^*(\omega_{FS})$ is a metric on X, which can be viewed as algebraic objects.

Let now h be a metric on L with positive curvature $\omega = -\frac{i}{2}\partial\bar\partial \log h > 0$. The Tian-Yau-Zelditch theorem asserts that the metric h on L and ω on X can be approximated asymptotically in k by the metrics $h(k)$ and $\omega(k)$, if the basis $\underline{s} = \{s_\alpha\}_0^{N_k}$ used to construct the Kodaira imbedding is an orthonormal basis of $H^0(X, L^k)$ with respect to the L^2 metric, $\|s\|^2 \equiv \int_X |s|^2 h^k \omega^n$. To see how this comes about, we write

$$\log \frac{h(k)}{h} = -\frac{1}{k} \log \rho_k(z), \quad \omega - \omega(k) = -\frac{i}{2k}\partial\bar\partial \log \rho_k(z) \tag{13.19}$$

where $\rho_k(z)$ is the Bergman kernel (or density of states, since it integrates to $\dim H^0(X, L^k)$),

$$\rho_k(z) \equiv \sum_{\alpha=0}^{N_k} |s_\alpha(z)|^2 h^k(z). \tag{13.20}$$

The desired statement follows then from an asymptotic expansion for $\rho_k(z)$ [**Z**]

$$\rho_k(z) = k^n \left(1 + \sum_{p=1}^{\infty} A_p(z) k^{-p}\right), \tag{13.21}$$

This expansion is itself a consequence of an asymptotic expansion obtained by Boutet de Monvel and Sjöstrand [**BS**] for the Szegö kernel on strongly pseudo-convex domains, refining an earlier expansion along the diagonal obtained by Fefferman [**?**]. The coefficient $A_1(z)$ has been shown by Lu [**L**] to be given by $\frac{1}{2}R(z)$, where R is the scalar curvature of ω. This turns out to provide a key link with the problem of constant scalar curvature metrics [**D01, D10**]. Other generalizations and applications of asymptotic expansions and of the Tian-Yau-Zelditch theorem can be found in [**MM, Wg**].

13.4. Semi-classical constructions. Let $\mathcal{K}_k$ denote the space $\{(\iota_{\underline{s}}(h_{FS}))^{\frac{1}{k}}\}$ of pull-backs of the Fubini-Study metrics, under Kodaira imbeddings defined by an arbitrary basis $\underline{s}$ of $H^0(X, L^k)$. Since h_{FS} is invariant under $SU(N_k+1)$, we have

$$\mathcal{K}_k = SL(N_k+1)/SU(N_k+1). \tag{13.22}$$

Note that the right hand side is a symmetric space with negative curvature. A suggestive consequence of the Tian-Yau-Zelditch theorem is that, in a pointwise sense, we have

$$\mathcal{K} = \lim_{k\to\infty}\mathcal{K}_k = \lim_{k\to\infty} SL(N_k+1)/SU(N_k+1). \tag{13.23}$$

It is a natural question whether this pointwise approximation can be extended to the approximation of more geometric properties, for example of extended geometric objects. A prime example is whether geodesics in $\mathcal{K}$ can be approximated by geodesics in $\mathcal{K}_k$. This translates precisely into whether the solutions of the homogeneous complex Monge-Ampère equations can be approximated by one-parameter subgroups of Bergman kernels [**PS06**]. We shall see below that the answer is affirmative, see [**PS06, PS07, PS09b**]. Some refinements of these approximations and their rate of convergence can be found in Berndtsson [**Be1, Be2**], and in [**SZ07, SZ10**] in the case of toric varieties. For toric varieties, a similar approximation has been extended to harmonic maps by Rubinstein and Zelditch [**RZ08**].

We provide now some details. Fix a metric h_0 on L with positive curvature ω_0. Let $\Phi(z,w)$ be a solution of the Dirichlet problem (13.7) with boundary value φ_0 when $|w|=1$. If we view it as either a geodesic segment or a geodesic ray in $\mathcal{K}$ emanating from the corresponding potential φ_0, then this geodesic segment or ray should be the limit of a sequence of one-parameter subgroups in $\mathcal{K}_k$, as $k\to\infty$. Let $B_k \in GL(N_k+1)$ be the infinitesimal generator of the one-parameter subgroup in $\mathcal{K}_k$, so that the subgroup is given by w^{B_k}, $w\in\mathbf{C}^\times$. We can assume that B_k is diagonal, with eigenvalues $\lambda_\alpha^{(k)}$, $0 \le \alpha \le N_k$. Let $\underline{s} = \{s_\alpha\}_0^{N_k}$ be a basis of $H^0(X, L^k)$ which is orthonormal with respect to the L^2 norm defined by h_0 and the volume form ω_0^n. The subgroup w^{B_k} acts on the basis $\{s_\alpha\}_0^{N_k}$ to produce the basis $w^{B_k} \cdot \underline{s} \equiv \{w^{\lambda_\alpha^{(k)}} s_\alpha\}_0^{N_k}$. The corresponding pull-backs of the Fubini-Study metrics can be written explicitly as

$$\Phi_k(z, w) = \frac{1}{k} \log \sum_{\alpha=0}^{N_k} |w|^{2\lambda_\alpha^{(N_k)}} |s_\alpha(z)|^2 h_0(z)^k \tag{13.24}$$

Thus the main problem is to choose the appropriate generators B_k and to show that the $\Phi_k(z, w)$ converge, in a suitable sense, to a solution $\Phi(z, w)$ of the homogeneous complex Monge-Ampère equation. Note that, by the Tian-Yau-Zelditch theorem, the functions $\Phi_k(z, w)$ converge to the correct boundary value when $|w| = 1$.

We address the problem of choosing the generators B_k.

Consider first the case of a generalized geodesic segment joining two points φ_0 and φ_1 in $\mathcal{K}$. There is in this case a natural choice of infinitesimal generator B_k: B_k is just the matrix of change of bases, from a basis $\{s_\alpha^{(0)}\}_0^{N_k}$ orthonormal with respect to the L^2 metric defined by h_0, ω_0^n, to a basis $\{s_\beta^{(1)}\}_0^{N_k}$ orthonormal with respect to the L^2 metric defined by h_1, ω_1^n.

Next, consider the case of a test configuration. The group action $\rho(w)$, $w \in \mathbf{C}^\times$ preserves the central fiber (X_0, L_0). Thus it induces a one-parameter subgroup w^{B_k} on the space of holomorphic sections,

$$w^{B_k} : \; H^0(X_0, L_0^k) \to H^0(X_0, L_0^k). \tag{13.25}$$

The generators B_k are the generators that we are looking for.

Once the generators B_k, and hence their eigenvalues $\lambda_\alpha^{(k)}$ have been chosen, we need a criterion for when the expressions (13.24) have the desired convergence properties. This is provided by the following lemma [**PS08**]:

Lemma 22. *Fix $h_0 \in \mathcal{K}$ as before. Consider the general Ansatz*

$$\Phi_k(z, w) = \frac{1}{k} \log \sum_{\alpha=0}^{N_k} |w|^{2\lambda_\alpha^{(k)}} |^2 |s_\alpha^{(k)}(z)|^2_{h_0^k} \tag{13.26}$$

where $\{s_\alpha^{(k)}(z)\}$ is an orthonormal basis for $H^0(X, L^k)$ with respect to the L^2 norm defined by (h_0^k, ω_0^n), and $\lambda_\alpha^{(k)}$ are real numbers, $0 \le \alpha \le N_k$ for each k. Then if

(1) *There exists a constant C independent of both α and k so that*

$$|\lambda_\alpha^{(k)}| \le C\, k \tag{13.27}$$

(2) *There exists a constant C_T independent of k so that*

$$\int\int_{X \times \{e^{-T} < |w| \le 1\}} \left(\pi^* \omega_0 + \frac{i}{2} \partial\bar\partial \Phi_k\right)^{n+1} \le C\, k^{-1}. \tag{13.28}$$

then $\Phi(z,w)=\lim_{k\to\infty}[\sup_{\ell\geq k}\Phi_k(z,w)]^*$ *is continuous at* $|w|=1$, *and satisfies in the sense of pluripotential theory*

$$\left(\pi^*\omega_0+\frac{i}{2}\partial\bar\partial\Phi\right)^{n+1}=0 \text{ on } X\times\{e^{-T}<|w|<1\},\quad \Phi(z,w)=0 \text{ for } |w|=1. \tag{13.29}$$

It can be shown that, in both choices of generators B_k for geodesic segments and for geodesic rays, the condition on the uniform growth of the eigenvalues $\lambda_\alpha^{(k)}$ is satisfied. The only non-trivial step remaining is to verify the condition on the decay of the masses of the Monge-Ampère measures of $\Phi_k(z,w)$. The key observation here is that these masses are essentially cohomological, and given by

$$\int\int_{X\times\{e^{-T}<|w|<1\}}\left(\pi^*\omega_0+\frac{i}{2}\partial\bar\partial\Phi\right)^{n+1}=\dot E(\varphi(\cdot,0))-\dot E(\varphi(\cdot,-T)) \tag{13.30}$$

where $E(\varphi)$ is the functional (11.1) we had encountered earlier as the anti-derivative of the Monge-Ampère measure ω_φ^n.

We can now state the main theorems of this section and complete their proofs. For geodesic segments, we have [**PS06**]:

THEOREM 26. *Let* $h_1\in\mathcal{K}$ *be another metric on* L *with* $\omega_1=-\frac{i}{2}\partial\bar\partial\log h_1>0$, $M\equiv X\times\{e^{-1}<|w|<1\}$. *Then the generalized solution* $\Phi(z,w)$ *of the Dirichlet problem (13.7) can be expressed as*

$$\Phi(z,w)=\lim_{k\to\infty}[sup_{\ell\geq k}\Phi_\ell(z,\log|w|)]^* \tag{13.31}$$

where the eigenvalues $\lambda_\alpha^{(k)}$ *defining* $\Phi_k(z,w)$ *are the eigenvalues of the matrix* B_k *of change of bases from an orthonormal basis of* $H^0(X,L^k)$ *with respect to* h_0,ω_0^n *to an orthonormal basis of* $H^0(X,L^k)$ *with respect to* h_1,ω_1^n.

PROOF. We compute explicitly the right hand side of (13.30). Denote the approximating Fubini-Study metrics in (13.19) for the metrics h_0 and h_1 as $h_0(k)=e^{-\varphi_0(k)}h_0$ and $h_1(k)=e^{-\varphi_1(k)}h_0$ respectively. Let $\omega_a(k)$ be their curvatures. Let $\{s_{a,\alpha}^{(k)}\}$ be orthonormal bases with respect to h_a,ω_a^n. Then we have, with $a=0,1$,

$$\int_X\dot\varphi_a(k)\omega_a(k)^n=\frac{2}{k^{n+1}}\int_X\sum_{\alpha=0}^{N_k}\lambda_\alpha^{(k)}|s_{a,\alpha}^{(k)}(z)|^2h_a(k)^k\omega_a(k)^n$$

$$=\frac{2}{k^{n+1}}\sum_{\alpha=0}^{N_k}\lambda_\alpha^{(k)}+O\left(\frac{1}{k^{n+2}}\right)N_k\max_\alpha|\lambda_\alpha^{(k)}|, \tag{13.32}$$

where we have applied the Tian-Yau-Zelditch theorem. Thus the leading terms cancel in the difference (13.30), giving the desired estimate. Q.E.D.

By the uniqueness of the solution of the Dirichlet problem, this solution must coincide with the solution obtained by Chen [**C00**] from the method of a priori estimates (see Theorem 24), so it must be $C^{1,\alpha}$. It has also been shown by Berndtsson [**Be2**] that the convergence described in [**PS06**] can actually be strengthened to uniform convergence.

For geodesic rays defined by a test configuration, we have [**PS07, PS09b**]:

THEOREM 27. *Let ρ be a test configuration for a positive line bundle $L \to X$ over a compact complex manifold X. Let h_0 be a metric on L, with $\omega_0 = -\frac{i}{2}\partial\bar\partial \log h_0 > 0$. Let $\Phi_k(z,w)$ be defined by (13.24), where the eigenvalues $\lambda_\alpha^{(k)}$ are the eigenvalues of the endomorphisms B_k on $H^0(X_0, L_0^k)$ induced by the group action ρ. Then*

$$\Phi(z,w) = \lim_{k\to\infty}[\sup_{\ell\geq k}\Phi_k(z,w)]^* \tag{13.33}$$

defines a generalized solution of the Dirichlet problem

$$\left(\pi^*\omega_0 + \frac{i}{2}\partial\bar\partial\Phi\right)^{m+1} = 0 \text{ on } X\times D^\times, \quad \Phi(z,w) = 0 \text{ for } |w| = 1. \tag{13.34}$$

The solution is actually of class $C^{1,\alpha}(X\times D^\times)$ for any $0<\alpha<1$. It is non-constant when the test configuration is non-trivial.

PROOF. As before, it remains only to prove the bound $O(k^{-1})$ on the mass of the Monge-Ampère measure on $M = X\times D^\times$. Consider the functions $\Phi_k^\#(z,w)$ defined by the same formula as $\Phi_k(z,w)$, but with the eigenvalues $\lambda_\alpha^{(k)}$ replaced by their traceless counterparts

$$\lambda_\alpha^{\#,(k)} = \lambda_\alpha^{(k)} - \frac{\mathrm{Tr}\, B_k}{N_k+1}. \tag{13.35}$$

Then $\Phi_k(z,w) = \Phi_k^\#(z,w)(z,w) + \frac{\mathrm{Tr}\, B_k}{k(N_k+1)}\log|w|^2$. Thus they have the same complex Hessian, and we can evaluate (13.30) with $\Phi_k(z,w)$ replaced by $\Phi_k^\#(z,w)$. The formula (13.30) gives then, with obvious notations,

$$\begin{aligned}\int_{X\times D^\times}\left(\pi^*\omega_0 + \frac{i}{2}\partial\bar\partial\Phi_k^\#\right)^{n+1} &= \lim_{T\to\infty}\int_X \dot\varphi(T)^\#\omega_k(T)^n \\ &\quad - \int_X \dot\varphi(0)^\#\omega_k(0)^n.\end{aligned} \tag{13.36}$$

Since the eigenvalues $\lambda_\alpha^{\#(k)}$ sum to 0, the leading term in the second expression on the right hand side is 0. As for the first expression, the lemma below shows that it is automatically $O(k^{-1})$. This establishes the fact that the limit $\Phi(z,w)$ satisfies the homogeneous complex Monge-Ampère equation. □

LEMMA 23. *Let $\mathcal{T}$ be a test configuration. Then we have*

$$\lim_{T\to\infty}\int_X \dot\varphi(T)\omega_k(T)^n = \frac{1}{k}F \tag{13.37}$$

where F is the Donaldson-Futaki invariant of $\mathcal{T}$, defined as the second term F in the following asymptotic expansion

$$\frac{\mathrm{Tr}\, B_k}{k(N_k+1)} = F_0 + F\,k^{-1} + O(k^{-2}). \tag{13.38}$$

We should say that this lemma was implicit in the paper of Donaldson [**D04**]. Its explicit statement and proof can be found in [**PS07**], Mabuchi [**M09**], and Donaldson [**D10**].

Unlike in the case of the geodesic segments, the regularity of the geodesic rays obtained from the above theorem does not follow as yet, since the behavior of the ray near $w=0$ has not been addressed. The following lemma gives a complete description of this behavior [**PS09b**]:

LEMMA 24. *Let $\mathcal{T}$ be a test configuration, and let p be an equivariant resolution of singularities as considered earlier in §13.2 . Then the function*

$$\Psi_k \equiv \Phi_k - \Phi_1 \tag{13.39}$$

extends as a smooth function over the whole of $\tilde{\mathcal{X}}_D$. Furthermore, it satisfies the following uniform estimate

$$\sup_{k\geq 1}\sup_{\tilde{\mathcal{X}}_D}|\Psi_k| \leq C < \infty. \tag{13.40}$$

With this lemma, we can show that the function

$$\Psi = \Phi - \Phi_1 \tag{13.41}$$

is a bounded solution of a homogeneous complex Monge-Ampère equation on $\tilde{\mathcal{X}}_D$ with a non-negative background form Ω_1. This can be expressed in turn as a homogeneous complex Monge-Ampère equation with a background form Ω which satisfies all the hypotheses of Theorem 14. Thus the solution must be $C^{1,\alpha}$, and the proof of the theorem is complete.

13.5. The toric case. In the toric case, the previous constructions of solutions of homogeneous complex Monge-Ampère equations as limits of Bergman metrics can be analyzed more precisely. We can obtain in this manner more detailed information on the approximating paths and their rates of convergence. A remarkable feature also emerges from this study, which is an unexpected relation between the previous semiclassical constructions and the theory of large deviations [**SZ07, SZ10**].

13.5.1. *Bergman geodesics.* Let X be an n-dimensional toric manifold and $L \to X$ be a positive toric line bundle over X. Let $\mathcal{K}_T$ be the space of positively curved smooth toric hermitian metrics on L which are invariant under the compact $(S^1)^n$ torus action. Let $h_0, h_1 \in \mathcal{K}_T$ and let h_t for $0 \le t \le 1$ be the Monge-Ampère geodesic between them as defined in Section §13.4. We define

$$\varphi_t(z) = \log(h_t(h_0)^{-1}). \tag{13.42}$$

Then $Ricci(h_t) = Ricci(h_0) + \frac{i}{2}\partial\bar\partial\varphi_t$.

The line bundle L is associated to a convex polytope P in $\mathbf{R}^n$ which coincides with the image of the moment map by any toric Kähler metric in $c_1(L)$. Each integral point $\alpha \in k\overline{P}$ corresponds to a holomorphic section in $H^0(X, L^k)$. In particular, $\{z^\alpha\}_{\alpha \in k\overline{P}}$ form a basis for $H^0(X, L^k)$ for $z \in (\mathbf{C}^*)^n$. Let $\{s_\alpha(z)\}_{\alpha \in k\overline{P} \cap \mathbf{Z}^n}$ be an orthonormal toric basis for $H^0(X, L^k)$ with respect to the L^2 norm defined by (h_0^k, ω_0^n), where $\omega_0 = -\frac{i}{2}\partial\bar\partial \log h_0$. Then we define the following paths of Bergman metrics,

$$h_{t,k}(z) = \sum_{\alpha \in k\overline{P} \cap \mathbf{Z}^n} \frac{1}{(Q_{h_0^k}(\alpha))^{1-t}(Q_{h_1^k}(\alpha))^t} |s_\alpha(z)|^2, \tag{13.43}$$

where

(13.44)

$$Q_{h_0^k}(\alpha) = \|s_\alpha\|^2_{h_0^k} = \int_X |s_\alpha(z)|^2_{h_0^k}\omega_0^n, \quad Q_{h_1^k}(\alpha) = \|s_\alpha\|^2_{h_1^k} = \int_X |s_\alpha(z)|^2_{h_1^k}\omega_1^n.$$

The corresponding potentials are given by

$$\varphi_k(t,z) = \frac{1}{k}\log(h_{t,k}h_0^{-k}). \tag{13.45}$$

They correspond to the potentials $\Phi_k(z,w)$ of Theorem 26 with $t = \log|w|$.

The following theorem [**SZ10**] shows that the potentials $\varphi_k(t,z)$ actually converge in C^2:

THEOREM 28. *We have*

$$\lim_{k\to\infty} \|\varphi_k(t,z) - \varphi_t(z)\|_{C^2([0,1]\times X)} = 0. \tag{13.46}$$

Theorem 28 is a considerable strengthening of both Theorem 26 and Berndtsson's result [**Be2**] in the toric case. The advantage of studying Bergman metrics on toric manifolds is that toric holomorphic sections are naturally orthogonal to each other and one can analyze the norming constants $Q_{h^k}(\alpha)$. On the other hand, toric geodesics φ_t are alway smooth [**Gd**]. Thus one may expect a higher order of convergence.

In [**RZ08**], Theorem 28 is generalized from geodesics of toric Kähler metrics to harmonic maps of a compact Riemannian manifold with boundary into the space of toric Kähler metrics. More precisely, such a harmonic map equation can always be solved, and the solution approximated by harmonic maps into the space of toric Bergman metrics in the C^2-topology. The case of geodesics corresponds to the case when the Riemannian manifold with smooth boundary is the interval $[0, 1]$.

13.5.2. *Geodesic rays and large deviations.* The approximation of the geodesic rays associated in Section §13.4 to a test configuration $\mathcal{T}$ for a polarization $L \to X$ can similarly be refined in the case of toric manifolds [**SZ07**]. An interesting observation is made in [**SZ07**], which relates the geodesic rays constructed in §13.4 on toric manifolds to the large deviations principle [**V**].

Let $L \to X$ be a very ample toric line bundle over a toric manifold X. We use the same notations as in Section §13.5.1. Let $h = e^{-\varphi}$ a smooth toric hermitian metric on L such that $\omega = \frac{i}{2}\partial\bar{\partial}\varphi \in c_1(L)$ is a toric Kähler metric on X. On $(\mathbf{C}^n)^*$,the potential φ can be identified with a smooth convex function ψ on $\mathbf{R}^n$, $\varphi(z) = \psi(\rho)$, with $\rho = \log |z|^2$. The Legendre transform of ψ defined by

$$u(x) = \sup_{\rho \in \mathbf{R}^n} (x \cdot \rho - \psi(\rho)) \tag{13.47}$$

is called the symplectic potential associated to ψ. The function $u(x)$ is a smooth convex function on P with appropriate boundary singularities.

Let $\psi_t(\rho)$ be the geodesic ray constructed in Theorem 27 in the toric setting. Then the symplectic potential associated to ψ_t is given by

$$u_t(x) = u_0(x) - t(R - f(x))$$

for some positive piecewise linear convex function $f(x)$ on P, and $R \in \mathbf{R}$ with $R - f(x) > 0$ on P. The piecewise function $f(x)$ is an alternative way of describing a test configuration $\mathcal{T}$ in the toric setting [**D02**].

Define the pair $(d\mu_k^\rho, I^\rho(x))$ by

$$\begin{aligned} d\mu_k^\rho(x) &= (\Pi_{h^k}(z,z))^{-1} \sum_{\alpha \in k\overline{P} \cap \mathbf{Z}^n} \frac{|s_\alpha|^2_{h^k}(z)}{Q_{h^k}(\alpha)} \delta_{\frac{\alpha}{k}}(x) \\ I^\rho(x) &= u_0(x) + \psi_0 - x \cdot \rho \end{aligned} \tag{13.48}$$

where $\Pi_{h^k}(z, z)$ is the Szegö kernel for (L^k, h^k) and $\delta_{\alpha/k}(x)$ is a delta function at α/k. Then the pair $(d\mu_k^\rho, I^\rho(x))$ satisfies the large deviation principle. The measure $d\mu_k^\rho$ is a probability measure on $\overline{P}$ and the function $I^\rho(x)$ is called the rate function associated to $d\mu_k^\rho$. Varadhan's lemma says that for each t and ρ,

$$\lim_{k \to \infty} \frac{1}{k} \log \int_P e^{kt(R-f(x))} d\mu_k^\rho(x) = \sup_{x \in P} (t(R - f(x)) - I^\rho(x)). \tag{13.49}$$

However, it turns out that

$$\psi_{t,k}(\rho) - \psi_0(\rho) = \frac{1}{k} \log \int_P e^{kt(R-f(x))} d\mu_k^\rho(x) \tag{13.50}$$

is exactly the Bergman geodesic ray constructed in Theorem 27, while

$$\begin{aligned}\sup_{x\in P}(t(R-f(x))-I^\rho(x)) &= \sup_{x\in P}\{x\cdot\rho-(u_0(x)-t(R-f(x))\}-\psi_0(\rho)\\ &= \psi_t(\rho)-\psi_0(\rho). \qquad (13.51)\end{aligned}$$

Therefore Varadhan's lemma immediately gives the pointwise convergence of $\psi_{k,t}$ to ψ_t. The uniform convergence in $C^1(X\times[0,1])$ topology is proved in [**SZ07**]. The toric geodesic ray ψ_t is also shown in [**SZ07**] to be $C^{1,1}(X\times[0,1])$, but not $C^2(X\times[0,1])$ in general.

13.5.3. *Counter-examples to regularity of higher order than* $C^{1,1}$. It is well-known that solutions of the homogeneous real Monge-Ampère equation may be only of class $C^{1,1}$ and not higher. Such counterexamples have been extended to the complex case by Gamelin and Sibony [**GS**]. A more general argument for why solutions cannot always be smooth has been given by Donaldson [**D**]. In this section, we would like to give a simple example which illustrates the fact that, even when geodesic segments may be smooth, as in the case of toric varieties, geodesic rays associated to a test configuration may be again at most $C^{1,1}$ [**SZ07**].

Consider the following simple example of a $C^{1,1}$ geodesic ray over $\mathbf{CP}^1$. We consider the standard Fubini-Study metric $g_{FS}=\frac{i}{2}\partial\bar\partial\varphi_0=\frac{i}{2}\partial\bar\partial\log(1+|z|^2)=\frac{i}{2}\partial\bar\partial\log(1+e^\rho)$ on $\mathbf{CP}^1$, where $\varphi_0=\log(1+e^\rho)$ and $\rho=\log|z|^2$. Then the moment map can be constructed by

$$x=\frac{\partial\log(1+e^\rho)}{\partial\rho}=\frac{e^\rho}{1+e^\rho}\in(0,1). \qquad (13.52)$$

The symplectic potential u_0 corresponding to φ_0 is given by

$$u_0(x)=x\rho-\varphi_0(\rho)=x\log x+(1-x)\log(1-x),\qquad x=\frac{e^\rho}{1+e^\rho}. \qquad (13.53)$$

Let $f(x)=|x-1/2|$ be a piecewise linear convex function. Then

$$u_t(x)=u_0(x)+tf(x)=x\log x+(1-x)\log(1-x)+t|x-1/2| \qquad (13.54)$$

induces a geodesic ray. Now we can calculate the Kähler potential $\varphi_t(\rho)$ corresponding to u_t.

By applying the Legendre transform to u_t, we can show by a straightforward calculations that

$$\varphi_t(\rho)=\rho x-u_t(x),\qquad \rho=\frac{\partial u_t}{\partial x}, \qquad (13.55)$$

and

$$\varphi_t(\rho)=\begin{cases}-\dfrac{t}{2}+\log(1+e^{\rho+t}), & \rho\in(-\infty,-t)\\ \dfrac{\rho}{2}+\log 2, & \rho\in(-t,t)\\ \dfrac{t}{2}+\log(1+e^{\rho-t}), & \rho\in(t,\infty)\end{cases} \qquad (13.56)$$

Let $\varphi(t,z) = \varphi_t(\rho) - \varphi_0$. Then φ is an ω_{FS}-psh function in $C^{1,1}(\mathbf{R}\times X)$, but it is not of class $C^2(\mathbf{R}\times X)$.

13.6. The Cauchy problem for the homogeneous Monge-Ampère equation. The construction of geodesic rays associated to a test configuration can be viewed as the solution of the Cauchy problem for the homogeneous complex Monge-Ampère equation, with the initial velocity provided, implicitly, by the test configuration. Even though the Cauchy problem is not well-posed in the sense of Hadamard, it is instructive to examine when there are solutions and when and how they break down. Such an analysis has been provided recently by Rubinstein and Zelditch [**RZ10a**] for convex solutions of the real Monge-Ampère equation.

Consider the Cauchy problem

$$
\begin{aligned}
&\det(\nabla^2\varphi) = 0 \text{ on } [0,T]\times \mathbf{R}^n,\\
&\varphi(0,x) = \varphi_0(x) \text{ on } \mathbf{R}^n, \quad \frac{\partial\varphi}{\partial t}(0,x) = \psi_0(x) \text{ on } \mathbf{R}^n.
\end{aligned}
\tag{13.57}
$$

It is shown in [**RZ10b**] that the above equation can be solved via Legendre transform until the Legendre transform of the solution stops being convex. More precisely, the Legendre transform of φ_t is given by

$$u_t = u_0 + tv. \tag{13.58}$$

The function u_t will stop being convex at a certain $T_{span} > 0$ if v is not convex, thus ψ_t stops solving the HRMA after T_{span}. A candidate solution is constructed in [**RZ10b**] for $t \geq T_{span}$, however, it does not solve the homogeneous real Monge-Ampère equation even in a weak sense and it is not differentiable in general. Before T_{span}, the solution can also be approximated by Toeplitz quantization of the Hamiltonian flow defined by the Cauchy data [**RZ10a**].

14. Envelopes and the Perron Method

The complex Monge-Ampère equation satisfies the comparison principle. Thanks to this, generalized solutions can be obtained by the Perron method, as envelopes of families of plurisubharmonic functions. In this section, we describe some results obtained in this manner, focusing on the homogeneous case.

14.1. Envelopes. The Perron method for complex Monge-Ampère equations was first developed by Bedford and Taylor [**BT76**] for degenerate complex Monge-Ampère equations on bounded domains in $\mathbf{C}^n$. A special case of their results of particular interest to our considerations is the following. Let $D \subset \mathbf{C}^n$ be a smooth, bounded strictly pseudoconvex domain in $\mathbf{C}^n$, and let $f \in C(\partial D)$. Define the following family of plurisubharmonic functions

$$\mathcal{E}_{D,f} = \{u \in PSH(D) | u|_{\partial D} \le f\} \tag{14.1}$$

and its upper envelope

$$\hat{u}(z) = \sup_{\mathcal{E}_{D,f}} u(z). \tag{14.2}$$

Then $\hat{u} \in \mathcal{E}_{D,f} \cap C(\overline{D})$, u is of class $C^{1,1}$ in the interior of D, and u is the unique solution of the Dirichlet problem for the homogeneous complex Monge-Ampère equation

$$\left(\frac{i}{2}\partial\bar{\partial}u\right)^n = 0 \text{ on } D, \quad u|_{\partial D} = f \text{ on } \partial D. \tag{14.3}$$

The local $C^{1,1}$ regularity follows from the $C^{1,1}$ regularity of the solution of the Dirichlet problem for the unit ball [**BT76**].

The Perron method has been widely applied since for the complex Monge-Ampère equations, and there has been considerable progress, thanks partly to the infusion of new techniques and improved approximation theorems for plurisubharmonic functions. In particular, the following results were obtained relatively recently by Berman and Demailly for certain homogeneous complex Monge-Ampère equations, generalizing the geodesic equations considered in §13.

Let (X, ω_X) be an n-dimensional compact Kähler manifold with a smooth Kähler form ω_X. Let Σ be a strictly pseudoconvex domain in $\mathbf{C}^m$ with ρ being a smooth strictly plurisubharmonic defining function for Σ and $\Sigma = \{\rho < 0\}$. We consider the product manifold $M = X \times \Sigma$ and let $\omega_M = \omega_X + \omega_\Sigma$ be a Kähler form on M, where $\omega_\Sigma = \frac{i}{2}\partial\bar{\partial}\rho$. Let $\pi_X : M \to X$ and $\pi_\Sigma : M \to \Sigma$ be the natural projection maps.

Let α be closed real $(1,1)$ form on M with bounded coefficients, with $\alpha_s \equiv \alpha|_{\{s\}\times X} \ge \varepsilon\omega_X$ for some $\varepsilon > 0$ and all $s \in \Sigma$. Let f be a continuous function on $\overline{M}$ such that for each $s \in \partial\Sigma$, $f|_{X_s} \in PSH(X, \alpha_s)$, where $X_s = \pi_X^{-1}(s)$. We define

$$\mathcal{E}_{M,\alpha,f} = \{u | u \in PSH(M, \alpha) \cap C(M), \ u \le f \text{ on } \partial M\}. \tag{14.4}$$

and the upper envelope of $\mathcal{E}_{M,\alpha,f}$ by

$$\varphi = \sup_{\mathcal{E}_{M,\alpha,f}} u. \tag{14.5}$$

The following theorem was proved in [**BD**];

THEOREM 29. *Let (X, ω_X), Σ, α, and $f \in C(\partial M)$ satisfy all the properties listed above, and define the family $\mathcal{E}_{M,\alpha,f}$ and its upper envelope φ as* *above. Then the function* φ *is the unique* α*-plurisubharmonic solution of the Dirichlet problem*

$$(14.6)\qquad \left(\alpha + \frac{i}{2}\partial\bar{\partial}\varphi\right)^{\dim M} = 0 \text{ on } M, \quad \varphi = f \text{ on } \partial M.$$

Furthermore, if $f \in C^{1,1}(\partial M)$*, then for any* $s \in \Sigma$*,* $i\partial\bar{\partial}\varphi|_{\{s\}\times X}$ *is locally bounded, uniformly in* $s \in \Sigma$*.*

The proof in [**BD**] depends on the Kiselman infimum principle and refined regularization techniques for plurisubharmonic functions. Here we discuss only the special case when $\alpha = \omega_X$ (more precisely, the pull-back of ω_X to M) and $f \in C^\infty(\partial M)$. In this case, the fiber-wise regularity can be obtained by the method of elliptic regularization used earlier for the geodesic equation and the standard C^2 estimates of Yau, as described in §6.

First we show that φ is the solution of the Dirichlet problem. Let $\omega_{X,f} = \omega_X + \frac{i}{2}\partial\bar{\partial}f$ after extending f to a smooth function on $\overline{M}$. The original problem is equivalent to the same problem formulated rather with

$$(14.7)\quad \mathcal{E}_{M,\omega_{X,f},0} = \{u | u \in PSH(M, \omega_{X,f}),\ u \leq 0 \text{ on } \partial M\}, \quad \varphi = \sup_{\mathcal{E}_{M,\omega_{X,f},0}} u.$$

We show that φ is continuous on ∂M. The form $\omega_{X,f} + A\frac{i}{2}\partial\bar{\partial}\rho$ is a Kähler form on M for sufficiently large $A > 0$. Hence $A\rho \in \mathcal{E}_{M,\omega_{X,f},0}$ and then $A\rho \leq \varphi$. On the other hand, for any $u \in \mathcal{E}_{M,\omega_{X,f},0}$, $u + A\rho$ is plurisubharmonic on $\pi_X^{-1}(z)$ for each $z \in X$, for sufficiently large $A > 0$ independent of the choice $z \in X$. Thus $u + A\rho \leq 0$ since $u + A\rho \leq 0$ on $\partial(\pi_X^{-1}(z))$. It easily follows that

$$(14.8)\qquad B\rho \leq \varphi \leq -B\rho$$

for some $B > 0$. In particular, φ is continuous on ∂M.

Next we show that φ is continuous in M. First we fix $A > 0$ with

$$(14.9)\qquad A\rho \leq \varphi \leq -A\rho.$$

For any compact subset K in M and any sufficiently small $\epsilon > 0$, we choose $\delta = (4A)^{-1}\epsilon$ so that $K \subset M_{4\delta}$ and $\varphi \leq \epsilon/4$ on $M \backslash M_{4\delta}$, where

$$(14.10)\qquad M_\delta = \{(z,s) \in M | \rho(z,s) < -\delta\}.$$

By Demailly's regularization techniques [**D**], there exists a decreasing sequence $\{u_j\} \subset PSH\left(M, \left(\frac{j+1}{j}\right)\omega_{X,f}\right) \cap C(M_{\delta/2})$ which converges to the φ^*, the upper semi-continuous envelope of φ. We define

$$(14.11)\qquad \tilde{u}_j(z,s) = \begin{cases} \max\left(\dfrac{j}{j+1}u_j - \epsilon, 2A\rho\right), & (z,s) \in M_\delta \\ 2A\rho & (z,s) \notin M_\delta. \end{cases}$$

On ∂M_δ, $\frac{j}{j+1}u_j - \epsilon \leq -\epsilon/2 \leq -2A\delta$. Hence $\tilde{u}_j \in \mathcal{E}_{M,\omega_{X,f},0}$. Furthermore, on K,

$$\frac{j}{j+1}u_j - \epsilon \geq \frac{j}{j+1}\varphi - \epsilon \geq \frac{j}{j+1}A\rho - 4A\delta \geq A\rho + A\rho = 2A\rho, \tag{14.12}$$

and so $\tilde{u}_j = \frac{j}{j+1}u_j - \epsilon$. It follows immediately that

$$\varphi^* \leq u_j = \frac{j+1}{j}(\tilde{u}_j + \epsilon) \leq \frac{j+1}{j}\varphi + \frac{j+1}{j}\epsilon \tag{14.13}$$

or $0 \leq u_j - \varphi \leq \frac{1}{j}|\varphi| + \epsilon$. Therefore u_j converges to φ uniformly in $L^\infty(K)$ and so φ^* is continuous in K. In conclusion, $\varphi = \varphi^* \in \mathcal{E}_{M,\omega_{X,f},0} \cap C(\overline{M})$.

Finally, we show that $(\frac{i}{2}\partial\bar{\partial}\hat{u})^{\dim M} = 0$. Fix any Euclidean ball B in M. There exists $\eta \in C^\infty(B)$ such that $\omega_{X,f} = \frac{i}{2}\partial\bar{\partial}\eta$ on B. Let

$$\mathcal{E}_{B,(\varphi+\eta)|_{\partial B}} = \{v | \varphi \in PSH(B), v|_{\partial B} = (\varphi + \eta)|_{\partial B}\} \tag{14.14}$$

and $\psi = \sup_{v \in \mathcal{E}_{B,(\varphi+\eta)|_{\partial B}}} v$. Then by the above Bedford-Taylor theorem for $\mathbf{C}^{\dim M}$, we have $(\frac{i}{2}\partial\bar{\partial}\psi)^{\dim M} = 0$ and $\varphi = \psi - \eta$ on $\overline{B}$. Hence $(\omega_{X,f} + \frac{i}{2}\partial\bar{\partial}\varphi)^{\dim M} = 0$ on B.

We turn to the proof of fiber wise regularity. Consider the following elliptic regularization of the homogeneous Monge-Ampère equation,

$$\left(\omega_X + \frac{i}{2}\partial\bar{\partial}\varphi_\varepsilon\right)^{n+m} = \varepsilon\,\omega_M^{n+m} \text{ on } M, \quad \varphi_\epsilon = f \text{ on } \partial M. \tag{14.15}$$

For any $\varepsilon > 0$, there exists a unique smooth solution φ_ϵ in $\overline{M}$. In fact, since $(\omega_X + \frac{i}{2}\partial\bar{\partial}\underline{u}_s)^{\dim M} > \epsilon\omega_M^{\dim M}$, the function $\underline{u}_s \equiv f + A\rho$ is a subsolution satisfying $\underline{u}_s|_{\partial M} = f|_{\partial M}$ for sufficiently large $A > 0$. Thus $\varphi_\varepsilon \geq \underline{u}_\varepsilon$. Furthermore $\varphi_\varepsilon \leq \varphi \in \mathcal{E}_{M,\omega_X,f}$, and we have

$$\|\varphi_\varepsilon\|_{C^0(M)} \leq C \tag{14.16}$$

uniformly in ε.

It suffices now to show that there exists $C > 0$ such that for any $\epsilon \in (0,1)$ and $s \in \overline{\Sigma}$,

$$\left.\left(\omega_X + \frac{i}{2}\partial\bar{\partial}\varphi_\varepsilon\right)\right|_{X_s} \leq C\omega_X|_{X_s}. \tag{14.17}$$

It would follow then that, for any $s \in \overline{\Sigma}$,

$$\left.\left(\omega_X + \frac{i}{2}\partial\bar{\partial}\varphi\right)\right|_{X_s} \leq C\omega_X. \tag{14.18}$$

since φ_ϵ is increasingly monotone as $\varepsilon \to 0$ and φ_ε converges to φ uniformly.

To establish (14.17), let $\omega_\varepsilon = \omega_X + \frac{i}{2}\partial\bar{\partial}\varphi_\varepsilon$. We denote g_X, g_Σ, g_M and g_ε be the Kähler metrics associated to ω_X, ω_Σ, ω_M and ω_ε. We always use the product coordinates for M, where $(z,s) = (z_1, \ldots, z_n, s_1, \ldots, s_m)$ and $z \in X$ and write $\omega_M = \frac{i}{2}(g_X)_{\bar{j}i}dz^i \wedge d\bar{z}^j + \frac{i}{2}(g_\Sigma)_{\bar{\beta}\alpha}ds^\alpha \wedge d\bar{s}^\beta$.

We define

$$H_\epsilon = \frac{\left(\omega_X + \frac{i}{2}\partial\bar\partial\varphi_\varepsilon\right) \wedge \omega_X^{n-1} \wedge \omega_\Sigma^m}{\omega_X^n \wedge \omega_\Sigma^m}. \tag{14.19}$$

Notice that H_ε is the trace of the relative endomorphism between $\omega_\varepsilon|_{X_s}$ and ω_X,

$$H_\varepsilon(z,s) = \mathrm{Tr}_{\omega_X}(\omega_\epsilon|_{X_s})(z) = \sum_{i,j=1}^{n} (g_X)^{i\bar j}(g_\varepsilon)_{\bar j i}, \tag{14.20}$$

where g_X is the Kähler metric associated to ω_X and g_ε is the Kähler metric associated to $\omega_X + \frac{i}{2}\partial\bar\partial\varphi_\varepsilon$. The same calculations as in Yau's Schwarz lemma **[Y78a]** or in Yau's second order estimates, see §6, show that there exists $C_1 > 0$ such that for all $\varepsilon \in (0,1)$, we have on M,

$$\Delta_\varepsilon \log H_\varepsilon \geq -C_1 \mathrm{Tr}_{\omega_\varepsilon}(\omega_M) - C_1, \tag{14.21}$$

where Δ_ε is the Laplacian operator on M with respect to ω_ε. We also have

$$\Delta_\varepsilon(-\varphi_\varepsilon + \rho) = -(n+m) + \mathrm{Tr}_{\omega_\varepsilon}(\omega_X + \omega_\Sigma) = -(n+m) + \mathrm{Tr}_{\omega_\varepsilon}(\omega_M). \tag{14.22}$$

Then there exist $C_2, C_3, C_4 > 0$ which are independent of $\epsilon \in (0,1)$ such that

$$\begin{aligned}\Delta_\varepsilon(\log H_\varepsilon - A\varphi_\varepsilon + A\rho) &\geq (A - C_1)\mathrm{Tr}_{\omega_\varepsilon}(\omega_M) - C_1 - A(n+m)\\ &\geq C_2(\mathrm{Tr}_{\omega_M}(\omega_\varepsilon))^{1/(n+m-1)}\left(\frac{\omega_M^{n+m}}{\omega_\varepsilon^{n+m}}\right)^{1/(n+m-1)} - C_2\\ &\geq C_3 H_\varepsilon - C_4.\end{aligned}$$

On the other hand,

$$H_\varepsilon|_{\partial M} = \left.\frac{\left(\omega_X + \frac{i}{2}\partial\bar\partial f\right) \wedge \omega_X^{n-1}}{\omega_X^n}\right|_{\partial M} \tag{14.23}$$

is uniformly bounded from above for all $\varepsilon \in (0,1)$. Applying the maximum principle, we obtain a constant $C_5 > 0$ such that for all $\varepsilon \in (0,1)$,

$$H_\epsilon \leq C_5 \tag{14.24}$$

since both u_ε and ρ are uniformly bounded in $C(\overline{M})$. The proof of the fiberwise regularity is complete.

14.2. Envelopes with integral conditions. In the previous section, we have seen how envelopes with pointwise Dirichlet conditions can produce solutions to the Dirichlet problem for complex Monge-Ampère equation. It would be interesting to determine whether envelopes with integral conditions can be effectively used to produce other solutions of Monge-Ampère equations, or other canonical metrics. We describe some examples of such envelopes below.

One example is the following hermitian metric defined by Tsuji [**Ts07**] on projective manifolds of general type, generalizing the metric introduced in [**NS**]. Let X be a smooth projective variety of general type. Fix a smooth hermitian metric h_0 on K_X and define

$$\varphi_{can}(z) = \sup\left\{\varphi(z) | Ric(h_0) + \frac{i}{2}\partial\bar{\partial}\varphi \geq 0, \quad \int_X e^{\varphi} h_0^{-1} = 1\right\} \tag{14.25}$$

and

$$h_{can} = e^{-\varphi_{can}} h_0. \tag{14.26}$$

It has been shown by Berman and Demailly [**BD**] that this metric h_{can} coincides with the metric $\tilde{h}_{can}$ defined instead by

(14.27)
$$\tilde{h}_{can}(z) = \inf_{m\in \mathbf{Z}^+} \inf \left\{ (\sigma \wedge \bar{\sigma}(z))^{-1/m} \middle| \int_X |\sigma \wedge \bar{\sigma}|^{1/m} = 1, \ \sigma \in H^0(X, mK_X) \right\}.$$

which is manifestly a birational invariant, $\tilde{h}_{can}$ is a birational invariant since $H^0(X, mK_X)$ is invariant under birational transformations.

Another example is the following. Let Ω be a bounded strictly pseudo-convex domain in $\mathbf{C}^n$. We define

$$\varphi_{can}(z) = \sup\left\{\varphi(z) \middle| \frac{i}{2}\partial\bar{\partial}\varphi \geq 0, \ \int_\Omega e^{\varphi}\left(\frac{i}{2}\partial\bar{\partial}|z|^2\right)^n = 1\right\}. \tag{14.28}$$

We also define $(h_{can})^{-1} = e^{\varphi_{can}}(\frac{i}{2}\partial\bar{\partial}|z|^2)^n$ to be the canonical measure on Ω.

LEMMA 25. *φ_{can} is a plurisubharmonic function on Ω.*

PROOF. First, we show that φ_{can} is bounded from above in any compact subset of Ω. Suppose not, then by taking a subsequence, there exist a sequence of points $z_j \to \hat{z} \in \Omega$ and a sequence of psh functions φ_j with $\int_\Omega e^{\varphi_j} = 1$ such that

$$\varphi_j(z_j) \to \infty.$$

Without loss of generality, we can assume that $B(z_j, r) \subset\subset \Omega$ for all j for some fixed $r > 0$. Then by the mean value inequality and Jensen's inequality, there exist positive constants C_1, C_2, C_3 independent on j such that

$$1 = \int_\Omega e^{\varphi_j} \geq C_1 e^{\int_\Omega \varphi_j} \geq C_2 e^{\int_{B(z_j,r)} \varphi_j} \geq C_3 e^{\varphi_j(z_j)} \to \infty. \tag{14.29}$$

This is a contradiction.

Next we have to show that $\varphi_{can} = (\varphi_{can})^*$. By the definition of φ_{can}, for any $\hat{z} \in \Omega$, there exists a sequence $z_j \to \hat{z}$ and psh φ_j with $\int_\Omega e^{\varphi_j} = 1$ such that

$$\varphi_j(z_j) \to (\varphi_{can})^*(\hat{z}). \tag{14.30}$$

By taking a subsequence, we can assume that φ_j converges to a psh function φ in $L^1(\Omega)$ and thus almost everywhere. In particular, $\int_\Omega e^\varphi \leq 1$ by Fatou's lemma. On the other hand,

$$\begin{aligned}\varphi_{can}(\hat{z}) \geq \varphi(\hat{z}) &= \lim_{r\to 0} \frac{1}{vol(B(\hat{z},r))} \int_{B(\hat{z},r)} \varphi \\ &= \lim_{r\to 0} \lim_{j\to\infty} \frac{1}{vol(B(z_j,r))} \int_{B(z_j,r)} \varphi_j \\ &\geq \lim_{j\to\infty} \varphi_j(z_j) = (\varphi_{can})^*(\hat{z}).\end{aligned}$$

The lemma is proved. Q.E.D.

The following theorem provides a more algebraic characterization of φ_{can}:

THEOREM 30. *$\varphi_{can} \in PSH(\Omega) \cap C(\Omega)$ and*

(14.31)
$$\varphi_{can}(z) = \sup_{m\in\mathbf{Z}^+} \sup \left\{ \frac{1}{m} \log |f|^2(z) | f \in \mathcal{O}(\Omega), \int_\Omega |f|^{\frac{2}{m}} \left(\frac{i}{2}\partial\bar{\partial}|z|^2\right)^n = 1 \right\}.$$

PROOF. Denote by $\varphi_{can,alg}$ the right-hand side of (14.31). It is easy to see that $\varphi_{can,alg} \leq \varphi_{can}$, so it suffices to show that $\varphi_{can,alg} \geq \varphi_{can}$.

By the Ohsawa-Takegoshi extension theorem, for any psh φ on Ω with $\int_\Omega e^\varphi = 1$ and any point $z \in \Omega$, there exists a holomorphic function f on Ω such that for any $m \in \mathbf{Z}^+$,

$$|f|^2 e^{-(m-1)\varphi}(z) = 1 \quad \text{and} \quad \int_\Omega |f|^2 e^{-(m-1)\varphi} \leq C, \tag{14.32}$$

where C does not depend on φ or m. By Hölder's inequality, we have

$$\int_\Omega |f|^{2/m} \leq \left(\int_\Omega |f|^2 e^{-(m-1)\varphi}\right)^{1/m} \left(\int_\Omega e^\varphi\right)^{(m-1)/m} \leq (C)^{1/m}.$$

Let $F = \frac{f}{(\int_\Omega |f|^{2/m})^m}$. Then $F \in \mathcal{O}(\Omega)$ with $\int_\Omega |F|^{2/m} = 1$, and

$$\frac{1}{m} \log |F|^2(z) \geq (m-1)\varphi(z) - \frac{1}{m}C = \varphi(z) - \frac{1}{m}(\varphi(z) + C). \tag{14.33}$$

For fixed z, $\varphi(z)$ and C are uniformly bounded from below. By letting $m \to \infty$, we get

$$\varphi_{can,alg}(z) \geq \varphi(z). \tag{14.34}$$

Since this is true for any z and any psh φ with $\int_\Omega e^\varphi = 1$, we have $\varphi_{can,alg}(z) \geq \varphi_{can}(z)$, and hence $\varphi_{can,alg} = \varphi_{can}$.

Now we can show that φ_{can} is continuous. It suffices to show that φ_{can} is lower semi-continuous. Suppose not. Then there exists $\hat{z} \in \Omega$ and $\varepsilon > 0$ and a sequence of points $z_j \in \Omega$ converging to $\hat{z}$ so that

$$\varphi_{can}(z_j) < \varphi_{can}(\hat{z}) - \epsilon. \tag{14.35}$$

Also there exist $f \in \mathcal{O}(\Omega)$ and $m \in \mathbf{Z}^+$ such that

$$\int_\Omega |f|^{2/m} = 1, \quad \frac{1}{m} \log |f|^2(\hat{z}) > \varphi_{can}(\hat{z}) - \frac{\epsilon}{2}. \tag{14.36}$$

Then there exist $r > 0$ such that for all $z \in B(\hat{z}, r)$,

$$\varphi_{can}(z) \geq \frac{1}{m} \log |f|^2(\hat{z}) > \varphi_{can}(\hat{z}) - \frac{\epsilon}{4}.$$

This is a contradiction. Q.E.D.

A natural question to ask is whether $g_{can} = \frac{i}{2}\partial\bar{\partial}\varphi_{can}$ defines a complete metric on a bounded strictly pseudoconvex domain in $\mathbf{C}^n$ and how it is related to other invariant metrics such as the Bergman, Carathéodory and Kobayashi metrics.

15. Further Developments

As we had acknowledged in the introduction, we could not cover all the possible recent developments, and this survey has not touched on many important topics. In this section, we would like to mention a few and provide some references, for readers who may be completely new to the subject.

A first major omission is a discussion of the important equation (2.1) with $F(z,\varphi) = e^{f(z)-\varphi}$, which corresponds to the open problem of Kähler-Einstein metrics on a compact Kähler manifold (X, ω_0) with $\omega_0 \in c_1(K_X^{-1})$. From the discussion of a priori estimates in §6 and §7, we see that the equation would be solvable if we can obtain a C^0 estimate. The problem is to link such an estimate to stability in GIT, as required by the conjecture of Yau [**Y93**]. Donaldson has recently laid out a program for achieving this [**D10, D11a, D11b, CDa, CDb**]. Prior to this program, Kähler-Einstein metrics with positive scalar curvature have been found in various geometric situations by Tian and Yau [**TY87**] and Tian [**T87**] using the α-invariant, and by Siu [**Si**] and Nadel [**N**] using multiplier ideal sheaves. Necessary conditions for Kähler-Einstein metrics have been obtained by Tian [**T97**]. It has been shown by Tian [**T90b**] that, for surfaces, the existence of Kähler-Einstein metrics is equivalent to the vanishing of the Futaki invariant. A full account of the arguments in [**T90b**] can be found in the paper of Tosatti [**T10**]. The same characterization of the existence of Kähler-Einstein metrics by the vanishing of the Futaki invariant has been established by Wang and Zhu [**WZ2**]. Their proof exploits the fact that the toric potentials on a toric variety satisfy a real Monge-Ampère equation, and the image of their gradients is the polytope of the variety. For a survey of some of these developments and the related issue of stability, see [**PS03, PS08**].

The related question of singularities for the Monge-Ampère equation when the manifold is unstable is of similar considerable interest, and even less explored. The case of holomorphic vector bundles has seen remarkable progress, with the recent works of G. Daskalopoulos and R. Wentworth for complex surfaces [**DW1, DW2**], and A. Jacob [**J1, J2**] for general Kähler manifolds, on the generalization to arbitrary dimensions of the Atiyah-Bott formula for complex curves. An analysis of a break-up of an unstable ruled surface by the Calabi flow has been given by G. Szekelyhidi [**Sz**].

Another major omission is parabolic complex Monge-Ampère equations, and particularly the Kähler-Ricci flow. As we had mentioned earlier, starting with the papers of Cao [**Cao**] and Tsuji [**Ts**], there has been a constant feedback between developments for the elliptic and for the parabolic Monge-Ampère equation. In fact, much of the material discussed in Sections 3, 6, 7 either arose from or are directly motivated by the study of the Kähler-Ricci flow on Fano manifolds (e.g. [**ST09, PS06a, PSSW1, CZ, MS, Yu, Zh**] and references therein) or on manifolds of general type (e.g. [**ST06, TZ, ST09, SW1**]). New powerful techniques have been introduced by Perelman (see [**ST**] for an account of Perelman's unpublished results on the Kähler-Ricci flow). We refer to the papers we listed for a fuller list of references. Related developments for the Sasaki-Ricci flow can be found in [**Co1, Co2, He1**].

The Monge-Ampère measure is uniquely defined by Bedford and Taylor [**BT76**] for locally bounded potentials. The largest classes of possibly unbounded potentials for which a well-behaved measure can be defined have been identified by Cegrell [**Ceg**] and Blocki [**B06**]. The Monge-Ampère measures can also be defined for unbounded potentials, if their singularity set is compact [**D, Sib**]. A prime example is the pluricomplex Green's function (see e.g. [**L, BD, Gb, B00, Ze1**] and references therein). As we saw in Section §9, a non-pluripolar definition can be given, and the range of the corresponding Monge-Ampère measures has been completely characterized by Guedj and Zeriahi [**GuZe**]. The investigation of Monge-Ampère measures which charge pluripolar sets is still in its infancy, see [**CG**] for examples on projective spaces and [**ACCH**] for some general results. It is an important direction for research.

In Section §13, we have seen how geodesics in the space of Kähler metrics lead to the homogeneous complex Monge-Ampère equation. Similarly, Donaldson [**D07**] has shown how geodesics in the space of volume forms on a Riemannian manifold lead to a non-linear equation now known as Donaldson's equation. He also showed how this equation can be interpreted as a PDE version of Nahm's equation in mathematical physics, and is closely related to well-known free boundary problems. The existence of $C^{1,\alpha}$ solutions of Donaldson's equation has been obtained by Chen and He [**CH**] and He [**H**]. The same questions of regularity and maximum rank arise for this equation as they do for the homogeneous complex Monge-Ampère equation. Some early results in low dimensions can be found in [**GPa, GPb**]. The existence and regularity of geodesics in the space of Sasaki metrics have also been investigated in [**GZ**].

We have seen in Section 3.2 how the most basic Alexandrov-Bakelman-Pucci estimates can be applied to the complex Monge-Ampère equations.

It would be interesting to find out whether this method can be carried out further. In this context, we would like to mention the recent remarkable ABP estimates on Riemannian manifolds obtained by Wang and Zhang [**WZ1**], building on earlier works of Cabré [**Ca1**].

Finally, we would like to mention viscosity methods. They have been very successful in the investigation of non-linear equations where no complex structure plays a particular role [**CIL**]. Even though the notion of plurisubharmonicity poses a number of difficulties, it may not be unreasonable to expect that viscosity methods can be developed and become of wider use for equations such as the complex Monge-Ampère equation. Some major steps in incorporating plurisubharmonicity in viscosity methods have been undertaken by Harvey and Lawson [**HL**], Eyssidieux, Guedj, and Zeriahi [**EGZ10**], and Wang [**W1**]. For example, a version of Theorem 12, establishing the existence and uniqueness of viscosity solutions to the Dirichlet problem for the equation (2.1) on domains in $\mathbf{C}^n$ for continuous data, has been established in [**W**].

A. Plurisubharmonic functions

We gather here for the convenience of the reader some basic properties of plurisubharmonic functions and of their Monge-Ampère measures.

A.1. The exponential estimate. Now the L^2 norm, in fact the L^p norm for any $p<\infty$, of any non-positive plurisubharmonic function φ is bounded by a constant depending only on the Kähler class of ω_0. This is a consequence of the following local estimate of Hörmander, extended to Kähler manifolds by [**T87, TY87, Ze**]

THEOREM 31. *Let ω_0 be a Kähler form. There exists a a strictly positive number α and a constant C depending only on ω_0 so that*

$$\frac{1}{[\omega_0^n]}\int_X e^{-\alpha(u-\sup_X u)}\omega_0^n \leq C \tag{A.1}$$

for all $u\in PSH(X,\omega_0)$.

Since $e^{\alpha t}\geq (\frac{\alpha}{p})^p t^p$ for all $t\geq 0$ and all $p>0$, it follows that $\|u-\sup_X u\|_{L^p}$ is bounded by a constant depending only on p and ω_0.

For recent advances on exponential estimates for plurisubharmonic functions, see [**DNS**].

A.2. Regularization of plurisubharmonic functions. The existence of approximations of ω_0-plurisubharmonic functions by monotone sequences of smooth ω_0-plurisubharmonic functions is much more delicate for Kähler manifolds than for domains in $\mathbf{C}^n$. Part of the difficulty resides in the conflicting roles of the differential geometric and the complex structure. An early approximation theorem with loss of ε-positivity is due to

Demailly [**D89**]. Many others are now available, including the recent ones of Demailly, Peternell, and Schneider [**DPS**] and of Demailly and Paun [**DP2**], which imply in particular the following statement: let (X,ω_0) be a Kähler manifold, and let γ be a continuous non-negative $(1,1)$-form. Then for any $\varphi \in PSH(X,\gamma)$ with Lelong numbers $\nu_\varphi(z)=0$ for all $z\in X$, and any subset $X'\subset X$ with compact closure, there exists a decreasing sequence $\varepsilon_j\downarrow 0$, and a sequence $\varphi_j\in PSH(X,\gamma+\varepsilon_j\omega_0)\cap C^\infty(X)$ with $\varphi_j\downarrow\varphi$ in a neighborhood of X'. This statement was given an independent proof by Blocki and Kolodziej [**BK**]. But particularly important for our purposes is the observation of Blocki and Kolodziej [**BK**] that, for X compact and $\gamma=\omega_0$, no loss of positivity is necessary:

THEOREM 32. *Let (X,ω_0) be a compact Kähler manifold. Then for every $\varphi\in PSH(X,\omega_0)$, there exists a sequence $\varphi_j\in PSH(X,\omega_0)\cap C^\infty(X)$ with $\varphi_j\downarrow\varphi$.*

For the convenience of the reader, we provide some details on how to derive Theorem 32 from the result of [**DPS, DP2**].

Let $\varphi\in PSH(X,\omega_0)$, $\varphi\le -1$ on X. For each j, the function $\max(\varphi,-j)$ is bounded, and hance has vanishing Lelong numbers. Thus the result of [**DPS, DP2**] implies the existence of a sequence of smooth functions $\psi_{jk}\in PSH(X,(1+\varepsilon_{j,k})\omega_0)$ with $\psi_{j,k}\downarrow\max(\varphi,-j)$ and $\varepsilon_{j,k}\downarrow 0$ as $k\to\infty$. By passing to a subsequence, we may assume that $\varepsilon_{j,k}\ge\varepsilon_{j+1,k}$ for all j,k. It suffices to show that there exists a sequence $k_1<k_2<\cdots$ so that $\tilde\varphi_j\equiv\psi_{j,k_j}+\frac{1}{2^j}$ is a decreasing sequence converging to φ. The sequence

$$\varphi_j\equiv\frac{\tilde\varphi_j}{1+\varepsilon_{j,k_j}} \tag{A.2}$$

is then a sequence of smooth functions in $PSH(X,\omega_0)$ with $\varphi_j\downarrow\varphi$.

We choose k_j inductively as follows. Fix j. Let

$$C_j=\left\{\psi_{j+1,k}\ge\psi_{l,k_l+j}+\frac{1}{2^{j+1}}\text{ for some } l\le j\right\}. \tag{A.3}$$

Since $\cap_k C_k=\emptyset$, we can choose k_{j+1} so that $C_{k_{j+1}}=\emptyset$. Note that

$$\psi_{j+1,k_j}\le\psi_{j,k_j}+\frac{1}{2^{j+1}}\quad\text{and so}\quad\tilde\varphi_{j+1}\le\tilde\varphi_j. \tag{A.4}$$

Now fix $x\in X$, $\varepsilon>0$, and assume that $\varphi(x)\ge -j_0$. Choose j so that

$$0\le\psi_{j_0,k_{j_0}+j}(x)-\varphi(x)<\varepsilon \tag{A.5}$$

Then $\psi_{j,k_j}-\varphi(x)<\varepsilon+\frac{1}{2^j}$, which implies $\tilde\varphi_j(x)-\varphi(x)<\varepsilon+\frac{1}{2^j}+\frac{1}{2^j}$. The argument is complete.

A.3. The comparison principle. The following is a useful version of the comparison principle. It follows from the standard arguments of Bedford and Taylor [**BT82**], using the above approximation theorem for plurisubharmonic functions on Kähler manifolds.

THEOREM 33. *Let* (X, ω_0) *be a compact Kähler manifold with smooth boundary* ∂X *and dimension* n, *and let* ω *be a smooth, non-negative, closed* $(1,1)$*-form. Then we have*

$$\int_{\{\varphi<\psi\}} \left(\omega + \frac{i}{2}\partial\bar{\partial}\psi\right)^n \leq \int_{\{\varphi<\psi\}} \left(\omega + \frac{i}{2}\partial\bar{\partial}\varphi\right)^n \tag{A.6}$$

for all $\varphi, \psi \in PSH(X, \omega) \cap L^\infty(X)$ *satisfying* $\liminf_{z \to \partial X}(\varphi(z) - \psi(z)) \geq 0$.

Acknowledgements

The first-named author would like to thank Professors Huai-Dong Cao and Xiaofeng Sun of Lehigh University for their very warm hospitality during his visit there during the 2010 conference in honor of Professor C.C. Hsiung. The authors would like to thank Professor Pengfei Guan for providing them with some invaluable references. They have also benefitted greatly from interactions with Professors O. Munteanu, G. Szekelyhidi, V. Tosatti, Ben Weinkove, and the participants of the informal Complex Geometry and PDE seminar at Columbia University.

References

[ACCH] Ahag, P., Cegrell, U., Czyz, R. and Hiep, P.H., *Monge-Ampère measures on pluripolar sets*, J. Math. Pures Appl. **92** (2009) 613–627.

[AT] Arezzo, C. and Tian, G., *Infinite geodesic rays in the space of Kähler potentials*, Ann. Sc. Norm. Super. Pisa Cl. Sci. (5) 2 (2003), no. 4, 617–630.

[A] Aubin, T., *Equations du type Monge-Ampère sur les variétés Kähleriennes compacts*, Bull. Sc. Math. 102 (1976), 119–121.

[B] Bedford, E., *Survey of pluripotential theory* , in *Several Complex Variables*, ed. J.E. Fornaess, Mittag-Leffler Institute, 1987/1988, Princeton University Press.

[BD] Bedford, E. and J.-P. Demailly, *Two counterexamples concerning the pluricomplex Green function in* $\mathbf{C}^n$, Indiana Univ. Math. J. **37** (1988), 865–867.

[BT76] Bedford, E. and Taylor, B.A. *The Dirichlet problem for the complex Monge-Ampère operator*, Invent. Math. 37 (1976), 1–44.

[BT82] Bedford, E. and Taylor, B.A., *A new capacity for plurisubharmonic functions*, Acta Math. 149 (1982), 1–40.

[BB] Berman, R. and Boucksom, S., *Growth of balls of holomorphic sections and energy at equilibrium*, Invent. Math. **181** (2010), no. 2, 337–394.

[BBGZ] Berman, R., Boucksom, S., Guedj, V., and Zeriahi, A. *A variatonal approach to complex Monge-Ampère equations*, arXiv:0907.4490.

[BD] Berman, R. and Demailly, J.P. *Regularity of plurisubharmonic upper envelopes in big cohomology classes*, preprint, arXiv:0905.1246.

[Be1] Berndtsson, B., *Positivity of direct image bundles and convexity on the space of Kähler metrics* J. Differential Geom. **81** (2009), 457–482.

[Be2] Berndtsson, B., *Probability measures associated with geodesics in the space of Kähler metrics*, preprint, arXiv:0907.1806.

[B00] Blocki, Z., *The $C^{1,1}$ regularity of the pluricomplex Green function*, Michigan Math. J. 47 (2000) 211–215.

[B06] Blocki, Z., *The domain of definition of the complex Monge-Ampère operator*, Amer. J. Math. **128** (2006) 519–530.

[B09a] Blocki, Z., *A gradient estimate in the Calabi-Yau theorem*, Math. Ann. **344** (2009), no. 2, 317–327.

[B09b] Blocki, Z., *On geodesics in the space of Kähler metrics*, to appear in the Proceedings of the "Conference in Geometry" dedicated to Shing-Tung Yau (Warsaw, April 2009).

[B11a] Blocki, Z., *On the uniform estimate in the Calabi-Yau theorem, II*, Science China Mathematics, **54** (2011), 1375–1377.

[B11b] Blocki, Z., *The Calabi-Yau theorem*, to appear in Lecture Notes in Mathematics as a part of the volume Complex Monge-Ampère equations and geodesics in the space of Kähler metrics (ed. V. Guedj).

[BK] Blocki, Z. and S. Kolodziej, *On regularization of plurisubharmonic functions on manifolds*, Proc. Amer. Math. Soc. **135** (2007), 2089–2093.

[BEGZ] Boucksom, S., Eyssidieux, P., Guedj, V. and Zeriahi, A., *Monge-Ampère equations in big cohomology classes* Acta Math. **205** (2010), no. 2, 199–262.

[BS] Boutet de Monvel, L. and Sjöstrand, J., *Sur la singularité des noyaux de Bergman et de Szegö*, Astèrisque 34–35 (1976), 123–164.

[Ca1] Cabre, X., *Nondivergent elliptic equations on manifolds with nonnegative curvature*, Comm. Pure Appl. Math., **50** (1997) 623–665.

[Ca] Caffarelli, L., *Interior $W^{2,p}$ estimates for solutions of Monge-Ampère equations*, Ann. of Math. **131** (1990), 135–150.

[CNS] Caffarelli, L., Nirenberg, L., and Spruck, J., *The Dirichlet problem for nonlinear second order elliptic equations. I. Monge-Ampère equations*, Comm. Pure Appl. Math. 37, 1984, pp. 369–402.

[CKNS] Caffarelli, L., Kohn, J.J., Nirenberg, L. and Spruck, J., *The Dirichlet problem for Nonlinear Second-Order Elliptic Equations. II. Complex Monge-Ampère, and Uniformly Elliptic, Equations*, Comm. Pure and Appl. Math. 38 (1985), 209–252.

[Ca2] Calabi, E., *Improper affine hyperspheres and generalization of a theorem by K. Jörgens*, Mich. Math. J. **5** (1958) 105–126.

[Cao] Cao, H.D., *Deformation of Kähler metrics to Kähler-Einstein metrics on compact Kähler manifolds*, Invent. Math. 81 (1985), no. 2, 359–372.

[CZ] Cao, H.D. and Zhu, M., *A note on compact Kähler-Ricci flow with positive bisectional curvature*, Math. Research Letters **16** (2009) 935–939.

[Cat] Catlin, D., *The Bergman Kernel and a Theorem of Tian*, Analysis and geometry in several complex variables (G. Komatsu et al., Eds.), Birkhauser, 1999, pp. 1–23.

[Ceg] Cegrell, U., *Pluricomplex energy*, Acta Math. **180** (1998), 187–217.

[CP] Cegrell, U. and Persson, P., *The Dirichlet problem for the complex Monge-Ampère operator: stability in L^2* , Michigan Math. J. **39** (1992) 145–151.

[C00] Chen, X.X., *The space of Kähler metrics*, J. Differential Geom. 56 (2000), no. 2, 189–234.

[C06] Chen, X.X., *"Space of Kähler metrics III: on the lower bound of the Calabi energy and geodesic distance"*, arXiv: math.DG/0606228.

[CDa] Chen, X.X. and Donaldson, S.K., *"Volume estimates for Kähler-Einstein metrics: the three dimensional case"*, arXiv:1104.0270

[CDb] Chen, X.X. and Donaldson, S.K., *"Volume estimates for Kähler-Einstein metrics and rigidity of complex structures"*, arXiv:1104.4331

[CH] Chen, X.X. and He, W., *The space of volume forms*, preprint, Int. Math. Res. Not. IMRM (2011), no. 5, 967–1009.

[CS] Chen, X.X. and Sun, S., *Calabi flow, geodesic rays, and uniqueness of constant scalar curvature Kähler metrics*, preprint, arXiv:1004.2018.

[CT08] Chen, X.X. and Tang, Y., *Test configuration and geodesic rays*, Astérisque No. 321 (2008), 139–167.

[CT] Chen, X.X. and Tian, G., *Geometry of Kähler metrics and foliations by holomorphic discs*, Publ. Math. Inst. Hautes Etudes Sci. No. 107 (2008), 1107.

[CW] Chen, Y.Z. and Wu, L.C., *Second order elliptic equations and elliptic systems*, Transl. Math. Monographs **174** (1998).

[CY80] Cheng, S.Y. and Yau, S.T., *On the existence of a complete Kähler-Einstein metric on non-compact complex manifolds and the regularity of Fefferman's equation*, Comm. Pure Appl. Math. **33** (1980) 507–544.

[CY86] Cheng, S.Y. and Yau, S.T., *Inequality between Chern numbers of singular Kähler surfaces and characterization of orbit space of discrete group of SU(2,1)*, Contemporary Math. **49** (1986) 31–44.

[Ch] Cherrier, P., *Équations de Monge-Ampère sur les variétés Hermitiennes compactes*, Bull. Sc. Math (2) **111** (1987), 343–385.

[Co1] Collins, T., *The transverse entropy functional and the Sasaki-Ricci flow*, preprint, arXiv:1103.5720.

[Co2] Collins, T., *Uniform Sobolev inequality along the Sasaki-Ricci flow*, preprint, arXiv:1104.1151.

[CG] Coman, D. and Guedj, V., *"Quasi plurisubharmonic Green's functions"*, J. de Mathématiques Pures et Appliquées Volume 92, Issue 5 (2009) 456–475.

[CIL] Crandall, M., Ishii, H and P.L. Lions, *Users guide to viscosity solutions of second order partia differential equations*, Bull. Amer. Math. Soc. **27** (1992) 1–67.

[DW1] Daskalopoulos, G. and Wentworth, R., *Convergence properties of the Yang-Mills flow on Kähler surfaces*, J. Reine Angew. Math. **575** (2004) 69–99.

[DW2] Daskalopoulos, G. and Wentworth, R., *On the blow-up set of the Yang-Mills flow on Kähler surfaces*, Math. Z. **256** (2007), no. 2, 301–310.

[D89] Demailly, J.P., *Potential theory in several complex variables*, Lecture notes, ICPAM, Nice 1989.

[D] Demailly, J.P., *Complex analytic and differential geometry*, book available on line at author's website.

[D1] Demailly, J.P., *Singular Hermitian metrics on positive line bundles*, in Complex algebraic varieties (Bayreuth, 1990), Lecture Notes in Math. 1507, Springer, Berlin, 1992, 87–104.

[DP] Demailly, J.P. and Pali, N., *Degenerate complex Monge-Ampère equations over compact Kähler manifolds*, Internat. J. Math. 21 (2010), no. 3, 357–405, arXiv:0710.51095.

[DP2] Demailly, J.P. and Paun, M., *Numerical characterization of the Kähler cone of a compact Kähler manifold*, Ann. of Math. (2) 159 (2004), no. 3, 1247–1274.

[DPS] Demailly, J.P., Peternell, T. and Schneider, M., *Pseudo-effective line bundles on compact Kähler manifolds*, Internat. J. Math. 12 (2001), 689–741.

[dH] den Hollander, F., *Large deviations*, Fields Institute Monographs, 14. American Mathematical Society, Providence, RI, 2000.

[Di09] Dinew, S., *Uniqueness in $\mathcal{E}(X,\omega)$*, J. of Functional Analysis, **256** (2009), 2113–2122.

[Di09a] Dinew, S., *An inequality for mixed Monge-Ampère measures*, Math. Z. **262** (2009), 1–15.

[DK] Dinew, S. and Kolodziej, S., *Pluripotenial estimates on compact Hermitian manifolds*, preprint, arXiv:0910.3937.

[DZ] Dinew, S. and Zhang, Z., *On stability and continuity of bounded solutions of degenerate complex Monge-Ampère equations over compact Kähler manifolds*, Adv. Math. 225 (2010), no. 1, 367–388.

[DZZ] Dinew, S., Zhang, X. and Zhang, X., "The $C^{2,\alpha}$ estimate of complex Monge-Ampère equation", arXiv:1006.4261.

[DNS] Dinh, T.C., Nguyen, V.A., and Sibony, N., *Exponential estimates for plurisubharmonic functions and stochastic dynamics*, arXiv:0801.1983.

[D87] Donaldson, S., *Infinite deteminants, stable bundles, and curvature*, Duke Math. J. **54** (1987) 231–247.

[D99] Donaldson, S.K., *Symmetric spaces, Kähler geometry and Hamiltonian dynamics*, Northern California Symplectic Geometry Seminar, Amer. Math. Soc. Transl. Ser. 2, 196, Amer. Math. Soc., Providence, RI, 1999, 13–33.

[D01] Donaldson, S.K., *"Scalar curvature and projective imbeddings I"*, J. Differential Geom. **59** (2001) 479–522.

[D90] Donaldson, S.K., *"Holomorphic disks and the complex Monge-Ampère equation"*, J. Symplectic Geometry 1 (2002) no. 2, 171–196.

[D02] Donaldson, S.K., *"Scalar curvature and stability of toric varieties"*, J. Differential Geom. **59** (2002) 289–349.

[D04] Donaldson, S.K., *"Scalar curvature and projective imbeddings II"*, Quarterly J. Math. **56** (2005) no. 3, 345–356.

[D07] Donaldson, S.K., *Nahm's equations and free-boundary problems*, The many facets of geometry, 71–91, Oxford Univ. Press, Oxford, 2010.

[D08] Donaldson, S.K., "Constant scalar curvature metrics on toric surfaces", J. Diff. Geo. **84** (2010), no. 3, 465–488.

[D10] Donaldson, S.K., *"Stability, birational transformations and the Kähler-Einstein problem"*, arXiv:1007.4220.

[D11a] Donaldson, S.K., *"Kähler metrics with cone singularities along a divisor"*, arXiv:1102.1196.

[D11b] Donaldson, S.K., *"b-Stability and blow-ups"*, arXiv:1107.1699.

[DE] Dupuis, P. and Ellis, R. S., *A weak convergence approach to the theory of large deviations*, Wiley Series in Probability and Statistics: Probability and Statistics, A Wiley-Interscience Publication. John Wiley & Sons, Inc., New York, 1997.

[EGZ] Eyssidieux, P., Guedj, V. and Zeriahi, A., *Singular Kähler-Einstein metrics*, J. Amer. Math. Soc., **22** (3) (2009) 607–639.

[EGZ08] Eyssidieux, P., V. Guedj and A. Zeriahi, *A priori L^∞-estimates for degenerate complex Monge-Ampère equations*, Int. Math. Res. Not. IMRN 2008, Art. ID rnn 070, 8 pp.

[EGZ10] Eyssidieux, P., Guedj, V. and Zeriahi, A., *Viscosity solutions to degenerate complex Monge-Ampère equations*, arXiv:1007.0076v1, 2010.

[F76] Fefferman, C., *"Monge-Ampère equations, the Bergman kernel, and geometry of pseudoconvex domains"*, Ann. of Math. **103** (1976) 395–416.

[FN] Fornaess, J. E. and Narasimhan, R., *The Levi problem on complex spaces with singularities*, Math. Ann. 248 (1980), no. 1, 47–72.

[GS] Gamelin, T. and Sibony, N., *Subharmonicity for uniform algebras*, J. Funct. Anal. **35** (1980), no. 1, 64–108.

[GT] Gilbarg, D. and Trudinger, N., *Elliptic partial differential equations of second order*, Classics in Mathematics. Springer-Verlag, Berlin, 2001. xiv + 517 pp. ISBN: 3-540-41160-7.

[Gm] Gill, M., *Convergence of the parabolic complex Monge-Ampère equation on compact hermitian manifolds*, arXiv:1009.5756.

[Gb] Guan, B., *The Dirichlet problem for complex Monge-Ampère equations and regularity of the pluricomplex Green's function*, Comm. in Anal. Geom., 8:213–218, 2000.

[Gb98] Guan, B., *"The Dirichlet problem for Monge-Ampère equations in non-convex domains and space-like hypersurfaces of constant Gauss curvature"*, Trans. Amer. Math. Soc. **350** (1998) 4955–4971.

[GL] Guan, B. and Li, Q., *Complex Monge-Ampère equations and totally real submanifolds*, Adv. Math. (2010), 1185–1223.

[GS] Guan, B. and Spruck, J., *"Boundary value problem on S^n for surfaces of constant Gauss curvature*, Annals of Math., **138** (1993), 601–624.

[Gd] Guan, D., *On modifed Mabuchi functional and Mabuchi moduli space of Kähler metrics on toric bundles*, Math. Res. Lett. 6 (1999), no. 5–6, 547–555.

[Gp] Guan, P., *A gradient estimate for complex Monge-Ampère equation*, 2008 preprint.

[GPa] Guan, P. and Phong, D.H., *A maximum rank problem for degenerate elliptic fully nonlinear equations*, preprint, arXiv:1009.3975.

[GPb] Guan, P. and Phong, D.H., *Partial Legendre transforms of non-linear equations*, preprint, arXiv:1010.2111.

[GZ] Guan, P. and Zhang, X., *"Regularity of the geodesic equation in the space of Sasakian metrics"*, arXiv:0906.5591.

[GuZe] Guedj, V., Zeriahi, A., *The weighted Monge-Ampre energy of quasiplurisubharmonic functions*, J. Funct. Anal. **250** (2007), 442–482.

[H] Han, Q., *Dirichlet problems of Monge-Ampère equations*, lecture notes.

[Ha] Hanani, A. *Equations du type de Monge-Ampère sur les variétés hermitiennes compactes*, J. Funct. Anal. **137** (1996), no. 1, 49–75.

[HL] Harvey, F.R. and Lawson, H.B., *Dirichlet duality and the nonlinear Dirichlet problem*, Comm. Pure Appl. Math., **62** (2009) 396–443.

[He1] He, W., *The Donaldson equation*, preprint, arXiv:0810.4123.

[He2] He, W., *Entire solutions of Donaldson's equation*, preprint, arXiv:1002.4851.

[Ho] Hörmander, L., *An introduction to complex analysis in several variables*, North-Holland, 1984.

[J1] Jacob, A., *The limit of the Yang-Mills flow on semi-stable bundles*, arXiv:1104.4767.

[J2] Jacob, A., *The Yang-Mills flow and the Atiyah-Bott formula for compact Kähler manifolds*, arXiv:1109.1550.

[K83] Kobayashi, R., *"Kähler-Einstein metrics on an open algebraic manifold"*, Osaka. J. Math. **21** (1984) 399–418.

[K98] Kolodziej, S., *The complex Monge-Ampère equation*, Acta Math. **180** (1998) 69–117.

[K03] Kolodziej, S., *The Monge-Ampère equation on compact Kähler manifolds*, Indiana Univ. Math. J. 52 (2003), no. 3, 667–686.

[K05] Kolodziej, S. *The complex Monge-Ampère equation and pluripotential theory*, Memoirs of AMS, Vol. 178, No. 840, 2005.

[KT] Kolodziej, S. and Tian, G., *A uniform L^∞ estimate for complex Monge-Ampère equations* Math. Ann. **342** (2008), 773–787.

[L] Lempert, L., *Solving the degenerate complex Monge-Ampère equation with one concentrated singularity*, Math. Ann. **263** (1983) 515–532.

[LV] Lempert, L. and Vivas, L., *Geodesics in the space of Kähler metrics*, preprint, arXiv:1105.2188.

[LY] Li, P., and Yau, S.T., *On the parabolic kernel of the Scrödinger operator*, Acta Math. **156** (1986), 153–201.

[LYZ] Loftin, J., Yau, S.T., Zaslow, E., *Affine manifolds, SYZ geometry and the "Y" vertex*, J. Differential Geom. **71** (2005), 129–158.

[L] Lu, Z., *On the lower order terms of the asymptotic expansion of Tian-Yau-Zelditch*, Amer. J. Math. **122** (2000) 235–273.

[MM] Ma, X., and Marinescu, G., *Generalized Bergman kernels on symplectic manifolds*, Adv. Math. **217** (2008), 1756–1815.

[M87] Mabuchi, T., *Some symplectic geometry on compact Kähler manifolds*, Osaka J. Math. **24** (1987) 227–252.

[M08] Mabuchi, T., *"K-stability of constant scalar curvature polarization"*, arXiv:0812.4093.

[M09] Mabuchi, T., *"A stronger concept of K-stability"*, arXiv:0910.4617.

[MY83] Mok, N. and Yau, S.T., *Completeness of the Kähler-Einstein metric on bounded domains and the characterization of domains of holomorphy by curvature conditions*, Proc. Symp. in Pure Math. **39** (1983) 41–59.

[MS] Munteanu, O. and Szekelyhidi, G., *"On convergence of the Kähler-Ricci flow"*, arXiv:0904.3505.

[N] Nadel, A., *Multiplier ideal sheaves and Kähler-Einstein metrics of positive scalar curva- ture*, Ann. of Math. 132 (1990) 549–596.

[NS] Narasimhan, M.S. and Simha, R.R. *Manifolds with ample canonical class*, Invent. Math. 5 (1968) 120–128.

[PS03] Phong, D.H. and Sturm, J., *"Stability, energy functionals, and Kähler-Einstein metrics"*, Commun. Anal. Geom. **11** (2003) 563–597.

[PS06] Phong, D.H. and Sturm, J., *The Monge-Ampère operator and geodesics in the space of Kähler potentials*, Inventiones Math. **166** (2006) 125–149.

[PS06a] Phong, D.H. and Sturm, J., *On stability and the convergence of the Kähler-Ricci flow*, J. Differential Geom. **72** (2006) no. 1, 149–168.

[PS07] Phong, D.H. and Sturm, J., *Test configurations and geodesic in the space of Kähler potentials*, J. Symplectic Geom. **5** (2007) no. 2, 221–247.

[PS07a] Phong, D.H. and Sturm, J., *On the regularity of geodesic rays associated to test configurations*, arXiv: 0707.3956.

[PS08] Phong, D.H. and Sturm, J., *"Lectures on Stability and Constant Scalar Curvature"*, in *Handbook of Geometric Analysis Vol III*, ALM 14, Higher Education Press and International Press, Beijing-Boston, 2009, arXiv: 0801.4179.

[PS09a] Phong, D.H. and Sturm, J., *"The Dirichlet problem for degenerate complex Monge-Ampère equations"*, Commun. Analysis and Geom. **18** (2010) 145–170, arXiv: 0904.1898.

[PS09b] Phong, D.H. and Sturm, J., *"Regularity of geodesic rays and Monge-Ampère equations"*, Proc. Amer. Math. Soc. **138** (2010), 3637–3650 arXiv: 0908. 0556.

[PS09c] Phong, D.H. and Sturm, J., *"On pointwise gradient estimates for the complex Monge-Ampère equation"*, arXiv:0911.2881, to appear in *Conference on Geometry in honor of S.T. Yau*, eds. S. Janeczko, J. Li, and D.H. Phong.

[PSS] Phong, D.H., Sesum, N. and Sturm, S., *Multiplier ideal sheaves and the Kähler-Ricci flow*, Commun. Anal. Geom. 15 (2007), no. 3, 613–632.

[PSSW] Phong, D.H., Song, J., Sturm, J. and Weinkove, B., *"The Moser-Trudinger inequality on Kähler-Einstein manifolds"*, Amer. J. Math. **130** No. 4 (2008) 1067–1085.

[PSSW1] Phong, D.H., Song, J., Sturm, J. and Weinkove, B., *"The Kähler-Ricci flow and the $\bar\partial$ operator on vector fields"*, J. Differential Geom. **81**, 3 (2009), 631–647.

[PSSW2] Phong, D.H., J. Song, J. Sturm, and Weinkove, B., *"On the convergence of the modified Kähler-Ricci flow and solitons"*, Comment. Math. Helv. **86** (2011), no. 1, 91–112. arXiv:0809.0941.

[RW] Ross, J. and Witt-Nystrom, D., *Analytic test configurations and geodesic rays*, arXiv:1101.1612.

[RZ08] Rubinstein, Y. and Zelditch, Z., *Bergman approximations of harmonic maps into the space of Kähler metrics on toric varieties*, J. Symplectic Geom. 8 (2010), no. 3, 239–265.

[RZ10a] Rubinstein, Y. and Zelditch, Z., *The Cauchy problem for the homogeneous Monge-Ampère equation, I. Toeplitz quantization*, preprint, arXiv:1008. 3577.

[RZ10b] Rubinstein, Y. and Zelditch, Z., *The Cauchy problem for the homogeneous Monge-Ampère equation, II. Legendre transform*, preprint, arXiv:1010.2442.

[Se] Semmes, S., *Complex Monge-Ampère equations and symplectic manifolds*, Amer. J. Math. 114 (1992) 495–550.

[ST09] Sesum, N., and Tian, G., *Bounding scalar curvature and diameter along the Kähler Ricci flow (after Perelman)*, J. Inst. Math. Jussieu **7** (2008), no. 3, 575–587.

[Sib] Sibony, N., *"Quelques problemes de prolongement de courants en analyse complexe"*, Duke Math. J. Volume 52, Number 1 (1985), 157–197.

[Si] Siu, Y.T., *The existence of Kähler-Einstein metrics on manifolds with positive anticanonical line bundle and a suitable finite symmetry group*, Ann. of Math. (2) 127 (1988), no. 3, 585–627.

[ST06] Song, J. and G. Tian, *The Kähler-Ricci flow on surfaces of positive Kodaira dimension*, Inventiones Math., **170** Number 3, 609–653.

[ST08] Song, J. and Tian, G. *Canonical measures and Kähler-Ricci flow*, preprint, J. Amer. Soc. **25** (2012), 308–353.

[SW] Song, J. and Weinkove, B., *On the convergence and singularities of the J-flow with appli- cations to the Mabuchi energy*, Comm. Pure Appl. Math. 61 (2008), no. 2, 210–229.

[SW1] Song, J. and Weinkove, B., *"Contracting exceptional divisors by the Kähler-Ricci flow"*, arXiv:1003.0718.

[SZ07] Song, J. and Zelditch, S., *Test configurations, large deviations and geodesic rays on toric varieties*, preprint, arXiv:0712.3599.

[SZ10] Song, J. and Zelditch, S., *Bergman metric and geodesics in the space or Kähler metrics on toric varieties*, Anal. PDE 3 (2010), no. 3, 295–358.

[Sz] Szekelyhidi, G., *The Calabi functional on a ruled surface*, arXiv: math/0703562.

[ST] Szekelyhidi, G. and V. Tosatti, *"Regularity of weak solutions of a complex Monge-Ampre equation"*, arXiv:0912.1808.

[T87] Tian, G., *On Kähler-Einstein metrics on certain Kähler manifolds with $C_1 > 0$*, Comm. Math. Phys. 112 (1987), no. 1, 175–203.

[T90a] Tian, G., *On a set of polarized Kähler metrics on algebraic manifolds*, J.Differential Geometry 32 (1990), 99–130.

[T90b] Tian, G., *On Calabis conjecture for complex surfaces with positive first Chern class*, Invent. Math. 101, no. 1 (1990), 101–172.

[T97] Tian, G., *Kähler-Einstein metrics with positive scalar curvature*, Invent. Math. **130** (1997), no. 1, 1–37.

[TY86] Tian, G. and Yau, S.T., *Existence of Kähler-Einstein metrics on complete Kähler manifolds and their applications to algebraic geometry*, in *Mathematical aspects of string theory*, ed. by S.T. Yau (1987) 575–627, World Scientific.

[TY87] Tian, G. and Yau, S.T., *Kähler-Einstein metrics on complex surfaces with $c_1(M)$ positive*, Comm. Math. Phys. 112 (1987) 175–203.

[TY90] Tian, G. and Yau, S.T., *"Complete Kähler manifolds with zero Ricci curvature I"*, J. Amer. Math. Soc. **3** (1990) 579–609.

[TY91] Tian, G. and Yau, S.T., *"Complete Kähler manifolds with zero Ricci curvature II"*, Inventiones Math. **106** (1991) 27–60.

[TZ] Tian, G. and Zhang, Z. *On the Kähler-Ricci flow on projective manifolds of general type*, Chinese Ann. Math. Ser. B 27 (2006), no. 2, 179–192.

[T10] Tosatti, V., *Kähler-Einstein metrics on Fano surfaces*, preprint, arXiv:1010.1500.

[TW1] Tosatti, V. and Weinkove, B., *Estimates for the complex Monge-Ampère equation on Hermitian and balanced manifolds*, Asian J. Math. 14 (2010), no. 1, 19–40.

[TW2] Tosatti, V. and Weinkove, B., "The complex Monge-Ampère equation on compact Hermitian manifolds", J. Amer. Math. Society **23** no. 4 (2010) 1187–1195.

[TWY] Tosatti, V., Weinkove, B. and Yau, S.T., *Taming symplectic forms and the Calabi-Yau equation*, Proc. Lond. Math. Soc. (3) 97 (2008), no. 2, 401–424.

[TrWa] Trudinger, N. and X.J. Wang, *"Boundary regularity for the Monge-Ampère and affine maximal surface equations"*, Ann. of Math. 167 (2008), 993–1028.

[Ts] Tsuji, H., *Existence and degeneration of Kähler-Einstein metrics on minimal algebraic varieties of general type*, Math. Ann. 281 (1988), 123–133.

[Ts07] Tsuji, H., *Canonical singular hermitian metrics on relative canonical bundles*, preprint, arXiv:0704.0566.

[UY] Uhlenbeck, K. and Yau, S.T., *On the existence of Hermitian Yang-Mills connections on stable vector bundles*, Commun. Pure Appl. Math. 39 (1986) 257–293.

[V] Varadhan, S.R.S., *Asymptotic probability and differential equations*, Comm. Pure Appl. Math. **19**, 261–286.

[Wg] Wang, X., *Canonical metrics on stable vector bundles*, Comm. Anal. Geom. **13** (2005), 253–285.

[W1] Wang, Y., *A viscosity approach to the Dirichlet problem for complex Monge-Ampère equations*, arXiv: 1010.1292.

[W2] Wang, Y., *A remark on $C^{2,\alpha}$-regularity of the complex Monge-Ampère equation*, arXiv:1111.0902.

[WZ1] Wang, Y. and Zhang, X., *Measure Estimates, Harnack Inequalities and Ricci Lower Bound*, arXiv:1102.5567.

[WZ2] Wang, X.J. and Zhu, X.H., *Kähler-Ricci solitons on toric manifolds with positive first Chern class*, Advances Math. **188** (2004) 87–103.

[W] Weinkove, B., *Convergence of the J-flow on Kähler surfaces*, Comm. Anal. Geom. 12 (2004), no. 4, 949–965.

[W08] Wu, D., *"Kähler-Einstein metrics of negative Ricci curvature on general quasi-projective manifolds"*, Commun. Analysis Geom. **16** (2008) 395–435.

[Y75] Yau, S.T., *Harmonic functions on complete Riemannian manifolds*, Commun. Pure Appl. Math. **28** (1975), 201–228.

[Y78] Yau, S.T., *On the Ricci curvature of a compact Kähler manifold and the complex Monge-Amp'ere equation, I*, Comm. Pure Appl. Math., **31** (1978), 339–411.

[Y78a] Yau, S.T., *A general Schwarz lemma for Kähler manifolds*, Amer. J. Math. 100 (1978), no. 1, 197–203.

[Y93] Yau, S.T., *Open problems in geometry*, Proc. Symposia Pure Math. 54 (1993) 1–28.

[Y94] Yau, S.T., *A survey of Kähler metrics*, Proc. Symp. in Pure Math. **41** (1984) 285–289.

[Y96] Yau, S.T., *"Review of Kähler-Einstein metrics in algebraic geometry"*, Israel Math. Conference Proc. **9** (1996) 433–443.

[Z] Zelditch, S., *Szegö kernels and a theorem of Tian*, Int. Math. Res. Notices, no.6 (1998), 317–331.

[Ze] Zeriahi, A., *Volume and capacity of sublevel sets of a Lelong class of plurisubharmonic functions*, Indiana Univ. Math. J. **50** (2001), 671–703.

[Ze1] Zeriahi, A., *Pluricomplex Green functions and the Dirichlet problem for the complex Monge-Ampère operator*, Michigan Math. J. **44** (1997) 589–596.

[Yu] Yuan, Y., *On the convergence of a modified Kähler-Ricci flow*, to appear in Math. Zeitschrift.

[Zh] Zhang, Z. *"Kähler Ricci flow on Fano manifolds with vanished Futaki invariants"*, arXiv:1010.5959, to appear in Math. Research Letters.

[ZZ] Zhang, X. and Zhang, X.W., *"Regularity estimates of solutions to complex Monge-Ampre equations on Hermitian manifolds"*, arXiv:1007.2627

Department of Mathematics, Columbia University, New York, NY 10027

Department of Mathematics, Rutgers University, Piscataway, NJ 08854

Department of Mathematics, Rutgers University, Newark, NJ 07102

Some speculations on pairs-of-pants decompositions and Fukaya categories

Paul Seidel

1. Introduction

This is partly a survey and partly a speculative article, concerning a particular question about Fukaya categories of symplectic manifolds. Namely, can we decompose a symplectic manifold into standard pieces, and then reconstruct its Fukaya category by gluing together categories depending only on the geometry of each piece, in a (loosely understood) sheaf-theoretic way? For this to work, some degree of control over pseudo-holomorphic curves is required, an issue which depends on the geometry of the decomposition under consideration. Sheaf-theoretic ideas have been successfully applied to the symplectic geometry of cotangent bundles, starting with the work of Fukaya-Oh [**17**], and followed by Kasturirangan-Oh [**26**, **39**] and Nadler-Zaslow [**35**, **36**] (for a survey of the last-mentioned work and related ideas of Fukaya-Smith, see [**19**]). Recently, Kontsevich [**28**] has proposed a generalization to Stein manifolds whose Lagrangian skeleta have certain singularities. However, that is not quite the direction we wish to take here.

Instead, we focus on situations inspired by homological mirror symmetry [**27**]. In its most naive form, the construction of the mirror via SYZ duality [**31**, **48**] is local. This leads to a sheaf-theoretic description of Fukaya categories of Lagrangian torus fibrations, proposed and partially proved by Kontsevich-Soibelman [**29**]; a closely related approach was pursued by Fukaya [**16**]. However, whenever the SYZ fibration has singularities, instanton corrections emanate from these and spread over the base [**22**, **30**], modifying the naive construction by non-local contributions. Our specific aim here is to look at some other situations where instanton corrections may be absent (as the word *may* indicates, this is strictly conjectural). In Section 2 we consider two instances, namely hypersurfaces in $(\mathbb{C}^*)^n$ and in an abelian variety. In both cases, the pieces of our decomposition are higher-dimensional pairs-of-pants [**32**], and the mirror description is in terms of categories of Landau-Ginzburg branes [**40**] (see also [**9**, **15**]). In Section 3 we make an attempt to merge this with [**29**], which means passing from algebraic to nonarchimedean analytic geometry. Actually, we will do so only in the lowest-dimensional case, so the outcome is a new conjectural model for the Fukaya categories of closed surfaces. This is obviously strongly influenced by the proposed homological mirror symmetry for surfaces of genus ≥ 2, due to Kapustin-Katzarkov-Orlov-Yotov [**24**] (some cases have now been proved [**14, 44**]).

Acknowledgments. This paper owes a somewhat larger than usual debt to others: Mohammed Abouzaid, Denis Auroux, Ludmil Katzarkov, and Nick Sheridan. In particular, Sheridan explained his unpublished work on signed affine structures [**46**] to me, which significantly influenced the material in Section 3. NSF provided partial financial support through grant DMS-0652620.

2. The Zariski topology

Hypersurfaces in the complex torus. We start by stating a version of the homological mirror symmetry conjecture for hypersurfaces in the complex torus $(\mathbb{C}^*)^n$. This amounts to using Mikhalkin's pair-of-pants decompositions [**32**] and the closely related theory of toric degenerations, for which see for instance [**37**]. Besides that, it is strongly influenced by other versions of mirror symmetry considered by Gross-Siebert [**20, 21**] and Abouzaid [**1, 2**].

Suppose that we have a function $\phi_{\mathbb{Z}} : \mathbb{Z}^n \to \mathbb{Z} \cup \{\infty\}$ which is finite at a finite nonempty set of points. To this and a choice of complex parameter $0 < |\epsilon| < 1$ one associates the Laurent polynomial

$$F(z_1, \dots, z_n) = \sum_{\phi_{\mathbb{Z}}(v) < \infty} \epsilon^{\phi_{\mathbb{Z}}(v)} z_1^{v_1} \dots z_n^{v_n}, \tag{2.1}$$

which defines a hypersurface $M = F^{-1}(0) \subset (\mathbb{C}^*)^n$. We will now make some further assumptions. First of all, the set of points where $\phi_{\mathbb{Z}} < \infty$ should be of the form $P_{\mathbb{Z}} = P \cap \mathbb{Z}^n$, where P is a bounded integer polytope with nonempty interior. Moreover, there should be a decomposition of P into simplices P_k which are integral and of volume $1/n!$, with the following properties. Extend $\phi_{\mathbb{Z}}$ to a function $\phi : P \to \mathbb{R}$ which is affine on each P_k. We ask that ϕ should be convex, and not differentiable at any point which lies on more than one P_k. Then, if we take ϵ small, M is smooth, and comes with a decomposition into $(2n-2)$-dimensional pairs-of-pants, one for each P_k [**32**, Theorem 4].

The Legendre transform of ϕ is the function $\psi : \mathbb{R}^n \to \mathbb{R}$ defined by

$$\psi(v) = \max_{w \in P} v \cdot w - \phi(w) = \max_{w \in P_{\mathbb{Z}}} v \cdot w - \phi_{\mathbb{Z}}(w). \tag{2.2}$$

This is again continuous and convex. There is a decomposition of $\mathbb{R}^n$ into finitely many integer polytopes Q_k, of which some will be noncompact, such that ψ is affine on each Q_k, and not differentiable at any point lying on more than one Q_k. One can use ψ to define a toric degeneration, whose generic fibres are isomorphic to $(\mathbb{C}^*)^n$, as follows. For any integer $k \geq 1$, let $\mathcal{R}^k$ be the complex vector space whose basis elements correspond to points $v \in \frac{1}{k}\mathbb{Z}^{n+1}$ such that $v_{n+1} \geq \psi(v_1, \ldots, v_n)$. These spaces form a commutative graded ring $\mathcal{R}$, where the product of generators $v \in \frac{1}{k}\mathbb{Z}^{n+1}$, $w \in \frac{1}{l}\mathbb{Z}^{n+1}$ is

$$\frac{kv + lw}{k + l} \in \tfrac{1}{k+l}\mathbb{Z}^{n+1}. \tag{2.3}$$

$\mathcal{R}$ is also naturally a free module over $\mathbb{C}[t]$, where t acts by mapping $v \in \frac{1}{k}\mathbb{Z}^{n+1}$ to $(v_1, \ldots, v_n, v_{n+1} + \frac{1}{k})$. Hence,

$$\mathcal{X} = \mathrm{Proj}(\mathcal{R}) \tag{2.4}$$

is a toric variety which comes with a flat morphism $W : \mathcal{X} \to \mathbb{A}^1 = \mathrm{Spec}(\mathbb{C}[t])$. Among the symmetries of that variety, we want to single out the $\mathbb{C}^*$-action for which the monomial corresponding to $v \in \frac{1}{k}\mathbb{Z}^{n+1}$ has weight kv_{n+1}. By definition, W is homogeneous with respect to this action. While these properties would be true in greater generality, our assumptions ensure that $\mathcal{X}$ is smooth, and covered by toric Zariski charts $\mathbb{C}^{n+1} \hookrightarrow \mathcal{X}$, one for each simplex in the original decomposition, such that $W(z) = -z_1 \ldots z_{n+1}$ in each chart (however, the weights of the $\mathbb{C}^*$-action can vary from one chart to the next).

EXAMPLE 2.1. *For $n = 2$, M is a curve whose genus is the number of integer points in the interior of P. For instance, take the following $\phi_{\mathbb{Z}}$ and its associated triangulation:*

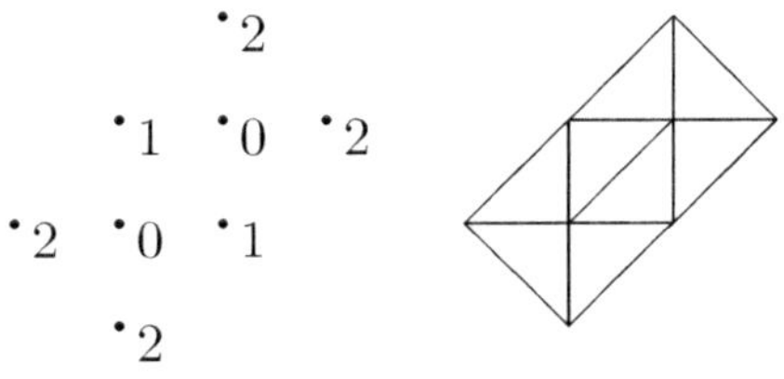

Then, M is a genus 2 curve with 6 points removed, decomposed into 8 pairs-of-pants. The Legendre dual yields the following decomposition:

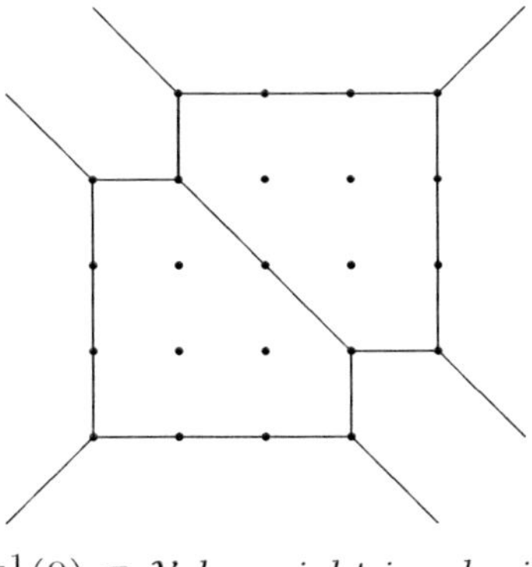

Hence, the fibre $X = W^{-1}(0) \subset \mathcal{X}$ has eight irreducible components, of which the two compact ones are isomorphic to $\mathbb{C}P^1 \times \mathbb{C}P^1$ blown up in one point. This can be considered as a degenerate version of the kind of pictures drawn

in [24], and it also appears naturally when one applies a version of the SYZ construction to obtain a mirror for $(\mathbb{C}^*)^2 \times \mathbb{C}$ *blown up at* $M \times \{0\}$ *[3].*

We now turn to categorical aspects. M is an exact symplectic manifold with a canonical (up to homotopy) trivialization of its canonical bundle. Let $\mathcal{F}(M)$ be its Fukaya category. Here, we restrict the class of objects to compact, exact, graded, and *Spin* Lagrangian submanifolds, so that $\mathcal{F}(M)$ is a $\mathbb{Z}$-graded A_∞-category defined over $\mathbb{C}$. On the other hand, there is a natural triangulated category $D^{\mathbb{C}^*}(W)$ associated to W, the category of equivariant Landau-Ginzburg branes [**41**, Section 3]. By using dg quotients à la [**12**], one can also define an underlying differential graded category $\mathcal{D}^{\mathbb{C}^*}(W)$.

CONJECTURE 2.2. *There is a cohomologically full and faithful embedding* $\mathcal{F}(M) \hookrightarrow \mathcal{D}^{\mathbb{C}^*}(W)$.

The algebro-geometric object $(\mathcal{X}, W)$ on the right hand side is glued together from pieces which follow an essentially unique standard model. In a suitable abstract sense, the same can be said of the associated category $\mathcal{D}^{\mathbb{C}^*}(W)$. One could therefore imagine an approach to Conjecture 2.2 where the Fukaya category is viewed in the same way, and glued together from standard pieces corresponding to pair-of-pants in our decomposition. A first step towards constructing restriction functors which would govern the gluing process was taken in [**4**]. For a partial analysis of the Fukaya category of a single pair-of-pants, see [**44**] (for $n = 1$) and [**45**] (for all n).

REMARK 2.3. *Let* $\mathcal{P}$ *be the fan in* $\mathbb{R}^{n+1}$ *consisting of the cones* $\mathcal{P}_k = \bigcup_{t \in \mathbb{R}^+} \{t\} \times tP_k$. *The associated toric variety is canonically isomorphic to* $\mathcal{X}$. *Projection to* $\mathbb{R}^+$ *yields a toric map* $\mathcal{X} \to \mathbb{A}^1$, *which is the same as our previously defined* W. *This description (for which see [**37**, Section 3] or [**23**, Section 1.2]) shows that* $W : \mathcal{X} \to \mathbb{C}$ *depends only on the decomposition* $P = \bigcup_k P_k$, *and not on the function* ϕ *inducing that decomposition (the additional information provided by* ϕ *amounts to fixing an ample line bundle on* $\mathcal{X}$*). Correspondingly, a construction of* M *which uses only the* P_k *is given in [**32**, Corollary 5.1 and Remark 5.2] (in this framework,* ϕ *determines an automorphism of* M, *namely the monodromy map obtained by moving the parameter* ϵ *around a small circle).*

REMARK 2.4. *One can change the* $\mathbb{C}^*$*-action on* $\mathcal{X}$ *by composing it with any one-parameter subgroup of the natural torus which leaves* W *invariant (which gives roughly a* $\mathbb{Z}^n$ *worth of choices), and such a modification would affect* $\mathcal{D}^{\mathbb{C}^*}(W)$. *There are corresponding variations of* $\mathcal{F}(M)$, *obtained by changing the homotopy class of the trivialization of the canonical bundle.*

REMARK 2.5. *The embedding from Conjecture 2.2 can't be an equivalence, for two different reasons. First,* $\mathcal{D}^{\mathbb{C}^*}(W)$ *is triangulated, and secondly (which is more important from a geometric perspective), it contains* *objects whose endomorphism spaces are infinite-dimensional. To improve the situation, one can enlarge $\mathcal{F}(M)$ to the so-called wrapped Fukaya category [4, **19**], which includes some noncompact Lagrangian submanifolds, and then pass to its triangulated closure. It is a possible, if somewhat hazardous, notion that the outcome might be actually equivalent to $\mathcal{D}^{\mathbb{C}^*}(W)$.*

Hypersurfaces in abelian varieties. There is a variant of the previous picture in which one considers infinite periodic triangulations. The objects encountered in this way are a little more interesting. In particular, on the symplectic geometry side, one gets closed (but nonsimply connected) manifolds.

Our starting point, in the viewpoint of [**33**, Section 5], is a polarized tropical abelian variety. Concretely, this will be given by specifying an integer lattice $\Gamma \subset \mathbb{Z}^n$ and a positive definite integral symmetric matrix g of size n, such that $\gamma \cdot g\gamma \in 2\mathbb{Z}$ for $\gamma \in \Gamma$. In addition, suppose that we have a function $\phi_{\mathbb{Z}} \colon \mathbb{Z}^n \to \mathbb{Z}$ such that $\phi_{\mathbb{Z}}(v) - \frac{1}{2}v \cdot gv$ is Γ-periodic; equivalently,

$$\phi_{\mathbb{Z}}(v+\gamma) = \phi_{\mathbb{Z}}(v) + v \cdot g\gamma + \tfrac{1}{2}\gamma \cdot g\gamma \tag{2.5}$$

for $v \in \mathbb{Z}^n$, $\gamma \in \Gamma$. Consider the entire function on $\mathbb{C}^n$ defined, for some $\epsilon > 0$, by

$$F(z) = \sum_{v \in \mathbb{Z}^n} \epsilon^{-2\pi i v \cdot z + 2\pi \phi_{\mathbb{Z}}(v)}. \tag{2.6}$$

Because it's a Fourier series, this is $\log(\epsilon)^{-1}\mathbb{Z}^n$-periodic, and by (2.5) it also satisfies

$$F(z + ig\gamma) = F(z)\epsilon^{2\pi i z \cdot \gamma - \pi \gamma \cdot g\gamma}. \tag{2.7}$$

Hence, its zero set descends to a hypersurface M in the torus $T = \mathbb{C}^n / \log(\epsilon)^{-1}\mathbb{Z}^n \oplus ig(\Gamma)$. If we equip M with the constant Kähler form

$$\omega = \log(\epsilon)\, d\mathrm{re}(z) \cdot g^{-1}\, d\mathrm{im}(z) \tag{2.8}$$

then the homology class of M is Poincaré dual to $[\omega]$ (since (2.6) is a linear combination of theta-functions for the factor of automorphy (2.7), see for instance [**6**, Chapter 2]). We will now impose further conditions. Namely, there should be a Γ-periodic tiling of $\mathbb{R}^n$ by integer simplices P_k with volume $1/n!$, with the following property. Let $\phi : \mathbb{R}^n \to \mathbb{R}$ be the unique extension of $\phi_{\mathbb{Z}}$ to a function which is continuous, and affine on each P_k. We ask that ϕ must be convex, and not differentiable at any point lying on two different simplices. Then, M will be smooth for sufficiently small values of ϵ.

Take the Legendre transform of ϕ, but where the standard scalar product in (2.2) is replaced by g. The resulting function ψ is convex, takes on integer values at integer points, and has the same periodicity property as ϕ. We associate to it a graded ring $\hat{\mathcal{R}}$, by modifying the previous construction for polytopes in the following way. Consider points $v \in \frac{1}{k}\mathbb{Z}^{n+1}$ such that $v_{n+1} \geq \psi(v_1, \dots, v_n)$, modulo the equivalence relation

$$v \sim \big(v_1 + \gamma_1, \dots, v_n + \gamma_n, v_{n+1} + (v_1, \dots, v_n) \cdot g\gamma + \tfrac{1}{2}\gamma \cdot g\gamma\big) \tag{2.9}$$

for $\gamma \in \Gamma$. Each such point has a height, which is $v_{n+1} - \psi(v_1, \dots, v_n)$, and this is unchanged under (2.9). When defining $\hat{\mathcal{R}}^k$, we allow infinite formal linear combinations of the v, as long as their heights go to infinity. The result is that $\hat{\mathcal{R}}^k$ is a finite rank free module over $\mathbb{C}[[t]]$, where t acts as before by raising the last coordinate by $1/k$. Given two equivalence classes which give rise to generators of $\hat{\mathcal{R}}$, we fix a point v representing the first one, then allow all possible representatives w of the second one under (2.9): the (infinite) sum of the resulting terms (2.3) defines the product of our generators. By taking an appropriate version of Proj, we can associate to $\hat{\mathcal{R}}$ a formal scheme $\hat{\mathcal{X}}$, which comes with a proper flat morphism $\hat{W} : \hat{\mathcal{X}} \to \hat{\mathbb{A}}^1$ to the formal disc. This construction has a long history in algebraic geometry, most of which is frankly outside the competence of this author (the starting point is [**34**]; for an expository account of recent developments, see [**8**]). Because of our assumptions, $\hat{\mathcal{X}}$ itself is smooth, and its zero fibre has toric components, which have normal crossings with each other.

EXAMPLE 2.6. *Take* $\Gamma = \mathbb{Z}$, $g = 2$, *and* $\phi_{\mathbb{Z}}(v) = v^2$. *After substituting* $\epsilon = \exp(\tau/2i)$ *and* $z = 2i\zeta/\tau$, *one sees that* F *is precisely the Jacobi theta function for* $\mathbb{C}/\mathbb{Z} \oplus \tau\mathbb{Z}$, *hence* M *will consist of a single point. The Legendre transform yields* $\psi = \phi$. *The quotient* $R = \hat{\mathcal{R}}/t\hat{\mathcal{R}}$ *has generators* $a \in R^1$, $b \in R^2$ *and* $c \in R^3$ *corresponding to points* $(0,0)$, $(\frac{1}{2}, \frac{1}{2})$ *and* $(\frac{1}{3}, \frac{1}{3})$, *respectively, and the relation*

$$abc = b^3 + c^2 = (\tfrac{1}{2}, \tfrac{1}{2}) + (\tfrac{1}{3}, \tfrac{1}{3}) \in R^6. \tag{2.10}$$

Therefore, the zero-fibre $W^{-1}(0) = \mathrm{Proj}(R)$ *of our degeneration is a rational curve with a node (lying at* $b = c = 0$ *in the coordinates introduced above).*

EXAMPLE 2.7. *Take* $\Gamma = 2\mathbb{Z} \times 2\mathbb{Z} \subset \mathbb{Z}^2$, $g = 4 \cdot \mathrm{Id}$, *and*

$$\phi_{\mathbb{Z}}(v) = \begin{cases} 2\|v\|^2 & \text{if } (v_1, v_2) = (0,0) \text{ mod } 2, \\ 2\|v\|^2 - 2 & \text{if } (v_1, v_2) = (0,1) \text{ or } (1,0) \text{ mod } 2, \\ 2\|v\|^2 - 3 & \text{if } (v_1, v_2) = (1,1) \text{ mod } 2. \end{cases} \tag{2.11}$$

This decomposes $\mathbb{R}^2/\Gamma$ *into eight triangles*

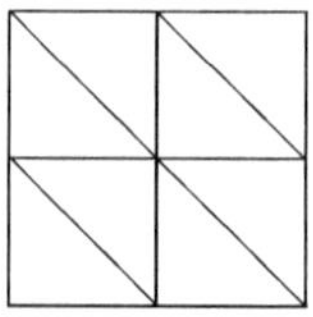

In this case, M *is a genus five curve, correspondingly decomposed into eight pairs-of-pants. The central fibre of the degeneration associated to the*

Legendre dual ψ has four irreducible components, each of which is a copy of $\mathbb{C}P^2$ blown up at three points.

By construction, all our hypersurfaces M have ample canonical bundle. More precisely, that bundle carries a connection whose curvature is proportional to ω. This allows one to define the Fukaya category $\mathcal{F}(M)$ as a $\mathbb{Z}/2$-graded category over $\mathbb{C}$, by appropriately restricting the class of Lagrangian submanifolds under consideration (compare [**44**]; for the more familiar monotone case which has the opposite sign of $c_1(M)$, see [**38, 51**]). On the other side, the main difference with respect to the previous situation is that there is no longer a $\mathbb{C}^*$-action on $\hat{\mathcal{X}}$ covering that on the base. Hence, we necessarily have to work with the ordinary differential graded category of Landau-Ginzburg branes $\mathcal{D}(\hat{W})$, which is $\mathbb{Z}/2$-graded.

CONJECTURE 2.8. *There is a cohomologically full and faithful embedding $\mathcal{F}(M) \hookrightarrow \mathcal{D}^\pi(\hat{W})$, where the superscript π stands for split-closure.*

REMARK 2.9. *The split-closure (also called idempotent or Karoubi completion) is necessary even in simple situations like Example 2.6. In that case, $\mathcal{D}^\pi(\hat{W})$ is generated by a single object, the structure sheaf at the singular point of the zero-fibre. The endomorphisms of that object form a two-dimensional Clifford algebra. Hence, in the split-closure, the object splits into two summands which are isomorphic up to a shift. Each summand corresponds to the point object in the Fukaya category $\mathcal{F}(M)$ (this problem would not have been present in the parallel case of Conjecture 2.2). More generally, passing to the idempotent completion is also a great computational simplification, since we know split-generators of $\mathcal{F}(M)$ in many cases [43].*

REMARK 2.10. *If we set $q = \log(\epsilon)$, then the torus $T \cong \mathbb{C}^n/\mathbb{Z}^n \oplus iqg(\Gamma)$ can be defined for complex values of q (in the left half-plane), and is invariant under $q \mapsto q+i$. The same holds for the hypersurface M, which therefore comes with a monodromy map (compare Remark 2.5). The same conclusion can be reached under slightly weaker assumptions than the ones used before: instead of asking for g to be integral, it is sufficient that $g(\Gamma) \subset \mathbb{Z}^n$. However, the Legendre transformed function ψ then no longer necessarily takes integer values on integers, so the construction of the mirror would have to be modified.*

3. The analytic topology

The local picture for torus fibrations. We start by reviewing some elementary material about the symplectic geometry of Lagrangian torus fibrations, see for instance [**13**]. Consider $M = T^n \times \mathbb{R}^n$ with its standard symplectic form. This is a Lagrangian torus fibration over $B = \mathbb{R}^n$, whose symplectic geometry is tied closely to the $\mathbb{Z}$-affine geometry of the base. For instance, take a function H on B, pull back it to M, and consider the associated Hamiltonian system. This system is one-periodic if and only if H is $\mathbb{Z}$-affine, which means of the form $H(b) = t + v \cdot b$ for some $t \in \mathbb{R}$ and $v \in \mathbb{Z}^n$. Similarly, consider $\mathbb{Z}$-affine automorphisms of B, which are maps of the form $\phi(b) = u + Ab$, where $u \in \mathbb{R}^n$ and $A \in GL_n(\mathbb{Z})$. Denote the group of such automorphisms by G. They then naturally lift to symplectic automorphisms of M. The last-mentioned observation has a more interesting converse:

THEOREM 3.1 (Benci [unpublished] for the convex case, Sikorav [**47**] in general). *Let $U, V \subset \mathbb{R}^n$ be connected open subsets with trivial first Betti number. If $T^n \times U$ is symplectically isomorphic to $T^n \times V$, there is a $\phi \in G$ such that $\phi(U) = V$.*

Let Λ_q be the Novikov field in one variable q, and with coefficients in $\mathbb{C}$. This is the field of formal series $\sum_r a_r q^r$, where $r \in \mathbb{R}$ and $a_r \in \mathbb{C}$, such that for each $C \in \mathbb{R}$, there are only finitely many $r \leq C$ with $a_r \neq 0$. It comes with an obvious real-valued valuation $\mathrm{val}_q : \Lambda_q \to \mathbb{R} \cup \{+\infty\}$. Kontsevich and Soibelman [**29**] introduced the following sheaf $\mathcal{O}_B$ of commutative Λ_q-algebras on B. For any connected open U, $\mathcal{O}_B(U)$ consists of series $f = \sum_{v \in \mathbb{Z}^n} c_v z_1^{v_1} \cdots z_n^{v_n}$, $c_v \in \Lambda_q$, such that the following property holds for each $b \in U$:

$$\text{(3.1)} \qquad \textit{given } C \in \mathbb{R}, \textit{ there are only finitely many } v \in \mathbb{Z}^n \textit{ such that } \mathrm{val}_q(c_v) - b \cdot v \leq C.$$

More canonically, the exponents v naturally belong to the lattice of integer cotangent vectors on B. This shows that the pairing in (3.1) is well-defined without reference to any basis.

EXAMPLE 3.2. *Take $n = 1$ and $U = (-\infty, 0)$. For a function $f = \sum_v c_v z^v$ to lie in $\mathcal{O}_B(U)$, the sequence $\mathrm{val}_q(c_v)$ must satisfy $\lim_{v \to -\infty} \mathrm{val}_q(c_v)/v \to -\infty$ (super-linear decay) and $\liminf_{v \to +\infty} \mathrm{val}_q(c_v)/v \geq 0$ (sub-linear growth). This is the nonarchimedean analogue of the notion of holomorphic functions on the punctured open unit disc in $\mathbb{C}$.*

Algebras of this kind are common in nonarchimedean geometry, but they can also be related to symplectic homology (references for symplectic homology are [**10**,**50**], or for instance the more recent [**11**]). Namely, let H be a function on B, and X_H the Hamiltonian vector field of its pullback to M, which is parallel to the torus fibres. Take the zero-section $L = \{0\} \times \mathbb{R}^n \subset M$. We are interested in length one chords, which are trajectories $x : [0,1] \to M$ of X_H with both endpoints on L. There is precisely one such chord x_b in the fibre over every point b where dH is an integral cotangent vector. The action functional for such chords is given by

$$\text{(3.2)} \qquad A_H(x_b) = H(b) - dH_b \cdot b.$$

Assume that all x_b are nondegenerate, which means that the Hessians $D^2 H_b$ are nonsingular. Then they have well-defined Maslov indices, which are just the Morse indices of H_b. One defines the Floer cochain space $CF^*(L, L; H)$ as follows. Its elements are formal series $\sum_{b,r} a_{b,r} q^r x_b$, where $a_{b,r} \in \mathbb{C}$, the x_b are chords of a given Maslov index, r runs over the real numbers, and the following condition holds: for each $C \in \mathbb{R}$, there are only finitely many nonzero $a_{b,r}$ such that $A_H(x_b) + r \leq C$. If there are only finitely many x_b overall, then $CF^*(L, L; H)$ is just the Λ_q-vector space generated by them. However, in general it is somewhat bigger, being complete for the q-adic topology.

Now take an open subset $U \subset B$ which is bounded and strongly convex. The last-mentioned condition means that it is convex, with smooth boundary, and that the Gauss map $N : \partial U \to S^{n-1}$ is a diffeomorphism. One can then find a continuous function $\bar{H} : \bar{U} \to \mathbb{R}$ which is constant on the boundary, restricts to a smooth and strictly convex function $H = \bar{H}|U$ on the interior, and such that the derivatives grow to infinity in normal direction near the boundary. More precisely, we want

$$\begin{aligned} &\lim_{b \to c} \|dH_b\| = \infty, \\ &\lim_{b \to c} dH_b / \|dH_b\| = N(c) \end{aligned} \tag{3.3}$$

uniformly for $c \in \partial U$ (a general reference for the construction of convex functions is [**7**], but our specific situation is actually much easier than the one considered in that paper). This in particular implies that $dH : U \to \mathbb{R}^n$ is a diffeomorphism. Temporarily ignoring the fact that the function is defined only partially, we can then consider the Floer cochain space $CF^*(L, L; H)$, which in this case is concentrated in degree zero. The chords x_b are naturally parametrized by $v = dH_b \in \mathbb{Z}^n$, so that we can write elements of $CF^0(L, L; H)$ as series $\sum_{v,r} a_{v,r} q^r z^v$. Due to (3.2) and the fact that H is bounded, the growth condition says that for each C, there are only finitely many nonzero $a_{v,r}$ such that $r - v \cdot b \leq C$. This is just the specialization of (3.1) to points b where the chords appear. However, because of (3.3) such points cluster everywhere near the boundary, which implies that

$$CF^0(L, L; H) \cong \mathcal{O}_B(U). \tag{3.4}$$

Hamiltonian functions such as H are tricky from the point of view of Floer cohomology, so one may have trouble using this approach directly (which would be necessary if one wanted to refine (3.4) to an isomorphism of rings rather than vector spaces, for instance). However, there are known workarounds where one first considers a certain class of functions which are constant outside U, and then passes to the limit, see again [**10, 11**].

Global torus fibrations. Again following [**29**], we now globalize the picture above. Namely, let B be a $\mathbb{Z}$-affine manifold. This structure can either be described through its sheaf $\mathcal{A}\!\mathit{ff}_B$ of $\mathbb{Z}$-affine functions (seen as a subsheaf of the continuous function sheaf $\mathcal{C}_B$), or more concretely by specifying an atlas with transition functions in G. There is a canonical symplectic manifold

M associated to B, defined as the quotient of T^*B by its fibrewise integer lattice $T^*_{\mathbb{Z}}B$, and a natural sheaf of Λ_q-algebras $\mathcal{O}_B$ in the base, locally isomorphic to the one introduced above for $B = \mathbb{R}^n$.

From now on, we assume that B is closed. Let $\mathcal{D}(\mathcal{O}_B)$ be the derived dg category of $\mathcal{O}_B$-module sheaves. This is a dg category over Λ_q, whose underlying cohomological category is the standard unbounded derived category, and is unique up to quasi-equivalence. On the symplectic side, consider the Fukaya category $\mathcal{F}(M)$, which is an A_∞-category over Λ_q. To precisely specify this category, we have to make at least two additional remarks. First of all, for any compatible almost complex structure on M, the canonical bundle $\mathcal{K}_M = \Lambda^{\mathrm{top}}_{\mathbb{C}} TM$ can be identified with the complexified pullback of $\Lambda^{\mathrm{top}}_{\mathbb{R}} TB$. In particular, $\mathcal{K}_M^2$ has a canonical trivialization up to homotopy, which we can use to define the gradings in the Fukaya category. Secondly, we can pull back the Stiefel-Whitney class $w_2(B)$ to M, and use Lagrangian submanifolds which are *Pin* relative to this class [**18**] to fix the signs. With this in mind, we have:

THEOREM 3.3 (Kontsevich-Soibelman [**29**]). *The full subcategory of $\mathcal{F}(M)$ consisting of Lagrangian sections of our torus fibration embeds in a cohomologically full and faithful way into $\mathcal{D}(\mathcal{O}_B)$.*

CONJECTURE 3.4 (Kontsevich-Soibelman [**29**]). *All of $\mathcal{F}(M)$ embeds into $\mathcal{D}(\mathcal{O}_B)$ in the same sense.*

The conjecture follows from the theorem above if one can prove that the Lagrangian sections are split-generators for the Fukaya category. This is known when B is a circle [**42**] or a two-dimensional square torus [**5**].

We also need a slight generalization. A *signed $\mathbb{Z}$-affine manifold* is a $\mathbb{Z}$-affine manifold B together with a torsor L^σ over the local system $T^*_{(1/2)\mathbb{Z}} B/T^*_{\mathbb{Z}}B \cong T^*_{\mathbb{Z}}B \otimes \mathbb{Z}/2$ (equivalently, one can express this additional datum as a lift of the transition maps from G to a finite extension $G^\sigma = (\mathbb{Z}/2)^n \rtimes G$). This naturally gives rise to a Lagrangian torus fibration $M \to B$ together with an antisymplectic fibrewise involution, whose fixed point set can be identified with L^σ. On the other hand, we can associate to our signed $\mathbb{Z}$-affine structure a sheaf of matrix algebras $\mathcal{O}_B^\sigma$ of size 2^n over $\mathcal{O}_B$. For $B = \mathbb{R}^n$ with the trivial torsor, rows and columns of the matrices are labeled by elements of $(\mathbb{Z}/2)^n$, and the (i,j) entries are elements of $z_1^{(i_1-j_1)/2} \cdots z_n^{(i_n-j_n)/2}\mathcal{O}_B$. In general, rows and columns are labeled by points in the fibre of L^σ. The analogue of Conjecture 3.4 says that $\mathcal{F}(M)$ should embed into the derived category of $\mathcal{O}_B^\sigma$-module sheaves. For the trivial torsor, this reduces to the original statement by way of Morita equivalence.

In the remainder of the paper, we will consider only the case when B is one-dimensional. In that case, $T^*_{(1/2)\mathbb{Z}}B/T^*_{\mathbb{Z}}B$ is canonically isomorphic to the constant sheaf $\mathbb{Z}/2$, so L^σ is just a double cover of B. We want to tweak the definition of $\mathcal{O}_B^\sigma$ slightly, by turning it into a sheaf of $\mathbb{Z}/2$-graded algebras, where the off-diagonal pieces $z^{1/2}\mathcal{O}_B$ are odd, and the others even. The covering transformation of L^σ acts by permuting the rows as well as the columns of $\mathcal{O}_B^\sigma$.

The pair-of-pants. Our next step is to adapt the mirror of the pair-of-pants from Section 2 to the nonarchimedean framework. Consider tropical affine three-space, which is $B = (\mathbb{R}^+)^3$ with the sheaf $\mathcal{O}_B$ of functions $f = \sum_{v\in\mathbb{Z}^3} c_v z^v$ satisfying the same condition as in (3.1), and with the following additional holomorphicity requirement: if an open connected subset $U \subset B$ contains a point b with $b_k = 0$ for some $k \in \{1,2,3\}$, then all $f \in \mathcal{O}_B(U)$ have $c_v = 0$ for all v such that $v_k < 0$. In particular, global sections consist of formal power series with rapidly decreasing Λ_q-coefficients. Take the superpotential $W = -z_1z_2z_3 \in \mathcal{O}_B(B)$, and its three obvious matrix factorizations, which are

$$E_1 = \{\cdots \mathcal{O}_B \xrightarrow{\quad z_1 \quad} \mathcal{O}_B \xrightarrow{\quad -z_2z_3 \quad} \mathcal{O}_B \cdots\} \tag{3.5}$$

and its cyclic permutations E_2, E_3. The endomorphisms of $E = E_1 \oplus E_2 \oplus E_3$ form a sheaf of $\mathbb{Z}/2$-graded differential algebras $\mathcal{A}_B = End(E)$. If we forget the differential and the grading, then this is the matrix algebra over $\mathcal{O}_B$ of rank 6.

Let $U_3 = (\mathbb{R}^+)^2 \times (0,\infty) \subset B$ be the complement of the third coordinate plane, $I_3 = (0,\infty)$ the open half-line, and $\iota_3 : I_3 \to U_3$ the embedding given by $\iota_3(b) = (0,0,b)$. Consider the trivial double cover of I_3 and the associated sheaf of $\mathbb{Z}/2$-graded algebras $\mathcal{O}_{I_3}^\sigma$.

LEMMA 3.5. *There is a quasi-isomorphism of sheaves of differential algebras,*

$$\phi_3 : (\iota_3)_*\mathcal{O}_{I_3}^\sigma \longrightarrow \mathcal{A}_B|U_3. \tag{3.6}$$

In informal language, the idea behind this statement is that on the region where $z_3 \neq 0$, the superpotential becomes a nondegenerate quadratic function of z_1, z_2. Matrix factorizations of such functions form a semisimple category, see for instance [**25**], which is described by the left hand side of (3.6). More concretely, one first uses z_3^{-1} to build a contraction of $E_3|U_3$, which shows that $\mathcal{A}_B|U_3$ is quasi-isomorphic to its subalgebra of endomorphisms of $E_1 \oplus E_2$. One then writes down an explicit quasi-isomorphism

$$(\iota_3)_*\mathcal{O}_{I_3}^\sigma \longrightarrow End(E_1 \oplus E_2)|U_3, \tag{3.7}$$

as follows. Elements on the left hand side are two-by-two matrices $(f_{ij}(z_3))$, where the diagonal entries are series with z_3^n exponents, while the off-diagonal entries have $z_3^{n+1/2}$ exponents. We take a function $f_{kk}(z_3)$ to the corresponding diagonal degree zero endomorphism of E_k, respectively. The off-diagonal entries go to the odd homomorphisms between E_1 and E_2, given respectively by

$$(3.8)\qquad \begin{array}{ccccc} E_1|U = \{\cdots \mathcal{O}_B & \xrightarrow{z_1} & \mathcal{O}_B & \xrightarrow{-z_2 z_3} & \mathcal{O}_B \cdots\} \\ & \searrow^{z_3^{-1/2} f_{12}(z_3)} & & \searrow^{z_3^{1/2} f_{12}(z_3)} & \\ E_2|U = \{\cdots \mathcal{O}_B & \xrightarrow[z_2]{} & \mathcal{O}_B & \xrightarrow[-z_3 z_1]{} & \mathcal{O}_B \cdots\} \end{array}$$

and the same formula in the other direction. Consider the transposition τ_{12} which exchanges the first two coordinates. Then we have a commutative diagram

$$(3.9)\qquad \begin{array}{ccc} \mathcal{O}^{\sigma}_{I_3} & \xrightarrow{\phi_3} & \mathcal{A}_B|U_3 \\ \downarrow & & \downarrow \tau_{12}^* \\ \mathcal{O}^{\sigma}_{I_3} & \xrightarrow{\tau_{12}^*(\phi_3)} & \tau_{12}^*\mathcal{A}_B|U_3, \end{array}$$

where the left hand $\downarrow$ is the involution of $\mathcal{O}^{\sigma}_{I_3}$ associated to the nontrivial covering transformation.

The *tropical pair-of-pants* is

$$(3.10)\qquad T = (\mathbb{R}^+ \times \{0\}) \cup (\{0\} \times \mathbb{R}^+) \cup \{(-b,-b) \,:\, b \geq 0\} \subset \mathbb{R}^2.$$

Take the embedding $\iota : T \to B$ given by $(b,0) \mapsto (b,0,0)$, $(0,b) \mapsto (0,b,0)$, $(-b,-b) \mapsto (0,0,b)$, and set

$$(3.11)\qquad \mathcal{O}^{\sigma}_T = \iota^* \mathcal{A}_B.$$

In general, as a pullback sheaf this would be defined as a direct limit, but Lemma 3.5 shows that in this case all the maps in the direct system are quasi-isomorphisms. Consider the boundary of a thickening of $T \subset \mathbb{R}^2$, together with its projection to T, which is a double cover except for the preimage of the vertex:

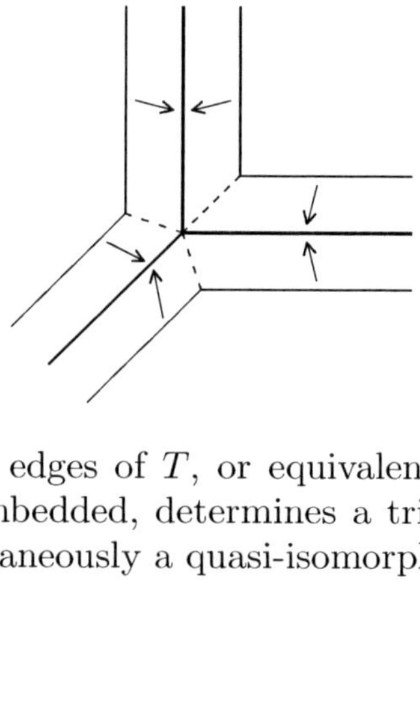

(3.12)

A cyclic ordering of the edges of T, or equivalently an orientation of the plane into which it is embedded, determines a trivialization of the double cover over I_3, and simultaneously a quasi-isomorphism $\mathcal{O}^{\sigma}_{I_3} \simeq \mathcal{O}^{\sigma}_T|I_3$.

Metrized graphs. Let C be a graph with trivalent vertices, but possibly with some semi-infinite edges. A *signed tropical structure* on C is given by a $\mathbb{Z}$-affine structure on each edge, together with a map $L^\sigma \to C$ from a topological one-manifold L^σ which is a double cover away from the vertices, and has the same local structure near those vertices as in (3.12). This restricts to give a signed $\mathbb{Z}$-affine structures on the edges, and it also allows us to glue together the resulting cylinders to form a symplectic surface M, in an essentially unique way (which means up to symplectic automorphisms which are Hamiltonian isotopic to the identity). The Fukaya category $\mathcal{F}(M)$ is a $\mathbb{Z}/2$-graded A_∞-category over Λ_q (if M is not a torus, there are smaller subcategories defined over $\mathbb{C}$, versions of which we used previously in Conjectures 2.2 and 2.8; however, those do not fit in naturally with the present argument).

On the other hand, one can associate to C a sheaf of differential $\mathbb{Z}/2$-graded algebras $\mathcal{O}_C^\sigma$, by using the standard matrix algebra sheaf over the edges, the model $\mathcal{O}_T^\sigma$ near each vertex, and the quasi-isomorphisms from Lemma 3.5 to glue the two together. The general theory of sheaves of dga's can be quite sophisticated, because one wants to allow homotopies between the gluing maps (compare for instance [**49**]), but the present case is much more straightforward, because of the particularly simple topology of the base. There is an associated dg category $\mathcal{D}(\mathcal{O}_C^\sigma)$ of sheaves of dg modules (appropriately derived), and our main idea is expressed by the following:

CONJECTURE 3.6. *$\mathcal{F}(M)$ embeds cohomologically fully and faithfully into $\mathcal{D}(\mathcal{O}_C^\sigma)$.*

REMARK 3.7. *It is helpful to compare this with previously proposed descriptions of the Fukaya category of closed higher genus surfaces. Several such constructions have appeared in the literature, starting with the pioneering [24] (and indeed, we've given another one in Section 2, for surfaces appearing as theta-divisors). In all such cases, the mirror is a Landau-Ginzburg model whose singular set is an algebraic curve consisting of rational components whose intersection models the pair-of-pants decomposition of the surface. However, the local algebro-geometric structure near that set varies, and even though one expects all of these descriptions to be equivalent (after idempotent completion), it is by no means obvious that this is true. One possible advantage of Conjecture 3.6 is that the model is constructed in a more hands-on way, with less extraneous information.*

References

[1] M. Abouzaid. Homogeneous coordinate rings and mirror symmetry for toric varieties. *Geom. Topol.*, 10:1097–1156, 2006.

[2] M. Abouzaid. Morse homology, tropical geometry, and homological mirror symmetry for toric varieties. *Selecta Math. (N.S.)*, 15:189–270, 2009.

[3] M. Abouzaid, D. Auroux, and L. Katzarkov. Work in progress.

[4] M. Abouzaid and P. Seidel. An open string analogue of Viterbo functoriality. *Geom. Topol.*, 14:627–718, 2010.
[5] M. Abouzaid and I. Smith. Homological mirror symmetry for the four-torus. *Duke Math. J.* 152:373–440, 2010.
[6] C. Birkenhage and H. Lange. *Complex abelian varieties (second edition)*. Springer, 2004.
[7] Z. Blocki. Smooth exhaustion functions in convex domains. *Proc. Amer. Math. Soc.*, 125:477–484, 1997.
[8] M. Brion. Compactification de l'espace des modules des variétés abéliennes principalement polarisées (d'après V. Alexeev). *Astérisque*, 311:Exp. No. 952, vii, 1–31, 2007.
[9] R.-O. Buchweitz. Maximal Cohen-Macaulay modules and Tate cohomology over Gorenstein rings. Unpublished manuscript, available at http://hdl.handle.net/1807/16682, 1986.
[10] K. Cieliebak, A. Floer, and H. Hofer. Symplectic homology II: a general construction. *Math. Z.*, 218:103–122, 1995.
[11] K. Cieliebak, U. Frauenfelder, and A. Oancea. Rabinowitz Floer homology and symplectic homology. *Ann. Sci. Ec. Norm. Super. (4)*, 43:957–1015, 2010.
[12] V. Drinfeld. DG quotients of DG categories. *J. Algebra*, 272:643–691, 2004.
[13] J.J. Duistermaat. On global action-angle coordinates. *Comm. Pure Appl. Math.*, 33:687–706, 1980.
[14] A. Efimov. Homological mirror symmetry for curves of higher genus. Preprint arXiv:0907.3903, 2009.
[15] D. Eisenbud. Homological algebra on a complete intersection, with an application to group representations. *Trans. Amer. Math. Soc.*, 260:35–64, 1980.
[16] K. Fukaya. Floer homology for families—a progress report. In *Integrable systems, topology, and physics (Tokyo, 2000)*, pages 33–68. Amer. Math. Soc., 2002.
[17] K. Fukaya and Y.-G. Oh. Zero-loop open strings in the cotangent bundle and Morse homotopy. *Asian J. Math.*, 1:96–180, 1998.
[18] K. Fukaya, Y.-G. Oh, H. Ohta, and K. Ono. *Lagrangian intersection Floer theory - anomaly and obstruction*. Amer. Math. Soc., 2010.
[19] K. Fukaya, P. Seidel, and I. Smith. The symplectic geometry of cotangent bundles from a categorical viewpoint. In *Homological Mirror Symmetry: New Developments and Perspectives*, volume 757 of *Lecture Notes in Physics*, pages 1–26. Springer, 2008.
[20] M. Gross and B. Siebert. Affine manifolds, log structures, and mirror symmetry. *Turkish J. Math.*, 27:33–60, 2003.
[21] M. Gross and B. Siebert. Mirror symmetry via logarithmic degeneration data. I. *J. Differential Geom.*, 72:169–338, 2006.
[22] M. Gross and B. Siebert. From real affine geometry to complex geometry. Preprint arXiv:math/0703822, 2007.
[23] M. Gross and B. Siebert. An invitation to toric degenerations. Preprint arXiv:math/0808.2749, 2008.
[24] A. Kapustin, L. Katzarkov, D. Orlov, and M. Yotov. Homological Mirror Symmetry for manifolds of general type. *Central European J. Math.*, 7:571–605, 2009.
[25] A. Kapustin and Y. Li. D-branes in Landau-Ginzburg models and algebraic geometry. *J. High Energy Phys.*, p. 005, 44 pp., 2003.
[26] R. Kasturirangan and Y.-G. Oh. Floer homology of open subsets and a relative version of Arnold's conjecture. *Math. Z.*, 236:151–189, 2001.
[27] M. Kontsevich. Homological algebra of mirror symmetry. In *Proceedings of the International Congress of Mathematicians (Zürich, 1994)*, pages 120–139. Birkhäuser, 1995.
[28] M. Kontsevich. Talk in Paris, 2009.

[29] M. Kontsevich and Y. Soibelman. Homological mirror symmetry and torus fibrations. In *Symplectic geometry and mirror symmetry*, pages 203–263. World Scientific, 2001.
[30] M. Kontsevich and Y. Soibelman. Affine structures and non-Archimedean analytic spaces. In *The unity of mathematics*, pages 321–385. Birkhäuser, 2006.
[31] N.C. Leung. Mirror symmetry without corrections. *Comm. Anal. Geom.*, 13:287–331, 2005.
[32] G. Mikhalkin. Decomposition into pairs-of-pants for complex algebraic hypersurfaces. *Topology*, 43:1035–1065, 2004.
[33] G. Mikhalkin and I. Zharkov. Tropical curves, their Jacobians and theta functions. In *Curves and abelian varieties*, pages 203–230. Amer. Math. Soc., 2008.
[34] D. Mumford. An analytic construction of degenerating abelian varieties over complete rings. *Compositio Math.*, 24:239–272, 1972.
[35] D. Nadler. Microlocal branes are constructible sheaves. *Selecta Math. (N.S.)*, 15:563–619, 2009.
[36] D. Nadler and E. Zaslow. Constructible sheaves and the Fukaya category. *J. Amer. Math. Soc.*, 22:233–386, 2009.
[37] T. Nishinou and B. Siebert. Toric degenerations of toric varieties and tropical curves. *Duke Math. J.*, 135:1–51, 2006.
[38] Y.-G. Oh. Floer cohomology of Lagrangian intersections and pseudo-holomorphic discs I. *Comm. Pure Appl. Math.*, 46:949–994, 1993.
[39] Y.-G. Oh. Naturality of Floer homology of open subsets in Lagrangian intersection theory. In *The Third Pacific Rim Geometry Conference (Seoul, 1996)*, pages 261–280. Intl. Press, 1998.
[40] D. Orlov. Triangulated categories of singularities and D-branes in Landau-Ginzburg models. *Proc. Steklov Inst. Math.*, 246:227–248, 2004.
[41] D. Orlov. Triangulated categories of singularities, and equivalences between Landau-Ginzburg models. *Mat. Sb.*, 197:117–132, 2006.
[42] A. Polishchuk and E. Zaslow. Categorical mirror symmetry: the elliptic curve. *Adv. Theor. Math. Phys.*, 2:443–470, 1998.
[43] P. Seidel. *Fukaya categories and Picard-Lefschetz theory*. European Math. Soc., 2008.
[44] P. Seidel. Homological mirror symmetry for the genus two curve. *J. Algebraic Geom.*, 20:727–769, 2011.
[45] N. Sheridan. On the homological mirror symmetry conjecture for pairs of pants. *J. Differential Geom.*, to appear.
[46] N. Sheridan. Unpublished notes, 2009.
[47] J.-C. Sikorav. Rigidité symplectique dans le cotangent de T^n. *Duke Math. J.*, 59:759–763, 1989.
[48] A. Strominger, S.T. Yau, and E. Zaslow. Mirror symmetry is T-duality. *Nucl. Phys. B*, 479:243–259, 1996.
[49] B. Toën and G. Vezzosi. From HAG to DAG: derived moduli stacks. In *Axiomatic, enriched and motivic homotopy theory*, pages 173–216. Kluwer Acad. Publ., 2004.
[50] C. Viterbo. Functors and computations in Floer homology with applications, Part I. *Geom. Funct. Anal.*, 9:985–1033, 1999.
[51] K. Wehrheim and C. Woodward. Functoriality for Lagrangian correspondences in Floer theory. *Quantum Topol.*, 1:129–170, 2010.

The k-Yamabe problem

Weimin Sheng, Neil S. Trudinger, and Xu-Jia Wang

1. Introduction

1.1. The k-Yamabe problem. The k-Yamabe problem is a higher order extension of the celebrated Yamabe problem for scalar curvature. It was initially proposed by Viaclovsky [**72**] and also arose in the study of Q-curvatures in [**11**]. Viaclovsky found that in a conformal metric, the resultant k-curvature equation can be expressed as an equation similar to the k-Hessian equation, which has been studied by many authors [**77**].

The k-Yamabe problem can be formulated as follows. Let $(\mathcal{M}^n, g)$ be a closed Riemannian manifold of dimension $n \geq 3$. It is well-known that there is an orthonormal decomposition of the Riemannian curvature tensor

$$Riem = W + A_g \odot g,$$

where W is the Weyl tensor and A_g is the Schouten tensor. The Weyl tensor is conformally invariant and Schouten tensor is given by

$$A_g = \frac{1}{n-2}(Ric_g - \frac{R_g}{2(n-1)}g),$$

where Ric_g and R_g are respectively the Ricci tensor and the scalar curvature. For $1 \leq k \leq n$, the k-Yamabe problem is to prove the existence of a conformal metric $\tilde{g}$ such that

$$\sigma_k(\lambda(A_{\tilde{g}})) = 1 \quad \text{on} \quad \mathcal{M}, \tag{1.1}$$

where $\lambda = (\lambda_1, ..., \lambda_n)$ are the eigenvalues of the Schouten tensor $A_{\tilde{g}}$ with respect to $\tilde{g}$, $\sigma_k(\lambda)$ is the k-th elementary symmetric polynomial,

---

2000 *Mathematics Subject Classification.* 53C21, 58J05, 35J60.

Key words and phrases. Yamabe problem, nonlinear elliptic equation, existence, compactness.

The first author was supported by NSFC 10771189 and 10831008. The second and third authors were supported by the Australian Research Council.

given by

$$\sigma_k(\lambda_1, ..., \lambda_n) = \sum_{1\le i_1<\cdots<i_k\le n} \lambda_{i_1}\lambda_{i_2}\cdots\lambda_{i_k}. \tag{1.2}$$

For convenience we call $\sigma_k(\lambda(A_{\widetilde{g}}))$ the k-scalar curvature, or simply k-curvature, of $(\mathcal{M}, \widetilde{g})$.

Let

$$\widetilde{g} = e^{-2w}g = v^{\frac{4}{n-2}}g, \quad (v = e^{-\frac{n-2}{2}w}). \tag{1.3}$$

By direct computation,

$$A_{\widetilde{g}} = \nabla^2 w + dw\otimes dw - \frac{1}{2}|\nabla w|^2 g + A_g. \tag{1.4}$$

Hence the k-Yamabe problem is equivalent to solving the following PDE

$$\sigma_k[\lambda(\nabla^2 w + dw\otimes dw - \frac{1}{2}|\nabla w|^2 g + A_g)] = e^{-2kw}. \tag{1.5}$$

When $k \ge 2$, this is a fully nonlinear partial differential equation closely related to the k-Hessian equation. Therefore techniques developed in the study of the Yamabe problem and the k-Hessian equation may be applied to the k-Yamabe problem.

1.2. The Yamabe problem. It is convenient to have a brief review of the classical Yamabe problem and the k-Hessian equation. The Yamabe problem, proposed by H. Yamabe in [**80**], is a higher dimensional extension of the famous uniformization theorem, namely any two dimensional Riemannian maniflod is conformally equivalent to one of constant Gauss curvature. Yamabe's paper [**80**] contains the following theorem on the existence of a conformal metric with constant scalar curvature.

THEOREM 1.1. *Let $(\mathcal{M}, g)$ be a compact Riemannian manifold of dimension $n \ge 3$ without boundary. Then there exists a conformal metric $\tilde{g} = v^{4/(n-2)}g$ of constant scalar curvature.*

Yamabe's proof is based on the variational approach. By direct computation, $\tilde{g} = v^{4/(n-2)}g$ is a solution to the Yamabe problem if v satisfies the equation

$$-\frac{4(n-1)}{n-2}\Delta_g v + R_g v = v^{\frac{n+2}{n-2}}, \tag{1.6}$$

or equivalently v is a critical point of the functional

$$\mathcal{F}_1(v) = \frac{\int_{\mathcal{M}} \left(\frac{4(n-1)}{n-2}|\nabla v|_g^2 + Rv^2\right) dvol_g}{\left(\int_{\mathcal{M}} v^{\frac{2n}{n-2}} dvol_g\right)^{\frac{n-2}{n}}}. \tag{1.7}$$

where Δ_g is the Laplace-Beltrami operator, R_g is the scalar curvature of g. However, Yamabe's proof is correct only when the Yamabe constant $Y_1 \le 0$. His proof contains a gap when $Y_1 > 0$, due to the lack of compactness in the Sobolev embedding $W^{1,2} \to L^{2n/(n-2)}$. The second author [**63**] found the gap and fixed it when $Y_1 < \varepsilon$ for some constant ε, (depending on the Sobolev constant). In [**1, 2**], Aubin proved the above theorem under the condition

$$Y_1(g) < Y_1(g_{S^n}), \tag{1.8}$$

where g_{S^n} denotes the standard metric of the n-sphere,

$$Y_1(g) = \inf\{\mathcal{F}_1(v) \mid v > 0, v \in C^\infty(\mathcal{M})\} \tag{1.9}$$

is the Yamabe constant. Aubin also verified (1.8) when $n \ge 6$ and $(\mathcal{M}, g)$ is not locally conformally flat. Schoen [**51**] verified (1.8) for the remaining cases by the positive mass theorem. The positive mass theorem implies a positive lower order term in the expansion of the fundamental solution, which is the key to verify the inequality (1.8). We refer the reader to [**40, 3, 56**] for a detailed account.

The resolution of the Yamabe problem was a milestone in differential geometry, and has stimulated great interest in the study of various prescribing curvature problems, including the k-Yamabe problem.

An interesting problem, raised by Schoen, is the compactness of the solution set when the manifold is not conformally equivalent to the unit sphere. The compactness problem has drawn much attention in recent years. For the 1-Yamabe problem, this problem has been completely resolved. Schoen first proved the compactness in the case when the manifold is locally conformally flat. He also developed a series of local estimates and fine local analysis for possible blow-ups of solutions to equation (1.6) [**52, 53, 54, 57**]. Some of his estimates can be found in [**35**]. His estimates were used by many researchers in the study of semilinear elliptic equations, including the compactness problem, and eventually led to the resolution of the compactness of the solution set for dimensions $n = 3$ in [**47**], $n = 4, 5$ in [**17**], $n = 6, 7$ in [**45, 49**], and also the compactness when $n \le 11$ under the assumption of positive mass theorem in [**46**]. Recently, Khuri, Marques and Schoen [**34**] established the compactness for dimensions $n \le 24$, and Brendle and Marques [**6, 7**] found the compactness does not hold when the dimension $n \ge 25$.

1.3. The k-Hessian equation. When $2 \le k \le n$, (1.5) is a fully nonlinear partial differential equation. It is closely related to the k-Hessian equation,

$$\sigma_k[\lambda(D^2 u)] = f \quad \text{in } \Omega, \tag{1.10}$$

where Ω is a bounded domain in $\mathbb{R}^n$, $D^2 u$ is the Hessian matrix of u, and λ denotes the eigenvalues of $D^2 u$. When $k = 1$, (1.10) is the Poisson equation, when $k = n$ it is the Monge-Ampere equation. For $2 \le k \le n$, the k-Hessian equation is elliptic when restricted to the space of admissible functions, just like the Monge-Ampere equation when restricted to convex functions.

The k-Hessian equation has been studied by many authors. The regularity of k-admissible solutions was established by Caffarelli, Nirenberg and Spruck [**10**]. See also Ivochkina [**33**] for some special cases, and Trudinger [**65**] for Hessian quotient equations. According to Caffarelli, Nirenberg and Spruck [**10**], a function is k-admissible if the eigenvalues of the Hessian matrix D^2u lie in the convex cone

$$\Gamma_k^+ = \{x \in \mathbb{R}^n \mid \sigma_j(x) > 0 \text{ for } 1 \le j \le k\}, \tag{1.11}$$

and equation (1.10) is elliptic when u is k-admissible. For some $k \ge 2$, $\mathbb{R}^n$ contains other cones in which the equation (1.10) is elliptic but in these cases the regularity is unclear at the moment.

The k-Hessian equation is of divergence form. A Sobolev type inequality for k-admissible functions was proved by Wang [**75**], and a Moser-Trudinger type inequality was proved by Tian and Wang [**61**]. The corresponding variational solutions, based on gradient flow approach, were obtained by Chou and Wang [**16**]. Furthermore, various potential theoretic results have also been obtained for the k-Hessian equation by Trudinger and Wang [**66, 67, 68**] and Labutin [**39**], just like those for the Laplacian operator. See also [**77**] for details.

1.4. Admissibility of metrics. Following Caffarelli, Nirenberg and Spruck [**10**], we say a metric $\tilde{g} \in [g]$ *k-admissible* (or simply *admissible*) if $\lambda(A_{\tilde{g}}) \in \Gamma_k^+$. Hence $\tilde{g}$ is 1-admissible iff $\sum \lambda_i > 0$ and $\tilde{g}$ is n-admissibe iff $\lambda_i > 0$ for all i. As in [**10**], one can show that equation (1.5) is elliptic if the eigenvalues $\lambda(A_{\tilde{g}}) \in \Gamma_k^+$. For simplicity we also call the functions u and w k-admissible if the metric $\tilde{g}$ in (1.3) is k-admissible. Denote

$$\begin{aligned} &[g] = \{\tilde{g} = v^{4/(n-2)}g \mid v > 0, v \in C^\infty(\mathcal{M})\}, \\ &[g]_k = \{\tilde{g} \in [g] \mid \tilde{g} \text{ is } k\text{-admissible}\}, \end{aligned} \tag{1.12}$$

where $[g]$ is the conformal class of $[g]$ and $[g]_k$ is the set of k-admissible metrics.

Based on the progress in the study of the k-Hessian equation, we wish to prove related results for the k-Yamabe problem, such as the existence of k-admissible solutions to the k-Yamabe problem when the initial metric g is k-admissible, and related variational properties. However, we would also like to draw the reader's attention to various differences between these two equations. See Remarks 3.1, 4.1 and 4.2.

1.5. Plan of this paper. Assuming the initial metric g is k-admissible for some $2 \le k \le n$, the k-Yamabe problem has been solved in the following cases:

- $(\mathcal{M}, g)$ is locally conformally flat, using the moment map of Schoen and Yau;
- $k = 2$, by the variational structure of equation (1.5) in this case;
- $k \geq \frac{n}{2}$, in which case, the Ricci curvature is nonnegative.

Resolution of the k-Yamabe problem in these cases is built upon the a priori estimates and the Liouville theorem. This paper is arranged as follows.

§2. Liouville theorem
The Liouville theorem for the case $k = 1$ was obtained in [**9**]. Under certain asymptotic assumptions at infinity, the Liouville theorem was proved in [**19**] and also follows from [**50**]. A. Li and Y. Li [**41, 42**] extended the Liouville theorem to the cases $2 \leq k \leq n$. The Liouville theorem was also obtained in [**13**] for $k = 2$ in lower dimensions.

§3. A priori estimates
By the regularity theory of Evans and Krylov for fully nonlinear, uniformly elliptic equations, it suffices to establish second derivative estimates for solutions of (1.5). If the solution is uniformly bounded, the second derivative estimate can be obtained by differentiating the equation (1.5) twice and applying the maximum principle [**73**]. But as with the Yamabe problem, a key issue is to establish the uniform estimate. For this purpose, one needs to prove the local gradient and second derivative estimates.

For equation (1.5), the local second derivative estimate can be obtained as in [**28, 76**]. Indeed after a proper change of the function w, the matrix (1.4) corresponds to that in the equation arising in the reflector design problem. The interior gradient estimate for (1.5) was first obtained by P. Guan and G. Wang [**25**]. Different proofs were later found in [**14, 78, 44**] and they apply to more general equations.

§4. Existence and compactness
This section is the main body of the paper. We divide it into four subsections.

§4.1. Existence on locally conformally flat manifolds.
Due to the characterization of locally conformally flat manifolds by Schoen and Yau, and the invariance of the matrix (1.4) under the Kelvin transform, one can apply the method of moving planes to get the uniform estimate. Therefore the existence of solutions can be obtained by either the elliptic method [**41**] or the parabolic method [**26**]. The existence of solutions has also been obtained by a variational approach in [**60**].

§4.2. Existence when the k-Yamabe problem is variational.
It is known that equation (1.5) is variational when $k = 2$ or $\mathcal{M}$ is locally conformally flat. In [**60**] we proved the existence of solutions when equation (1.5) is variational, namely it is the Euler equation of a functional. Our

result extended the one in [**11, 12**] where the existence was proved when $k = 2$ and $n = 4$; and that in [**26, 41**] on locally conformally flat manifolds. Note that our proof applies to the cases $k \geq 3$ for any manifolds, provided the equation (1.5) is variational. However, Branson and Gover [**5**] proved that when $k \geq 3$, equation (1.5) is variational if and only if the manifold $\mathcal{M}$ is locally conformally flat.

§4.3. Existence for $k > n/2$.
When $k > n/2$, the existence of solutions was proved by Gursky and Viaclovsky in [**30**]. They also proved the set of solutions is compact. However in the cases $k > n/2$, much stronger results were proved by the authors in [**70**]. It was proved in [**70**] that the set of admissible metrics, after normalization such that the underlying manifold is of constant volume, is compact. Therefore one can prove the existence of solutions for more general nonlinear right hand side term in (1.5). Some other interesting results on the singularity profile for admissible functions were also established.

§4.4. Existence for $k = n/2$.
The cases $k = 2$ and $k = \frac{n}{2}$ are probably the most interesting cases for the k-Yamabe problem (except $k = 1$). In the case $k = n/2$, we proved the solution set is compact and so also the existence of solutions [**71**]. As in [**70**], our proof is based on the analysis of a minimal radial function of the solution and also relies on the Liouville theorem of A. Li and Y. Li. A key fact for the cases $k \geq n/2$ is that if g is k-admissible, then the Ricci curvature of the manifolds is nonnegative, so that the Bishop′s volume growth theorem is applicable.

For the compactness of the solution set of the k-Yamabe problem, when $k < \frac{n}{2}$, nothing has been proved, but we expect similar results as the 1-Yamabe problem, namely the solution set is compact in lower dimensions and not otherwise. But when $k \geq \frac{n}{2}$, the solution set is compact [**30, 70, 71**]. Unlike the case $k = 1$, the compactness for $k \geq \frac{n}{2}$ is built upon the fact that a k-admissible metric must have nonnegative Ricci curvature [**24, 71**].

§5. Remarks.
We include in §5.1 a result proved by the first author in [**59**], which says that if a k-Yamabe constant is positive, then there is a k-admissible metric. In §5.2 we explain that the results obtained for equation (1.5) apply to a larger class of comformally invariant equations satisfying certain structural conditions. Finally we mention some unsolved problems.

2. Liouville theorem

Let $\tilde{g} = v^{\frac{4}{n-2}} g$. Then the Schouten tensor of $\tilde{g}$ is given by

$$A_{\tilde{g}} = \frac{2}{n-2}\frac{\nabla_c^2 v}{v} + A_g$$

and the k-Yamabe problem (1.1) becomes

$$\sigma_k(\lambda(\nabla_c^2 v + \tfrac{n-2}{2} v A_g)) = (\tfrac{n-2}{2})^k v^{k\frac{n+2}{n-2}}. \tag{2.1}$$

where

$$\nabla_c^2 v = -\nabla^2 v + \frac{n}{n-2}\frac{dv \otimes dv}{v} - \frac{|\nabla v|^2}{v}\frac{g}{n-2} \tag{2.2}$$

is the conformal Hessian matrix.

The following Liouville type theorem is due to Caffarelli, Gidas and Spruck [**9**] (see also [**35**]) for $k = 1$ and by A. Li and Y. Y. Li [**41, 42**] for $k \geq 2$. The Liouville theorem has been used in the proof for the existence of solutions to the k-Yamabe problem for $k = \frac{n}{2}$ [**71**]. It has also been used to establish the local gradient estimate for the equation (2.1) [**44, 78**].

THEOREM 2.1. *Let $v \in C^2(R^n)$ be a positive, k-admissible solution of*

$$\sigma_k(\lambda(\nabla_c^2 v) = (\frac{n-2}{2})^k v^{k\frac{n+2}{n-2}-\varepsilon} \tag{2.3}$$

in the entire space $\mathbb{R}^n$, where $n \geq 3$, $1 \leq k \leq n$, and $0 \leq \varepsilon < k\frac{n+2}{n-2}$. Then $\varepsilon = 0$ and there exist a constant $a > 0$ and a point $\overline{x} \in \mathbb{R}^n$ such that

$$v(x) = c(n,k)(\frac{a}{1 + a^2|x - \overline{x}|^2})^{\frac{n-2}{2}}, \tag{2.4}$$

where $c(n,k) = 2^{\frac{n-2}{4}} \binom{n}{k}^{\frac{n-2}{4k}}$.

When $k = 1$ and $\varepsilon = 0$, equation (2.3) becomes

$$-\Delta v = \frac{n-2}{2} v^{\frac{n+2}{n-2}} \quad \text{on } \mathbb{R}^n. \tag{2.5}$$

The above Liouville theorem asserts that a positive C^2 solution of (2.5) must be of the form

$$v(x) = (2n)^{\frac{n-2}{4}}(\frac{a}{1 + a^2|x - \overline{x}|^2})^{\frac{n-2}{2}}. \tag{2.6}$$

Under the decay hypothesis $v(x) = O(|x|^{2-n})$, the result was proved in [**19**] and also follows from Obata [**50**]. Obata proved the uniqueness of solutions (up to conformal diffeomorphism) for the Yamabe problem on the unit sphere.

Obata's result was extended to $k \geq 2$ by Viaclovsky [**72, 74**]. By the stereographic projection, Viaclovsky's uniqueness result implies the following: If v is a k-admissible solution to (2.3) and if the function $\tilde{v}(x) := |x|^2 v(\frac{x}{|x|^2})$ is smooth and positive at the origin, then v must be given by (2.4). This result can be proved by the method of moving planes as the equation (2.3) is invariant under Kelvin transform [**74**]. In the case $k = 2$ and

$n = 4, 5$, Theorem 2.1 was also proved by Chang, Gursky and Yang [**12, 13**] by different argument, where some further results were also obtained.

The proof of Theorem 2.1 by A. Li and Y. Li [**41, 42**] is also based on the Kelvin transform and the moving planes technique. But they need to establish some local estimates, such as a Harnack type inequality of Schoen, and prove an asymptotic behavior of the solution at infinity. Their initial proof [**41**] used the local gradient estimate in [**25**] but later they were able to avoid the a priori estimate by an observation on the behavior of isolated singularities. The Liouville theorem was extended to weak entire solutions in [**43**], which in turn can be used to establish the interior gradient estimate [**78, 44**]. A. Li and Y. Li have also extended the Liouville theorem to much more general fully nonlinear equations [**42**].

3. Local estimates

In this section we consider a priori estimates for k-admissible solutions to the equation

$$\sigma_k[\lambda(\nabla^2 w + dw \otimes dw - \frac{1}{2}|\nabla w|^2 g + A_g)] = f(x)e^{-2kw}. \tag{3.1}$$

Equation (3.1) is a fully nonlinear partial differential equation, analogous to the k-Hessian equation (1.10). Assume that $f > 0$ and is $C^{1,1}$ smooth. To obtain the $C^{3,\alpha}$ estimate for solutions to equation (3.1), by the regularity theory of Evans [**18**] and Krylov [**36**] it suffices to prove the second derivative estimate, so that the equations become uniformly elliptic.

The second derivative estimate on compact manifolds without boundary can be obtained easily by differentiating equation (3.1) twice and using the maximum principle [**73**]. For Yamabe type problems such as (3.1), the L^∞ estimate is usually a key issue, and we need interior gradient and second derivative estimates.

The interior second derivative estimate for (3.1) can be obtained similarly as in [**76, 28**]. Indeed, in the case $k = n$, equation (3.1) on sphere (with a different zero order term in the matrix) also arises in the reflector design [**76, 28**]. The interior gradient estimate for (3.1) seems more complicated and was first obtained in [**25**]. Different proofs for the interior gradient estimate were later found in [**14, 78, 44**].

The proof by Guan and Wang [**25**] of the interior gradient estimate relies on the algebraic structure of the operator σ_k. Chen's proof [**14**] combines the estimates for gradient and second derivatives in a single argument. The proof in [**78, 44**] is based on the Liouville theorem [**43**]. (The third author would like to point out that before his paper [**78**] was published in early 2006, he was not aware that Yanyan Li was trying to prove the regularity by his Liouville theorem, as there was no indication of it in the early version of [**43**]).

As pointed out in [**78**], Chen's proof is the favorable one. It is simpler than those in [**25, 78, 44**]. The proofs in [**14, 25**] also apply to corresponding parabolic equations. Below we adopt the proof of Chen [**14**].

THEOREM 3.1. *Let $w \in C^3(\mathcal{M}^n)$ be a k-admissible solution to (3.1) in the geodesic ball $B(x_0, r)$. Then there is a constant C such that*

$$\max_{B(x_0,r/2)} (|\nabla w|^2 + |\nabla^2 w|) \leq C(1 + \max_{B(x_0,r)} e^{-2w}). \tag{3.2}$$

PROOF. The proof also applies to more general equations of the form

$$F(W) = f(x, w),$$

where

$$W = A_g + \nabla^2 w + dw \otimes dw - \frac{1}{2}|\nabla w|^2 g.$$

We assume that $F = F(\lambda(W))$ is a function of the eigenvalues λ of the matrix W, satisfying the following conditions:

(i) $F(\lambda)$ is symmetric in λ,
(ii) F is homogeneous of degree 1,
(iii) F is concave, and
(iv) F is positive in the cone $\{\lambda \in \mathbb{R}^n \mid \lambda_1 > 0, \ldots, \lambda_n > 0\}$.

Note that the above conditions imply that $F(\lambda(W))$ is elliptic in w. In the proof here we assume

$$F = [\sigma_k(\lambda(W))]^{1/k}.$$

First, since $W \in \Gamma_k^+$, we have $tr_g W > 0$. Hence

$$tr_g A_g + \Delta w - \frac{n-2}{2}|\nabla w|^2 > 0.$$

Therefore we have

$$|\nabla w|^2 < C(\Delta w + 1). \tag{3.3}$$

This means Δw is bounded from below, and an upper bound of Δw implies an upper bound of $|\nabla w|^2$. Since $k \geq 2$, we also have

$$0 < \sigma_2(W) = \frac{1}{2}((tr_g W)^2 - |W|^2),$$

i.e.,

$$|\nabla^2 w|^2 \leq C((\Delta w)^2 + 1). \tag{3.4}$$

Hence an upper bound of Δw implies an upper bound for $|\nabla^2 w|^2$.

To prove the interior estimates, we introduce the auxiliary function

$$H = \eta(\Delta w + |\nabla w|^2) =: \eta L,$$

where $0 \le \eta \le 1$ is a cutoff function satisfying $\eta = 1$ in $B_{r/2}$, $\eta = 0$ outside B_r, $|\nabla \eta| < C(\sqrt{\eta}/r)$ and $|\nabla^2 \eta| < C/r^2$. For simplicity we assume $r = 1$. Suppose x_0 is the maximal point of H. We may assume Δw is positive at x_0. Take a local coordinates such that $g_{ij}(x_0) = \delta_{ij}$. Then at x_0 we have

$$
\begin{aligned}
&0 = \nabla_j H = \eta_j(\Delta w + |\nabla w|^2) + \eta(w_{kkj} + 2w_k w_{kj}),\\
&0 \ge \nabla_i \nabla_j H = (\eta_{ij} - \frac{2\eta_i\eta_j}{\eta})L + \eta L_{ij}.
\end{aligned} \tag{3.5}
$$

Hence

$$
\begin{aligned}
0 \ge F^{ij} H_{ij} &= F^{ij}((\eta_{ij} - \frac{2\eta_i\eta_j}{\eta})L + \eta L_{ij})\\
&\ge -C(\Sigma F^{ii})L + \eta F^{ij} L_{ij},
\end{aligned} \tag{3.6}
$$

where $F^{ij} = \frac{\partial F}{\partial r_{ij}}$ is positive definite. Now we compute

$$
\begin{aligned}
F^{ij}L_{ij} &= F^{ij}(w_{kkij} + 2w_{ki}w_{kj} + 2w_k w_{kij})\\
&\ge F^{ij} w_{ijkk} + 2F^{ij}(w_{ki}w_{kj} + w_k w_{ijk}) - C(\Sigma F^{ii})(1 + |\nabla^2 w|^{3/2})\\
&= I + II - C(\Sigma F^{ii})(1 + |\nabla^2 w|^{3/2}).
\end{aligned} \tag{3.7}
$$

Notice that

$$
W_{ij,kk} = w_{ijkk} + 2(w_{ik}w_{jk} + w_{ikk}w_j) - (|\nabla^2 w|^2 + w_l w_{lkk})\delta_{ij} + A_{ij,kk}.
$$

We have by (3.5)

$$
\begin{aligned}
I &= F^{ij} w_{ijkk}\\
&= F^{ij}(W_{ij,kk} - 2(w_{ik}w_{jk} + w_{ikk}w_j) + (|\nabla^2 w|^2 + w_l w_{lkk})\delta_{ij} - A_{ij,kk})\\
&\ge F^{ij}W_{ij,kk} - 2F^{ij}w_{ik}w_{jk} + 2F^{ij}w_j(\frac{\eta_i}{\eta}L + 2w_k w_{ki})\\
&\quad + (\Sigma F^{ii})(|\nabla^2 w|^2 - w_l(\frac{\eta_l}{\eta}L + 2w_k w_{kl})) - C(\Sigma F^{ii})(1 + |\nabla^2 w|^{3/2})\\
&\ge F^{ij}W_{ij,kk} + F^{ij}(4w_k w_j w_{ki} - 2w_{ik}w_{jk} + |\nabla^2 w|^2\delta_{ij} - 2w_k w_l w_{kl}\delta_{ij})\\
&\quad - C\eta^{-1/2}(\Sigma F^{ii})|\nabla w|L - C(\Sigma F^{ii})(1 + |\nabla^2 w|^{3/2}).
\end{aligned}
$$

Since

$$
W_{ij,k} = w_{ijk} + w_{ik}w_j + w_i w_{jk} - w_{lk}w_l\delta_{ij} + A_{ij,k},
$$

we have

$$
\begin{aligned}
II &= 2F^{ij}w_{ki}w_{kj} + 2w_k F^{ij}(W_{ij,k} - w_{ik}w_j - w_i w_{jk} + w_{lk}w_l\delta_{ij} - A_{ij,k})\\
&= 2w_k F^{ij} W_{ij,k} + F^{ij}(2w_{ki}w_{kj} - 4w_i w_k w_{kj} + 2w_k w_{lk} w_l \delta_{ij})\\
&\quad - C(\Sigma F^{ii})(1 + |\nabla^2 w|^{3/2}).
\end{aligned}
$$

Substituting I and II into (3.7) we have

$$\begin{aligned} F^{ij}L_{ij} \geq{}& F^{ij}W_{ij,kk} + 2w_kF^{ij}W_{ij,k} + (\Sigma F^{ii})|\nabla^2 w|^2 \\ &- C\eta^{-1/2}(\Sigma F^{ii})|\nabla w|L - C(\Sigma F^{ii})(1 + |\nabla^2 w|^{3/2}). \end{aligned}$$

Hence from (3.6) we get

$$\begin{aligned} 0 \geq \eta F^{ij}H_{ij} \geq{}& \eta^2 F^{ij}W_{ij,kk} + 2\eta^2 w_k F^{ij}W_{ij,k} + \eta^2(\Sigma F^{ii})|\nabla^2 w|^2 \\ &- C(\Sigma F^{ii})(1 + (\eta|\nabla^2 w|) + (\eta|\nabla^2 w|)^{3/2}). \end{aligned} \tag{3.8}$$

By (3.3) and (3.4),

$$C^{-1}L \leq |\nabla^2 w| \leq C(L+1) \tag{3.9}$$

for some constant C. Hence by (3.8),

$$\begin{aligned} 0 \geq{}& \eta^2 F^{ij}W_{ij,kk} + 2\eta^2 w_k F^{ij}W_{ij,k} + (\eta/C)^2(\Sigma F^{ii})L^2 \\ &- C(\Sigma F^{ii})(1 + \eta L + (\eta L)^{3/2}). \end{aligned} \tag{3.10}$$

By the concavity of F,

$$\begin{aligned} \Delta F(W) &= F^{ij,rs}W_{ij,k}W_{rs,k} + F^{ij}W_{ij,kk} \\ &\leq F^{ij}W_{ij,kk}. \end{aligned}$$

Hence by equation (3.1),

$$\begin{aligned} F^{ij}W_{ij,kk} &\geq \Delta[f(x)e^{-2w}] \\ &\geq -C(\Sigma F^{ii})(1 + e^{-2w})L. \end{aligned} \tag{3.11}$$

Here we used the property $(\Sigma F^{ii}) \geq 1$. Similarly from equation (3.1) we have

$$\begin{aligned} w_k F^{ij}W_{ij,k} &= \nabla_k F(W) w_k \\ &= \nabla_k(f(x)e^{-2w})w_k \\ &\geq -C(\Sigma F^{ii})(1 + e^{-2w})L. \end{aligned} \tag{3.12}$$

Inserting (3.11) and (3.12) into (3.10), we get

$$H \leq C(1 + \max_{B_r} e^{-2w})$$

for $x \in B_{r/2}$. □

Once the second derivative is bounded, equation (3.1) becomes uniformly elliptic, and Evans-Krylov's regularity theory implies higher order regularity for equation (3.1).

REMARK 3.1.

(i) Equation (3.1) is also elliptic when the eigenvalues

$$\lambda(A_{\tilde{g}}) \in -\Gamma_k^+. \tag{3.13}$$

If w is a solution to (3.1) with eigenvalues $\lambda(A_{\tilde{g}}) \in -\Gamma_k^+$, then $\tilde{w} = -w$ satisfies the equation

$$\sigma_k[\lambda(\nabla^2 \tilde{w} - d\tilde{w} \otimes d\tilde{w} + \frac{1}{2}|\nabla \tilde{w}|^2 g - A_g)] = e^{2k\tilde{w}}. \tag{3.14}$$

However there is no second derivative estimate for solutions to (3.14), a counterexample was given in [**60**]. See also the Heinz-Lewy counter-example in [**58**], and [**76, 48**]. The difference is due to the sign of the first derivative term $|\nabla w|^2$ in the matrices in (3.1) and (3.14).

(ii) For the k-Hessian equation (1.10), regularity for solutions u with eigenvalues $\lambda(D^2 u) \in \Gamma_k$ also hold for solutions with $\lambda(D^2 u) \in (-\Gamma_k)$. But the pure interior second derivative estimate (3.2) does not hold for equation (1.10). When the solution vanishes on the boundary, a Pogorelov type estimate was proved in [**16**].

(iii) In addition to the convex cones Γ_k and $-\Gamma_k$, equations (1.10) and (3.1) may have other elliptic branches, depending on k and n. Namely there are (nonconvex) cones $\Gamma \subset \mathbb{R}^n$ such that equation (1.10) (and also (3.1)) is elliptic if the corresponding eigenvalues $\lambda(D^2 u) \in \Gamma$. In this case, the existence and continuity of viscosity solutions have been proved by Harvey and Lawson [**31**], but the regularity is unclear at the moment.

4. Existence and compactness

In this section we always assume

(a) $(\mathcal{M}, g)$ is a compact Riemannian manifold of dimension $n \geq 3$, not conformally equivalent to the unit sphere;
(b) the metric g is k-admissible, or equivalently the class $[g]_k$ is not empty.

As mentioned in the introduction, the existence of solutions to the k-Yamabe problem (1.1) for $k \geq 2$ has been proved for

(i) locally conformally flat manifolds [**26, 41, 60**];
(ii) the case $k = 2$ [**60**];
(iii) the case $k > n/2$ [**30, 70**]; and
(iv) the case $k = n/2$ [**71**].

In the following we address the above four cases separately.

4.1. Existence on locally conformally flat manifolds. The existence of solutions to the k-Yamabe problem on locally conformally flat manifolds in [**26, 41**] is based on the characterization of locally conformally flat manifolds by Schoen and Yau [**55**], and that in [**60**] relies on the variational structure of the equation (2.1).

When $\mathcal{M}$ is locally conformally flat, there is a conformal immersion of the universal covering of $\mathcal{M}$ into the unit sphere S^n. By the stereographic projection, the solution v satisfies equation (2.1) in a domain Ω in the Euclidean space $\mathbb{R}^n$. Moreover, if $\Omega \neq \mathbb{R}^n$, then $v(x) \to +\infty$ as $x \to \partial\Omega$ [**55**]. Therefore by the moving plane argument, v is locally uniformly bounded in Ω, which in turn implies that v is uniformly bounded on the manifold $\mathcal{M}$. Once the solution is proven uniformly bounded, the solution can be obtained by a degree argument [**41**]. Instead of the degree argument, the existence can also be obtained by a parabolic equation approach [**26**], just as in [**81**] for the case $k = 1$. Note that once the solutions are proven uniformly bounded, then the solution set is compact.

4.2. Existence when the k-Yamabe problem is variational. It was verified that the k-Yamabe problem is variational when $k = 2$ or the manifold $\mathcal{M}$ is locally conformally flat [**72, 8, 60**]. In [**60**] we proved the existence of solutions to the k-Yamabe problem when equation (2.1) is variational. When $k = 2$ and $n = 4$, the existence was previously obtained in [**12**]. As with the 1-Yamabe problem, our proof consists of two steps. The first one is to show that the Yamabe problem is solvable if a Yamabe constant satisfies an inequality like (1.8). The second step is to verify the inequality for manifolds not conformally equivalent to the unit sphere S^n.

PROPOSITION 4.1. ([**72, 8, 60**]) *The k-Yamabe problem is variational if $k = 2$ or if the manifold $\mathcal{M}$ is locally conformally flat.*

A more general result can be found in [**60**]. Indeed, an operator $F[v] = F[\nabla^2 v, \nabla v, v, x]$ is variational if and only if its linearized operator is self-adjoint. If F is not homogeneous of degree -1, the functional is given by

$$\mathcal{F}[v] = \int_{\mathcal{M}} G[v],$$

where

$$G[v] = \int_0^1 vF[tv].$$

When F is homogeneous of degree -1, such as the case $k = \frac{n}{2}$ in (4.1), we may consider v as a composite function $v = \phi(w)$ and consider the operator $\widetilde{F}[w] = \phi'(w)F(\phi(w))$. Hence the corresponding functional is given by

$$\mathcal{F}_{n/2}(w) = \int_{\mathcal{M}} \int_0^1 wF[tw].$$

In particular if $v = e^{-\frac{n-2}{2}w}$, then we obtain the functional in [**8**]

$$\mathcal{F}_{n/2}(w) = -\int_{\mathcal{M}}\int_0^1 w\sigma_{n/2}(\lambda(A_{g_t})),$$

where $g_t = e^{-2tw}g_0$.

When $k = 2$ and $n = 4$, by the Gauss-Bonnet-Chern formula,

$$8\pi^2\chi(\mathcal{M}) = \int_{\mathcal{M}} |W|_g^2 dVol_g + \mathcal{F}_2(g).$$

The Weyl tensor is invariant under conformal change of metrics. Hence $\mathcal{F}_2(g)$ is a conformal invariant [**11**]. For $k = n/2$ and $n > 4$, it is shown in [**72**] that $\mathcal{F}_{n/2}$ is also a constant in $[g]$ when $\mathcal{M}$ is locally conformally flat.

Let $(\mathcal{M}, g_0)$ be a Riemannian manifold. If $g = v^{\frac{4}{n-2}}g_0$ is a solution of the k-Yamabe problem (1.1), then the Schouten tensor is given by $A_g = \frac{2}{(n-2)v}V$, and v satisfies the equation

$$L[v] := v^{(1-k)\frac{n+2}{n-2}}\sigma_k(\lambda(V)) = v^{\frac{n+2}{n-2}}, \tag{4.1}$$

where

$$V = -\nabla^2 v + \frac{n}{n-2}\frac{\nabla v \otimes \nabla v}{v} - \frac{1}{n-2}\frac{|\nabla v|^2}{v}g_0 + \frac{n-2}{2}vA_{g_0}. \tag{4.2}$$

As with the Yamabe problem, we introduce the k-Yamabe constant for $2 \le k \le \frac{n}{2}$,

$$Y_k(\mathcal{M}) = \inf\{\mathcal{F}_k(g) \mid g \in [g_0]_k, \mathrm{Vol}(\mathcal{M}, g) = 1\}, \tag{4.3}$$

where

$$\begin{aligned} \mathcal{F}_k(g) &= \int_{\mathcal{M}} \sigma_k(\lambda(A_g))d\,vol_g \\ &= \int_{\mathcal{M}} v^{\frac{2n}{n-2} - k\frac{n+2}{n-2}}\sigma_k(\lambda(V))\,d\,vol_{g_0}. \end{aligned} \tag{4.4}$$

Note that we have ignored a coefficient $(\frac{2}{n-2})^k$ in the second equality. We have the following.

THEOREM 4.1. *([**60**]) Assume $2 \le k \le \frac{n}{2}$, the set $[g_0] \neq \emptyset$, and the operator L is variational. Then there is a k-admissible solution to the k-Yamabe problem (1.1).*

REMARK 4.1.

(i) The condition $[g_0] \neq \emptyset$ is essential for the k-Yamabe problem. By this condition we can work in the class of k-admissible functions, such that the equation (4.1) is elliptic.

(ii) When a solution is k-admissible, we have the regularity of solutions established in Section 3. When the eigenvalues lie in the negative cone $-\Gamma_k$, there is no interior regularity and no existence is known.

(iii) A surprising result, proved by Branson and Gover [**5**], is that equation (4.1) is variational if and only if $k = 2$ or $\mathcal{M}$ is locally conformally flat. So if $k \geq 3$, then (4.1) is variational only if $\mathcal{M}$ is locally conformally flat.

Our proof follows the strategy of Aubin, which consists of two steps. In the first step (Lemma 4.1) we show that if $Y_k(\mathcal{M}) < Y_k(S^n)$, then there is a solution to (4.1). In the second step we verify the inequality $Y_k(\mathcal{M}) < Y_k(S^n)$ (Lemma 4.2).

In the first step one needs to use a gradient flow. One should be cautious with parabolic equations of the form

$$u_t = F[u] + \phi\left[u, \int g(D^2u)\right],$$

where F is an elliptic operator, g is a function of D^2u, ϕ is a given function. Under proper conditions one can establish the estimate $|\partial_t u| + |\partial_x^2 u| \leq C$. The estimate implies that the function $\phi[u, \int g(D^2u)]$ is bounded and measurable in t, it does not mean $\phi[u, \int g(D^2u)]$ is Lipschitz or Hölder continuous in t. Hence one cannot conclude $\|u\|_{C^{2,1}} \leq C$ by Krylov's regularity theory. In the general case the higher regularity was recently obtained in [**62**].

LEMMA 4.1. *Under the assumptions of Theorem 4.2, if*

$$Y_k(\mathcal{M}) < Y_k(S^n), \tag{4.5}$$

then there is a k-admissible solution to (4.1).

PROOF. A solution of the k-Yamabe problem is a min-max type critical point of the corresponding functional. As we need to restrict ourselves to k-admissible functions, we cannot directly use variational theory (such as the Ekeland variational principle). Instead we study a descent gradient flow of the functional and investigate its convergence. We need to choose a special gradient flow for which the a priori estimates can be established.

As with the original Yamabe paper [**80**], we first study the approximating problems

$$L(v) = v^p, \tag{4.6}$$

where $1 < p \leq \frac{n+2}{n-2}$. (4.6) is the Euler equation of the functional $J_p(v) = J_p(v; \mathcal{M})$,

$$J_p(v) = \frac{n-2}{2n-4k}\int_{(\mathcal{M},g_0)} v^{\frac{2n}{n-2}-k\frac{n+2}{n-2}}\sigma_k(\lambda(V)) - \frac{1}{p+1}\int_{(\mathcal{M},g_0)} v^{p+1}. \tag{4.7}$$

Let $\phi_1 = \varepsilon$ and $\phi_2 = \varepsilon^{-1}$, where $\varepsilon > 0$ is a small constant. Then when $k < \frac{n}{2}$, $J_p(\phi_1) \to 0$ and $J_p(\phi_2) \to -\infty$ as $\varepsilon \to 0$. Let P denote the set of paths in Φ_k connecting ϕ_1 and ϕ_2, namely

$$P = \{\gamma \in C([0,1], \Phi_k) \mid \gamma(0) = \phi_1, \gamma(1) = \phi_2\}, \tag{4.8}$$

where Φ_k denote the set of k-admissible functions. Denote

$$c_p[\mathcal{M}] = \inf_{\gamma \in P} \sup_{s \in [0,1]} J_p(\gamma(s); \mathcal{M}). \tag{4.9}$$

Then (4.5) is equivalent to

$$c_p[\mathcal{M}] < c_p[S^n] \tag{4.10}$$

with $p = \frac{n+2}{n-2}$. By (4.10) we can prove that J_p has a min-max critical point v_p with $J_p(v_p) = c_p[\mathcal{M}]$, in the sub-critical case $p < \frac{n+2}{n-2}$. By a blow-up argument, we can also prove that v_p converges to a solution of (4.1) under the assumption (4.10).

As in (1.3), let

$$w = -\frac{2}{n-2} \log v. \tag{4.11}$$

Our gradient flow is then given by

$$F[w] - w_t = \mu(f(x, w)), \tag{4.12}$$

where

$$F[w] := \mu(\sigma_k(\lambda(A_g)))$$

and $g = e^{-2w} g_0$. When $f(x, w) = e^{-2kw}$, a stationary solution of (4.12) is a solution to the k-Yamabe problem. We choose

$$\mu(t) = \begin{cases} t^{1/k} & t \geq 10, \\ \log t & t \in (0, \frac{1}{10}), \end{cases}$$

and

$$(t-s)(\mu(t) - \mu(s)) \geq c_0(t-s)(t^{1/k} - s^{1/k})$$

for some constant $c_0 > 0$ independent of t. We also assume that μ is monotone increasing and satisfies

$$\lim_{t \to 0^+} \mu(t) = -\infty.$$

This condition ensures the solution is k-admissible at any time t.

We say a solution w to the parabolic equation (4.12) is k-admissible if for any fixed t, w is k-admissible as a function of x. Denote $Q_r = B_r(0) \times (0, r^2]$.

To prove Lemma 4.1, we establish the following interior a prori estimates for w [**60**]:

$$|\nabla_x w(0, r^2)| \leq C_1, \tag{4.13}$$

$$|\nabla_x^2 w(0, r^2)| \leq C_2, \tag{4.14}$$

$$|w_t(0, r^2)| \leq C_3, \tag{4.15}$$

where C_1 is independent of $\sup w$ (if $f = \kappa(x)e^{-pw}$ for some constant $p > 0$ and smooth, positive function κ), C_2 and C_3 depend only on n, k, r, μ, $\inf w$, $\sup w$, and $\|A_{g_0}\|_{C^2}$.

Estimates (4.13)-(4.15) can be established similarly as for the elliptic equation (3.1), using the arguments in [**25**] or [**14**]. For the parabolic equation (4.12), to prove the L^∞ estimate we use a blow-up argument and we need to show the estimates (4.13) and (4.14) also hold for the equation

$$\frac{1}{a}\mu(a^k\sigma(\lambda(W))) - w_t = \frac{1}{a}\mu(a^k f),$$

where $a > 0$ is a constant, and the constants C_1, C_2 are independent of a if $a \geq 1$.

By estimates (4.13)–(4.15), equation (4.12) becomes locally uniformly parabolic. By Krylov's regularity theory, we obtain the $C^{4,2}$ a priori estimate and also the local existence. To prove Lemma 4.1, we first consider the existence of k-admissible solutions for $2 \leq k < n/2$ to equation (4.6) in the subcritical growth case $1 < p < \frac{n+2}{n-2}$. As a solution to (4.6) is a critical point of the functional $J = J_p$, we write the parabolic equation (4.12) in terms of v as

$$F[v] + \frac{v_t}{v} = \mu(f(v)), \tag{4.16}$$

where $f(v) = v^{\frac{4k}{n-2}-\varepsilon}$, $F[v] = \mu(\sigma_k(\lambda(\frac{V}{v})))$, and $\varepsilon = \frac{n+2}{n-2} - p$. Equation (4.16) is a descent gradient flow of the functional $J(v)$. By a blow-up argument and the Liouville Theorem 2.1, we can prove that there is an initial function such that the solution to (4.16) is uniformly bounded and converges to a solution v_p of (4.6) with

$$J_p(v_p) = c_p > 0, \tag{4.17}$$

where c_p is the min-max critical value given in (4.9). Moreover, the set of solutions of (4.6) is compact.

Let $p \nearrow \frac{n+2}{n-2}$. By condition (4.5), the Liouville theorem 2.1, and (4.17), we can prove that the solution v_p is uniformly bounded. Hence by the a priori estimates, v_p converges smoothly along a subsequence to a solution of (4.1). Hence Lemma 4.1 is proved for $2 \leq k < \frac{n}{2}$.

For the case $k = \frac{n}{2}$, we can verify that the functional $\mathcal{F}_{n/2}(g)$ is a constant, and by (4.5), the constant is strictly smaller than $Y_{n/2}(S^n)$. Hence by the Liouville theorem in [**41**], one can prove that the set of solutions of (4.1) is compact. The existence of the solutions can be obtained by a degree argument. □

For the existence of min-max critical points of fully nonlinear elliptic functionals, a logarithm gradient flow was first used by Chou [**15**]. More general gradient flows were later used in [**75**] to prove the Sobolev type inequality (4.20) below and in [**16**] to prove the existence of min-max critical points of the functional corresponding to the k-Hessian equation (1.10). Our argument above was inspired by the argument in [**15, 75, 16**].

LEMMA 4.2. *The critical inequality (4.5) holds for any compact manifold not conformally equivalent to the unit sphere* S^n.

PROOF. We reduce the proof of the lemma for $k \geq 2$ to the case $k = 1$, namely (1.8). Let v_0 be the solution to the classic Yamabe equation

$$-\Delta v + \frac{n-2}{4(n-1)} R v = n(n-2) v^{\frac{n+2}{n-2}}.$$

Let v_k be the solution of

$$\sigma_k(\lambda(V)) = C_{n,k} v_0^{k\frac{n+2}{n-2}} \quad \text{on } \mathcal{M},$$

where $C_{n,k} = \frac{n!(n-2)^k}{k!(n-k)!}$. Then

$$-\Delta v_k + \frac{n-2}{4(n-1)} R v_k \geq n(n-2) v_0^{\frac{n+2}{n-2}}.$$

By comparison principle, $v_k \geq v_0$. By direct computation, $g = v_k^{4/(n-2)} g_0$ satisfies

$$Y_k(\mathcal{M}) \leq \mathcal{F}_k(g) < Y_k(S^n).$$

This finishes the proof of Lemma 4.2. □

As a consequence of our argument, we have the following Sobolev type inequality. For locally conformally flat manifolds, it was first proved in [**27**].

COROLLARY 4.1. *Let* $2 \leq k < \frac{n}{2}$. *Then there exists a constant* $C > 0$ *such that the inequality*

$$(Vol(\mathcal{M}_g))^{\frac{n-2}{2n}} \leq C \left[\int_{\mathcal{M}} \sigma_k(\lambda(A_g)) d\, vol_g \right]^{\frac{n-2}{2n-4k}} \tag{4.18}$$

holds for any conformal metric $g = v^{\frac{4}{n-2}} g_0 \in [g_0]_k$.

REMARK 4.2. Inequality (4.18) can be written in the form

$$\|v\|_{L^{2n/(n-2)}(\mathcal{M})} \leq C \left[\mathcal{F}_k[v]\right]^{\frac{n-2}{2n-4k}}. \tag{4.19}$$

where $\mathcal{F}_k$ is given in (4.4). When $k = 1$, we recover the Sobolev inequality

$$\|v\|_{L^{2n/(n-2)}(\mathcal{M})} \leq C \left[\int |Dv|^2 \right]^{1/2}.$$

In [**27**], Guan and Wang also proved the following inequality for locally conformally flat manifolds,

$$\left[\int_{\mathcal{M}} \sigma_k(\lambda(A_g)) d\, vol_g \right]^{\frac{n-2}{2n-4k}} \leq \left[\int_{\mathcal{M}} \sigma_m(\lambda(A_g)) d\, vol_g \right]^{\frac{n-2}{2n-4m}}$$

for $1 \leq k < m \leq \frac{n}{2}$ (the opposite inequality was proved when $\frac{n}{2} \leq k < m \leq n$ in [**27**]).

The above inequalities are similar to those for the k-Hessian equation, proved in [**75, 69**]. Let Ω be a bounded, $(k-1)$-convex domain in $\mathbb{R}^n$, i.e. for any point $x \in \partial\Omega$, $\sigma_{k-1}(\kappa(x)) > 0$, where $\kappa(x) = (\kappa_1(x), \ldots, \kappa_{n-1}(x))$ are the principal curvatures of $\partial\Omega$ at x. Denote by $\Phi_0^k(\Omega)$ the set of all k-admissible functions vanishing on $\partial\Omega$ and

$$\|u\|_{\Phi_0^k} = \left[\int_{\Omega} (-u) \sigma_k[\lambda(D^2 u)] dx \right]^{\frac{1}{k+1}}$$

One easily verifies that $\|\cdot\|_{\Phi_0^k}$ is a norm in Φ_0^k. The Sobolev type inequality

$$\|u\|_{L^p(\Omega)} \leq C \|u\|_{\Phi_0^k} \tag{4.20}$$

was proved in [**75**], where $u \in \Phi_0^k(\Omega)$, $p \leq k^* := \frac{n(k+1)}{n-2k}$ and $C = C(n, k, p, |\Omega|)$ if $1 \leq k < \frac{n}{2}$; $p < \infty$ and $C = C(n, p, \text{diam}(\Omega))$ if $k = \frac{n}{2}$; and $p = \infty$ and $C = C(n, k, \text{diam}(\Omega))$ if $\frac{n}{2} < k \leq n$. In [**69**] the following inequality was proved,

$$\|u\|_{\Phi_0^k} \leq C \|u\|_{\Phi_0^m},$$

where $u \in \Phi_0^m(\Omega)$, $1 \leq k < m \leq n$. See also [**77**] for details.

If $1 \leq k < \frac{n}{2}$, the exponent k^* in (4.20) is optimal. In [**75**] it was proved that the best constant in (4.20) is achieved by radially symmetric functions. Therefore when $p = k^*$ and $k < \frac{n}{2}$, by the classical Sobolev embedding, the best constant C is attained when $\Omega = \mathbb{R}^n$ by the function

$$u(x) = [1 + |x|^2]^{(2k-n)/2k}.$$

When $k = \frac{n}{2}$, a Moser-Trudinger type inequality has also been proved in [**61**]. The reader should note the different powers in inequalities (4.19) and (4.20).

In the case $\frac{n}{2} < k \leq n$, (4.20) can be strengthened. It was shown [**67**] that a k-admissible function to the k-Hessian equation (1.10) is Hölder continuous with exponent $\alpha = 2 - \frac{n}{k}$. This observation gives a guidance for us to prove the results in the next two sub-sections.

4.3. The cases for $k > n/2$. The existence of solutions and the compactness of the solution set for the k-Yamabe problem have been proved in [**30**] for $k > n/2$ and [**70, 71**] for $k \geq n/2$. In the case $k > n/2$, much stronger results were proved in [**70**], i.e., not only the solution set, but the set of all k-admissible metrics with fixed volume, is compact. These results are based on the following observation.

PROPOSITION 4.2. *If the metric g is k-admissible for some $k \geq \frac{n}{2}$, then the Ricci curvature is positive if $k > \frac{n}{2}$ and nonnegative if $k = \frac{n}{2}$.*

Proposition 4.2 was first observed in the case $k = 2$, $n = 4$ in [**11**] and proved in the general case in [**24**]. Our proof of Proposition 4.2 is based on a simple inequality in [**67**]. Note that the Ricci curvature $\mu = (\mu_1, \ldots, \mu_n)$ is given by

$$\mu_i = (n-2)\lambda_i + \sum \lambda_j.$$

Hence the Ricci curvature $\mu_i \geq 0$ for all $1 \leq i \leq n$ if and only if $P_{\frac{1}{n-2}}(\lambda) \geq 0$, where

$$P_\delta(\lambda) = \min \lambda_i + \delta \sum \lambda_i$$

is the Pucci operator. For any $\lambda \in \Gamma_k$, let $f(x) = \frac{1}{2}\sum \lambda_k x_k^2$. By direct computation, $\Delta_p f \geq 0$ for $p \leq 2 + \frac{n(k-1)}{n-k}$ (this is the case $l = 1$ in Lemma 4.2 [**67**]). Hence

$$\sum \lambda_i + \frac{n(k-1)}{n-k}\lambda_j \geq 0$$

for every j. When $k = \frac{n}{2}$, $\frac{n(k-1)}{n-k} = n-2$. Hence $Ric_g \geq 0$ if $\lambda(A_g) \in \Gamma_{n/2}$, and $Ric_g > 0$ if $\lambda(A_g) \in \Gamma_k$ for $k > \frac{n}{2}$.

By Proposition 4.2, one can consider the existence of conformal metrics such that the k-curvature is equal to a prescribed function, namely the existence of conformal metrics $g \in [g_0]$ such that

$$\sigma_k(\lambda(A_g)) = f, \tag{4.21}$$

where f is a given positive smooth function on $\mathcal{M}$. As noted above, when $k > n/2$, problem (4.21) was first solved by Gursky and Viaclovsky [**30**], who proved the compactness of the solution set. Their proof involved a detailed analysis of the growth rate of solutions at isolated singular points.

In [**70**], we prove a sharp Harnack inequality for k-admissible metrics when $(\mathcal{M}, g_0)$ is not conformally equivalent to the unit sphere S^n. As a consequence we obtained the compactness of the set of all k-admissible metrics with fixed volume. When $(\mathcal{M}, g_0)$ is the unit sphere we prove there is a unique admissible metric with singularity. As a consequence we prove an existence theorem for the equation (4.21), thereby recovering Gursky and Viaclovsky's results [**30**].

Our proof needs to introduce the minimal radial function and also involves nonsmooth k-admissible metrics, which are limit of smooth k-admissible functions in L^p norm for some $p > 0$.

Minimal radial function. A main new idea in our proof is the introduction of the minimal (maximal, resp) radial function for lower (upper, resp) semi-continuous functions. For any given lower semi-continuous function v in a geodesic ball $B_R(x_0) \subset \mathcal{M}$, (here x_0 can be any fixed point and $R > 0$ can be any constant), the maximal radial function of v, also defined in $B_R(x_0)$, is a rotationally symmetric function $\tilde{v}$ satisfying $\tilde{v} \le v$, and $\tilde{v} \ge \phi$ for any radial function ϕ satisfying $\phi \le v$. Explicitly it can be given by

$$\tilde{v}(x) = \inf\{v(y) \mid y \in \partial B_r(x_0),\ r = d(x, x_0)\}.$$

Obviously $\tilde{v}$ is also lower semi-continuous, and for any $r \in (0, R)$, there is a point $x_r \in \partial B_r(x_0)$ such that $\tilde{v}(x_r) = v(x_r)$. If v is superharmonic, then $\tilde{v}$ is also superharmonic, and as a function of $d(x, x_0)$, is monotone decreasing.

Similarly for an upper semi-continuous function w, the maximal radial function of w is given by

$$\tilde{w}(x) = \sup\{w(y) \mid y \in \partial B_r(x_0),\ r = d(x, x_0)\}.$$

When the function w is subharmonic (with respect to an elliptic operator), $\tilde{w}$ is monotone increasing, and it can also be equivalently defined as follows. For any $h \in \mathbb{R}$, denote $\Omega_h = \{x \in \mathcal{M} \mid \ w(x) < h\}$, which is open as w is upper semi-continuous. For any given point x_0, the minimal radial function $\tilde{w}$ (with respect to x_0) can also be given by

$$\widetilde{w}(r) = \inf\{h \mid \operatorname{dist}(x_0, \partial\Omega_h) > r\}.$$

Note that $\widetilde{w}$ is radial, so it is a function of $r = d(x, x_0)$, and can be regarded as a function of one variable. Obviously $\widetilde{w} \ge w$ and $\widetilde{w} \le \phi$ for any radial function ϕ satisfying $\phi \ge w$.

Nonsmooth k-admissible metric. Our argument involve limit of k-admissible metrics. Therefore we allow singular metrics and call a metric $g = \chi g_0$ k-admissible if $\chi : \mathcal{M} \to (0, \infty]$, χ is lower semi-continuous, $|\{x \in \mathcal{M} \mid \chi(x) = \infty\}| = 0$, and there exists a sequence of k-admissible metrics $g_m = \chi_m g_0$, $\chi_m \in C^2(\mathcal{M})$, such that $\chi_m \to \chi$ almost everywhere in $\mathcal{M}$. If g is k-admissible, then the function $v = \chi^{(n-2)/4}$ is subharmonic with respect to the operator

$$\Box := -\Delta_{g_0} + \frac{n-2}{4(n-1)} R_{g_0}$$

and hence by the weak Harnack inequality [**20**], the set $\{\chi = \infty\}$ has measure zero.

We first prove the classification of the possible singularities of k-admissible metrics on $\mathbb{R}^n$.

THEOREM 4.2. *Let g be k-admissible metric on $\mathbb{R}^n$ with $\frac{n}{2} < k \le n$. Then either*

$$g(x) = \frac{C}{|x - x_0|^4} g_0(x) \tag{4.22}$$

for some point $x_0 \in \mathbb{R}^n$ *and positive constant* C*, or the conformal factor* χ *is Hölder continuous with exponent* $\alpha = 2 - \frac{n}{k}$*, where* g_0 *is the standard metric on* $\mathbb{R}^n$*.*

PROOF. Let $g = e^{-2w} g_0$. Let $\widetilde{w}$ be the maximal radial function of w with respect to a point x_0, which we take as the origin 0. Then $\widetilde{w}$ is k-admissible in $\mathbb{R}^n \backslash \{0\}$. Since $\widetilde{w}$ is radial, it satisfies an ode. When $k > n/2$, by $\sigma_k(\lambda(W)) \geq 0$ and from the ode we show that either $\widetilde{w}$ is Hölder continuous in $\mathbb{R}^n$ with exponent $\alpha = 2 - \frac{n}{k}$, or

$$\widetilde{w}(x) = 2 \log |x| + C \tag{4.23}$$

for some constant C.

In the latter case, observe that the function $2 \log |x| + C$ is harmonic in $\mathbb{R}^n \backslash \{0\}$ and $\widetilde{w}$ is subharmonic. But by our definition of $\widetilde{w}$, $w - \widetilde{w}$ contains interior local maximum points. By the maximal principle, we conclude that $w(x) = 2 \log |x| + C$. □

By Theorem 4.2, we see that when $k > \frac{n}{2}$, there is only a unique (up to conformal diffeomorphism) singular k-admissible function in $\mathbb{R}^n$ and on S^n. This result is analogous to Obata's uniqueness result for the constant scalar curvature. But note that our uniqueness is not just for metrics with prescribed k-curvature, but for *all* k-admissible metrics with singularity. This result implies the set of all k-admissible function is quite "small". Theorem 4.2 has some other interesting consequences, see [**70**]. Next we prove the following Harnack inequality.

THEOREM 4.3. *If* $(\mathcal{M}, g_0)$ *is not conformally equivalent to the unit sphere* S^n *and* $\frac{n}{2} < k \leq n$*, then* $[g_0]_k$ *is compact in* C^0 *and satisfies the following Harnack inequality, namely for any* $g = \chi g_0 \in [g_0]_k$*,*

$$\max_{x,y \in \mathcal{M}} \frac{\chi(x)}{\chi(y)} \leq \exp(C|x-y|^{2-\frac{n}{k}}) \tag{4.24}$$

for some fixed constant C *depending only on* n, k *and* $(\mathcal{M}, g_0)$*, where* $|x - y|$ *denotes the geodesic distance in the metric* g_0 *between* x *and* y*.*

PROOF. First we note that if u is k-admissible to the k-Hessian equation (1.10) and if $k > \frac{n}{2}$, then $u \in C^\alpha$ with $\alpha = 2 - \frac{n}{k}$ [**67**]. For k-admissible functions to the k-Yamabe problem, using the above Hölder continuity for the k-Hessian equation (1.10), we can prove that

$$\frac{u(x) - u(y)}{|x-y|^\alpha} \leq C \int_{\mathcal{M}} u, \tag{4.25}$$

where $\alpha = 2 - \frac{n}{k}$ and C is independent of u.

Next, as in the proof of Theorem 4.2, the minimal radial function of w is either Hölder continuous or satisfies (4.23). In the first case, we are through.

In the latter case, noting that the function $2\log|x| + C$ is harmonic in $\mathbb{R}^n\backslash\{0\}$, and $w - \widetilde{w}$ contains interior local maximum points. By the maximal principle, a blow-up limit of $w - \widetilde{w}$ must be a constant. Hence we have

$$w(x) = 2\log|x| + o(1). \tag{4.26}$$

By a similar blow-up argument we can prove the singularities are isolated. Therefore there are at most finitely many singular points.

From the singularity behavior (4.26), as in [**30**] we can prove that w is C^∞ smooth away from the singular points. Therefore the manifold equipped with the metric $g = e^{-2w}g_0$ is a complete manifold with nonnegative Ricci curvature. By Bishop's volume comparison theorem, the quantity

$$Q(r) = \frac{Vol_g(B_r(y))}{r^n}$$

is monotone decreasing in r. So $Q(r) \le \lim_{r\to 0} Q(r) = \frac{1}{n}|S^{n-1}|$, where $|S^{n-1}|$ is the volume of the standard unit $(n-1)$-sphere. By the asymptotic estimate (4.26), we have

$$Q(r) \to \frac{m}{n}|S^{n-1}| \quad \text{as } r \to \infty,$$

where m is the number of singular points. Therefore we have either $m = 0$, or $m = 1$ and $Q(r) \equiv \frac{1}{n}|S^{n-1}|$. If $m = 0$, we are through. If $m = 1$, w has a unique singularity $\{0\}$. One can show that $R_g \equiv 0$. Hence $(\mathcal{M}\backslash\{0\}, g)$ is a complete manifold with $Ric_g \ge 0$. So it is isometric to the Euclidean space $\mathbb{R}^n$, which in turn implies that $(\mathcal{M}, g_0)$ is conformally equivalent to S^n. This contradicts with the assumption in Theorem 4.3.

We have thus proved that there is no singularity in k-admissible metrics, all conformal factors are Hölder continuous. One can furthermore prove that

$$\begin{aligned} &\sup_{\mathcal{M}} w - \inf_{\mathcal{M}} w \le K, \\ &|w(x) - w(y)| \le K|x - y|^{2-\frac{n}{k}} \end{aligned} \tag{4.27}$$

for some constant $K > 0$. The first estimate in (4.27) can be proved by contradiction. If it is not true, there exists a sequence of k-admissible functions w_m such that $\sup_{\mathcal{M}} w_m = 0$ and $\inf_{\mathcal{M}} w_m \to -\infty$. Suppose that $w_m(0) \to -\infty$. By the Hölder continuity, we may assume that e^{w_m} converges locally uniformly to e^w in $\mathcal{M}\backslash\{0\}$ and $\lim_{x\to 0} w(x) = -\infty$. By a weak Harnack inequality, the set $\{w = -\infty\}$ has measure zero. So w is k-admissible. Hence by the above argument, 0 is the unique singular point of w, w is C^∞-smooth away from 0, and $(\mathcal{M}, g_0)$ is conformally equivalent to the unit sphere S^n, which contradicts with our assumption. The second estimate in (4.27) follows from (4.25). □

From the Harnack inequality in Theorem 4.3 and by the Arzela-Ascoli theorem, we have the following compactness.

Corollary 4.1. *Suppose $(\mathcal{M}, g_0)$ is not conformally equivalent to the unit sphere S^n. Then the set of all k-admissible metrics of fixed volume is compact.*

By the Harnack inequality in Theorem 4.3, we can not only prove the existence of solutions to equation (4.21) for positive and smooth f, but to more general equations. For simplicity we state the existence for a special right hand side. We refer the reader to [**70**] for more general existence results.

THEOREM 4.4. *Let $(\mathcal{M}, g_0)$ be a compact n-manifold not conformally equivalent to the unit sphere S^n. Suppose $\frac{n}{2} < k \le n$ and $[g_0]_k \neq \varnothing$. Then for any smooth positive function f and any constant $p \neq k$, there exists a positive admissible solution to the equation*

$$\sigma_k(\lambda(V)) = f(x)v^p. \tag{4.28}$$

The solution is unique if $p < k$. When $p = k$, then there exists a unique constant $\theta > 0$ such that

$$\sigma_k(\lambda(V)) = \theta f(x)v^k \tag{4.29}$$

has a solution, which is unique up to a constant multiplication.

The proof is by a degree argument as in [**79**]. In [**79**] the existence of nontrivial solution to the Monge-Ampere equation was proved for very general superlinear right hand side. The proof in [**79**] relies on the convexity of solutions but for equation (4.29), the Hölder continuity of solutions, implied by the Harnack inequality (4.24), is sufficient for the degree argument.

As a special case of Theorem 4.4, letting $p = k\frac{n+2}{n-2}$, we obtain the existence of the solutions to (4.21) for $n/2 < k \le n$, which is the k-Yamabe problem when $f \equiv 1$, proved in [**30**]. Note that the compactness of the set of k-admissible functions automatically implies that the set of solutions is compact.

4.4. The case $k = n/2$. When $k = 2$ and $n = 4$, the existence and the compactness of the solution set for the $\frac{n}{2}$-Yamabe problem on compact Riemannian manifolds were proved by Chang, Gursky and Yang ([**11, 12**]). The paper [**11**] also contains various geometric applications of the problem. We have the following result for all even $n \ge 4$ [**71**].

THEOREM 4.5. *Let $(\mathcal{M}, g_0)$ be a compact Riemannian manifold of even dimension $n \ge 4$. Suppose g_0 is $\frac{n}{2}$-admissible and $\mathcal{M}$ is not conformally equivalent to the unit sphere. Then the set of solutions to the $\frac{n}{2}$-Yamabe problem is compact. In particular, there is a $\frac{n}{2}$-admissible solution to (4.1).*

We point out that the compactness is true not only for the k-Yamabe problem with $k = \frac{n}{2}$, but also for the more general equation (4.21). About the proof of Theorem 4.5, we cannot prove that the maximal radial function

$\widetilde{w}$ satisfies the asymptotic behavior (4.23) for any k-admissible function function w. But by the Liouville theorem 2.1, we can prove (4.23) if $g = e^{-2w} g_0$ is a solution to the k-Yamabe problem. Once (4.23) is proved, by the same argument as in the case $k > \frac{n}{2}$ in §4.3, we can prove w has the asymptotic behavior (4.26). Once (4.26) holds, the proof in §4.3 also applies to the case $k = \frac{n}{2}$ and we are through.

Therefore in the case $k = \frac{n}{2}$, it suffices to show that if $g = e^{-2w} g_0$ is a solution to the k-Yamabe problem, then (4.23) holds.

The proof of (4.23) uses several results on subharmonic functions and the Liouville Theorem 2.1, which we list as follows.

(a) Let $v \geq 0$ and satisfy

$$-\Delta v + \frac{n-2}{4(n-1)} R_{g_0} v \geq 0. \tag{4.30}$$

Then we have the weak Harnack inequality [**20**]

$$\|v\|_{L^p(\mathcal{M})} \leq C \inf_{\mathcal{M}} v, \tag{4.31}$$

where $1 \leq p < \frac{n}{n-2}$ and C is a positive constant depending on n, p, and g_0.

(b) If $v \in B_1(x_0)$ satisfy (4.30) (in the weak sense), then $v \in W^{1,p}(B_{1/2})$ for $1 \leq p < \frac{n}{n-1}$, and [**32**]

$$\|v\|_{W^{1,p}(B_{1/2})} \leq C \|v\|_{L^1(B_1)}, \tag{4.32}$$

where C is a positive constant depending on n, p, and g_0.

From (4.31) and (4.32) it follows that

$$\|v\|_{W^{1,p}(\mathcal{M})} \leq C \inf_{\mathcal{M}} v, \tag{4.33}$$

where $1 \leq p < \frac{n}{n-1}$, and C depends on n, p, and g_0. Therefore if v_j is a sequence of k-admissible functions satisfying $\inf_{\mathcal{M}} v_j = 1$, then v_j is uniformly bounded in $W^{1,p}(\mathcal{M})$ for any $p < \frac{n}{n-1}$, and v_i subconverges to a (generalized) k-admissible function. In particular, if v satisfies (4.30), then the maximal radial function $\tilde{v}$ is locally uniformly Hölder continuous away from 0, with Hölder exponent $\alpha \in (0, 1/n)$.

(c) Let v be a radially symmetric, super-harmonic function in $B_1(0) \subset \mathbb{R}^n$, namely $\Delta v \leq 0$ in $B_1(0)$. Then

$$v(x) = \frac{c_0 + o(1)}{|x|^{n-2}}, \tag{4.34}$$

where c_0 is a nonnegative constant.

(d) Suppose $\{v_j\}$ is a sequence of solutions with local maximum $v_j(0) \to \infty$. Denote

$$w_j(x) = -\frac{2}{n-2} \log \frac{v_j(x)}{m_j}$$

where $m_j = \inf_{\mathcal{M}} v_j$. Let $\widetilde{w}_j$ be the maximal radial function of w_j. Then $\widetilde{w}_j \to \widetilde{w}$ and $\widetilde{w}$ is a k-admissible function. By Liouville Theorem 2.1, the corresponding radial function $\widetilde{v}$ (by (4.11), $e^{\widetilde{w}} = \widetilde{v}^{-2/(n-2)}$) satisfies (4.34) for some positive constant $c_0 > 0$. Hence

$$\widetilde{w}(r) \le 2\log r + C.$$

On the other hand, as $\widetilde{v}$ is k-admissible for some $k \ge n/2$, it satisfies an ode and from which we have

$$\widetilde{w}(r) \ge 2\log r - C'.$$

Therefore we have

$$\widetilde{w}(r) = 2\log r + C + o(1).$$

namely (4.23) holds.

5. Remarks

5.1. Admissibility of the initial metric. In §4 we assume the initial metric g is k-admissible, or equivalently the set $[g]_k \neq \emptyset$. This is a pointwise condition. For the Yamabe problem (1.6), this condition can be replaced by the Yamabe constant $Y_1 > 0$, namely if $Y_1 > 0$, then there exists a metric $\widetilde{g} \in [g]$ such that the scalar curvarture $R_{\widetilde{g}} > 0$.

In the case $n = 4$ and $k = 2$, Chang, Gursky and Yang [**11**] proved that $[g]_2 \neq \emptyset$ if the Yamabe constants $Y_1 > 0$ and $Y_2 > 0$. See also [**29**] for a different proof. The Yamabe constants Y_1 and Y_2 are defined as follows.

$$Y_1 = \inf_{\widetilde{g}\in[g]} vol(\widetilde{g})^{-\frac{n-2}{n}} \int_{\mathcal{M}} \sigma_1(\widetilde{g})dvol_{\widetilde{g}},$$
$$Y_2 = \inf_{\widetilde{g}\in[g]} vol(\widetilde{g})^{-\frac{n-4}{n}} \int_{\mathcal{M}} \sigma_2(\widetilde{g}^{-1}A_{\widetilde{g}})dvol_{\widetilde{g}}.$$

By Chern-Gauss-Bonnet Formula, Y_2 is a conformal invariant when $n = 4$.

Denote

$$\hat{Y}_k([g]) = \begin{cases} \inf\{\mathcal{F}_k(\widetilde{g}) \mid \widetilde{g} \in [g]_{k-1}, vol(\mathcal{M},\widetilde{g}) = 1\} & \text{if } [g]_{k-1} \neq \emptyset, \\ -\infty, & \text{if } [g]_{k-1} = \emptyset. \end{cases} \tag{5.1}$$

where $\mathcal{F}_k$ is given in (4.4). Then we have

THEOREM 5.1. *Let $(\mathcal{M}, g)$ be a compact n-dimensional manifold. Assume that $\hat{Y}_k([g]) > 0$ for $k \ge 2$. Then there exists a k-admissible conformal metric $\widetilde{g} \in [g]$.*

This result was proved by the first author in [**59**]. For locally conformally flat manifolds, it was proved by Guan, Lin and Wang [**23**].

PROOF. Consider a conformal metric $g = e^{-2u}g_0$ satisfying

$$\sigma_k^{1/k}(g_0^{-1}A_g^t) = f(x)e^{2u}, \tag{5.2}$$

where

$$A^t = \frac{1}{n-2}(Ric_g - \frac{t}{2(n-1)}R_g \cdot g),$$

$t \leq 1$, and $f(x) > 0$ is any smooth function. The idea of introducing the family of equations is such that the equation becomes uniformly elliptic when $t < 1$. This idea was used by Krylov [**37**] and Trudinger [**64**]. One may easily show that the linearized operator of this equation is invertible. Choosing a proper function f, we can establish the a priori estimates for solutions of (5.2), and by the continuity method, there exists a solution of (5.2) for $t = 1$. □

5.2. Existence of solutions for more general nonlinear curvature functions. In the case when $\mathcal{M}$ is locally conformally flat, as explained in §4.1, the proof is based on the a priori estimates and the moving plane argument. Therefore it applies to more general nonlinear curvatures. Similarly the compactness results in §4.3 and 4.4 relies essentially on the condition $k \geq \frac{n}{2}$, and they can also be extended to more general curvature functions. More specifically, these results can be extended to the following more general equation

$$\sigma(\lambda(A_{\tilde{g}})) = \phi, \tag{5.3}$$

where ϕ is a positive, smooth function on $\mathcal{M}$, σ is a nonlinear function defined on an open convex cone $\Gamma \subset \mathbb{R}^n$. We assume that σ satisfies

($\mathbb{C}_1$) $\sigma > 0$ in Γ and $\sigma = 0$ on $\partial\Gamma$;
($\mathbb{C}_2$) σ is concave;
($\mathbb{C}_3$) σ is invariant under exchange of variables;
($\mathbb{C}_4$) σ is homogeneous of degree 1.

We also assume that Γ satisfies

($\mathbb{G}_1$) If $\lambda \in \Gamma$, then any permutation of λ also lies in Γ;
($\mathbb{G}_2$) $\Gamma_n \subset \Gamma \subset \Sigma_{\frac{1}{n-2}}$, where Γ_n is the positive cone and

$$\Sigma_\theta = \{\lambda \in \mathbb{R}^n \mid P_\theta(\lambda) > 0\},$$

and P_θ is the Pucci operator given after Proposition 4.2.

Then the existence results on locally conformally flat manifolds in §4.1 can be extended to equation (5.3) if $\mathbb{C}_1$-$\mathbb{C}_4$ and $\mathbb{G}_1$ hold; and the existence and compactness results in Theorem 4.5 holds in $\mathbb{C}_1$-$\mathbb{C}_4$ and $\mathbb{G}_1$-$\mathbb{G}_2$ hold.

Recently there have also been some interesting works on the singularity analysis for k-admissible functions and on the a priori estimates and existence of solutions for the k-Yamabe problem on manifolds with boundary. However, research in these directions are in progress and we will not discuss these aspects in this work.

5.3. Open problems. We finish this paper by mentioning a few unsolved problems.

1. A natural question is the existence of solutions to the k-Yamabe problem when $2 < k < n/2$ and $\mathcal{M}$ is not locally conformally flat.
2. The compactness of the solution set to the k-Yamabe problem for $k = 1$ and $k \geq \frac{n}{2}$ has been completely solved. One would also like to know the answer for the cases $2 \leq k < \frac{n}{2}$, in particular the case $k = 2$.
3. The full k-Yamabe problem [**38**]. In the introduction of the k-Yamabe problem we have dropped the Weyl tensor in the decomposition Riem $= W + A_g \odot g$. Labbi [**38**] proposed to study the existence of a conformal metric such that the quantity $s_k := c^{2k}\text{Riem}^k$ is equal to a constant, where c is the standard contraction operator, and Riem^k denotes the Kulkarni-Nomizu product. When $k = 1$, s_k is exactly the scalar curvature of $(\mathcal{M}, g)$. When $k = \frac{n}{2}$ and n is even, s_k is the Lipschitz-Killing curvature. As shown by Labbi, this quantity s_k is always variational and if $\mathcal{M}$ is a hypersurface, s_k is the $(2k)^{th}$ mean curvature of $\mathcal{M}$, namely the $(2k)^{th}$ elementary symmetric polynomial of the principal curvatures of $\mathcal{M}$. When M is locally conformally flat, then the Weyl curvature vanishes, and $s_k = \sigma_k(\lambda(A_g))$ for $k = 1, 2, \ldots, [\frac{n}{2}]$ (up to a constant).

References

[1] T. Aubin, Équations différentielles non linéaires et problème de Yamabe concernant la courbure scalaire, J. Math. Pures Appl. 55 (1976), 269–296.

[2] T. Aubin, Problèmes isopérimetriques et espaces de Sobolev, J. Diff. Geom. 11 (1976), 573–598.

[3] T. Aubin, Some nonlinear problems in Riemannian geometry, Springer, 1998.

[4] A.L. Besse, Einstein manifolds, Springer-Verlag, Berlin, 1987.

[5] T. Branson and R. Gover, Variational status of a class of fully nonlinear curvature prescription problems, Calc. Var. PDEs 32 (2008), 253–262.

[6] S. Brendle, Blow-up phenomena for the Yamabe PDE in high dimensions, J. Amer. Math. Soc. 21 (2008), 951–979.

[7] S. Brendle and F. C. Marques, Blow-up phenomena for the Yamabe problem II, J. Diff. Geom. 81(2009), 225–250.

[8] S. Brendle and J. Viaclovsky, A variational characterization for $\sigma_{n/2}$, Calc. Var. PDEs 20(2004), 399–402.

[9] L.A. Caffarelli, B. Gidas and J. Spruck, Asymptotic symmetry and local behavior of semilinear elliptic equations with critical Sobolev growth, Comm. Pure Appl. Math., 42 (1989), 271–297.

[10] L.A. Caffarelli, L. Nirenberg, and J. Spruck, Dirichlet problem for nonlinear second order elliptic equations III, Functions of the eigenvalues of the Hessian, Acta Math. 155(1985), 261–301.

[11] A. Chang, M. Gursky, P. Yang, An equation of Monge-Amère type in conformal geometry, and four-manifolds of positive Ricci curvature, Ann. of Math. (2) 155(2002), 709–787.

[12] A. Chang, M. Gursky, P. Yang, An a priori estimate for a fully nonlinear equation on four-manifolds, J. Anal. Math. 87 (2002), 151–186.

[13] A. Chang, M. Gursky, and P. Yang, Entire solutions of a fully nonlinear equation, Lectures on partial differential equations, 43–60, Inter. Press, 2002.
[14] S. Chen, Local estimates for some fully nonlinear elliptic equations, Int. Math. Res. Not., 55 (2005), 3403–3425.
[15] K.S. Chou (K. Tso), On a real Monge-Amperé functional, Invent. Math. 101(1990), 425–448.
[16] K.S. Chou and X.-J. Wang, A variational theory of the Hessian equation, Comm. Pure Appl. Math. 54 (2001), 1029–1064.
[17] O. Druet, Compactness for Yamabe metrics in low dimensions, Int. Math. Res. Not., 23 (2004), 1143–1191.
[18] L. C. Evans, Classical solutions of fully nonlinear, convex, second-order elliptic equations, Comm. Pure Appl. Math., 35 (1982), 333–363.
[19] B. Gidas, W. M. Ni and L. Nirenberg, Symmetry and related properties via the maximum principle, Comm. Math. Phys., 68(1979), 209–243.
[20] D. Gilbarg and N.S. Trudinger, Elliptic partial differential equations of second order, Springer, Berlin, 1983.
[21] B. Guan, The Dirichlet problem for Hessian equations on Riemannian manifolds, Calc. Var. PDEs, 8 (1999), 45–69.
[22] B. Guan, Conformal metrics with prescribed curvature functions on manifolds with boundary, Amer. J. Math. 129 (2007), 915–942.
[23] P. Guan, C. S. Lin and G. Wang, Application of the method of moving planes to conformally invariant equations, Math. Z. 247 (2004), 1–19.
[24] P. Guan, J. Viaclovsky and G. Wang, Some properties of the Schouten tensor and applications to conformal geometry, Trans. Amer. Math. Soc. 355 (2003), 925–933.
[25] P. Guan and G. Wang, Local estimates for a class of fully nonlinear equations arising from conformal geometry, Int. Math. Res. Not. (2003), 1413–1432.
[26] P. Guan and G. Wang, A fully nonlinear conformal flow on locally conformally flat manifolds, J. Reine Angew. Math. 557 (2003), 219–238.
[27] P. Guan and G. Wang, Geometric inequalities on locally conformally flat manifolds, Duke Math. J. 124 (2004), 177–212.
[28] P. Guan and X.-J. Wang, On a Monge-Ampère equation arising in geometric optics, J. Diff. Geom., 48(1998), 205–223.
[29] M. Gursky and J. Viaclovsky, A fully nonlinear equation on four-manifolds with positive scalar curvature, J. Differential Geom. 63 (2003), 131–154.
[30] M. Gursky and J.Viaclovsky, Prescribing symmetric functions of the eigenvalues of the Ricci tensor, Ann. of Math. (2) 166 (2007), 475–531.
[31] F.R. Harvey and H.B. Lawson, Dirichlet duality and the nonlinear Dirichlet problem, Comm. Pure Appl. Math. 62 (2009), 396–443.
[32] J. Heinonen, T. Kilpelainen, O. Martio, Nonlinear potential theory of degenerate elliptic equations, Oxford Univ. Press, 1993.
[33] N. Ivochkina, Solution of the Dirichlet problem for certain equations of Monge-Ampère type (Russian), Mat. Sb. (N.S.) 128 (1985), 403–415.
[34] M.A. Khuri, F.C. Marques and R. M. Schoen, A compactness theorem for the Yamabe problem, J. Diff. Geom. 81 (2009), 143–196.
[35] N. Korevaar, R. Mazzeo,F. Pacard, R. Schoen, Refined asymptotics for constant scalar curvature metrics with isolated singularities, Invent. Math. 135 (1999), 233–272.
[36] N. V. Krylov, Boundedly inhomogeneous elliptic and parabolic equations in a domain, Izv. Akad. Nauk SSSR Ser. Mat., 47 (1983), 75–108.
[37] N.V. Krylov, Nonlinear elliptic and parabolic equations of second order, Reidel, 1987.
[38] M.-L. Labbi, Variational properties of the Gauss-Bonnet curvatures, Calc. Var. PDEs 32 (2008), 175–189.
[39] D. Labutin, Potential estimates for a class of fully nonlinear elliptic equations, Duke Math. J. 111 (2002), 1–49.

[40] J.M. Lee and T.H. Parker, The Yamabe problem, Bull. Amer. Math. Soc. 17(1987), 37–91.
[41] A. Li and Y.Y. Li, On some conformally invariant fully nonlinear equations, Comm. Pure Appl. Math. 56 (2003), 1416–1464.
[42] A. Li and Y.Y. Li, On some conformally invariant fully nonlinear equations II, Liouville, Harnack, and Yamabe, Acta Math. 195 (2005), 117–154.
[43] Y.Y. Li, Degenerate conformally invariant fully nonlinear elliptic equations, Arch. Ration. Mech. Anal. 186 (2007), 25–51.
[44] Y.Y. Li, Local gradient estimates of solutions to some conformally invariant fully nonlinear equations, C. R. Math. Acad. Sci. Paris 343 (2006), 249–252; Comm. Pure Appl. Math. 62 (2009), 1293–1326.
[45] Y.Y. Li and L. Zhang, Compactness of solutions to the Yamabe problem II, Calc. Var. PDE, 24 (2005), 185–237.
[46] Y.Y. Li and L. Zhang, Compactness of solutions to the Yamabe problem III ,J. Funct. Anal., 245 (2007), 438–474.
[47] Y.Y. Li and M. Zhu, Yamabe type equations on three-dimensional Riemannian manifolds, Commun. Contemp. Math. 1 (1999), 1–50.
[48] X.N. Ma, N.S. Trudinger, and X.-J. Wang, Regularity of potential functions to the optimal transportation problem, Arch. Ration. Mech. Anal. 177 (2005), 151–183.
[49] F.C. Marques, A priori estimates for the Yamabe problem in non-locally conformal flat cases, J. Diff. Geom. 71 (2005), 315–346.
[50] M. Obata, the conjecture on conformal transformations of Riemannian manifolds, J. Diff. Geom. 6 (1971), 247–258.
[51] R. Schoen, Conformal deformation of a Riemannian metric to constant scalar curvature, J. Diff. Geom. 20(1984), 479–495.
[52] R. Schoen, Variational theory for the total scalar curvature functional for Riemannian metrics and related topics. Topics in Calculus of Variations, Lectures Notes in Math. 1365 (1989), pp. 120–154.
[53] R. Schoen, Lecture notes on differential geometry, 1988 at Stanford Univ, 1989 at Courant Inst.
[54] R. Schoen, On the number of constant scalar curvature metrics in a conformal class, Differential geometry, 311–320, Pitman Monogr. Surveys Pure Appl. Math. 52, 1991.
[55] R. Schoen and S.T. Yau, Conformally flat manifolds, Kleinian groups and scalar curvature, Invent. Math. 92 (1988), 47–71.
[56] R. Schoen and S.T. Yau, Lectures on Differential geometry, International Press, 1994.
[57] R. Schoen and D. Zhang, Prescribed scalar curvature on the n-sphere, Calc. Var. PDEs 4(1996), 1–25.
[58] F. Schulz, Regularity theory for quasi-linear elliptic systems and Monge-Ampère equations in two dimensions, Lecture Notes in Math. vol. 1445, Springer, 1990.
[59] W.M. Sheng, Admissible metrics in the σ_k Yamabe problem, Proc. Amer. Math. Soc. 136 (2008), 1795–1802.
[60] W.M. Sheng, N.S. Trudinger and X.-J. Wang, The Yamabe problem for higher order curvatures, J. Diff. Geom. 77(2007), 515–553.
[61] G. Tian and X.-J. Wang, Moser-Trudinger type inequalities for the Hessian equation, J. Funct. Anal., 259 (2010), 1974–2002.
[62] G. Tian and X.-J. Wang, A priori estimates for fully nonlinear parabolic equations, preprint, 2011.
[63] N.S. Trudinger, Remarks concerning the conformal deformation of Riemannian structures on compact manifolds, Ann. Scuola Norm. Sup. Pisa (3) 22(1968), 265–274.
[64] N.S. Trudinger, The Dirichlet problem for the prescribed curvature equations, Arch. Rat. Mech. Anal. 111 (1990), 153–179.
[65] N.S. Trudinger, On the Dirichlet problem for Hessian equations, Acta Math., 175(1995), 151–664.

[66] N.S. Trudinger and X.-J. Wang, Hessian measures I, Topol. Methods Nonlinear Anal. 10 (1997), 225–239.
[67] N.S. Trudinger and X.-J. Wang, Hessian measures II, Ann. of Math. (2) 150 (1999), 579–604.
[68] N.S. Trudinger and X.-J. Wang, Hessian measures III, J. Funct. Anal. 193 (2002), 1–23.
[69] N.S. Trudinger and X.-J. Wang, A Poincaré type inequality for Hessian integrals, Calc. Var. PDEs 6(1998), 315–328.
[70] N.S. Trudinger and X.-J. Wang, On Harnack inequalities and singularities of admissible metrics in the Yamabe problem, Calc. Var. PDEs 35 (2009), 317–338.
[71] N.S. Trudinger and X.-J. Wang, The intermediate case of the Yamabe problem for higher order curvatures, Int. Math. Res. Not. Vol. 2010, No. 13, pp. 2437–2458.
[72] J. Viaclovsky, Conformal geometry, contact geometry, and the calculus of variations, Duke Math. J. 101 (2000), 283–316.
[73] J. Viaclovsky, Estimates and existence results for some fully nonlinear elliptic equations on Riemannian manifolds, Comm. Anal. Geom. 10 (2002), 815–846.
[74] J. Viaclovsky, Conformally invariant Monge-Ampere equations: global solutions, Trans. Amer. Math. Soc., 352 (2000), 4371–4379.
[75] X.-J. Wang, A class of fully nonlinear elliptic equations and related functionals, Indiana Univ. Math. J. 43 (1994), 25–54.
[76] X.-J. Wang, On the design of a reflector antenna, Inverse Problems, 12(1996), 351–375.
[77] X.-J. Wang, The k-Hessian equation, Geometric analysis and PDEs, 177–252. Lecture Notes in Math., vol.1977, Springer, 2009.
[78] X.-J. Wang, A priori estimates and existence for a class of fully nonlinear elliptic equations in conformal geometry, Chinese Ann. Math. Ser. B 27 (2006), 169–178.
[79] X.-J. Wang, Existence of multiple solutions to the equations of Monge-Ampère type, J. Differ. Eqs., 100 (1992), 95–118.
[80] H. Yamabe, On a deformation of Riemannian structures on compact manifolds, Osaka Math. J. 12(1960), 21–37.
[81] R. Ye, Global existence and convergence of Yamabe flow, J. Diff. Geom. 39 (1994), 35–50.

Centre for Mathematics and Its Applications, The Australian National University, Canberra, ACT 0200, Australia; and, Department of Mathematics, Zhejiang University, Hangzhou, 310027, P.R. China.
E-mail address: weimins@zju.edu.cn

Centre for Mathematics and Its Applications, The Australian National University, Canberra, ACT 0200,
Australia.
E-mail address: Neil.Trudinger@anu.edu.au, Xu-Jia.Wang@anu.edu.au

Topics on Geometric Analysis

Shing-Tung Yau

I recall that about twenty years ago, Peter Li told me that the National Science Foundation has a program called geometric analysis. I was curious about what that was until Peter explained to me. I believe this is a real contribution of NSF to mathematics, on top of the funding that we have enjoyed all these years.

Just like any subject in mathematics, the roots of geometric analysis dates to ancient times and also to the not so ancient contributions of many modern mathematicians. And as many scientists like to say, we stood on the shoulders of giants, who laid down the basic tools and concepts and made important progress. But I believe only starting in the 1970s, did we see a systematic development based on analysis and nonlinear differential equations to solve important problems in geometry and topology. This development is vigorous and it has had strong feedback to the development of differential equations and analysis in general.

I remember that as a graduate student, I read a great deal of Russian literature on surface theory based on the Alexandrov school and works of Pogorelov. I also noticed the great works of Morrey, Nash, Nirenberg, de Giorgi and others. But they were more interested in analysis than geometry. Problems in geometry are good testing grounds for developments of partial differential equations. This is still so. But we started to develop into other directions.

When I entered the subject, it seemed to me that Riemannian geometry could benefit from the introduction of techniques beyond the geodesic methods which were almost exclusively used at that time. Indeed, the great power of the Atiyah-Singer Index Theorem had shown prevailing influence in the subject of geometry, topology and algebraic geometry already. On the other direction, the deep insights offered by integration by part and

This article is based on the C.-C. Hsiung Lecture that the author gave on the International Symposium on Geometry and Topology in Memory of C.-C Hsiung at Lehigh University on May 28, 2010.

by the maximum principle in partial differential equations work like magic, once the right quantity is found.

I was convinced that there are much for geometers to learn from analysis and partial differential equations. With many friends with great strengths in both analysis and geometry, we started to look into various subjects such as minimal submanifolds, harmonic maps, Monge-Ampère equations, and Ricci flows.

I shall select a few of these topics and discuss their prospects. And I should stress that the subject matter of geometric analysis covers much more than the topics that I am going to discuss here.

Let me quote my good friend Hamilton: "The subject has just begun."

It will be a pity if young bright geometers are misled by ignorant comments that geometric analysis is dead. If we want to construct an interesting and deep geometric structure over a manifold that is not known to us in any concrete manner, differential equations seem to be the only way that I know of. This was indeed the guiding principle to prove the Poincaré conjecture or the geometrization conjecture using Ricci flow.

1. Hamilton's Theory of Ricci Flow

I recall that I met Hamilton in Cornell in 1979, who told me that he was working on the Ricci flow. At the same period, I was talking with Rick Schoen then in a similar direction of adapting the ideas of harmonic maps to study deformation of metrics. It was indeed a very natural concept. I may also say that the physicist Friedan also had came up with the equation in quantum gravity as a renormalization group flow. I recall that many analytic approaches to the solution of the Poincaré conjecture were proposed by geometers. But we all know that practically all of these other approaches led to ashes.

Friedan never really thought in terms of using the flow for understanding the structure of geometry or topology per say. The later claims of how string theory or physics led to the break through in Ricci flow, is just a myth. It did not really occur in that way.

In any case, the key to the development of Ricci flow is of course what we can do with such equations. I thought that Hamilton was just toying around with superficial claims until in 1980, he called me explaining what he could do with the Ricci flow for three-manifolds with positive Ricci curvature [**26**]. Immediately I realized that to be a major breakthrough for the whole subject of flows for geometric structures.

I had three of my graduate students to work on it. I instructed Bando [**1**] to work on Kähler-Ricci flow. He gave the first major result there: that positivity of bisectional curvature is preserved in complex three dimensions. Then I told Mok [**51**] to generalize it, which he did.

I told Cao and Chow [**8**] to use the flow to reprove the Frankel conjecture. Unfortunately, despite many attempts by my later students who claimed to have success, the problem is still not solved - basically the flow proof of the Frankel conjecture still depends on the argument that Siu and I [**63**] used in 1979.

The news media has created an image that Hamilton did not do anything since the paper in 1983 until the work of Perelman [**55, 56, 57**]. This is far from true. In fact, great progress was made in those 20 years. Creative new techniques were introduced by Hamilton, most of which revolutionized the subject of geometry. The deep understanding of Ricci soliton to understand singularities was made by Hamilton [**27**]. The profound technology of metric surgery was introduced by him [**29**]. He was able to give a deep and precise description of the metric when it is close enough to the singularity [**28**]. This includes statements on how a metric looks like when it starts to break up or acquire a singularity. In particular, it includes a true understanding of the Mostow Rigidity Theorem when the metric is only close to the hyperbolic space in a rough sense [**30**].

Hamilton [**27**] introduced a great deal of analysis beyond the classical study of partial differential equations. He managed to generalize the work of Li-Yau [**43**] on understanding the behavior of the heat equation. He adapted the argument to the Ricci flow in an ingenious manner. Note that Perelman [**55**] proved a noncollapsing estimate by proving Li-Yau's inequality on backward solutions to the adjoint heat equation, which he used to rule out the cigar singularity, which was anticipated by Hamilton. The generalization by Hamilton of the Li-Yau estimate is fundamental to several important points for understanding the singularities of the Ricci flow. All of these works are employed in the study of the Poincaré conjecture and the geometrization conjecture.

Let me say that the great prowess of Ricci flow goes way beyond the Poincaré conjecture, despite what some so-called experts said in the news media, that the method of the proof has no other future directions. On the contrary, I like to mention a few examples here.

We saw the success of Brendle-Schoen [**3**] using Ricci flow to prove the quarter pinching conjecture of the differentiable sphere theorem: a conjecture that seems to be far beyond what classical comparison theorems for geodesics can achieve. One sees that new techniques in partial differential equations are created and new problems are waiting to be solved.

There are natural open problems in metric geometry that should be able to be solved by Ricci flow. I would like to encourage our young geometric analysts to spend time on them:

A. A flow proof of the Gromoll-Meyer-Cheeger-Perelman [**11, 54**] soul theorem of complete open manifolds with non-negative curvature. Perhaps a flow argument can also be used to prove the converse: that the total space of any vector bundle over a compact manifold with positive curvature admits a complete metric with non-negative curvature?

B. A proof of Gromov's theorem [**24**] of a bound on the Betti numbers of manifolds with positive sectional curvature depending only on dimension, settling the conjecture that the torus is the worst case. It is known that this follows from the statement that the Betti numbers of the loop space of the manifold grow at most polynomially. Perhaps it is instructive to compare the approach of Micallef-Moore [**50**] on the proof of the topological pinching theorem by studying the index of the Sacks-Uhlenbeck spheres. Can one use minimal surface techniques to settle this problem?

Of course, one would like to go beyond what is known in the area of metric geometry. For example, the following topics may be interesting:

Prove that there is only a finite number of manifolds with positive sectional curvature for each dimension when the dimension are greater than a large number. The number can be 25. In fact, would all these be compact symmetric spaces? If so, Ricci flow should be very useful as we are now trying to flow it to some canonical models.

The problem of understanding Hamilton's flow on manifolds of negative curvature has not been explored as much as it can be. Perhaps some modifications are needed. The structure of the fundamental group of such manifold needs to be understood. For example:

C. Thurston made the conjecture that each hyperbolic three dimensional manifold is covered by another compact manifold which admits an embedded incompressible surface with genus $g > 1$. Can one prove this statement based on analysis?

D. Given a compact manifold with negative curvature, can one find a "canonical" embedding of its universal cover into an open subset of some compact manifold with positive Ricci curvature so that all the deck transformations can be extended to be homeomorphisms of the larger manifold which may satisfy additional properties such as quasi-conformality? Hopefully, this can be simpler if the manifold is Kähler and in that case, we like to have everything to be holomorphic. The boundary of the open set could be complicated, but can be understood by looking at the set of the domain of discontinuity of the group.

E. For most structures that are canonically defined by the metric, it should be compatible with the Ricci flow. In those cases, the Ricci flow should give powerful information.

That was one of the reasons that I asked my former students to work on Kähler-Ricci flow. Unfortunately, besides the important advances made by Donaldson [**15**], the progress has been slow, since the thesis of Cao [**5**]. The most recent estimates of Perelman [**55**] should shed more light on how to proceed. The major question is how to understand the question of stability of the complex structure through the flow. After all, the conjecture that I [**70**] made on the equivalence of manifold stability with the Einstein condition should certainly show up in the study of the limit of the flow. One should study the limit of the Kähler-Ricci flow carefully to find the right definition of stability of Fano manifolds.

F. Kähler-Ricci flow has been used to help answer my conjecture [**69**] that a complete non-compact Kähler manifold with positive bisectional curvature is biholomorphic to $\mathbb{C}^n$. So far the best result on this is by Chau-Tam [**10**], who proved the conjecture in the case of bounded curvature and maximal volume growth. The work of Chau-Tam and others that use Käher-Ricci flow to address my conjecture make use of the important Li-Yau-Hamilton inequality, which was proved to hold for Kähler-Ricci flow by Cao [**6, 7**]. It should be possible to complete the program in the near future.

G. For manifolds with non-positive Chern classes, the problem of metric singularities near base points of the canonical map is very important for applications to algebraic geometry. The same question applies to non-compact Kähler-Einstein manifolds with non-positive Chern class. One would like to study the structure of the points that compactify the manifold, and we hope to find the ansatz or structure of the metric at infinity. We also like to be able to calculate the residue of Chern classes at the singular points or at infinity. Hence a good description of the metric should include such information. The first paper in this direction was described by me in 1978 [**71, 72, 73, 74, 75**], where I proved that for algebraic manifold M with a divisor D, that M-D admits a canonical Kähler-Einstein metric if $K + D > 0$. The singularity of the metric near D is well understood.

Once the structure of those singular metrics are understood, the application to algebraic geometry should follow easily. This should include the minimal model theory, the abundance conjecture and the structure of the moduli space of algebraic manifolds. In my original paper on the Calabi conjecture, I made the first attempt to understand the singularity of metrics. But there was not enough knowledge of algebraic geometry to help then and I hope our knowledge is matured enough to tackle such problems now.

H. The full theory of Ricci flow with boundary is not well developed yet. It will be nice to see how to deform such manifolds to Einstein manifolds whose boundary has constant mean curvature. We also need to develop a theory of Ricci flow for complete non-compact manifolds with no upper bound on curvature. In particular, prove the theorem of Schoen-Yau [**60**] that every complete non-compact three dimensional manifolds with positive Ricci curvature is diffeomorphic to Euclidean space.

I. Many years ago, I [**70**] proposed to study the groups of exotic spheres in the following way: An exotic spheres admit metrics with positive scalar curvature iff it bounds a spin manifold. An exotic spheres admit metrics with positive Ricci curvature iff it bounds a parallelizable manifold. The first statement was proved by the work of Stolz [**64**] based on the surgery result of Schoen-Yau [**59**] and Gromov-Lawson [**25**]. The second assertion can be refined by replacing parallelizable manifolds with manifolds where some of their Pontryagin classes are zero, while we only require some of the eigenvalue of the Ricci curvature to be positive. Perhaps we can achieve this by flows on compact manifolds with boundary.

J. Can one find a reasonable canonical bounding manifold, similar to those appear in AdS/CFT [**68**] theory where we require the bounding manifold to admit an Einstein metric that is asymptotic to hyperbolic space near the boundary?

K. Can one characterize those exotic spheres that admit metrics with positive sectional curvature? Should they admit a fiber structure where both fiber and base admit metrics with positive curvature so that the fiber bounds a manifold with positive curvature in a smooth fashion?

L. It is well known that an evolution equation very close to Hamilton's Ricci flow is the mean curvature flow, where several people, starting from Gage [**19**], Hamilton [**20**], Grayson [**23**], and Huisken [**34, 35**], made important contributions. It remains to be seen how to use the mean curvature flow, or other submanifold flows, to handle the famous problem of the Schoenflies conjecture [**17**], that any smoothly embedded three-sphere in $\mathbb{R}^4$ bounds a four-ball.

2. Manifolds with Special Holonomy Groups

Can flows generate manifolds with G_2 or $Spin(7)$ holonomy groups? Compact manifolds of these types were constructed by Joyce [**37, 38, 39**]. But unfortunately, Joyce's manifolds are based on singular perturbation methods, and it is difficult to use the method to study the global moduli problem, which is important for the M-theory linked to the manifold. It will be great to find a flow or other global elliptic method to construct such manifolds.

There are some obvious obstructions for a manifold to admit G_2 structures, in terms of nontriviality of the third cohomology and positive definiteness of the quadratic form associated to the first Pontryagin form, as can be found in the book of Joyce [**40**]. To the best of my knowledge, these are the only known obstructions. Can one prove it?

Many construction of such manifolds are related to constructions of Calabi-Yau manifolds, and there are proposals to relate them to a pair of Calabi-Yau manifold plus a special lagrangian cycle in this manifold. It was proposed that some total space of a circle fiber space over the manifold degenerate along the cycle may do the job. Such constructions work better for non-compact manifolds and the technology is far from mature.

Special holonomy manifolds play an important chapter in differential geometry and physics. However, there are other manifolds of interest that do not have special holonomy group. The most notable of such manifolds are Einstein manifolds. Major construction of such manifolds either come from Kähler geometry, from reduction of dimension by group actions or from the Wick rotation from Lorentzian Einstein manifolds [**21**]. It will be interesting to see how general such constructions are.

This is especially true for the Wick rotation construction: it is possible to start out from a singular Lorentzian manifold and obtain a non-singular Einstein manifold. Under what conditions can we reverse this procedure to produce interesting solutions to the Einstein equation which have physical interest?

When we study cone singularities for manifolds with a special holonomy group, we can find new structures, such as Sasaki-Einstein manifolds [**4**] or nearly-Kähler manifolds [**22**]. They themselves do not have special holonomy group and yet they are constructed from such manifolds by reduction of dimension. If we look at the category of all manifolds generated by manifolds with special holonomy group by doing reduction of this sort or manifolds with a foliation whose leaves are given by special holonomy group, how big a class of manifolds can we generate? How do we characterize them?

3. Four Manifolds

A very important problem regarding structures on manifolds relate to four dimensional manifolds where few structures are known. But there is an important one, besides the space of Einstein metrics. These are the self-dual or anti-self-dual metrics. They are naturally defined for four-dimensional manifolds and their existence says a lot about the manifolds. For one thing, the twistor space of anti-self-dual manifolds admits an integrable complex structure and the anti-self-dual bundles lift to holomorphic bundles. It therefore has a natural connection to the theme of complex geometry.

Hitchin [**31**] showed that the twistor space is not Kähler except for a couple of obvious examples. However, they can have a lot of meromorphic functions and naturally the function field gives us natural invariants of the manifolds. There is also the concept of Kodaira dimension of the complex manifold that may give some understanding for the anti-self dual metrics.

While there are many constructions of self-dual manifolds, especially the spectacular works of Taubes on the existence of such metrics after stabilization by connected sums with complex projective plane, the general existence theorem is still not known.

But in order to understand such metrics, we need to be able to give direct proof of existence without using singular perturbation method (Perhaps one can use flows or elliptic techniques). We need to understand the global moduli space of such metrics. In particular, do they have an infinite number of components once we fix the diffeomorphic type of the manifold. This is of course, very relevant to the question of the topology of four dimensional manifolds.

Similar questions can be addressed for Einstein metrics. Our understanding on Einstein manifolds is also rather poor. We are not even sure whether there can be more than two distinct Einstein metrics on the four sphere.

4. Six Dimensional Manifolds

Six dimensional geometries motived by string theory have pointed to new directions to look for geometric structures. The first important question is to give a topological classification of all Calabi-Yau manifolds. But it is also interesting to study the non-Kähler complex manifolds. I believed that in six dimensions, any almost complex manifold admits an integrable complex structure. If my conjecture is true, the problem will be very different from four dimensions. Twistor space of anti-self-dual four manifolds already give plenty of such manifolds which may be considered as supporting evidence.

An important class of examples come from Strominger's [**65**] study of the heterotic string where a hermitian metric is coupled with a stable holomorphic bundle: Given a complex manifold which admits a no-where zero holomorphic three form, a balanced metric and a stable holomorphic bundle whose first Chern class is zero, and whose second Chern form is equivalent to the second Chern form of the manifold up to $\partial\overline{\partial}$ class, then we study a coupled system of elliptic equations. The system admits interesting parallel spinors and was introduced by Strominger.

Examples were constructed by Li-Yau [**42**] and Fu-Yau [**18**]. For the twistor space mentioned above, if the anticanonical divisor exists as an effective divisor with normal crossings, we can form a branched cover of the complex manifold to kill the canonical divisor. The resulting manifold should satisfy the conditions mentioned above. Hence we expect a close relation between anti-self-dual four dimensional manifolds with such six dimensional manifolds

5. Metric Cobordism

The famous works by physicists on the holographic principle [**44**] gave the AdS/CFT correspondence, which says that if an Einstein manifold which is asymptotic to the hyperbolic metric with conformal boundary given by another manifold M with positive scalar curvature, the quantum gauge theory on the boundary is isomorphic to the quantum theory of gravity in the bulk.

It opens up many interesting questions for geometry. Which compact manifold with positive scalar curvature can be written as the conformal boundary of hyperbolic Einstein manifolds? See [**68**]. How do we describe the moduli space of such manifolds?

A remarkable holographic principle appears in classical general relativity that is related to the positive mass conjecture and quasi-local mass. For a space-like two dimensional surface in a spacetime which has space-like mean curvature vector and satisfies the local energy condition, a suitably defined quasilocal mass, which only depends on the boundary data attached to the two dimensional surface, vanishes only if it bounds a space-like three dimensional hypersurface that can be embedded into the Minkowski spacetime.

All these theories indicate that there should be interesting metric discussion on cobordism theory. Many people, including Hopkins-Singer [**33**], Dan Freed [**16**], Simons-Sullivan [**62**] and others have studied K-theory with connections. Such a theory should be compatible with metric cobordism and the AdS/CFT correspondence. A full understanding of this will be important for the future of geometry.

6. Affine Metrics

This problem arose in studying the construction of Calabi -Yau metrics. Suppose we have a compact manifold with a codimensional two subcomplex C such that outside C, there is an affine structure which means that it is covered by coordinate charts whose coordinate transformations are affine transformations with determinant equal to one. The affine structure has natural monodromy group around C. Suppose on each chart, we have a convex function whose Hessian defines a global metric on the complement of C.

We are interested in classifying those metrics, which Cheng and I [**12**] called affine Kähler metric, whose volume form is the given volume form attached to the affine structure. The moduli of the affine structure and the metrics and also the possible structure of C and the monodromy group around C needs to be classified.

This is a subject which is much related to the subject of affine spheres in the theory of affine geometry. For two dimensional affine spheres in three dimensional affine space, some of them are invariant under the affine group and the quotient is a nice Riemann surface equipped with a holomorphic cubic form. It is not clear whether there is a higher dimensional analogue of such a construction.

In this regard, there is a study of Einstein's equation in $2+1$ formulation by studying the constraint equations on Riemann surfaces. Since it satisfies the Einstein equation, the spacetime that evolves is flat due to dimensional consideration. The universal cover of the surface can be embedded into Minkowski spacetime as a hypersurface. The movement of the Riemann surface due to the Einstein flow is not trivial. It moves the conformal structure along some direction in the Teichmüller space. Moncrief and I tried to prove that this movement will deform the conformal structure to the Thurston boundary.

7. Isometric Embeddings

The question of local or global isometric embeddings of a Riemannian manifold to Euclidean space is still far away from being answered. Most of the interesting progress has been made for two dimensional surfaces, where existence and uniqueness are relatively easier to understand. The uniqueness is a big mess in higher dimension because the optimal embedding dimension is $n(n+1)/2$ and this makes the codimension too high for uniqueness, as long as the present technology is concerned. When $n > 3$, the local embedding problem is a hyperbolic system. But little is known about them.

Perhaps we can try some more restricted question:

Assume a closed manifold can be conformally embedded into Euclidean space as a hypersurface, can it be deformed to be an isometric embedding? And can we prove it is unique? Perhaps some assumption such as the positivity of scalar curvature or other stronger curvature assumption should be made?

Isometric embedding into Lorentizian space time can be easier and interesting due to applications for general relativity. In many cases, we are interested to isometrically embed a spacelike hypersurface into Minkowski spacetime or some other homogenous spacetime as codimensional two spacelike hypersurface.

In the works that I did with Mu-Tao Wang, we can minimize energy among such embeddings to find some canonical one. And they can exhibit beautiful properties. Problems in higher dimensions need to be overcome.

A simple question that I encountered about thirty years ago is the following:

Is it true that a complete manifold with Ricci curvature bounded from below can be isometrically embedded into Euclidean space with bounded mean curvature?

The problem of isometric deformation of manifolds is still far from being solved. I remember I gave lectures on this subject in Berkeley in 1977. Besides the surfaces with nonnegative curvature and some surfaces with rotational symmetry, really not much is known about the problem. But even for surfaces of rotation, it is interesting enough. For in some very special cases, the problem is related to the problem of the spectrum of operators.

8. Minimal Submanifolds

The subject of minimal submanifolds should be considered as one of the major foundation of non-linear elliptic theory, where major techniques were developed.

In two dimensions, conformal geometry brought in beautiful technology. I am glad to see great developments led by Meeks [**46**], Colding and Minicozzi [**13**], Hoffman [**32**] and Rosenberg [**49**], and others. Under their hands, we see an almost complete picture of complete minimal surface theory in Euclidean space.

There are still open problems that need to be solved though. A notable one is what some people call the Embedded Calabi-Yau Problem. The major contributors are Meeks [**47, 48**], Martin [**45**], Nadirashvilli [**53**], Perez and Ros. It says that for a surface to admit a complete bounded minimal embedding into the Euclidean three space, it is necessary and sufficient that every end of it has infinite genus.

Also, much more need to be said about minimal surfaces with constant mean curvature in other simply connected homogeneous three manifolds. For example, it will be useful to know whether there are only a finite number of closed embedded minimal surfaces with a fixed genus in the three sphere. (Due to the work of Choi-Schoen [**9**], we know the space is compact. The question is whether a continuous family of minimal surfaces exists or not.) I believe that it may be finite. And in that case, what is the number of such embedded minimal surfaces for each genus. It looks like the spectrum of Laplacian of these surfaces is very special. But concrete computations are needed before one can make any sensible statements.

There are plenty of works on minimal surfaces in general three dimensional manifolds. The most notable development starts from Morrey [**52**], and then Sacks-Uhlenbeck [**58**]. The bubbling of spheres and its applications to geometry are tremendous. The argument was made precise in the proof by Siu-Yau [**63**] of the Frankel conjecture. The proof of the topological theorem of Micallef-Moore [**50**] also depends on this argument. The developments of J-holomorphic curves in symplectic geometry owed a great deal to the work of Sacks-Uhlenbeck beyond what some authors would like to acknowledge.

There is also the great achievement of Simon-Smith [**61**] on the existence of embedded spheres with index less than one. Jost [**36**] generalized the work of Simon-Smith to the existence of two spheres in any manifold diffeomorphic to the three sphere. Colding-Minicozzi [**14**] studied such minimal surfaces and gave a very important application to the extinction time theorem for Hamilton's flow on simply connected three manifolds.

There is still much to be learned about how minimal surfaces are created and disappear when the metric of the three manifold changes. This is related to the appearance of black holes in general relativity.

As for higher dimensional minimal submanifolds, regularity theory has been a central problem, mostly related to geometric measure theory.

For minimal submanifolds of manifolds with special holonomy group, there are many special submanifolds that are of great interest. They are part of the calibrated submanifolds discussed by Harvey-Lawson. In many ways, they are analogues of complex submanifolds of Kähler manifolds where the restrictions of the Kähler form raised to a suitable power becomes the volume form. For manifolds with special holonomy groups, there are such forms that are parallel. The most notable is the holomorphic n-form of the Calabi-Yau manifolds. They define a special class of submanifolds that minimize volume in their homology class.

Such calibrated submanifolds were rediscovered by Becker-Becker-Strominger [2] in their study of cycles that preserve some supersymmetry of the ambient manifolds. They are special cases of branes in string theory. As such, they play important roles in the consideration of supersymmetry, which gives beautiful duality with holomorphic bundles. Many analogues with holomorphic bundles can be drawn.

Richard Thomas and I [67] looked into various properties of such cycles. We conjectured that classes of lagrangian cycles with zero Maslov index should admit special lagrangian cycles and that they can be obtained by the mean curvature flow where long time existence should hold. And if some stability condition holds, it will converge to special lagrangian cycles. It is important to check that special lagrangian cycles automatically satisfies such stability conditions. The mirror symmetry between these cycles with Hermitian-Yang-Mills bundles is really fascinating.

An important question that I raised many years ago, is the mirror of the Hodge conjecture. The question here is that every middle dimensional cycles of the odd dimensional Calabi-Yau manifold should be written as a (rational) linear combination of special lagrangian cycles. Note that special lagrangian cycles mean cycles that can be calibrated by possibly different choices of $(n, 0)$-forms with norm one, and also the coefficients can have different signs.

If we insist that the coefficients be positive, then there is an interesting question: When will a homology class be representable by an irreducible special lagrangian cycle?

Once we have obtained a special lagrangian cycle, we would like to know its topology. For a Calabi-Yau manifold, what kind of topology can a special lagrangian cycle admit? Can a real hyperbolic manifold occur? When will it be a torus? When it is a torus, what is its deformation space? Can we fill in the Calabi-Yau manifold with special lagrangian tori so that the moduli space can be identified with some three-manifold with an affine structure that has singularity along some codimensional two set. This is related to what is called the Strominger-Yau-Zaslow [66] fibration and the mirror conjecture.

References

[1] S. Bando, *On the classification of three-dimensional compact Kaehler manifolds of nonnegative bisectional curvature.* J. Differential Geom. **19**(1984), no. 2, 283–297.

[2] K. Becker, M. Becker, Melanie and A. Strominger, *Fivebranes, membranes and non-perturbative string theory.* Nuclear Phys. B 456 (1995), no. 1–2, 130–152.

[3] S. Brendle and R. Schoen, *Manifolds with 1/4-pinched curvature are space forms.* J. Amer. Math. Soc. **22**(2009), no. 2, 287–307.

[4] C. P. Boyer and K. Galicki, *Sasakian geometry*, Oxford Mathematical Monographs, Oxford University Press (2007).

[5] H. D. Cao, *Deformation of Kähler metrics to Kähler-Einstein metrics on compact Kähler manifolds.* Invent. Math. **81**(1985), no. 2, 359–372.

[6] H. D. Cao, *On Harnack inequalities for the Kähler-Ricci flow.* Invent. Math. **109**(1992), 247–263.
[7] H. D. Cao, *Limits of solutions to the Kähler-Ricci flow.* J. Differential Geom. **45**(1997), no. 2, 257–272.
[8] H.D. Cao and B. Chow, *Compact Kähler manifolds with nonnegative curvature operator.* Invent. Math. **83**(1986), 553–556.
[9] H. I. Choi and R. Schoen, *The space of minimal embeddings of a surface into a three-dimensional manifold of positive Ricci curvature.* Invent. Math. **81**(1985), no. 3, 387–394.
[10] A. Chau and L.-F. Tam, *On the complex structure of Kähler manifolds with non-negative curvature.* arXiv:math.DG/0504422.
[11] J. Cheeger and D. Gromoll, *On the structure of complete manifolds of nonnegative curvature.* Ann. of Math., **96** (1978) no. 3, 413–443.
[12] S. Y. Cheng and S.-Y. Yau, *The real Monge-Ampère equation and affine flat structures*, in Proceedings of the 1980 Beijing Symposium on Differential Geometry and Differential Equations, Science Press, Beijing, China (1982), Gordon and Breach Science Publishing Company, New York , pp. 339–370.
[13] T. H. Colding and W. P. Minicozzi II *The space of embedded minimal surfaces of fixed genus in a 3-manifold.* Ann. of Math. (2) **160**(2004), I., 27–68; II., 69–92; III., 523–572; IV., 573–615.
[14] T. H. Colding and W. P. Minicozzi II *Estimates for the extinction time for the Ricci flow on certain 3-manifolds and a question of Perelman.* J. Amer. Math. Soc., **18** (2005) no. 3, 561–569.
[15] S. K. Donaldson, *Scalar curvature and projective embeddings.* I. J. Differential Geom. **59**(2001), no. 3, 479–522.
[16] D. S. Freed, *Pions and generalized cohomology.* J. Differential Geom. **80**(2008), 45–77.
[17] M. H. Freedman and F. Quinn, *Topology of 4-manifolds*, Princeton Mathematical Series, 39., Princeton University Press, Princeton, NJ, 1990.
[18] J. X. Fu and S.-T. Yau, *The theory of superstring with flux on non-Kahler manifolds and the complex Monge-Ampere equation.* J. Differential Geom. 78(2008), 369–428.
[19] M. E. Gage, *On an area-preserving evolution equation for plane curves*, pp. 51–62. in Nonlinear problems in geometry (Mobile, Ala., 1985), edited by D. M. DeTurck, Contemporary Mathematics 51, Amer. Math. Soc., Providence, 1986.
[20] M. E. Gage and R. S. Hamilton, *The heat equation shrinking convex plane curves.* J. Differential Geom., **23** (1986), 69–96.
[21] G. W. Gibbons, S. W. Hawking, and M. J. Perry, *Path integrals and the indefiniteness of the gravitational action.* Nuclear Phys. B, 138:1 (1978), 141–150.
[22] A. Gray, *Nearly Kahler manifolds.* J. Differential Geom. 4 (1970), 283–309
[23] M. A. Grayson, *The heat equation shrinks embedded plane curves to round points.* J. Differential Geom., **26** (1987), 285–314.
[24] M. Gromov, *Curvature, diameter and Betti numbers.* Comment. Math. Helv. **56** (1981), 179–195.
[25] M. Gromov and H. B. Lawson, *The classification of simply connected manifolds of positive scalar curvature.* Ann. of Math.,**111** (3) (1980), 423–434.
[26] R. S. Hamilton, *Three-manifolds with positive Ricci curvature.* J. Differential Geom. **17**(1982), no. 2, 255–306.
[27] R. S. Hamilton, *The Harnack estimate for the Ricci flow.* J. Differential Geom. **37**(1993), no. 1, 225–243.
[28] R. S. Hamilton, *The formation of singularities in the Ricci flow.* Surveys in differential geometry, Vol. II (Cambridge, MA, 1993), 7–136, Internat. Press, Cambridge, MA, 1995.

[29] R. S. Hamilton, *Four-manifolds with positive isotropic curvature.* Comm. Anal. Geom. **5**(1997), no. 1, 1–92.
[30] R. S. Hamilton, *Non-singular solutions of the Ricci flow on three-manifolds.* Comm. Anal. Geom. **7**(1999), no. 4, 695–729.
[31] N. J. Hitchin, *Kählerian twistor spaces.* Proc. London Math. Soc. **43** (1981), 133–150.
[32] D. Hoffman, *Computing minimal surfaces. Global theory of minimal surfaces,* 259–282, Clay Math. Proc., 2, Amer. Math. Soc., Providence, RI, 2005.
[33] M. J. Hopkins and I. M. Singer, *Quadratic functions in Geometry, Topology, and M-theory.* J. Differential Geom. **70**(2005) 329–452.
[34] G. Huisken, *Flow by mean curvature of convex surfaces into spheres.* J. Differential Geom. **20**(1984), no. 1, 237–266.
[35] G. Huisken, *Contracting convex hypersurfaces in Riemannian manifolds by their mean curvature.* Invent. Math. **84**(1986), no. 3, 463–480.
[36] J. Jost, *Embedded minimal surfaces in manifolds diffeomorphic to the three-dimensional ball or sphere.* J. Differential Geom. **30**(1989), no. 2, 555–577.
[37] D. D. Joyce, *Compact* 8*-manifolds with holonomy* Spin(7). Invent. Math. **123**(1996), no. 3, 507–552.
[38] D. D. Joyce, *Compact Riemannian* 7*-manifolds with holonomy* G_2*. I, II.* J. Differential Geom. **43**(1996), no. 2, 291–328, 329–375.
[39] D. Joyce, *A new construction of compact 8-manifolds with holonomy* Spin(7). J. Differential Geom. **53**(1999), no. 1, 89–130.
[40] D. Joyce, *Compact Riemannian manifolds with special holonomy,* 436 pages, Oxford University Press, 2000.
[41] H. B. Lawson Jr. and S.-T. Yau, *Scalar curvature, non-abelian group actions, and the degree of symmetry of exotic spheres.* Comment. Math. Helv. **49**(1974), 232–244.
[42] J. Li and S.-T. Yau, *The Existence of Supersymmetric String Theory with Torsion.* J. Differential Geom. **70**(2005), 143–181.
[43] P. Li and S.-T. Yau, *On the parabolic kernel of the Schrödinger operator.* Acta Math. **156**(1986), no. 3-4, 153–201.
[44] J. M. Maldacena, *The Large N limit of superconformal field theories and supergravity.* Adv. Theor. Math. Phys. **2**(1998), 231–252.
[45] F. Martin and S. Morales, *Complete proper minimal surfaces in convex bodies of* $\mathbb{R}^3$. Duke Math. J. **128** (2005), no. 3, 559–593.
[46] W. H. Meeks, III, *Applications of minimal surfaces to the topology of three-manifolds.* Current developments in mathematics, 2004, 95–109, Int. Press, Somerville, MA, 2006.
[47] W. H. Meeks, III, J. Perez and A. Ros, *The geometry of minimal surfaces of finite genus. I. Curvature estimates and quasiperiodicity.* J. Differential Geom. **66**(2004), 1–45.
[48] W. H. Meeks, III, J. Perez and A. Ros, *Liouville-type properties for embedded minimal surfaces.* Comm. Anal. Geom. **14**(2006), no. 4, 703–723.
[49] W. H. Meeks, III and H. Rosenberg, *The minimal lamination closure theorem.* Duke Math. J. **133**(2006), 467–497.
[50] M. Micallef and J.D. Moore, *Minimal two-spheres and the topology of manifolds with positive curvature on totally isotropic two-planes.* Ann. of Math. (2) **27**(1988), 199–227.
[51] N. Mok, *The uniformization theorem for compact Kähler manifolds of nonnegative holomorphic bisectional curvature.* J. Differential Geom. **27**(1988), no. 2, 179–214.
[52] C. B. Morrey, *The problem of Plateau in a Riemannian manifold.* Ann. of Math. **49**(1948), 807–851.
[53] N. Nadirashvili, *Hadamard's and Calabi-Yau's conjectures on negatively curved and minimal surfaces.* Invent. Math. **126** (1996) 457–465.

[54] G. Perelman, *Proof of the soul conjecture of Cheeger and Gromoll.* J. Differential Geom **40**(1994), no. 1, 209–212.
[55] G. Perelman, *The entropy formula for the Ricci flow and its geometric applications,* arXiv:math.DG/0211159.
[56] G. Perelman, *Ricci flow with surgery on three-manifolds,* arXiv:math/0303109.
[57] G. Perelman, *Finite extinction time for the solutions to the Ricci flow on certain three-manifolds,* arXiv:math.DG/0307245.
[58] J. Sacks and K. Uhlenbeck, *The existence of minimal immersions of 2-spheres.* Ann. of Math. (2) **113**(1981), no. 1, 1–24.
[59] R. Schoen and S.-T. Yau, *On the structure of manifolds with positive scalar curvature, Manuscripta Math.* **28** (1979), 159–183.
[60] R. Schoen and S.-T. Yau, *Complete three-dimensional manifolds with positive Ricci curvature and scalar curvature.* Seminar on Differential Geometry, pp. 209–228, Ann. of Math. Stud., **102**, Princeton Univ. Press, Princeton, N.J., 1982.
[61] L. Simon and F. Smith, *On the existence of embedded minimal 2-spheres in the 3-sphere, endowed with an arbitrary metric,* unpublished.
[62] J. Simons and D. Sullivan, *Structured vector bundles define differential K-theory.* arXiv:0810.4935.
[63] Y. T. Siu and S.-T. Yau, *Compact Kähler manifolds of positive bisectional curvature.* Invent. Math. **59**(1980), no. 2, 189–204.
[64] S. Stolz, *Simply connected manifolds of positive scalar curvature.* Ann. of Math. **136** (3) (1992), 511–540.
[65] A. Strominger, *Superstrings with torsion.* Nuclear Phys. B **274**(1986), no. 2, 253–284.
[66] A. Strominger, S.-T. Yau and E. Zaslow, *Mirror symmetry is T-duality.* Nuclear Phys. B **479**(1996), no. 1-2, 243–259.
[67] R. P. Thomas and S.-T. Yau, *Special Lagrangians, stable bundles and mean curvature flow.* (English. English summary) Comm. Anal. Geom. **10**(2002), no. 5, 1075–1113.
[68] E. Witten and S.-T. Yau, *Connectedness of the boundary in the AdS/CFT correspondence.* Adv. Theor. Math. Phys. **3**(1999), no. 6, 1635–1655 (2000).
[69] S.-T. Yau, *A review of complex differential geometry.* Proc. Symp. Pur Math. **52**(1991), 619-625.
[70] S.-T. Yau, *Open problems in Geometry.* Proc. Symp. Pure Math. **54**(1993) 1–28.
[71] S.-T. Yau, *Métriques de Kähler-Einstein sur les variétés ouvertes; Premiére Classe de Chern et courbure de Ricci: Preuve de la conjecture de Calabi.* Séminaire Palaiseau, Astérisque, 58 (1978), 163–167.
[72] Shiu Yuen Cheng and Shing-Tung Yau, *On the existence of a complete Kähler metric on noncompact complex manifolds and the regularity of Fefferman's equation.* Comm. Pure Appl. Math. 33 (1980), no. 4, 507–544.
[73] Shiu Yuen Cheng and Shing-Tung Yau, *Inequality between Chern numbers of singular Kähler surfaces and characterization of orbit space of discrete group of $SU(2,1)$; Complex differential geometry and nonlinear differential equations.* (Brunswick, Maine, 1984), 31–44, Contemp. Math., 49, Amer. Math. Soc., Providence, RI, 1986.
[74] Gang Tian and Shing-Tung Yau, *Complete Kähler manifolds with zero Ricci curvature, I.* J. Amer. Math. Soc. 3 (1990), no. 3, 579–609.
[75] Gang Tian and Shing-Tung Yau, *Complete Kähler manifolds with zero Ricci curvature, II.* Invent. Math. 106 (1991), no. 1, 27–60.

Department of Mathematics, Harvard University, One Oxford Street, Cambridge, MA 02138, USA
E-mail address: yau@math.harvard.edu